ANNUAL REVIEW OF BIOPHYSICS AND BIOENGINEERING

EDITORIAL COMMITTEE (1980)

R. M. ARTHUR
D. M. ENGELMAN
J. GERGELY
W. A. HAGINS
L. J. MULLINS
C. NEWTON
J. H. WANG
G. WEBER

Responsible for the organization of Volume 9
(Editorial Committee, 1978)

D. M. ENGELMAN
W. A. HAGINS
L. J. MULLINS
C. NEWTON
J. NICHOLLS
H. W. SHIPTON
J. H. WANG
G. WEBER
M. MORALES (Guest)

Production Editor	J. L. COHEN
Indexing Coordinator	M. A. GLASS
Subject Indexer	S. M. SORENSEN

ANNUAL REVIEW OF BIOPHYSICS AND BIOENGINEERING

L. J. MULLINS, *Editor*
University of Maryland

WILLIAM A. HAGINS, *Associate Editor*
National Institute of Arthritis and Metabolic Diseases

CAROL NEWTON, *Associate Editor*
University of California, Los Angeles

GREGORIO WEBER, *Associate Editor*
University of Illinois, Urbana

VOLUME 9

1980

ANNUAL REVIEWS INC.
Palo Alto, California, USA

COPYRIGHT © 1980 BY ANNUAL REVIEWS INC., PALO ALTO, CALIFORNIA, USA. ALL RIGHTS RESERVED. The appearance of the code at the bottom of the first page of an article in this serial indicates the copyright owner's consent that copies of the article may be made for personal or internal use, or for the personal or internal use of specific clients. This consent is given on the condition, however, that the copier pay the stated per-copy fee of $1.00 per article through the Copyright Clearance Center, Inc. (P. O. Box 765, Schenectady, NY 12301) for copying beyond that permitted by Sections 107 and 108 of the US Copyright Law. The per-copy fee of $1.00 per article also applies to the copying, under the stated conditions, of articles published in any Annual Review serial before January 1, 1978. Individual readers, and nonprofit libraries acting for them, are permitted to make a single copy of an article without charge for use in research or teaching. This consent does not extend to other kinds of copying, such as copying for general distribution, for advertising or promotional purposes, for creating new collective works, or for resale.

REPRINTS The conspicuous number aligned in the margin with the title of each article in this volume is a key for use in ordering reprints. Available reprints are priced at the uniform rate of $1.00 each postpaid. The minimum acceptable reprint order is 5 reprints and/or $5.00 prepaid. A quantity discount is available.

International Standard Serial Number: 0084-6589
International Standard Book Number: 0-8243-1809-9
Library of Congress Catalog Card Number: 79-188446

Annual Reviews Inc. and the Editors of its publications assume no responsibility for the statements expressed by the contributors to this Review.

PRINTED AND BOUND IN THE UNITED STATES OF AMERICA

PREFACE

We continue to be amazed at the diversity of topics that make up the fields of Biophysics and Bioengineering. A comparison of Volume 1 with Volume 9 shows that a majority of the topics in this volume are quite new. This is surely because the field is developing strongly, and we intend to seek out for review those areas where development is the most rapid. Our annual editorial meetings are marked by lively argument as to just where the field is going—yet this is the only way to fathom the kinds of materials we need to include.

As in the past, we urge readers to help us by suggesting topics for review that we ought to include. Some of our best reviews are generated this way. We are eternally grateful to our authors for their splendid efforts in producing the reviews you will read; we also wish to thank our production editor, Joan Cohen, for her untiring efforts to get this volume into print.

THE EDITORS AND EDITORIAL COMMITTEE

SOME RELATED ARTICLES APPEARING IN OTHER *ANNUAL REVIEWS*

From the *Annual Review of Biochemistry*, Volume 49 (1980)
Proteins Containing 4Fe-4S Clusters: An Overview, William V. Sweeney and Jesse C. Rabinowitz
Energy Transduction in Chloroplasts: Structure and Function of the ATPase Complex, Noun Shavit
On the Mechanism of Action of Folate- and Biopterin-Requiring Enzymes, Stephen J. Benkovic
The Proton-Translocating Pumps of Oxidative Phosphorylation, Robert H. Fillingame

From the *Annual Review of Genetics*, Volume 13 (1979)
Regulatory Sequences Involved in the Promotion and Termination of RNA Transcription, Martin Rosenberg and Donald Court
Regulation of Initiation of DNA Replication, Roberto Kolter and Donald R. Helinski

From the *Annual Review of Microbiology*, Volume 33 (1979)
Dynamics of the Macrophage Plasma Membrane, Steven H. Zuckerman and Steven D. Douglas
Concentration of Viruses from Water by Membrane Chromatography, Craig Wallis, Joseph L. Melnick, and Charles P. Gerba

From the *Annual Review of Pharmacology and Toxicology*, Volume 20 (1980)
The Electrogenic Na^+, K^+-Pump in Smooth Muscle: Physiologic and Pharmacologic Significance, William W. Fleming

From the *Annual Review of Physical Chemistry*, Volume 30 (1979)
Coherent Transient Effects in Optical Spectroscopy, R. L. Shoemaker
Hemoglobin and Myoglobin Ligand Kinetics, Lawrence J. Parkhurst

From the *Annual Review of Physiology*, Volume 42 (1980)
Co- and Counter-Transport Mechanisms in Cell Membranes, Robert B. Gunn
Receptors for Amino Acids, Ernest J. Peck, Jr.
Role of Cyclic Nucleotides in Excitable Cells, Irving Kupfermann
Biophysical Analysis of the Function of Receptors, Charles F. Stevens

CONTENTS

RECENT DEVELOPMENTS IN SOLUTION X-RAY SCATTERING, *Vittorio Luzzati and Annette Tardieu*	1
PHOTOACOUSTIC SPECTROSCOPY, *Allan Rosencwaig*	31
STIMULUS-RESPONSE COUPLING IN GLAND CELLS, *B. L. Ginsborg and C. R. House*	55
X-RAY DIFFRACTION STUDIES OF THE HEART, *I. Matsubara*	81
THE OPTICAL ACTIVITY OF NUCLEIC ACIDS AND THEIR AGGREGATES, *Ignacio Tinoco, Jr., Carlos Bustamante, and Marcos F. Maestre*	107
MODULATION OF IMPULSE CONDUCTION ALONG THE AXONAL TREE, *Harvey A. Swadlow, Jeffery D. Kocsis, and Stephen G. Waxman*	143
TRANSFER RNA IN SOLUTION: SELECTED TOPICS, *Paul R. Schimmel and Alfred G. Redfield*	181
NERVE GROWTH FACTOR: MECHANISM OF ACTION, *Stanley Vinores and Gordon Guroff*	223
THE STRUCTURE OF PROTEINS INVOLVED IN ACTIVE MEMBRANE TRANSPORT, *Ann S. Hobbs and R. Wayne Albers*	259
MAGNETIC CIRCULAR DICHROISM OF BIOLOGICAL MOLECULES, *John Clark Sutherland and Barton Holmquist*	293
RADIOIMMUNOASSAY, *Rosalyn S. Yalow*	327
SPECIAL TECHNIQUES FOR THE AUTOMATIC COMPUTER RECONSTRUCTION OF NEURONAL STRUCTURES, *I. Sobel, C. Levinthal, and E. R. Macagno*	347
BIOPHYSICAL APPLICATIONS OF NMR TO PHOSPHORYL TRANSFER ENZYMES AND METAL NUCLEI OF METALLOPROTEINS, *Joseph J. Villafranca and Frank M. Raushel*	363
MACHINE-ASSISTED PATTERN CLASSIFICATION IN MEDICINE AND BIOLOGY, *Ching-Chung Li and King-Sun Fu*	393
CERTAIN SLOW SYNAPTIC RESPONSES: THEIR PROPERTIES AND POSSIBLE UNDERLYING MECHANISMS, *JacSue Kehoe and Alain Marty*	437
COMPARATIVE PROPERTIES AND METHODS OF PREPARATION OF LIPID VESICLES (LIPOSOMES), *Francis Szoka, Jr. and Demetrios Papahadjopoulos*	467
DISPLAY AND ANALYSIS OF FLOW CYTOMETRIC DATA, *J. W. Gray and P. N. Dean*	509
BIOMATHEMATICS IN ONCOLOGY: MODELING OF CELLULAR SYSTEMS, *Carol M. Newton*	541
MEDICAL INFORMATION SYSTEMS, *Eugene M. Laska and Scott G. Abbey*	581
INDEXES	
Author Index	605
Subject Index	625
Cumulative Index of Contributing Authors, Volumes 5-9	638
Cumulative Index of Chapter Titles, Volumes 5-9	640

ANNUAL REVIEWS INC. is a nonprofit corporation established to promote the advancement of the sciences. Beginning in 1932 with the *Annual Review of Biochemistry*, the Company has pursued as its principal function the publication of high quality, reasonably priced Annual Review volumes. The volumes are organized by Editors and Editorial Committees who invite qualified authors to contribute critical articles reviewing significant developments within each major discipline.

Annual Reviews Inc. is administered by a Board of Directors whose members serve without compensation. The Board for 1980 is constituted as follows:

Dr. J. Murray Luck, Founder and Director Emeritus of Annual Reviews Inc.
  Professor Emeritus of Chemistry, Stanford University
Dr. Joshua Lederberg, President of Annual Reviews Inc.
  President, The Rockefeller University
Dr. James E. Howell, Vice President of Annual Reviews Inc.
  Professor of Economics, Stanford University
Dr. William O. Baker, *President, Bell Telephone Laboratories*
Dr. Robert W. Berliner, *Dean, Yale University School of Medicine*
Dr. Sidney D. Drell, *Deputy Director, Stanford Linear Accelerator Center*
Dr. Eugene Garfield, *President, Institute for Scientific Information*
Dr. William D. McElroy, *Chancellor, University of California, San Diego*
Dr. William F. Miller, *President, SRI International*
Dr. Colin S. Pittendrigh, *Director, Hopkins Marine Station*
Dr. Esmond E. Snell, *Professor of Microbiology and Chemistry,*
  University of Texas, Austin
Dr. Harriet A. Zuckerman, *Professor of Sociology, Columbia University*

The management of Annual Reviews Inc. is constituted as follows:
John S. McNeil, Chief Executive Officer and Secretary-Treasurer
William Kaufmann, Editor-in-Chief
Charles S. Stewart, Business Manager
Sharon E. Hawkes, Production Manager
Ruth E. Severance, Promotion Manager

Annual Reviews are published in the following sciences: Anthropology, Astronomy and Astrophysics, Biochemistry, Biophysics and Bioengineering, Earth and Planetary Sciences, Ecology and Systematics, Energy, Entomology, Fluid Mechanics, Genetics, Materials Science, Medicine, Microbiology, Neuroscience, Nuclear and Particle Science, Pharmacology and Toxicology, Physical Chemistry, Physiology, Phytopathology, Plant Physiology, Psychology, Public Health, and Sociology. In addition, five special volumes have been published by Annual Reviews Inc.: *History of Entomology* (1973), *The Excitement and Fascination of Science* (1965), *The Excitement and Fascination of Science, Volume Two* (1978), *Annual Reviews Reprints: Cell Membranes, 1975–1977* (published 1978), and *Annual Reviews Reprints: Immunology, 1977–1979* (published 1980). For the convenience of readers, a detachable order form/envelope is bound into the back of this volume.

RECENT DEVELOPMENTS IN SOLUTION X-RAY SCATTERING

♦9139

Vittorio Luzzati and Annette Tardieu

Centre de Génétique Moléculaire, Centre National de la Recherche Scientifique, 91190 Gif-sur-Yvette, France

INTRODUCTION

We are concerned here with the structural information that it is possible to retrieve from the X-ray scattering study of macromolecules in solution. The general law governing X-ray scattering phenomena expresses the angular dependence of the scattered intensity as a function of the space distribution of the interatomic distances in the scatterer and of the nature of the atoms involved. When the scatterer is isotropic, the intensity is a function only of the moduli of the interatomic vectors, not of their orientation. Let $r^2 p(r)$ be this isotropic distribution of the interatomic distances (a precise definition is given below). For a dilute solution of macromolecules the function $p(r)$ can be expected to display sharp short-range fluctuations (1 to 5 Å) corresponding to pairs of neighboring atoms, followed by more damped fluctuations if middle-range regularities are present—for example in the 5 to 10 Å region for α-helical proteins. Beyond approximately 10 Å, the number and variety of the interatomic vectors increases very rapidly with increasing r, and the function $p(r)$ gradually smoothes out and slowly decreases. When r exceeds the maximal dimensions of the macromolecule, $p(r)$ becomes uniform, since all the vectors involve only the solvent.

Because the scattered intensity $i(s)$ and the distribution $p(r)$ are related by a Fourier transformation, such a separation of the fluctuations of $p(r)$ into two classes entails the presence in $i(s)$ of two distinct regions. One, at s small, contains the information relevant to the long-range (macromolecular) organization; the other, at s large, mirrors the structural features at the molecular level. The separation between the two regions can be set at approximately the inverse of the position of the boundary between the short- and the long-range fluctuations of $p(r)$, namely $s = 2\sin\theta/\lambda \sim (10\text{ Å})^{-1}$ (2θ is the scattering angle, λ the wavelength). Since the wavelength used in X-ray scattering experiments is restricted for practical reasons to the vicinity of 1.5 Å, the scattering angles corresponding to $s < (10\text{ Å})^{-1}$ are smaller than 10°. These angles are "small" in the sense that for $2\theta < 10°$, the approximations $\sin 2\theta = 2\theta$ and $\cos 2\theta = 1$ are acceptable. Thus the operational distinction between "small" and "large" angles acquires a fairly precise physical meaning; this is the reason why the expression "small-angle X-ray scattering" is often used as a synonym for "solution X-ray scattering" in the study of macromolecular systems. This expression is sometimes extended to neutron scattering studies, in spite of the fact that the wavelength is commonly much larger for neutrons than for X rays and thus the scattering angles corresponding to macromolecular dimensions are not always "small."

The beginning of an explicit concern for the small-angle properties of the X-ray scattering curves dates from 1938, when Guinier (8) established the well-known theorem relating the curvature of the intensity curve at the origin of s to the radius of gyration of the solute. Biophysicists, whose main concern in those days was the molecular nature of proteins, incorporated small-angle X-ray scattering into the panoply of hydrodynamic techniques that they were then developing (sedimentation, diffusion, etc); hence it became customary to present small-angle X-ray scattering as one of the physical properties of macromolecules, rather than as one aspect of diffraction theory. This connection with macromolecular physics was of little consequence as long as the scope of solution X-ray scattering studies was a coarse morphological analysis based essentially upon measurements of radii of gyration. With increasing sophistication, the conceptual framework provided by macromolecular physics became inadequate; theoretical support was then sought in diffraction theory, Fourier transformation providing the formalism.

Over the last three decades, the focus of structural studies of biological interest has shifted from the macromolecular to the atomic level, with a much stronger emphasis on crystallographic high resolution analyses than on solution scattering studies. Yet, when a biological system is not amenable to crystallization, solution scattering often provides one of the best tools for gaining structural information. Besides, solution techniques offer the possibility of varying chemical and physical conditions and of exploring structural transitions that are difficult to induce and to study in crystalline preparations.

For a long time solution X-ray scattering studies suffered from severe technical limitations; the experiments were cumbersome and time consuming, the results often disappointingly inaccurate. Around 1972 a major technical breakthrough took place, namely the introduction of a position-sensitive X-ray detector that cut down exposure times by a factor larger than 100 (6). The new detector turned out to be ideally suited for solution scattering studies and was immediately adopted for that purpose (see examples below). More recently, the use of synchrotrons and storage rings as X-rays sources (28), as well as the design of improved one- and two-dimensional position sensitive detectors, has speeded up X-ray scattering experiments by another large factor. The overall effect is that in a span of a few years the exposure times have been shortened by a factor larger than 10^5. At almost the same time, the use of high flux reactors (11) has also transformed neutron scattering into a powerful tool for structural studies of macromolecules in solution. Since neutron and X-ray scattering are very similar phenomena, the wealth and accuracy of the data made available by the new techniques have justified more sophisticated analyses, which in turn have prompted a revived interest in some theoretical aspects of solution scattering studies.

We focus this review on recent developments in solution X-ray scattering studies that have followed the introduction of position sensitive detectors; previous work in this area has been reviewed by other authors (13, 14). As a point of semantics, we find it appropriate to use the expression "small angle scattering," in a mathematical rather than in a physical sense, to specify the properties of the scattered intensity near the origin of s and to designate as "solution scattering" the general approach to the structure of macromolecules in solution. In the first part of this paper we discuss some experimental aspects of solution X-ray scattering studies. In the second part we summarize the theoretical background, with special emphasis on the contrast variation approach and the invariant volume hypothesis, namely, the systematic variation of the electron density of the solvent and the assumption that a volume is associated to each particle in solution, inside of which the electron density distribution is invariant with respect to the density of the solvent. This approach has played a fundamental role in most of the recent solution scattering studies, both by X rays and by neutrons. Finally, we illustrate the technical and theoretical parts with one example.

TECHNIQUES

A small-angle X-ray scattering camera consists essentially of a set of elements—slits, sample holder, beam stop, detector—supported by a bench. The X-ray source can be a conventional or a rotating anode tube, a synchrotron, or a storage ring. The incident beam, collimated by slits, mirrors, or monochromators, must be reasonably monochromatic and can be parallel or can converge to a focus. Other slits stop the radiation scattered by the collimating system. The quality of the slits—mechanical settings; material, shape, and polishing of the jaws—is one of the limiting factors in solution X-ray scattering technology.

The geometry of the beam dictates the size and shape of the sample. Often the cross section of the beam at the sample is narrow and long (typically a few tenths of a millimeter by a few millimeters); in this case the most convenient sample cell is a low absorption capillary tube of diameter 0.5 to 1.0 mm.

Conventional electronic detectors, for example scintillation or proportional counters, require mechanical scanning of the scattered beam. Position sensitive detectors record both number and position of the photons and thus avoid the use of mechanical scanning. For example, the detector designed by Gabriel & Dupont (6) and used in our laboratory since 1972, specifies the position of the photons in the direction parallel to its wire with a resolution of approximately 0.20 mm; since its length is 5 cm, one detector is equivalent to approximately $5/0.20 = 250$ conventional detectors. This factor defines the gain in exposure time; it is thus clear why position-sensitive detectors have replaced conventional detectors in solution X-ray scattering technology.

Blank scattering is another limiting factor; indeed scattering intensities are generally weak and the signal/noise ratio must be as high as possible. Air scattering can be eliminated by operating in vacuo; in practice, however, it is difficult to avoid interrupting the vacuum path in the vicinity of the sample. It is essential that the most rigorous precautions be taken in order to keep the residual scattering originating from this region as low as possible; the two windows near to the sample must be made of monocrystalline material (for example mica), the air path must be short, the volume of irradiated air "seen" by the detector must be extremely small. For the sake of an illustration we can mention that, according to the routine of our laboratory, a camera is satisfactorily set when the whole of the blank scattering—from slits, air, windows, sample holder, electronic noise—is not larger than the scattering of 1 mm of distilled water (with the exception of a very narrow region around the beam stop where the blank scattering increases sharply).

One problem that is difficult to eliminate completely is collimation distortions. Most often the incident beam is very narrow—infinitely narrow for practical purposes—but rather long. A variety of mathematical procedures can be used to correct these distortions; yet, it is worthwhile to stress that the corrections are all the more accurate, the shorter the beam. The same also applies to polychromatism distortions.

A typical solution scattering experiment unfolds as follows. First, the adjustable parameters of the camera (sample-detector distance, slit width and position, etc) are chosen according to the requirements of the experiment, and the various elements (slits, beam stop, etc) are set. Then the intensity scattered by the solution, by the solvent, as well as that of blank scattering are recorded. Each of these measurements is performed with a defined exposure time, and in each case the energy of the incident beam is measured; this is done either by recording the incident beam properly attenuated by a set of calibrated filters or by using a calibrated scatterer. The thickness of the sample is measured either directly or by its X-ray absorption. All these data lead to the determination of the intensity scattered by the solute; correction for collimation and chromatic distortions finally yield the intensity curves used in subsequent operations (see below).

THEORY

We present the theoretical treatment within the framework of a set of ideal conditions, chosen in accordance with the experimental conditions of the example discussed below.

The sample is homogeneous and isotropic; its shape is a planar slab, oriented at right angle to the direction of the incident beam and intercalated over the entire cross section of the beam; the thickness of the slab is negligible with regard to the distance to the detector. The sample is supposed to be a solution of macromolecules and all the scattered intensity is assumed to be restricted to small scattering angles —namely in the angular range where the approximations $\sin 2\theta = 2\theta$ and $\cos 2\theta = 1$ are acceptable. The X-ray beam is strictly monochromatic; multiple and incoherent scattering are disregarded, as well as absorption (the absorption factor, independent of the scattering angle, is the same for the incident and the scattered beam). We also assume that the collimation distortions are corrected.

We use the following notation (see 17, 19, 20, 26):

$F(\mathbf{r}) * G(\mathbf{r}) = \int F(\mathbf{r}-\mathbf{R})G(\mathbf{R})dv_\mathbf{R}$: convolution of the functions $F(\mathbf{r})$ and $G(\mathbf{r})$;

$\langle F(\mathbf{r}) \rangle$: spherical average of $F(\mathbf{r})$;

$\mathbf{r}, \mathbf{s}$: vectors that specify position in real and reciprocal space; r and s are the moduli; the units are Å for r, Å^{-1} for s ($s = 2\sin\theta/\lambda$, where 2θ is the scattering angle, λ is the wavelength, in Å);

E_o, η, ν: energy E_o of the incident beam, thickness η of the sample (in e cm^{-2}), a physical constant $\nu = \lambda^2 \times 7.9 \times 10^{26}$;

$I(s)$: experimental intensity curve, corrected for collimation distortions;

V, N: volume of the sample (Å^3), number of solute particles in V;

$\rho(\mathbf{r}), \bar{\rho}$: distribution of the electron density in the sample (e Å^{-3}) and its average value;

$p(r) = \langle \rho(\mathbf{r}) * \rho(-\mathbf{r}) \rangle$: autocorrelation function of $\rho(\mathbf{r})$;

$i(s) = (2/s) \int_0^\infty rp(r) \sin 2\pi rs\, dr$: intensity curve relevant to the electron density distribution $\rho(\mathbf{r})$;

d_o, ρ_o: density (g cm^{-3}) and electron density (e Å^{-3}) of the solvent, assumed to be uniform;

c, c_e, c_v, c_{ev}: concentration (solute : solution) respectively in g g^{-1}, e e^{-1}, g cm^{-3}, e Å^{-3};

m, M, μ: number of electrons and molecular weight (in Daltons) of the macromolecular solute, $\mu = m/M$;

$\bar{v}, \psi$: partial specific volume (cm g^{-3}) and electron partial specific volume (Å e^{-3}).

It is convenient to express the scattered intensity normalized to one electron of solution:

$$i_n(s) = I(s)/E_o \eta \nu = i(s)/V\bar{\rho}. \qquad 1.$$

In this expression the experimental information (grouped together in the second term) is clearly separated from the structural parameters (gathered together in the third term).

The Sample Is an Ideal Solution of Identical Globular Particles

A solution is ideal when the distribution of the solute particles is random, without correlations in position and orientation. The scattered intensity is in this case the sum of the intensity scattered by the individual particles:

$$i(s) = \sum_{k=1}^{N} i_k(s) \qquad 2.$$

where (see notation)

$$i_k(s) = (2/s) \int_0^\infty r p_k(r) \sin 2\pi r s \, dr; \qquad 3.$$

$$p_k(r) = \langle \Delta\rho_k(\mathbf{r}) * \Delta\rho_k(-\mathbf{r}) \rangle; \qquad 4.$$

$$\Delta\rho_k(\mathbf{r}) = \rho_k(\mathbf{r}) - \rho_o; \qquad 5.$$

$\Delta\rho_k(\mathbf{r})$ is the electron density contrast, with respect to the solvent, associated to the kth particle.

Replacing Equation 2 in Equation 1 we obtain:

$$i_n(s) = \sum_{k=1}^{N} i_k(s)/V\bar{\rho} = c_e \left[\sum_{k=1}^{N} i_k(s) / \sum_{k=1}^{N} m_k \right]. \qquad 6.$$

If, moreover, all the particles are identical Equation 6 becomes:

$$i_n(s)/c_e = i_1(s)/m \qquad 7.$$

(the suffix 1 refers to one particle). Equation 7 is another convenient normalized expression of the intensity in which the experimental information is clearly separated from the structural parameters.

If the particles are globular—more precisely if all of their dimensions are smaller than $(s_{\min})^{-1}$ ($s_{\min}$ is the lower limit of the experimental range of s)—Equation 7 can be expanded as follows:

$$i_n(s)/c_e = (\Delta m/m)^2 m \left[1 - (4/3)\pi^2 R^2 s^2 + \cdots \right] \qquad 8.$$

where

$$\Delta m = \int \Delta\rho_1(\mathbf{r}) \, dv_r; \qquad 9.$$

$$R^2 = \int r^2 \Delta\rho_1(\mathbf{r}) \, dv_r / (\Delta m). \qquad 10.$$

In Equation 10 the origin of $\mathbf{r}$ is chosen to be the center of mass of $\Delta\rho_1(\mathbf{r})$. Equation 8 is known as Guinier's law (8) where R is the radius of gyration of $\Delta\rho_1(\mathbf{r})$.

If the solvent contains one component, $\Delta m/m$ takes the form (17):

$$\Delta m/m = 1 - \rho_o \psi \qquad 11.$$

and Equation 8 becomes:

$$i_n(s)/c_e(1-\rho_o\psi)^2 = m \left[1 - (4/3)\pi^2 R^2 s^2 + \cdots \right]. \qquad 12.$$

We discuss below the more general case of a solvent containing more than one component.

It is worthwhile to point out that the determination of the radius of gyration does not require knowledge of concentration and intensity on an absolute scale. If absolute scales and partial specific volumes are known, then it is possible, in addition, to determine the molecular weight of the solute using Equation 12 at $s=0$.

The Electron Density Is Uniform Throughout the Particle

A hypothesis often made in solution X-ray scattering studies is that the electron density is approximately constant throughout the solute particles. This hypothesis, whose validity should always be tested, is involved, implicitly or explicitly, whenever shape is discussed.

The intensity scattered by an object with uniform electron density has the following asymptotic form (17, 25):

$$8\pi^3 \lim_{s \to \infty} \left[s^4 i_n(s)/c_e \right] = S(\rho_1 - \rho_o)^2/m \qquad 13.$$

where S and ρ_1 are the surface area and the electron density of the particle. An additional term B takes into account the short-range electron density fluctuations around the average value (21):

$$\lim_{s \to \infty} \overline{s^4 i_n(s)} = A + Bs^4. \qquad 14.$$

Thus the validity of the hypothesis of quasi-uniform electron density distribution requires the plot $s^4 i_n(s)$ versus s^4 to become linear as s increases. If this is the case the term B can be determined and subtracted from $i_n(s)$ for all subsequent operations. Moreover it becomes possible in this case to determine the volume v_1 of the particle. Indeed, if $i_n(s)$ is known from $s=0$ to infinity (Equations 12 and 14 provide the extrapolations to the two ends), we may write (see Equations 3 and 8):

$$i_n(0)/c_e = (\Delta m)^2/m = v_1^2(\rho_1 - \rho_o)^2/m; \qquad 15.$$

$$(4\pi/c_e)\int_0^\infty s^2 i_n(s)\, ds = p_1(0)/m = v_1(\rho_1 - \rho_o)^2/m. \qquad 16.$$

The ratio of these two equations yields:

$$v_1 = \left[i_n(0)/c_e \right] \Big/ \left[(4\pi/c_e)\int_0^\infty s^2 i_n(s)\, ds \right]. \qquad 17.$$

Therefore, in the case of an ideal solution of identical globular particles with quasi-uniform electron density, it is possible to determine three morphological parameters: the radius of gyration R, the volume v_1 and the surface area S, even if intensities and concentrations are not known on an absolute scale. These three parameters can be used to analyze the shape and size of the particle. A widespread procedure is to postulate that the shape of the particle is a simple solid (sphere, cylinder, ellipsoid, etc) and to determine its dimensions according to the values of the morphological parameters. An alternative procedure is to compare the whole of the experimental curve $I(s)$ with the scattering curve calculated for model objects. It is worth noting that this procedure is not necessarily more accurate than use of the three morphological parameters (see below the discussion of the information content of the scattering curves).

If the solvent contains one component, if concentration and intensities are measured on absolute scales, and if the partial specific volume is also known, it is then possible to determine m (Equation 2), ρ_1 (Equations 15, 16), as well as the solvation α, namely the ratio of the number of electrons (solvent/solute) contained in the volume v_1:

$$\alpha = (v_1 - m\psi)\rho_o/m. \qquad 18.$$

Variable Contrast Experiments: The Invariant Volume Hypothesis

In solution X-ray scattering studies it is most rewarding to perform experiments at variable solvent electron density. The density variation is generally achieved by adding an electron dense component—for example sucrose or salt in the case of aqueous solvents. In interpreting these experiments it is necessary to take into account the possible preferential associations of the different solvent components with the macromolecular solute.

In the following analysis of this problem we make a few simplifying assumptions: the effects of electrical charges (Donnan equilibrium, volume of electroneutrality, etc) are disregarded; the systems contains one impermeant macromolecular component, water, and another freely permeant component; the solution is ideal and all the molecules of the impermeant component are identical. In this case the electron density contrast associated with one macromolecule involves the macromolecule itself and the perturbations of the distribution of the diffusible components in the vicinity of the macromolecule. The electron density ρ_o of the regions of the solution which are not perturbed by the solute is equal to the electron density of the solvent at dialysis equilibrium with the solution; therefore "solution" and "solvent" must be defined at dialysis equilibrium.

When the solvent contains more than one component it is still possible to determine m using Equation 8 (at $s = 0$), provided the correct expression of $\Delta m/m$ is used. In this case the electron density contrast $\Delta m(\rho_o)$ associated with one particle is equal to $V(\bar{\rho} - \rho_0)/N$ (see Equation 9), where $\bar{\rho}$ and ρ_o are the average electron densities of solution and solvent. Therefore we have

$$\Delta m(\rho_o)/m = [V(\bar{\rho}-\rho_o)/N]/[Vc_{ev}/N] = (\bar{\rho}-\rho_o)/c_{ev} = \delta\rho/\delta c_{ev}. \quad 19.$$

It is possible, although not straightforward, to determine $(\delta\rho/\delta c_{ev})$ experimentally; this involves measuring the specific density increment $[(\bar{d}-d_o)/c_v]$ (at constant chemical potential) as well as the elementary composition of solution and solvent. If this is done, m can be determined using the following equation:

$$m = [i_n(0,\rho_o)]/[c_e(\delta\rho/\delta c_{ev})^2]. \quad 20.$$

Equation 20 is as far as general thermodynamic arguments can lead; it is hardly possible to retrieve additional structural information without reference to models. One model, particularly useful in solution X-ray (and neutron) scattering studies at variable contrast, is specified by the *invariant volume hypothesis*, which is often formulated as follows (1, 20, 29, 31): a volume is associated with each macromolecule in solution, inside of which the electron density distribution is invariant with respect to the density of the solvent. In fact the mathematical formulation generally adopted,

$$\Delta\rho(\mathbf{r},\rho_o) = \rho(\mathbf{r},\rho_o) - \rho_o = \Delta\rho(\mathbf{r},\rho_x) - (\rho_o - \rho_x)v_1(\mathbf{r}), \quad 21.$$

is based upon a less restrictive hypothesis, namely that the electron density distribution inside that volume is linearly dependent upon the electron density of the solvent (19). In Equation 21 ρ_x is an arbitrary reference electron density level, $\Delta\rho(\mathbf{r},\rho_x)$ is the distribution of the electron density contrast at $\rho_o = \rho_x$ (this function is independent of ρ_o, but not of ρ_x), and $v_1(\mathbf{r})$ is a function independent of both ρ_o and $\bar{\rho}_1$. The more restrictive hypothesis is equivalent to limiting the values of $v_1(\mathbf{r})$ to 1 inside the particle and 0 outside. Letting $v_1(\mathbf{r})$ take values different from 0 or 1 means that the volume, at any point $\mathbf{r}$, is not restricted to being either totally permeable or totally impermeable to the diffusible component; in this case $v_1(\mathbf{r})$ defines the volume fraction at the point $\mathbf{r}$ excluded to the permeant component. From the physical viewpoint, this situation is not implausible. For example, in neutron scattering studies at variable D_2O/H_2O ratios, the presence of exchangeable protons in the macromolecules means that $v_1(\mathbf{r})$ necessarily takes values intermediate between 1 and 0. Let us consider first the general case, without a priori restrictions on $v_1(\mathbf{r})$. The integral

$$v_1 = \int v_1(\mathbf{r}) dv_r \quad 22.$$

defines the total excluded volume, v_1.

It is convenient to set the reference electron density level at the "buoyant" electron density, namely at the particular solvent electron density, $\bar{\rho}_1$, at which $\Delta m(\rho_1)$ vanishes. In this case we write (see Equations 8, 9, 19, 21):

$$\Delta\rho(\mathbf{r},\rho_o) = \Delta\rho(\mathbf{r},\bar{\rho}_1) - (\rho_o - \bar{\rho}_1)v_1(\mathbf{r}); \qquad 23.$$

$$\Delta m(\rho_o) = \int \Delta\rho(\mathbf{r},\rho_o) dv_r = -(\rho_o - \bar{\rho}_1)v_1; \qquad 24.$$

$$\{i_n(0,\rho_o)/c_e\}^{1/2} = \Delta m(\rho_o)m^{-1/2} = -(\rho_o - \bar{\rho}_1)v_1 m^{-1/2}; \qquad 25.$$

$$\delta\rho/\delta c_{ev} = \Delta m(\rho_o)/m = -(\rho_o - \bar{\rho}_1)v_1/m. \qquad 26.$$

Therefore, if the invariant volume hypothesis is fulfilled, the functions $[i_n(0,\rho_o)/c_e]^{1/2}$ versus ρ_o and $(\delta\rho/\delta c_{ev})$ versus ρ_o are straight lines.

It is possible to determine m using Equation 12 (at $s=0$) if accurate X-ray and densimetry experiments are available in water. Otherwise, and more generally, other measurements can be used for the same purpose. For example, the specific density increment is in this case a linear function of d_o (see Equations 19, 26 and References 1, 19, 31):

$$\delta d/\delta c_v = \Delta M(d_o)/M = \left[(\bar{d}_1 - d_o)v_1 0.602\right]/M$$

$$= (\bar{d}_1 - d_o)(v_1/m)0.602\mu. \qquad 27.$$

($\bar{d}_1$ is the buoyant density of the particle). If the linear relationships of Equations 25 and 27 are fulfilled, the ratio of the slopes of the two straight lines provides an accurate expression of m (1):

$$m = (0.602\mu)^2 \left\{\delta[i_n(0,\rho_o)/c_e]^{1/2}/\delta\rho_o\right\}^2 / \left[\delta(\delta d/\delta c_v)/\delta d_o\right]^2 \qquad 28.$$

Moreover, the straight line of Equation 25 defines the values of $\bar{\rho}_1$ and of $v_1/m^{1/2}$; if m is known, v_1 may be determined.

The invariant volume hypothesis specifies the mathematical form of the ρ_o dependence of the scattered intensity at any point s. We write (see notation):

$$p_1(r,\rho_o) = \langle \Delta\rho(\mathbf{r},\rho_o) * \Delta\rho(-\mathbf{r},\rho_o) \rangle = p_\rho(r) + (\rho_o - \bar{\rho}_1)p_{\rho v}(r)$$

$$+ (\rho_o - \bar{\rho}_1)^2 p_v(r) \qquad 29.$$

where:

$$p_\rho(r) = \langle \Delta\rho(\mathbf{r},\bar{\rho}_1) * \Delta\rho(-\mathbf{r},\bar{\rho}_1) \rangle; \qquad 30.$$

$$p_{\rho v}(r) = -\langle [\Delta\rho(\mathbf{r},\bar{\rho}_1) * v_1(-\mathbf{r})] + [\Delta\rho(-\mathbf{r},\bar{\rho}_1) * v_1(\mathbf{r})] \rangle; \qquad 31.$$

$$p_v(r) = \langle v_1(\mathbf{r}) * v_1(-\mathbf{r}) \rangle. \qquad 32.$$

Thus we obtain:

$$i_n(s,\rho_o)/c_e = \left[i_\rho(s) + (\rho_o - \bar{\rho}_1)i_{\rho v}(s) + (\rho_o - \bar{\rho}_1)^2 i_v(s)\right]m^{-1}. \qquad 33.$$

The functions $i_\rho(s)$, $i_{\rho v}(s)$, and $i_v(s)$ are called the characteristic functions (29). Therefore, if the invariant volume hypothesis is fulfilled, at each value of s the experimental points $i_n(s,\rho_o)/c_e$, plotted as a function of ρ_o, must fit to a parabola. If this is the case, and if the value of $\bar{\rho}_1$ is known (see above), the three characteristic functions can be determined. These functions contain all the information which can be retrieved from the whole family of X-ray scattering experiments at variable solvent density. A region of special interest is the vicinity of the origin of s. The two first terms of the series expansion of the intensity take the form (see Equation 8):

$$i_n(s,\rho_o)/c_e = v_1^2(\rho_o - \bar{\rho}_1)^2 m^{-1}[1 - (4/3)\pi^2 R^2(\rho_o)s^2 + \cdots] \qquad 34.$$

where (see 19):

$$R^2(\rho_o) = R_v^2 - a(\rho_o - \bar{\rho}_1)^{-1} - b(\rho_o - \bar{\rho}_1)^{-2}; \qquad 35.$$

$$R_v^2 = v_1^{-1} \int r^2 v_1(\mathbf{r}) dv_r; \qquad 36.$$

$$a = m_{2\rho}/v_1 = v_1^{-1} \int r^2 \Delta\rho(\mathbf{r},\bar{\rho}_1) dv_r; \qquad 37.$$

$$b = v_1^{-2} \left| \int \mathbf{r} \Delta\rho(\mathbf{r},\bar{\rho}_1) dv_r \right|^2. \qquad 38.$$

In Equation 37 the origin is taken at the centre of mass of $v_1(\mathbf{r})$; Equation 38 is independent of the origin of $\mathbf{r}$. R_v is the radius of gyration of $v_1(\mathbf{r})$. The terms a and b depend on the internal structure of the particle: a takes into account the relative distribution of the high and the low density regions in the particle; b is a function of the symmetry of the internal structure, or more precisely $b(\rho_o - \bar{\rho}_1)^2$ is the distance between the centers of mass of $\Delta\rho(\mathbf{r},\rho_o)$ and of $v_1(\mathbf{r})$.

All the results up to this point (Equations 21–38) involve the validity of Equation 21 without restrictions on $v_1(\mathbf{r})$, and therefore cannot provide any information about the form of $v_1(\mathbf{r})$. However the values of the autocorrelation functions at $r=0$ may yield some information of $v_1(\mathbf{r})$. The following expressions can be used (see Equations 30–33):

$$p_\rho(0) = \int \{\Delta\rho(\mathbf{r},\bar{\rho}_1)\}^2 dv_r = 4\pi \int_0^\infty s^2 i_\rho(s) ds; \qquad 39.$$

$$p_{\rho v}(0) = -2 \int \Delta\rho(\mathbf{r},\rho_1) v_1(\mathbf{r}) dv_r = 4\pi \int_0^\infty s^2 i_{\rho v}(s) ds; \qquad 40.$$

$$p_v(0) = \int \{v_1(\mathbf{r})\}^2 dv_r = 4\pi \int_0^\infty s^2 i_v(s) ds \qquad 41.$$

The experimental values of these parameters can be determined when the intensity curves are known from $s=0$ to infinity (see above). If $v_1(\mathbf{r})$ takes only two values, 1 or 0, Equations 39–41 become:

$$p_\rho(0)/m = v_1 \overline{\{\Delta\rho(\mathbf{r},\bar{\rho}_1)\}^2}/m; \qquad 42.$$

$$p_{\rho v}(0)/m = 0; \qquad 43.$$

$$p_v(0)/m = v_1/m. \qquad 44.$$

When m and v_1 are known (see above), Equation 44 can be tested. The solvent density dependence of the density increments provides also a test of Equation 44 (see Equation 27 and Reference 1):

$$v_1/m = -\frac{1}{0.602\mu}\frac{\delta}{\delta d_o}\left[\frac{\delta d}{\delta c_v}\right] = \frac{4\pi}{m}\int_0^\infty s^2 i_v(s)\,ds. \qquad 45.$$

Therefore the hypothesis that the particle volume is totally impenetrable to the permeant component, which is particularly interesting in solution X-ray scattering experiments, can be tested using Equation 43, which involves only the scattering data, and Equations 44–45, which involve other measurements besides the scattering data (for example, molecular weight or specific density increments).

The invariant volume hypothesis does not require the content of the volume v_1 to be specified. If v_1 contains m electrons of the macromolecule, $\beta_w m$ electrons of water, $\beta_s m$ electrons of the third component, if ψ, ψ_w and ψ_s are the electron partial specific volumes, and if the β's and the ψ's are independent of ρ_o, one can write:

$$\bar{\rho}_1 = (1+\beta_w+\beta_s)/(\psi+\beta_w\psi_w+\beta_s\psi_s); \qquad 46.$$

$$v_1/m = \psi + \beta_w\psi_w + \beta_s\psi_s. \qquad 47.$$

Since $\bar{\rho}_1$ and v_1/m are experimental parameters (see Equations 25 and 26) β_w and β_s can be determined if the ψ's are known. If, moreover, the amount of the third component contained in the volume v_1 is negligible (this proposition can be tested if experiments in water are available), $\beta_s = 0$ and the hydration can be calculated from Equations 46 and 47 (see also Equation 18 and Reference 16):

$$\alpha = \beta_w = (\bar{\rho}_1 v_1/m) - 1. \qquad 48.$$

Although we deal in this Section with a three-component system, it would not be difficult to extend most of the results to a system containing a larger number of components.

Information Content of Solution Scattering Experiments

A satisfactory discussion of the information content of solution scattering experiments requires a careful analysis of the accuracy of the experimental data and of the distortions introduced by the mathematical operations (background subtraction, correction of collimation distortions, etc). Without attempting a rigorous treatment of these problems we may try to estimate the maximal and ideal number of independent structural parameters that one can hope to determine by the analysis of solution X-ray scattering experiments (see References 18 and 19). A simple answer is at hand for a particle whose largest dimension—namely the maximal distance between two of its points—is bounded by an upper limit, D_{max}. In this case the autocorrelation function vanishes beyond D_{max}; consequently (sampling theorem) the entire scattering curve is defined by the values of the intensity at the points of a lattice $s_h = h/2D_{max}$, and each of these intensities is equivalent to one independent structural parameter. In theory, intensity is scattered over an infinite range of s; in practice, the signal/noise ratio decreases steadily as s increases, and most often the only information retrieved beyond some value of s ($s > s_{max}$) are the terms A and B of the asymptotic form (see Equation 14). In this case the number J of independent parameters is equal to the number of lattice points in the range $0 < s < s_{max}$, plus the two parameters A and B:

$$J = 2D_{max}s_{max} + 2. \qquad 49.$$

An example can illustrate this equation. Consider a protein of spherical shape, molecular weight M, 30% hydration, with $\bar{v} = 0.74\ cm^3 g^{-1}$. The diameter of the sphere is $D_{max} = 1.78\ M^{1/3}$; in this case J is (see Equation 49):

$$J = 3.56\ M^{1.3}s_{max} + 2 \qquad 50.$$

In solution X-ray scattering experiments on globular proteins s_{max} is often in the vicinity of $(25\ \text{Å})^{-1}$. Thus is this case the information which can be retrieved is equivalent, at best, to the four morphological parameters—molecular weight, radius of gyration, volume, surface area—plus the parameter B related to the short range fluctuations, when $M \sim 10,000$ daltons. The information content increases (rather slowly, in fact) with molecular weight, more significantly with s_{max} and when the particles are anisometric (D_{max} increases in this case).

For a complete density contrast study, which leads to the determination of the three independent characteristic functions, the maximal number of parameters is $3J$.

Strategy for Structure Analysis

We have defined above a variety of theoretical frameworks to which one can refer the interpretation of solution X-ray scattering experiments. The choice of the proper framework is a matter of great practical importance.

Ideality can be approached by performing scattering experiments at variable concentration and extrapolating to infinite dilution. Generally, correlation effects are a problem only near $s=0$ and can be eliminated by linear extrapolations, as a function of c_e, of the various parameters: $i_n(0)$, R, etc.

The presence of density heterogeneities, which might have serious consequences on the X-ray scattering study (26), could be tested using the scattering experiments, although other techniques (for example density gradient sedimentation) provide easier tests.

Morphological monodispersity cannot be tested satisfactorily using X-ray scattering experiments, since in the presence of heterogeneities the equations established above remain formally correct, although some of the parameters are replaced by proper average values (see 26). Other techniques can be used for this purpose: chemical analysis, electron microscopy, sedimentation, etc.

The determination of molecular weight involves concentration and specific density increments, which in turn require a good chemical characterization and a densimetric study of the system. Le Maire, Tardieu & Vachette (16) have recently discussed the correlations of X-ray scattering, densimetry, and equilibrium sedimentation experiments with a variety of chemical and thermodynamic parameters: concentration scale, molecular weight, partial specific volume, interaction parameters. It is safe to conclude that a sound interpretation of solution X-ray scattering experiments requires some tests to be provided by other techniques.

The hypothesis of a uniform electron density distribution inside the particle entails the decay of the scattering intensity at large s, according to Equation 14. This asymptotic law should be tested before engaging a morphological analysis of the particle in terms of volume, surface area, ellipsoids, etc.

We have discussed above the consequences of the invariant volume hypothesis. Some of these provide criteria for testing the validity of the hypothesis. As far as the X-ray scattering experiments are concerned, the tests are the following:

1. The relationship $\{i_n(0,\rho_o)/c_e\}^{1/2}$ versus ρ_o must be linear (Equation 25);

2. at each value of s the scattered intensity must display a parabolic dependence on ρ_o (Equation 33);
3. the ρ_o dependence of the radius of gyration must follow Equation 35.

Other experiments can provide additional tests. For example:

4. The specific density increments (measured by densimetry and/or equilibrium sedimentation) must be a linear function of the density of the solvent (Equation 27);
5. if the tests 1 to 4 above are fulfilled, Equation 45 provides an additional and very sensitive verification and shows, moreover, whether the volume associated with one particle is totally, or partly, excluded to the high density permeant component.

The value of the maximal dimension D_{max} is defined by the point r beyond which the autocorrelation function vanishes; the value of s_{max} depends on the experimental conditions. Once D_{max} and s_{max} are known, it is possible to estimate the maximal number of independent structural parameters that can ideally be retrieved from the data; this figure provides useful guidance against over- and under-interpretation.

AN EXAMPLE: A LOW DENSITY SERUM LIPOPROTEIN

Serum lipoproteins are lipid-rich particles, present in vertebrate blood, which play an important role in lipid transport and energy metabolism; these particles are also implicated in some pathological disorders, for example atherosclerosis (7, 12, 24). Serum lipoproteins can be fractionated into a variety of particles, some of which are well characterized according to weight and chemical composition. One of these classes, called low density serum lipoprotein (abbreviated LDL) contains particles which are fairly large—molecular weight in the range 2–3 10^6—and contain approximately 20% protein and 80% lipid. LDL are ideal objects for X-ray scattering studies at variable contrast, given the wide range of the electron density fluctuations inside the particles (0.28 e $Å^{-3}$ in the hydrocarbon regions to 0.41 e $Å^{-3}$ in the protein regions). The ability of lipoproteins to withstand high NaBr concentrations permits variation of the solvent electron density from 0.334 e $Å^{-3}$ (0.1 M NaBr) to 0.446 e $Å^{-3}$ (6 M NaBr). It is also worth mentioning that LDL particles undergo a reversible temperature-induced transition that is known to involve the organization of the lipids (3, 4).

LDL has been the subject of several solution X-ray scattering studies over the last few years (2, 15, 22, 32, see also 30 for a neutron scattering study). The most recent of these (19) provides an excellent illustration of the latest developments of solution scattering studies. The samples used were extracted from hyperlipidemic rhesus monkeys (5) and were studied both below and above the transition temperature; only the low temperature (21°C) results are discussed here. The chemical composition is 19% protein, 24% phospholipid, 9% free cholesterol, 47% cholesterylesters, and 1% triglycerides. Previous hydrodynamic and electron microscope studies had shown that the preparation is homogeneous and that the particles are quasi-spherical in shape. The X-ray scattering experiments were performed using a rotating anode generator and a position sensitive linear detector constructed by A. Gabriel. The scattering experiments were carried out over an extended range of NaBr concentrations; the curves were recorded from $s \sim (900$ Å$)^{-1}$ to $\sim (20$ Å$)^{-1}$.

Specific density increments of LDL solutions were also measured at dialysis equilibrium in the presence of variable amounts of NaBr (1). The experimental points, plotted in Figure 1, are in excellent agreement with the straight line predicted by the invariant volume hypothesis (see Equation 27).

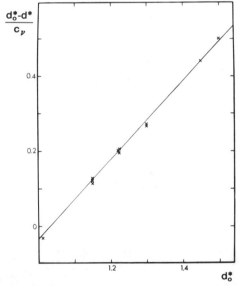

Figure 1 Results of densimetry experiments (see 1). d^* and d_o^* are the densities of "solution" and "solvent" at dialysis equilibrium. $(d_o^* - d^*)/c_v = \delta d/\delta c_v$. According to Equation 27 the experimental points fit to a straight line, whose slope and intercept define the volume of the hydrated particle and its water content: $v_1/m = 3.18$ Å^3 e^{-1}, $a = 0.0897$ g_{H_2O}/g_{LDL}.

Each experimental X-ray scattering curve, corrected for collimation distortions, yields a normalized function $i_n(s,\rho_o)/c_e$ when absolute scale and LDL concentration are taken into account (see Equations 1 and 7). The extrapolation to $s=0$ (see Equations 8–10) defines the values of $\{i_n(0,\rho_o)/c_e\}$ and of $R^2(\rho_o)$. Figure 2 presents the values of $[i_n(0,\rho_o)/c_e]^{1/2}$ plotted as a function of ρ_o; the points fit to a straight line (see Equation 25) that defines the values of $\bar{\rho}_1$ and $v_1 m^{-1/2}$. Similarly the plot $(\rho_o-\bar{\rho}_1)R^2(\rho_o)$ versus $(\rho_o-\bar{\rho}_1)$ yields a straight line (see Figure 3), in agreement with the invariant volume hypothesis and with the notion that the position of the center of mass of the particle is independent of the electron density of the solvent (see Equation 35); the straight line defines the values of R_v and of $m_{2\rho}$. Figure 4 shows the position of the experimental points $i_n(s,\rho_o)/c_e$ with respect to the functions defined by Equation 33; the excellent fit provides a justification for determining the three characteristic functions according to Equation 33. These results lead to the first conclusion:

1. all the tests illustrated in Figures 1–4 are consistent with the validity of the invariant volume hypothesis.

Moreover, in keeping with Equation 45, the values of $-(0.602\mu)^{-1} \cdot \delta(\delta d/\delta c_v)/\delta d_o$ ($=3.18$, see Figure 1) and of $m^{-1}\int_0^\infty s^2 i_v(s)\,ds$ ($=3.41$) are almost equal (see 1). Thus we can conclude:

2. the particle volume appears to be totally impenetrable to NaBr (in other words the function $v_1(\mathbf{r})$ is equal to 1 or to 0).

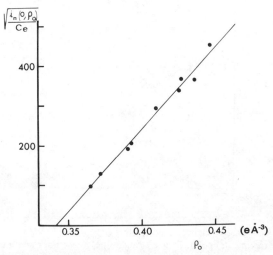

Figure 2 Plot $\{i_n(0,\rho_o)/c_e\}^{1/2}$ versus ρ_o (see Equation 25). The experimental points fit to a straight line, whose slope and intercept define the values of $v_1 m^{-1/2}$ and of $\bar{\rho}_1$.

SOLUTION X-RAY SCATTERING 19

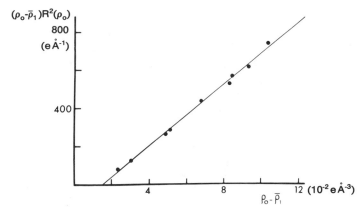

Figure 3 Plot $(\rho_o - \bar{\rho}_1) R^2(\rho_o)$ versus $(\rho_o - \bar{\rho}_1)$; the value of $\bar{\rho}_1$ is defined in Figure 2. According to Equation 35 the points fit to a straight line, whose slope and intercept define the values of R_v and of $m_{2\rho}/v_1$.

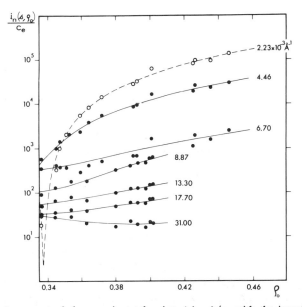

Figure 4 Agreement of the experimental points $i_n(s, \rho_o)/c_e$ with the invariant volume hypothesis. At a few values of s (the values are labeled in the figure), the experimental points are plotted as a function of ρ_o. The curves are the parabolae obtained by least squares analysis of Equation 33.

Table 1 Experimental values of some morphological parameters

Parameter	Reference	Value
$\bar{\rho}_1$	Equation 25; Figure 2	0.3417 e Å^{-3}
$v_1 m^{-1/2}$	Equation 25; Figure 2	4.177×10^3 Å^3 e$^{-1/2}$
R_v	Equations 35, 36; Figure 3	89.8 Å
$m_{2\rho}/v_1$	Equations 35, 37; Figure 3	118.6 e Å^{-1}
v_1/m	Equation 27; Figure 1	3.18 Å^3 e^{-1}
m	Equation 28	1.726×10^6 e
v_1	Equations 27, 28	5.489×10^6 Å^3
α	Figure 1	6.47×10^{-2} e$_{H_2O}$/e$_{LDL}$

The results presented in Figures 1–3 lead to the determination of the morphological parameters reported in Table 1: This is traditionally the first step, and most often the only reliable one, in the analysis of solution scattering experiments at variable solvent contrast. The data of Table 1 lead to the following conclusions:

3. the positive sign of $m_{2\rho}$ (see Equation 37) confirms one well-known feature of LDL, namely the preferential location of the high density region in the periphery of the particle;
4. the ratio R_v^3/v_1 shows that the particles are fairly isometric but that their shape departs markedly from that of a sphere [$R_v^3/v_1 = 0.132$ (see Table 1), corresponds to ellipsoids with axial ratios 1:1:0.576 and 1:1:1.636; for a sphere $R_v^3/v_1 = 0.110$].

The next step involves an analysis of the autocorrelation function. This function—more precisely the values of $r^2 p_1(r,\rho_o)/m$ in the space $\{r,\rho_o\}$—calculated using the three characteristic functions, is plotted in Figure 5. The observation of extended regions with a negative value of $p_1(r,\rho_o)$ reflects the presence of conspicuous electron density fluctuations inside the particle. More precisely, if at any solvent density ρ_o the function $p_1(r,\rho_o)$ displays negative values, then somewhere in the particle the electron density is bound to be higher, somewhere else lower, than ρ_o. Thus the electron density of the particle must take values equal to, or larger than, the upper ρ_o limit of the negative regions of $p_1(r,\rho_o)$, as well as values equal to, or smaller than, the lower ρ_o limit of the negative regions. These limits (see Figure 5) are $\bar{\rho}_1 + 0.062$ and $\bar{\rho}_1 - 0.012$ e Å^{-3}. Moreover the tails of the autocorrelation function—namely the region at $r > 200$ Å in Figure 5—can provide some information about the structure of the outer layer of the particle. It is convenient to analyze this information with references to Figures 5 and 6.

First, note that the three curves of Figure 6 drop to zero quite sharply and almost at the same point. This leads to the conclusion:

5. the length of the maximal chord D_{max} of the particle is 295 ± 5 Å.

Such a large value of D_{max} is not compatible with a spherical shape. Indeed, if the particle were a sphere, D_{max} would be its diameter; now the volume and radius of gyration of a sphere of diameter 295 Å are 1.344×10^7 Å^3 and 114.2 Å, much larger than the observed values ($v_1 = 5.489 \times 10^6$ Å^3 and $R_v = 89.8$ Å, see Table 1). This leads to the next conclusion:

6. the particle cannot be a smooth sphere, even to a first approximation: the volume of the spherical envelope is at least 2.4 times the volume v_1 of the particle.

It is generally agreed that in LDL the high density components are located preferentially in the periphery of the particle. It is possible to make and test a more specific assumption, namely that the outer region of the particle is occupied by a medium of high and uniform electron density $\rho_b = \bar{\rho}_1 + \Delta\rho_b$. If this is the case, the outer region is matched in a solvent of density ρ_b; the autocorrelation function $p_1(r, \rho_b)$ is thus bound

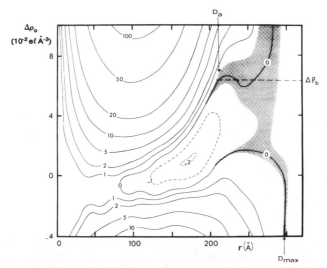

Figure 5 Plot of the function $r^2 p_1(r, \Delta\rho_o)/m$ as a function of $\Delta\rho_o$. $p_1(r, \Delta\rho_o)$ is calculated using the three characteristic functions (see Equation 29). Each contour line corresponds to one value of the function; the units are e Å^{-1}. The region of uncertainty around the 0 contour line is hatched (see Reference 19 for a discussion of accuracy). The uncertainty regions around the other contour lines are very narrow. See Figure 6 for the definition of $\Delta\rho_b$, D_{max}, and D_a.

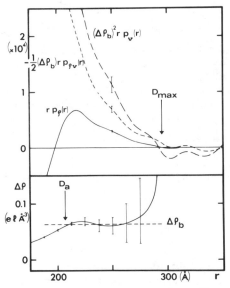

Figure 6 Tails of the autocorrelation functions. *Lower frame.* The root of Equation 51, which defines the electron density level of the high density outer region. The vertical bars are the errors (see Reference 19). The value adopted is shown by the dotted line $\Delta\rho_b$. *Upper frame.* The three autocorrelation functions; in this representation the errors, measured by the vertical bars, are independent of r. Note that the maximal chord of the particle D_{max} is defined with good accuracy (± 5 Å), as well as the point D_a where $\Delta\rho$ departs from $\Delta\rho_b$. Note also that from D_{max} inwards $(\Delta\rho_b)^2 \overline{p}_v(r) > -(1/2)(\Delta\rho_b)p_{\rho v}(r) > p_\rho(r)$. $\Delta\rho_b = 0.0624$ e Å^{-3}; $D_{max} = 295$ Å; $D_a = 208$ Å.

to vanish at all values of r larger than the longest vector D_a joining two points of electron density different from ρ_b (see Figure 7). This situation is equivalent in Figure 5 to the zero contour line becoming horizontal; this is indeed consistent with the figure, account being taken of the experimental errors. A more accurate procedure for testing this assumption and for determining the values of $\Delta\rho_o$ is to solve the equation (see Equation 29):

$$p_1(r,\Delta\rho) = p_\rho(r) + (\Delta\rho)p_{\rho v}(r) + (\Delta\rho)^2 p_v(r) = 0 \qquad 51.$$

for $\Delta\rho$, at different values of r. The highest root is plotted in Figure 6; $\Delta\rho_b$ is reasonably constant and equal to approximately 0.0624 e Å^{-3} over an extended range of r. It must be noted that $\bar{\rho}_1 + 0.0624 = 0.4041$ e Å^{-3} is the electron density of a hydrated protein. These results lead to the following conclusion:

7. the outer region of the particle is occupied by a medium of high and uniform electron density, most likely a hydrated protein.

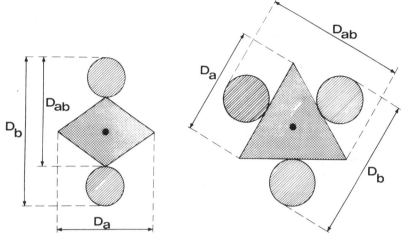

Figure 7 Effect of a center of symmetry on the tails of the autocorrelation functions. Two-dimensional representation of particles consisting of an outer region of high and uniform electron density, ρ_b (*hatched*) and of a core of density $\rho_a(\mathbf{r})$ (*cross-hatched*). $\rho_a(\mathbf{r})$ is not bound to be uniform; $\rho_b > \rho_a(\mathbf{r})$. In agreement with the properties of LDL the position of the center of gravity is independent of the density of the solvent. D_a, D_b, D_{ab} are the longest vectors joining, respectively, two points of the core, two points of the outer region, one point of the core, and one of the outer region. R_a and R_b are the maximal distances to the center of mass of the particle of any point in the two regions. D_{max} is the maximal chord of the particle. In general the following conditions are fulfilled: $D_a \leqslant D_{max}$; $D_b \leqslant D_{max}$; $D_{ab} \leqslant D_{max}$; $D_{ab} \leqslant R_a + R_b$. (*left*) Centrosymmetric case: $R_a = D_a/2$; $R_b = D_b/2$; $D_{ab} \leqslant (D_a + D_b)/2$. (*right*) Non centrosymmetric case: $R_a > D_a/2$; $R_b > D_b/2$. D_a and D_b set no restriction on D_{ab}.

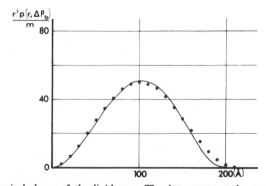

Figure 8 Spherical shape of the lipid core. The dots represent the experimental points $r^2 p_1(r, \rho_b)/m$, namely for a solvent whose density matches the outer protein region. The full line is the function calculated for a uniform sphere of radius 97.9 Å and of electron density $(\Delta\rho_b - \Delta\rho_a) = 8.62 \times 10^{-2}$ e Å^{-3}.

If its electron density is uniform, the outer layer can be "erased" by setting $\rho_o = \rho_b$; the function $r^2 p_1(r, \rho_b)/m$ is plotted in Figure 8. The edge of this function—which corresponds in Figure 6 to the inner point at which $\Delta\rho_b$ departs from a constant—defines the maximal chord D_a of the core of the particle ($D_a = 208 \pm 5$ Å). Besides, in spite of the electron density fluctuations present in the lipids, the curve $p_1(r, \rho_b)$ turns out to fit quite closely the autocorrelation function of a uniform sphere whose electron density is close to the average electron density of the lipid moiety of LDL (see Figure 8). Therefore we can conclude:

8. the core of the particle is occupied by a medium of low average density; its maximal chord is 208 ± 5 Å;
9. the core is approximately spherical in shape and it appears to consist mainly of lipids.

The presence of an almost spherical core surrounded by a markedly nonspherical protein shell indicates that:

10. the shape of the particle is grossly spherical, the external surface is deeply corrugated, and the outer layer is sparsely occupied by a hydrated protein.

The tails of the autocorrelation functions can also provide some information on the symmetry of the particle. In a qualitative way this is illustrated (see Figure 5) by a remarkable property of the autocorrelation functions, namely that over a wide range of $\Delta\rho_0$ (0.02 to 0.06 e Å^{-3}), $p_1(r, \rho_o)$ takes negative values in the region bordering upon $D_{\max}$; this observation shows that the longest vectors of the particle join regions with different electron density and thus that the particle is not centrosymmetric. It is possible to discuss this point in a more rigorous way with the help of Figure 6 and of a two-dimensional example sketched in Figure 7. If D_{ab} (see Figure 7) is the longest vector joining one point of the outer region to one point of the core, all the vectors in the range $D_{\max} > r > D_{ab}$ join two points of density ρ_b, those in the range $D_{ab} > r > D_a$ join one point of density ρ_b and one of density either ρ_b or ρ_a. For all points in the interval $D_{ab} < r < D_{\max}$, one can substitute $\Delta\rho_1(\mathbf{r}) = \Delta\rho_b v_1(\mathbf{r})$. For the points contributing in the interval $D_a < r < D_{ab}$, it is on the average true that $\Delta\rho_1(\mathbf{r}) < \Delta\rho_b v_1(\mathbf{r})$, since on the average $\rho_a < \rho_b$. Therefore, taking into account the definitions of $p_\rho(r)$, $p_{\rho v}(r)$ and $p_v(r)$ (see Equations 30–32) we obtain:

$$p_\rho(r) = -1/2(\Delta\rho_b) p_{\rho v}(r) = (\Delta\rho_b)^2 p_v(r) \qquad 52.$$

if $D_{ab} < r < D_{\max}$, and

$$p_\rho(r) \leqslant -1/2(\Delta\rho_b) p_{\rho v}(r) \leqslant (\Delta\rho_b)^2 p_v(r) \qquad 53.$$

if $D_a < r < D_{ab}$. It is clear (see Figure 7) that the presence of a center of symmetry sets a lower limit to D_{ab}:

$$D_{ab} \leqslant (D_a + D_b)/2. \qquad \qquad 54.$$

No such restriction exists in the absence of a center of symmetry. Therefore the presence of a center of symmetry entails that, in the range $(D_a + D_{max})/2 < r < D_{max}$, Equation 52 is fulfilled. Figure 6 shows that the three functions are far from coincident in the expected range of r; instead it is rewarding to find that the inequalities of Equation 53 are fulfilled, in agreement with the notion that the vectors joining the outer region to the core come into play at a distance quite close to D_{max}. It is thus possible to draw the conclusion:

11. the structure of the particle is not centrosymmetric. The symmetry in question here is a coarse one, at a resolution as low as 100 Å, without reference to molecular properties.

In summary, the particle appears to contain an almost spherical, predominantly lipid core and an outer protein region, sparsely occupied and devoid of a coarse center of symmetry. It is hard to visualize how these conditions can be met unless the protein of the outer region is condensed in a small number of lumps, widely separated from each other. Thus the final conclusion drawn from the analysis of the autocorrelation functions is:

12. the protein is distributed very unevenly in the outer region, most likely in a small number of globules.

This set of conclusions is as far as a straightforward analysis of the solution scattering data can lead; it is hardly possible to pursue the analysis beyond this point without making specific assumptions on the structure of the particles. In the case of LDL the information content of the data (see sections on Theory) can be estimated to be equivalent to approximately 78 independent structural parameters ($3J = 78$ for $D_{max} = 295$ Å and $s_{max} = (25$ Å$)^{-1}$, see Equation 49). This figure, surprisingly large for a solution scattering study, stimulated a more elaborate analysis of the structure of LDL, which is described in Reference 19. Without entering into the details of this last step of the analysis, which would take us beyond the scope of this article, it is worthwhile to point out that the structural assumptions were focused on symmetry; on the basis of a variety of arguments it was shown that LDL particles are likely to display symmetry and that the most probable symmetry is tetrahedral (point group 23). Given this symmetry and taking into account the previous conclusions, it was possible to construct a model consisting of a spherical lipid core surrounded by 4 protein globules located in tetrahedral positions, which was shown to be in excellent agreement with the physical and the chemical properties of LDL. A freeze etching electron microscope study (10) provided essential information for the construction of the model. From the viewpoint of the information content, the model proposed is defined by 8 independent parameters; 3 additional parameters were introduced to take into account the internal organization of the lipid core.

CONCLUSIONS

We wish to stress once more that the main novelties in solution scattering studies are the position sensitive detectors and the high flux sources: the revived interest for some theoretical aspects of this technique and the concern for more sophisticated analyses of the data are the consequences of the experimental improvements. It is worthwhile noting that since the above reported experiments on LDL were carried out (summer 1976)—a rotating anode generator and a linear detector were used then, with a gain of approximately 250 over a conventional instrument—storage ring radiation has become routinely available at Laboratoire pour l'Utilisation du Rayonnement Electromagnétique (LURE) (Orsay, France), with an additional gain of 100 to 500.

We have also emphasized the variable contrast approach—namely the systematic variation of the solvent electron density—especially if the invariant volume hypothesis is fulfilled. We have discussed extensively the theoretical aspects of that approach and how the validity of the invariant volume hypothesis can be tested.

LDL has provided the most convenient illustration of the wealth and accuracy of the data made available by the new techniques and of the unusual degree of sophistication involved in the analysis of the data. Lack of space has precluded a discussion of other examples of recent solution X-ray scattering studies performed since the introduction of position sensitive detectors; we can mention bovine rhodopsin (26), ribosomal proteins (9), chromatin particles (27), and allosteric transition of aspartate transcarbamylase (23). Several solution X-ray scattering studies have been performed lately at LURE (sarcoplasmic reticulum Ca^{2+} ATPase, *Escherichia coli* ribosomal subunits, chromatin core particles, allosteric transition of phosphofructokinase, ribosomal proteins); the interpretation of the experiments is still underway (J. Dijk, A. M. Freund, A. Gulik, M. Le Maire, M. Lemonnier, F. Rousseaux, F. Seydoux, L. Sperling, A. Tardieu, P. Vachette, personal communications).

The example of LDL illustrates the various steps of the Strategy for Structure Analysis discussed above. The chemical characterization of the sample and especially the determination of homogeneity should always precede the X-ray scattering study; indeed the assumption of monodispersity, involved in the interpretation of the data, cannot be tested by the scattering experiments (19, 26). Concentration must also be known on an absolute scale. As mentioned above, it is possible to bring the solutions close to thermodynamic ideality by extrapolating to vanishing concentration; usually this extrapolation does not set serious problems, at least in the absence of strong electrostatic interactions. A variety of physico-chemical techniques—densimetry, sedimentation, light scattering, etc—provide useful information with regard to homogeneity and also allow one to determine the preferential association of the different diffusable components (water, salt, sucrose, etc) with the macromolecular component and thus to test the invariant volume hypothesis (1, 16). All the tests of this hypothesis, provided by the X-ray scattering as well as by the densimetry experiments, are fulfilled in the case of LDL. Once the invariant volume hypothesis is adopted, an analysis of the solvent density dependence of the scattering functions leads to the determination of the characteristic functions; besides, the properties of the scattering functions at and near to the origin of s specify the values of the morphological parameters—molecular weight, radius of gyration, volume, etc. The three characteristic functions contain all the information of a complete X-ray scattering study at variable contrast. The amount and accuracy of the structural information that it is possible to retrieve from the three functions depends upon the properties of the system under investigation. In the case of LDL the analysis is facilitated by the conspicuous electron density fluctuations due to the presence of protein and hydrocarbon regions, and by the clear-cut separation of the low density core from the high density outer region. The electron microscope observations, as well as the hypotheses relevant to the symmetry of the particle, also play an essential role.

One may wonder to what extent the approach adopted for LDL is of more general use. The first of the two conditions mentioned above—wealth and accuracy of the data and validity of the invariant volume hypothesis—is improving dramatically with the use of position sensitive detectors and of synchrotron radiation. The problem of varying the electron density of the solvent without altering the structure of the solute has been solved quite easily in number of systems. Yet, it must be noted that even if the invariant volume hypothesis is not strictly fulfilled, a wealth of information could still be extracted from the scattering data. With regard to scattering density contrast, neutrons are more favorable than X-rays since the D_2O/H_2O ratio is much easier to alter than the electron density of the solvent; nevertheless, it is questionable whether the wealth and accuracy of the neutron scattering data presently available are up to the standard of the best X-ray scattering studies.

The problems mentioned above illustrate the most promising areas for solution scattering studies. Several of the examples—membranes proteins, ribosomal subunits, serum lipoproteins, chromatin particles —are objects of obvious biological interest most of which are not (yet?) amenable to crystallization. The others—allosteric transitions of aspartate transcarbamylase and of phosphofructokinase—deal with structural transitions that would be difficult to study in the crystalline state (22a). Besides, it is worthwhile mentioning that the new experimental techniques are opening the way to the kinetic study of structural transitions.

ACKNOWLEDGMENTS

Many of the recent developments described in this article are the result of collaborations with Lon Aggerbeck, Leonardo Mateu, and Patrice Vachette. We are also grateful to Patrice Vachette for stimulating discussions, and to him and Linda Sperling for a critical reading of the manuscript. We are greatly indebted to Angelo Scanu and his colleagues for their long-lasting collaboration, to Heinrich Stuhrmann for discussions and encouragements, and to Marc Le Maire, Christian Sardet, and Linda Sperling for their participation in the solution X-ray scattering studies. We owe to André Gabriel the position sensitive detectors, and we thank Jean-Claude Dedieu, Bernard Jullemier, and Claude Pontillon for their skillful technical assistance.

The work described in this article was supported mainly by the Centre National de la Recherche Scientifique and partly by grants and fellowships from the Délégation Générale à la Recherche Scientifique et Technique, the National Science Foundation, the European Molecular Biology Organization, the Philippe Foundation, and the Simone and Cino del Duca Foundation.

Part of the text and all the figures relevant to the low-density serum lipoprotein are reproduced from an article published in the *Journal of Molecular Biology* (19).

Literature Cited

1. Aggerbeck, L., Yates, M., Tardieu, A. Luzzati, V. 1978. *J. Appl. Crystallogr.* 11:466–72
2. Atkinson, D., Deckelbaum, R. J., Small, D. M., Shipley, G. G. 1977. *Proc. Natl. Acad. Sci. USA.* 74:1042–46
3. Deckelbaum, R. J., Shipley, G. G., Small, D. M. 1977. *J. Biol. Chem.* 252:744–54
4. Deckelbaum, R. J., Shipley, G. G., Small, D. M., Less, R. S., George, P. K. 1975. *Science* 190:392–94
5. Fless, G. M., Wissler, R. W., Scanu, A. M. 1976. *Biochemistry* 15:5799–805
6. Gabriel, A., Dupont, Y. 1972. *Rev. Sci. Instrum.* 43:1600–3
7. Goldstein, J. L., Brown, M. S. 1977. *Ann. Rev. Biochem.* 46:897–930

8. Guinier, A. 1939. *Ann. Phys. Paris* 12:161–237
9. Gulik, A., Freund, A. M., Vachette, P. 1978. *J. Mol. Biol.* 119:391–97
10. Gulik-Krzywicki, T., Yates, M., Aggerbeck, L. P. 1979. *J. Mol. Biol.* 131:475–84
11. Ibel, K. 1976. *J. Appl. Crystallogr.* 9:296–309
12. Jackson, R. L., Morrisett, J. D., Gotto, A. M. 1976. *Physiol. Rev.* 56:259–316
13. Kratky, O., Pilz, I. 1972. *Q. Rev. Biophys.* 5:481–537
14. Kratky, O., Pilz, I. 1978. *Q. Rev. Biophys.* 11:39–70
15. Laggner, P., Müller, K., Kratky, O., Kostner, G., Holasek, A. 1976. *J. Colloid Interface Sci.* 55:102–8
16. Le Maire, M., Tardieu, A., Vachette, P. In preparation
17. Luzzati, V. 1960. *Acta Crystallogr.* 13:939–45
18. Luzzati, V. 1979. Imaging processes and coherence in physics. In *Lecture Notes in Physics*. Heidelberg:Springer In press
19. Luzzati, V., Tardieu, A., Aggerbeck, L. 1979. *J. Mol. Biol.* 131:435–73
20. Luzzati, V., Tardieu, A., Mateu, L., Stuhrmann, H. B. 1976. *J. Mol. Biol.* 101:115–27
21. Luzzati, V., Witz, J., Nicolaieff, A. 1961. *J. Mol. Biol.* 3:367–78
22. Mateu, L., Tardieu, A., Luzzati, V., Aggerbeck, L., Scanu, A. M. 1972. *J. Mol. Biol.* 70:105–16
22a. Monaco, H. L., Crawford, J. L., Lipscomb, W. N. 1978. *Proc. Natl. Acad. Sci. USA* 75:5276–80
23. Moody, M. F., Vachette, P., Foote, A. M. 1979. *J. Mol. Biol.* 133:517–32
24. Morrisett, J. D., Jackson, R. L., Gotto, A. M. Jr. 1975. *Ann. Rev. Biochem.* 44:183–207
25. Porod, G. 1951. *Kolloid. Z.* 124:83–114
26. Sardet, C., Tardieu, A., Luzzati, V. 1976. *J. Mol. Biol.* 105:383–407
27. Sperling, L., Tardieu, A. 1976. *FEBS Lett.* 64:89–91
28. Stuhrmann, H. B. 1978. *Q. Rev. Biophys.* 11:71–98
29. Stuhrmann, H. B., Kirste, R. G. 1965. *Z. Phys. Chem. Frankfurt* 46:247–50
30. Stuhrmann, H. B., Tardieu, A., Mateu, L., Sardet, C., Luzzati, V., Aggerbeck, L. P., Scanu, A. M. 1975. *Proc. Natl. Acad. Sci. USA.* 72:2270–73
31. Tardieu, A. 1979. *Eur. J. Biochem.* 96:621–24
32. Tardieu, A., Mateu, L., Sardet, C., Weiss, B., Luzzati, V., Aggerbeck, L., Scanu, A. M. 1976. *J. Mol. Biol.* 101:129–53

PHOTOACOUSTIC SPECTROSCOPY

♦9140

Allan Rosencwaig

University of California, Lawrence Livermore Laboratory, Livermore, California 94550

INTRODUCTION

In its broadest sense, spectroscopy can be defined as the study of the interaction of energy with matter. As such it is a science encompassing many disciplines and many techniques. The energy used in the oldest form of spectroscopy, optical spectroscopy, is in the form of photons, with wavelengths ranging from less than one Angstrom in the X-ray region, to more than a hundred microns (10^6 Angstroms) in the far-infrared. Because of its versatility, range, and nondestructive nature, optical spectroscopy remains a widely used and most important tool for investigating and characterizing the properties of matter.

Conventional optical spectroscopic techniques tend to fall into two major categories. In the first category, one studies the optical photons that are transmitted through the material of interest, that is, those photons that have not interacted with the material. In the second category, one studies the light that is scattered or reflected from the material, that is, those photons that have undergone some interaction with the material. Almost all conventional optical methods are variations on these two basic techniques. As such, they are distinguished not only by the fact that optical photons constitute the incident energy beam but also by the fact that the data is obtained by detecting some of these photons after the beam has interacted with the matter or material under investigation. It should be noted that these optical techniques preclude the detection and analysis of those photons that have undergone an absorption, or annihilation, interaction with the material—the process that is often of most interest to the investigator.

Optical spectroscopy has been a major scientific tool for over a century, and it has proven invaluable in studies on reasonably clear media such as solutions and crystals, and on clean specularly reflective surfaces. There are, however, several instances when conventional spectroscopy is inadequate, even for clear, transparent materials. Such a situation arises when one is attempting to measure a very weak absorption, which in turn involves the measurement of a very small change in the intensity of the strong, essentially unattenuated, transmitted signal. In addition to weakly absorbing materials, there are a great many nongaseous substances, both organic and inorganic, that are not readily amenable to the conventional transmission or reflection forms of optical spectroscopy. These materials are usually highly light-scattering, such as powders, amorphous solids, tissues, gels, smears, and suspensions. Other difficult materials are those that are optically opaque, with dimensions that far exceed the penetration depth of the photons. In the former case, the optical signal is composed of a complex combination of specularly reflected, diffusely reflected, and transmitted photons, which makes the analysis of the data extremely difficult. In the latter case, the absorptive properties of the material are difficult, if not impossible, to determine, since essentially no photons are transmitted. Over the years, several techniques have been developed to permit optical investigation of highly light-scattering and opaque substances. The most common of these are diffuse reflectance, attenuated total reflection or internal reflection spectroscopy, and Raman scattering. All of these techniques have proven to be very useful. Yet, each suffers from serious limitations. In particular, each method is useful for only a relatively small category of materials, each has a limited wavelength range, and the data obtained is often difficult to interpret.

During the past few years, another optical technique has been developed to study those materials that are unsuitable for the conventional transmission or reflection methodologies (7). The technique, called photoacoustic spectroscopy (PAS), is distinguished from the conventional techniques chiefly by the fact that, even though the incident energy is in the form of optical photons, the interaction of these photons with the sample under investigation is studied not through subsequent detection and analysis of some of these photons but rather through a direct measure of the energy absorbed by the material because of its interaction with the photon beam.

Although more is said about experimental methodology later in this chapter, a brief description here might be appropriate. In photoacoustic spectroscopy, or PAS, the sample to be studied is placed in a closed cell or chamber. In the case of gases and liquids the sample generally fills the entire chamber. For solids, the sample fills only a portion of the chamber, with the rest of the chamber filled with a nonabsorbing gas such as air. In addition to the sample, the chamber also contains a sensitive microphone. The sample is illuminated with monochromatic light that either passes through an electromechanical chopper or is intensity-modulated in some other fashion. If any of the incident photons are absorbed by the sample, internal energy levels within the sample are excited. Upon subsequent deexcitation of these energy levels, all or part of the absorbed photon energy is then transformed into heat energy through nonradiative deexcitation processes. In a gas this heat energy appears as kinetic energy of the gas molecules, whereas in a solid or liquid it appears as vibrational energy of ions or atoms. Since the incident radiation is intensity-modulated, the internal heating of the sample is periodic.

The periodic heating of a solid or liquid sample from the absorption of the optical radiation results in a periodic heat flow from the sample to the nonabsorbing gas. This in turn produces pressure and volume changes in the gas, which are detected by the condensor microphone. This method is quite sensitive, especially for samples with large surface: volume ratios such as powders, and it is capable of detecting temperature rises of 10^{-6} to 10^{-5}C in the sample, or thermal power generation of about 10^{-6} calories/cm$^3 \cdot$ sec.

There are several advantages to photoacoustic spectroscopy. Since *absorption* of optical or electromagnetic radiation is required before a photoacoustic signal can be generated, light that is transmitted or elastically scattered by the sample is not detected and hence does not interfere with the inherently absorptive PAS measurements. This is of crucial importance when working with essentially transparent media such as pollutant-containing gases that have few absorbing centers. This insensitivity to scattered radiation also permits the investigator to obtain optical absorption data on highly light-scattering materials such as powders, amorphous solids, gels, colloids, etc. Another advantage is the capability of obtaining optical absorption spectra on materials that are completely opaque to transmitted light. Coupled with this, is the capability, unique to photoacoustic spectroscopy, of performing nondestructive depth-profile analysis of absorption as a function of depth into a material.

Furthermore, since the sample itself constitutes the electromagnetic radiation detector, no photoelectric device is necessary; thus studies over a wide range of optical and electromagnetic wavelengths are possible without the need to change detector systems. The only limitations are that the source be sufficiently energetic (at least 10 μwatts/cm^2) and that the window of the photoacoustic cell be reasonably transparent to the radiation. Finally, the photoacoustic effect results from a radiationless energy conversion process and is therefore complementary to radiative and photochemical processes. Thus PAS may itself be used as a sensitive, though indirect, method for studying the phenomena of fluorescence and photochemistry in matter.

PHOTOACOUSTIC THEORY

The mathematical analysis of a photoacoustic signal is usually fairly laborious and complex. For a full description of photoacoustic theory we refer the reader to the review by Rosencwaig (7) and references therein. Here, we present a simplified photoacoustic theory that can be derived from basic physical insights.

When a sample absorbs intensity-modulated optical radiation, the sample will undergo periodic heating. This periodic heating results in a periodic heat flow from the sample to the gas at the sample-gas boundary. A thin layer of gas near the boundary is then cyclically heated by this heat flow. The thickness of the boundary layer is determined by the thermal diffusion length in the gas, μ', which is given by

$$\mu' = \left(\frac{2\alpha}{\omega}\right)^{1/2} \qquad 1.$$

where ω is the radial frequency at which the light is intensity-modulated and α is the thermal diffusivity defined as

$$\alpha = \kappa/\rho C \qquad 2.$$

where κ is the thermal conductivity, ρ the density and C the specific heat. Generally, the thermal diffusion length for most gases runs in the range of 25–500 μm for the frequencies usually used in photoacoustic spectroscopy (1000 Hz–5 Hz). The thermal diffusion length represents the distance where the temperature rise due to conduction is e^{-1} that at the origin of the heating, which for our case is the sample surface.

The localized heating of a layer of fluid or gas can be viewed as producing a localized stress that is then rapidly transmitted through the rest of the enclosed gas in the photoacoustic cell. The local pressure or stress that is generated within a thermal diffusion length of the sample surface can be approximated by the expression

$$p_\mu \simeq B'\alpha'_t(1/2\Theta_0) \qquad 3.$$

where B' is the bulk modulus and α'_t the volume expansion coefficient of the gas, and where $1/2\Theta_0$ is approximated as the average temperature within this thermal diffusion length, if Θ_0 is taken as the temperature at the sample-fluid interface. The pressure at the microphone, a distance l'

away, is given by

$$p = p_\mu\left(\frac{\mu'}{l'}\right) = 1/2 B' \alpha_t' \Theta_0 \left(\frac{\mu'}{l'}\right) = \frac{\gamma P_0}{2 T_0 a' l'} \Theta_0, \qquad 4.$$

since for a gas, $B' = \gamma P_0$ where γ is the ratio of specific heats and P_0 is the pressure, and $\alpha_t' = 1/T_0$ where T_0 is the temperature and $a' = 1/\mu'$.

When the gas or fluid is completely constrained at its borders, then the pressure p is the same everywhere in the cell, as long as the cell dimensions are much smaller than the acoustic wavelength.

The temperature Θ_0 at the sample-gas interface can be approximated by

$$\Theta_0 \simeq H_{abs}/M_{th} \qquad 5.$$

where H_{abs} is the amount of heat absorbed per unit time within the first thermal diffusion length μ in the sample and M_{th} is the thermal mass of this region of the sample. For the case where the thermal diffusion length in the sample is smaller than the sample thickness ($\mu < l$) and where A is the area illuminated by the light,

$$H_{abs} = \frac{I_0 A (1 - e^{-\beta\mu})}{\omega}, \qquad 6.$$

with I_0 being the light intensity and β the optical absorption coefficient

$$M_{th} = \rho C \mu A. \qquad 7.$$

This then gives

$$\Theta_0 = \frac{I_0(1 - e^{-\beta\mu})}{\rho C \omega \mu}. \qquad 8.$$

Similarly, when $\mu > l$, then

$$H_{abs} \simeq \frac{I_0(1 - e^{-\beta l})}{\omega} \qquad 9.$$

and

$$M_{th} \simeq \rho'' C'' \mu'' A \qquad 10.$$

where the unprimed symbols represent the sample and the double-primed symbols represent the parameters of the material directly behind the sample.

Thermal conduction introduces a phase lag in the PAS signal. There is a $\pi/4$ phase lag due to conduction in the gas and an additional phase lag due to conduction in the sample. The sample phase lag is approximated by $\phi \simeq 1/\beta\mu$. When $\mu < 1/\beta$, ϕ reaches a maximum value of $\pi/4$ as well.

Combining these expressions for the magnitude of the PAS signal with the above remarks on phase, we can reconstruct all six of the special photoacoustic cases treated in the more detailed Rosencwaig-Gersho theory (8). We find that aside from a factor of 1/2, all six cases are properly given by the above arguments.

EXPERIMENTAL METHODOLOGY

As in other forms of spectroscopy, a photoacoustic spectrometer is composed of three main parts: a source of incoming radiation, the experimental chamber, and the data acquisition system. A typical photoacoustic spectrometer is shown in block diagram fashion in Figure 1.

Radiation Sources

The most common and most versatile sources of optical radiation in the ultraviolet, visible, and infrared regions are provided by the conventional light sources. These are the arc lamp for the ultraviolet-visible, the incandescent lamp for the visible and near infrared, and the glow-bar for the mid- to far-infrared regions. All three light sources provide strong, broadband optical radiation, and they must therefore be used in conjunction with suitable monochromators. Since the signal-to-noise ratio in photoacoustic spectroscopy increases linearly with the amount of light falling on the sample, one desires an intense light source and a high light throughput (i.e. low f-number) monochromator. These light

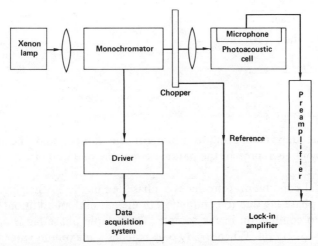

Figure 1 A block diagram of a single-beam photoacoustic spectrometer [Rosencwaig (6)].

sources generally operate in a continuous mode, and thus a light chopper, usually electromechanical in nature, must be used.

Another source of optical radiation that can be used in photoacoustic spectroscopy of solids is the laser. A laser requires no monochromator and, if operated in a pulsed mode, would also require no chopper. In the visible wavelength region, dye lasers provide an intense, highly monochromatic light, readily tunable over a fairly large wavelength range. Dye lasers can also be used with reasonable intensity in the ultraviolet region with the aid of frequency-doubling crystals. In the infrared there are currently no continuously tunable lasers that cover a wide spectral range, although if the experiment can be performed over a narrow wavelength range, then a discrete infrared laser (e.g. the CO or CO_2 laser) or a tunable spin-flip Raman laser can be used to great advantage to provide intense, highly monochromatic radiation.

Experimental Chamber

The experimental chamber is the section containing the photoacoustic cell or cells and all the required optics. The actual design of this chamber will vary, depending on whether one is using a single-beam system employing only one photoacoustic cell or a double-beam system containing two cells, with appropriate beam-splitting optics. The photoacoustic cells generally incorporate a suitable microphone with its preamplifier. Both conventional condenser microphones with external biasing and electret microphones with internal self-biasing provided from a charged electret foil are good microphones to use.

Some criteria governing the actual design of the photoacoustic cell are: (*a*) acoustic isolation from the outside world; (*b*) minimization of extraneous photoacoustic signal arising from the interaction of the light beam with the walls, windows, and the microphone in the cell; (*c*) microphone configuration; (*d*) acoustical means for maximizing the acoustic signal within the cell; (*e*) the requirements set by the samples to be studied and the type of experiments to be performed.

The above criteria are considered in detail in the review chapter by Rosencwaig (7).

Data Acquisition

The tasks of acquiring, storing, and displaying the data can be performed in many ways. However certain basic procedures should be followed. For example, the signal from the microphone preamplifier should be processed by an amplifier tuned to the chopping frequency in order to maximize the signal-to-noise ratio. If phase as well as signal amplitude is desired, then a phase-sensitive lock-in amplifier should be used.

For a single-beam spectrometer, provision must generally be made to remove from the photoacoustic spectrum any spectral structure resulting from the lamp, monochromator, and optics of the system. This normalization can be done conveniently by digitizing the analog signal from the tuned amplifier and then performing a point-by-point normalization (i.e. division) with either a power meter reading or a previously recorded photoacoustic spectrum obtained with a black absorber. In a double-beam spectrometer, normalization can be performed in analog real-time fashion by dividing the analog output from the tuned amplifier processing the sample signal, with the output derived from a reference signal. This reference output may be from a power meter or from a second photoacoustic cell.

With regard to the storage and display of the data, there are of course many possible schemes ranging from the relatively inexpensive chart recorder to the sophisticated minicomputer.

STUDIES IN BIOLOGY

One of the most promising areas for the use of the photoacoustic method is in the study of biological systems, for here most of the materials to be studied are often in a form that makes them difficult if not impossible to study by any other optical technique. Although many biological materials occur naturally in a soluble state, many others are membrane-bound or part of bone or tissue structure. These materials are insoluble and function biologically within a more or less solid matrix. Optical data on these materials are usually difficult to obtain by conventional techniques, since these materials are generally not in a suitable state for conventional transmission spectroscopy and if solubilized are often significantly altered. Photoacoustic spectroscopy, through its capability of providing optical data on intact biological matter, even on matter that is optically opaque, holds great promise, both as a research and a diagnostic tool in biology and medicine.

Hemoproteins

One of the first illustrations of the capabilities of PAS in biology was an experiment on cytochrome-C (6), with the results shown in Figure 2. Since cytochrome-C is readily soluble in water, its optical absorption spectrum is well known. Figure 2(a) shows the optical absorption spectrum of the oxidized and reduced forms of cytochrome-C in aqueous solution obtained with a conventional spectrophotometer. Figure

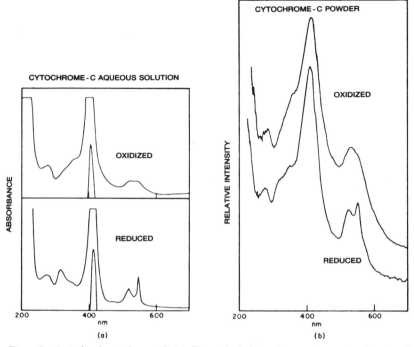

Figure 2 A study of cytochrome-C. (*a*) The optical absorption spectrum of oxidized and reduced cytochrome-C in aqueous solution. (*b*) The photoacoustic spectra of oxidized and reduced cytochrome-C in powder form [Rosencwaig (6)].

2(*b*) shows the PAS spectra of oxidized and reduced cytochrome-C in lyophilized or powder form. The photoacoustic spectra are qualitatively very similar to the optical absorption spectra. In particular, all of the differences between the oxidized and reduced forms visible in the absorption spectra are also visible in the photoacoustic spectra. This experiment indicates that it is possible with photoacoustic spectroscopy to obtain spectra on biological material in solid form that are comparable to those obtained in solution. This capability becomes extremely important when the compound to be studied is insoluble.

The oxygen-carrying protein of red blood cells, hemoglobin also displays a strong and characteristic heme absorption spectrum. Conventional absorption spectroscopy on whole blood, even when the blood is diluted in a suitable buffer, does not produce satisfactory results. The inadequacy of the conventional technique arises from the strong light-scattering properties of whole blood, due primarily to the presence of the other protein and lipid material in the plasma and the red blood cells. In the conventional process, one first extracts the hemoglobin from the whole blood and then obtains an absorption spectrum of an aqueous solution of the extracted hemoglobin. The PAS spectrum of a smear of whole blood in Figure 3, exhibits the characteristic spectrum of oxyhemoglobin as clearly as in the PAS spectrum of the red blood cells and even of the extracted hemoglobin itself (6). The presence of the other protein and lipid material in whole blood, which causes so much

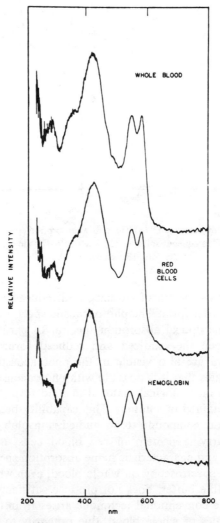

Figure 3 Photoacoustic spectra of smears of whole blood, red blood cells, and hemoglobin [Rosencwaig (7)].

difficulty in conventional spectroscopy from light scattering, creates no problem in photoacoustic spectroscopy. Thus, it is now possible to study the hemoglobin directly in whole blood, that is, in situ, without resort to extraction procedures.

Plant Matter

The unique capabilities of photoacoustic spectroscopy enable it to be used to obtain optical absorption data on even more complex biological systems such as green leaf and other plant matter. A PAS spectrum on an intact green leaf in Figure 4 clearly shows all of the optical characteristics of the chloroplasts in the leaf matter—the Soret band at 420 nm, the carotenoid band structure between 450 and 550 nm, and the chlorophyll band between 600 and 700 nm. PAS can thus be used to study intact plant matter, even living plant matter, and to obtain valuable information about normal and abnormal plant processes and pathology.

Since one can study intact plant matter, it is clear that photoacoustic spectroscopy can be used as a quick and efficient screening tool for those natural products with strong specific absorbance in some region of the optical spectrum. Using the PAS spectrum obtained from only a few milligrams of a natural source, such as a plant, animal, or microorganism, one can readily determine the types and relative concentrations of secondary metabolites present, that is, the chemical byproducts

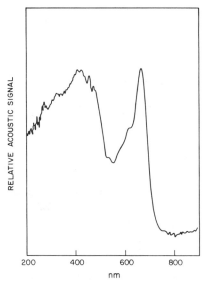

Figure 4 A photoacoustic spectrum of an intact green leaf [Rosencwaig (7)].

of the organism's metabolism; these compounds may prove to have important physiological or biomedicinal value. At present such information can be arrived at only after laborious extraction and analysis procedures of the natural source.

Applications of PAS in natural products chemistry are illustrated with a study on marine algae, a promising source of new and unusual natural products. The photoacoustic spectra shown in Figure 5 were taken on the algae after they had been air-dried, using less than 10 mg of material in each case (6). Although there is a considerable amount of spectral detail in all the PAS spectra, the main region of interest is around 320 nm, where the sought-after aromatic or highly conjugated secondary metabolites could be expected to have an absorption band. The PAS spectra indicated that only two of the algae would yield any quantity

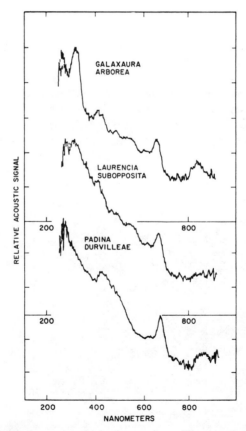

Figure 5 Photoacoustic spectra obtained on a few milligrams of dried marine algae [Rosencwaig (7)].

of these aromatic compounds. Conventional extraction and analysis procedures have fully confirmed the photoacoustic data. Photoacoustic spectroscopy can thus provide, in minutes, information about the compounds present in complex biological systems; moreover, it is capable of doing this nondestructively, since it requires only milligrams of material and no special sample preparation.

Another illustration of the use of photoacoustic spectroscopy in biology is in the study of marine phytoplankton illustrated in Figure 6 (5). In this study the spectra of five phylogenetically disparate algal species were determined in replicate by PAS and by conventional

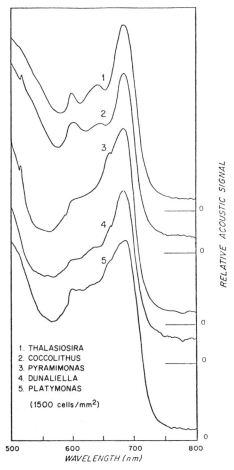

Figure 6 Photoacoustic spectra of representative marine phytoplankton [Ortner & Rosencwaig (5)].

spectrophotometric analysis. Monospecific cultures were obtained of *Coccolitus huxleyii, Thalassiosira pseudonana, Dunaliella tertiolecta, Pyraminmonas* sp., and *Platymonas* sp. Cells from these cultures were then collected on Whatman glass fiber filters and examined by both PAS and spectrophotometric methods. In comparing the PAS and optical absorption spectra, it was found that the PAS spectra exhibited more detail than the corresponding absorption spectra.

The extra sensitivity of the photoacoustic technique is due to the fact that the photoacoustic signal is dependent not only on the absorption characteristics of the specimen but also on its deexcitation mechanism (fluorescence and phosphorescence) and on its thermal properties, whereas absorption spectra depend only on the absorption characteristics.

Besides obtaining spectral information on bulk biological matter, one can also use PAS to investigate the macroscopic structure of the sample.

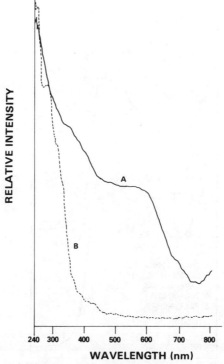

Figure 7 Photoacoustic spectra on intact apple peel. The dashed line spectrum was taken at 220 Hz, showing absorption only within the waxy top layer. The solid line spectrum was taken at 33 Hz, showing absorption within the red peel beneath the waxy layer [Rosencwaig (7)].

This can be done in several ways. For example, Figure 7 shows the PAS spectrum obtained on a piece of apple peel for two chopping frequencies (7). At the higher frequency, the PAS signal arises solely from the waxy layer at the surface. This layer exhibits mainly a UV absorption from its protein content. At the lower frequency, the thermal diffusion length extends below the waxy layer and a PAS signal is obtained from the biological material beneath the waxy layer, such as the carotenoids and chlorophyll compounds in the peel itself. By changing the chopping or modulation frequency in a continuous fashion, one is thus able to perform a depth-profile analysis of biological specimens. Another method for obtaining depth-profile information is to record the PAS spectrum at different phase angles, as demonstrated by Kirkbright and his co-workers (1).

Photosynthesis

The photoacoustic effect measures the nonradiative deexcitation processes that occur in a system after it has been optically excited. This selective sensitivity of the PAS technique to the nonradiative deexcitation channel can be used to great advantage in the study of biological systems that exhibit photochemistry. In biology, one of the most important manifestations of photochemistry is the process of photosynthesis, both in green plants and in certain bacterial organisms. Photochemical processes like photosynthesis compete with the photoacoustic process and thus can be studied with PAS.

The normalized photoacoustic signal p can be written as (3)

$$p = q\left\{1 - \sum_i (\phi_i \Delta E_{p,i}/Nh\nu) - (\phi_l \nu_l/\nu)\right\}. \qquad 11.$$

Here q is the fraction of light absorbed by the sample; ϕ_i is the quantum yield for photochemical reaction; $\Delta E_{p,i}$ is the molar internal energy change per product formation in this reaction; ϕ_l is the quantum yield for luminescence; ν_l and ν are the frequencies of radiated and absorbed light respectively; N is avogadro's number; and h is Planck's constant.

If one neglects the role of radiative deexcitation, it is clear that any photochemical process will result in a decrease of the PAS signal, with increasing quantum yield for photochemistry. Furthermore, by varying the modulation frequency and analyzing the phase of the PAS signal, information can be obtained on kinetic parameters and on the energy content of intermediates.

Cahen et al (3) have studied photosynthesis in lettuce chloroplast membranes by means of photoacoustic spectroscopy. Figure 8 shows the PAS spectra obtained for photosynthetically active membranes and for DCMU-poisoned [DCMU = 3-(3,4-dichlorophenyl)-1,1-dimetheylurea, an electron transport inhibitor] chloroplast membranes. The spectra are normalized at 440 nm where little photosynthetic activity is to be expected. At the 680 nm band of chlorophyl, the DCMU-poisoned chloroplasts give a signal that is 10% stronger than that of the active chloroplasts, indicating the lack of photosynthetic activity. As the modulation frequency is decreased, the difference between the spectra of normal and poisoned membranes increases. Cahen et al explained this observation by hypothesizing that the inhibitor activity of the DCMU did not reach completion at the higher modulation frequencies and shorter time periods.

Another experiment performed by Cahen et al (2) investigated the photosynthetic process in the purple membrane of *Halobacterium halobium*. This membrane contains a single protein, bacteriorhodopsin, covalently bonded to a retinal molecule. Absorption of light by the retinal brings about a cyclic photochemical process that drives the translocation of protons from one side of the membrane to the other through conformational changes of the protein. Figure 9 shows the PAS spectra of dried purple membrane fragments with and without strong side illumination. The side illumination causes an accumulation of the photointermediates absorbing at 415 nm and 660 nm, and a decrease of the population absorbing at 565 nm.

In lyophilized purple membranes, the absorbed light drives the photocycle only, whereas in the whole cell, part of the absorbed energy is stored as ATP and ion gradients. From Equation 11, the photo-

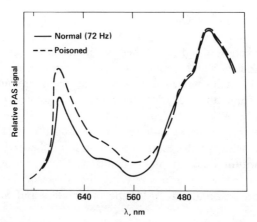

Figure 8 Photoacoustic spectra of photosynthetically active lettuce chloroplast membranes, and for DCMU-poisoned membranes [Cahen et al (3)].

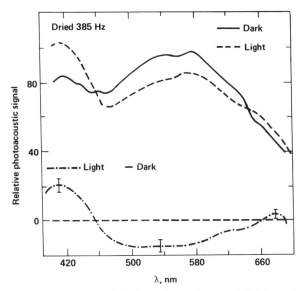

Figure 9 Photoacoustic spectra of dried purple membranes of Halobacterium halobium, with and without strong side illumination [Cahen et al (2)].

acoustic signal normalized to the absorbed energy is given by

$$p/q = 1 - \frac{\phi \Delta E_p \lambda}{\text{const}}.$$ 12.

If $\phi \Delta E_p = 0$, as would be the case if there was no photochemical reaction ($\phi = 0$) or if there were a cyclic reaction with $\Delta E_p = 0$, then p/q would be independent of λ. However, if some of the absorbed energy is stored in the products sensed by PAS, a valley-shaped curve would be obtained, with the lowest point in the region of the highest energy storage.

The results of Cahen et al (2) appear to demonstrate this situation. In the freeze-dried purple membrane fragments, where presumably no energy is stored (on the time scale of the experiment) in the steady state, p/q is found to be constant within experimental error. However, for intact cells p/q passes through a minimum in the 540–620 nm region, indicating energy storage in this wavelength region.

These results indicate that photoacoustic spectroscopy appears to be a sensitive tool for the study of photosynthesis. Furthermore, PAS can provide information on the energetics of the process by clearly distinguishing between samples in which part of the absorbed energy is stored or used to do work and those in which all of the absorbed energy is dissipated in the photocyle as heat.

STUDIES IN MEDICINE

Perhaps the most exciting area for photoacoustic studies in biophysics lies in the field of medicine. One can use PAS to obtain optical data on medical specimens that are not amenable to conventional study because of excessive light-scattering; one can study specimens that are completely opaque to transmitted radiation; and one can perform depth-profile analysis by phase or frequency adjustments. Furthermore, it is possible to construct a photoacoustic cell that can be used to perform all of the above measurements in an in-vivo mode.

Bacterial Studies

An example of the use of photoacoustic spectroscopy in medical studies is in the identification of bacterial states. Although conventional light-scattering methods can be used to monitor bacterial growth on various substrates, the light-scattering makes it most difficult to obtain spectroscopic data on bacteria and thus to identify them. However, bacterial samples are quite suitable for study by PAS. Thus the PAS spectrum of *Bacillus subtilis* var. *Niger*, a common airborne bacteria, contains a strong absorption band at 410 nm when this bacteria is in its spore state (10). This band is absent when the bacteria is in its vegetative state. Thus, photoacoustic spectroscopy enables the detection and discrimination of bacterial states and allows one to monitor and detect bacteria in various stages of development.

Drugs in Tissues

Another medical use for PAS is in the study of animal and human tissues, both hard tissues such as teeth and bone, and soft tissues such as skin, muscle, etc. An example of a PAS experiment on soft tissue (7) is illustrated in Figure 10 where photoacoustic spectra of guinea pig epidermis are shown. The top spectrum is of an epidermis that has been treated with a 2% solution of tetrachlorosalicylanilide (TCSA) in ethanol. This compound is known to be a highly effective antibacterial agent but also causes photosensitivity and other skin problems. The central spectrum is of control guinea pig epidermis. The bottom spectrum is the difference spectrum found by subtracting the control spectrum from the TCSA spectrum. The difference spectrum is thus the absorption spectrum of TCSA bound within the epidermis. From this spectrum one can establish the state of the TCSA compound in situ, when it is incorporated into the skin, and thus learn more of its action on and within the skin, under various conditions.

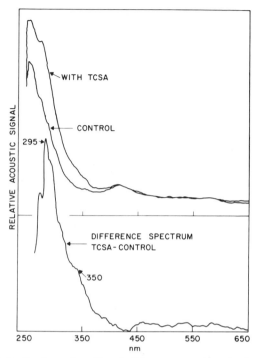

Figure 10 A study of guinea pig epidermis. The upper spectrum is of epidermis treated with tetrachlorosalicylanilide (TCSA); the middle spectrum is of untreated epidermis; and the bottom spectrum is the difference spectrum showing the absorption bands of the TCSA compound within the epidermis [Rosencwaig (6)].

Another example of a drug-in-tissue experiment has been reported by Campbell et al (4). In this study the presence of topically applied tetracycline (TCN) was detected by working at a wavelength of 380 nm where TCN is fairly strongly absorbing, whereas skin is not. The experimental results are shown in Figure 11 where the PAS signal is substantially greater for the TCN-treated samples than for the untreated sample and the signal amplitude is roughly proportional to TCN concentration.

These examples illustrate the potential usefulness of photoacoustic spectroscopy in the study of natural, medical, and cosmetic compounds present in and on human tissue.

Human Eye Lenses

Figure 12 depicts PAS spectra taken on intact human eye lenses (7). These experiments were conducted to study the processes by which

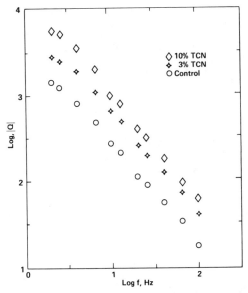

Figure 11 Photoacoustic signal dependence on chopping frequency for a control human stratum corneum and for stratum corneum topically treated with tetracycline (TCN) [Campbell et al (4)].

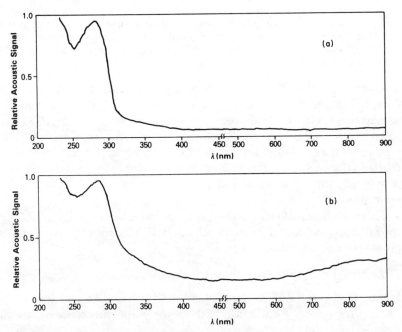

Figure 12 A study on intact human eye lenses. (*a*) The PAS spectrum of a normal lens; (*b*) the spectrum of a lens with a brunescent cataract [Rosencwaig (7)].

human eye cataracts are formed. Little is known about this ancient disease except that it is probably due to a photo-oxidative process in which the tryptophan and tyrosine residues in the protein matter of the lens form complexed compounds that are either highly light-scattering (cortical cataracts) or else are colored (brunescent or nuclear cataracts). Cataract studies are severely hampered by the fact that few spectroscopic investigations can be conducted on the intact lenses. In general these lenses must be solubilized, and except under the strongest solubilizing agents, only about one half of the lens material goes into solution and is suitable for study. Unfortunately the stronger solubilizing agents do too much damage to the inherent protein to be useful.

The PAS spectra shown in Figure 12 were obtained from intact human eye lenses. The upper spectrum is of a normal lens, whereas the lower spectrum is of a lens with a brunescent cataract. Both spectra exhibit the characteristic peak at 280 nm owing mainly to absorption by the tryptophan and tyrosine protein residues. The cataractous lens shows a much broader 280 band than does the normal lens. This is in agreement with the theory that the cataract is a result of conjugated tryptophan and tyrosine compounds. In addition, however, the PAS spectra indicate that the cataractous lens also exhibits an increased infrared absorption. This unexplained feature suggests that cataractous formation not only impairs vision in the blue end of the spectrum but actually impairs vision throughout the visible region, with broad absorption bands moving in from both the ultraviolet and the infrared ends of the visible spectrum.

Tissue Studies

The first extensive PAS study on mammalian tissue was on the stratum corneum, the outermost layer of the epidermal tissue. The stratum corneum, a translucent membrane in the visible region, is a highly effective light scatterer, especially in the ultraviolet region, and thus cannot be studied readily by conventional optical absorption techniques.

The amount of water present and the role this water plays in mammalian tissues strongly determines the physiochemical properties of tissues like the stratum corneum. Since the photoacoustic effect is sensitive to the presence of water because of its dependence on the thermal diffusivity of the sample, it has been possible to apply photoacoustic spectroscopy for moisturization (hydration) studies, which are related to the role of water in the stratum corneum (9). Figure 13 shows the photoacoustic signal at the tryptophan-tyrosine "protein" band at 285 nm as a function of the water content in the stratum corneum, as determined gravimetrically. Identical results were obtained at other

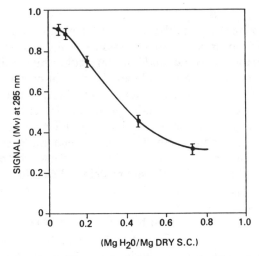

Figure 13 Photoacoustic signal strength at 285 nm as a function of water content for intact newborn rat stratum corneum [Rosencwaig & Pines (9)].

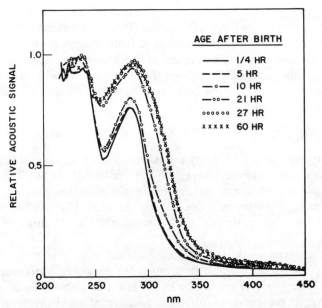

Figure 14 Photoacoustic spectra of a series of newborn rat stratum corneum during the postpartum maturation period [Rosencwaig & Pines (9)].

spectral wavelengths. There is at first very little change in the photoacoustic signal, then a fairly rapid, and then a slower decrease, with the curve apparently approaching a limiting value at high water content. The shape of this curve can be explained by the changes that occur in the thermal properties of the stratum corneum when water content increases. Such data has supplied information on the presence of free and "bound" water in stratum corneum.

Figure 14 shows a series of photoacoustic spectra obtained from newborn rat stratum corneum membrane during the initial 60 hr maturation period after birth (9). The times shown represent the postpartum age of the rats at the time of sacrifice. The 280 nm band undergoes a major change, particularly in the 10–30 hour postpartum period. Identical results were obtained on all sets of rat litters studied, irrespective of the period between the time of harvesting of the stratum corneum and the time the photoacoustic spectra were actually obtained.

Mammalian stratum corneum obviously serves quite different roles in the pre- and postpartum periods. Major and rapid biochemical and structural changes can be expected during the initial postpartum maturation period when the stratum corneum matrix undergoes alteration to develop its so-called "barrier" functions and adapt to its new and strikingly different environment. Rosencwaig & Pines (9) have postulated that the observed changes in the PAS spectrum of Figure 14 reflect this maturation process of the stratum corneum. In particular, they have postulated that the tyrosine residues in the α-keratin protein of stratum corneum undergo a major molecular modification, probably by enzymatic action. This modification results in increased inter-chain hydrogen-bonding, which can play a crucial role in stabilizing the coiled-coil structure of the α-keratin in the stratum corneum, thus enabling the stratum corneum to take on its postpartum barrier functions.

Although there is as yet no direct evidence for the appearance of a modified tyrosine, such as a hydroxyl-tyrosine, during the maturation or keratinization period, this hypothesis is attractive because it bears a close analogy to the known developmental processes of collagen, the structural protein in muscle tissue. It has recently been determined that, as collagen is being formed, some of the proline residues are enzymatically modified to hydroxyproline, and that the extra hydroxyl group of the hydroxyproline contributes significantly to the stability of the triple helix of collagen by additional hydrogen bonds. Rosencwaig & Pines believe, therefore, that the possibility of an enzymatic molecular change in the tyrosine of a postpartum stratum corneum merits further investigation (9).

FUTURE TRENDS

Photoacoustic spectroscopy of solids and liquids is still in its formative stages, yet its potential, both as a research and analytical tool, is already impressive. Until the development of this technique, many materials, both natural and synthetic, could not be readily investigated by conventional optical methodologies, since these materials occur in the form of powders or amorphous solids or as smears, gels, oils, suspensions, and so on. With photoacoustic spectroscopy, optical absorption data on virtually any solid and liquid material can now be obtained.

In this chapter we have reviewed recent experiments with the photoacoustic effect in the fields of biology and medicine—experiments that indicate some of the possible applications of this technique in biophysics and bioengineering.

As photoacoustic spectroscopy becomes better known, investigators with many different problems and orientations will adapt this technique to their own uses. Furthermore, since this is a spectroscopic technique which is not detector-limited, it will surely be extended into the far ultraviolet and infrared regions of the optical spectrum, and possibly into other regions of the electromagnetic spectrum as well.

In the near future, it is quite likely that photoacoustic spectroscopy will become a common and useful analytical and research tool in many scientific laboratories. Its ease of operation and versatility can only increase its area of applications. In the field of biology the PAS technique will be used to study intact biological systems, both in the laboratory and out in the field, providing data that can now be obtained only after extensive wet chemical procedures. In medicine, photoacoustic spectroscopy offers the opportunity for extending the exact science of noninvasive spectral analysis to intact medical substances such as tissues, with the possibility that by such noninvasive techniques new light might be shed on the diseases that afflict mankind.

The next few years promise to be an exciting period of growth for the rediscovered science of photoacoustic spectroscopy.

Literature Cited

1. Adams, M. J., Beadle, B. C., King, A. A., Kirkbright, G. F. 1976. *Analyst London* 101:553–61
2. Cahen, D., Garty, H., Caplan, S. R. 1978. *FEBS Lett.* 91:131–34
3. Cahen, D., Malkin, S., Lerner, E. I. 1978. *FEBS Lett.* 91:339–42
4. Campbell, S. D., Yee, S. S., Afromowitz, M. A. 1977. *J. Bioeng.* 1:185–88
5. Ortner, P. B., Rosencwaig, A. 1977. *Hydrobiologia* 56:3–6
6. Rosencwaig, A. 1975. *Anal. Chem.* 47:A592–604
7. Rosencwaig, A. 1978. *Adv. Electron. Electron Phys.* 46:207–311
8. Rosencwaig, A., Gersho, A. 1976. *J. Appl. Phys.* 47:64–69
9. Rosencwaig, A., Pines, E. 1977. *Biochim. Biophys. Acta* 493:10–23
10. Somoano, R. B. 1978. *Angew. Chem. Int. Ed. Engl.* 17:238–45

STIMULUS-RESPONSE COUPLING IN GLAND CELLS

♦9141

B. L. Ginsborg
Department of Pharmacology, University of Edinburgh, Edinburgh, Scotland

C. R. House
Department of Veterinary Physiology, University of Edinburgh, Edinburgh, Scotland

INTRODUCTION

Secretion essentially means the transport of matter from the inside of a cell to the outside. Some gland cells secrete only small quantities of highly active substances, a task which could be accomplished entirely by exocytosis. Others, which secrete into ducts, require carrier fluid to be transported as well. This could be achieved by the activation of ion pumps and the entrainment of passive water flow. The nervous or humoral stimuli which give rise to secretion also produce electrical and morphological effects, and a central problem is the relationship between the various observed changes. Current hypotheses for stimulus-response coupling will be discussed and illustrated in relation to a number of different secreting tissues. Among exocrine glands, the cockroach salivary gland, mammalian salivary and lacrimal glands and the pancreas have been selected. The pancreatic β-cell and the adrenal chromaffin cell are used as examples for endocrine secretion. Other tissues which have been successfully studied include the blowfly salivary gland [see Fain & Berridge (49)], molluscan salivary and pedal glands [see Kater (88)], the liver [see Burgess et al (23)] and the pituitary (see 48), but they have been excluded for lack of space. Many of the topics discussed here are reviewed comprehensively in references (137) (salivary secretion), (26) (pancreatic secretion), and (117) (electrophysiology).

EXOCRINE GLANDS: STIMULI AND RESPONSES

Figure 1 illustrates a scheme which encompasses many current hypotheses. The upper part deals with agonists other than those activating β-adrenergic receptors, which, for acinar cells, are established only in mammalian salivary glands. Details will be discussed in separate sections. It is generally agreed that the agonist-receptor interaction gives rise to A, an increase in ionized calcium in the cytosol, and B, an increase in conductance to one or several environmental ions. It is not clear which effect comes first or if one is the cause of the other; thus C1 and D1 are seen as alternatives and not as a positive feedback loop. Nor is there agreement about the source of the calcium; some assume a

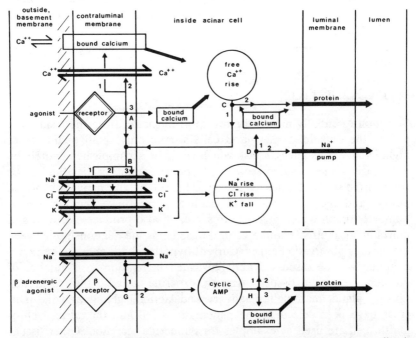

Figure 1 Scheme illustrating some current hypotheses for stimulus-response coupling in exocrine tissue. *Upper half*, agonist-receptor interaction leads either to the release of calcium from a store associated with the membrane (A1) or an internal structure (A3) or to a trans-membrane influx (A2). The rise in cytosolic Ca^{2+} leads to membrane conductance increases (one or several of B1, 2, and 3) and to exocytosis of protein. Fluid secretion is the result of an active extrusion of Na^+ across the luminal membrane (D2). A variant is that the agonist-receptor interaction leads first to membrane conductance changes (A4); the increase in intracellular sodium then releases calcium from a store (D1). *Lower half* applies to mammalian salivary glands and is discussed in section on cyclic nucleotides in salivary secretion.

transmembrane flux as the only source [e.g. pancreas (86)], some, release from intracellular stores [e.g. pancreas (27)] and others, a combination of the two [e.g. pancreas (125)]. Most, although not all (see section on exocytosis and protein secretion), are agreed that the increased cytosolic calcium concentration is responsible for initiating protein secretion by exocytosis. Whether or not they are primary or secondary, the changes in ion conductance (B, Figure 1), which vary both in extent and in the identity of the ions, depending on the gland (see section on electrical responses), will generally have two immediate consequences: (a) a change in potential difference across the contraluminal membrane, and (b) transmembrane ion fluxes. Although the conductance changes involved will determine whether the electrical response is hyperpolarizing or depolarizing and the extent of the ion exchanges, their direction will be expected to be dependent only on the electrochemical gradients; thus an inflow of sodium and chloride and an outflow of potassium would be anticipated. If, as seems to be the case, the resting sodium conductance is high, there may be a significant gain in sodium (notwithstanding the agonist-induced conductance change). Thus it seems reasonable to suppose that fluid secretion is a consequence of a corresponding sodium efflux across the luminal membrane. There is however no significant additional evidence to add to that given by Schneyer et al in 1972 (137).

Preparations and Stimuli

In vivo all three mammalian glands are activated by parasympathetic cholinergic nerves acting on "muscarinic" receptors. The salivary glands are also activated by sympathetic adrenergic nerves acting both via α- and β-adrenoceptors, and the exocrine pancreas is activated by the peptide hormones cholecystokinin (CCK, sometimes called pancreozymin, PZ, or cholecystokinin-pancreozymin, CCK-PZ), secretin and possibly others. In vitro preparations, which have provided much of the information to be reviewed, include tissue fragments, tissue slices, and dispersed cells. Since these are clearly not accessible to nerve stimulation, responses have been obtained by the addition of cholino- or sympathomimetic substances or the peptide hormones to the bathing fluid. Salivary glands have also been found to respond to the peptides substance P and eledoisin, lacrimal glands to α-adrenergic agonists, and the pancreas to a variety of "foreign" peptides, including caerulin and bombesin, which have structural similarities with the "native" hormones. The receptors for cholinergic, α-adrenergic, β-adrenergic, and peptidergic agonists are distinct as illustrated by the fact that responses to each of the first three kind of agonist may be blocked selectively by atropine, phentolamine, and propranolol.

Exocrine gland cells are inexcitable in the sense that they do not produce any kind of action potential [but see (88) for excitable molluscan examples]. Until recently it appeared that depolarization was not an effective stimulus. For example, pancreatic tissue is not activated by a tenfold increase in external potassium concentration (e.g. 125). An increase in external concentration of potassium to 75 mM has however now been reported to release prelabelled protein from guinea-pig pancreatic lobules (136).

In the study of exocrine secretion products of acinar cells, perfused tissue has the disadvantage that the primary secretion is modified by its passage through ducts; superfused tissue may suffer not only this disadvantage but also does not allow what enters the superfusate via the luminal membrane to be distinguished from what enters via the contraluminal membrane. Dispersed single cells have the disadvantages (see 26) that (a) drastic procedures are required to produce them and their sensitivity to agonists is low, and (b) their behavior may not be the same as when they are coupled to one another in an acinus. A new preparation of functional isolated pancreatic acini has recently been described (144).

Responses

EXOCYTOSIS AND PROTEIN SECRETION Evidence exists that protein release occurs by exocytosis in all three mammalian tissues under consideration (for references see also Table 1). A number of reports, however, suggest that exocytosis is not the only mechanism for release. In salivary glands where there appear to be two separate pathways for amylase release (see section on cyclic nucleotides in salivary secretion), it has been reported that the ratio of the amount of degranulation to the amount of amylase released varied with the method of stimulation (54) and that the enzymes peroxidase and amylase were not secreted in parallel (55) and may therefore have been released from different pools. In the pancreas, although changes in the shape and cross sectional area of the lumen provide strong evidence for exocytosis of the large zymogen granules, protein secretion continues, during prolonged stimulation, after their disappearance. This continued secretion has been attributed rather than demonstrated to be due to the exocytosis of smaller granules. To quote Jamieson & Palade (83), "Discharge of small storage granules to the acinar lumen by membrane fusion-fission has not yet been observed in the stimulated condition but their proximity to the apical plasma-lemma and the fact that they carry labeled secretory proteins which are ultimately secreted into the duct space, strongly suggest that they function in the same manner as mature zymogen

Table 1 Responses of exocrine glands

Gland	Stimulus	Response
Cockroach salivary gland	Nerve; dopamine (not on α- or β-adrenoceptors)	Hyperpolarization: reversal potential E_K; due to increase in g_K sometimes followed by depolarization due to increase in g_{Na} (56) NaCl and fluid (141)
Mammalian salivary glands	Agonists acting on either acetylcholine (muscarinic) receptors or α-adrenoceptors	Hyperpolarization or depolarization followed by hyperpolarization; reversal potential close to resting potential due to increase in g_K and g_{Na} (132). For electrical response to nerve stimulation see (85, 52) K-loss from cells (124) K-gain by medium (138, 126) Exocytosis and secretion of amylase (122, 97), kallikrein (1), sialic acid (19)
	Agonists on β-adrenoceptors	Very slow depolarization (119, 132) Exocytosis (97, 126, 138) and enzymes as above
	Nerve stimulation	As above (55, 84)
	Substance P and eledoisin (peptides) not via acetylcholine or adrenoceptors	Amylase and K-release (135)

Table 1 (Continued)

Gland	Stimulus	Response
Lacrimal gland	Acetylcholine; epinephrine (on α-adrenoceptors	Hyperpolarization: reversal potential—60 mV due to increase in g_K and g_{Na} (78)
	Intracellular Ca injection	Response as above (79) and uncoupling (80)
	Carbachol; epinephrine (on α-adrenoceptors); A23187 and Ca	K-release (115)
	Carbachol; α-adrenergic agonists	Enzyme (peroxidase) release (70, 127)
Pancreas	Acetylcholine	Depolarization; reversal potential—15 mV; due to increase in g_{Na}, g_K and g_{Cl} (76); uncoupling (77); for nerve stimulation see (36)
	Peptides: CCK and analogue; bombesin	Responses as above (118, 120)
	Intracellular Ca injection	Responses as above (74, 118)
	A23187 and Ca	Depolarization and amylase secretion (125)
	Increased external K (75 mM)	Protein release (136)
	For secretion details see (26)	

granules." The concept of exocytosis from pancreatic acinar cells has been challenged because the proportions of different proteins secreted at the same time have been found to vary with different stimuli (134; see also 26). However a recent report by Dagorn & Estival (33) suggests that the variation depends only upon the overall rate of secretion and can be explained by postulating two different protein pools of constant proportion. This might be expected on morphological grounds (see 26).

Fluid secretion Fluid secretory mechanisms of acinar cells are particularly difficult to study since perfused tissue must be used and, as already mentioned, the composition of the primary fluid is modified by duct cells. A vexatious interaction of a different kind may also occur between protein release and fluid secretion in superfused fragments of tissue (e.g. 121). Since fluid secretion is necessary to wash protein out of the lumen into the superfusate, it is difficult to decide if an observed inhibition of protein release is primary or due to lack of fluid.

Electrical Responses of Acinar Cells

Electrical responses have been recorded with intracellular electrodes to nerve stimulation and/or ionophoretically applied agonists from acinar cells of the cockroach and mammalian salivary glands and from the mammalian lacrimal gland and pancreas. In all four tissues, cells in a single acinus appear to be closely coupled in normal conditions (56, 80, 131).

SITE OF RESPONSES It is generally assumed that intracellular electrodes signal only events occurring at the contraluminal (basolateral) membrane. This is equivalent to supposing that the luminal membrane is separated by a resistance much higher than its own from the indifferent (bath) electrode in the external solution or, in other words, that LR/R' in Figure 2 is large. There is no direct experimental evidence to support this idea but electrotonic potentials produced by current pulses passed between the inside of the cell and the external solution have been reported to be unaffected during intense amylase secretion (82). This presumably involves a large increase in luminal membrane area (see 138). Since this would be expected to cause an increase in membrane capacity and reduction in resistance, the absence of a change in electrotonic potential indirectly supports a large value for LR/R'.

Site of receptors Agonists have been applied ionophoretically both extra- and intracellularly to cockroach salivary gland cells and to pancreatic cells. In the former (72), catecholamines were found to be approximately equipotent in both sites. The response to intracellular agonist did not have a smaller latent period, nor was it faster than the

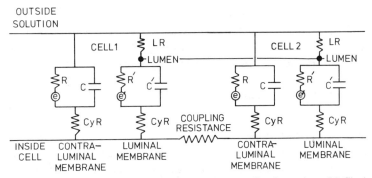

Figure 2 Simplified equivalent circuit of two contiguous cells of an acinus. R', C', e' refer to luminal membrane, CyR is cytoplasmic resistance, R, C, e refer to contra-luminal membrane, and LR is resistance between lumen and outside solution. If LR/R' is sufficiently large, changes in e', R', and C' will not influence the PD between intracellular electrode and the outside solution (cf reference 133).

response to extracellular agonist. The results therefore do not provide evidence that *extracellularly* applied agonists act on internal receptors. In the pancreas, acetylcholine (75) and peptide agonists (123) were found to be ineffective when applied intracellularly.

NATURE OF RESPONSES Responses of the four tissues have a number of features in common (Table 1). With the exception of the effect of isoprenaline acting on β-adrenoceptors of mammalian salivary glands, all agonists induce increase in potassium conductance. This is the first and predominant effect seen in the cockroach salivary gland in response to either nerve stimulation or applied dopamine (17) and leads to a hyperpolarization (Figure 3A) which may be "inverted" (Figure 3C) by setting the membrane potential to a level more negative than the potassium equilibrium potential (56); there is probably also a subsequent increase in sodium conductance that can outlast the increase in potassium conductance and give rise to a depolarizing phase of the response (Figure 3B). In salivary glands an increase in sodium conductance overlaps and sometimes precedes the potassium conductance increase so that the responses may be biphasic, the initial phase being depolarizing. The different time courses of the sodium and potassium conductances suggest that the ions may pass through separate channels. The responses to cholinergic and α-adrenergic agonists are closely similar. However β-agonists produce a greatly delayed and prolonged depolarization, which has not been analysed in detail. The response in the lacrimal gland to cholinergic and α-adrenergic agonists is hyperpolarizing but involves increase in both sodium and potassium conductance; there is no response to β-adrenergic agonists. The responses of

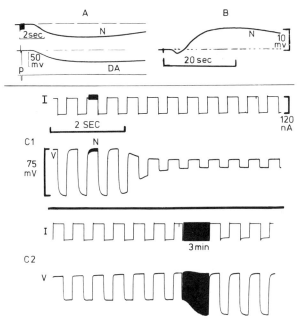

Figure 3 Cockroach salivary gland. Intracellular electrical records, hyperpolarization downwards. *A*. N, response to nerve stimulation; DA to an ionophoretic dopamine pulse, p (reference 17). *B*. Response showing large prolonged depolarizing phase: *C1*. Reduction in resistance and "reversal" of response to nerve stimulation, N. *Upper traces*: current pulses through one electrode; *lower traces*: voltage from second electrode in a closely coupled cell. *C2*. Recovery from stimulation (reference 56).

the pancreas to a variety of peptides are all closely similar to the response to acetylcholine and have the same reversal potential. The acetylcholine response has been studied in detail and shown to be due to a simultaneous increase in potassium, sodium, and chloride conductances, the last two being dominant.

ION CONCENTRATION CHANGES   As has been pointed out the changes in conductance may lead to significant changes in transmembrane flux. In the salivary and lacrimal glands a significant efflux of potassium occurs; this has been demonstrated both by the increase in potassium concentration in the solution that incubates tissue slices [which shows that a net loss of intracellular potassium occurs (see 138)] and by an increased efflux of ^{42}K (or ^{86}Rb) from tissue slices. A fall in intracellular potassium concentration has also recently been demonstrated (124) by direct measurement with potassium selective intracellular electrodes. No such fall occurred from pancreatic acinar cells; this is consistent with the conclusion from direct investigations that the dominant conductance changes are to sodium and potassium.

UNCOUPLING By simultaneous recording from two or three cells of the electrotonic potentials caused by current passed through one of them, Iwatsuki & Petersen (e.g. 80) have succeeded in demonstrating that agonists can cause uncoupling of previously closely coupled cells in the same acinus in lacrimal glands and in the pancreas. (The observed effects were those expected from an increase in the coupling resistance in Figure 1). Fairly intense stimuli were required but not larger than required to produce maximal responses.

LATENCIES OF ELECTRICAL RESPONSES All the responses recorded so far have been found to have long latent periods. In the cockroach salivary gland where typical values are between about 0.5 and 1 sec at room temperature, a rise in temperature of 10°C reduces the period by a factor of about 3 (17). This result among others (18) makes it unlikely that the latency is primarily due to a long diffusion time. In mammalian salivary glands the minimum latency for the response to ionophoretically applied acetylcholine was reported to be 210 msec and for epinephrine, 420 msec. Similar results were obtained by Creed & Wilson (31) for responses to parasympathetic (273 msec) and sympathetic stimulation (420 msec). The difference between the latencies to the responses mediated via cholinergic and α-adrenergic receptors is of particular interest in view of the possibility that they operate a common mechanism (see 126). In the pancreas the latency of the response to vagal stimulation is about 1 sec (39); ionophoretic application of acetylcholine and various peptides to pancreatic acinar cells is reported to give rise to similar minimum latencies of 0.5 sec. Iwatsuki & Petersen (74) have shown that it is possible to mimic the action of acetylcholine by the intracellular injection of calcium. If the response is coupled to receptor activation by an increase in cytosolic calcium (C1 in Figure 1), a knowledge of the latency of the response to calcium will clearly be of great importance. Latencies of the values reported are not unique to exocrine tissue. They are also found in the responses of excitable cells to muscarinic agonists. The very much larger latency [>4 sec (132)] for the action of β-adrenergic agonists on salivary glands is also characteristic of the β-adrenergic effects on the heart [see (71, 109) for discussion and further references].

CALCIUM AND EXOCRINE RESPONSES

Procedures which have been used to test the idea that responses are calcium dependent are indicated in Tables 2 and 3.

Table 2 Evidence for Ca-dependence

Tissue	Response	Agonist	Test
Cockroach salivary gland	Electrical	Dopamine	Ca-withdrawal[a]
		—	Ca-readmission (107)
		—	A23187[b]
	Fluid	—	Ca-readmission (107)
Mammalian salivary gland	Amylase	Acetylcholine epinephrine (α)	Ca-withdrawal (97, 122)
		peptides	Ca-withdrawal (135)
	Exocytosis	Acetylcholine, epinephrine (α)	Ca-withdrawal (97)
	K-release	Acetylcholine, epinephrine (α)	Ca-withdrawal (126, 138)
		Peptides	Ca-withdrawal (135)
		—	A23187 (138)
Lacrimal gland	Electrical	—	Intracellular Ca injection (79)
	Uncoupling	Acetylcholine (80)	—
	Peroxidase exocytosis	Carbachol	Ca-withdrawal, lanthanum (127)
	K-release	Carbachol, epinephrine	Ca-withdrawal, A23187 (115)
Pancreas	Electrical	Acetylcholine (prolonged application)	Ca-withdrawal (95)
		—	Intracellular Ca injection (74, 118)
		—	A23187 (125)
	Uncoupling	Acetylcholine (80)	Ni, Co (77)
		Peptides (80, 81)	—
	Amylase	Acetylcholine, CCK-like polypeptides	Ca-withdrawal (121)
		—	Ca-readmission (125, 121)
	Acinar fluid	Acetylcholine, CCK-like polypeptides	Ca-withdrawal (87, 143)

[a]Unpublished experiments, B. L. Ginsborg, C. R. House and M. R. Mitchell.
[b]Unpublished experiments, A. R. Martin and M. R. Mitchell.

Table 3 Evidence not in favor of Ca-dependence

Tissue	Response	Agonist	Test
Cockroach salivary gland	Fluid	Dopamine	Ca-withdrawal[a]
Mammalian salivary gland	Electrical (hyper-polarization)	Acetylcholine (only brief application)	Ca-withdrawal (110)
		α-agonists (only brief application)	Ca-withdrawal (119)
	(Depolarization)	β-agonists	Ca-withdrawal (119)
	Amylase	β-agonists	Ca-withdrawal (97, 122, 138; but see 128)
	Exocytosis	β-agonists	Ca-withdrawal (97, 138; but see 128)
Lacrimal gland	K-release	Phenylephrine (α-agonist)	Ca-withdrawal (127)
Pancreas	Bicarbonate, fluid	Secretin	Ca-withdrawal (87, 143)

[a]Unpublished experiments, C. R. House and R. K. Smith.

Calcium Withdrawal, Intracellular Injection, and Readmission

ELECTRICAL RESPONSES In the cockroach salivary gland (Figure 4) and in the pancreas, electrical responses to agonists have been found to be reduced and eventually abolished by calcium withdrawal. The effect is however much slower and less dramatic than on responses which are established as calcium dependent (see section on ionic requirements for adrenal secretion). Furthermore, even after prolonged exposure to calcium-free solutions a large amount of agonist evokes a response. The results are difficult to explain if one accepts the hypothesis that responses depend directly on a transmembrane influx of calcium (Figure 1, A2). A peculiar diffusion barrier might be postulated but the idea seems unattractive, at least for the cockroach salivary gland, since in the same experiment calcium withdrawal abolishes the response to nerve stimulation before having any effect on the response to ionophoretically applied agonist (B. L. Ginsborg, C. R. House, M. R. Mitchell, unpublished). Even prolonged calcium withdrawal has been reported to have little effect on responses to isolated brief applications of agonist to mammalian salivary glands (Table 3). A reduction was observed in responses of pancreatic cells, but only in the absence of *any* divalent

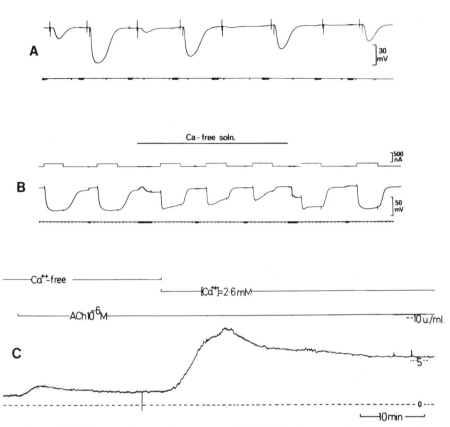

Figure 4 Calcium involvement in responses. *A, B*. Electrical responses from acinar cells of cockroach salivary gland. Time marks, 10 sec. *A*. Effect of calcium withdrawal, 3 min before start of trace, on responses to two dopamine pulses of different duration. *B. Upper trace*: ionophoretic currents. Effect of calcium withdrawal on maximal responses to prolonged dopamine ejections (unpublished experiments of B. L. Ginsborg, C. R. House, and M. R. Mitchell). *C*. Effect of calcium lack on amylase secretion induced by acetylcholine [from (122), with permission].

cation (75). Responses were obtained in calcium-free solutions containing manganese, which inhibits calcium influx (63) in excitable cells. In the pancreas also, there is no compelling evidence that ^{45}Ca uptake is increased in parallel with the amplitude of the response (see 53). The evidence described thus far does not support the idea that a transmembrane influx of calcium is obligatory for the electrical response. That calcium is essential however, at least for the cockroach salivary gland and pancreas, is shown when repeated or prolonged stimuli are applied in its absence. In the cockroach (B. L. Ginsborg, C. R. House, M. R. Mitchell, cf Figure 4B) "strong" ionophoretic applications of dopamine evoke only a few successively smaller responses and then become ineffective; in the pancreas sustained stimulation by acetylcholine has been recently shown to produce only a transient response in the absence of calcium.

The effect of calcium withdrawal on electrical responses to repeated or sustained stimulation has not been tested for mammalian salivary or lacrimal glands but it has been tested on a counterpart, namely agonist induced release of ^{42}K or ^{86}Rb from preloaded tissue slices (see section on ion concentration changes). Whether calcium is present or not, agonists induce an immediate increase in efflux rate, but in the absence of calcium this subsides in 3 or 4 min, whereas in the presence of calcium the decline is considerably slower.

All the results described would be consistent with the idea that extracellular calcium was not an immediate requirement for agonists to evoke membrane conductance changes but that it was required to replenish a limited pool of calcium, which was emptied in some way during a sustained response. The "pool calcium" might be envisaged as bound to the outside of the luminal membrane or within it or to its inside surface, but in equilibrium with the external solution. In the terms of Figure 1, the electrical response is generated by A1 followed by C1. In the presence of external calcium, the pool is continuously replenished; in its absence it is gradually exhausted. This differs somewhat from previous schemes which propose that *sustained* responses are generated by a transmembrane influx of calcium produced by an increase in the calcium conductance of the contraluminal membrane (e.g. 115) and that these responses require two separate mechanisms for raising cytosolic calcium—one for brief and another for sustained responses.

Intracellular calcium injection There is now no doubt that an increase in cytosolic calcium does produce responses similar to those induced by agonists. Iwatsuki & Petersen (see 80) have given a direct demonstration by ionophoretic injection of brief pulses of calcium applied intracellularly to lacrimal and pancreatic acinar cells. Responses were obtained that were closely similar to those evoked by extracellularly applied agonists.

Since there is evidence that uncoupling of cells is associated with an increase in cytosolic calcium, the uncoupling of lacrimal and pancreatic acinar cells (see section on uncoupling), also demonstrated by Iwatsuki & Petersen, provides strong, albeit indirect, evidence that agonists can increase cytosolic calcium in these cells. Moreover the demonstration that lacrimal gland uncoupling is associated with hyperpolarization (80) disposes of the possibility that depolarization rather than calcium was the essential stimulus to uncoupling (see 140).

In summary there is now direct evidence that intracellular injection of calcium imitates the application of agonists and strong indirect evidence that agonists cause a rise in cytosolic calcium for two representative exocrine glands.

Two tests listed in Tables 2 and 3 which have provided evidence that responses might be due to an increase in cytosolic calcium are calcium readmission and A23187. In the first test, the tissue is exposed to a calcium-free solution for a period and then to a calcium-containing solution. The appearance of a response (in the absence of any agonist and in the presence of a blocking agent to inhibit the action of any neurotransmitter that might be released) is taken to imply that it was induced by an influx of calcium into the cytosol. The same conclusion is reached if the antibiotic A23187, which acts as an ionophore for divalent cations (51), evokes a response in the presence of calcium and not in its absence. Neither test of course provides evidence about the origin of the agonist-induced responses.

SECRETORY RESPONSES Sufficiently prolonged and severe reduction in calcium concentration abolishes or greatly reduces most agonist-induced secretory responses. Exceptions are those from mammalian salivary glands to β-adrenergic agonists and from the pancreas to secretin. Both of these are currently believed to be mediated by cyclic AMP (see section on cyclic nucleotides in salivary secretion). More unexpected is the insensitivity to calcium withdrawal of dopamine-induced secretion from the cockroach salivary gland (C. R. House and R. K. Smith, unpublished) and phenylephrine-induced potassium release from lacrimal gland tissue slices (Table 3).

CYCLIC NUCLEOTIDES IN SALIVARY SECRETION

The evidence that cyclic AMP may mediate bicarbonate-rich fluid secretion by pancreatic duct cells is reviewed by Case (26). Another secretion possibly mediated by cyclic AMP is the β-adrenoceptor-activated enzyme secretion from mammalian salivary glands. On the basis of a series of experiments made mainly on tissue slices between 1969 and 1974 by Schramm and Selinger and their colleagues, it was originally proposed that enzymes were secreted exclusively by such a process and that cholinergic and α-adrenergic activation leads only to a calcium-mediated secretion of fluid and potassium ions (138). It is now clear, however, that cholinergic and α-adrenergic receptor activation can also release amylase without producing a concomitant increase in cyclic AMP (97, 145). The same is true for the undecapeptides, substance P and eledoisin (135); these activate receptors that are different from cholinergic and adrenergic ones, since the peptides are not antagonized by atropine or phentolamine or propranolol. Cholinergic and α-adrenoceptor activation leads to an increase in cyclic GMP (145), which has been proposed as an alternative mediator to cyclic AMP. However, the peptides do not increase cyclic GMP (135) and furthermore cyclic GMP can be increased by sodium azide without a concomitant release of amylase (66). In summary it thus seems that amylase secretion brought about by cholinergic, α-adrenergic, and peptide receptor activation is probably not mediated by either cyclic AMP or cyclic GMP.

The evidence in favor of cyclic AMP mediating the β-receptor-activated amylase secretion (Figure 1, H2) is stronger, and there is general agreement that when isoproterenol causes an increase in cyclic AMP, amylase is released. However, an increase in cyclic AMP does not always parallel an increase in amylase release; (+)-propranolol carbachol, and α-adrenergic agonists inhibit cyclic AMP production by isoproterenol (24, 25, 66, 112) but none reduces the amount of amylase it releases (66). As Harper & Brooker point out (66), it is reasonable to regard cyclic AMP mediation as a useful working hypothesis rather than as an established fact. Cyclic AMP was originally regarded as acting independently of calcium, since amylase release was thought not to be affected by calcium withdrawal [(138) but see Figure 3 in (122)]. However, prolonged incubation of tissue slices in calcium-free media with high concentrations of EGTA do lead to a reduction of isoproterenol-induced amylase release without affecting the amount of cyclic AMP produced. Putney et al (128) have therefore suggested that cyclic AMP is not the final mediator of β-adrenoceptor-induced amylase release but an intermediate, which causes release of calcium from an intracellular store with a special relationship to the luminal membrane (see Figure 1, H3).

ENDOCRINE GLAND CELLS

Adrenal Chromaffin Cell

The chromaffin cell inspired much of the present interest in stimulus-secretion coupling and, indeed, the term itself (45). The early results (45–47) reviewed by Douglas (43) in 1968 soon led to the idea that the essential link was the entry of calcium, which then promoted exocytosis of the chromaffin granule contents in much the same way as it promotes the release of acetylcholine from motor-nerve terminals (see 89, 90).

Experimentally, catecholamine secretion can be evoked by the application of acetylcholine, which is the "natural agonist" and is released from cholinergic axons in vivo, and by a number of amines and peptides (see 43). Secretion can also be evoked by a rise in external potassium (e.g., 10, 46) and even by a relatively small increase in its concentration; in this it differs from exocrine glands (see section on preparations and stimuli) and, like other endocrine secretory cells, resembles the motor nerve terminal.

IONIC REQUIREMENTS FOR CATECHOLAMINE RELEASE   No direct test (see section on intracellular calcium injection) has yet shown either that an increase in cytosolic calcium is a sufficient stimulus for secretion or that agonists cause such an increase; nevertheless, a formidable array of results (reviewed in 43, 44) supports both concepts. The rapid inhibition of agonist-induced catecholamine release, which occurs when calcium is withdrawn from the external solution (by contrast with the slow effect on exocrine glands), suggests that secretagogues must cause a transmembrane influx to be effective. A striking "readmission response" is indirect evidence that a calcium influx is a sufficient stimulus. A clear-cut calcium dependent response to A21387 has not been obtained (28); an opposite anomaly is that secretion is stimulated by isotonic sucrose in the absence of external calcium (94, 129). Clearly neither result, however, prejudices the calcium hypothesis for agonist action (see also 142).

Acetylcholine-induced calcium influx   The way in which acetylcholine causes calcium to enter chromaffin cells is not known. One possibility is that acetylcholine, which is known to depolarize the cells, causes a nonspecific increase in their membrane permeability; this would explain, simply, how acetylcholine evokes secretion in the absence of cations other than calcium (see 43). An alternative view has been put forward by Kidokoro et al (91) who have found that acetylcholine-induced epinephrine release from the perfused rat adrenal medulla was (*a*) abolished by cobalt and (*b*) reduced by tetrodotoxin. Kidokoro et al argue that, by analogy with the action of cobalt on chromaffin cell action potentials (21) and elsewhere (see 8–9, 63) (*a*) shows that the calcium influx is entirely via voltage-sensitive calcium channels and is thus due to the acetylcholine-induced depolarization (rather than via channels opened by acetylcholine acting on receptors directly). Observation (*b*) implies a proportion of these channels are opened by the larger depolarization resulting from chromaffin cell sodium-dependent action potentials (16, 21) (see section on chromaffin cell action potentials). Earlier work had already suggested the existence of a sodium-influenced component of catecholamine release (11–12, 94, see also 142). It has also recently been reported (92) that veratridine, a substance that enhances sodium permeability of excitable cells (e.g. 111), evokes a calcium-dependent catecholamine release from perfused adrenal glands.

A model system, the clonal cell line derived from rat phaeochromocytoma (59), has also recently been used for studies on stimulus secretion coupling (41, 58, 116). In this preparation secretion is reported to be insensitive to tetrodotoxin (130).

Chromaffin cell action potentials Spontaneous action potentials have been recorded from chromaffin cells in culture with an extracellular suction electrode (21); evoked action potentials have been recorded in response to both electrotonic depolarization (Figure 5) and ionophoretically applied acetylcholine (16, 21) with intracellular electrodes. Action potentials have also been recorded from cells in fresh tissue slices; they are likely therefore to have a physiological significance (16). In the absence of sodium the action potential no longer overshoots, but signs of a regenerative response remain (Figure 5B). The large amplitude "spike" has been interpreted as a sodium action potential since it is blocked by tetrodotoxin but not by cobalt (see 9, 63). The small amplitude response which could be enhanced by replacing calcium with barium (see 50, 63) was therefore interpreted as a calcium response. With external recording it has been shown that acetylcholine causes a large increase in frequency of the action potentials (21), and it has been suggested that this corresponds to acetylcholine-induced catecholamine secretion.

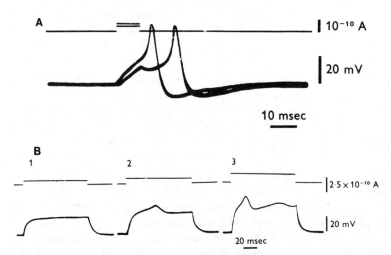

Figure 5 Intracellular records of responses of chromaffin cell to outward current pulses. *A*. Regenerative responses to brief pulses delivered to a cell bathed in standard saline. Resting potential, -45 mV. *B*. Depolarizing responses to pulses delivered to a cell bathed in sodium-free saline. The current intensity was increased threefold. Resting potential, -70 mV [from (21), with permission].

Pancreatic β-cell

Table 4 lists a number of stimulants of insulin secretion. All except D-galactose also produce changes in membrane potential. Since D-galactose is chemically close to D-glucose its comparative lack of activity is taken to imply that insulin secretion is not elicited by the interaction of glucose with a conventional "membrane" receptor. There must however be some target for glucose (or some metabolite) and it is convenient to call it the glucoreceptor.

NATURE OF THE GLUCORECEPTOR   Two findings favor the idea, which is widely held (e.g. 40, 61), that a glucose metabolite, rather than glucose, is the true agonist: (a) several monosaccharides not metabolized in the β-cell evoke no activity; and (b) mannoheptulose, which inhibits the formation of glucose-6-phosphate by hexokinase, also specifically prevents glucose-induced activity, since it does not affect tolbutamide stimulation (30, 40). Two metabolites which have been suggested as trigger molecules are glucose-6-phosphate (4) and D-glyceraldehyde (40).

In favor of the view that glucose acts via a membrane glucoreceptor (e.g. 93) Pace et al (114) found that even after glucose metabolism had been blocked by iodoacetate, glucose could still stimulate β-cells in the presence of 5 mM pyruvate as an alternative energy source. Pyruvate itself neither reverses idodacetate inhibition of glycolysis nor stimulates β-cells (100). Insulin release is biphasic, the initial transient reaching a peak within 2 min of the start of glucose stimulation (32) or less when tolbutamide is the stimulant (15). It is possible that the first phase is mediated via a membrane receptor, whereas the second sustained phase is related to an intracellular action. However, biochemical attempts to resolve the question of the site(s) of the glucoreceptor have as yet not been successful (see also 73, 101).

An interesting in vitro insulin-releasing system, described by Davis & Lazurus (35, see also 96), consists of β-cell membrane incubated with insulin granules, and it does not metabolize glucose. Nevertheless in the presence of ATP and calcium, glucose evokes the release of a very large fraction of the insulin within 10 min. Moreover glucose-6-phosphate also acts as a stimulant. The authors have therefore proposed that there is a membrane glucoreceptor mediating secretion and that it is not specific for glucose.

CALCIUM AND INSULIN RELEASE   Indirect evidence that a calcium influx is responsible for insulin release (see e.g. 98–99) resembles that for catecholamine release (see section on ionic requirements for catecholamine release). Thus glucose-induced secretion is rapidly abolished

Table 4 Chemical stimulants of insulin secretion and electrical activity in pancreatic beta cells

Stimulant	Insulin secretion	Electrical activity
D-glucose	Biphasic response (32)	Slow depolarizing waves with superimposed spikes (37) Spike frequency shows biphasic characteristics (14)
D-galactose[a]	Small multiphasic response, which can be potentiated by theophylline (93)	No effect (40)[b]
Tolbutamide	Initial rapid release followed by very small maintained secretion (32)	Depolarization and spikes for up to 40 min, then block developed (but no repolarization) (40)
Ouabain	Time course of secretory response depends on concentration (65)	Gradual depolarization followed by continuous spikes (104)
High $[K]_o$	Initial rapid release (57)	Depolarization but no spikes (6, 38)

[a]No effect observed in rabbit pancreas by Coore & Randle (30) who cite additional evidence, however, that D-galactose is effective in the dog.
[b]Conditions of experiment different from that of Landgraf et al (93).

by calcium withdrawal (60-62, 65), in contrast to the slow effect of sodium withdrawal (64), and by the addition of cobalt (69). The stimulant action of A23187 in the absence of calcium has been taken to imply that the release of calcium from intracellular binding sites suffices to promote insulin release (but see 3). In view of the recent investigations of Atwater and her colleagues, to be described below (see section on electrical activity of β-cells), the fact that quinine stimulates insulin release in the absence of glucose (68) and that TEA potentiates the action of glucose but does not release insulin by itself (see 67) may also be regarded as indirect evidence for calcium as an intermediate in insulin release.

Sodium and insulin secretion It seems likely that sodium is involved in insulin release not directly, but, as in the case of catecholamine release from chromaffin cells, by increasing the calcium influx during the β-cell action potential. More speculatively, it has been suggested that the increase in intracellular sodium that results from an influx of sodium leads to an increase in cytosolic calcium by (*a*) a transmembrane exchange of sodium for calcium and (*b*) by release of calcium from binding sites (see 42).

Cyclic AMP and insulin release The evidence consistent with cyclic AMP acting as an obligatory intermediate in insulin release is scanty and consists in the augmentation of glucose-induced release by theophylline and by dibutyryl cyclic AMP (22). In opposition to this idea however are the following findings: (*a*) glucose neither increases cyclic AMP (108) nor activates adenylate cyclase (34), and (*b*) caffeine and theophylline increase β-cell cyclic AMP but they do not release insulin (e.g. 29).

ELECTRICAL ACTIVITY OF β-CELLS In spite of the small size of β-cells a number of investigators have succeeded in impaling them with an intracellular microelectrode and studying the changes in membrane potential induced by glucose and other agonists (e.g. 5-7, 13, 14, 37, 38, 40, 102, 105, 106, 113). The first results were reported by Dean & Matthews (37, 38).

The response to glucose may be summarized as follows: when its concentration exceeds about 5 mM, β-cells depolarize by about 10 mV from their resting potential (-30 to -50 mV) in a series of "slow waves" with superimposed "spikes" (Figure 6), which are about 20 mV in amplitude and 50-200 msec in duration. At a glucose concentration of 2.8 mM, no spikes occur and at about 17 mM, the slow waves fuse and the train of spikes becomes continuous, the frequency increasing as the glucose concentration is raised (37, 38, 40, 106). The spikes appear to be generated mainly by calcium inflow, since (*a*) their amplitude varies with external calcium but not with sodium concentration (38), (*b*) they are not blocked or greatly reduced in amplitude by tetrodotoxin (106), but (*c*) they are blocked by D-600 (see 102). The implication of this result is that the β-cell membrane has a well-developed voltage-sensitive calcium system (see 63), such that depolarization leads to a calcium influx. Recent investigations by Atwater and her colleagues (5–7) go some way to show how the slow waves are generated. In essence their results are in agreement with the suggestion of Matthews that the potassium permeability is determined by the intracellular calcium concentration [see Meech (103) for review] and that glucose reduces either the cytosolic calcium or the sensitivity of the calcium-potassium channel interaction. Evidence already existed (20, 67, 139) from studies of potassium efflux and concentration that glucose could produce a reduction in potassium permeability. The effect of glucose is thus to shut off part of the potassium conductance, which in turn leads to a depolarization that opens the voltage-sensitive calcium channels and leads to the spiking activity and calcium entry. The intracellular calcium concentration rises and reopens the calcium-sensitive potassium channels so that the cell repolarizes and the cycle starts once again. As

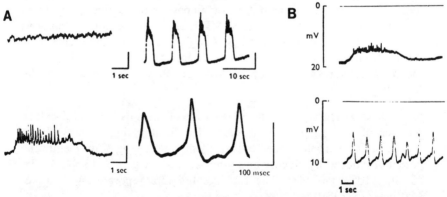

Figure 6 Spikes induced in islet cells by D-glucose. *A*. Intracellular recordings in presence of normal calcium concentration. Control record (*top left*) in the presence of 2.8 mM glucose. Rhythmic bursts of spikes superimposed on slow waves (*top right*) elicited by 16.6 mM glucose. Record of a single burst (*lower left*) in presence of 16.6 mM glucose. Record (*lower right*) of individual spikes. In all records vertical calibrations, 2 mV. *B*. Effect of absence of external calcium on spikes induced in islet cells by D-glucose. A normal burst of spikes (*top*) induced by 11.1 mM glucose in presence of 2.56 mM calcium. Continuous firing of spikes (*bottom*) in zero calcium and 11.1 mM glucose [from (37, 38), with permission].

already mentioned, the actions of quinine and TEA on insulin release are easily explained on the basis of this hypothesis and their actions on the electrical events support it. Quinine which is known to "shut" calcium-sensitive potassium channels (2) would be expected to act in a similar way to glucose and was found to do so. On the other hand, TEA inhibits the opening of voltage-sensitive potassium channels. It would therefore be expected to prolong and enhance the "spikes" (since repolarization is brought about by an increase in potassium permeability), with a resulting increase in calcium flux, but not to initiate them, and this too was the result obtained.

CONCLUSIONS

Electrophysiological studies of secretory cells have made significant progress in the last few years. Evidence has been obtained for the existence of conductance changes induced in the cell membrane of exocrine cells by neurotransmitters and for the excitability of endocrine cells. It has generally been agreed that changes in intracellular ion concentrations, especially that of calcium, serve as a stimulus for secretion, but how they do so remains mysterious. A major obstacle to progress is that there is no method for monitoring secretory changes fast enough to match electrical recording, so that even the order of events between stimulus and secretion is uncertain.

Literature Cited

1. Albano, J., Bhoola, K. D., Heap, P. F., Lemon, M. J. C. 1976. *J. Physiol. London* 258:631–58
2. Armando-Hardy, M., Ellory, J. C., Ferreira, H. G., Fleminger, S., Lew, V. L. 1975. *J. Physiol. London* 250: 32–33P
3. Ashby, J. P., Speake, R. N. 1975. *Biochem. J.* 150:89–96
4. Ashcroft, S. J. H., Hedeskov, C. H., Randle, P. J. 1970. *Biochem. J.* 118:143–47
5. Atwater, I., Dawson, C. M., Ribalet, B., Rojas, E. 1979. *J. Physiol. London* 288:575–88
6. Atwater, I., Ribalet, B., Rojas, E. 1978. *J. Physiol. London* 278:117–39
7. Atwater, I., Ribalet, B., Rojas, E. 1979. *J. Physiol. London* 288:561–74
8. Baker, P. F., Hodgkin, A. L., Ridgway, E. B. 1971. *J. Physiol. London* 218:709–55
9. Baker, P. F., Meves, H., Ridgway, E. B. 1973. *J. Physiol. London* 231:511–26
10. Baker, P. F., Rink, T. J. 1975. *J. Physiol. London* 253:593–620
11. Banks, P. 1967. *J. Physiol. London* 193:631–37
12. Banks, P., Biggins, R., Bishop, R., Christian, B., Currie, N. 1969. *J. Physiol. London* 200:797–805
13. Beigelman, P. M., Ribalet, B., Atwater, I. 1977. *J. Physiol. Paris* 73:201–17
14. Beigelman, P. M., Thomas, L. J., Shu, M. J., Bessman, S. P. 1976. *J. Physiol. Paris* 72:721–28
15. Bennett, L. L., Curry, D. L., Curry, K. 1973. *Proc. Soc. Exp. Biol. Med.* 144:436–39
16. Biales, B., Dichter, M., Tischler, A. 1976. *J. Physiol. London* 262:743–53
17. Blackman, J. G., Ginsborg, B. L., House, C. R. 1979. *J. Physiol. London* 287:67–80
18. Blackman, J. G., Ginsborg, B. L., House, C. R. 1979. *J. Physiol. London* 287:81–92
19. Bogart, B. I., Picarelli, J. 1978. *Am.*

20. Boschers, A. C., Kawazu, S., Duncan, G., Malaisse, W. J. 1977. *FEBS Lett.* 83:151–54
21. Brandt, B. L., Hagiwara, S., Kidokoro, Y., Migazaki, S. 1976. *J. Physiol. London* 262:417–40
22. Brisson, G. R., Malaisse-Lagne, F., Malaisse, W. J. 1972. *J. Clin. Invest.* 51:232–41
23. Burgess, G. M., Claret, M., Jenkinson, D. H. 1979. *Nature* 279:544–46
24. Butcher, F. R., Goldman, J. A., Nemerovski, M. 1975. *Biochim. Biophys. Acta* 392:82–94
25. Butcher, F. R., Rudich, L., Emler, C., Nemerovski, M. 1976. *Mol. Pharmacol.* 12:862–70
26. Case, R. M. 1978. *Biol. Rev. Cambridge Philos. Soc.* 53:211–354
27. Chandler, D. E., Williams, J. A. 1978. *J. Cell. Biol.* 76:371–85
28. Cochrane, D. E., Douglas, W. W., Mouri, T., Nakazato, Y. 1975. *J. Physiol. London* 252:363–78
29. Cooper, R. H., Ashcroft, S. J. H., Randle, P. J. 1973. *Biochem. J.* 134:599–605
30. Coore, H. G., Randle, P. J. 1964. *Biochem. J.* 93:66–78
31. Creed, K. E., Wilson, J. A. P. 1969. *Aust. J. Exp. Biol. Med. Sci.* 47:135–44
32. Curry, D. L., Bennett, L. L., Grodsky, G. M. 1968. *Endocrinology* 83:572–84
33. Dagorn, J. C., Estival, A. 1979. *J. Physiol. London* 290:51–58
34. Davis, B., Lazarus, N. R. 1972. *Biochem. J.* 129:373–79
35. Davis, B., Lazarus, N. R. 1976. *J. Physiol. London* 256:709–29
36. Davison, J. S., Pearson, G. T. 1979. *J. Physiol. London.* 295:45P
37. Dean, P. M., Matthews, E. K. 1970. *J. Physiol. London* 210:255–64
38. Dean, P. M., Matthews, E. K. 1970. *J. Physiol. London* 210:265–75
39. Dean, P. M., Matthews, E. K. 1972. *J. Physiol. London* 225:1–13
40. Dean, P. M., Matthews, E. K., Sakamoto, Y. 1975. *J. Physiol. London* 246:459–78
41. Dichter, M. A., Tischler, A. S., Greene, L. A. 1977. *Nature* 268:501–4
42. Donatsch, P., Lowe, D. A., Richardson, B. P., Taylor, P. 1977. *J. Physiol. London* 267:357–76
43. Douglas, W. W. 1968. *Br. J. Pharmacol.* 34:451–74
44. Douglas, W. W. 1977. *Ciba Found. Symp.* 54:61–87
45. Douglas, W. W., Rubin, R. P. 1961. *J. Physiol. London* 159:40–57
46. Douglas, W. W., Rubin, R. P. 1963. *J. Physiol. London* 167:288–310
47. Douglas, W. W., Rubin, R. P., 1964. *J. Physiol. London* 175:231–41
48. Douglas, W. W., Taraskevich, P. S. 1978. *J. Physiol. London* 285:171–84
49. Fain, J. N., Berridge, M. J. 1979. *Biochem. J.* 178:45–69
50. Fatt, P., Ginsborg, B. L., 1958. *J. Physiol. London* 142:516–43
51. Foreman, J. C., Mongar, J. L., Gomperts, B. D. 1973. *Nature* 245:249–51
52. Gallacher, D. V. 1979. *J. Physiol. London.* 295:46P
53. Gardner, J. D. 1979. *Ann. Rev. Physiol.* 41:55–66
54. Garrett, J. R., Thulin, A. 1975. *Cell. Tissue Res.* 159:179–93
55. Garrett, J. R., Thulin, A., Kidd, A. 1978. *Cell Tissue Res.* 188:235–50
56. Ginsborg, B. L., House, C. R., Silinsky, E. M. 1976. *J. Physiol. London* 262:489–500
57. Gomez, M., Curry, D. L. 1973. *Endocrinology* 92:1126–32
58. Greene, L. A., Rein, G. 1977. *Brain Res.* 138:521–28
59. Greene, L. A., Tischler, A. S. 1976. *Proc. Natl. Acad. Sci. USA* 73:2424–28
60. Grodsky, G. M. 1972. *Diabetes* 21:Suppl. 2, pp. 584–93
61. Grodsky, G. M., Batts, A. A., Bennett, L. L., Vealla, C., McWilliams, N. B., Smith, D. F. 1963. *Am. J. Physiol.* 205:638–51
62. Grodsky, G. M., Bennett, L. L. 1966. *Diabetes* 15:910–13
63. Hagiwara, S. 1975. In *Membranes: Lipid Bilayers and Biological Membranes. Dynamic Properties*, ed. G. Eisenmann, pp. 359–81. New York: Dekker
64. Hales, C. N., Milner, R. D. G. 1968. *J. Physiol. London* 194:725–43
65. Hales, C. N., Milner, R. D. G. 1968. *J. Physiol. London* 199:177–87
66. Harper, J. F., Brooker, G. 1978. *Mol. Pharmacol.* 14:1031–45
67. Henquin, J. C. 1978. *Nature* 271:271–73
68. Henquin, J. C., Horemans, B., Nenquin, M., Verniers, J., Lambert, A. E. 1975. *FEBS Lett.* 67:371–74
69. Henquin, J. C., Lambert, A. E. 1975. *Am. J. Physiol.* 228:1669–77
70. Herzog, V., Sies, H., Miller, F. 1976. *J. Cell. Biol.* 70:692–706
71. Hill-Smith, I., Purves, R. D. 1978. *J. Physiol. London* 279:31–54

72. House, C. R., Ginsborg, B. L., Mitchell, M. R. 1978. *Nature* 275:663
73. Idahl, L. A. 1973. *Diabetologia* 9:403-12
74. Iwatsuki, N., Petersen, O. H. 1977. *Nature* 268:147-49
75. Iwatsuki, N., Petersen, O. H. 1977. *J. Physiol. London* 269:723-33
76. Iwatsuki, N., Petersen, O. H. 1977. *J. Physiol. London* 269:735-51
77. Iwatsuki, N., Petersen, O. H. 1978. *J. Physiol. London* 274:81-96
78. Iwatsuki, N., Petersen, O. H. 1978. *J. Physiol. London* 275:507-20
79. Iwatsuki, N., Petersen, O. H. 1978. *Pfluegers Arch.* 377:185-87
80. Iwatsuki, N., Petersen, O. H. 1978. *J. Cell. Biol.* 79:533-45
81. Iwatsuki, N., Petersen, O. H. 1978. *J. Clin. Invest.* 61:41-46
82. Iwatsuki, N., Petersen, O. H. 1979. *J. Physiol. London* 292:81-82P
83. Jamieson, J. D., Palade, G. E. 1971. *J. Cell. Biol.* 50:135-58
84. Jirakulsomchok, D., Schneyer, C. A. 1979. *Am. J. Physiol.* 236:E371-85
85. Kagayama, M., Nishiyama, A. 1974. *J. Physiol. London* 242:157-72
86. Kanno, T., Saito, A., Sato, Y. 1977. *J. Physiol. London* 270:9-28
87. Kanno, T., Yamamoto, M. 1977. *J. Physiol. London* 264:787-99
88. Kater, S. B. 1977. In *Approaches to the Cell Biology of Neurons*, ed. W. M. Cohen, J. A. Ferrendelli, pp. 195-215. Bethesda: Soc. Neurosci.
89. Katz, B. 1969. *The Release of Neural Transmitter Substances* Liverpool: Liverpool Univ. Press. 60 pp.
90. Katz, B., Miledi, R. 1967. *J. Physiol. London* 189:535-44
91. Kidokoro, Y., Ritchie, A. K., Hagiwara, S. 1979. *Nature* 278:63-65
92. Kirkepar, S. M., Prat, J. C. 1979. *Proc. Natl. Acad. Sci. USA* 76:2081-83
93. Landgraf, R., Kotler-Brajtburg, J., Matschinsky, F. M. 1971. *Proc. Natl. Acad. Sci. USA* 68:536-40
94. Lastowecka, A., Trifaró, J. M. 1974. *J. Physiol. London* 236:681-705
95. Laugier, R. 1979. *J. Physiol. London.* 295:43P
96. Lazarus, N. R., Davis, B., O'Connor, K. J. 1976. *J. Physiol. Paris* 72:787-94
97. Leslie, B. A., Putney, J. W., Sherman, J. M. 1976. *J. Physiol. London* 260:351-70
98. Malaisse, W. J. 1973. *Diabetologia* 9:167-73
99. Malaisse, W. J., Herchuelz, A., Devis, G., Somers, G., Boschero, A. C., Hutton, J. C., Kawazu, S., Sener, A., Atwater, I. J., Duncan, G., Ribalet, B., Rojas, E. 1978. *Ann. NY Acad. Sci.* 307:562-82
100. Matschinsky, F. M., Ellerman, J. 1973. *Biochem. Biophys. Res. Commun.* 50:193-99
101. Matschinsky, F. M., Ellerman, J. E., Krzanowski, J., Kotler-Brajtburg, J., Landgraf, R., Fertel, R. 1971. *J. Biol. Chem.* 246:1007-11
102. Matthews, E. K., Sakamoto, Y. 1975. *J. Physiol. London* 246:421-37
103. Meech, R. W. 1978. *Ann. Rev. Biophys. Bioeng.* 7:1-8
104. Meissner, H. P. 1976. *J. Physiol. Paris* 72:757-67
105. Meissner, H. P., Atwater, I. 1976. *Horm. Metab. Res.* 8:11-15
106. Meissner, H. P., Schmelz, H. 1974. *Pfluegers Arch.* 351:195-206
107. Mitchell, M. R., Ginsborg, B. L., House, C. R. 1979. *Experientia.* In press
108. Montague, W., Cook, J. R. 1971. *Biochem. J.* 122:115-20
109. Niedergerke, R., Page, S. 1977. *Proc. R. Soc. London Ser. B* 197:333-62
110. Nishiyama, A., Petersen, O. H. 1974. *J. Physiol. London* 242:173-88
111. Ohta, M., Narahashi, T., Keeler, R. F. 1973. *J. Pharmacol. Exp. Ther.* 184:143-54
112. Oron, Y., Kellogg, J., Larner, J. 1978. *Mol. Pharmacol.* 14:1018-30
113. Pace, C. S., Price, S. 1974. *Endocrinology* 94:142-47
114. Pace, C. S., Stillings, S. N., Hover, B. A., Matschinsky, F. M. 1975. *Diabetes* 24:489-96
115. Parod, R. J., Putney, J. W. 1978. *J. Physiol. London* 281:371-81
116. Patrick, J., Stallcup, W. B. 1977. *Proc. Natl. Acad. Sci. USA* 74:4689-92
117. Petersen, O. H. 1979. *Electrophysiology of Gland Cells* London: Academic. In press
118. Petersen, O. H., Iwatsuki, N. 1978. *Ann. NY Acad. Sci.* 307:599-615
119. Petersen, O. H., Pedersen, G. L. 1974. *J. Membr. Biol.* 16:353-62
120. Petersen, O. H., Philpott, H. G. 1979. *J. Physiol. London* 290:305-15
121. Petersen, O. H., Ueda, N. 1976. *J. Physiol. London* 254:583-606
122. Petersen, O. H., Ueda, N., Hall, R. A., Gray, T. A. 1977. *Pfluegers Arch.* 372:231-237
123. Philpott, H. G. 1979. *Pfluegers Arch.* In press

124. Poulsen, J. H., Oakley, B. 1979. *Proc. R. Soc. London Ser. B* 204:99–104
125. Poulsen, J. H., Williams, J. A. 1977. *J. Physiol. London* 264:323–39
126. Putney, J. W. 1978. *Am. J. Physiol.* 235:C180–87
127. Putney, J. W., VandeWalle, C. M., Leslie, B. A. 1978. *Am. J. Physiol.* 235:C188–98
128. Putney, J. W., Weiss, S. J., Leslie, B. A., Marier, S. H. 1977. *J. Pharmacol. Exp. Ther.* 203:144–55
129. Rink, T. J. 1977. *J. Physiol. London* 266:297–325
130. Ritchie, A. K. 1979. *J. Physiol. London* 286:541–61
131. Roberts, M. L., Iwatsuki, N., Petersen, O. H. 1978. *Pfluegers Arch.* 376:159–67
132. Roberts, M. L., Petersen, O. H. 1978. *J. Membr. Biol.* 39:297–312
133. Rose, R. C., Schultz, S. G. 1971. *J. Gen. Physiol.* 57:639–63
134. Rothman, S. S. 1977. *Ann. Rev. Physiol.* 39:373–89
135. Rudich, L., Butcher, F. R. 1976. *Biochim. Biophys. Acta* 444:704–11
136. Scheele, G. A., Haymovits, A. 1978. *Ann. NY Acad. Sci.* 307:648–52
137. Schneyer, L. H., Young, J. A., Schneyer, C. A. 1972. *Physiol. Rev.* 52:720–77
138. Schramm, M., Selinger, Z. 1975. *J. Cycl. Nucl. Res.* 1:181–92
139. Sehlin, J., Taljidal, I. B. 1975. *Nature* 253:635–36
140. Sheridan, J. D. 1978. *Intercellular Junctions and Synapses*, ed. J. Feldman, N. B. Gilula, J. D. Pitts, pp. 39–59. London: Chapman & Hall
141. Smith, R. K., House, C. R. 1977. *Experientia* 33:1182–83
142. Trifaró, J. M. 1977. *Ann. Rev. Pharmacol. Toxicol.* 17:27–47
143. Ueda, N., Petersen, O. H. 1977. *Pfluegers Arch.* 370:179–83
144. Williams, J. A., Korc, M., Dormer, R. L. 1978. *Am. J. Physiol.* 235:E517–24
145. Wojcik, J. D., Grand, R. J., Kimberg, D. V. 1975. *Biochim. Biophys. Acta* 411:250–62

X-RAY DIFFRACTION STUDIES OF THE HEART

♦9142

I. Matsubara

Department of Pharmacology, Tohoku University School of Medicine, Seiryo-machi, Sendai, Japan 980

USE OF X-RAY DIFFRACTION IN PHYSIOLOGY OF HEART MUSCLE

The contractile apparatus of vertebrate heart muscle is similar to that of vertebrate skeletal muscle when viewed in the electron microscope (52, 61). The sarcomeres contain regular, overlapping arrays of thick and thin filaments as in skeletal muscle, and the length of these filaments are about the same as those of skeletal myofilaments. The amounts of the major myofibrillar proteins (myosin, actin, tropomyosin, and troponin) are comparable to those found in skeletal muscle (34). Based on such structural and biochemical similarities, the contractile mechanism of heart muscle is believed to be essentially the same as that of skeletal muscle. According to currently accepted models of skeletal muscle contraction (14, 20, 21, 25), the myosin heads that project from the backbone of the thick filament react with actin molecules aligned on the thin filament. A relative sliding force between the two types of filaments is produced in the course of the reaction, leading to contraction of muscle. These models have also been applied to heart muscle to explain the molecular basis of its mechanical properties.

However, heart muscle has important mechanical properties that are not shared by skeletal muscle. A typical example is the staircase phenomenon in which the contractile tension of heart muscle increases with the frequency of contraction (4). The twitch tension of skeletal muscle, in contrast, is relatively independent of the frequency. Many such differences between the two types of muscle have been ascribed to differences in the events responsible for excitation-contraction coupling.

In skeletal muscle, excitation of the surface membrane causes an increase in the myoplasmic calcium concentration by releasing calcium from an intracellular store (the sarcoplasmic reticulum) (10). On the other hand, in heart muscle there are at least two mechanisms by which the myoplasmic calcium is increased: calcium influx across the surface membrane and calcium release from a store (for recent reviews see 11, 31). The contribution of these two components of calcium to contractile activity depends on a number of factors, including the heart rate. Also in other respects, myofilaments in heart muscle do differ from those in skeletal muscle, for example in the speed of tension development. These differences probably contribute to specific features of the contractile performance of the heart. Therefore, simple application of models derived from studies with skeletal muscle is unlikely to lead to a better understanding of cardiac function. For these reasons, detailed study of the functional behavior of the contractile proteins will be necessary.

Various mechanical experiments have been carried out to assess the state of contractile proteins in heart muscle during contraction. These experiments involve, for instance, measurements of tension responses to quick stretch and release, or measurements of the maximum shortening speed of muscles released during isometric contraction (for review see 5). Analysis of such measurements is usually based on mechanical models that consist of three independent components: a contractile component, a series elastic component, and a parallel elastic component. Two types of models, which differ in the positions of the series and parallel elastic components relative to one another, have been used: i.e. the Maxwell model, in which the two elastic components are connected in parallel, and the Voigt model, in which these are connected in series. The analysis gives the force and the length of each component as a function of time. Some of these quantities, for example the time course of the force produced by the contractile component at a constant length (i.e. the intensity of the active state) (19), may provide insight concerning the time course of molecular events underlying contraction. However, serious limitations of the models have been pointed out in both skeletal and heart muscles, suggesting the difficulties in studying the molecular events by means of such models (55, 59). Some of the limitations are as follows: (*a*) the elastic components do not seem to be independent of the contractile component, since the muscle compliance changes as the tension develops (22); (*b*) it is difficult to relate each component to a separate structural component in muscle (32); (*c*) correct choice between the Maxwell and Voigt models is often difficult in heart muscle and, indeed, varies from one type of experiment to the other (3). These limitations indicate that mechanical analysis must incorporate structural information and that a more direct method of studying the molecular events is required.

X-ray diffraction is suitable, in principle, for direct observation of molecules in living muscle. This technique can be regarded as microscopy using X rays instead of light or electrons (37). In optical and electron microscopes a magnified image of an object is obtained, first by scattering light or electrons with the object, and then by converging these with lenses to form an image. With X rays it is not practicable to form an image from scattered X rays, since there is no lens to converge X rays. Instead, an image is obtained through calculations (Fourier transform) based on measurements of the scattered X rays. In other words, X-ray diffraction is microscopy in which the function of lenses is carried out by computation. Unfortunately the computation cannot be as efficient as lenses, since vital information about scattered X rays is lost in the course of measurements; the phases of the scattered X rays cannot be measured. This "phase problem," in addition to the low resolution presently available, makes X-ray diffraction a rather imperfect type of microscopy.

However, X-ray diffraction does have important advantages when applied to biological materials. First, unlike electrons, X rays can penetrate thick layers of water and tissue, allowing study of the internal structure of living muscle in physiological solutions. Secondly, unlike light, X rays have wavelengths short enough to resolve structures at the molecular level. Recently, significant improvements have been made in the intensity of X-ray beams and the speed of recording diffraction patterns. By combining these improvements it has become possible to record diffraction patterns from skeletal and heart muscles on a very short time basis (27, 47). This has made feasible the study of movements of contractile proteins during contraction under various physiological conditions.

Thus X-ray diffraction provides a direct means for clarifying the molecular mechanisms underlying characteristic features of heart muscle. However, the X-ray diffraction method can only examine certain aspects of these mechanisms. Contraction of heart muscle is a complex phenomenon, indeed a chain of events: excitation of the cell membrane—an increase in myoplasmic calcium concentration—calcium binding to the regulatory proteins—interaction between contractile proteins—tension production. X-ray diffraction gives information about the event, directly linked to tension production, and therefore may clarify immediate causes of characteristic mechanical phenomena in heart muscle. When such an immediate cause is found, the next step is to investigate whether it originates from peculiar properties of cardiac contractile proteins or from an earlier event in the chain.

MUSCLES IN THE QUIESCENT STATE AND IN RIGOR

Heart muscle preparations suitable for X-ray diffraction are papillary and trabecular muscles in which fibers are to a large extent aligned parallel to one another. These muscles are usually dissected from the right ventricle of a freshly killed animal. Hearts of various mammals have been studied, including dogs, cats, guinea pigs, goats, and rabbits (41, 43). Rabbit hearts give poorer diffraction patterns, but otherwise little difference has been found between the species (43).

In the first myocardial X-ray studies the muscles were held in a specimen chamber filled with Tyrode solution and the diffraction patterns were recorded on X-ray film (42, 43). As the exposure time required (30 min–12 hr) was considerably longer than the duration of contraction, patterns could be obtained only from muscles under static conditions. Two types of static preparation were used: (*a*) Living resting muscles with little or no spontaneous activity. This state may be called "quiescent state" to distinguish it from the resting state between cyclic contractions, i.e. the diastolic state. (*b*) Muscles in rigor. These are either dead muscles or glycerol-extracted muscles.

Meridional Patterns From Quiescent Muscles

Diffraction patterns of muscles are often divided into the meridional and equatorial patterns. The word *meridional* is used in a broad sense; the meridional pattern consists of reflections on the meridian, which is the pattern's center line parallel to the muscle axis, and those in the near-meridional region. These reflections are associated with the periodic structure along the muscle axis, for example regular packing of protein molecules in each myofilament. The equatorial pattern consists of reflections on the equator, which is perpendicular to the muscle axis. The equatorial reflections are associated with regular side-to-side packing of myofilaments in each sarcomere.

The meridional pattern recorded from quiescent heart muscle shows the reflections associated with myosin, as well as those associated with actin (Figure 1*a*). The most prominent feature of the pattern is a series of reflections on the meridian at orders of 42.6 nm. The second (21.3 nm), third (14.2 nm), and sixth (7.1 nm) order reflections are strong,

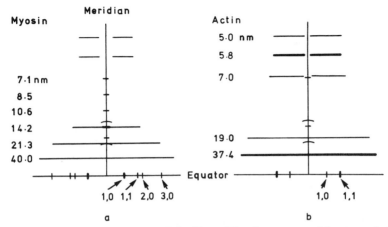

Figure 1 Schematic representation of the X-ray diffraction patterns of heart muscle (*a*) in the quiescent state and (*b*) in rigor. Only main reflections are shown. The arcs on the meridian represent collagen reflections at 21.3 and 12.8 nm.

whereas the fourth (10.6 nm) and fifth (8.5 nm) order reflections are weak. Each meridional reflection is accompanied by near-meridional reflections, forming a layer line at each latitude (for the first order layer line, see below). In patterns recorded from vertebrate skeletal muscle in the living resting condition, the same meridional reflections and layer lines are seen at orders of 42.9 nm and have been ascribed to a helical arrangement of myosin projections along the thick filament. Three different models of the helical arrangement are known to be consistent with the meridional pattern of skeletal muscle, although it is not as yet possible to distinguish between them; these are models with two-stranded (Figure 2*a*), three-stranded, and four-stranded helices formed by myosin molecules (29, 60). The following features are common to all these models: (*a*) myosin projections emerge from the backbone of the thick filament at 14.3-nm intervals along the filament axis; (*b*) the orientation of the projections repeats itself every 42.9 nm (=3×14.3 nm). In heart muscle it is also impossible to choose between the three models. However, the diffraction pattern indicates that the helical arrangement of the projections in heart muscle shares the above two features, except that the repeats are slightly smaller (14.2 and 42.6 nm).

Although the myosin-related diffraction patterns of heart and skeletal muscles are very similar, detailed comparison reveals some differences (43): (*a*) The intensities of all layer lines (relative to the equatorial reflections) in the heart muscle pattern are much weaker than those in the skeletal muscle pattern. (*b*) In the skeletal muscle pattern the first order layer line occurs at the axial position (i.e. the latitude) expected from the myosin periodicity (42.9 nm). On the other hand, in the heart muscle pattern the first layer line occurs at the axial position of 40.0 nm, which is significantly different from that expected from the myosin periodicity (42.6 nm).

The weakness of the layer lines suggests that the myosin projections are less well ordered in heart muscle than in skeletal muscle. Two explanations for the poor ordering have been proposed. First, some muscle fibers in the preparation might have gone into rigor during the 6–12 hr exposure required for recording the meridional pattern; when fibers go into rigor the myosin projections attach to actin, resulting in poor ordering (see section on meridional patterns from muscles in rigor). Secondly, some actin-myosin interaction may normally be present, even in the quiescent state. To date, it is not clear which is the correct explanation. This ambiguity might be resolved by using either an intense X-ray source or a more sensitive detector; these will considerably shorten the time required for recording the meridional pattern,

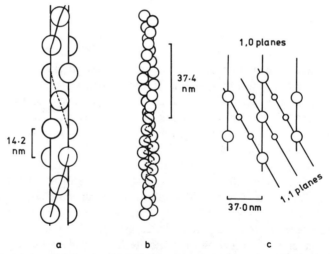

Figure 2 (*a*) A model showing the helical arrangement of myosin molecules in the thick filament. In this model myosin projections form a two-stranded helix of the type proposed by Huxley & Brown (29); two myosin projections emerge every 14.2 nm. Each projection is represented by a sphere in the same manner as in Figure 3. (*b*) A model showing the arrangement of actin in the thin filament (58). Actin subunits form two right-handed strands of pitch 74.8 nm, wound helically around one another with a cross-over repeat of 37.4 nm. The alternative description, a left-handed primitive helix of 5.8-nm pitch, is shown in the lower half of the model. (*c*) A model showing the hexagonal arrangement of myofilaments in the A-band at a sarcomere length 2.2 μm. The large and small circles represent the cross-sections of thick and thin filaments respectively. Note that the 1,0 planes consist only of thick filaments, whereas the 1,1 planes consist of thick and thin filaments with a ratio 1:2.

thereby reducing the possibility of fibers going into rigor during the exposure.

The apparent shift of the first layer line in the heart muscle pattern has been attributed to an overlap of the actin layer line at 37.4 nm (Figure 1b) with the myosin layer line at 42.6 nm; the overlapping layer lines would give the appearance of a reflection at an intermediate axial position. This explanation is based on an assumption that the actin layer lines are intensified by attachment of some myosin projections to the thin filament.

In addition to the myosin reflections, the pattern includes weak near-meridional reflections at 5.0 and 5.8 nm (Figure 1a). The same reflections are seen in the patterns recorded from skeletal muscle (29) and have been ascribed to the primitive helices of actin molecules in the thin filament (Figure 2b). The heart muscle pattern also includes a weak meridional reflection at 18.5 nm. This seems to represent the second order of a 37.0-nm repeat, since faint meridional reflections are occasionally seen at the first, fourth, fifth, and sixth orders of this spacing. In the skeletal muscle pattern, the corresponding reflections are seen at orders of 38.3 nm and have been associated with a regular arrangement of troponin and tropomyosin in the thin filament (29, 57).

Meridional Patterns From Muscles in Rigor

The pattern from heart muscle in rigor is similar to that from skeletal muscle in rigor (29, 43) (Figure 1b). The pattern is characterized by strong reflections arising from the regular alignment of actin in the thin filament. The innermost layer line occurs at 37.4 nm and the second layer line is at 19.0 nm. The near-meridional reflections at 5.8 and 5.0 nm are greatly increased in intensity, compared with those in the "quiescent" pattern. Other reflections from actin periodicities are seen in the near-meridional region at axial spacings of 7.0 and 4.5 nm.

The pattern indicates that the arrangement of actin molecules in heart muscle is almost the same as that in skeletal muscle. Namely, the thin filament consists of two long strands of actin subunits wound helically around one another (Figure 2b). The 37.4-nm layer line arises from the distance between cross-over points of the two strands, indicating that the helical pitch of each strand is 74.8 nm, twice the cross-over distance. The same helical arrangement of actin may be described alternatively by either of two primitive helices (left-handed, pitch 5.8 nm or right-handed, pitch 5.0 nm) linking all subunits. Since the long pitch of the actin strand corresponds approximately to 13 turns of the 5.8-nm helix or 15 turns of the 5.0-nm helix, the near-meridional reflections at 5.8 and 5.0 nm can be regarded as 13th and 15th orders of the long pitch.

The other near-meridional reflections at 7.0 and 4.5 nm can be indexed as 11th and 17th orders of the same pitch.

No myosin layer lines can be seen in the rigor pattern. The only reflections arising from myosin are weak meridional reflections at the second (21.4 nm), third (14.3 nm), and sixth (7.1 nm) orders of 42.9 nm. This repeat is slightly longer than that in quiescent muscle (42.6 nm) (43). A similar change in the myosin periodicity has been observed with skeletal muscle when it goes from the living resting state to rigor (16, 29). A possible mechanism of this change has been discussed by Haselgrove (16).

The fact that heart muscle in rigor gives only weak myosin-related reflections but strong actin-related reflections has been interpreted to indicate that myosin projections attach to actin when the muscle goes into rigor (29, 43). According to this interpretation the projections move out of the helical positions characteristic of the quiescent state, thus weakening the myosin-related reflections. The intensification of actin reflections in the rigor pattern has been attributed to "labeling" of actin subunits by myosin projections. Thus the change in the meridional pattern on going into rigor has been taken as evidence supporting the idea that the same cross-bridge mechanism as postulated in skeletal muscle operates also in heart muscle.

Reflections Arising From Collagen

Both quiescent and rigor patterns show other reflections that are not attributable to myofilaments (43). These can be indexed as orders of a 64.0-nm periodicity and are characteristic reflections of the meridional diffraction pattern from collagen filaments (2). The third (21.3 nm) and fifth (12.8 nm) orders are strong. The collagen filaments must be aligned in parallel to the muscle axis, since the reflections are co-linear with the meridional reflections from the myofilaments and probably are at least part of the structural basis for the high resting tension in heart muscle.

Equatorial Patterns From Quiescent Muscles

Quiescent heart muscles give a pair of reflections on the equator that can be indexed as the 1,0 and 1,1 reflections from a hexagonal lattice (Figure 2c) (43). These reflections are much stronger than reflections in the meridional pattern, and an exposure of 30 min–1 hr (in a low-angle camera combined with a rotating-anode X-ray generator) is sufficient for recording them on film. The 2,0 and 3,0 reflections from the lattice require a longer exposure (Figure 1a). The equatorial pattern is similar to that of skeletal muscle (23) and is consistent with the hexagonal arrangement of the thick and thin filaments seen in cross-sections of heart muscle in the electron microscope.

The side spacing between myofilaments can be calculated from the equatorial pattern. For example, the center-to-center distance between neighboring thick filaments corresponds to $2/\sqrt{3}$ times the spacing between the 1,0 lattice planes ($d_{1,0}$) obtained from the positions of the 1,0 reflections. The side spacing is dependent on the sarcomere length; the spacing decreases as the sarcomere length increases, so that the lattice volume is constant ($d_{1,0}^2 \times$ the sarcomere length = constant) (43). Constant lattice volume has also been found in skeletal muscle (9, 23) and is attributed to the constant fiber volume maintained by an osmotic balance across the sarcolemma (40). The lattice spacing is also sensitive to the osmolarity of the bathing solution (56). Although no direct comparison has been made between the spacings in heart and skeletal muscles in solutions of the same osmolarity, the following observation suggests that the two types of muscle bathed in appropriate physiological solutions have approximately the same interfilament spacing: the 1,0 spacing of feline heart muscle in Tyrode solution is 37.0 nm at a sarcomere length of 2.2 μm (43), whereas that of frog skeletal muscle in Ringer solution at the same sarcomere length is 36.7 nm (40).

In skeletal muscle the sarcomere length affects also the intensities of the equatorial reflections (9). The ratio of the integrated intensities of the 1,0 and 1,1 reflections ($I_{1,0}/I_{1,1}$) is often used to express the equatorial intensity distribution. In resting skeletal muscle the 1,0 reflection is stronger than the 1,1 reflection; the intensity ratio is about 2.0 at a sarcomere length of 2.1 μm at which the thick and thin filaments fully overlap (17). When the muscle is stretched, the 1,1 intensity rapidly decreases and the ratio becomes nearly 10 at a sarcomere length of 2.8 μm (9). Two effects of stretch on the filament lattice are responsible for this phenomenon: (a) Electron micrographs of skeletal muscle show that the thin filaments are aligned in the hexagonal lattice only in the A-band; they are poorly ordered in the I-band. Upon stretch, the thin filaments are withdrawn from the A-bands, reducing the X-ray scattering mass of the 1,1 planes where the thin filaments are located (Figure 2c) and thus decreasing the 1,1 intensity (9). (b) Stretch causes disorder in the hexagonal lattice (15). Such disorder weakens all the equatorial reflections, but the effect is greater at higher angles of diffraction. Therefore, the 1,1 intensity decreases more than the 1,0 intensity and the ratio $I_{1,0}/I_{1,1}$ increases.

In quiescent heart muscle the 1,0 reflection is stronger than the 1,1 reflection; at a sarcomere length of 2.1 μm the intensity ratio is approximately 3.0 (41), which is much greater than in skeletal muscle (2.0). This difference seems to arise from the fact that the ordering of the thin filaments in heart muscle is not as good as in skeletal muscle. When heart muscle is stretched, the intensity ratio increases in a manner similar to that in skeletal muscle. Although no systematic study has been done on this phenomenon, it seems reasonable to assume that the change in the ratio is caused by the mechanisms discussed above.

Equatorial Patterns From Muscles in Rigor

Glycerol-extracted muscles and dead muscles also give the reflections from the hexagonal lattice (43). The lattice spacing in glycerol-extracted muscles at a sarcomere length of 2.1 μm is about 40 nm (43), considerably greater than that in living quiescent muscle (37.0 nm). The increase in spacing is due to loss of the osmotic control of cell volume by the sarcolemma, which limits expansion of the lattice. The lattice spacing is variable in dead muscles but generally greater than that in living muscles. In both glycerol-extracted and dead muscles the relative intensity of the 1,0 and 1,1 reflections is reversed as compared with that in living quiescent muscle, and the ratio $I_{1,0}/I_{1,1}$ is 0.37 at a sarcomere length of 2.1 μm (41). This change in the relative intensity is very similar to that found in skeletal muscle going into rigor (24).

The interpretation of changes in the intensity ratio is controversial, so an explanation of the situation is required before the ratio can be used as an index of molecular events for heart muscle. The basic problem is that the ratio represents the intensity distribution between only two reflections, and therefore provides rather poor resolution in analyzing the structure. This makes it possible to interpret an observed change in the intensity ratio in various ways and to postulate different types of molecular changes. Moreover, it is difficult to improve the resolution, since higher order equatorial reflections in muscles in rigor and during contraction are weak and therefore hard to record with the present technique.

The most widely accepted interpretation is the one proposed by Huxley (24). He has shown that the change in the intensity ratio observed in skeletal muscle between the resting state and rigor can be interpreted as the result of radial transfer of myosin projections from the vicinity of the thick filament to the thin filament to attach to actin. To justify this interpretation Huxley has made the following points: (*a*) the decrease in the intensity ratio observed when a resting muscle goes into rigor can be explained quantitatively by assuming that all the myosin projections are transferred radially to the thin filament region; (*b*) electron micrographs show the occurrence of such radial transfer.

An alternative interpretation has been proposed by Lymn (38). Based on the equatorial reflections including higher orders (up to the 3,1 and 1,3 reflections in the resting pattern), he has proposed a model in which myosin projections in the resting condition are stretched further away from the backbone of the thick filament than in Huxley's interpretation. The change in the intensity ratio when the muscle goes into rigor or during activation can be explained if the projections move azimuthally, rather than radially, to attach to actin. However, this view has been criticized on the following grounds (18): (*a*) the predicted decrease in the 1,0 intensity accompanying the azimuthal movement is much smaller than the actual decrease; (*b*) analysis of the equatorial pattern including the higher order reflections shows that the myosin projections in resting muscle are more likely to be located in the vicinity of the thick filaments. Therefore, it is unlikely that the projections move only azimuthally; the movement must have a radial component to account for the changes of the intensities of the reflections. However, the above criticism does not deny the possibility of azimuthal movement and a contribution to changes in the intensity ratio.

It is also important to note that changes in the intensity ratio do not necessarily represent attachment or detachment between myosin projections and actin. Any configurational change of projections that accompanies a shift of mass between the 1,0 and 1,1 planes could affect the ratio. For example, a modeling study by Lymn (39) has shown that the ratio could change if myosin projections that are attached to actin tilt relative to the axis of the thin filament. Thus, the observed change in the intensity ratio may result from a configurational change, as well as from attachment or detachment. In other words, a change in the ratio may represent more than one molecular process.

In this review the interpretation proposed by Huxley is followed; changes in the intensity ratio are attributed to radial transfer of myosin projections, which is presented schematically in Figure 3. In this model the myosin projections form two-stranded helices along the thick filament as proposed by Huxley & Brown (29). The projections appear to emerge in six directions (rather than two) from the backbone of the filament, but this is due to the fact that the diagram represents a view of the lattice projected onto the plane perpendicular to the muscle axis. In the living resting state (Figure 3*a*) myosin projections, which are represented by spheres (18, 38), are in contact with the backbone of the thick filament. When muscle goes into rigor (Figure 3*b*), all the projections move radially to attach to thin filaments (62).

The equatorial patterns recorded from heart muscle indicate that the intensity ratio changes from 3.07 to 0.37 as the muscle shifts from the living quiescent state to rigor (40). Using each intensity ratio, the electron density distribution in the filament lattice projected onto the plane perpendicular to the muscle axis can be calculated. In the calculation, which is a Fourier transform, the phases of the 1,0 and 1,1 reflections are assumed to be 0. This phase assignment has been made

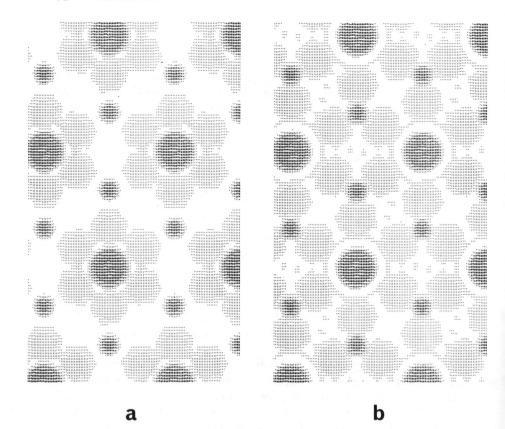

Figure 3 A model of the myofilament lattice in skeletal muscle (*a*) in the resting state and (*b*) in rigor. This model has been favored in a modeling study (64, 66) in which various arrangements of myosin projections (29, 38, 60) were tested. In the modeling the shaft of the thick filament was assumed to be cylindrical and each myosin projection was assumed to be spherical (18, 38). The thin filament was assumed either to be cylindrical or to have the composite helical structure (50) consisting of actin, along with tropomyosin. The diameter of the cylinder(s) and the sphere were varied to find a set of values that can explain the observed intensities (see below) within two standard deviations. Myosin projections (diameter 10.4–14.0 nm) form two-stranded helices along the thick filament shaft (diameter 9.2–14.8 nm). This model can account for (*a*) the intensity distribution among the equatorial reflections (up to the 3,0 reflection) recorded from frog resting muscle at a sarcomere length of 2.2 μm, (*b*) the changes in the 1,0 and 1,1 intensities occurring when muscle goes into rigor, and (*c*) the intensity changes of the equatorial reflections upon stretching sarcomeres. The mass of the projection and the shaft in this model suggest that four myosin molecules occur per 14.3 nm interval along the thick filament, supporting the biochemical data by Morimoto & Harrington (51) and Pepe & Drucker (53).

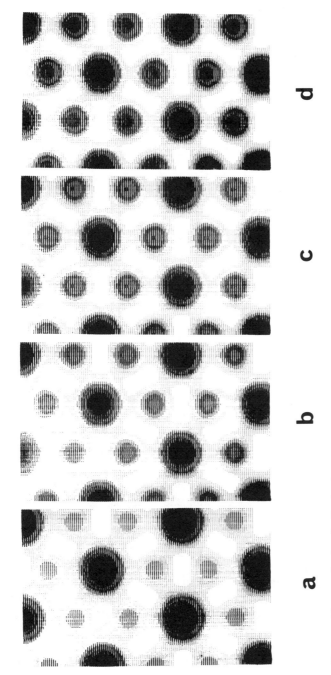

Figure 4 Electron density distribution in the myofilament lattice of heart muscle projected onto the plane perpendicular to the muscle axis: (*a*) quiescent state, (*b*) diastolic state, (*c*) systolic state, (*d*) rigor. The larger peaks in each diagram represent the thick filaments and the smaller peaks represent the thin filaments [from (44) by permission of the publisher].

on the following grounds: (a) the phases of the equatorial reflections are either 0 or π, since the hexagonal arrangement of the myofilaments has a center of symmetry (at least at the resolution available); (b) the hexagonal arrangement cannot be reproduced in the map if the phase of either the 1,0 or 1,1 reflection is π. Figure 4a represents the density distribution map for the quiescent state, whereas Figure 4d represents the one for rigor. In each diagram, the mass associated with the thick and thin filaments can be calculated using reasonable assumptions about the level of background intensity: either the lowest density in the map or the lowest density along the line connecting the centers of the thick and thin filaments can be taken as the background level. Usually the calculation is quoted for each of these assumptions (17). In the quiescent state the mass associated with the thick filament is much greater than that associated with the thin filament. When the muscle goes into rigor, the thick filament mass decreases and the thin filament mass increases. These changes in the mass are interpreted as caused by radial transfer of myosin projections from the vicinity of the thick filament to that of the thin filament. The mass so transferred is referred to as "rigor transfer" in the next section.

SYSTOLIC AND DIASTOLIC PATTERNS

Patterns From Skeletal Muscle During Contraction

Diffraction patterns of contracting skeletal muscle were first obtained with a stroboscopic technique (8, 30). This technique was to make the muscle contract many times and superimpose the diffraction patterns from all the contractions on one film by using a shutter synchronized with each contraction. The principal features of the meridional pattern thus recorded are as follows: (a) the myosin layer lines are considerably weaker than in the resting pattern, indicating that the helical arrangement of the myosin projections along the thick filament becomes less regular; (b) much less enhancement of the actin reflections occurs compared with that of the rigor pattern, suggesting that the myosin projections associated with actin during contraction are not uniform in configuration (26).

The equatorial pattern from contracting muscle is intermediate between the resting and rigor patterns. The 1,0 reflection is weaker, and the 1,1 reflection is stronger, than in the resting pattern. The intensity ratio is smaller than the resting value (2.63 at a sarcomere length of 2.2 μm) but not as small as in rigor (0.26). In early stroboscopic experiments that required a great number of contractions for recording the patterns, the intensity ratio during contraction was 0.78 (17). Using this ratio the density distribution map was obtained, and the mass associated with the thick and thin filaments was calculated. The mass associated with the thick filament was smaller, and that associated with the thin filament was greater, than in the resting state. A mass transfer amounting to 45–58% of the rigor value would explain these changes. Since almost all myosin projections are supposed to attach to actin in skeletal muscle on going into rigor (62), this result was interpreted to suggest that 45–58% of the myosin projections are attached to actin (or in the vicinity of the thin filaments) at any particular moment during contraction. More recent stroboscopic experiments, which required a smaller number of contractions and therefore caused less fatigue to muscle, showed a greater change in the equatorial intensity ratio (54, 63, 65). In an experiment in which the equatorial pattern was obtained from a single isometric tetanus of 10-sec duration, the intensity ratio was 0.51, very close to the rigor value (49). This led to the conclusion that 81–89% of the myosin projections are in the vicinity of the thin filaments during a maximum tetanus.

The number of projections near the thin filaments does not necessarily represent the number of projections attached to actin. However, it is often assumed that the two numbers agree or parallel each other. Plausible support for such an assumption seems to come from the following observations: (a) stiffness measurements on skinned muscle fibers (13) have suggested that the number of actin-myosin links during a maximum isometric contraction roughly agrees with the number of projections found in the vicinity of the thin filaments in the more recent X-ray study (49); (b) at the onset of contraction, the number of myosin projections in the vicinity of the thin filaments increases roughly in parallel to the tension development (28, 45). On the other hand, the following observations showing lack of correlation between the number and the tension do not support the equivalence of mass transfer and cross-bridge attachment: (a) the equatorial intensity ratio does not return to its resting value until some time after the tension has fallen to zero (63); (b) the lack of correlation is more pronounced in heart muscle during the diastolic state (see below). Thus at present it seems reasonable to suggest that although the intensity ratio may indicate the number of myosin projections near the thin filaments, it does not necessarily give the number of attached projections.

Patterns From Heart Muscle During Cyclic Contraction

The stroboscopic technique has been applied to heart muscle contracting cyclically with systolic (contracted) and diastolic (relaxed) phases (41, 44). Only the equatorial patterns have been studied to date. The preparations most suitable for such studies are canine papillary muscles perfused with blood through their blood vessels. The perfusion improves oxygen supply to individual fibers and minimizes deterioration of the muscle during repeated contractions. In all the experiments reviewed in this section the muscle was held isometrically with one end connected to a tension transducer, and a shutter was coupled to the transducer. To record the systolic pattern, the shutter was opened as the tension exceeded two thirds of peak tension and closed as it fell below half of the peak tension. To record the diastolic pattern, the shutter was opened when the tension was at the steady diastolic level. Since the equatorial pattern is sensitive to the sarcomere length, the systolic and diastolic patterns had to be compared at approximately the same sarcomere length. For recording the systolic pattern, the muscle was held at the optimal length (L_{max}), which gave a sarcomere length of about 2.1 μm at the peak of contraction (36, 44). For recording the diastolic pattern, the muscle was held at a shorter length in order to adjust the sarcomere length to 2.1 μm during the diastolic phase.

The first preparation used in the stroboscopic experiments was the papillary muscle perfused with diluted blood and stimulated 12 times per min (41). This produced a peak systolic tension of 135 g/cm^2 at L_{max}. The intensity ratio ($I_{1,0}/I_{1,1}$) was 2.66 during the diastolic phase and 2.19 during the systolic phase, which suggests that the myosin projections are transferred from the thick to the thin filaments as the heart muscle shifts from the diastolic to the systolic phase. The diastolic intensity ratio was smaller than the quiescent ratio (3.07) at the same sarcomere length, indicating that the diastolic state is different from the quiescent state. A certain number of myosin projections seem to remain in the vicinity of the thin filaments during the diastolic phase.

If an estimate of the number of projections associated with the thin filament is to be made from the intensity ratios during the cardiac cycle, the values for the number near the thin filaments must be known for the quiescent and rigor states. The weakness of the myosin layer lines in the quiescent pattern suggests that there may be some projections associated with actin even in a quiescent muscle. The number attached in rigor is likewise subject to doubt. Therefore, a reliable estimate of the number of projections associated with actin cannot be made from the data obtained so far, and the results in various conditions are expressed better as the fraction of the rigor transfer.

The analysis based on the intensity ratio obtained from the above experiments suggested that the average mass transfer was 7–9% of the rigor transfer during the diastolic phase and 18–20% during the systolic phase. The lower values (7 and 18%) were obtained when the lowest density in the density distribution map was chosen as the background, and the higher values (9 and 20%) when the background level was taken as the lowest density along the line connecting thick and thin filaments. The transfer during the systolic phase was considerably smaller than that in skeletal muscle during the maximum tetanus. This was attributed to the fact that the peak systolic tension of the preparation was only 10–20% of the maximum tension that mammalian heart muscle can produce (600–1500 g/cm^2) (5, 36).

A greater amount of transfer was observed in a type of preparation that produced a larger systolic tension (44). The preparation was a papillary muscle remaining attached to a cross-perfused heart beating with its spontaneous frequency (approximately 115/min). The muscle produced a peak tension of about 600 g/cm^2 at L_{max}. The intensity ratio was 1.19 in the diastolic phase and 0.79 in the systolic phase. Figure 4b,c show the density distribution map calculated from these ratios. The average mass transfer was 51–52% during the diastolic phase (Figure 4b) and 70–71% during the systolic phase (Figure 4c). This result suggests that a large number of myosin projections remain near the thin filaments during the diastolic phase. The following experiment was performed to determine if these projections produce contractile force during the diastolic phase (33). The systolic and diastolic tensions were recorded from a papillary muscle in a cross-circulated heart. Then the same muscle was isolated from the heart and perfused with diluted blood to simulate the earlier type of preparation. During the isolation the muscle was kept connected to a transducer so that there was no change in the muscle length. The muscle was stimulated at the frequency of 12/min and the systolic and diastolic tensions were measured. The result indicated that, although the peak systolic tension fell to 20–25% of that of the cross-circulated preparation, the steady level of the diastolic tension did not change significantly. This suggests that most of the myosin projections found near the thin filaments in the cross-circulated preparation during the diastolic phase are not producing significant force.

The transfer during the systolic phase in relation to the systolic tension is discussed later.

Staircase Phenomenon

A third type of preparation was used in studying the effect of contraction frequency on the equatorial diffraction pattern (46). The preparation was an isolated papillary muscle cross-circulated with arterial blood of a donor dog. The peak systolic tension at L_{max} was 287 g/cm^2 when the muscle was stimulated at a frequency of 120/min. The systolic intensity ratio was 1.34, suggesting that the systolic transfer was 44–47% of the rigor transfer (the word *transfer*, rather than implying movement of material during the onset of muscular activity as such, is used to express the steady level of mass transfer from thick to thin filaments during some particular state of activation *relative to the quiescent state*; e.g. systolic transfer then reflects the proportion of projections associated with the thin filaments during the systolic phase). When the frequency was decreased to 60/min, the peak systolic tension decreased. This was accompanied by a decrease in the systolic transfer to 28–31% of the rigor transfer. When the frequency was increased from 120/min to 180/min, the peak systolic tension increased. This was accompanied by an increase in the systolic transfer to 58–60% of the rigor transfer. Thus the systolic transfer is greater at a higher frequency of contraction, and this may be the basis of the concomitant increase in the peak tension.

Two mechanisms can be proposed as to how an increase in the frequency leads to an increased systolic transfer:

1. Using the calcium sensitive photoprotein aequorin, Allen & Blinks (1) have studied the transient rise of the intracellular calcium concentration on contraction of heart muscle. They have shown that, when the contraction frequency is raised, the amplitude of the transient rise of the calcium increases. This is likely to lead to a greater number of projections attached to actin.
2. The decrease in the number of transferred projections after the peak tension is rather slow (see below), and continues even after the tension has fallen to the steady diastolic level. At a higher frequency, the diastolic interval is shorter and preliminary observations suggest that a greater number of projections remain associated with the thin filaments at the end of the diastolic phase (48). As a result of this, as will be proposed later, the number of projections available for attachment in the systolic phase is increased.

Relation Between Systolic Transfer and Systolic Tension

In Figure 5 the fraction of projections associated with actin in each type of heart muscle preparation during the systolic phase is plotted against the peak systolic tension. The graph suggests that a rise in peak tension, caused either by increasing the frequency of contraction or by improving the quality of the preparation, is accompanied by an increase in the number of myosin projections in the vicinity of the thin filaments. One interesting feature of Figure 5 is that the systolic transfer in papillary

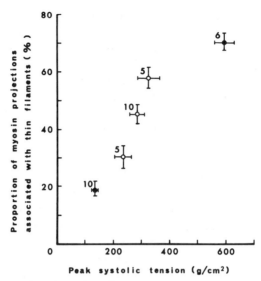

Figure 5 The proportion of myosin projections found in the vicinity of the thin filaments during the systolic phase plotted against the peak systolic tension. The mass transferred from thick to thin filaments when a quiescent muscle goes into rigor is assumed to represent the total mass of myosin projections (see text). The open and filled circles represent mean values (±SE of the mean) obtained from three different types of heart muscle preparations, including one type contracting at three different frequencies (open circles). The numbers beside the circles indicate the number of observation.

muscle attached to a cross-circulated heart (70–71% of the rigor transfer) is almost as great as that found in skeletal muscle during maximum tetanus (81–89%). If the latter represents the maximum possible transfer during contraction, the systolic transfer in this type of heart muscle preparation may be regarded as rather close to the upper limit. Then, it is not surprising that the staircase phenomenon is less prominent in a similar type of cross-circulated preparation (35).

It is not clear what the inotropic states of the various types of preparations are, relative to that of the normal heart muscle in vivo. The physiological state of the muscle of a cross-circulated heart might be assumed to approximate to that in vivo. It would then follow that the muscle in vivo is nearly maximally activated during each systolic phase, as far as the number of cross-bridges is concerned.

DYNAMIC STUDY

Use of a position sensitive X-ray detector instead of film has greatly shortened the exposure time required to obtain the diffraction patterns. Combination of such a detector with a digital memory circuit has allowed changes in the diffraction patterns during contraction to be followed with high time resolution (12, 27, 54). This was achieved by synchronizing the memory circuit with the stimulus and separately storing the data from different periods during contraction. Contractions were repeated until reasonable counting statistics were obtained. Both one- and two-dimensional detectors are available (7). The one-dimensional detector is most suitable for studying the equatorial patterns, and the time resolution obtained in such a study has been 7.5 msec in skeletal muscle (45) and 32 msec in heart muscle (47).

Time-Resolved Diffraction Studies of Contracting Skeletal Muscle

In frog skeletal muscle at the onset of isometric tetanus, the 1,0 intensity decreases and the 1,1 intensity increases as the tension develops (17, 27, 45). When the intensity ratio $(I_{1,0}/I_{1,1})$ is plotted against time, it forms an approximate mirror image of the tension curve (45). However, a detailed comparison of intensity-ratio and tension curves has revealed that the change in the X-ray intensities precedes the tension change; at 10°C the change in the intensity ratio is half complete within 20 msec after the onset of stimulation, a time when the tension has hardly developed (27). The intensity ratio reaches the minimum before the maximum tension (27, 45), and stays at the minimum level during the steady tetanic contraction (65). At the end of tetanus, the return of the intensity ratio to its resting level occurs in two stages; a rapid return is followed by a slow return. The initial rapid return occurs almost simultaneously with the fall of tension. The slow return continues after the tension has fallen to zero (63).

Measurements during repeated single-stimulus twitches have yielded the following results (45): (a) Before the latency relaxation (an indication of the onset of contraction) there is no change in the equatorial intensities. (b) The intensity ratio starts to change at the very beginning of tension rise (8–14 msec after stimulus at 0°C). (c) The change in the intensity ratio, once initiated, proceeds faster than the tension change. At 0°C the intensity ratio reaches the minimum level 60–70 msec after the stimulus, 90 msec before the peak of twitch tension. The minimum level is approximately the same as that reached during the maximum isometric tetanus. (d) The intensity ratio does not reach the resting level until 0.7–1.1 sec after the tension has fallen to zero.

The above results from muscles during tetanus and twitch have been interpreted as follows: (a) A certain period of time is needed for myosin projections to exert their full strength after their arrival in the vicinity of the thin filaments. (b) The maximum number of myosin projections near the thin filaments during twitch (at 0–4°C) does not differ significantly from the number obtained during the maximum isometric tetanus. (c) Myosin projections do not return to their resting positions promptly after contraction. The same conclusion has been reached by a study of the myosin layer lines after tetanus (26); the layer lines do not resume their resting intensities until several seconds after the tension has fallen to zero.

Time-Resolved Diffraction Study of Heart Muscle During Cyclic Contraction

Figure 6 shows the results obtained from a cross-circulated papillary muscle contracting at a frequency of 1/sec (47). The intensity ratio decreases during the systolic phase, forming an approximate mirror image of the tension curve. This decrease is due not only to the movement of myosin projections but also to internal shortening of the muscle, even though this was held at a constant length. However, the contribution of this shortening has been shown to be relatively small; when the intensity-ratio change at a constant sarcomere length was estimated by making a correction for the effect of internal shortening, the resulting curve had almost the same time course as that shown in Figure 6 (47).

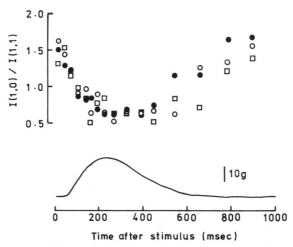

Figure 6 Changes in the equatorial intensity ratio in a papillary muscle during the cardiac cycle. The 1-sec cycle was divided into 16 phases, and the ratio was measured for each phase. The same set of diffraction patterns was measured by three people; different symbols represent the measurements by different people. The tension record was an average of all the contractions needed for obtaining the diffraction patterns [from (47) by permission of the publisher].

The data suggest that the number of myosin projections in the vicinity of the thin filaments reaches a maximum approximately at the time of peak tension, at about 250 msec after the stimulus at 37°C. This is much slower than in skeletal muscle in which the number reaches a maximum 60–70 msec after the stimulus at 0°C, and even earlier at higher temperature. The slow transfer of myosin projections in heart muscle is probably the basis of the slow development of the active state in heart muscle (5).

Figure 6 suggests that the return of myosin projections to the thick filaments continues after the tension has fallen to a steady diastolic level. Even at the end of the diastolic phase a considerable number of myosin projections, corresponding to 30–40% of the rigor transfer, remain near the thin filaments.

Paired-Pulse Stimulation

When heart muscle is stimulated with a pair of pulses spaced at a suitable interval, the muscle responds with a stronger contraction (6). The rate of tension development and the peak systolic tension are greater than those obtained with single-pulse stimulation. A diffraction study comparing single- and paired-pulse stimulation has shown that the mass transfer is increased in both systolic and diastolic phases by paired stimulation (47). The increased systolic transfer is likely to be the basis of the strengthening of contraction. On the other hand, the increase in the mass transfer of the diastolic phase is not accompanied by any change in the steady level of diastolic tension. This observation supports the idea that the myosin projections found in the vicinity of the thin filaments during the diastolic phase do not produce a significant contractile force.

Possible Physiological Significance of Myosin Projections Remaining Near the Thin Filaments During the Diastolic Phase

Both stroboscopic and time-resolved studies have shown that the systolic transfer changes with the systolic tension, whereas the diastolic transfer (i.e. the number of projections remaining near the thin filaments during the diastolic phase) is not related to the diastolic tension. The simplest interpretation of these results is that the amount of radial transfer affects the tension only when the thin filaments are activated. One possibility is that the tension is roughly proportional to the product of the radial transfer and the degree of thin filament activation (i.e. the number of active actin sites). A preparation with a greater systolic transfer would then produce a greater systolic tension (Figure 5).

However, according to the above idea, the tension would not be linearly related to the systolic transfer, since tension would depend also on the extent of thin filament activation, which may differ in the different preparations used in Figure 5. The diastolic tension would be unrelated to the diastolic transfer, since most of thin filaments would be inactive during the diastolic phase.

Although the projections found in the vicinity of the thin filaments during the diastolic phase are apparently not producing tension, they may be primed to produce tension on activation of thin filaments more readily, compared to other projections. The rate of tension development and the peak systolic tension may be greater when the thin filaments are activated with the projections in the "ready" state. Such effects should be more significant in heart muscle than in skeletal muscle, since the transfer of projections from thick to thin filaments on activation is slower in heart muscle. This idea is supported by (a) comparing different types of heart muscle preparation, and (b) examining a given preparation subjected to different patterns of stimulation, either with single pulses of variable frequency or with paired-pulses (see previous sections). In all these experiments the rate of tension development and the peak systolic tension were greater when the diastolic transfer was increased. Thus the number of myosin projections remaining near the thin filaments during the diastolic phase seems to be related to the magnitude of the subsequent contraction.

SUMMARY

X-ray diffraction studies on mammalian heart muscle have shown that the molecular packing of myosin and actin in the myofilaments is the same as in skeletal muscles. The changes in the diffraction patterns occurring when heart muscle contracts or goes into rigor are almost the same as those in skeletal muscle. These findings suggest that a crossbridge mechanism, similar to that in skeletal muscle, operates in heart muscle.

Both in skeletal and heart muscles the intensities of the principal equatorial reflections change as the contractile tension develops. These intensity changes have been interpreted as indicating a transfer of myosin projections from the vicinity of the thick filaments to that of the thin filaments. The time course of the transfer is much slower in heart muscle, and this may be the basis of the slow development of the active state in myocardium. When the peak systolic tension is increased in heart muscle by improving the quality of the preparation, by increasing the frequency of contraction, or by stimulating the muscle with paired pulses, the systolic transfer of myosin projections increases. This suggests that the inotropic effect (i.e. the tension increase) is caused by an increase in the number of cross-bridges at the peak of contraction.

In cyclically contracting heart muscle, the array of myosin projections does not go into the fully rested state during the diastolic phase. A considerable number of myosin projections remain near the thin filaments during this phase, but they do not seem to produce significant force, since a change in the mass transfer does not affect diastolic tension. However, a change in the mass transfer during diastole is associated with changes in the rate of tension development and the peak tension of the subsequent contraction. This suggests that the myosin projections which remain near the thin filaments during the diastolic phase may produce tension in the subsequent contraction more readily than those which have returned to the vicinity of the thick filaments. Thus important physiological alterations of cardiac contractility may result from changes in the distribution of projections in the diastolic interval.

ACKNOWLEDGMENTS

I am grateful to Drs. Pauline Bennett, Arthur Elliott, Yale Goldman, Rolf Niedergerke, Edward O'Brien, Gerald Offer, Sally Page, and Mr. Roger Starr for helpful comments, and to the Wellcome Trust for support.

Literature Cited

1. Allen, D. G., Blinks, J. R. 1978. *Nature* 273:509–13
2. Bear, R. S. 1944. *J. Am. Chem. Soc.* 66:1297–1305
3. Blinks, J. R., Jewell, B. R. 1972. In *Cardiovascular Fluid Dynamics*, ed. D. H. Bergel, 1:225–60. London/New York: Academic
4. Bowditch, H. P. 1871. *Ber. Sachs. Ges. (Akad.) Wiss.* pp. 652–89
5. Brady, A. J. 1968. *Physiol. Rev.* 48:570–600
6. Brutsaert, D. L. 1966. *Arch. Int. Physiol. Biochim.* 74:642–64
7. Charpak, G. 1977. *Nature* 270:479–82
8. Elliott, G. F., Lowy, J., Millman, B. M. 1965. *Nature* 206:1357–58
9. Elliott, G. F., Lowy, J., Worthington, C. R. 1963. *J. Mol. Biol.* 6:295–305
10. Endo, M. 1977. *Physiol. Rev.* 57:71–108
11. Fabiato, A., Fabiato, F. 1979. *Ann. Rev. Physiol.* 41:473–84
12. Faruqi, A. R. 1975. *IEEE Trans. Nucl. Sci.* NS-22:2066–73
13. Goldman, Y. E. Simmons, R. M. 1977. *J. Physiol. London* 269:55–57P
14. Hanson, J., Huxley, H. E. 1955. *Symp. Soc. Exp. Biol.* 9:228–64
15. Haselgrove, J. C. 1970. *X-ray diffraction studies on muscle*. PhD thesis. Univ. Cambridge, England
16. Haselgrove, J. C. 1975. *J. Mol. Biol.* 92:113–43
17. Haselgrove, J. C., Huxley, H. E. 1973. *J. Mol. Biol.* 77:549–68
18. Haselgrove, J. C., Stewart, M., Huxley, H. E. 1976. *Nature* 261:606–8
19. Hill, A. V. 1949. *Proc. R. Soc. London Ser. B* 136:399–420
20. Huxley, A. F. 1957. *Prog. Biophys. Chem.* 7:255–318
21. Huxley, A. F., Simmons, R. M. 1971. *Nature* 233:533–38
22. Huxley, A. F., Simmons, R. M. 1973. *Cold Spring Harbor Symp. Quant. Biol.* 37:669–80
23. Huxley, H. E. 1953. *Proc. R. Soc. London Ser. B* 141:59–62
24. Huxley, H. E. 1968. *J. Mol. Biol.*

25. Huxley, H. E. 1969. *Science* 164:1356-66
26. Huxley, H. E. 1972. *Cold Spring Harbor Symp. Quant. Biol.* 37:361-76
27. Huxley, H. E. 1975. *5th Int. Biophys. Congr. Copenhagen* S53
28. Huxley, H. E. 1975. *Acta Anat. Nippon.* 50:310-25
29. Huxley, H. E., Brown, W. 1967. *J. Mol. Biol.* 30:383-434
30. Huxley, H. E., Brown, W., Holmes, K. C. 1965. *Nature* 206:1358
31. Jewell, B. R. 1977. *Circ. Res.* 40:221-30
32. Jewell, B. R., Wilkie, D. R. 1958. *J. Physiol. London* 143:515-40
33. Kamiyama, A., Matsubara, I., Suga, H. 1976. *J. Physiol. London* 256:15-16P
34. Katz, A. M. 1970. *Physiol. Rev.* 50:63-158
35. Kavaler, F., Harris, R. S., Lee, R. J., Fisher, V. J. 1971. *Circ. Res.* 28:533-44
36. Krueger, J.W., Pollak, G. H. 1975. *J. Physiol. London* 251:627-43
37. Lipson, H. 1972. In *Optical Transforms*, ed. H. Lipson, pp. 1-25. London/New York: Academic
38. Lymn, R. W. 1975. *Nature* 258:770-72
39. Lymn, R. W. 1978. *Biophys. J.* 21:93-98
40. Matsubara, I., Elliott, G.F. 1972. *J. Mol. Biol.* 72:657-69
41. Matsubara, I., Kamiyama, A., Suga, H. 1977. *J. Mol. Biol.* 11:121-28
42. Matsubara, I., Millman, B. M. 1973. *J. Physiol. London* 230:62-63P
43. Matsubara, I., Millman, B. M. 1974. *J. Mol. Biol.* 82:527-36
44. Matsubara, I., Suga, H., Yagi, N. 1977. *J. Physiol. London* 270:311-20
45. Matsubara, I., Yagi, N. 1978. *J. Physiol. London* 278:297-307
46. Matsubara, I., Yagi, N., Endoh, M. 1978. *Nature* 273:67
47. Matsubara, I., Yagi, N., Endoh, M. 1979. *Nature* 278:474-76
48. Matsubara, I., Yagi, N., Endoh, M. *Eur. J. Cardiol.* In press
49. Matsubara, I., Yagi, N., Hashizume, H. 1975. *Nature* 255:728-29
50. Miller, A., Tregear, R. T. 1972. *J. Mol. Biol.* 70:85-104
51. Morimoto, K., Harrington, W. F. 1974. *J. Mol. Biol.* 83:83-97
52. Page, S. G. 1974. *Ciba Found. Symp.* 24:13-25
53. Pepe, F. A., Drucker, B. 1979. *J. Mol. Biol.* 130:379-93
54. Podolsky, R. J., St. Onge, R., Yu, L., Lymn, R. W. 1976. *Proc. Natl. Acad. Sci. USA* 73:813-17
55. Pringle, J. W. S. 1960. *Symp. Soc. Exp. Biol.* 14:41-68
56. Rome, E. M. 1968. *J. Mol. Biol.* 37:331-44
57. Rome, E. M., Hirabayashi, T., Perry, S. V. 1973. *Nature New Biol.* 244:154-55
58. Selby, C. C., Bear, R. S. 1956. *J. Biophys. Biochem. Cytol.* 2:71-85
59. Simmons, R. M., Jewell, B. R. 1974. In *Recent Advances in Physiology*, ed. R. J. Linden, pp. 87-147. London: Livingstone
60. Squire, J. M. 1972. *J. Mol. Biol.* 72:125-38
61. Stenger, R. J., Spiro, D. 1961. *Prog. Cardiovasc. Dis.* 7:295-335
62. Thomas, D. D., Cooke, R. 1979. *Biophys. J.* 25:19a
63. Yagi, N., Ito, M. H., Nakajima, H., Izumi, T., Matsubara, I. 1977. *Science* 197:685-87
64. Yagi, N., Matusbara, I. 1976. *Kagaku Tokyo* 46:330-38 (In Japanese)
65. Yagi, N., Matsubara, I. 1977. *Pfluegers Arch.* 372:113-14
66. Yagi, N., Matsubara, I. 1978. In *Diffraction Studies of Biomembranes and Muscles*, ed. T. Mitsui, pp. 142-52. Japan: Taniguchi Found.

THE OPTICAL ACTIVITY OF NUCLEIC ACIDS AND THEIR AGGREGATES

Ignacio Tinoco, Jr. and Carlos Bustamante
Chemistry Department and Laboratory of Chemical Biodynamics, University of California, Berkeley, California 94720

Marcos F. Maestre
Donner Laboratory, Division of Medical Physics, University of California, Berkeley, California 94720

INTRODUCTION

This review includes experimental and theoretical work on absorption and scattering of light by chiral macromolecules. All molecules absorb light, but only chiral (handed) molecules show a preferential absorption for right or left circularly polarized light. This phenomenon of circular dichroism (CD) has been very useful in characterizing any chiral aggregate of chromophores—including proteins, nucleic acids, and their complexes. We discuss theoretical methods that relate the absorption and circular dichroism of a polymer or aggregate to the optical properties of its constituent parts. The experimental data reviewed is limited essentially to nucleic acids, for lack of space.

Two new experimental methods that are particularly useful for macromolecules or for systems that scatter a significant fraction of the incident light are fluorescence-detected circular dichroism and circular intensity differential scattered light. Fluorescence-detected circular dichroism (FDCD), as the name implies, used the intensity of the fluorescence emitted to monitor the intensity of the light absorbed (167). This method provides two main advantages: (*a*) The spectrum of a complex system containing many chromophores, but only a few fluorophores is greatly simplified, and (*b*) scattering artifacts, which plague circular dichroism studies using transmitted light measurements, are partly avoided by use of fluorescence detection. The theory (159, 160) and practice (108, 166) of fluorescence-detected circular dichroism have recently been discussed, so we will not review them further. The intensity of light scattered by chiral systems will depend on whether the incident light is right or left circularly polarized. This is known as the circular intensity differential of scattered light. Studies of differential light scattering for chiral systems are very recent, but we think that they will become increasingly important. Circular intensity differential scattered light is most informative when periodicities in the system are of the order of the wavelength of the incident light.

Other recent reviews which provide a broader picture have been published by Woody (176), Johnson (92), and Bayley (8). The proceedings of a NATO Institute on Optical Activity and Chiral Discrimination held in 1978 (115) discuss topics such as vibrational circular dichroism, Raman optical activity, and circular polarization of luminescence (155), which are omitted here.

CIRCULAR DICHROISM: THEORY

We want to be able to interpret a measured optical property for a macromolecule or, in general, a large aggregate of chromophores, in terms of structure. That is, we want to know how the chromophores are arranged relative to each other. For example, in chromatin we want to know what is the secondary structure of the DNA (A, B, or C-like) and what is its tertiary structure (pitch and radius of super helix)? In membranes, organelles, cells, etc larger partially ordered arrays may be of interest. It is clear that useful theoretical methods will have to be approximate; one cannot treat the system as a many electron molecule. Therefore, the necessary approximation in practical theories of aggregate optical properties is that the subunits do not exchange or transfer electrons. With this constraint the steps used to devise a useful theory are:

1. Divide the system into subunits.
2. Decide how the subunits interact. Usually only two subunits are considered to interact at one time; the influence of other subunits is included as an effective dielectric constant. Both dynamic and static electric interactions can be considered.
3. Write the optical properties of the aggregate in terms of the properties of the subunits and their interactions. The structure of the aggregate determines the magnitude of the interactions.

It is thus clear that from the (assumed) known properties of the subunits, a proposed structure of the aggregate, and the theory, one can calculate an optical property. Comparison with experiment will show which proposed structures are possible. If enough optical data are measured, it may be possible to severely limit the number of possible structures. Ignoring scattering effects, we can measure absorption, circular dichroism or ellipticity, refraction, and optical rotation or circular birefringence. The wider the wavelength range used, the better; also, studies of oriented systems provide more information.

We can roughly estimate the amount of experimental information available from all the measurements. The absorption and circular dichroism spectra can provide the number and positions of transitions present in the wavelength region. For an oriented system we could measure up to three different absorption spectra (three directions of polarization) and three different circular dichroism spectra (three directions of incidence). For each transition we could thus get six experimental quantities: three dipole strengths measuring absorption and three rotational strengths measuring circular dichroism. For a system with only one axis of orientation such as an electrically oriented or flow-oriented system, only two independent directions (parallel and perpendicular to the axis) are present. Each transition can be characterized by two dipole strengths and two rotational strengths. For an unoriented system there is one average dipole strength and one average rotational strength for each transition. The circular dichroism and circular birefringence (or optical rotation) are related by Kronig-Kramers transforms, as are the absorption and refraction. Therefore, if we knew the circular dichroism at all wavelengths, it would be redundant to also measure the optical rotation. However, for a finite wavelength region, measurement of both circular dichroism and optical rotatory dispersion can provide new information such as the location of transitions outside the accessible wavelength region (24).

The most useful type of theory—which has not yet been developed as far as we know—is one that would directly give some structural information from the measured spectrum. The Patterson function in X-ray diffraction is an example of this type of method. It would be good to be able to use the measured polymer spectrum and subunit spectra to calculate a function which depended only on the interactions between subunits.

Polymers

We will consider a polymer, or aggregate, made up of subunits which have no electron overlap between them. We must now decide what kinds of interactions between subunits are pertinent. Simpson & Peterson (149) deduced some very general conclusions about this and introduced the idea of strong and weak coupling. Strong coupling occurs when the largest interaction between subunits is greater than the half-width at half height of the absorption band involved in the interaction. Weak coupling occurs when the largest interaction between subunits is smaller than the half-width at half height. The reason for making the distinction is that different approximations are appropriate for these two cases. In strong coupling one should consider interactions between electronic transitions; in weak coupling, interactions between individual vibronic transitions of an electronic band are important. In principle one should decide if the polymer corresponds to weak or strong coupling, and then one should apply the appropriate theory. In practice, as we shall see, useful information can be obtained by applying either strong or weak coupling methods to the same polymer. Furthermore, it is possible that a polymer will show weak or strong coupling depending on the wavelength region.

Most of the biological polymers have fairly broad absorption bands and thus show weak coupling. For example the nucleic acid bases have absorption widths at half height of 3000 cm^{-1} to 5000 cm^{-1}. The most quoted example of strong coupling is found in aggregates of cyanine dyes in aqueous solution (122).

STRONG COUPLING If the subunits have line spectra (absorption bands have zero width), then any interaction gives strong coupling. Therefore, the simplest theory for treating polymers is to approximate the subunit spectra by lines and to calculate a line spectrum for the polymer. This is the exciton model (8, 156).

The polymer Hamiltonian is a sum of subunit Hamiltonians plus an interaction potential V_{ij}.

$$H = \sum_i H_i + \sum_i \sum_{j>i} V_{ij}. \qquad 1.$$

V_{ij} is Coulomb's Law between all particles in the two subunits i and j; an effective dielectric constant is sometimes used. The polymer electronic wavefunction, ψ, is expanded in a basis set of products of subunit electronic wavefunctions, ϕ_{ia}, where i numbers the subunit and a labels the state (ground, first excited, etc). From the Hamiltonian and the basis set the energy matrix (Table 1) is obtained. The eigenvalues and eigenvectors of this matrix characterize the polymer spectra. Let us consider the types of matrix elements that appear in the energy matrix.

The subunit electronic wavefunctions are orthonormal, real wavefunctions so $\langle \phi_{io} | \phi_{io} \rangle = 1$, $\langle \phi_{io} | \phi_{ia} \rangle = 0$ and $\langle \phi_{ia} | H | \phi_{ia} \rangle = E_{ia}$, the energy of the i subunit in its a excited state. The matrix elements of the

Table 1 The energy matrix in strong coupling[a]

	Singly excited states			Doubly excited states		Triply excited states	
	$\phi_{10}\phi_{20}\cdots\phi_{N0}$	$\phi_{1a}\phi_{20}\cdots\phi_{N0}$	$\phi_{10}\phi_{2b}\cdots\phi_{N0}$	...	$\phi_{1a}\phi_{2b}\cdots\phi_{N0}$	...	$\phi_{1a}\phi_{2b}\cdots\phi_{Nc}$	...	Etc.
$\phi_{10}\phi_{20}\cdots\phi_{N0}$	$\sum_{ij} V_{ioo,joo}$	$\sum_j V_{10a,joo}$	$\sum_j V_{20b,joo}$	...	$V_{10a,20b}$	...	$\phi_{1a}\phi_{2b}\cdots\phi_{Nc}$ 0	...	...
$\phi_{1a}\phi_{20}\cdots\phi_{N0}$		E_{1a} $+\sum'_j V_{1aa,joo}$ $+\sum'_{ij} V_{ioo,joo}$	$V_{10a,20b}$	...	$V_{1aa,20b}$ $+\sum'_j V_{20b,joo}$	...	$V_{20b,Noc}$	...	...
$\phi_{10}\phi_{2b}\cdots\phi_{N0}$			E_{2b} $+\sum_j V_{2bb,joo}$ $+\sum'_{ij} V_{ioo,joo}$	...	$V_{10a,2bb}$ $+\sum'_j V_{10a,joo}$	...	$V_{10a,Noc}$	...	...
$\phi_{1a}\phi_{2b}\cdots\phi_{N0}$					$E_{1a}+E_{2b}$ $+V_{1aa,2bb}$ $+\sum'_j V_{1aa,joo}$ $+\sum'_j V_{2bb,joo}$ $+\sum'_{ij} V_{ioo,joo}$	...	$V_{1aa,Noc}$ $+V_{2bb,Noc}$ $+\sum'_j V_{Noc,joo}$	...	...
$\phi_{1a}\phi_{2b}\cdots\phi_{Nc}$							$E_{1a}+E_{2b}+E_{Nc}$ $+$ static field terms		

[a] The Hamiltonian is $H=\sum_i H_i + \sum_{ij} V_{ij}$ and the zero of energy is the sum of the ground state energies of the subunits, $\sum_i E_{i0}$. The matrix is symmetric so only the upper half is given. A prime on a sum means that a term in the summation is omitted.

interaction potential are designated as follows:

$$\langle \phi_{io}\phi_{io}|V_{ij}|\phi_{jo}\phi_{jo}\rangle \equiv V_{ioo,joo}$$
$$\langle \phi_{io}\phi_{ia}|V_{ij}|\phi_{jo}\phi_{jo}\rangle \equiv V_{ioa,joo}$$
$$\langle \phi_{ia}\phi_{ia}|V_{ij}|\phi_{jo}\phi_{jo}\rangle \equiv V_{iaa,joo} \qquad 2.$$
$$\langle \phi_{io}\phi_{ia}|V_{ij}|\phi_{jo}\phi_{jb}\rangle \equiv V_{ioa,job}.$$

$V_{ioo,joo}$ represents the interaction between the static charge densities in the ground state of subunits i and j (139). $V_{ioa,joo}$ is the interaction between the transition density of i and the static ground state of j. $V_{iaa,joo}$ is the interaction between the static charge density of excited state a of subunit i and the static ground state of j. $V_{ioa,job}$ is the interaction between transition charge densities of i and j. Usually the V's are calculated by either a dipole-dipole approximation (100) or a monopole-monopole approximation (81). Experimentally available data such as the permanent electric dipole moment in the ground state, μ_{oo}, the permanent dipole in the excited state, μ_{aa}, and the transition electric dipole moment, μ_{oa}, are used to calculate the dipole-dipole interactions. In a monopole approximation wavefunctions for the subunits are needed. Each atom, except hydrogen atoms, is assigned a permanent, ρ_{oo}, or transition charge, ρ_{oa}. The monopole-monopole interaction involves a sum over all charges.

Once values for the elements of the energy matrix are obtained, the matrix can be diagonalized to obtain eigenvalues and eigenvectors. To reduce the size of the matrix and to simplify diagonalization, further approximations are sometimes made. One simplification is to omit all static field terms; all terms involving static charge densities are set equal to zero. The rationale for omitting these terms is that the largest contribution to shifts in energy levels between subunits in a polymer is the interactions between transition charge densities, which couple singly excited states of the same or similar energies. These interactions give rise to the so-called exciton splittings in polymers made of identical subunits. To further reduce the size of the matrix, triply excited and higher excited states are omitted.

The desired polymer properties are calculated from the eigenvalues and the eigenfunctions of the energy matrix. We are usually only interested in the ground and the singly excited states. The polymer optical spectra are characterized by the positions and magnitudes of the absorption, or circular dichroism bands. The positions are

$$\nu_K = (E_K - E_o)/h. \qquad 3.$$

The magnitudes are calculated as dipole strengths for absorption

$$D_K = \langle O|\boldsymbol{\mu}|K \rangle \cdot \langle K|\boldsymbol{\mu}|O \rangle, \qquad 4.$$

and rotational strengths for circular dichroism

$$R_K = \text{Im}\langle O|\boldsymbol{\mu}|K \rangle \cdot \langle K|\mathbf{m}|O \rangle. \qquad 5.$$

The symbols O and K refer to the ground and excited states of the polymer (the eigenstates of the energy matrix). The electric dipole moment operator for the polymer is $\boldsymbol{\mu}$ and $\mathbf{m}$ is the magnetic dipole moment operator; they are both written as a sum over subunit operators ($\boldsymbol{\mu}_i$ and $\mathbf{m}_i$). Because the magnetic dipole operator depends on choice of origin, the rotational strength includes a term which depends on the distances ($\mathbf{R}_j - \mathbf{R}_i$) between subunits. The subunit parameters are the subunit transition frequencies, ν_{mn}, electric dipole transition moments, $\boldsymbol{\mu}_{jmn}$, and magnetic dipole transition moments, $\mathbf{m}_{jnm}$.

The experimental polymer data to compare with the calculated polymer properties are the positions of the bands (ν_K), the integrated intensity of the absorption bands (D_K) and the integrated intensity of the circular dichroism bands (R_K). The experimental problem of resolving a measured spectrum into bands is not trivial, particularly for the circular dichroism where much cancellation of positive and negative regions can occur.

The expressions for the dipole strength and the rotational strengths were derived in 1928 (133) with the assumption that the wavelength of incident light was large compared to dimensions of the molecule. For macromolecules this is not necessarily true, but the expressions are valid as long as the wavelength is large compared to the distance between significantly interacting subunits. For shorter wavelengths or for significant interactions over longer distances, the dipole approximation of spectroscopy cannot be made. The correct expressions involve matrix elements of the dot product of the vector potential of the light with the linear momentum operator (135). The absorption and circular dichroism is then written in terms of the following operator (161, 162)

$$\langle O|T|K \rangle = \langle O|\exp[i\mathbf{k}\cdot\mathbf{r}]\mathbf{p}|K \rangle \qquad 6.$$

Here $\mathbf{r}$ is the position operator for an electron, $\mathbf{p}$ is its momentum operator, $\mathbf{k}$ is the wave vector of the light and a sum over all electrons in the system is implied.

We have written various equations and the notation is, as usual, arbitrary and unfamiliar to many readers. However, the main points should be clear. In strong coupling only the electronic states are important. Products of electronic states for the subunits form a basis set for obtaining polymer electronic (exciton) states. The most general method to calculate the polymer states is to diagonalize the energy matrix. Once the polymer states are obtained, the (line) absorption or circular dichroism spectrum of the polymer can be calculated using either the dipole approximation or the full vector potential for the interaction with the light.

WEAK COUPLING   In weak coupling the interaction is between individual vibronic components of the subunits. A variety of alternative, but equivalent, methods have been used to obtain polymer spectra from monomer spectra. The quantum mechanical self-consistent field methods are called time-dependent Hartree or random phase approximations (78, 131, 132). The equivalent classical method is the local field approximation (40, 41). These methods include explicitly only the dynamic coupling (transition charge densities) between subunits. The shape of the absorption band of each subunit is used to calculate the shape of the polymer spectrum. Electrostatic effects can be introduced by using a shifted absorption band for each subunit. The shift is obtained by calculating the effect of the electrostatic fields of the rest of the polymer on each subunit.

We describe the classical method (40, 41) here, because it is easiest to understand. Scientists tend to downgrade classical approximations relative to quantum mechanical methods. However, the classical and quantum mechanical methods used in weak coupling are essentially equivalent.

The subunit absorption and circular dichroism can be written in terms of subunit electric and magnetic polarizabilities. The polarizabilities are frequency-dependent, complex tensors. The real parts characterize the refractive index and optical rotation; the imaginary parts characterize the absorption and circular dichroism. In the polymer each subunit experiences the incident electromagnetic field of the light plus the induced fields produced by the light incident on all the other subunits. The difference in the local field that each subunit senses in the polymer is what causes the polymer optical properties to be different from the sum of the subunit properties.

The polymer polarizability (which gives the polymer optical properties) is obtained by solving a set of Nr linearly dependent equations. Here N is the number of subunits, each of which has r bands. The equations depend on the complex subunit polarizabilities and on their interactions. Thus, the shapes of the polymer spectra are obtained as a function of the shapes of the subunit spectra and the all-order interactions between subunits. The equations are solved at each wavelength by matrix inversion.

Applequist et al (1) have shown that if all subunits are assumed to have Lorentzian band shapes of the same width, then the polymer spectra can be obtained as a function of frequency with only one matrix inversion. This is a large saving in computer time and makes the weak coupling computation (one matrix inversion) no more time-consuming than the strong coupling computation (one matrix diagonalization).

The weak coupling and strong coupling approximations are very different in the assumptions made and in the methods used to obtain polymer spectra. However, for some presumably weak-coupled systems the results are nearly the same. Figure 1 shows a comparison between two calculations (26) for the circular dichroism of the dinucleoside phosphate adenyl-3′,5′-adenosine (ApA). The weak coupling calculation used the resolved shape of the adenosine spectrum for the subunit shape; a dimer spectrum was then calculated. The strong coupling calculation gave a line spectrum for the dimer shown as the vertical bars; Gaussian bands with 1500 cm^{-1} bandwidths were superimposed to obtain a dimer spectrum. In both calculations static fields were ignored and the same subunit interactions between transition charge densities were used. The calculations agree with each other and both agree well with experiment above 230 nm.

ANY COUPLING There have been "exact" treatments available for considering simple, model polymers for some time (52, 116). Identical subunits with only one excited electronic state and a simple vibrational progression are common restrictions for these methods. One important

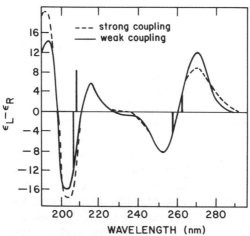

Figure 1 Calculated CD for the dinucleoside phosphate ApA using a weak coupling assumption and a strong coupling assumption. The interesting finding is that these two very different approximations give similar results.

application of these methods has been to test the more practical methods to find when their approximations become inadequate.

A recent method which is approximate, but which applies to weak, strong, and intermediate coupling has been developed by Hemenger et al (83, 84). It is called the degenerate ground state model, because an average vibrational energy is used for the ground states of the subunits. So far only calculations for dimers and for infinite linear polymers have been presented. Identical subunits with only one excited state were treated. However, further developments may make this method a useful, practical one for treating real polymers.

HELICES Symmetry can be very useful in calculating any polymer spectra using any method. For large polymers ($n \rightarrow \infty$) use of symmetry is necessary to obtain a solution at all. The symmetry of the polymer (helical or other symmetry) is used to obtain selection rules so that only a few eigenstates and eigenfunctions need to be calculated. The first application of helical symmetry to polymer circular dichroism spectra was incorrect (117). Selection rules for absorption of an infinite polymer can be obtained using the usual spectroscopic assumption that the wavelength of light is large compared to the size of the molecule. However, for circular dichroism this cannot be done, because of the contribution from a term linear in the distance between subunits. To obtain the correct selection rules, the finite wavelength of light must be acknowledged; the vector potential of the light, not just its linear expansion is necessary. Selection rules for infinite, helical polymers have been discussed for both the quantum mechanical (39, 109, 130, 158) and classical (103) methods.

CIRCULAR DICHROISM: EXPERIMENT

Nucleic acids are optically active not only at the primary level (the nucleic acid bases with their attached sugars and phosphates), but also at the secondary structural levels (sequence dependence and dependence on geometrical organization) (16, 17, 65, 113, 157, 165, 172). There is increasing evidence that tertiary structural organization also has a strong influence on the optical activity of large assemblies of nucleic acids, nucleoproteins, viruses, and such biologically significant structures as chromatin, chromosomes, and indeed whole cells (44, 45, 70, 74–77, 87, 105–107, 110, 145, 146, 163). Remarkable optical activities have been reported for nucleohistone aggregates (50), DNA condensed in buffer-alcohol solvents, and the condensates produced by two phase separation methods such as the so called Giannoni particles or "crystals" (47–49, 54, 55, 66, 95).

We shall present a selected review of the pertinent papers demonstrating each of these effects, i.e. primary effects, secondary effects, and tertiary effects. There will be necessarily some overlap, since for most systems more than one effect may be present.

Primary Structure Effects

Nucleic acid bases have no optical activity since they have a plane of symmetry. However, once they are combined with the sugar phosphate moieties, this basic symmetry is broken and they develop enough activity to be measurable. When incorporated into polymers, the base-base interactions produce large changes in the magnitudes and shapes of the CD spectra, which reflect the types of nucleotides, their sequences along the strands, the number of strands (single, double, or triple helices) and the geometries of the strands (16, 17, 20, 21, 53, 63–65, 90, 104, 129, 153, 154, 157, 170).

At present the most successful explanation of the optical activity of these polymers is based on interactions between $\pi - \pi^*$ transitions in the pyrimidine and purine bases. With these transitions as a basis, attempts have been made to calculate the CD spectra of a variety of single- and double-stranded polymers with some degree of success (25–27, 93, 94). Qualitative agreement with experiment is usually found, although large discrepancies sometimes occur. However, it is apparently not necessary to postulate $n - \pi^*$ transitions in this region of the spectrum (220–300 nm) to explain the optical activity of nucleic acid polymers. Also, base-sugar and base-phosphate interactions do not seem to be necessary (118, 119). Further discussion of these and other theories of the optical activity of nucleic acids has been recently presented by Woody (176).

Secondary Structure Effects

Regardless of base sequence or whether the nucleic acid polymers are single or double stranded the most striking differences are those measured between the CD of RNA and of DNA (104, 136, 175). The CD spectrum of a DNA molecule in aqueous solutions of 0.1 M ionic strength, pH 7.0 is mainly "conservative". This means that the total area (rotational strength) of the first positive band (maximum at 270–275 nm, zero at 257–259 nm) is approximately equal in absolute value to that of the negative band (minimum at 245–248 nm, zero at approximately 228 nm). Thus the sum of the rotational strengths for the polymer in this region is equal to zero, hence the name conservative (63–68, 88, 89, 104, 157, 177). The magnitudes for the CD of DNA are as follows: $\Delta \varepsilon = 2.2$ to 2.4 liter/mole-cm for the maximum and -2.4 to as low as -5.5 liter/mole-cm for the minimum. The CD spectrum of

RNA (both double-stranded and so called single-stranded RNA) has quite a different shape. It has a maximum near 260 nm, with values ranging from 6 to 10 (liter/mole-cm), a zero at around 240 nm, and a minimum at 210 nm with a variation in magnitude of -2 to -10 liter/mole-cm. This minimum, which is always present in all the measured CD of RNA, has not been given enough emphasis in the literature and indeed most of the time is not even reported in the published RNA CD spectra. However, this is an important part of the characterization of the CD of nucleic acids whose geometry is similar to that of RNA. It is particularly important for the analysis of DNA in various stages of condensation or aggregation as we will discuss later (55, 165).

At present the evidence shows that the difference in CD spectra between RNA and DNA is due to the differences in the geometries of the polymers. Roughly, A geometry corresponds to 11 base pairs per turn, a tilt of the base plane of 70 degrees to the helix axis, a dihedral angle of 16 degrees between base pairs and a rotation per residue of 32.7 degrees. B geometry has 10 bases per turn, the base plane is perpendicular to the helix axis and the rotation per residue is 36 degrees. The C geometry is very closely related to the B geometry with about 9.3 bases per turn. RNA is presumably always found in an A-type geometry (A, A' family of geometries, 2), whereas DNA can be found in any of a variety of geometries as determined from X-ray analysis of fibers in different relative humidities. DNA thus shows widely differing CD spectra depending on the temperature, solvent, degree of hydration, type of aggregation or condensation, and type and amount of complexation with polypeptides and proteins (4, 15–19, 20, 21, 53, 54, 66, 67, 70, 72–77, 88, 89, 110, 112, 120, 128, 129, 146, 164, 165, 177–179). Attempts have been made to relate the measured changes in CD to corresponding changes in geometry. RNA with a reasonably constant A structure is one reference and films of DNA, which have been equilibrated against relative humidity conditions analogous to those used to produce the A, B, and C-type geometries found in fibers, are other references (2, 128, 165). The CD of films of DNA at different relative humidities was measured by Tunis-Schneider & Maestre (165), and the spectra obtained were compared with the CD spectra of RNA and DNA in a variety of solvent conditions. The DNA films with presumably A-type geometry showed a nonconservative CD similar to that of RNA in solution. The B-type film conditions gave a conservative CD spectrum similar to that of fully hydrated DNA in low ionic strength, neutral pH buffers. In these two cases there was agreement between the geometries expected from the CD spectra and those predicted from X-ray DNA fiber analysis. A further example is the work of Gray & Ratliff (64) in which the CD of poly [d(A-C)·d(G-T)] is compared with its ribose analog poly [r(A-C)·r(G-U)] and with the hybrids poly [d(A-C)·r(G-U)] and poly [r(A-C)·d(G-T)]. These duplex polymers with essentially identical base sequences in each of the strands gave A-type CD spectra for the ribose polymers, A-type for the ribose-deoxyribose hybrids, and B-type CD spectra for the deoxyribose duplex in phosphate buffers. However, at the same salt concentration and 60% (w/w) ethanol concentration the deoxyribose polymer CD spectrum became essentially the same as that of the ribose polymers. This is a clear demonstration that DNA can take the same geometry as RNA under proper conditions of lowered water activity (dehydration). For these polymers in the A geometry there is a maximum at 260 nm with $\Delta\varepsilon = 8$, a crossover at 240 nm and a negative band with minimum at 210 nm ($\Delta\varepsilon = -7.5$).

A third type of CD spectrum for films of DNA in LiCl was found at low humidity (165). By analogy with X-ray data for fibers of Li DNA under the same relative humidity the CD was assigned to a C-type geometry. Similar CD spectra had been reported for DNA in a variety of solvents such as ethylene glycol, methanol, ethanol, and high ionic strength solutions. All these solvents have the common property that they lower the water activity (32, 34, 69, 120, 164). This interpretation that the geometry of DNA in these solvents tends to C geometry is controversial. Ivanov and his group (88) and Brahms and co-workers have shown that there is a generalized change in geometry towards C forms as function of dehydration of the DNA molecule (15, 17). Hanlon and co-workers have done extensive work on the relations between CD spectra and the winding angle of DNAs in aqueous electrolyte solutions (28, 77). They first obtained a set of reference CD spectra for the A, B, and C geometries and studied the effect of the winding angle of the duplex on the measured CD spectra. Their conclusion that the magnitude of the positive band above 260 nm decreases in a linear fashion as the duplex winding angle increases is consistent with the interpretation that the DNA moves smoothly from a B-like geometry to a C-like geometry (28). Theoretical computations by Cech & Tinoco (25–27) also show that for a C-type geometry there is a general decrease of positive magnitude in the first CD band and an increase in absolute (negative) magnitude for the second CD band of some of the polymers when compared with a B-type geometry. In opposition to the above interpretation are the conclusions drawn by Maniatis et al (114) in which they claim that X-ray scattering studies of gels of DNA in high salt solutions do not show the scattering patterns associated with the computed C-type geometry but rather are more closely related to those computed from B-type geometry. Similar conclusions were obtained by Baase & Johnson (4) from their studies on the CD of super-coiled DNA as a function of winding angle.

It is important to remember that a basic assumption in all these arguments has been that the B-type geometry as determined by fiber X-ray scattering is the one found in physiological ionic strength buffers. However, Wang (171) has recently determined that the average DNA in dilute and moderate ionic strength solutions has 10.4 bases per turn instead of the 10 bases per turn given by the X-ray fiber data. Therefore the association of film CD spectra with the measured values has to be modified.

DNA in Complexes and Aggregates

The CD spectra of nucleic acids also change when the nucleic acids interact with a variety of biological materials such as polypeptides, histones, polyamines, and proteins (21–23, 30, 35, 36, 44, 50, 51, 72–74, 90, 97, 105–107, 123–125, 137, 146, 152, 173). These changes in general show a reduced magnitude of the first CD band of DNA corresponding to a purported B to C geometry transition (74, 105). However, when the complexes are prepared in a particular manner, remarkable changes in the CD spectra of the DNA occur. Very large negative or positive rotations are seen (23, 50, 51, 147, 149). Since the interpretation of the reduction of the first peak in the CD spectra in terms of specific changes in structure of the DNA duplex is a controversial one at present, it should be approached with caution. The workers in this field are divided into two general camps: those who interpret the CD changes as a manifestation of secondary geometry changes, and those that claim that the CD changes are primarily a manifestation of tertiary structure. If tertiary structure is the explanation, it must surely be a manifestation of some long-range chiral ordering of the type that occurs in liquid crystals (29, 37). In this sense it is a type of scattering phenomena which we will discuss in the next section.

Recent work by Baase & Johnson (4) has determined the change in the average rotation of the bases about the helix axis as DNA is transferred from solutions containing 0.05 M NaCl to 3.0 M CsCl, 6.2 M LiCl, and 5.4 M NH_4Cl. This work allowed them to relate the intensity of the CD of DNA at 275 nm directly to the change in the number of base pairs per turn. The geometrical change on going from low salt to high salt is of order of $-0.22\pm.02$ base pairs per turn. Baase & Johnson applied these results to the CD studies of Cowman & Fasman (36) on nucleosome structure. Their conclusions are that there is a change in secondary structure corresponding to 0.31 (±0.03) bases per turn of DNA. This change in secondary structure can be combined with the change in linking number of −1.25 turns per nucleosome (96, 148) to give about −1.6 turns about the histone core or about 90 bases per turn.

Scattering Particles

When circular dichroism techniques are applied to structures whose size is comparable to the wavelength of light, several complications arise that affect the CD signals in sometimes quite surprising fashion. CD studies have now been applied quite extensively to complicated biological macrostructures. Much CD work has been published on the optical behavior of DNA in chromosomes (21), chromatin (38, 146, 152), bacteria (163), sperm (150, 151), and related material such as reconstituted nucleohistones (22, 23, 124, 125) and other model systems for DNA-protein complexes (82, 144–146). DNA condensates in ETOH solvents also show CD scattering components (66, 67). The structure of bacteriophages has been studied with CD (43–45, 110, 111) and the conformations of membrane-incorporated proteins (58, 102, 140, 142, 168, 169), whole red blood cell, and cell ghosts (56, 57, 140, 142, 143) have also motivated considerable effort in recent years.

All these objects due to their large size and particulate nature tend to form light-scattering suspensions, some of which scatter quite intensely. The effect on CD measurements is still not well understood, although it has been extensively discussed (10–13, 58–61, 141, 142, 168). Several possible artifacts of CD measurements have been proposed, among them: absorption or Duysens flattening (46, 140, 168), concentration-obscuring effects (140, 142, 143, 168, 169), differential light scattering (43–45), and liquid crystal behavior (86, 87). Corrections for these effects have been attempted in a variety of ways, both by computational approximations and theoretical applications of classical scattering theories modified for optically active particles (10–13, 59–61, 87). Instrumental corrections have been attempted with varying degrees of success by Dorman & Maestre (44–45), Gregory & Raps (71), and M. Maestre & C. Reich (in preparation).

In general, the presence of possible optical scattering perturbations on CD signals are manifested by the following: (a) Long CD tails—CD signals away from the absorbance bands; (b) extremely large CD signals —factors up to 10^3 times the intrinsic CD of the component material; (c) abnormally low values for CD bands and gross distortion of shapes (Duysens flattening effects); (d) position-sensitive CD spectra that depend strongly on the position of the sample with respect to the detector; this last condition must be differentiated from linear dichroism or linear birefringence of the material, which will produce very large apparent CD and which can be detected by rotating the sample about the axis of the light beam (42, 165); (*e*) apparent CD signals that vary with the acceptance angle of the detector or that are detected at an angle to the incident beam (43–45); these signals can be measured by beam goniometer instruments or instruments that measure selected sectors of the scattering pattern (see Figure 2); (*f*) signals that vary slowly over a period of time in nonreproducible fashion; these are

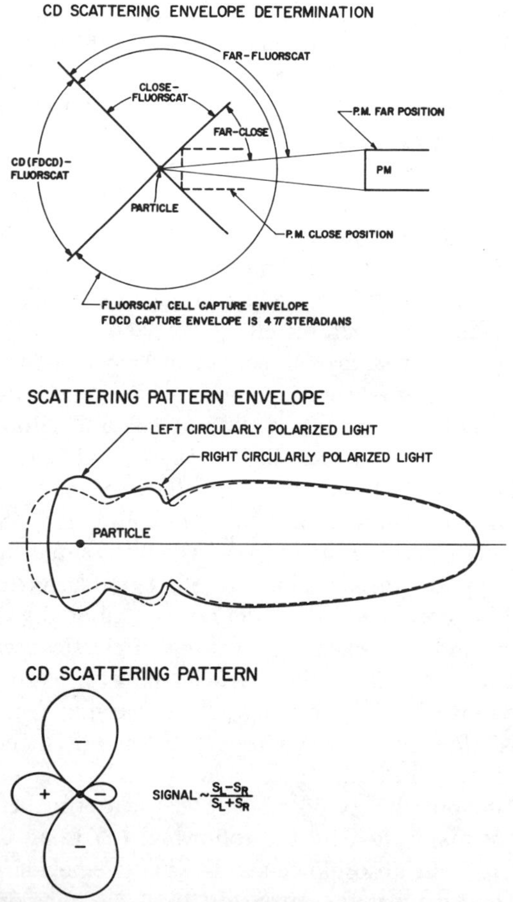

CD SCATTERING ENVELOPE DETERMINATION

SCATTERING PATTERN ENVELOPE

CD SCATTERING PATTERN

$$\text{SIGNAL} \sim \frac{S_L - S_R}{S_L + S_R}$$

Figure 2 A diagrammatic representation of the CD scattering envelope and its measurement for a hypothetical particle. Different sectors are measured by computing the appropriate differences between measuring methods. For example far-close means the difference is calculated between the signals measured with the photomultiplier far from and close to the cell.

usually associated with precipitation or condensation of nucleic acids from solutions, or complexation with polypeptides or histones. In general some of these condensates produce CD signals that vary widely with preparation history and are hard to reproduce.

Historically, the problem of scattering was first discussed by Ji & Urry (91) in their explanation for the distorted CD spectra of proteins in membranes. They explained the reduction of magnitudes and the red-shifted minima as a manifestation of some type of absorbance flattening equivalent to that postulated by Duysens for the distortion in absorbance spectra of very large particles. Similar interpretations of the distorted CD spectra of red blood cell membranes were proposed by Schneider et al (140, 141), as correction could be obtained by sonication of the membranes.

The CD of bacteriophages T2, T4, T6, showed CD tails that extended far into the visible range of the spectrum (111). The CD tails were effectively eliminated by the use of variable acceptance detectors and the use of an integrating device (fluorscat cell, Dorman et al, 1973) (43–45) that covered all regions of the scattering envelope except the back region. These measurements presumably have corrected for that part of the spectrum which depends on differential scattering, that is, preferential scattering of right and left circularly polarized light. In Figure 2 we can see the different degrees of correction as the angle of acceptance of the detector is gradually increased.

Similar instrumentation was used to study the CD of nucleohistone aggregates (B. Dorman and M. Maestre, unpublished data). Figure 3 shows the separation of the CD spectrum of an H1 histone-DNA complex into a differential scattering component and a complexation-aggregation CD signal, which is presumably an intrinsic property of the complex. Notice how the differential scattering component contains all the long CD tails properties. The CD scattering component can be shown to be very similar to the Kramers-Kronig transform of the intrinsic signal, i.e., it comes from the dispersive part of the complex index of refraction for the nucleohistone aggregate.

The type of CD spectra shown above is similar to the so-called psi (ψ) type spectra as described by Jordan et al (95). Similar spectra were obtained by Carrol (22, 23) for complexes of DNA and RNA with polypeptides; polylysine-DNA complexes gave very large positive and negative psi-type spectra. Evdokimov et al (47–49) have reported on the very large psi-type CD (both positive and negative) for DNA particles precipitated from aqueous salt solutions containing various concentrations of polyethylene glycol. Again all the reported CD spectra show very large values and long CD tails away from the main absorbance bands into the visible region of the spectrum. Shin & Eichhorn (147)

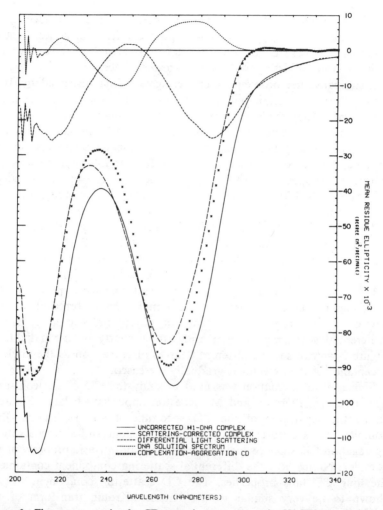

Figure 3 Fluorscat correction for CD scattering components for H1-DNA nucleohistone complexes. The differential scattering component accounts for the CD tails outside the main absorbance bands of the complex. The nucleohistone CD is at least one order of magnitude larger than for DNA in solution.

were able to show a reversible change of a negative psi spectrum into a positive psi spectrum for DNA-polylysine complexes upon the addition and removal of metal ions. They proposed that a reversible change in tertiary structure was involved. Similar types of psi-type spectra have been reported by Fasman and co-workers for the optical properties of reconstituted nucleohistones (50, 51). Other types of structures such as intact chloroplasts show CD spectra that are extremely variable, depending on the acceptance angle of the detector and also on its angular orientation relative to the incident beam (71, 128).

Not all large nucleic acids show such remarkable CD scattering properties. Gosule & Schellman (62) have described a DNA-spermidine aggregate that shows no change at all in the CD spectra of the DNA as compared to the solution spectra. Maestre (unpublished) has measured the CD spectra of RNA viruses of the f2 family in which the CD is the same as the solution spectra; also in DNA bacteriophage α there is no change in the DNA CD spectrum relative to the solution spectrum.

As pointed out by Holzwarth et al (87) the CD spectra of the DNA inside the head of the T-even bacteriophage show many of the characteristics of the optical activity of liquid crystals, i.e. strong CD signals with absorbance band shapes and long CD tails with "hooks" close to the absorbance bands. Furthermore, C. Reich et al (in preparation) found that the scattering properties of DNA condensates in ethanolic solutions showed some striking resemblances to the scattering patterns of liquid crystals. Using a novel application of fluorescence-detected circular dichroism they developed a way of measuring the total differential scattered CD components in all possible directions, or sectors, of the scattering pattern. This will measure all light that is not absorbed. The result is depicted in Figure 4. Reich et al found that for this type of psi-type CD spectra (very large in magnitude and positive) there is very little forward differential scattering. Most of the differential scattering is perpendicular to the light beam, but some differential scattering occurs in the backward direction. Even more remarkable was the fact that for some of the particles correction for differential scattering produced larger psi-type CD spectra.

The above results illustrate the futility of only measuring one direction of differential scattering with the hope that it will provide insight into the total CD scattering of the particle. Extrapolations such as performed by the computational approximation of Nicolini et al (121) and Gitter-Amir et al (56, 57) are fraught with possible misunderstandings about the optical behavior of such large particles. This is because these workers are assuming that CD scattering behaves as ordinary, unpolarized Rayleigh-Debye or Mie-type scattering. As pointed out by Maestre and Reich (in preparation) CD measurement is a ratio of the intensity difference of left-minus right-polarized light divided by the sum of such intensities. This signal can obviously have a very large value even if the total scattered intensity is very small, and conversely, the ratio can be very small if the total intensity scattered is very large.

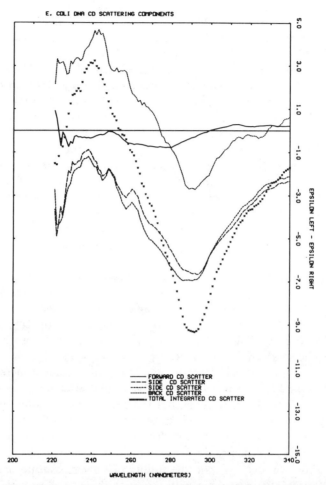

Figure 4 The CD scattering components of DNA condensates in 80% EtOH-buffer (w/w) solutions as measured in different directions. An important result for this case is that the component in the forward direction is very small as compared to the side-scattering and back-scattering components. For this type of particle the CD scattering is mainly in the side directions.

Maestre and Reich (in preparation) have measured the CD of DNA films with induced twists produced by shearing the film between quartz plates and have obtained CD values of extremely large magnitudes (Figure 5). They also have shown that the sense of twist of the films determines the sign of the measured CD. Moreover, they measured the back scattering of these films and found that the polarization of the scattered light is the same as that of the incoming beam; this is one of the main characteristics of cholesteric liquid crystal reflections (29, 54,

85, 86). Measurement of psi-type particles, i.e. polylysine-T7 phage DNA aggregates and ethanolic condensates, shows behavior similar to the scattering behavior of the films. All the psi-type CD are probably manifestation of some cholesteric or twisted nematic type of arrangement of the DNA molecules in the films and particles (43, 45, 49, 54, 82, 95). Maestre and Reich (in preparation) postulated the following explanations. (*a*) Psi-type spectra are mainly a manifestation of resonance-scattering phenomena with associated anomalous transmission in the absorbing bands. (*b*) The apparent CD bands outside the absorbing regions (CD tails) can be eliminated by the use of devices such as fluorscat cells and FDCD methods (C. Reich et al, in preparation). This apparent CD is caused by differential scattering of right and left circularly polarized light; it is not only a function of the bulk index of refraction, but also reflects the size and shape of the particle. (*c*) Some of the scattering particles have optically active behavior such as found in cholesteric liquid crystals or twisted nematic crystals. In such cases the sense of twist reflects the helical organization of the packing of the DNA molecule; that is, a left-handed helical array gives positive

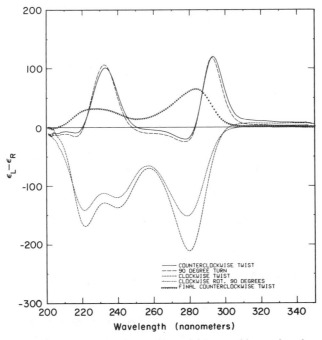

Figure 5 CD of DNA films measured in a vertical CD assembly as a function of twist of film and orientation of film about the optical axis of light beam. Twisting occurred in the sequential order described in curve label.

CD. The wavelength at which the large CD occurs is determined by the Bragg law as applied to liquid crystal as follows (29, 37): $n\lambda = 2L\cos r$. Here n is the order, λ is the wavelength, L is the periodicity of the helical organization, and r is an internal reflection angle (obtained from Snell's Law). (d) It is obvious that for this kind of structure it matters little what the intrinsic CD of the material is, since the CD produced by the super organization is so large as to completely mask the underlying optical activity. Indeed as in liquid crystals the component molecules do not have to be optically active at all, but they must have absorbance bands at or near the critical wavelength, with linearly dichroic behavior (29, 37).

Theoretical Models

As seen from the above discussion it is reasonable to expect that any attempt to explain the source of scattering perturbations by the use of ordinary scattering theory is doomed to failure. The most serious attempts at applying scattering theory have been done by Gordon & Holzwarth (60), Gordon (59), Holzwarth et al (87), and Bohren (10-13). The result of the theories, which involve modification of the Mie scattering theory for symmetrical objects, is that only small corrections are necessary to obtain the intrinsic CD of these aggregates. Indeed, when the Mie scattering case was solved for a spherical particle as a model for the CD of an intact bacteriophage (87), the theory failed to compute the scattering components as measured by Dorman & Maestre (45).

The main weakness of the above computations is that a very symmetric model is assumed in which the distribution of optically active components is essentially uniform; furthermore, at present only the forward direction is computed because of mathematical difficulties. As implied by the experimental data the scattering systems show a highly asymmetric distribution of materials with internal organization. In the following section we present a model based upon a helix; it provides results which are much more consistent with experiments.

CIRCULAR INTENSITY DIFFERENTIAL SCATTERING

Some media interact differentially with right and left circularly polarized light. These media are characterized macroscopically as being circularly birefringent, i.e. having a different refractive index for right and left circularly polarized light, and circularly dichroic, i.e. having a different absorption coefficient for right and left circularly polarized light. The phenomenon of optical rotatory dispersion (ORD) is directly related to the real parts of the refractive indices, whereas circular dichroism (CD) is directly related to the imaginary parts of the refractive indices (the absorption coefficients). The refractive index can in turn be considered as a manifestation of a forward scattering process. It is experimentally observed that chiral molecules differentially scatter right and left circularly polarized light; it is possible to relate such measurements to the molecular dissymmetry of the medium. It has been suggested [Atkins & Barron (3) and Barron et al (5)] that this effect be called Circular Intensity Differential Scattering (CIDS).

An appropriate measure of CIDS is the ratio of the difference of scattered intensities between left and right circularly polarized light to its sum:

$$\text{CIDS} = (I_L - I_R)/(I_L + I_R)$$

where I_L and I_R are the scattered intensities for the two polarizations of the incident light. This definition of CIDS can apply to both elastic (Rayleigh) and inelastic (Raman) scattering.

The useful features that this method can provide as an experimental tool include: (a) Since the signal can be measured for 4π radians around the sample, it is possible to get information about the system that is not obtainable from forward scattering measurements; (b) it is possible to detect different orders of molecular regularity by changing the wavelength of the incident radiation; (c) measurements of CIDS should give additional information regarding the dissymmetry of molecules and their geometry.

Some very general theoretical treatments of scattering of circularly polarized light have appeared in the literature [see (7) for a review]. The main purpose of this article is not, however, to provide an extensive review of these theoretical works, but instead to communicate some of the preliminary results that have been obtained in our laboratory on the CIDS of helical molecules. Here we will only mention some of the more important contributions to the theory in this field. Atkins & Barron (3) have given general expressions for the differential scattering of light both for Rayleigh and Raman processes, from a quantum mechanical point of view. These authors were able to obtain explicit expressions by constructing the scattering matrix and relating it to molecular parameters. In their treatment the wavelength of light is large compared to the dimensions of the molecule and no absorptive phenomena were taken into account. Later Barron & Buckingham (6) presented both a classical and quantum mechanical analysis for the elastic and inelastic differential scattering of circularly polarized light. Harris & McClain (79, 80) have presented a theory for the polarization of light scattered by polymers emphasizing its applicability for wavelengths shorter than the polymer dimensions. They construct the Perrin matrix for the scattering of light according to Stokes' formalism (127) and relate this phenomenological treatment to molecular parameters. The wavelength is assumed to be large compared to the size of the monomers, but not necessarily large relative to the polymer. The treatment is general; it does not explicitly consider any particular geometry.

CIDS of Helical Molecules

We present here a classical treatment of the differential elastic scattering of a continuous helical structure in which all the optical properties of the helix are described by a polarizability tensor. The derivation is obtained within the dipole radiation approximation. In what follows we make explicit reference to the approximations involved, although in many cases the details of the derivation must be omitted for lack of space. We treat successively three cases in order of increasing complexity of the model. First, we deal with a helix of point scatterers having noninteracting spherically symmetric polarizabilities. It will be shown that no differential response to circularly polarized light exists in this case and that CIDS is therefore zero for all scattering directions. An asymmetrical polarizability tensor is then treated. General expressions for the CIDS as a function of the helical parameters and the scattering angle are obtained and some calculations presented. Finally, we consider the case in which dipole-dipole interactions between different parts of the helix are included. In all cases the theories are valid for all wavelengths of the incident radiation.

General Theory

The classical solution to the interaction of radiation with matter is described by Maxwell's equations in material media. From these expressions a differential wave equation can be easily obtained by assuming periodic solutions for the fields:

$$(\nabla^2 + k^2)\mathbf{E}(\mathbf{r}) = -[\varepsilon(\mathbf{r}) - 1]k^2\mathbf{E}(\mathbf{r}) + \nabla\nabla \cdot \mathbf{E}(\mathbf{r}). \qquad 7.$$

Here ε is the dielectric tensor of the medium and $\mathbf{k}$ is the wave vector of the electric field $\mathbf{E}$. In principle, any scattering problem could be solved by obtaining the solution of this differential equation, subjected to adequate boundary conditions, at the surface of the scatterer and at infinity. This approach is rather cumbersome if the geometry of the scatterer on which we must specify the boundary conditions is as complicated as a helical array. A solution is obtained by transforming the differential wave equation into an integral equation and using the

Born-Kirchoff approximation (98, 138):

$$E(\mathbf{r}') = \mathbf{F} \cdot \int_V \exp[i(\mathbf{k}-\mathbf{k}_0)\cdot\mathbf{r}][(\varepsilon-1)/4\pi]\cdot\mathbf{j}\,dV,$$

$$\mathbf{F} = (1-\mathbf{kk}/k^2)(k^2/r')E_0\exp(i\mathbf{k}\cdot\mathbf{r}').$$

8.

Here $E(\mathbf{r}')$ is the far field approximation to the electric field (126) at the point in space $(\mathbf{r}')$ where the scattered signal is measured; E_0 is the incident field with polarization direction $\mathbf{j}$; $\mathbf{r}$ is the position of the scatterer; $\mathbf{k}$ and $\mathbf{k}_0$ are the scattered and incident wave vectors, respectively. The integral extends over the positions of all the scatterers. The physical relation between all these quantities is schematically shown in Figure 6a for a segment of a helix. Equation 8 incorporates through $\mathbf{r}$ all the boundary conditions dependent on the geometry of the scatterers. Furthermore, since $(\varepsilon-1)/4\pi$ is proportional to the polarizability tensor, α, associated with the scatterer, we see that Equation 8 gives the electric field at a distance r' from an origin as the coherent sum (integral) of all the contributions of dipole radiation induced by the incident field.

HELICAL ARRAY WITH A SPHERICALLY SYMMETRICAL POLARIZABILITY
The equation of a helix can be written in parametric form as:

$$\mathbf{r} = \mathbf{e}_1 a\cos\theta + \mathbf{e}_2 a\sin\theta + \mathbf{e}_3(P\theta/2\pi)$$

9.

where a is the radius of the helix, P its pitch and $\mathbf{e}_1$, $\mathbf{e}_2$, $\mathbf{e}_3$ are orthogonal unit vectors. The helix is assumed to be infinitely thin. In order to

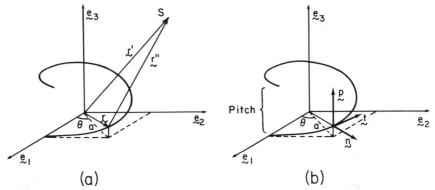

Figure 6(a) A segment of the helix of radius a, showing the relations between $\mathbf{r}$, the vector position of the scatterer; $\mathbf{r}'$, the vector position of the point of observation S; and $\mathbf{r}''$, the distance from the scatterer to the observation point. In the far field approximation $|\mathbf{r}'-\mathbf{r}|$ is approximately r'. (b) The same segment of the helix shows the local orthogonal coordinate system in terms of which the dielectric tensor is defined. The parameter θ is shown also. See text for the definitions of $\mathbf{t}$, $\mathbf{p}$, $\mathbf{n}$.

describe adequately the polarizability of the helix, we define a new cartesian coordinate system along orthogonal directions at each point on the helix, as is shown in Figure 6b. The parametric form of each of these unit vectors is:

$$\mathbf{n} = \mathbf{e}_1 \cos\theta + \mathbf{e}_2 \sin\theta,$$

$$\mathbf{t} = d\mathbf{r}/d\theta = -\mathbf{e}_1(a/M)\sin\theta + \mathbf{e}_2(a/M)\cos\theta + \mathbf{e}_3(P/2\pi M), \qquad 10.$$

$$\mathbf{p} = \mathbf{n} \times \mathbf{t} = \mathbf{e}_1(P/2\pi M)\sin\theta - \mathbf{e}_2(P/2\pi M)\cos\theta + \mathbf{e}_3(a/M),$$

with $M^2 = (a^2 + P^2/4\pi^2)$ a normalization constant. Now, since the polarizability is in general a second rank tensor, we can use dyadic notation and write it as:

$$\boldsymbol{\alpha} = \alpha_{tt}\mathbf{tt} + \alpha_{tn}\mathbf{tn} + \alpha_{tp}\mathbf{tp} + \alpha_{nt}\mathbf{nt} + \cdots. \qquad 11.$$

We see that the case of spherically symmetric polarizability corresponds to a diagonal tensor of the type

$$\boldsymbol{\alpha} = \alpha(\mathbf{nn} + \mathbf{tt} + \mathbf{pp}) \equiv \alpha \mathbf{1}, \qquad 12.$$

since $\mathbf{nn} + \mathbf{tt} + \mathbf{pp} \equiv \mathbf{1}$. Thus a helix with a spherically symmetric polarizability, which can be thought of as a helical array of spherically symmetric point scatterers, will not alter the polarization of the incident light. Therefore, although scattering occurs, the scattered intensities for right and left circularly polarized light are the same and no differential scattering (CIDS) is predicted for this model.

HELIX WITH AN ASYMMETRICAL POLARIZABILITY We choose (Bustamante, Maestre, Tinoco, unpublished) the helix to be polarizable only along its tangential direction: $\boldsymbol{\alpha} = \alpha \mathbf{tt}$. This approximation would become exact in the limit of an infinitely thin helical structure.[1] With this assumption, the scattered electric vector (Equation 8) can be written:

$$\mathbf{E}(\mathbf{r}') = \alpha \mathbf{F} \cdot \int_V \exp[i(\mathbf{k} - \mathbf{k}_0) \cdot \mathbf{r}] \mathbf{tt} \cdot \mathbf{j} \, dV. \qquad 13.$$

Now dV equals the cross sectional area of the helix, A, times dL, the differential of arc length of the helix. We can then explicitly write Equation 8 as

$$\mathbf{E}(\mathbf{r}') = \alpha B \mathbf{F} \cdot \int_0^{lP} \exp[i(aQ_x \cos(2\pi z/P)$$

$$+ aQ_y \sin(2\pi z/P) + z\Delta k_z)] \mathbf{tt} \cdot \mathbf{j} \, dz \qquad 14.$$

$$B = A[(4\pi^2 a^2/P^2) + 1]^{\frac{1}{2}}.$$

---

[1]This restriction is not as strong as it may appear. In fact the derivation above will be valid (and the results for CIDS will be the same) if we take the polarizability to be: $\alpha_1 \mathbf{tt} + \alpha_2(\mathbf{pp} + \mathbf{nn})$, since this is equivalent to: $\alpha_2 \mathbf{1} + (\alpha_1 - \alpha_2)\mathbf{tt}$. Since, as was shown before, $\mathbf{1}$ does not contribute to CIDS, our choice is equivalent to an ellipsoidal polarizability.

l is the number of turns of the helix and $Q=(Q_x^2+Q_y^2)^{\frac{1}{2}}$ is the projection of the vector $\mathbf{k}-\mathbf{k}_0$ in the xy plane (see Figure 7). After writing explicitly the expression for $\mathbf{t}$ in Equation 14 and choosing a polarization for the incident light $\mathbf{j}$, we can perform the integration and obtain for incidence along y and polarization along z:

$$\mathbf{E}(\mathbf{r}')=(\alpha BP/4\pi M^2)\mathbf{F}\cdot\sum_{n=-\infty}^{\infty}J_n(Qa)\exp[in(\psi^*+\pi/2)](-1)^{l(n+1)}$$
$$\times\{-2iaS_n\mathbf{e}_1+2aPT_nS_n\mathbf{e}_2+(-1)^l(P/\pi^2T_n)\mathbf{e}_3\}\sin(\Delta k_z lP/2),$$
$$T_n=(\Delta k_z/2\pi)-(n/P),\qquad\qquad\qquad\qquad\qquad 15.$$
$$S_n=\{\pi P[(\Delta k_z/2\pi-n/P)^2-1/P^2]\}^{-1},$$
$$J_n=n\text{th order Bessel function}.$$

This equation is completely general and is valid for all ratios of pitch/wavelength. For the case of $P \geqslant$ wavelength, Equation 15 can be simplified:

$$\mathbf{E}(\mathbf{r}')=B\alpha(P/4\pi M^2)\mathbf{F}\cdot\{ia[J_{n-1}(Qa)X_{n-1}-J_{n-3}(Qa)X_{n-3}]\mathbf{e}_1$$
$$+a[J_{n-1}(Qa)X_{n-1}+J_{n-3}(Qa)X_{n-3}]\mathbf{e}_2$$
$$+(P/\pi)J_{n-2}(Qa)X_{n-2}\mathbf{e}_3\},\qquad\qquad 16.$$
$$X_n=\exp[in(\psi^*+\pi/2)];n=(\Delta k_z P/2\pi)+2.$$

In this last equation we obtain the result that if the wavelength of light is smaller or equal to the pitch of the helix, the scattering pattern is not

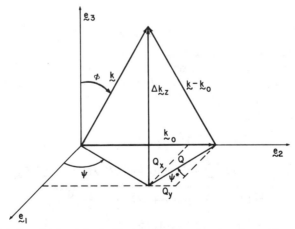

Figure 7 Relations between the incident and scattered vectors $\mathbf{k}_0$ and $\mathbf{k}$ are shown for light incident along $\mathbf{e}_2$; Q is the projection of the vector $\mathbf{k}-\mathbf{k}_0$ onto the plane $\mathbf{e}_1$, $\mathbf{e}_2$ ($x-y$ plane). ψ^* is the angle between Q and the $\mathbf{e}_1$-axis. The third component of $\mathbf{k}-\mathbf{k}_0=\Delta k_z$ is also shown in the figure. The experimentally observed angle of scattering, ψ, is shown.

continuous, but discrete. This appears in the equation as a selection rule that specifies the allowed values that the $\mathbf{e}_3$ component (z component) of the $\mathbf{k}-\mathbf{k}_0$ vector can take (see Figure 7). Cochran, Crick & Vand (33) obtained a similar result for unpolarized light. The scattering pattern of a helix with light incident perpendicular to the helix axis is therefore (for $P \gg \lambda$) composed of layer lines, the maximum number of which can be easily shown to be:

$$n_{\max} = \{P/\lambda\} + 1 \qquad 17.$$

where $\{P/\lambda\}$ means the integral part of P/λ. Equations 16 and 17 are valid only for $P \gg \lambda$. For $P < \lambda$, the more complicated Equation 15 must be used. In this case the scattering field is continuous and no layer structure is associated with it. Similar results can be obtained for any other direction of incidence and possible polarization including circular polarization. From the field strengths, the intensity of the scattered radiation can be calculated; in this way we have obtained an expression for CIDS for $P \gg \lambda$ and incidence along y:[2]

$$\frac{I_L - I_R}{I_L + I_R} = \frac{(2aP/\pi)\{Ma^2 + NP^2/2\pi^2\}\cos(\psi^* + \pi/2)}{(a^2P^2/2\pi^2)U + (P^4/2\pi^4)V + a^4W}$$

where:

$$M = J_{n-1}(J_n - J_{n-2}) + J_{n-3}(J_{n-2} - J_{n-4}),$$

$$N = J_{n-2}(J_{n-1} - J_{n-3}),$$

$$U = 3J_{n-3}^2 + 3J_{n-1}^2 - 2J_{n-3}J_{n-1}\cos 2(\psi^* + \pi/2), \qquad 18.$$

$$V = J_{n-2}^2,$$

$$W = J_n^2 + 2J_{n-2}^2 + J_{n-4}^2 - 2(J_n + J_{n-4})J_{n-2}\cos 2(\psi^* + \pi/2),$$

and the argument of the Bessel functions is Qa. The same equation is obtained for the incidence along x direction, as expected. CIDS for light incident along the helix axis (z direction), can also be obtained in a closed form for the case of $P \gg \lambda$ to give (including the terms for transversality of the field):

$$\frac{I_L - I_R}{I_L + I_R} = \frac{a^2[(S_- + X_-)\sin^2\phi - 2S_-] + \left(\frac{aP}{\pi}\right)T_- \sin(2\phi) - (P^2/\pi^2)G_- \sin^2\phi}{a^2[(O_+ + X_+)\sin^2\phi + 2O_+] + \left(\frac{aP}{\pi}\right)T_+ \sin(2\phi) + (P^2/\pi^2)G_+(1 + \cos^2\phi)}, \qquad 19.$$

---

[2] We have not included in Equation 18 the terms that account for the transversality of the scattered field. In the calculations and discussion, however, the complete expression has been used.

$$S\pm = (J_n^2 \pm J_{n-4}^2),$$
$$X\pm = 2J_{n-2}(J_n \pm J_{n-4}),$$
$$G\pm = (J_{n-1}^2 \pm J_{n-3}^2),$$
$$T\pm = (2/Qa)((n-3)J_{n-3}^2 \pm (n-1)J_{n-1}^2),$$
$$O_+ = 2J_{n-2}^2 + J_n^2 + J_{n-4}^2.$$

There are two observations to be made about Equations 18 and 19. First, we notice that in Equation 19 all the dependence of angle ψ^* has disappeared; thus the CIDS for light incident along the z axis is cylindrically symmetric, as expected. This result is of course not true along the other two directions of incidence. Secondly, we notice that all the constants appearing in front of Equations 15 and 16 cancel when the CIDS ratio is taken. The main result is that this ratio is independent of the actual value of the polarizability along the tangential direction. This means that this simple model is unable to take into account possible absorptive phenomena, damping or band shape effects. This is related to the fact that we have not allowed off-diagonal terms in the polarizability tensor. In this context, since there is no interaction between different parts of this helix we can refer to these results as *form* CIDS (9, 174). Equations 17 and 18 have been used to calculate CIDS for different layer lines and for all angles ψ. In Figure 8 polar graphs of the zero layer line pattern of CIDS for ratios of P/λ of 10, 1 and, 0.1 are shown. Also shown is the normalized total $(I_L + I_R)$ scattered intensity. We outline the general results only in connection with these calculations. (*a*) There is no differential forward scattering in the zero layer line. This is true for any ratio of P/λ. For all other layer lines the differential forward scattering is not zero. (*b*) The pattern of layer lines is symmetric about the $x-y$ plane (zero layer line). (*c*) The differential scattered light, both for the backward direction in the zeroth layer line and perpendicular to the $x-y$ plane, passes through a minimum changing the sign of its polarization for a given relation of pitch/wavelength and pitch/radius. These results are related to those described for the sign of polarization of the radiation scattered by helical antennas of definite handedness (101) and might also be connected to the results described in the scattering experiments with liquid crystals (29, 37) and DNA films (Maestre and Reich, work in progress). (*d*) The number of lobes of the scattering pattern for the zero layer line is independent of the ratio of pitch/wavelength, but it is an increasing function of the ratio of radius/wavelength. The lobes alternate in sign as indicated in Figure 8. (*e*) The number of lobes decreases in going from the zero to the higher order layer lines. (*f*) The envelope of the patterns has a complicated behavior: maxima and minima can be found as a function

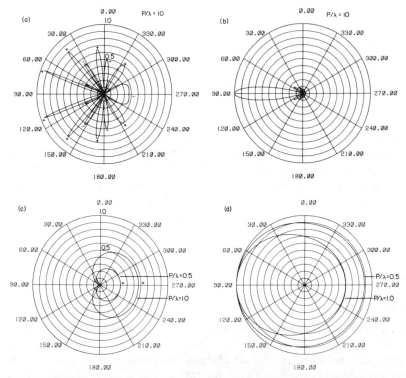

Figure 8 Polar plots of the scattering of light incident perpendicular to the helix axis of a helix of radius = 1 and pitch = 10. The light is incident from the right of each figure. The differential scattering is shown on the left hand side of the figure. The absolute magnitude of $(I_L - I_R)/(I_L + I_R)$ is plotted vs angle; the sign of each lobe is given. In Figure 8a the wavelength = 1 ($P/\lambda = 10$); in Figure 8b the wavelength = 10 ($P/\lambda = 1$) and 20 ($P/\lambda = 0.5$). The total scattering is shown on the right hand side of the figure. The normalized magnitude of $(I_L + I_R)$ is plotted vs angle for the same wavelengths.

of the ratio P/λ. The minima seem to be equivalent to the "invisible" particles described by Kerker (99) and Chew & Kerker (31) for regular scattering. (g) As the wavelength grows relative to the dimensions of the helix (pitch, radius), the differential scattering decreases and increasingly resembles the functional dependence of Rayleigh scattering; for wavelengths much larger than the pitch or radius the differential scattering is zero, whereas the total intensity approaches that of Rayleigh (spherical) scattering (Figure 8).

SELF-INTERACTING HELIX WITH AN ANISOTROPIC POLARIZABILITY Here we construct a solution to the scattering of radiation by a helix in which interactions between different parts of the scatterer are included. In fact by allowing a coupling between the electric dipoles induced along the helix we are actually describing the field at a given point as the sum of the initial field plus the field caused by the electric dipoles that were induced somewhere else along the helix by the same initial field. The end result of this correlation between different parts of the helix is, therefore, to go from an external field to a "local" field description. We allow only dipole-dipole interactions (134) and in this way the effective field at a given polarizable point j in the helix can be written as a first approximation:

$$\mathbf{E}_0 \exp(-i\mathbf{k}_0 \cdot \mathbf{r}_j) - \sum_l \mathbf{T}_{jl} \cdot \mathbf{E}_0 \exp(-i\mathbf{k}_0 \cdot \mathbf{r}_l) \qquad 20.$$

where the summation runs over all the other dipoles induced on the helix where $\mathbf{T}_{jl}$ is the dipole interaction tensor (134). Although we will not write down the equations here, one can obtain an expression for the scattered electric field from a self-interacting continuous helix. The scattering field contains two terms. The first one corresponds to the incident field and the second one is the correction for the dipole-dipole interactions.

As it turns out, following a similar analysis to that for the noninteracting helix, one arrives at equations that cannot be integrated analytically. Therefore, the contribution to the CIDS due to the dipole-dipole interaction term must be evaluated numerically. However, for light incident along the z axis it is possible to obtain a closed expression. The result obtained resembles the equations of the noninteraction case. Of course because the total scattered field is: $E_{\text{incident}} + E_{\text{interaction}}$, the CIDS will involve cross terms that again cannot in general be obtained analytically. To this point we have done no calculations with the equations obtained for the interacting case. Further research to determine in what way the presence of the correlation might change the observed scattering patterns is being carried out in our laboratory. There are a few remarks to make at this point. Due to the nonlinear form of the fields (Equation 20), the polarizabilities will not be cancelled in the CIDS and, as expected, absorptive phenomena and band shapes will be relevant in this model. What has been accomplished by allowing interaction is equivalent to the macroscopic, ad hoc introduction of off-diagonal terms in the polarizability to describe absorptive phenomena. Although we have not yet made calculations with this model, we might predict a priori that two different sources of differential scattering are now operative: first, what we have previously called *form* scattering, which is due to the anisotropy of the polarizabilities involved; and a second contribution that we might call *intrinsic* CIDS, which is due to the anisotropic coupling between the radiation dipoles. This, as is well known, is the way classical electrodynamics takes account of the optical activity of matter (14).

Future research will involve a better understanding of the preliminary results that have been communicated here, as well as the relative importance of both intrinsic and form contributions to the observed differential scattering patterns within and outside of absorption bands. We are currently working on a quantum mechanical description of this model.

Acknowledgments

The research described here was supported in part by National Institutes of Health Grants GM 10840 (IT) and AI 08247 (MFM) and by the Environmental Research and Development Division of the US Department of Energy under contract No. W-7405-ENG-48. We also thank Professor Robert Harris for many helpful discussions on scattering theory.

Literature Cited

1. Applequist, J., Sundberg, K. R., Olson, M. L., Weiss, L. C. 1979. *J. Chem. Phys.* 70:1240–46
2. Arnott, S. 1970. In *Progress in Biophysics and Molecular Biology*, ed. T. A. V. Butler, D. Noble, 21:265–319. Oxford/New York: Pergamon
3. Atkins, P. W., Barron, L. D. 1969. *Mol. Phys.* 16:453–66
4. Baase, W. A., Johnson, W. C. 1979. *Nucleic Acids Res.* 6:797–814
5. Barron, L. D., Bogaard, M. P., Buckingham, A. D. 1973. *J. Am. Chem. Soc.* 95:603–5
6. Barron, L. D., Buckingham, A. D. 1971. *Mol. Phys.* 20:1111–19
7. Barron, L. D., Buckingham, A. D. 1975. *Ann. Rev. Phys. Chem.* 26:381
8. Bayley, P. M. 1973. *Prog. Biophys. Biophys. Chem.* 27:3–76
9. Bearden, J. Jr., Bendet, I. J. 1972. *J. Cell. Biol.* 55:489
10. Bohren, C. F. 1975. *Light scattering by optically active particles.* PhD thesis. Univ. Arizona, Tucson. 99 pp.
11. Bohren, C. F. 1975. *J. Chem. Phys.* 62:1566
12. Bohren, C. F. 1974. *Chem. Phys. Lett.* 29:458
13. Bohren, C. F. 1977. *J. Theor. Biol.* 65:755–67
14. Born, M. 1918. *Ann. Phys. Leipzig.* 55:177–240
15. Brahms, J., Pilet, J., Lan, T. T. P., Hill, L. R. 1973. *Proc. Natl. Acad. Sci. USA* 70:3352–55
16. Brahms, J., Mommaerts, W. F. H. M. 1964. *J. Mol. Biol.* 10:73
17. Brahms, S., Brahms, J., Van Holde, K. E. 1976. *Proc. Natl. Acad. Sci. USA* 73:3453–57
18. Bram, S. J. 1971. *J. Mol. Biol.* 58:277–88
19. Bram, S. J. 1971. *Nature New Biol.* 233:161
20. Cantor, C. R., Tinoco, I. Jr. 1965. *J. Mol. Biol.* 13:65–77
21. Cantor, K. P., Hearst, J. E. 1969. *J. Mol. Biol.* 49:213–29
22. Carroll, D. 1970. PhD thesis. *Physical studies of polylysine-polynucleotide complexes.* Univ. Calif., Berkeley, 200 pp.
23. Carroll, D. 1972. *Biochemistry* 11:426–33
24. Cassim, J. Y., Yang, J. T. 1970. *Biopolymers* 9:1475–1501
25. Cech, C. L., Tinoco, I. Jr. 1976. *Nucleic Acids Res.* 3:399–404
26. Cech, C. L. 1975. *Polynucleotide circular dichroism calculations.* PhD thesis. Univ. Calif., Berkeley. 262 pp.
27. Cech, C. L., Hug, W., Tinoco, I. Jr. 1976. *Biopolymers* 15:131–52
28. Chan, A., Kilkuskie, R., Hanlon, S. 1979. *Biochemistry* 18:84–91
29. Chandrasekhar, S. 1974. *Liquid Crystals.* London: Cambridge Univ. Press. 342 pp.
30. Cheng, S. M., Mohr, S. C. 1975. *Biopolymers* 14:663–74
31. Chew, H., Kerker, M. 1976. *J. Opt. Soc. Am.* 66:445
32. Chung, S.-Y., Holzwarth, G. 1975. *J. Mol. Biol.* 92:449–66
33. Cochran, W., Crick, F. H. C., Vand, V. 1952. *Acta Crystallogr.* 5:581–86

34. Cohen, P., Kidson, C. 1962. *J. Mol. Biol.* 35:241–43
35. Cowman, M. K., Fasman, G. D. 1978. *Biophys. J.* 21:A96
36. Cowman, M. K., Fasman, G. D. 1978. *Proc. Natl. Acad. Sci. USA* 75:4759–63
37. DeGennes, P. G. 1974. *The Physics of Liquid Crystals.* Oxford: Clarendon. 347 pp.
38. DeMurcia, G., Das, G. C., Erard, M., Daune, M. 1978. *Nucleic Acids Res.* 5:523–35
39. Deutsche, C. S. 1970. *J. Chem. Phys.* 52:3703–14
40. DeVoe, H. 1964. *J. Chem. Phys.* 41:393–400
41. DeVoe, H. 1965. *J. Chem. Phys.* 43:3199–208
42. Disch, R. L., Sverdlik, D. I. 1967. *J. Chem. Phys.* 47:2137
43. Dorman, B. P., Hearst, J. E., Maestre, M. F. 1973. In *Methods in Enzymology,* ed. C. H. W. Hirs, S. N. Timasheff, 27:767–96. New York: Academic
44. Dorman, B. P. 1973. *Ultraviolet absorption and circular dichroism measurements on light-scattering biological specimens.* PhD thesis. Univ. Calif. Berkeley. 195 pp.
45. Dorman, B. P., Maestre, M. F. 1973. *Proc. Natl. Acad. Sci. USA* 70:255–59
46. Duysens, L. M. N. 1956. *Biochim. Biophys. Acta* 19:1
47. Evdokimov, Y. M., Pyatigorskaya, T. L., Kadykov, V. A., Polyvtsev, O. F., Doskocil, J., Koudelka, J., Varshavsky, Y. M., 1976. *Nucleic Acids Res.* 3:1533–47
48. Evdokimov, Y. M., Platonov, A. L., Tikhonenko, A. S., Varshavsky, Y. M. 1972. *FEBS Lett.* 23:180–84
49. Evdokimov, Y. M., Pyatigorskaya, T. L., Polyvtsev, O. F., Akimenko, N. M., Kadykov, V. A., Tsvankin, D. Y., Varshavsky, Y. M. 1976. *Nucleic Acids Res.* 3:2353–66
50. Fasman, G. D., Schaffhausen, B., Goldsmith, L., Adler, A. 1970. *Biochemistry* 9:2814–22
51. Fasman, G. D., Cowman, M. K. 1978. *The Cell Nucleus Chromatin,* Pt. B, H. Busch, pp. 55–57. New York: Academic
52. Fulton, R. L., Gouterman, M. 1964. *J. Chem. Phys.* 41:2280–86
53. Gennis, R. B., Cantor, C. R. 1972. *J. Mol. Biol.* 65:381–99
54. Giannoni, G., Padden, F. J. Jr., Keth, H. D. 1969. *Proc. Natl. Acad. Sci. USA* 63:964–71
55. Girod, J. C., Johnson, W. C., Huntington, S. K., Maestre, M. F. 1973. *Biochemistry* 12:5092–96
56. Gitter-Amir, A., Rosenheck, K., Schneider, A. S. 1976. *Biochemistry* 15:3131
57. Gitter-Amir, A., Schneider, A. S., Rosenheck, K. 1976. *Biochemistry* 15:3138
58. Glaser, M., Singer, S. J. 1971. *Biochemistry* 10:1780
59. Gordon, D. J. 1972. *Biochemistry* 11:413
60. Gordon, D. J., Holzwarth, G. 1971. *Proc. Natl. Acad. Sci. USA* 68:2365
61. Gordon, D. J., Holzwarth, G. 1971. *Proc. Natl. Acad. Sci. USA* 68:2365–69
62. Gosule, L. C., Schellman, J. A. 1976. *Nature* 259:333–35
63. Gray, D. M., Lee, S. C., Skinner, D. M. 1978. *Biopolymers* 17:107–14
64. Gray, D. M., Ratliff, R. L. 1975. *Biopolymers* 14:487–98
65. Gray, D. M., Tinoco, I. Jr. 1970. *Biopolymers* 9:223
66. Gray, D. M., Taylor, T. N., Lang, D. 1978. *Biopolymers* 17:145–47
67. Gray, D. M., Edmonson, D. P., Lang, D., Vaughn, M. R., Nave, C. 1979. *Nucleic Acids Res.* 6:2089–107
68. Gray, D. M., Hamilton, F. D., Vaughn, M. R., Nave, C. 1978. *Biopolymers* 17:85–106
69. Green, G., Mahler, H. R. 1971. *Biochemistry* 10:2200
70. Green, G., Mahler, H. R. 1968. *Biopolymers* 6:1509
71. Gregory, R. P. T., Raps, S. 1974. *Biochem. J.* 142:193
72. Greve, J., Maestre, M. F., Moise, H., Hosoda, J. 1978. *Biochemistry* 17:887–93
73. Greve, J., Maestre, M. F., Moise, H., Hosoda, J. 1978. *Biochemistry* 17:893–98
74. Hanlon, S., Johnson, R. S., Wolf, B., Chan, A. 1972. *Proc. Natl. Acad. Sci. USA* 69:3263–67
75. Hanlon, S., Johnson, R. S., Chan, A. 1974. *Biochemistry* 13:3963–71
76. Hanlon, S., Johnson, R. S., Chan, A. 1974. *Biochemistry* 13:3972–81
77. Hanlon, S., Brudno, S., Wu, T. T., Wolf, B. 1975. *Biochemistry* 14:1648–60
78. Harris, R. A. 1965. *J. Chem. Phys.* 43:959–70
79. Harris, R. A., McClain, W. M. 1977. *J. Chem. Phys.* 67:269
80. Harris, R. A., McClain, W. M. 1977. *J. Chem. Phys.* 67:265–68
81. Haugh, E. F., Hirschfelder, J. O.

82. Haynes, M., Garrett, R. A., Gratzer, W. B. 1970. *Biochemistry* 9:4410–16
83. Hemenger, R. P. 1978. *J. Chem. Phys.* 68:1722–28
84. Hemenger, R. P., Kaplan, T., Gray, L. J. 1979. *J. Chem. Phys.* 70:3324–32
85. Holzwarth, G., Chabay, I., Holzwarth, N. A. W. 1973. *J. Chem. Phys.* 58:4816–19
86. Holzwarth, G., Holzwarth, N. A. W. 1973. *J. Opt. Soc. Am.* 63:324–31
87. Holzwarth, G., Gordon, D. G., McGinness, J. E., Dorman, B. P., Maestre, M. F. 1974. *Biochemistry* 13:126–32
88. Ivanov, V. I., Minchenkova, L. E., Schyolkina, A. K., Poletayev, A. I. 1973. *Biopolymers* 12:89–110
89. Ivanov, V. I., Minchenkova, L. E., Minyat, E. E., Frank-Kamenetskii, M. D., Schyolkina, A. K. 1974. *J. Mol. Biol.* 87:817–33
90. Jensen, D. E., Kelly, R. C., von Hippel, P. H. 1976. *J. Biol. Chem.* 251:7215–28
91. Ji, T. H., Urry, D. W. 1969. *Biochem. Biophys. Res. Commun.* 34:404
92. Johnson, W. C. Jr. 1978. *Ann. Rev. Phys. Chem.* 29:93–114
93. Johnson, W. C. Jr., Tinoco, I. Jr. 1969. *Biopolymers* 7:727–49
94. Johnson, W. C. Jr. 1980. In *Origins of Optical Activity in Nature*, ed. D. C. Walker. New York: Elsevier
95. Jordan, C. F., Lerman, L. S., Venable, J. H. Jr. 1972. *Nature New Biol.* 236:67–70
96. Keller, W. 1975. *Proc. Natl. Acad. Sci. USA* 72:4876–80
97. Kelly, R. C., Jensen, D. E., von Hippel, P. H. 1976. *J. Biol. Chem.* 251:7240–50
98. Kerker, M. 1969. *The Scattering of Light and Electromagnetic Radiation*. New York, NY: Academic. 666 pp.
99. Kerker, M. 1975. *J. Opt. Soc. Am.* 65:376
100. Kirkwood, J. G. 1937. *J. Chem. Phys.* 5:479–91
101. Kraus, J. D., 1950. *Antennas*. New York, NY: McGraw-Hill. 553 pp.
102. Lenard, J., Singer, S. J. 1966. *Proc. Natl. Acad. Sci. USA* 56:1828
103. Levin, A. I., Tinoco, I. Jr. 1977. *J. Chem. Phys.* 66:3491–97
104. Lewis, D. G., Johnson, W. C. 1974. *J. Mol. Biol.* 86:9–96
105. Li, H. J., Chang, C., Weiskopf, M. 1973. *Biochemistry* 12:1763–72
106. Li, H. J., Brand, B., Rotter, A., Chang, C., Weiskopf, M. 1974. *Biopolymers* 13:1681–97
107. Li, H. J., Herlands, L., Santella, R., Epstein, P. 1975. *Biopolymers* 14:2401–15
108. Lobenstein, E. W., Turner, D. H. 1979. *J. Am. Chem. Soc.* 101: In press
109. Loxom, F. M. 1969. *J. Chem. Phys.* 51:4899–905
110. Maestre, M. F., Tinoco, I. Jr. 1967. *J. Mol. Biol.* 23:323–35
111. Maestre, M. F., Gray, D. M., Cook, R. B. 1971. *Biopolymers* 10:2537–53
112. Maestre, M. F., Greve, J., Hosoda, J. 1977. In *Excited States in Organic Chemistry and Biochemistry*, ed. B. Pullman, N. Goldblum, pp. 99–111. Dordrecht, Holland: Raidel.
113. Maestre, M. F. 1970. *J. Mol. Biol.* 52:543–56
114. Maniatis, T., Venable, J. H., Lerman, L. S. 1974. *J. Mol. Biol.* 84:37–64
115. Mason, S. F. 1979. *Optical Activity and Chiral Discrimination*. Boston: Reidel. 368 pp.
116. Merrifield, R. E. 1963. *Radiat. Res.* 20:154–58
117. Moffitt, W. 1956. *J. Chem. Phys.* 25:476–78
118. Moore, D. S., Wagner, T. E. 1973. *Biopolymers* 12:201–27
119. Moore, D. S., Wagner, T. E. 1974. *Biopolymers* 13:977–86
120. Nelson, R. G., Johnson, W. C. 1970. *Biochem. Biophys. Res. Commun.* 41:211–16
121. Nicolini, C., Baserga, R., Kendall, F. 1976. *Science* 192:796–98
122. Norden, B. 1977. *J. Phys. Chem.* 81:151–59
123. Olins, D. E. 1969. *J. Mol. Biol.* 43:439–60
124. Olins, D. E., Olins, A. L. 1972. *J. Cell. Biol.* 53:715–36
125. Olins, D. E., Olins, A. L. 1974. *Science* 183:330–32
126. Pauli, W. 1978. In *Pauli Lectures on Physics 2* ed. C. P. Enz. Cambridge, Mass./London, England: MIT Press
127. Perrin, F. 1942. *J. Chem. Phys.* 10:415
128. Philipson, K. D., Sauer, K. 1973. *Biochemistry* 12:3454–58
129. Pilet, J., Brahms, J. 1972. *Nature New Biol.* 236:99–100
130. Rhodes, W. 1961. *J. Am. Chem. Soc.* 83:3609–17
131. Rhodes, W., Chase, M. 1967. *Rev. Mod. Phys.* 348–61
132. Rhodes, W., Redmann, S. M. 1977. *Chem. Phys.* 22:215–20
133. Rosenfeld, L. 1928. *Z. Phys.* 52:161–74
134. Rosenfeld, L. 1965. *Theory of Elec-*

trons, New York, NY: Dover
135. Sakurai, J. J. 1967. *Advanced Quantum Mechanics.* Reading, Mass: Addison-Wesley. 336 pp.
136. Samejima, T., Hashizume, H., Imahori, K., Fujii, I., Miura, K. 1968. *J. Mol. Biol.* 34:39–48
137. Santella, R. M., Li, H. J. 1977. *Biopolymers* 16:1879–94
138. Saxon, D. S. 1955. *Lectures on the Scattering of Light.* Sci. Rep. No. 9, Contract AF19(122)-239. Dep. Meterol. Univ. Calif., Los Angeles. 100 pp.
139. Scheraga, H. A. 1968. *Adv. Phys. Org. Chem.* 6:103–84
140. Schneider, A. S., Schneider, M. J., Rosenheck, K. 1970. *Proc. Natl. Acad. Sci. USA* 66:793
141. Schneider, A. S. 1971. *Chem. Phys. Lett.* 8:604–8
142. Schneider, A. S. 1973. In *Methods in Enzymology*, 27D, Chap. 29, p. 751. New York: Academic. 1063 pp.
143. Schneider, A. S., Harmatz, D. 1976. *Biochemistry* 15:4158–62
144. Schwartz, A., Fasman, G. 1979. *Biopolymers* 18:1013–45
145. Shapiro, J. T., Leng, M., Felsenfeld, G. 1969. *Biochemistry* 8:3219
146. Shih, T. Y., Fasman, G. D. 1970. *J. Mol. Biol.* 52:125–29
147. Shin, Y. A., Eichhorn, G. L. 1977. *Biopolymers* 16:225–30
148. Shure, M. Vinograd, J. 1976. *Cell* 8:215–26
149. Simpson, W. T., Peterson, D. L. 1957. *J. Chem. Phys.* 26:588–93
150. Sipski, M. L., Wagner, T. E. 1977. *Biol. Reprod.* 16:428–40
151. Sipski, M. L., Wagner, T. E. 1977. *Biopolymers* 16:573–82
152. Sponar, J., Fric, I. 1972. *Biopolymers* 11:2317–30
153. Sprecher, C. A., Johnson, W. C. 1977. *Biopolymers* 16:2243–64
154. Sprecher, C. A., Johnson, W. C. 1979. *Biopolymers* 18:1009–19
155. Steinberg, I. Z. 1978. *Ann. Rev. Biophys. Bioeng.* 7:113–37
156. Tinoco, I. Jr. 1962. *Adv. Chem. Phys.* 4:113–60
157. Tinoco, I. Jr., Cantor, C. R. 1970. *Methods Biochem. Anal.* 18:81–203
158. Tinoco, I. Jr., Woody, R. W., Bradley, D. F. 1963. *J. Chem. Phys.* 38:1317–25
159. Tinoco, I. Jr., Turner, D. H. 1976. *J. Am. Chem. Soc.* 98:6453–56
160. Tinoco, I. Jr., Ehrenberg, B., Steinberg, I. Z. 1977. *J. Chem. Phys.* 66:3491–97
161. Tinoco, I. Jr. 1979. *Int. J. Quantum Chem.* 16:111–17
162. Tobias, I., Brocki, T. R., Balazs, N. L. 1975. *J. Chem. Phys.* 62:4181–83
163. Torten, M., Schneider, A. S. 1973. *J. Infect. Dis.* 127:319–20
164. Tunis, M. J. B., Hearst, J. E. 1968. *Biopolymers* 6:1325–44
165. Tunis-Schneider, M. J. B., Maestre, M. F. 1970. *J. Mol. Biol.* 52:521–41
166. Turner, D. H. 1978. *Methods Enzymol.* 49G:199–214
167. Turner, D. H., Tinoco, I. Jr., Maestre, M. F. 1974. *J. Am. Chem. Soc.* 96:4340–42
168. Urry, D. W. 1972. *Biochim. Biophys. Acta* 265:115–68
169. Urry, D. W., Krivacic, J. 1970. *Proc. Natl. Acad. Sci. USA* 65:845–52
170. Vandegrift, V., Serra, M., Moore, D. S., Wagner, T. E. 1974. *Biochemistry* 13:5087–92
171. Wang, J. 1979. *Proc. Natl. Acad. Sci. USA* 76:200–3
172. Warshaw, M. W., Tinoco, I. Jr. 1965. *J. Mol. Biol.* 13:54–64
173. Weiskopf, M., Li, H. J. 1977. *Biopolymers* 16:669–84
174. Wiener, O. 1926. *Kolloidchem. Beth.* 23:189
175. Wolf, B., Hanlon, S. 1975. *Biochemistry* 14:1661–70
176. Woody, R. W. 1977. *J. Polym. Sci. Macromol. Rev.* 12:181–321
177. Yang, J. T., Samejima, T. 1969. *Prog. Nucleic Acid Res. Mol. Biol.* 9:223–300
178. Zimmer, C., Luck, G. 1973. *Biochim. Biophys. Acta* 312:215–27
179. Zimmer, C., Luck, G. 1974. *Biochim. Biophys. Acta* 361:11–32

MODULATION OF IMPULSE CONDUCTION ALONG THE AXONAL TREE

♦9144

Harvey A. Swadlow
Department of Psychology, University of Connecticut, Storrs, Connecticut 06268; Department of Neurology, Harvard Medical School, Beth Israel Hospital, Boston, Massachusetts 02215

Jeffery D. Kocsis and Stephen G. Waxman
Department of Neurology, Stanford University Medical School, Veterans Administration Hospital, Palo Alto, California 94304

INTRODUCTION

As early as 1935, the experiments of Barron & Matthews (9) suggested that axons might not always conduct impulses faithfully but might, in fact, function in some systems as variable filters modulating the spatial and temporal relations between impulses. Subsequent advances in our understanding of axonal membrane mechanisms underlying the initiation and conduction of the action potential, along with advances in our understanding of the mechanisms of synaptic transmission and integration led to a relative neglect of the role of the axon in models of neural integration. Thus, the classical view of impulse conduction along the axon has been that of a transmission line, in which impulses initiated near the soma are faithfully distributed along the axonal arbors and reproduced at the axon terminals, and in which the temporal relations between impulses are maintained throughout the axonal tree.

It is clear from cable theory, however, that geometrical constraints play a role in defining the characteristics of impulse conduction through a region of axon (see below). Thus, morphological inhomogeneities such as branchpoints and dilations may result in regions of low safety factor (i.e. regions where impulse conduction is likely to fail). In recent years it has become clear that in some systems such regions of morphological inhomogeneity are dynamically involved in the modulation of neural activity. Thus, impulse conduction through branchpoints has been shown to be (*a*) dependent on the history of impulse conduction along the axon and (*b*) subject to modulation from synaptic and other external factors. In the present chapter we review the static characteristics of axons which result in regions of low safety factor and the dynamic axonal and extraaxonal influences on impulse conduction through such regions.

SOME THEORETICAL CONSIDERATIONS

Early quantitative descriptions of the propagating nerve impulse (where partial differential equations were solved to describe membrane current and voltage changes during generation of the action potential) assumed a uniform axon geometry, with the result that shape and velocity of the action potential were uniform (53). More recently, non-uniformities in axon geometry have been introduced into simulation studies of the propagating nerve impulse. The introduction of structural non-uniformities may influence not only conduction properties such as shape and velocity of the impulse but also the ability of the impulse to propagate without block. In the present section we review simulation studies dealing with non-uniform axonal morphology, including regions of axonal branching. Such studies have typically made the simplifying assumptions of uniform specific membrane properties, extracellular isopotentiality, and constant specific axoplasmic conductance.

One of the first changes in axon structure to be modeled was a step increase in fiber diameter (15, 42, 63, 64). If an action potential is initiated at a thin diameter axon segment and propagates toward the point of step increase in diameter, the velocity of the action potential slows and the amplitude is reduced as the point of step diameter increase is approached (see Figure 1). The degree of slowing is dependent on the amount of diameter increase. If the step increase is large enough, the action potential will fail. Note that failure occurs before the point of step increase. For a larger step increase in diameter, the point of action-potential failure will be at a greater distance from the point of diameter change.

If failure of the impulse does not occur as the action potential approaches the point of step diameter increase, slowing of conduction will occur, and the impulse will propagate with delay through the region of diameter change. The conduction velocity of the impulse in the larger diameter segment will, of course, be greater than in the smaller segment. When the delay through the region of diameter change is longer than

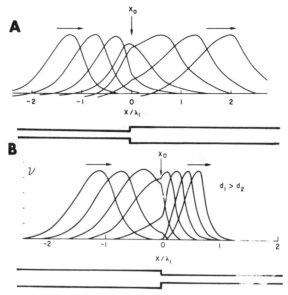

Figure 1 Computed action potentials for step increase (*A*) and for step decrease (*B*) in diameter at point X_0. Sections through the simulated axons are shown below the action potentials. The action potentials are plotted as voltage versus distance and they are propagating from left to right (*arrows*). Each action potential is plotted at an instant of time and the impulses are separated by equal time intervals [based on Goldstein, 1978 (41a)].

the refractory period, impulse reflection (i.e. reverse propagation) would be expected to occur (e.g. see 42, 79).

Another nonhomogeneous condition that has been modeled is a step reduction in axon diameter (42). Conduction into a region of step diameter reduction is very secure, and failure will not occur. The conduction velocity of the propagating impulse increases as the impulse approaches the point of step reduction, but after passing into the region of smaller axon diameter the impulse propagates at a lesser velocity (see Figure 1).

In order to understand the biophysical basis of the alterations in the action potential for step diameter change, one must consider the influences of axonal geometry and the electrical properties of the fiber on impulse initiation and propagation. Briefly, action potential initiation occurs when enough current is supplied to the axon membrane to charge its capacitance to a voltage level above threshold. The action potential will supply current to, and excite, adjacent membrane regions. Larger diameter axons have a greater surface area per unit length and less input resistance; therefore, more total current is necessary to drive a larger diameter fiber to threshold. Once excited, however, the larger diameter fiber generates more current than a smaller fiber. If then, the specific membrane properties are uniform, the propagation of the action potential from a small to a large diameter axon region is vulnerable, since a relatively low current generator (the action potential in the small diameter region of the fiber) must provide current to a region of increased surface area with a decreased input resistance. These changes (and the resultant impedance mismatch) lead to a reduction in the current density across the axon membrane. As the action potential approaches the step increase in axon diameter it is marked either by a slowing in conduction velocity (since more time would be required for threshold to be reached) or by conduction failure (if the increased electrical load of the larger diameter axon segment is sufficient to prevent attainment of threshold). When an impulse is traveling toward a step reduction in diameter, more secure transmission through the junction occurs, since a relatively large current generator is supplying current to a region of decreased surface area and increased input resistance. Therefore, threshold is attained more quickly as the point of diameter reduction is approached.

Changes in axon diameter may, in some cases, be quite gradual. Conduction through flaring and tapering axon lengths has been simulated (15, 42, 79). Goldstein & Rall (42) have demonstrated that in flaring core conductors (continuously increasing diameter), conduction may block if the flare is large or conduction may continue if the flare is gradual. If conduction continues, the impulse will gradually increase in velocity, but the velocity at any given point along the axon will be less than that of a uniform fiber with an equivalent diameter.

Computer simulations of myelinated nerve fiber have revealed that conduction failure may occur at the junction between a parent myelinated fiber and the nonmyelinated preterminal axon (100) and at regions of focal demyelination (66, 147). Such sites represent an inhomogeneity of the fiber; the input resistance of the fiber is reduced because of the increased surface area of the nonmyelinated portion of the fiber. The behavior of the impulse while traveling through such a region is similar to that seen at a step increase in diameter.

Early mathematical treatments of current and potential distributions along branching neurites dealt with dendritic branches that were assumed to be inexcitable (85, 86, 87). The branch pattern of the particular dendritic tree was represented by a set of equivalent cylinders, which in turn were transformed into compartments. Each compartment had similar electrical properties, thus simplifying the analysis of electrical properties within the dendritic tree.

Active impulse propagation through axonal branching regions has also been studied with simulation techniques (16, 42, 81, 82, 153). The membrane of axons in axonal branching networks differs from the passive membrane described in the above dendritic studies in that it is capable of sustaining regenerative action potentials. Since the passive spread of current generated by the action potential is responsible for propagation of the impulse (52), simulation studies of passive dendritic trees are relevant to models of conduction through branching axonal networks. A useful concept applied to studies of passive cable properties (85–87) and excitable branch networks (42) was a geometric ratio, also referred to as a load ratio (153). In this case, the conduction properties through a branching cable were modeled via transformations into an idealized equivalent single unbranched axonal element. This single nonbranched fiber can be treated similarly to a single unbranched fiber that has a step change in diameter. In simulation studies of both passive (85–87) and active (42, 81) branching networks it has proved useful to compare the input resistance (Ri) of the fiber carrying an action potential toward a branchpoint to the equivalent input resistance (Req) of the total post-branch network. The ratio of these input resistances, Ri/Req, is an important determinant of conduction near the branch point, just as was the case for propagation through a region of step diameter change.

If uniform specific membrane resistances and specific core resistances are assumed throughout the branch network, the ratio Ri/Req can be simplified and described in terms of diameters of the individual branches. This result can be obtained from first looking at the well-known equation for Ri (e.g. see Reference 88 for derivation).

$$R_i = \frac{2\sqrt{(RmRc)}}{\pi} \cdot d_a^{-3/2} \qquad 1.$$

where Rm and Rc are the specific membrane resistance and axoplasmic resistance, respectively, and d_a is the diameter of the fiber. Westerfield et al (153) have given a lucid account of the extrapolation of this type of model to branching axons.

Since Rm and Rc are assumed to be constant:

$$R_i = K \cdot d_a^{-3/2} \qquad 2.$$

where

$$K = \frac{2\sqrt{(RmRc)}}{\pi}.$$

Since the input resistances of the branches are in parallel, their conductances will add:

$$\frac{1}{Req} = \sum_j \frac{1}{R_j} \qquad 3.$$

where R_j equals the input resistance of the jth branch of the network. Expressing R_j in terms of axon diameter:

$$\frac{1}{Req} = \sum_j \frac{1}{K \cdot d_j^{-3/2}} \qquad 4.$$

where d_j equals the diameter of the jth branch. The equivalent input resistance is then:

$$Req = K \cdot \sum_j d_j^{-3/2}. \qquad 5.$$

The load ratio (153) or the geometric ratio (GR) (42) is determined by dividing Equation 2 by Equation 5:

$$\frac{Ri}{Req} = GR = \frac{K \cdot d_a^{-3/2}}{K \cdot \sum d_j^{-3/2}} = \frac{\sum d_j^{3/2}}{d_a^{3/2}} \qquad 6.$$

where d_a is the diameter of the axon carrying the action potential into the branching network. When the ratio is greater than 1, the action potential propagating toward the branch network faces a region of lowered input resistance; changes in the action potential as it approaches the branchpoint are similar to those described above for an impulse approaching a point of step increase (i.e. slowing, possible failure of invasion, or impulse reflection). Conversely, when GR is less than 1, changes in the approaching action potential are similar to those of a step decrease in diameter. If the ratio equals one, there is no change in conduction of the action potential as it approaches the branchpoint and the impulse propagates into all of the respective branches. After initiation of the impulses in the branches, the conduction velocities are dependent upon the individual branch diameters.

Just as with a step diameter increase, if impulse failure occurs, the point of failure may be at a region proximal to the branch junction. It is important to note that in this model the failure or success of propagation into the branches, *will apply to all of the branches*. Preferential conduction or routing of impulses into individual branches, as has been described experimentally (see below), is not accounted for.

No model of a branch network has demonstrated impulse routing or filtering through branches of differing diameters. One can imagine that if certain properties, which are usually assumed to be constant in the models, were varied, the experimental observation of impulse routing might be present in the simulations. If, for example, active membrane properties such as ionic conductances and metabolic pump activities were assumed to differ from one branch to another, one might expect differential filtering effects in these branches. Different time courses of the refractory period have also been demonstrated for large and small diameter fibers (76, 128), and this may also contribute to the filtering capabilities of a branched network.

The role of extracellular potassium accumulation in the branch network has been modeled by Adelman & Fitzhugh (1). Utilizing Hodgkin and Huxley membrane and allowing for potassium accumulation in the extracellular space, they demonstrated membrane depolarization and conduction block during high frequency impulse activation along a homogeneous axon.

Parnas et al (79) also allowed for extracellular potassium accumulation in their model, but they used an inhomogeneous axon model. An approximation of a flaring core conductor was introduced into the modeled fiber by a series of step diameter increases. They found that although single spikes could pass through the region of diameter increase, subsequent spikes could fail and even small changes in diameter of axon segments of the taper could alter the conduction properties so as to change from unimpeded conduction to impulse failure. Since the depolarizing effects of potassium accumulation following impulse activity may be greater for small diameter fibers than for large fibers (see below), smaller diameter branches could be more susceptible to depolarization block from potassium accumulation, and at certain frequencies impulse block could possibly occur in the smaller diameter branch, but not the larger.

ACTIVITY-DEPENDENT FLUCTUATIONS IN AXONAL EXCITABILITY AND CONDUCTION VELOCITY

There is significant evidence which indicates that many axons demonstrate variations in conduction velocity and excitability which can be attributed to the history of impulse conduction along the axon. Such history dependence of conduction properties has been found in both invertebrates and vertebrates, peripheral nervous system and central nervous system, and in both myelinated and nonmyelinated axons. Since such phenomena may interact significantly with the geometrical constraints on impulse conduction through branch points, they are reviewed below.

The Refractory Period

Following an action potential, an interval ensues during which a second, propagated action potential cannot be elicited. This interval is referred to as the absolute refractory period and, according to the ionic theory of Hodgkin & Huxley (53), is due to Na^+ inactivation and high K^+ conductance. The relative refractory period follows the absolute refractory period and consists of a period of decreased axonal excitability and conduction velocity. In myelinated and nonmyelinated axons of both the peripheral (76) and central (128) nervous system, there is an approximate inverse relationship between the conduction velocity of the axon at rest and the duration of the absolute refractory period. In these studies, absolute refractory period varied in a continuous manner with conduction velocity, and no discontinuity in the duration of the absolute refractory period of myelinated and nonmyelinated axons was observed. Since axon diameter is positively correlated with conduction velocity in both myelinated and nonmyelinated axons (e.g. see 146), it follows that the absolute refractory period of small diameter axons will be greater than that of larger axons.

The Supernormal Period

In many axonal systems, the relative refractory period is followed by a period of increased axonal excitability and conduction velocity. These observations were first made early in this century (2), and subsequent early investigations led to the erroneous conclusion that these phenomena were due to the deteriorated physiological condition of the preparations under study (43). The experiments of Bullock (22), however, in which the physiological condition of several types of invertebrate axons was carefully controlled, clearly showed a period of supernormal conduction velocity which followed the relative refractory period of a single impulse and lasted for 100–300 msec. The period of supernormal conduction velocity was roughly correlated with a period of supernormal axonal excitability. Bullock also demonstrated supernormal conduction velocity and excitability following the relative refractory period of sciatic nerve of frog. The magnitude of these variations was small following a single conditioning impulse, but was augmented following several conditioning impulses.

More recently, Raymond, Lettvin, and their colleagues (74, 94, 95) have systematically studied the effects of single and multiple impulses on the subsequent excitability of axons in sciatic nerve of frog. These workers have found that supernormal axonal excitability is maximal at 7–20 msec following a single conditioning impulse, and that supernormal axonal excitability is highly correlated with supernormal conduction velocity. Supernormal axonal excitability slowly decreases and usually reattains control values at 500 to 1500 msec following a single impulse. The left side of Figure 2 presents supernormal excitability curves obtained by Raymond and Lettvin. Following several closely spaced conditioning impulses, the magnitude of the supernormal period was greater than that following a single impulse, but maximal effects were found following 4–10 impulses. The right side of Figure 2 is discussed below.

Other axonal systems in vertebrate peripheral nervous system in which supernormal axonal excitability or conduction velocity have been demonstrated include olfactory nerve of tortoise (19), median (14,40) and ulnar (14) nerve of humans, and mammalian A and C fibers (38, 46). The super-normal excitability found in human median nerve following a single conditioning impulse lasted 15–60 msec, a value faster than that found in frog sciatic nerve but comparable to that found in the faster visual callosal axons of the rabbit brain (see below).

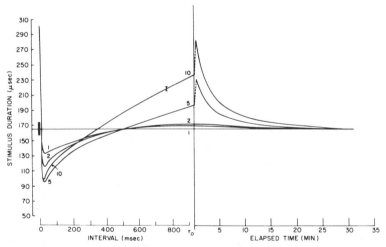

Figure 2 Threshold curves following bursts of impulses. The figure was traced directly from a colored multiple graph generated by the threshold hunter and x-y plotter. Additional results from later plots were used to sketch out the last 20 min of the two longest recovery periods. To the left of τ_0 the x-axis gives the interval between the last conditioning stimulus at 0 and the hunting stimulus. At τ_0 conditioning stimuli were turned off and threshold during recovery was hunted every 2 sec for approximately 20 min. The dark bar (*at left*) shows the duration change required to vary probability of response from nearly 0 to nearly 1. The range mark on the curve associated with bursts of 10 impulses shows the extent of variation in the duration of the threshold hunter with respect to the smooth traced line. Approximately 400 tests for threshold were made to compile each trace during the 900 msec interval after conditioning pulses. Bursts were repeated every 2 sec; pH 7.45, temperature 17°C, conduction velocity of axon used was 16 meters/sec [figure and caption from Raymond & Lettvin, 1978 (95)].

In vertebrate central nervous system, supernormal axonal conduction velocity was first reported to follow activity in nonmyelinated parallel fibers of cat cerebellum (3). Using the evoked potential technique, conduction velocity was observed to increase by approximately 18% at an interval of 22 msec following a prior impulse. Conduction velocity decreased to control values after 150 to 200 msec Increases in axonal excitability (up to 40%) were found to be concomitant with increases in conduction velocity. Parallel fibers, however, are somewhat atypical of central axons. Conduction velocities are very low (<0.5 m/sec), packing density is high, and axon diameter varies every few micrometers as the axons pass through the dendritic fields of Purkinje cells, giving off synaptic contacts (34).

Visual callosal axons of the rabbit have proved useful in the study of impulse conduction in central nervous system. The cells of origin of this neocortical system are located primarily in supragranular layers of binocular visual area I and in visual area II of the cerebral cortex (125, 130). The axons of these cells traverse the splenium of the corpus callosum to terminate in the contralateral visual cortex. The splenium of the corpus callosum contains both myelinated and nonmyelinated axons (151) and their morphology is rather typical of that of central axons. We have studied the aftereffects of impulse activity of both myelinated and nonmyelinated elements of this system (124, 127, 128, 152). In order to insure that the physiological condition of these axons was optimal, we have studied conduction properties in the unanesthetized, unparalyzed rabbit (123, 128). Our general strategy has been to study the conduction velocity and excitability of callosal axons by measuring threshold and latency to antidromic activation via stimulating electrodes located at one or several sites along the course of the callosal axon. Following the identification of antidromically activated neurons, latency to a single test pulse is determined at various intervals following either single conditioning pulses or trains of conditioning pulses.

In Figure 3 the antidromic latency to a test stimulus is shown as a function of conditioning stimulus–test stimulus interval for three units with control conduction velocities of 0.3, 1.3, and 4.1 m/sec, respectively. In Figure 3 A_1 the slowest callosal axon that we have studied is shown, with a control antidromic latency of 37.5 msec (conduction velocity=0.3 m/sec). At intervals of 10 and 17 msec, latency decreased to approximately 34.7 msec. Latency slowly increased to reattain the control value at an interval of approximately 170 msec. In Figure 3 B_1 and C_1, latency decreases are shown for two faster conducting units. It is clear that the duration of the decrease in latency was greater for the slower conducting axons. In Figure $3A_2$–C_2, variations in threshold are plotted for the same three units shown in Figure $3A_1$–C_1, respectively.

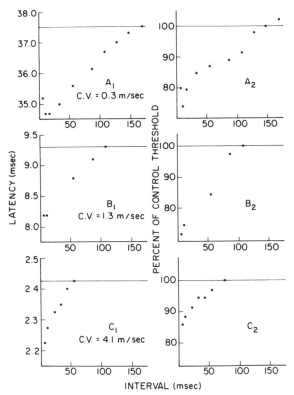

Figure 3 A_1-C_1: The latency to antidromic activation to a test stimulus as a function of conditioning stimulus–test stimulus interval for three rabbit visual callosal neurons with axon conduction velocities of 0.3, 1.3, and 4.1 meters/sec. Points show the antidromic latency at each conditioning stimulus–test stimulus interval. Pairs of stimuli were delivered at a rate of 1/3.3 sec. Each conditioning stimulus is presented at 1.5 times threshold intensity, and the test stimulus is presented at 1.2 times the threshold at each conditioning stimulus–test stimulus interval. A_2-C_2: Variations in threshold to antidromic test volley following a single antidromic conditioning volley for the same three units presented in A_1-C_1. The changes in threshold follow a similar time course to the changes in latency [from Swadlow & Waxman, 1976 (128)].

Variations in threshold follow roughly the same time course as do variations in latency.

We have found an approximately inverse relationship between the duration of the increase in conduction velocity and the control conduction velocity of the axon. Figure 4 shows this relationship for callosal axons of the rabbit. The sample contains both myelinated axons (conduction velocities > 3.4 m/sec) and nonmyelinated axons (conduction velocities < 0.8 m/sec). Axons with conduction velocities of between 0.8 and 3.4 m/sec could not be classified unequivocably as being either

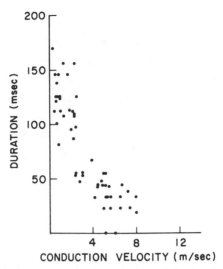

Figure 4 Relationship between conduction velocity and the duration of the decrease in latency for callosal axons. Each point represents a single unit [from Swadlow & Waxman, 1976 (128)].

primarily myelinated or nonmyelinated (see Reference 128 for criteria for identifying myelinated and nonmyelinated axons on the basis of conduction velocity). Note that the duration of the increase in conduction velocity appears to vary continuously with control conduction velocity and that there is no abrupt discontinuity that might indicate a qualitative difference between myelinated and nonmyelinated axons. We have also found a roughly inverse relationship between the control conduction velocity of the axon and the magnitude (%) of the increase in conduction velocity that occurs during the supernormal period in myelinated callosal axons of rabbit (128) and macaque monkey (126).

In callosal axons of both rabbit and monkey, increasing the number of conditioning pulses does not result in an increase in either the magnitude or the duration of the supernormal period. In contrast, in sciatic nerve of frog, both Bullock (22) and Raymond et al (74, 94, 95) have found an augmentation of supernormality with an increase in the number of conditioning pulses (from 1 to 4–10 pulses). Similarly, Zucker (158) has found an augmentation in the supernormal excitability of terminals of crayfish motor neurons with increases in the number of conditioning pulses.

Since both the duration and the magnitude of the supernormal period may be greater in slowly conducting axons and since conduction velocity is constrained by axon diameter, one would suspect that axonal branches (which are usually of smaller diameter than the parent axon) would demonstrate a longer-lasting and possibly larger supernormal period than would the larger parent axon. Similarly, two daughter branches of different diameter would be expected to demonstrate different aftereffects of impulse activity. A direct approach to this question is possible in larger vertebrate and invertebrate axons (see below) by recording directly from axon trunk and branches. Indirect approaches, however, must be used in many central axonal systems, especially in the mammalian brain, where the main axonal trunk may be 1 μm or less in diameter and axonal branches may be 0.1-0.2 μm. One indirect approach to the comparative analysis of the conduction properties of elements of the axonal tree (which we have used in fine-diametered callosal axons of rabbit) involves the use of collision techniques for the identification of the main axon and its branches (112, 128). An example of one such axon is provided by a neuron that could be antidromically activated at three separate stimulation sites. Site "a" was located in the

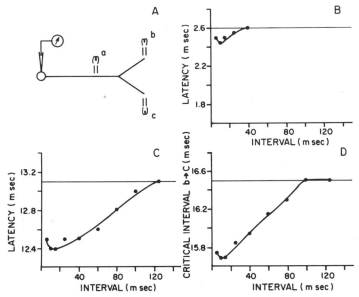

Figure 5 Conduction properties of axon trunk and preterminal branches of a callosal efferent neuron, which was antidromically activated at several locations along the axonal tree. *A.* Schematic illustration of the position of stimulating electrodes relative to axonal structures as revealed by collision techniques. *B.* Supernormal period of increased conduction velocity that occurred along the length of the axon between the soma and "a," as measured by conditioning and test stimuli delivered at "a." Points represent the latency to the test stimulus at each conditioning stimulus–test stimulus interval. *C.* Supernormal period that occurred along the length of the axon between the soma and "c" when conditioning and text stimuli were delivered at "c." *D.* Supernormal conduction along the preterminal axon between site "a" and site "c." (See text for explanation.)

splenium of the corpus callosum, just contralateral to the midline, whereas sites "b" and "c" were located approximately 2 mm apart, in the contralateral visual cortex. Antidromic latency via "a," "b," and "c" was 2.6, 8.6, and 13.1 msec respectively. Antidromic collision tests (112, 128) established that, for this cell, "b" and "c" are located on two separate branches, whereas "a" lies on the parent axonal trunk. Figure 5a is a schematic illustration of the position of stimulating electrodes relative to axonal structures as revealed by collision techniques. Figure 5b shows the supernormal period of increased conduction velocity that occurred along the length of the axon between the soma and "a," as measured by conditioning and test stimuli delivered at "a." Estimated conduction velocity along this segment of axon is approximately 5 m/sec. Note that the duration of the supernormal period is approximately 40 msec. In Figure 5c, both the conditioning stimulus and test stimulus are presented at "c." Note the increase in the duration of the supernormal period that results from the contribution of the more slowly conducting segment of axon between "a" and "c."

It is possible to study the conduction properties of the preterminal axonal branches by a method that totally eliminates the contribution of the main axon trunk and soma. For the above cell, the minimal interval following stimulation at "b," during which a stimulus applied at "c" yields an antidromic spike at the soma (the critical interval "b"→"c"), was 16.5 msec. At intervals shorter than 16.5 msec, the stimulus applied at "c" failed to elicit a response at the soma (owing to collision of impulses). For this cell it was determined that this critical interval was due to a conduction time of 15 msec between "b" and "c" and a refractory period of 1.5 msec at "c." Any change in the conduction velocity of the axonal branches between "b" and "c" would alter this critical interval. Thus, if two pulses were presented at "b" at an interval of 10 msec (i.e. the second impulse was presented near the peak of the supernormal period following the first impulse), one would expect the velocity of the second impulse to be increased relative to control values, yielding a shorter conduction time between "b" and "c." This would result in a correspondingly shorter critical interval "b"→"c."[1] In Figure 5d, two stimuli are presented at "b" at varying intervals. Points represent the critical interval following the second stimulus at "b," during which no spike resulted from a stimulus presented at "c." This curve represents supernormal conduction along preterminal axonal branches, which is independent of the main axon trunk. The magnitude and duration of

---

[1]Variations in the critical interval "b"→"c" may also reflect variations in the duration of the refractory period at "c," which in this case were shown to be negligible.

the supernormal period derived in this manner is similar to that seen in Figure 5c, which is derived from the entire length of axon between the soma and "c." This is not surprising, since both results, though derived by totally different and independent methods, reflect the conduction properties of the segment of axon between the branchpoint and "c." Note that the duration of the supernormal period demonstrated by the axon branches is significantly greater than that shown by the main axon trunk (Figure 5b).

In addition to parallel fibers of cat cerebellum (37, 71) and visual callosal axons of rabbit (124, 127, 128, 152) and monkey (126, 129) cortex, supernormal conduction velocity and excitability have recently been described following single impulses in a number of axonal systems of mammalian central nervous system. Such systems include caudatofugal axons of cat (65), hypothalamic efferent neurons of rats (98), somato-sensory callosal axons of rabbit (128), and Lissauer tract axons in cat (71). Furthermore, in our laboratory, we have recently found that neurons in visual area I of rabbit cerebral cortex, which project to either visual area II or to the ipsilateral lateral geniculate nucleus, have conduction properties that are essentially similar to those of visual callosal axons in the rabbit.

It is important to note, however, that in the cells which we have studied there is significant variability in both the magnitude and the duration of supernormality, which cannot be accounted for by differences in the control conduction velocity of the axon. Some cells, in fact, demonstrate no supernormal period (e.g. the two cells in Figure 4 with conduction velocities of 5.0 and 6.1 m/sec). Furthermore, the observed variability in supernormality cannot be accounted for by differences in the physiological state of the preparation, since neighboring cells in the same penetration showed typical supernormality. In addition, axons of some *systems* demonstrate only minimal or no supernormality. Corticotectal axons of the rabbit visual cortex, for example, are found in the same microelectrode penetrations as callosal efferent neurons, neurons projecting to ipsilateral visual area II, and neurons projecting to the ipsilateral lateral geniculate nucleus. Approximately 40% of corticotectal cells, however, show no supernormal conduction velocity, whereas the remainder show only minimal increases in conduction velocity, (i.e. most show a maximal increase in conduction velocity of from 0.5–2.0% while the duration of the supernormal period lasts from 10–20 msec.) Although this may be partially attributable to the relatively high conduction velocities of many corticotectal axons, even the slower corticotectal axons demonstrate a supernormal period of much less magnitude and shorter duration than axons *of comparable conduction velocity* in the callosal or cortico-geniculate system.

The Subnormal Period

Following a single impulse in axons of sciatic nerve of frog, a very slight decrease in excitability and conduction velocity is observed to follow the initial supernormal period (74, 94, 95). Increases in the number of conditioning stimuli result in an increase in both the magnitude and the duration of the subnormal period. The right side of Figure 2 shows the subnormality that develops in sciatic nerve of frog after 1–10 conditioning pulses, and the subsequent recovery of threshold to control values.

Similarly, following a single impulse in callosal axons of both rabbit (124, 127, 128) and monkey (126), a very small period of subnormal conduction velocity and excitability sometimes is observed to follow the initial supernormality. Following several impulses, however, subnormality is always seen to develop. The recovery of control levels of axonal conduction velocity is, however, faster in callosal axons than is the case in frog sciatic nerve. Following 20 prior impulses, for example, control values are generally reattained in rabbit after 1–2 sec, whereas following a 30–60 sec tetanus (33 pulses/sec), control values are reattained after 1–2 min.

Mechanism of Supernormal and Subnormal Periods

It is tempting to speculate that variations in axonal conduction velocity and excitability are due to variations in membrane potential. It can be demonstrated, however, that there is no *necessary* relationship between membrane potential and either conduction velocity or excitability. Both conduction velocity and axonal excitability are controlled by a multiplicity of interacting variables, each of which may be related in a different way to membrane potential. Lorente de No (69) and more recently Bennett et al (13) have pointed out the multiplicity of factors constraining the threshold voltage for an action potential, and Raymond & Lettvin (95) have shown that, making reasonable assumptions, one can construct a model in which either membrane hyperpolarization or depolarization results in superexcitability.

Nevertheless, one hypothesis which has received some experimental verification is that a depolarizing (or negative) after-potential follows the relative refractory period and is responsible for both supernormal excitability and conduction velocity. Such an after-potential has, in fact, been observed to follow the relative refractory period of mammalian A (38) and C (46) fibers and is associated with increased axonal excitability (39). One mechanism proposed to account for the after-depolarization is an increase in extracellular potassium, which follows the impulse

(35, 44, 109, 110). An increase in the concentration of extracellular potassium would result in a decrease in the potassium equilibrium potential and would therefore result in a decrease in resting membrane potential. Frankenhaeuser & Hodgkin (35) have studied the afterdepolarization found in giant axons of the squid and have calculated the accumulation of extracellular potassium following repetitive activity. In order to account for the magnitude and other phenomenological aspects of the afterdepolarization, they postulated that potassium was free to diffuse only 300 Å from the axon membrane. If no diffusion barrier was postulated, the amount of extracellular potassium was not sufficient to account for the observed effects.

Supernormal excitability has been found by Zucker (158) in axon terminals of crayfish motor neurons, but only occasionally in preterminal branches and never in the main axon. Changes in the extracellular concentration of sodium, manganese, and calcium had very little effect on the supernormal period. Variations in extracellular potassium, however, led to changes in supernormality that were consistent with the hypothesis of a transient post-impulse increase in the concentration of extracellular potassium. In support of this passive "potassium hypothesis," Zucker found that strophanthedin, an inhibitor of the sodium-potassium pump, had no effect on supernormal excitability.

There is no evidence regarding the mechanism of either super- or subnormality of axons in mammalian brain. If one assumes that supernormality in both sciatic nerve of frog and callosal axons of the rabbit and monkey is due to a transient increase in extracellular potassium, several factors might account for the differences in the time course of supernormality in these species. (a) The temperature difference between frog sciatic nerve preparation and rabbit callosal axons is approximately 21–22°C. Frog sciatic nerve was recorded at 17°C whereas the awake rabbit maintains a temperature of nearly 39°C. Zucker (158) has found the decay rate of superexcitability in crayfish terminals to have a Q_{10} of about 2.5, whereas the depolarizing after-potential in mammalian peripheral nerve has a Q_{10} of approximately 2.0 (44). In addition (b) the diffusion barrier around the membrane may be tighter, and/or the glial environment different, in central than in peripheral axons, resulting in a greater concentration of extracellular potassium following a single impulse, and faster subsequent diffusion back into the axon. A tighter diffusion barrier around central axons due, for example, to differences in the axon-glial relationship, could also result in a lack of augmentation of supernormality with increases in the number of impulses, since the functional extracellular space might become saturated with potassium after only a single impulse. As noted above, some axonal systems demonstrate only minimal supernormality. It would be interesting to compare the ultrastructure of the axon-glial relationship and the periaxonal space of such systems with that of axonal systems which demonstrate a marked supernormal period.

The mechanisms underlying the subnormal period of axonal excitability and conduction velocity have received far more experimental attention than those of the supernormal period. There is a body of evidence indicating that a hyperpolarization follows repetitive stimulation of a variety of axons (25, 102, 103, 121). In mammalian C fibers, this effect is associated with a decrease in axonal conduction velocity (20). Rang & Ritchie (89) have presented evidence that the post-tetanic hyperpolarization of mammalian C fibers reflects, at least in part, activity of a sodium pump. They found the post-tetanic hyperpolarization to be blocked by ouabain (which inhibits the sodium pump) and lithium and to be activated by a variety of cations known to activate the sodium pump. Two hypotheses have been put forth as to how the sodium pump could result in membrane hyperpolarization: (a) if the pump were a coupled sodium-potassium pump, extrusion of sodium would be coupled with uptake of potassium and could result in a depletion of potassium in the extracellular space immediately surrounding the membrane (103). Alternatively, (b) the extrusion of sodium might not be directly coupled with an influx of cations and would therefore be electrogenic (i.e. the extrusion of sodium from the cell with no concomitant cation influx would directly generate a membrane hyperpolarization (25, 121). Rang & Ritchie (89) have presented evidence that in mammalian C fibers the sodium pump is, at least in part, electrogenic. They show that, following stimulation of a nerve in potassium-free solution, a hyperpolarization results if potassium is added to the solution, rather than the usual depolarization. From these and other data they argue "...this potential is generated directly by the sodium pump, by the extrusion of sodium ions in excess of the number of potassium ions that are captured." Rang & Ritchie were careful to point out, however, that their data suggest the presence of coupled, as well as electrogenic pumping of sodium in C fibers, and in fact, suggest (90) that under normal conditions electrogenic pumping may play a lesser role than coupled pumping in maintaining the ionic balance in C fibers.

A post-tetanic hyperpolarization is also seen in myelinated vertebrate axons (e.g. 25, 103, 121) and also appears to be due to a sodium pump. Both Connelly (25) and Straub (121) have presented evidence that an uncoupled, electrogenic sodium pump may be the main cause of this hyperpolarization. That the post-tetanic, pump-related hyperpolarization is the cause of the subnormal phase of axonal excitability and conduction velocity is suggested by experiments (95) in which ouabain was found to abolish the subnormal phase of axonal excitability in fibers of frog sciatic nerve. Further, it was found that the time course of recovery of excitability is greater when the magnitude of subnormal excitability is greater. Similarly, Rang & Ritchie found the time course of the post-tetanic hyperpolarization in C fibers to be increased by increasing the duration of the tetanus, suggesting that similar mechanisms may be involved in the two processes (89).

THE FAILURE OF IMPULSES AT REGIONS OF LOW SAFETY FACTOR

It is reasonable to assume that the majority of axons are so designed that, under some conditions, the safety factor for a single impulse at all points along the axon is greater than 1. Thus, under ideal conditions, an impulse will travel from the soma, past inhomogeneities and branch points, to finally invade the terminal arbors. As described above, a number of workers have simulated the effects of increases in the diameter of modified Hodgkin-Huxley axons on impulse conduction (42, 63). Small increases were shown to reduce safety factor and to result in a slowing of conduction proximal to the inhomogeneity, whereas greater increases in diameter result in a failure of conduction (i.e. safety factor was reduced to less than 1). Whether axonal branchpoints can be considered to be formally equivalent to a decrease, no change or an increase in axon diameter will depend on the relative diameter of parent axon and daughter branches (42, 153; see Equation 5).

In the previous section, history-dependent effects of impulse conduction were described which will interact with passive axonal properties in determining the characteristics of impulse conduction through nonhomogeneous regions of the axon. The interaction between passive and dynamic axonal properties becomes more complex in regions of structural inhomogeneity and may, for example, result in routing of impulses down one or another branch at axonal bifurcations (45, 138, 155). In the present section we review empirical studies of impulse conduction through regions of axonal inhomogeneity.

Effects in Invertebrate Nervous System

The rapid increase in our understanding of the axon membrane, which resulted from the introduction of the giant axon of the squid (156), made clear the advantages of the relatively large diameter and accessible invertebrate axons. Careful dissection of some invertebrate systems allows the exposure of axonal trunk, preterminal branches, and postsynaptic elements. In many cases, intracellular and/or extracellular recordings can be obtained from each of these regions.

Conduction along branched axons emerging from pleural and abdominal ganglia of the mollusk *Aplysia* has been studied by Tauc & Hughes (133). They recorded intracellularly from the cell body, and they antidromically activated various axonal branches with electrical stimuli. It was found that antidromic volleys initiated in one axonal branch may differentially invade the other branch or the soma (via the main axonal trunk), depending on the frequency of antidromic stimulation. These authors suggested that impulse initiation in these cells may ordinarily occur at trigger zones near axonal branch points and that synaptic input onto the axon may modulate the invasion of successive branches.

Evidence that different branches of the same axon may act as filters with different frequency characteristics has been found by several authors. Bittner (17) has studied single axons whose branches innervate different components of the crayfish opener muscle. He presents evidence that the probability of invasion of preterminal branches of the same axon differs, depending on the frequency of impulses traveling along the main axon. This result has been confirmed by Hatt & Smith (51) who, in the same system, studied the depression of synaptic transmission that occurs during repetitive activation. Extracellular recordings from the preterminal axon near points of bifurcation revealed conduction block at or near the branchpoints along the length of the presynaptic nerve. Conduction block of one branch was seen to occur while the other branch continued to conduct. Smith (115) has shown that this conduction block can be relieved by hyperpolarizing currents and by reducing the concentration of extracellular potassium, and that it is associated with a reduced sodium inward current. He concludes that the conduction block is due to a *depolarization* of the axon, possibly resulting from an accumulation of extracellular potassium. Following even more prolonged stimulation of these axons, Smith & Hatt (116) described a second type of conduction block (a failure of alternate impulses) that occurred along a region of axon surrounded by dense connective tissue.

Parnas (78) has studied another crustacean system in which the branches of single axons innervate different muscles. In this system one axonal branch innervates the deep extensor abdominalis medialis (DEAM), whereas the other branch innervates the deep extensor abdominalis lateralis (DEAL). Utilizing intracellular recording from the muscles, extracellular recordings for the preterminal branches, and stimulation of the main axon, it was found that the branches innervating DEAM and DEAL respond differently to varying frequencies.

Ordinarily, single impulses to DEAM yield large EPSPs while those to DEAL elicit small EPSPs. Terminals leading to DEAM, however, are blocked at frequencies of 40–50/sec, whereas those innervating DEAL can sustain frequencies of 80–100/sec. The response of DEAL was, in fact, facilitated at high frequencies. The author concluded that the frequency-specific invasion of axon terminals probably occurs at the branchpoints.

Grossman et al (45) have studied the same system that was studied by Parnas (78). They utilized intracellular recording from the main axon just proximal to the bifurcation, and from the larger branch just distal to the bifurcation (which leads to DEAM), as well as extracellular recording from the smaller branch leading to DEAL. Their results corroborate Parnas's conclusion that the block of DEAM at high frequencies is due to a block of conduction into the branch leading to DEAM, and indicate that the point of failure was close to the branch point. The branch leading to DEAM was always larger than the branch leading to DEAL, but the relative diameter of the main axon and its two branches was such that if membrane properties were constant, no impedance mismatch would occur at the branchpoint. Grossman, et al. suggest that the frequency-specific channeling of impulses in this system may therefore be due to frequency-specific relative changes in the membrane properties of the two branches and that these changes may, in part, be due to a buildup in the concentration of extracellular potassium following impulse activity.

A series of studies by Nicholls and his collaborators (10, 11, 12, 59, 75) have led to an understanding of the mechanism by which conduction block at branchpoints results in certain adaptation phenomena in sensory fibers of the leech *Hirudo medicinalis* (138, 155). Impulse activity in these neurons was shown to result in both an increase in extracellular potassium and a membrane hyperpolarization. Following a single impulse, increases in extracellular potassium of 0.8 mM/1 were found, which decreased exponentially with a time constant of about 100 msec. Concentrations of up to 8 mM/1 (double the normal concentrations) of extracellular potassium resulted from higher frequency stimulation. Such stimulation was within the normal range of activity of these cells. The membrane potential of these cells was relatively insensitive to changes in the concentration of extracelluar potassium and did not react in a Nernstian fashion. Repetitive impulse activity in these cells led in fact to a marked hyperpolarization (11, 59), the magnitude of which was dependent on the rate of activity. Several findings indicated that the hyperpolarization was due in part to the activity of an electrogenic sodium pump. Among these were the observation that the hyperpolarization (*a*) was eliminated by ouabain and reversibly blocked by strophanthedin; (b) was intitiated following iontophoretic intracellular injection of sodium, but not potassium; (c) was inhibited by cooling and (d) was dependent upon the presence of extracellular potassium. This latter finding has been shown to be the case for the sodium pump in many excitable cells (73, 89, 117, 135). Although an increase in potassium conductance contributes to the hyperpolarization of some sensory cells in leech, hyperpolarization in touch cells, which are discussed below, appears to be mainly due to the activation of the electrogenic sodium pump (59). Following repetitive activity, IPSPs were shown to reverse and EPSPs were shown to be facilitated, indicating that the hyperpolarization also occurs at distal dendrites within the neurophile (11) and may affect synaptic transmission. Furthermore, hyperpolarization induced by current injection was shown to block conduction along some axonal branches (75).

In the same preparation, Van Essen (138) has elaborated on the finding of hyperpolarization-induced conduction block and has demonstrated the relevance of this phenomenon to sensory adaptation in touch-sensitive cells. These mechanoreceptors send processes that branch multiply before terminating in the skin (75). Intracellular and extracellular recordings were obtained from the soma and axonal branches respectively. The cell could be activated via intracellular current pulses or by mechanical stimulation of its receptive field on the skin. Two types of adaptation were shown to follow mechanical stimulation of the skin. Sustained pressure on a discrete portion of the receptive field of the cell led to a rapid adaptation of the response which was restricted to the portion of the receptive field being stimulated. More prolonged mechanical stimulation, or activation evoked by intracellular stimulation, resulted in a longer lasting rise in threshold throughout the entire receptive field of the cell. During adaptation resulting from continual stimulation, conduction block of some branches developed, and in some cells it was observed that one branch of a process conducted impulses continually while another failed intermittently. The adaptation phenomena and conduction block were apparently due to the post-activity hyperpolarization found in these cells. Conduction block was reduced by strophanthodin and mimicked by intracellular injection of positive current.

Intracellular injection of horseradish peroxidase into the touch sensitive mechanoreceptor in the leech has revealed that, in addition to the arborization of the fiber in the body segment associated with its ganglion, a very fine-diameter process extends to the immediately adjacent segment and then arborizes (155). Physiological studies indicated that conduction block was much easier to obtain following stimulation of adjacent segments than following stimulation of the main, central segment. Yau points out that the relative diameter of the process leading impulses from the adjacent segment and the first branch encountered was such that the predicted impedance mismatch would lead, in theory, to susceptibility to conduction block of even a single impulse (155). In fact, the amount of impulse activity sufficient to produce conduction block in these fibers was very small. In addition, this portion of the fiber acted as a low-pass filter, since the refractory period for conduction of impulses through the branch was greatly increased.

Parnas, Spira, and their co-workers (80, 119, 120) have examined impulse conduction along giant axons of the cockroach *Periplaneta americana*. These axons run continuously through the length of the nervous system (118) and constrict at the level of the third thoracic ganglia to send off collateral branches and form an isthmus (23, 118). The diameter of the axon is approximately 30–45 μm and 40–60 μm anterior and posterior of the constriction respectively, and the diameter at the isthmus is generally 20–40 μm (23). Following high frequency stimulation, conduction block occurred at the constricted portion of the axon. Higher frequencies were necessary for blockage of impulses in ascending than in descending direction. In contrast to the finding in sensory cells of leech (138, 155), where conduction block is caused by a hyperpolarization, both ascending and descending conduction block was associated with a depolarization of the axon membrane, which was thought to result from an increase in membrane conductance. Hyperpolarizing the axon with intracellular inward current injection was shown to relieve the conduction block. Ultrastructural studies (23) showed that the axon received synaptic input at the site of the constriction. Conduction block was facilitated by this synaptic input but still occurred in its absence. A theoretical model (79) based on modified Hodgkin-Huxley axon enabled computer simulation of impulse conduction along an axon similar to that studied in the cockroach. The model included the variations in extracellular potassium and was successful in predicting the shape of the conducted spikes, the frequency-specific conduction block, and the increase in the refractory period that occurred at the site of axonal constriction.

Effects in Vertebrate Nervous System

Afferent impulses entering the spinal cord must first traverse a bifurcation, one branch of which leads to the somata of bipolar cells in the dorsal root ganglion. As early as 1876 (154), a delay in impulse propagation past this branchpoint was reported. Subsequent studies over the next 60 years either confirmed (31) or denied (32) the existence of this delay. More recently Dun (29) has studied impulse propagation past dorsal root ganglia of the frog and found a delay of approximately .09, .15, and .23 msec in alpha, beta, and gamma fibers respectively. This delay was found to be present only when the preparation was in good physiological condition, thus explaining some of the earlier discrepant results. The composition of the bathing fluid was found to be critical, and the delay could not be found following mechanical trauma to the ganglion. Ito & Takahashi (57) have shown that the slowing of conduction results, in part, from a decrease in internodal length just proximal to the branchpoint, which appears to partially compensate for the impedance mismatch at the bifurcation. Dun (29) also found, as did earlier workers (3, 21), that a second impulse, traveling early in the relative refractory period of a prior impulse, would be blocked altogether as it passed the branchpoint. He explained this in terms of the additional electrical load of the branchpoint on the second impulse which, since it was traveling in a relatively refractory axon, already had a lowered safety factor.

Barron & Matthews reported anatomical (8) and physiological (9) evidence indicating that, after entering the spinal cord via the dorsal roots, some afferent fibers travel rostrally in the dorsal columns and send out a branch that exits via the neighboring dorsal roots and travels peripherally. In the first systematic study of intermittent conduction, Barron & Matthews (9) compared, in cat and frog, the patterns of orthodromic conduction of impulses entering the spinal cord at one dorsal root with the pattern of antidromic impulses which, after traversing at least one central branchpoint in the dorsal columns, were leaving the cord at other dorsal rootlets. They found the pattern of orthodromic impulses entering the cord to be very similar to that found in peripheral nerve. They found, however, that the burst of antidromic impulses was intermittently interrupted and that "...at fairly regular intervals it stops abruptly without any change in frequency and restarts a moment later equally abruptly." A similar intermittence was found when the orthodromic input was elicited by natural stimuli (i.e. a slowly-adapting stretch receptor). Furthermore, symmetrical effects were found when stimulating and recording electrodes were reversed. Barron & Matthews also found intermittent conduction in ascending fibers within the dorsal column. The intermittent conduction was subject to modulation by temperature and by activity in adjacent fibers. "Often a single drop of cold ringer would suffice to release the block when applied to a particular point on the column between the entrant and emergent fibers. A second drop of warm ringer on the same spot would reestablish the block immediately." The effects of temperature on conduction block at branchpoints are more fully discussed below. Conduction block was also reduced by cutting the dorsal roots on the contralateral side and it was modulated by afferent input from the contralateral side of the body. Furthermore, the block could be removed by undercutting the dorsal columns and (presumably) severing the branch leading to the grey matter. Barron & Matthews suggested that the intermittence was due to electrotonic effects originating at the termination of the collateral within the grey matter; they further proposed that this phenomenon may provide a mechanism whereby "...the grey matter is able to influence the discharges passing by it in the long conducting pathways of the spinal cord."

The findings of Barron & Matthews were challenged on both anatomical (7) and physiological (50, 136, 137) grounds. The discrepant anatomical findings may be explained in part by radiotracer studies (93) which indicated that the intraaxonal components of only about 1% of the fibers in adjacent dorsal roots are in continuity (as contrasted to the 32% proposed by Barron & Matthews). Alternatively, these effects may be due to electrotonic coupling between axons. Anatomical evidence suggesting possible electrical interaction between dorsal column fibers has in fact been found (105). Whatever its nature, the existence of intermittent conduction was confirmed by Fuortes (36). Wall et al (139) found blockage of ascending impulses of dorsal columns while finding continuous conduction in both sciatic nerve and dorsal root. They noted that safety factor for high frequency transmission was higher in sciatic nerve or dorsal root than in the sciatic nerve–dorsal column channel or dorsal root–dorsal column channel. Raymond and his colleagues (24, 93) have repeated the observations of Barron & Matthews and have developed an explanatory model based on activity-dependent shifts in excitability at regions of low safety factor along the axon. This model, based on work in sciatic nerve of frog (74, 94, 95), spinal cord (24, 93), and simulation studies (96, 97), explains the effects of stimulus rate on the periodicity of intermittent conduction in terms of the interaction of supernormal and subnormal phases of impulse conduction. When stimuli are delivered at a high rate (e.g. 20/sec), each stimulus falls within the supernormal period of the preceding stimulus, and a greater depression must develop before conduction will fail. Thus, under these conditions long periods of impulse conduction will alternate with long periods of failure. At lower stimulation rates, neither supernormality nor subnormality is well developed and periods of impulse conduction and failure are more closely spaced. Evidence for ephaptic modulation of impulse conduction was also found. Chung et al (24) have summarized their findings as follows:

> ... Thus, the probability that a particular impulse will invade a particular branch seems to be conditional on the time lag between that impulse and ones that preceded it in time or happened to be close to it in space. This implies that the subset of daughter terminals invaded by an impulse can be considered to depend on the temporal relations of the impulse to others in the discharge train of the axon. Temporal patterns such as those found in optic nerve fibers may thus be transformed into spatial ones at the axon terminals which can be distinguished by the cells on which these terminals end.

Krnjević & Miledi (67) presented evidence for two types of conduction failure in axons of phrenic nerve of the rat. In one, a cyclic pattern of impulses results from continuous high frequency stimulation, similar to the "intermittent conduction" described by Barron & Matthews (9). This pattern was ascribed to alternate periods of failure and conduction in the main nerve trunk that were thought to result from the interaction of super- and subnormal excitability of the main axonal trunk. A second type of conduction failure was thought to occur at preterminal branchpoints within the body of the muscle. One result of this presynaptic failure was that the failure of various muscle fibers of the same motor unit occurred asynchronously during repetitive stimulation of the main axon. Both types of conduction failure were frequency dependent. The preterminal failure was very sensitive to temperature and the effects of anoxia. This susceptibility to anoxia led Krnjevic & Miledi to suggest that under normal conditions hypoxia resulting from the reduced blood supply to contracting muscles may lead to presynaptic failure of conduction and that this may be partially responsible for the limited duration of sustained contraction found in certain muscles (67).

FURTHER COMMENTS ON THE MECHANISM OF IMPULSE CONDUCTION BLOCK

The above considerations indicate that a host of dynamic variables interact with static variables such as axonal geometry, stable membrane characteristics, and the morphology of the extracellular space in determining whether an impulse will invade or fail to invade a given segment of axon. Such dynamic variables include the history of impulse activity along the axon itself, axo-axonal synaptic effects, ephaptic effects, and the effects of temperature and anoxia.

As reviewed above, the geometry of the axon and its static membrane characteristics define specific regions of low safety factor where failure of impulse conduction is most likely to occur. Changes in the morphology of the extracellular space may also result in regions of low safety factor along a length of homogeneous axon. Smith & Hatt (116) have shown a failure of alternate impulses in axons innervating the opener muscle of the crayfish following prolonged stimulation. The block was suggested to occur at a region where the axon passed through a zone of dense connective tissue. Smith & Hatt suggested that the block was the result of membrane depolarization and consequent sodium inactivation due to the increase in the concentration of extracellular potassium in the reduced extracellular space. The size of the extracellular space surrounding the axon has been shown theoretically (35, 44, 111) to be important in controlling the dynamic properties of the axon membrane. Empirically, Baylor & Nicholls (10) have shown that removing the glial investment around pressure-sensitive mechanoreceptor cells in the leech effectively eliminates the increase in extracellular potassium that results from impulse activity in these cells. They point out that glial cells may act as spatial barriers around these neurons, thereby preventing the diffusion of extracellular potassium.

In myelinated axons, glial cells define not only many of the parameters of the extracellular space but also the internodal conduction distance. There is some evidence that the axon-glial relationship may vary along the length of the axon in a manner that facilitates impulse conduction past regions of low safety factor. In dorsal root ganglia Ito & Takahashi (57) have demonstrated a decrease in internodal length, just proximal to the branchpoint, which appears to compensate for the impedance mismatch at the bifurcation. Computer simulation studies of impulse conduction from myelinated axon into nonmyelinated terminal (100) indicate that the impedance mismatch resulting from the increased area of nonmyelinated membrane may be reduced by decreasing the length of the internode proximal to the terminal and by decreasing the diameter of the terminal nonmyelinated segment. Such decreases in internodal length have, in fact, been observed in preterminal regions of both peripheral nerve (157) and central axons (140).

Under some conditions the internodal length along a given segment of axon may vary over time. In teleost nervous system, compensatory increases in internodal length appear to follow the increases in axon diameter that occur as the animal grows (134). Decreases in internodal distance have been observed proximal to demyelinated regions resulting from multiple sclerosis (41, 122). Computer simulation studies have suggested that such decreases in internodal distance might facilitate impulse conduction past demyelinated regions (147).

As discussed above, two relatively long-lasting ionic consequences of impulse activity have been shown to influence conduction properties or to result in conduction block. Impulse activity may result in an increase in the concentration of extracellular potassium (e.g. see 10, 35, 44, 68, 79, 109, 110, 114, 116, 120, 158). The consequence of such an increase may be membrane depolarization and a concomitant increase in the excitability of the axon. When greater concentrations of extracellular potassium result, conduction failure may occur, presumably due to sodium inactivation. Impulse activity may also result in membrane hyperpolarization, which in many systems and species (e.g. 11, 25, 29, 73, 89, 121, 135) has been shown to be due, at least in part, to a noncoupled, electrogenic sodium pump. Observed results of this hyperpolarization have been decreased axonal excitability and conduction velocity and an increase in the probability of conduction block. The conduction block observed at regions of low safety factor along the axon may thus result from either excessive depolarization or hyperpolarization.

The changes in membrane potential and resultant conduction block produced at such regions as a result of prior impulse activity may be modulated by axo-axonic synaptic input. In giant axons of the cockroach (120) the modulation of spike trains, which results from synaptic input onto regions of low safety factor, apparently is due to the increased membrane conductance and resultant shunting of current. Similarly, Dudel (27, 28), has shown that in the crayfish leg, presynaptic inhibition is due to an axo-axonic synapse from which GABA is released, shunting the membrane. In touch-sensitive mechanoreceptors of the leech, synaptic contacts are found on some axonal branches that are subject to conduction block due to the hyperpolarizing effects of impulse activity (11). Van Essen (138) has made the interesting observation that the activation of some axo-axonic synapses may result in either facilitation or inhibition of impulse transmission, according to the history-dependent membrane potential of the axon. Thus, during a period of quiescence, one synaptic input had an inhibitory (hyperpolarizing) effect. The same input, however, during a period of activity-induced hyperpolarization and conduction block, had a depolarizing effect that in fact relieved the conduction block. This latter facilitatory effect resulted because, following activity, the membrane had been hyperpolarized beyond the equilibrium potential of the IPSP. This observation suggests the possibility that some axo-axonic synapses near regions of low safety factor may be classified as neither excitatory nor inhibitory. Their effects will depend on the prior impulse activity of the postsynaptic axon.

Neighboring neural structures may alter asynaptically the conduction properties and the probability of conduction failure at regions of low safety factor. Such nonsynaptic or "ephaptic" interaction between axons has been recognized for some time (6, 61, 70, 99). One system in which presynaptic excitability changes occur in the apparent absence of synaptic input is the mammalian lateral geniculate nucleus. Increased excitability of optic tract terminals with concomitant depolarization is found in the lateral geniculate nucleus of cat and monkey following electrical stimulation of the visual cortex (5, 58), mesencephalic reticular formation (4, 83), and following eye movement (18, 62). No anatomical evidence, however, for axo-axonic synapses has been found (47, 48, 60, 131). Singer & Lux (114), using potassium-sensitive microelectrodes, have shown that an increase in extracellular potassium occurs concomitantly with the evoked increase in terminal excitability and depolarization and that the latency and time course of these phenomena are very similar. Singer & Lux speculate that the increase in extracellular potassium is due to stimulation-evoked activity of neurons in the lateral geniculate nucleus and that the resultant depolarization may facilitate the invasion of presynaptic terminals entering this structure.

During interictal epileptic discharge in mammalian cortex, action potentials are initiated in the terminals of some thalamo-cortical (49, 107, 108) and callosal (106) terminals. These impulses travel antidromically to invade the cell body. It has been suggested (49) that impulse initiation in axon terminals is due in part to a depolarization that results from an increase in extracellular potassium in epileptic cortex. Such an increase has been observed using potassium-sensitive electrodes, but it was found during both ictal and interictal events (72). The above data thus suggest that the extracellular potassium liberated following neural activity may modulate the excitability of axon terminals and in some instances may even result in the initiation of action potentials. Since neurons within structures such as cortex and lateral geniculate nucleus are not activated uniformly under natural conditions, one might expect that the excitability (and therefore the invadability) of different preterminal branches of the same parent axon would vary with the neural activity in their immediate surroundings and would result in the differential routing of impulses.

Impulse conduction at regions of low safety factor is quite sensitive to anoxia (67) and to temperature variations. In squid giant axons, Westerfield et al (153) have utilized computer simulation and electrophysiological recordings to analyze the effects of temperature variations at axonal branchpoints. They found that at a given temperature there is a "critical ratio" of postbranch to prebranch diameters, which, if exceeded, results in failure of impulse propagation. The value of this critical ratio was shown to be very sensitive to temperature and was smaller at higher temperatures. The failure of impulse conduction that resulted when this value was exceeded was shown to be due primarily to the temperature-dependent change in the width of the action potential. An increase in the probability of conduction failure, with increases in temperature, was also shown by Barron & Matthews (9). Krnjević & Miledi (67) found mixed effects of temperature variations on presynaptic conduction block, which may have been partially due to increased oxygen utilization at higher temperatures. Impulses were found to conduct at higher frequencies at higher temperatures, but total block occurred earlier. At lower temperatures conduction occurred for a longer period but only lower frequencies could be maintained. We would point out that, given the previous arguments indicating that conduction block may be due to either excessive depolarization or hyperpolarization, the relationship between temperature and conduction block may not be simple but may depend on the etiology of the conduction block involved.

The finding of increased conduction block at higher temperatures is consistent with clinical observations on patients suffering from multiple sclerosis. Many such patients show a significant aggravation of symptoms (113) with an increase in body temperature of only 1° or 2°C. This effect is presumably due to increased conduction block in marginally conducting fibers. Rasminsky (91) has shown that conduction block in demyelinated fibers may occur with temperature increases of as little as 0.5°C and that this block is rapidly reversible when temperature is lowered.

The above discussion has emphasized regions of low safety factor along the axon, such as branchpoints and terminals, as areas where *conduction block* may occur. We would point out, however, that if an impulse does succeed in invading a given terminal, many of the influences on impulse conduction, which were described above, may result in *modulation* of transmitter release (e.g. see 54, 55, 132).

Finally, we consider the question of whether an axonal branchpoint may function as a 2, 3, or 4 position switch. If one assumes uniform properties of the extracellular space, and constant static and dynamic membrane characteristics, then any change in the safety factor observed at a branchpoint will depend solely on the geometrical considerations discussed in section 1. Thus, if the geometry of main axon and daughter branches is such that no impedance mismatch occurs, the safety factor at the branchpoint will essentially be the same as along the main axon.

If the branches are sufficiently small, safety factor may even increase at the branchpoint. The velocity of an impulse invading the branches will simply take on a new value, appropriate to the characteristics of the daughter branches.[2] If an impedance mismatch does result, however, then safety factor may be lowered. It has been shown (42, 64) that, assuming constant static membrane properties, an impedance mismatch of sufficient magnitude will result in failure of both branches in an all-or-none fashion. Thus, in this view, the branchpoint is considered a two position switch.

Several experiments, however, have now demonstrated that impulses may fail to invade one daughter branch while they successfully invade the other branch, thus forming a three position switch (45, 51). Several factors could account for these findings. First, static membrane characteristics may not be constant and branches may differ, for example, in their specific membrane properties. A large body of evidence indicates that the plasma membranes of axons are not spatially homogeneous in terms of structure and physiological properties but, on the contrary, exhibit a high degree of regional differentiation (141, 143). This is probably most strikingly illustrated by the membrane of normal myelinated fibers, in which sodium channel density is significantly higher at the nodes of Ranvier than in the internodal axon membrane beneath the myelin sheath (101, 104, 145, 149). Electrocyte axons in the gymnotid *Sternarchus albifrons* also show a high degree of regional specialization and exhibit two types of nodes of Ranvier along single fibers: Type I excitable nodes, with a dense cytoplasmic undercoating subjacent to the axolemma; and Type II inexcitable nodes, which do not possess the dense undercoating (148). On the basis of comparative arguments, it appears likely that the dense undercoating observed in electron micrographs subjacent to the axolemma at nodes of Ranvier and axon initial segments represents a morphological correlate of a very high density of ionic channels (150). In this regard, it has been noted that the cytoplasmic dense undercoating is not present beyond the last heminode where the myelin of preterminal fibers ends (84), or for more

---

[2]It should be pointed out that if safety factor at the branchpoint is equal to that of the parent axon, the characteristics of the branches must still accommodate the change in temporal characteristics of impulses in more slowly conducting axons (76, 77). Thus, internodal conduction time is not constant but is less in smaller diameter axons (26). This is probably a consequence of the decrease in spike rise time in smaller diameter axons. Moreover, refractory period is longer in smaller diameter fibers (76, 128).

than 1 μm along nonmyelinated collaterals of central myelinated fibers (142). Rather than assuming, however, that there are two types of axon membrane (high and low ionic channel density), it seems reasonable to speculate that the axon membrane may exhibit a *spectrum* of ionic channel densities and possibly of other physiological properties. It should be noted in this context that Ito et al (56) have shown that morphologically distinct branches leading from frog muscle spindle differ in their ability to generate and conduct single impulses.

Secondly, the dynamic membrane characteristics may vary, for example, as a result of differences in the diameters of the daughter branches (24). In the callosal system we have shown that the duration of the supernormal period of conduction velocity and axonal excitability is greater in slower conducting axons (Figure 4) and may be substantially shorter in daughter branches than in parent axon (Figure 5). Since the diameters of the branches at a bifurcation are often very different (144), one would expect that at some intervals following impulse activity the excitability of one branch would be supernormal while that of the other might have entered the subnormal phase.

In mammalian central nervous system many nonmyelinated axons have diameters of 0.1–0.2 μm. For axons of such small diameter, the aftereffects of impulse activity on daughter branches will be magnified. If one assumes a constant extracellular space of 200 Å surrounding the axon, the ratio of intracellular to surrounding extracellular space (neglecting the axon membrane) is approximately 125:1, 12:1, and 1:1 for axons with a diameter of 10.0, 1.0, and 0.1 μm respectively. Thus, in small fibers an increase in extracellular potassium following an impulse will be accompanied by a significant decrease in intracellular potassium, amplifying any Nernstein shift in membrane potential. Similarly, intracellular sodium concentrations following an impulse will be greater in small fibers. This would both reduce spike amplitude and, since the activity of the electrogenic sodium pump may depend on sodium concentrations, result in a greater afterhyperpolarization in the smaller branches. Thus, branches of different diameter may demonstrate very different aftereffects of impulse activity. For similar reasons, small differences in the size of the extracellular space around two branches would result in differences in the dynamics of ion flow following impulse activity. In small fibers with a high packing density, extracellular concentrations of potassium may increase by 4 mM/l per impulse (30). The above empirical and theoretical considerations indicate that some branchpoints may act as a three position switch. Although there have been no empirical demonstrations of branchpoints behaving as a 4 position switch, one could make minimal assumptions and model the branchpoint as such.[3]

FINAL REMARKS

The above considerations indicate that in some systems the temporal relationship between impulses and the connectivity between elements of the axonal tree varies dynamically with the history of impulse conduction along the axon and with the history of impulse activity (both synaptic and ephaptic) in the immediate microenvironment. Thus, although impulse conduction is sensitive to axonal history, that history is viewed in the context of the local environment. As reviewed above, the relevant time frame for such effects may extend from milliseconds to minutes. This view is quite different from the model of the axon as a simple transmission line that is faithful in both time and space.

A question of crucial interest concerns the functional significance of the temporal and spatial transformations of impulses that may occur along axonal trees. One distinct possibility is that such transformations simply introduce noise into the system and result in limitations in the information-carrying capacity of the axon. Alternatively, variations in the instantaneous conduction velocity of an axon, in the distribution of impulses along its arbors, and in the excitability of the terminals may be effectively utilized by some systems. In touch fibers of the leech, for example, activity-dependent conduction block is responsible for one type of adaptation of these cells to sensory stimuli (138, 155) whereas in spiny lobster (45, 78), frequency-specific conduction block gains a single axon differential control over two muscle systems. An important and difficult step in assessing the functional significance of such phenomena will be to relate the observed modulation of impulse conduction along individual axonal trees to changes in the behavior of the organism.

It is a matter for speculation as to whether impulse conduction along the axonal tree is subject to long-term modification. It is interesting to note in this regard that small structural variations in the preterminal axon (e.g. changes in diameter or taper) or changes in membrane (e.g. sodium channel density) may result in characteristic changes in the

---

[3]Routing of impulses down one or another subset of terminal arbors can occur whether the branchpoint is considered a 2, 3, or 4 position switch. Assuming an axonal tree consisting of a series of cascaded bifurcations, if each branchpoint is considered a 2 position switch, the number of possible combinations of invaded terminals will equal $2^{N/2}$ (where N equals the number of terminals), whereas if each branchpoint is considered a 4 position switch, the number of possible combinations will equal 2^N.

probability of invasion of the terminal region and in the amount of transmitter released when terminals are invaded. As discussed above, variations in the internodal length in preterminal axons or in the size or the functional extracellular space will also influence the invadability of the terminal. Thus, a plasticity in the axon-glia relationship could result in long-term modulation of the probability of invasion of various segments of the axonal tree. We bring up these speculations in order to emphasize the possibilities that emerge when the axon is viewed as being active in the integration as well as in the transmission of neural activity.

ACKNOWLEDGMENTS

It is a pleasure to thank Dr. S. A. Raymond for helpful comments regarding this manuscript and Mrs. Doris Walters for expert secretarial assistance. Research in the authors' laboratories was supported by grants from the National Institutes of Health, the Research Foundation of the University of Connecticut, the Kroc Foundation, the Paralyzed Veterans of America and by the United States Veterans Administration.

Literature Cited

1. Adelman, W. J., Fitzhugh, R. 1975. Fed. Proc. 34
2. Adrian, E. D. 1920. J. Physiol. London 54:1–31
3. Amberson, W. R., Downing, A. C. 1929. J. Physiol. London 68:1–18
4. Angel, A., Magni, F., Strata, P. 1965. Arch. Ital. Biol. 103:668–93
5. Angel, A., Magni, F., Strata, P. 1967. Arch. Ital. Biol. 105:104–17
6. Arvanitakl, A. 1942. J. Neurophysiol. 5:89–108
7. Barron, D. H. 1940. J. Neurophysiol. 3:403–6
8. Barron, D. H., Matthews, B. H. C. 1934. J. Physiol. London 85:104–8
9. Barron, D. H., Matthews, B. H. C. 1935. J. Physiol. London 85:73–103
10. Baylor, D. A., Nicholls, J. G. 1969 J. Physiol. London 203:555–69
11. Baylor, D. A., Nicholls, J. G. 1969 J. Physiol. London 203:571–89.
12. Baylor, D. A., Nicholls, J. G. 1969 J. Physiol. London 203:591–609
13. Bennett, M. V. L., Hille, B., Obara, S. 1970. J. Neurophysiol. 33:585–94
14. Bergmans, J. 1973. In *New Developments in Electromyography and Clinical Neurophysiology*, ed. J. E. Desmedt, 2:89–127. Basel: Karger
15. Berkinblit, M. B., Vvedenskaya, N. D., Gnedenko, L. S., Kovalev, S. A., Kholopov, A. V., Fomin, S. V., Chailakhyan, L. M. 1970. Biophysics 15:1121–31
16. Berkinblit, M. B., Vvedenskaya, N. D., Gnedenko, L. S., Kovalev, S. A., Kholopov, A. V., Fomin, S. V., Chailakhyan, L. M. 1971. Biophysics 16:105–13
17. Bittner, G. D. 1968. J. Gen. Physiol. 51:731–58
18. Bizzi, E. 1965. Physiologist 81:113
19. Bliss, T. V. P., Rosenberg, M. E. 1974. J. Physiol. London 239:60–61
20. Brown, G. L., Holmes, O. 1956. Proc. R. Soc. London Ser. B 145:1–14
21. Brücke, E. T. von, Early, M., Forbes, A. 1941. J. Neurophysiol. 4:80–91
22. Bullock, T. H. 1951. J. Physiol. London 114:89–97
23. Castel, M., Spira, M. E., Parnas, I., Yarom, Y. 1976. J. Neurophysiol. 39:900–8
24. Chung, S., Raymond, S. A., Lettvin, J. Y. 1970. Brain Behav. Evol. 3:72–101
25. Connelly, C. M. 1959. Rev. Mod. Phys. 31:475–84
26. Coppin, C. M. L., Jack, J. J. B. 1972. J. Physiol. London 222:91P
27. Dudel, J. 1965. Pflügers Arch. Gesamte Physiol. 282:323–37
28. Dudel, J. 1965. Pflügers Arch. Gesamte Physiol. 282:66–80
29. Dun, F. T. 1955. J. Physiol. London 127:252–64
30. Eccles, J. C., Dorn, H., Taborikova, H., Tsukahara, N. 1969. Brain Res.

15:276–80
31. Erlanger, J., Bishop, G. H., Gasser, H. S. 1929. *Am. J. Physiol.* 78:574–91
32. Erlanger, J., Blair, E. A. 1938. *Am. J. Physiol.* 121:431–53
33. Forbes, A., Ray, L. H., Griffith, F. R. 1923. *Am. J. Physiol.* 66:553–617
34. Fox, C. A., Hillman, D. E., Seigesmund, H. A., Dutta, C. R. 1967. In *Progress in Brain Research*, ed. C. A. Fox, R. S. Snyder, 25:174–225. Amsterdam: Elsevier
35. Frankenhaeuser, B., Hodgkin, A. L. 1956. *J. Physiol. London* 131:341–76.
36. Fuortes, M. G. F. 1950. *J. Physiol. London* 112:42
37. Gardner-Medwin, A. R. 1972. *J. Physiol. London* 222:357–71
38. Gasser, H. S. 1938. *J. Appl. Phys.* 9:88–96.
39. Gasser, H. S. 1939. *J. Neurophysiol.* 2:361–69
40. Gilliatt, R. W., Willison, R. G. 1963. *J. Neurol. Neurosurg. Psychiatry* 26:136–43
41. Gledhill, R. F., Harrison, B. M., McDonald, W. I. 1973. *Nature* 244:443–44
41a. Goldstein, S. S. 1978. In *Physiology and Pathobiology of Axons*, ed. S. G. Waxman. New York: Raven
42. Goldstein, S. S., Rall, W. 1974. *Biophys. J.* 14:731–58
43. Graham, H. T., Lorente de No, R. 1938. *Am. J. Physiol.* 123:326–40
44. Greengard, P., Straub, R. W. 1958. *J. Physiol. London.* 144:442–62
45. Grossman, Y., Spira, M. E., Parnas, I. 1973. *Brain Res.* 64:379–86
46. Grundfest, H., Gasser, H. S. 1938. *Am. J. Physiol.* 123:307–18
47. Guillery, R. W. 1969. *Z. Zellforsch.* 96:1–38
48. Guillery, R. W. 1969. *Z. Zellforsch.* 96:39–48
49. Gutnick, M. J., Prince, D. A. 1972. *Science* 176:424–26
50. Habgood, J. S. 1953. *J. Physiol. London* 121:264–74
51. Hatt, H., Smith, D. O. 1976. *J. Physiol. London* 259:367–93.
52. Hodgkin, A. L. 1937. *J. Physiol. London* 90 (2):183–210
53. Hodgkin, A. L., Huxley, A. F. 1952. *J. Physiol. London* 117:500–44
54. Hubbard, J. I., Willis, W. D. 1962. *J. Physiol. London* 163:115–37
55. Hubbard, J. I., Willis, W. D. 1968. *J. Physiol. London* 194:381–405
56. Ito, F., Kanamori, N., Kuroda, H. 1974. *J. Physiol. London* 241:389–405
57. Ito, M., Takahashi, I. 1960. In *Electrical Activity of Single Cells*, ed. Y. Katsuke, pp. 157–59. Tokyo: Ikagu Shoin
58. Iwama, K., Sakakura, H., Kasamatsu, T. 1965. *Jpn. J. Physiol* 15:310–322
59. Jansen, J. K. S., Nicholls, J. G. 1973. *J. Physiol. London* 229:635–55
60. Jones, E. G., Powell, T. P. S. 1969. *Proc. R. Soc. London Ser. B* 172:153–71
61. Katz, B., Schmitt, O. H. 1940. *J. Physiol. London* 97:471–88
62. Kawamura, H., Marchiafava, P. L. 1966. *Brain Res.* 2:213–15
63. Khodorov, B. I., Timin, E. N., Vilenkin, S. Ya., Gul'ko, F. B. 1969. *Biophysics USSR* 14:323–35
64. Khodorov, B. I., Timin, E. N. 1975. *Prog. Biophys. Mol. Biol.* 30:145–84
65. Kocsis, J. D., VanderMaelen, C. P. 1979. *Exp. Brain Res.* 36:381–86
66. Koles, Z. J., Rasminsky, M. 1972. *J. Physiol. London* 227:351–64
67. Krnjević, K., Miledi, R. 1959. *J. Physiol. London* 149:1–22
68. Krnjević, K., Morris, M. E. 1974. *Can. J. Physiol. Pharmacol.* 52:852–71
69. Lorente de No, R. 1947. *A Study of Nerve Physiology*, Pts. I, II Studies from The Rockefeller Institute. Vols. 131, 132. New York: Rockefeller Inst.
70. Marrazzi, A. S., Lorente de No, R. 1944. *J. Neurophysiol.* 7:83–101
71. Merrill, E. G., Wall, P. D., Yaksh, T. L. 1978. *J. Physiol. London* 284:127–45
72. Moody, W. J. Jr., Futamachi, K. J., Prince, D. A. 1974. *Exp. Neurol.* 42:248–63
73. Nakajima, S., Takahashi, K. 1966. *J. Physiol. London* 187:105–27
74. Newman, E. A., Raymond, S. A. 1971. *Q. Prog. Rep. MIT Res. Lab. Electr.* 102:165–87
75. Nicholls, J. G., Baylor, D. A. 1968. *J. Neurophysiol.* 31:740–56
76. Paintal, A. S. 1966. *J. Physiol. London* 184:791–811
77. Paintal, A. S. 1967. *J. Physiol. London* 193:523–33
78. Parnas, I. 1972. *J. Neurophysiol.* 35:903–14
79. Parnas, I., Hochstein, S., Parnas, H. 1976. *J. Neurophysiol.* 39:909–23
80. Parnas, I., Spira, M. E., Werman, R., Bergmann, F. 1969. *J. Exp. Biol.* 50:635–49
81. Pastushenko, V. F., Markin, V. S., Chizmadzhev, Yu. A. 1969. *Bio-*

physics 14:929-37
82. Pastushenko, V. F., Markin, V. S., Chizmadzhev, Yu. A. 1969. *Biophysics* 14:1130-38
83. Pecci-Saavedra, J., Wilson, P. D., Doty, R. W. 1965. *Nature* 210:740-42
84. Peters, A. 1966. *Q. J. Exp. Physiol.* 51:229-36
85. Rall, W. 1959. *Exp. Neurol.* 1:491-527
86. Rall, W. 1962. *Ann. NY Acad. Sci.* 96:1071-92
87. Rall, W. 1964. In *Neurol Theory and Modeling*, ed. R. Reiss, pp. 73-97. Stanford: Stanford Univ. Press
88. Rall, W. 1977. In *Handbook of Physiology*, ed. J. M. Brookhart, V. B. Mountcastle, 1(Pt. 1):39-97. Bethesda, Md: Am. Physiol. Soc.
89. Rang, H. P., Ritchie, J. M. 1968. *J. Physiol. London* 196:183-221
90. Rang, H. P., Ritchie, J. M. 1968. *J. Physiol. London* 196:223-36
91. Rasminsky, M. 1973. *Arch. Neurol.* Chicago 28:287-92
92. Rasminsky, M., Sears, T. A. 1972. *J. Physiol. London* 227:323-50
93. Raymond, S. A. 1969. Physiological influences on axonal conduction and distribution of nerve impulses. PhD thesis. MIT, Cambridge, Mass.
94. Raymond, S. A. 1979. *J. Physiol. London*. In press
95. Raymond, S. A., Lettvin, J. Y. 1978. In *Physiology and Pathobiology of Axons*, ed. S. G. Waxman, pp. 203-25. New York: Raven
96. Raymond, S. A., Pangaro, P. 1975. *Q. Prog. Rep. MIT Res. Lab. Electr.* 116:273-81
97. Raymond, S. A., Pangaro, P. 1975. *Nerve threshold and intermittent conduction.* Res. Lab. Electr. MIT, Cambridge, Mass. (Color film)
98. Renaud, L. P., Hopkins, D. A. 1977. *Brain Res.* 121:201-13
99. Renshaw, B. 1946. *Am. J. Physiol.* 146:443-
100. Revenko, S-V., Timin, Y. N., Khodorov, B. I. 1973. *Biophysics.* 18:1140-45
101. Ritchie, J. M., Rogart, R. B. 1977. *Proc. Natl. Acad. Sci. USA* 74:221-15
102. Ritchie, J. M., Straub, R. W. 1956. *J. Physiol. London* 134:698-711
103. Ritchie, J. M., Straub, R. W. 1957. *J. Physiol. London* 136:80-97
104. Rosenbluth, J. 1976. *J. Neurocytol.* 5:731-45
105. Scheibel, M. E., Scheibel, A. B. 1969. *Brain Res.* 13:417-33
106. Schwartzkroin, P. A., Futamachi, K. J., Noebels, J. L., Prince, D. A. 1975. *Brain Res.* 99:59-68
107. Schwartzkroin, P. A., Mutani, R., Prince, D. A. 1975. *J. Neurophysiol.* 38:795-811
108. Scobey, R. P., Gabor, A. J. 1975. *J. Neurophysiol.* 38:383-94
109. Shanes, A. M. 1949 *J. Gen. Physiol.* 33:57-73
110. Shanes, A. M. 1949. *J. Gen. Physiol.* 33:75-102
111. Shanes, A. M., Grundfest, H., Freygang, W. 1953. *J. Gen. Physiol.* 37:39-51
112. Shinoda, Y., Arnold, A. P., Asanuma, H. 1976. *Exp. Brain Res.* 26:215-34
113. Simons, D. J. 1937. *Bull. Neurol. Inst. NY* 6:385-86
114. Singer, W., Lux, H. D. 1973. *Brain Res.* 64:17-33
115. Smith, D. O. 1978. *Neurosci. Abstr.* p. 391.
116. Smith, D. O., Hatt, H. 1976. *J. Neurophysiol.* 39:794-801
117. Sokolove, P. G., Cooke, I. M. 1971. *J. Gen. Physiol.* 57:125-63
118. Spira, M. E., Parnas, I., Bergmann, F. 1969 *J. Exp. Biol.* 50:615-27
119. Spira, M. E., Parnas, I., Bergmann, F. 1969. *J. Exp. Biol.* 50:629-34
120. Spira, M. E., Yarom, Y., Parnas, I. 1976. *J. Neurophysiol.* 39:882-99
121. Straub, R. W. 1961. *J. Physiol. London* 159:19-20
122. Suzuki, K., Andrews, J. M., Waltz, J. M., Terry, R. D. 1969 *Lab. Invest.* 20:444-54
123. Swadlow, H. A. 1974. *Exp. Neurol.* 43:424-44
124. Swadlow, H. A. 1974. *Exp. Neurol.* 43:445-51
125. Swadlow, H. A. 1977. *Exp. Neurol.* 57:516-31
126. Swadlow, H. A., Rosene, D. L., Waxman, S. G. 1978. *Exp. Brain Res.* 33:455-67
127. Swadlow, H. A., Waxman, S. G. 1975. *Proc. Natl. Acad. Sci. USA* 72:5156-59
128. Swadlow, H. A., Waxman, S. G. 1976. *Exp. Neurol.* 53:128-50
129. Swadlow, H. A., Waxman, S. G., Rosene, D. L. 1978. *Exp. Brain Res.* 32:439-43
130. Swadlow, H. A, Weyand, T. G., Waxman, S. G. 1978. *Brain Res.* 156:129-34
131. Szentagothai, J., Hamori, J., Tombol, T. 1966. *Exp. Brain Res.* 2:283-301
132. Takeuchi, A., Takeuchi, N. 1962. *J.*

Gen. Physiol. 45:1181–93
133. Tauc, L., Hughes, G. M. 1963. *J. Gen. Physiol.* 46:533–49
134. Thomas, P. K., Young, J. Z. 1949. *J. Anat.* 83:336–51
135. Thomas, R. C. 1969. *J. Physiol. London* 201:495–514
136. Toennies, J. F. 1938. *J. Neurophysiol.* 1:378–90
137. Toennies, J. F. 1939. *J. Neurophysiol.* 2:515–25
138. Van Essen, D. 1973. *J. Physiol. London* 230:509–34
139. Wall, P. D., Lettvin, J. Y., McColloch, W. S., Pitts, W. H. 1956. In *Information Theory. 3rd London Symp.*, ed. C. Cherry, pp. 329–44 London: Butterworth
140. Waxman, S. G. 1971. *Brain Res.* 27:189–201
141. Waxman, S. G. 1972. *Brain Res.* 47:269–88
142. Waxman, S. G. 1974. *Brain Res.* 65:338–42
143. Waxman, S. G. 1975. *Int. Rev. Neurobiol.* 18:1–40
144. Waxman, S. G. 1975. *Neurosci. Lett.* 1:251–56
145. Waxman, S. G. 1977. *Arch. Neurol. Chicago* 34:585–90
146. Waxman, S. G., Bennett, M. V. L. 1972. *Nature New Biol.* 238:217–19
147. Waxman, S. G., Brill, M. H. 1978. *J. Neurol. Neurosurg. Psychiatry* 41:408–17
148. Waxman, S. G., Pappas, G. D., Bennett, M. V. L. 1972. *J. Cell Biol.* 53:210–24
149. Waxman, S. G., Quick, D. C. 1977. *J. Neurol. Neurosurg. Psychiatry* 40:379–86
150. Waxman, S. G., Quick, D. C. 1978. *Brain Res.* 144:1–10
151. Waxman, S. G., Swadlow, H. A. 1976. *Exp. Neurol.* 53:115–28
152. Waxman, S. G., Swadlow, H. A. 1976. *Brain Res.* 113:179–87
153. Westerfield, M., Joyner, R. W., Moore, J. W. 1978. *J. Neurophysiol.* 41:1–10
154. Wundt, W. 1876. *Untersuchungen zur Mechanik der Nerven und Nervencentren. 2. Abteilung: Ueber den Reflexvorgang und das Wesen der centralen Innervation.* Stuttgart: Enke
155. Yau, K. W. 1976. *J. Physiol. London* 263:513–38
156. Young, J. Z. 1936. *Cold Spring Harbor Symp. Quant. Biol.* Vol. 4
157. Zenker, W. 1964. *Z. Zellforsch.* 62:531–45
158. Zucker, R. S. 1974. *J. Physiol. London* 241:111–26

TRANSFER RNA IN SOLUTION: SELECTED TOPICS

♦9145

Paul R. Schimmel
Department of Biology, Massachusetts Institute of Technology, Cambridge, Massachusetts 02139

Alfred G. Redfield
Departments of Biochemistry and Physics, Brandeis University, Waltham, Massachusetts 02254

INTRODUCTION

Ever since the elucidation of the primary structure of a specific tRNA (45), there have been a progression of physical studies aimed at determining the detailed structure and solution behavior of tRNA. Some of this work culminated in the three-dimensional crystal structure determination (65, 90), which provided a concrete foundation on which to design future experiments. These experiments have aimed at establishing the relationship between the crystal structure and that of the molecule in solution (particularly with respect to the important tertiary interactions), the relationship between ion binding (e.g. Mg^{2+}, spermine, etc.) and conformation, and the conformational mobility of tRNA, especially with respect to the structural changes that occur while carrying out the complex process of protein synthesis. The challenge has been to formulate the ideas in specific, molecular terms, and to abandon the phenomenological and vague concepts that permeated the field in its earlier years. And to an increasing degree, this objective is being achieved.

Interest in tRNA has remained high over the years. This is because the molecule itself is inherently intriguing and complex from a physical standpoint, and therefore always offers many important challenges. It is also because of the great interest in specific protein–nucleic acid interactions involving tRNA and a variety of proteins (101), and the growing awareness that tRNA is involved in physiological processes other than protein synthesis and in systems where its protein synthetic function is coupled to regulation of gene transcription by RNA polymerase (3, 21, 33, 89, 115, 130, 131). It is these circumstances that have enabled the field to maintain its momentum, even though many of tRNA's basic properties were established years ago.

There have been several reviews of tRNA structure and physical properties, particularly in many of the articles contained in recent tRNA monographs (1, 102, 115). In view of this situation, we have chosen to select only a few topics that are of current interest and have not been extensively reviewed elsewhere or where we believe that previous analyses need correction or clarification. Our review concentrates on: use of tritium to probe tRNA structure and interactions in

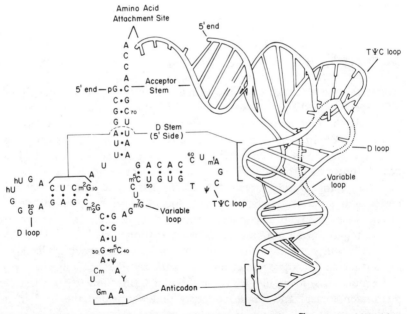

Figure 1 Sequence and cloverleaf structure (82) of yeast tRNA^{Phe} side-by-side with a schematic illustration of three-dimensional crystal structure (66). Major landmarks on the tRNA structure are indicated. In the cloverleaf diagram on the left, every tenth base is numbered. In the three-dimensional structure illustration, the ribose phosphate backbone is drawn as a continuous tube, cloverleaf hydrogen bonds are designated by crossbars, and thin lines join bases that have tertiary interactions.

solution; interaction of ions with tRNA; and recent applications of NMR to tRNA conformation and dynamics.

Tritium labeling of carbon 8 positions in purines is an example of the chemical probe approach to questions of nucleic acid structure. Unlike other probes of this type, tritium is of an atomic size; many sites may be simultaneously probed without causing structural alterations in the molecule under investigation. The interaction of ions with tRNA has been an area of considerable activity and controversy and is now at the stage where specific ion sites are being identified in X-ray crystallographic analyses. Also, we believe that enough confusion still exists in this area so as to warrant a brief summary and interpretation of some of the earlier work, as well as of more recent investigations. And, finally, NMR continues to be the major physical approach to tRNA structure and dynamics in solution, for which many new applications are steadily brought forth.

Figure 1 gives the sequence and cloverleaf structure of yeast tRNA^{Phe}, side-by-side with a schematic illustration of the three-dimensional crystal structure. Familiar landmarks (acceptor stem, dihydrouridine (D) loop, etc.) are indicated. In the case of the three-dimensional structure, the backbone is represented as a continuous tube and cloverleaf base pairs are designated by open crossbars. Bases that associate through tertiary interactions are joined by a thin line, and the dotted lines show segments that can vary in length in the various tRNA species. It is helpful to refer to this figure during the discussion of the material given below.

PROBING SPECIFIC SITES IN TRANSFER RNA WITH TRITIUM

An inherent problem associated with determining macromolecular structure in solution is the difficulty of obtaining physical information on specific sites in the polymer. This particularly true in the case of nucleic acids where, for the most part, there are only four basic repeating units. Therefore, many techniques, such as those based on optical properties, cannot resolve residues of a given kind (such as adenine) which occupy different positions in the sequence. It is for this reason that special site-specific probes have been developed as a means to probe the local environment about a particular residue in the sequence. For example, the reagent kethoxal specifically reacts with guanine residues that are exposed; by reacting a nucleic acid with kethoxal and determining which residues have reacted, one obtains information on the exposed residues in a nucleic acid structure

(39, 71, 72). Likewise, the presence of 4-thiouridine in certain tRNA's enables one to attach specific probes to this residue (such as a spin label) and thereby monitor the local environment in this region (17).

A major difficulty is that only a few residues can usually be studied by site-specific probes. Moreover, the introduction of a spin label, or a chemical modification, introduces a perturbation into the structure; this means that, for example, a spin label may give signals that are unique to a nonnative conformation rather than the desired native structure. Similarly, in a chemical modification study, one worries that, after the first base is modified, subsequent reactions with other bases may really be more indicative of a distorted structure generated by the first modification rather than be an accurate portrayal of the conformation of the native structure.

What is needed is a probe that does not introduce structural alterations and that is broadly applicable to a wide variety of sites, rather than to a select few such as unusual bases (e.g. thiouridine) or exposed residues. The smallest atomic size probe is hydrogen and its isotopes. Because hydrogen atoms are widely distributed in all nucleotide units, it is clear that any method that relies on a hydrogen exchange reaction (such as exchanging tritium or deuterium for hydrogen) can overcome the major objections associated with the usual site-specific probes and chemical modification reactions. This is because all sites, in principle, can be probed and because the substitution of tritium (or deuterium) for hydrogen in the nucleic acid structure should have a minor effect on conformation. It is for these reasons that isotopes of hydrogen are attractive probes of structure.

One of the earliest applications of hydrogen exchange took advantage of the relatively rapidly exchanging hydrogens associated with hydrogen-bonding protons (25, 26, 128). Depending on experimental conditions, these protons typically exchange in times of the order of seconds to minutes. In the usual applications, these rates are too rapid to enable the study of exchange rates of distinct sites distributed throughout the molecule. Instead, the overall exchange for the entire macromolecule is monitored, and from this certain useful structural information can be deduced. Of course, specific exchanging sites may be monitored by NMR if one is able to determine the exact resonance positions of a given residue in the spectrum. But, in general, this kind of application is limited by the resolving power of the NMR system and by the ability to unambiguously identify specific resonances.

There are two other exchange reactions associated with nucleic acid bases that, because of their extreme slowness, enable one to monitor exchange rates at specific loci distributed throughout the structure. These are the H-8 exchange reaction of purines (23, 24, 31,

32, 73, 76, 103, 118) and the H-5 exchange of pyrimidines (14, 28, 40, 50, 60, 100, 108, 112, 122, 123). Because these reactions have time constants that are on the order of days, at room temperature, it is possible to do a direct analysis of tritium incorporation into specific sites distributed throughout the molecule. Of the two slow exchange reactions, the H-8 exchange is more useful because it has a rate that is estimated to be about ten times more rapid than that associated with H-5 exchange (105). The low exchange rate of the latter reaction means that longer incubation times and higher specific activities of isotopic water need to be used in order to achieve meaningful incorporation in a reasonable period of time. It is for this reason that the H-8 exchange reaction has been preferred for studies that have been done thus far. This means that only purine sites can be analyzed, but because roughly half of all bases in a typical nucleic acid are purines, this is not a serious limitation.

After incubation for a few hours in tritiated water of high specific activity, sufficient incorporation occurs into purine C-8 positions to be detected by standard scintillation counting. Although the amount of tritium incorporation may only be a fraction of 1% of the maximum possible value (associated with full exchange), the low rate of exchange ensures that the incorporated radioactivity is exchanged out very slowly. What this means is that, after the nucleic acid has been exposed to tritium for a few hours, it may be isolated from free and loosely bound tritium and subjected to a fingerprint analysis, without loss of a significant amount of tritium during the fingerprinting procedure (31, 32). Through fingerprinting and subsequent analyses, it is possible to determine the extent of tritium labeling of specific purines distributed throughout the structure. In this way, from one experiment one is able to probe simultaneously many sites in the molecule with an atomic size reagent.

There have been three principal applications of the H-8 labeling reaction to transfer RNA. These are: investigation of tRNA structure in solution (31, 32, 106); studies of conformational changes in tRNA upon interaction with ribosomes (C. Farber and C. Cantor, in preparation); and mapping contact points in a complex of tRNA with an aminoacyl tRNA synthetase (107). Some of the essential features of these investigations are considered below.

Basic Aspects of H-8 Exchange

H-8 exchange follows simple first order kinetics. Therefore, the incorporation of ^{3}H into purine P is given by

$$-\frac{d\Delta P^*}{dt} = \frac{\Delta P^*}{\tau} \qquad \qquad 1.$$

or

$$\Delta P^* = \Delta P^{0*} e^{-t/\tau} \qquad 2.$$

where ΔP^* is the deviation in the concentration of tritium-labeled purine from the value achieved at isotopic equilibrium; τ is the time constant associated with exchange, and ΔP^{0*} is the value of ΔP^* at $t=0$.

The rate constants for H-8 exchange are small and somewhat different for adenine and guanine nucleotides. For example, at 37°, around neutral pH, the rate constant for 5'-AMP is 0.083% h^{-1}; that for 5'-GMP is 0.17% h^{-1} (32). The 3'-nucleotides and the nucleosides have similar rate constants (30).

The approximately twofold difference in rates for AMP and GMP holds over a relatively wide temperature range (25 to 90°C). In both cases, the exchange is quite temperature dependent and has an activation energy of about 22 kcal/mole^{-1} (32). Because exchange is so slow at 37° and because the activation energy for exchange is relatively high, it is currently impractical to do experiments with tRNA's much below 37°C.

There are several reports that the reaction proceeds by an ylid intermediate (23, 24, 118). It is envisioned that protonation first occurs at N-7 and is then followed by removal of the proton from C-8 to give the ylid. Subsequent protonation at C-8 (with hydrogen from the solvent) and dissociation of the proton from N-7 completes the reaction cycle. Clearly, any environment that affects the acidity of N-7 or of C-8 or that affects the accessibility of solvent to the imidazole part of the purine ring should affect the rate constants for H-8 exchange.

It is convenient to compare the exchange rate of a specific nucleotide unit within a nucleic acid with that of the corresponding free nucleotide. This ratio is designated R_i and is defined by (32)

$$R_i = \tau_0 / \tau_i \qquad 3.$$

where τ_0 is the time constant for H-8 exchange of the free nucleotide and τ_i is that for the same nucleotide unit at the i-th position in the chain. Therefore, R_i measures the rate for a specific residue normalized to what it would be were it like the unperturbed free mononucleotide unit. With this definition, values of R greater than unity correspond to a retardation in exchange. In all cases studied to date, values of R for a particular base within a nucleic acid are found to be equal to or greater than unity; that is, accelerated exchange, relative to the mononucleotide unit, has not been observed (31, 32, 106, 107). There is some indication, although by no means definitive proof, that retarded exchange rates result mostly from a decrease in accessibility of the exchanging site to the solvent (32, 107).

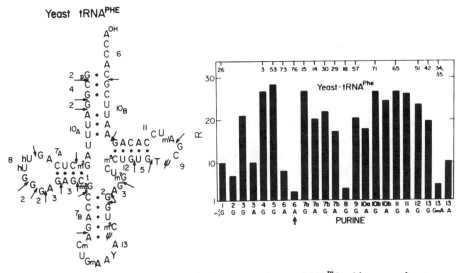

Figure 2 (*Left*) Sequence and cloverleaf structure of yeast tRNA^{Phe}, with arrows denoting cleavage points of T1 ribonuclease. The fragments are numbered in accordance with their positions on a chromatogram. Lower case numbers designate the position of every fifth base in the sequence. (*Right*) R values for specific bases in yeast tRNA^{Phe} incubated in tritiated water at pH 6.5, 10 mM Mg^{2+}, 37°. Horizontal axis gives the purines (A or G) from numbered T1 fragments; numbered tick marks across the top denote positions of bases in the sequence. An arrow indicates the 3′-terminal A. GmA is 2′-0-methyl GpA. Adapted with permission from Reference 32.

Applications to tRNA Structure in Solution

In order to measure tritium incorporation rates into specific sites in tRNA, it is necessary to have a method for isolating specific parts of the molecule after they have been labeled with tritium. This is accomplished by taking advantage of nucleic acid fingerprinting techniques which enable the isolation of specific nucleotide residues. Figure 2 gives the sequence and cloverleaf structure of yeast tRNA^{Phe} (82). Arrows designate positions in the structure that are cut by T1 ribonuclease. This nuclease cuts after G's and generates a series of fragments that are numbered from 1 to 13. These fragments can all be separated on a two-dimensional chromatogram. After separation on a chromatogram, individual fragments may be eluted and subject to further digestion with T2 ribonuclease, which splits after every base. The T2 digestion of each T1 fragment is also run on a chromatogram in order to separate the G's from the A's. This final separation enables one to obtain specific bases from defined positions in the sequence (32). For example, from the T1 and T2 digestions, A14 and G15 are individually obtained from fragment 7A.

No more than 1 G is obtained from each fragment. However, some fragments may contain more than one A; in these cases, the different A's from the same fragment will mix together on the chromatogram. Alternatively, other types of nuclease digestion procedures may be used in order to fractionate out the specific A residues. Also, some fragments are redundant; for example, fragment 3 (a dinucleotide) corresponds to A-G, which comes from 3 different sections of the sequence. Here again, in order to resolve the different bases, a different set of nuclease digestion procedures must be used. Nevertheless, utilizing just T1 and T2 ribonucleases, in conjunction with the appropriate chromatograms, many specific sites in the molecule can be monitored.

The basic procedure is to expose tRNA to tritiated water (1 to 3 Ci/ml) for a period of a few hours. The tRNA is then removed from the free and loosely bound tritium and subjected to the digestions and chromatographies described above. These steps do not result in a significant amount of exchange-out (from C-8) of the previously incorporated tritium.

The spectrum of tritium exchange rates at 37° (pH 6.5, 50 mM Na^+, 10 mM Mg^{2+}) for yeast $tRNA^{Phe}$ is also shown in Figure 2. This figure plots R values versus the T1 fragment number. In addition, the A's and G's associated with each fragment are indicated separately. In those cases where there are no ambiguities associated with the redundancies, the specific position of the nucleotide residue within the sequence is indicated across the top of the figure.

From Figure 2, it is clear that the tritium exchange rates show a remarkable variation from site to site within the structure. Several conclusions can be drawn from these and related data.

1. Bases in cloverleaf helical sections show remarkably high R-values, indicating a participation in helical structures. The purines in cloverleaf helices include those from fragments 4, 5, 7B, 10B, and 11, as well as the guanosines from fragments 10A, 12, and 13. The R values for these 12 purines fall in the range of approximately 17 to 29. In studies with the poly A:poly U duplex it was found that, at 37°C, R-values for the helical A's are about 17 (32). On the other hand, when poly A:poly U is melted out, or when tRNA is melted, the R values drop to 1.0. These results with poly A:poly U show that helix formation results in strongly retarded H-8 exchange rates, and certainly the same is found for cloverleaf helical bases in tRNA.

2. Several bases in single-stranded sections of the cloverleaf have high R values and these correspond to ones that participate in tertiary interactions, according to the crystal structure. For example, an important tertiary interaction in the crystal structure is the bonding

between G15 and C48 (65,90). The R value of G15 is among the highest in the molecule, with a value over 25. Another example is G57, which has a value of $R \approx 20$; in the crystal structure, this base interacts with the dihydrouridine loop. Also, A14 has $R \approx 20$, and this base is believed to participate in a hydrogen-bonded triple base pair interaction with U8 and A21. Finally, a more detailed analysis of the data suggests that A9 has $R > 13$; this base participates in a hydrogen-bonded triplet with U12 and A23. Therefore, a number of bases in cloverleaf single-stranded sections have high R values and each of these participates in tertiary interactions, according to the X-ray structural model (65,90).

3. A76 is evidently the most rapidly exchanging site in the molecule. This base is designated by an arrow in Figure 2. It is the 3'-terminal adenosine and, as far as is known, participates in no structural interactions. On the other hand, the exchange rate for A73 is about three times lower than that for A76, even though both residues are in close proximity within the same single-stranded segment of the tRNA. This is most likely due to the stacking interaction of A73 with the end of the acceptor helix (see Figure 2); A76 is further removed from this helix and will not be as strongly biased into a stacked conformation. In other studies, it was found that stacked residues in single-stranded poly A have retarded exchange rates, whereas unstacked residues do not have retarded rates (32). As expected, the tritium-labeling data suggest that the degree of stacking of A73 is greater than that of A76.

4. One of the more rapidly exchanging sites is G18 ($R = 3.7$) (32). Its rapid exchange rate suggests that this base does not participate in any strong interaction with another part of the structure. However, X-ray data suggest this base interacts by intercalation with the TψC loop (65,90). It is possible that the interactions of this base in the crystal structure are broken down under the conditions of the tritium-labeling experiment (pH 6.5, 37°). Alternatively, the interactions of this base may not produce a perturbation of the tritium exchange. But this seems unlikely, because of the good correlation between retarded exchange rates and structural interactions associated with other bases (see above). Moreover, other lines of evidence indicate that, with tRNA^{Phe} in solution, this base is freely accessible to the solvent.

In summary, the tritium-labeling approach has been shown to be remarkably sensitive to local structural effects in the RNA molecule. It is clear that bases in cloverleaf single-stranded sections can have markedly retarded exchange rates and that these bases have been shown by X-ray methods to be involved with tertiary interactions, such as hydrogen-bonded base triplets. In general, the method has provided a means to probe specific bases in solution and to test the applicability of the X-ray model to the molecule in solution. With the exception noted above, the labeling data on yeast tRNA^{Phe} in solution can be harmonized with its crystal structure.

However, the tritium-labeling data cannot be interpreted in a molecular sense, but only in terms of whether or not a base is "perturbed." It is not possible, at present, to assign a particular labeling rate to a particular orientation or interaction of a base, although studies with model compounds containing bases in rigidly fixed orientations may eventually improve this situation. But in spite of this limitation, the method does clearly identify bases that are involved in some kind of structural interaction. Furthermore, if we deal with nucleic acids of known sequence, but unknown secondary or tertiary structure, then the method should be able to eliminate some structural models and to support others. This is because RNA bases that are hydrogen bonded invariably show markedly retarded labeling rates. For example, C. Farber and C. Cantor (in preparation) have recently found that tritium-labeling data on native 5S RNA eliminate many of the proposed 5S RNA structural models and give support to only one.

The tritium-labeling analysis has also been applied to *Escherichia coli* tRNA^{Ile} and *E. coli* tRNA^{Tyr}; although data on these tRNA's corroborate many of the major conclusions obtained with yeast tRNA^{Phe}, they also suggest subtle structural differences may exist between the three tRNA's in solution (106). These subtle structural variations may be of importance for proteins such as aminoacyl tRNA synthetases, which must discriminate between tRNA species.

Interaction of tRNA with Aminoacyl tRNA Synthetase and with Ribosomes

The tritium-labeling method is ideally suited for exploring protein-tRNA interactions and also for searching for structural changes in tRNA that occur in particular situations. This is because the method is gentle and therefore can be done without fear of introducing denaturation into the materials (such as an aminoacyl tRNA synthetase) under investigation. It is also particularly useful for studying macromolecular complexes because most physical methods are inherently unable to give precise structural information on these systems in solution. For example, in the case of NMR, transverse relaxation times for a protein-nucleic acid complex are usually so short that spectral lines are too broad to be interpreted in a meaningful way (113).

One synthetase-tRNA complex has been studied in detail by the tritium-labeling method (107). In this case, five bases in the tRNA structure were found to be strongly perturbed by the bound synthetase. In each case, tritium incorporation into the base was blocked by the bound enzyme. All five of these bases were in sections of the structure that other studies had identified as important for synthetase-tRNA interactions.

In a more recent investigation, C. Farber and C. Cantor (in preparation) have investigated structural alterations in tRNA upon interaction with the ribosomes. One of the major unanswered questions is whether, and how, the tRNA molecule undergoes conformational changes during the course of protein synthesis. These authors have found that interaction of tRNA with ribosomes is accompanied by tRNA conformational changes, as elucidated by the tritium-labeling approach. Specifically, G57 (Figure 2), which (as mentioned above) in the free tRNA has a strongly retarded labeling rate owing to its participation in tertiary interactions, has an approximately twofold faster H-8 exchange in the presence of the 70S ribosome particle than as an isolated free tRNA species. Other data show that exchange is retarded, relative to free tRNA, at the 3'-end (Fragment 6), anticodon region (Fragment 13), and TψC stem area (Fragments 11 and 12). They have interpreted their findings as due, in part, to a structural alteration in tRNA that occurs upon interaction with the ribosome (C. Farber and C. Cantor, in preparation). Further studies of this nature will no doubt shed more light on the flexibility of tRNA as it participates in various physiological interactions.

INTERACTION OF IONS WITH TRANSFER RNA

The conformation of tRNA is well known to be sensitive to the concentration of cations (11, 12, 79). For this reason, considerable effort has been directed at understanding the effect of ion concentration on tRNA conformation. While it is obvious that positive counter ions must be present to stabilize the folded structure of a negatively charged polyelectrolyte, the challenge has been to understand how cations stabilize the particular folded form of tRNA that is adopted, and also to understand the rather unusual ion binding characteristics that are observed in solution. These two issues are closely related and considerable progress has been made in understanding both of them.

Experimental Results on tRNA in Solution

Studies of ion interactions have been typically done with Mg^{2+}, Mn^{2+}, spermine, and spermidine (5, 10, 15, 16, 44, 52, 69, 74, 75, 81, 84, 88,

95, 99, 109, 110, 116, 121, 124, 125). In addition, the interactions of monocovalent cations such as Na^+ and of other ions have also been explored (11, 12, 59, 63, 96, 126). As explained below, examination of the literature suggests significant discrepancies between the various studies. However, these apparent discrepancies are due in large part to the great sensitivity of the ion equilibria (with tRNA) in the conditions employed, and there have now been a sufficient number of investigations to give a fairly consistent picture of the ionic interactions.

A convenient way to catalog the metal ion-binding data is to make use of the standard Scatchard analysis (100a). If there are two classes of "identical," independent binding sites, with n_1 sites in Class 1 and n_2 sites in Class 2, then the moles of ion bound per mole of tRNA (ν) is related to the free ligand concentration C by the equation

$$\frac{\nu}{C} = \frac{\frac{n_1}{k_1}}{1 + \frac{C}{k_1}} + \frac{\frac{n_2}{k_2}}{1 + \frac{C}{k_2}} \qquad 4.$$

where k_1 and k_2 are microscopic dissociation constants for Class 1 and Class 2 sites, respectively. We arbitrarily define Class 1 sites as the strongest binding sites on the molecule. Obviously, additional terms could be added corresponding to additional classes of independent sites. This equation can be used to describe the absorption of ions to tRNA in certain situations.

However, in other studies one group of sites is found to be interacting and, therefore, not independent. These are invariably the strongest binding sites and, by the convention we follow here, are designated as Class 1 sites. (We use a prime on n_1 and k_1 to denote that these parameters now apply to interacting sites.) In these instances, Equation 4 is replaced by an alternate expression (109)

$$\frac{\nu}{C} = \frac{\frac{n_1'^{\alpha}}{k_1'^{\alpha}} C^{\alpha-1}}{1 + \frac{C^{\alpha}}{k_1'^{\alpha}}} + \frac{\frac{n_2}{k_2}}{1 + \frac{C}{k_2}} + \cdots \qquad 5.$$

where n_1' is the number of interacting sites and the apparent dissociation constant $k_1'^{\alpha}$ is given by

$$k_1'^{\alpha} = \frac{C^{\alpha}(n_1' R_0 - C_b)}{C_b} \qquad 6.$$

In this expression, C_b represents the concentration of bound ligand and

R_0 is the total tRNA concentration. The exponent α is in the range $1 \leq \alpha \leq n_1'$; if the system is infinitely cooperative, then $\alpha = n_1'$; otherwise $\alpha < n_1'$. In this situation, the strongest binding sites are responsible for a cooperative association of the first few ions bound to tRNA and the remaining ions bind to a second class of identical and independent sites and possibly even a third class of sites which is also noninteracting (109).

It should be emphasized that these equations are, essentially, semiempirical and as such provide little insight into the actual mechanism of the binding equilibria. Nevertheless, in order to catalog the information and to compare results of various investigations, they provide a useful common framework.

Table 1 summarizes results obtained by a variety of investigators. Two or three classes of sites are generally observed. In many cases, binding in the Class 1 category is highly cooperative and this is indicated in the table. It is apparent that for the ions Mg^{2+} and Mn^{2+} at least 25 sites are generally observed. A smaller number of sites are observed with spermine and spermidine, and this is consistent with the larger charges on these cations (approximately $+3$ and $+4$, respectively).

What is striking is the variable results on the nature of the binding to the Class 1 sites. Of those investigations which report cooperativity in Class 1 for binding of Mn^{2+} or Mg^{2+}, there is generally a concensus that the number of such sites is 4 to 6. The dissociation constant for this cooperative association tends to fall in the range of $1-10^2$ μM, with the range of values undoubtedly reflecting the different ionic conditions employed (see below). In cases where no cooperativity is associated with Class 1 binding, there are typically 1-6 such sites identified.

In situations where cooperativity is observed, care has generally been taken to use buffers of low ionic strength and, additionally or alternatively, buffers that have a cation form which is bulky and therefore should associate weakly with phosphate groups. This is the case with a quaternary amine such as triethanolamine. In this case, the ion under investigation (Mg^{2+} or Mn^{2+}) is faced with much less competition from ions in the buffer solution. In such situations, one studies an ion binding to tRNA amidst a background where binding of endogenous buffer species has been suppressed.

There is good evidence that the cooperative ion binding associated with Class 1 is coupled to a conformational change in the tRNA and that after the Class 1 cooperative sites are filled, the native tertiary structure has been assembled. This was first shown by equilibrium and kinetic studies, which utilize a site-specific probe to follow structural

Table 1 Parameters for ion binding to transfer RNA

tRNA	Conditions	Ion	Class 1 sites			Class 2 sites			Comments[c]	Ref.
			n_1	$k_1, \mu M$		n_2	$k_2, \mu M$			
E. coli unfractionated	50 mM Tris-Cl, 50 mM KCl, pH 7.5, 22°	Mn^{2+}	6	50		~18	—		C	10
E. coli unfractionated[a]	0.1 M triethanol-amine, pH 7.8, 25°	Mn^{2+}	6	2.6		~25	200		C	15
Yeast Phe	5 mM Na phosphate, 5 mM NaCl, pH 7.2, 25°	Mg^{2+}	4	1		20	91		NC	88
Yeast Phe	0.1 M triethanol-amine, pH 8.15, 0°	Mn^{2+}	5	3.7		12[b]	16[b]		C	109
Yeast Phe	0.1 M triethanol-amine, pH 8.35, 0°	Spermidine	5	4.8		~11	19		C	109
Yeast Phe	0.1 M triethanol-amine, pH 8.35, 0°	Spermine	4	0.27		~9	1.4		C	109
Yeast Phe	10 mM Na cacodylate, 22 mM NaCl, pH 6, 30°	Mg^{2+}	6.5	11		17	167		NC	95
E. coli Met$_n$	170 mM Na^+, phospate-cacodylate buffer, pH 7.0, 4°	Mg^{2+}	1	34		26	2400		NC	116
E. coli Glu$_2$ (Mixture of "denatured" and "active" conformers)	20 mM Na cacodylate, 80 mM NaCl, pH 7.0, 34°	Mg^{2+}	4	150		36	1800		C	5
E. coli Glu$_2$ ("Native" conformer)	20 mM Na cacodylate, 80 mM NaCl, pH 7.0, 4°	Mg^{2+}	1	13.3		36	1200		NC	5

[a] Data also obtained on specific tRNA's.
[b] A third class (of weaker sites) was also observed.
[c] C = Cooperativity; Nc = No Cooperativity.

changes in the tRNA associated specifically with cooperative ion association (74,75). From these investigations and from other studies involving tRNA fragments and tRNA molecules that had chain breaks (109), there seems to be little doubt that cooperativity is associated with structural changes needed to produce the native tertiary structure. Further support for these ideas was obtained in subsequent investigations (5,95). For example, studies with *E. coli* tRNA$_2^{Glu}$ showed that in 0.1 M Na$^+$ there is no cooperativity associated with binding of Mg^{2+} to the native structure (5). However, if a mixture of the "denatured" conformer (an ordered structure different from the native one) and "native" conformer is studied, cooperativity in Mg^{2+} binding can be observed. Experiments of a somewhat different kind, but leading to the same conclusion, have also been done (95).

Thus, cooperativity is associated with the assembly of the native tertiary structure; the starting material is probably one in which cloverleaf helical hydrogen bonds are intact, but the precise and specific interactions between many of the single-stranded regions (particularly in the dihydrouridine loop and TψC loop) that generate the L-shaped native tertiary structure are absent. It is likely that many of the transitions from alternate structures to the proper one (with its various tertiary interactions) will be associated with cooperative ion binding. This is because proper folding depends critically upon the binding (in concert) of ions to specific sites which, when occupied, strongly stabilize the native tertiary structure.

There are several explanations for the failure to observe cooperativity in certain studies. First, if counter ions such as Na$^+$ are present in concentrations as high as 0.1 M or more, it is likely that the tRNA structure will be stabilized in its native conformation even in the absence of a divalent cation such as Mg^{2+} or Mn^{2+} (11, 12, 74). In these cases it is doubtful that cooperativity can be observed. A second contributing factor is the hydrogen ion concentration. Careful studies of the effect of pH on cooperative Mg^{2+} binding have shown that, in low ionic strength buffers, Mg^{2+} binding is weaker and less cooperative as the pH is lowered (75). For example, at approximately 40 mM Na$^+$, cooperative Mg^{2+} binding can be observed at pH 7.5 but cannot be seen at pH 6. Moreover, the binding is at least an order of magnitude stronger at pH 7.5. This is apparently due to abnormal base protonations in the tRNA caused by low ionic strengths (75). The notion is that, as a compensation for the high density of negative charges on the macromolecule, sites such as N-3 on cytidine can have highly elevated pKs. The result of these protonations is to make the magnesium-induced conformational change associated with the breakdown of the low salt structure somewhat less facile and, consequently, both the cooperativity and affinity of Mg^{2+} binding are affected (75).

Finally, even when conditions should be favorable for observing cooperative association, it can be missed if the Scatchard analysis is not carried out to experimental values of ν close to 0 (109). This is because, in the Scatchard analysis, a humped ν/C versus C plot is obtained at low values of ν when cooperative binding occurs. If the lowest value of ν examined is 3 or 4, then the peak in the Scatchard plot, which generally occurs in this area, can be missed. This possibly explains why one of the low ionic strength studies shown in Table 1 did not observe cooperative binding.

In summary, the apparently disparate results of various investigators, with respect to the issue of cooperativity in the Class 1 sites, can be rationalized. It is clear that cooperativity is associated with a crucial structural change that involves the binding of ions to sites which must be occupied in order for the proper tertiary structure to form. The establishment of this point has been a major benefit of the studies conducted under conditions where cooperativity can be observed.

The occurrence of structural changes in concert with cooperative ion association underscores the semiempirical nature of the Scatchard analysis of the binding. A more accurate mechanistic picture has been derived by utilizing kinetic data, in conjunction with equilibrium titrations, so that the cooperative ion association has been partially subdivided into sequential elementary steps (74). An analysis incorporating a Monod-Wyman-Changeaux type of "allosterism" has also been proposed (5).

Class 2 sites are invariably noninteracting and, to a significant extent, are probably associated with the binding of ions to sites in helical regions (109). Some of the quantitative differences observed in various investigations can probably be ascribed to effects of ionic strength and pH. Finally, it should be mentioned that the temperature dependence of the binding (as indicated by ΔH) is estimated to be small (68, 80, 88).

Location of Ion-Binding Sites by X-ray Crystallographic Analysis

In recent X-ray crystallographic analyses, several ion-binding sites have been clearly identified (44, 52, 81). In particular, in the monoclinic form, three magnesium-binding sites have been clearly identified (52), and four have been studied in the orthorhombic form (44, 81). The three sites identified in the monoclinic form are the same as three of those found in the orthorhombic crystal. It should be mentioned that, at this point, it is not known whether these sites are the same as the strong binding sites studied in solution (Table 1).

The magnesium ion has six coordination positions and has an octrahedral geometry. The location of the four sites studied in the orthorhombic crystal are illustrated schematically by solid circles in Figure 3 (81). The magnesium ions are hydrated with four, five, or six water molecules; in cases where the coordination with water is less than 6, the remaining positions are linked to phosphate oxygens. In addition, in certain cases the coordinated water molecules are hydrogen bonded to phosphate oxygens or to hydrogen-bonding sites on bases.

One magnesium ion is located at the sharp turn that occurs in traveling sequentially along residues 8, 9, 10, 11, and 12. This turn is at the juncture of the amino acid acceptor helix and dihydrouridine helix. This specific magnesium is believed to be coordinated in all six positions with water and to use these water molecules as bridges to form hydrogen bonds with the phosphates of residues 8, 9, 11, and 12. Such a network of hydrogen bonds could provide firm stabilization of the crucial sharp corner associated with the native tertiary structure. As discussed below, NMR studies show evidence of Mg^{2+} binding in this region.

A second Mg^{2+} is in the dihydrouridine loop where it is directly coordinated to the phosphate oxygen of G19 and hydrogen bonded, through the coordinated water molecules, to the bases G20, U59, and C60. In this way a linkage is established between the dihydrouridine and TψC loops, which is another important feature of the native tertiary structure.

Another Mg^{2+} is also located in the dihydrouridine loop and appears to be directly coordinated to two phosphate oxygens of residues 20 and 21. And finally, there is an additional magnesium located in the anticodon loop. This is believed to be directly coordinated to the phosphate oxygen of Y37, and hydrogen bonded through water molecules to bases C32, Y37, A38, and ψ39. These interactions can impart some rigidity to the anticodon loop.

In the orthorhombic form of yeast tRNA^{Phe}, Mg^{2+} and spermine are simultaneously bound to the tRNA molecule (44, 81). The location of two of the spermine sites is also given in Figure 3. One of these is in the deep groove of the double helix formed by the D stem and anticodon stem. The spermine appears to hydrogen bond to four different phosphate residues located on both sides of the deep groove. This interaction appears to bring together phosphates on opposite sides of the major

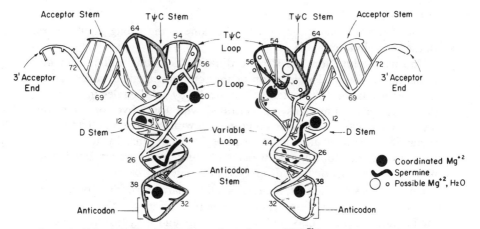

Figure 3 Schematic illustration of two views of yeast tRNA^{Phe} crystal structure (compare with Figure 1). Crossbars denote base pairs and abbreviated bars are single bases. Solid black symbols are bound Mg^{2+} and spermine positions; open symbols are possible Mg^{2+} or H$_2$O binding positions. Adapted with permission from Reference 81.

groove to a distance perhaps as much as 3 Å closer than the analogous distances in the deep groove of the amino acid acceptor-TψC helix. This interaction is believed to stabilize the approximately 25° deviation of the axis of the D stem from that of the anticodon stem, and also serves to impart rigidity to the anticodon end of the molecule.

The second spermine appears to be located at the sharp bend that occurs in the region of residues 8–12. In this area, there is also a rather close approach between the phosphate group in this region with those in the extra loop. For example, phosphate 10 is only 7.2 Å from phosphate 47. There is a string of negative charges in this area, which spermine appears to neutralize as it interacts with phosphates 9, 10, and 11 and phosphates 45, 46, and 47. Thus, the polyamines stabilize an interaction between two widely separated (in the sequence) parts of the ribose-phosphate backbone.

These studies show that Mg^{2+} and spermine play a crucial role in stabilizing the tertiary structure. It is striking that 3 of the 4 Mg^{2+} sites that have been identified occur in single-stranded loops, whereas the 4th is involved in making a sharp turn between two helical axis. Although there is no way unequivocally to identify these sites with those studies by equilibrium-binding measurements in solution, the conclusion (from solution studies) that the cooperative sites are associated with the tertiary interactions is certainly consistent with the picture provided by the crystallographic analysis.

NMR STUDIES

As with many other physical techniques applied to macromolecules, NMR can help confirm solution state structure as inferred from X-ray structure; can indicate new structural features, modes of flexibility and of interaction with cofactors and other macromolecules; and can yield kinetic information. The particular advantage of NMR is that signals are obtainable from reporters on many parts of the molecule simultaneously, and that several different kinds of information are forthcoming, at least in principle, from each group. The number of identifiable signals in tRNA's as compared to proteins is unusually large and they are spread out over the molecule. Furthermore, although NMR is notoriously insensitive, this disadvantage should not be overemphasized; quality spectra are obtainable in, for example, five minutes with ~ 150 nanomoles (5 mg) of material, and useable spectra in less than one minute.

The ultimate objective of the nearly ten years of study of tRNA by NMR has been to obtain biochemically and structurally new information. To some extent this has been achieved, because many features of the structure suggested by sequence, genetic, and X-ray studies have been confirmed. However, much of the NMR work to date has been concerned with identification of specific resonances with specific nuclei in the structure, and more biophysically relevant information has been a byproduct of such studies. It is certain that this technical phase of the subject will soon give way to more problem-oriented experiments, because we now have identified, or have methodology for identifying, a large number and variety of markers throughout the molecule. Furthermore, suitable instrumentation is becoming more widely available and there will soon be a few instruments operating at nearly double the magnetic field recently available, with a modest increase in sensitivity.

Many excellent reviews on different aspects of NMR in tRNA have appeared (8, 12, 41, 57, 64, 78, 85, 86, 91, 93). Spin label studies have also been reviewed (22). We will not recount the entire history of the field, but stress those developments not fully covered in these reviews, with emphasis on kinetics. Because of their particular applicability to tRNA, we also describe (in the next section) several technical aspects of NMR methodology, even though these have been reviewed elsewhere.

NMR Methods

We discuss methods in the context of proton NMR (Figure 4) because most work in tRNA has been limited to this nucleus, owing to the difficulty of procuring samples containing sufficient quantities of other suitable nuclei.

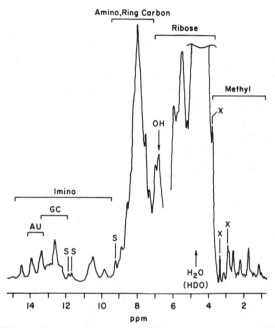

Figure 4 Composite NMR spectrum of yeast tRNA^{Phe} obtained at the Brandeis University 270 MHz spectrometer. The section upfield of 6.5 ppm was run in D$_2$O buffer and that below 6 ppm was run in H$_2$O. Slightly higher quality spectra of the imino and methyl regions have been obtained elsewhere at 360 MHz and can be found in many of the references. Spectra are slightly sensitive to Mn^{++}, as well as to temperature; temperature was 30°C and samples were dialyzed into 5 mM MgCl$_2$ buffer. X's are known instrumental artifacts or trace EDTA; S's are single spin resonances.

The large number of ring carbon, amino, and sugar resonances (7) in the region 9 to 4 ppm has so far rendered these protons unidentifiable and therefore useless; however, selected protons in this region can now be picked out by nuclear Overhauser effect (below) and may be useful in future studies. The region from 4 ppm to 0 ppm is relatively simple and contains methyl resonances that are often identifiable, especially in tRNA's containing few methylated bases (19, 61, 94). All these resonances (except amino resonances and the rapidly exchanging C-8 proton of m^{7}G) are observed by standard Fourier transform (FT) methods.

Spectroscopy of the region below 9 ppm, pioneered by Kearns and Shulman, (8, 12, 41, 64, 78, 91, 93) is especially interesting because this region is relatively uncluttered; it contains a single resonance from each A-U (-13.1 to -14 ppm) and G-C (-11.8 to 13.3 ppm) base pair, and resonances from many, perhaps all, of the non-Watson-Crick (W-C) imino nitrogen protons. Identifications of many of these will be discussed in the next section. Almost all these resonances disappear within minutes when tRNA is dissolved in $D_2O(54)$, so that the buffer must be nearly pure H_2O in order to observe them. This presents a technical problem for conventional FT NMR the 55 molar concentration of H_2O makes detection of the $\gtrsim 10^{-3}$ molar tRNA difficult from an engineering point of view.

Thus the spectrum is best observed by techniques that do not stimulate the H_2O resonance; these have been reviewed elsewhere (83). The method often used in the past was developed by J. Dadok and is called correlation NMR. It consists of excitation by a moderately strong radio frequency field, whose frequency is swept rapidly across the interesting spectral region and not through water, so that the water protons are not much excited or saturated. This mode of excitation produces a complicated signal, but the signal can be disentangled by computer analysis to produce a normal spectrum. More recently (57, 83), a special pulse has been used that flips over the interesting region of the spectrum while not affecting the region close to water. The most useful such pulse yet developed, called a 214 pulse, has been shown (86) to be several times more sensitive than correlation NMR for ordinary spectroscopy, as well as being more useful for kinetic studies.

KINETIC STUDIES Many NMR studies of tRNA have been simple observations of spectra, but kinetic and/or double irradiation methods are becoming more important. Most kinetic studies have measured the rate of exchange of the imino protons with solvent protons, in the hope that this rate reflects the rate of unfolding of local structure and is thus a measure of the relative instability of local structure. Several methods are summarized in Table 2; specific applications will be discussed later.

Table 2 Summary of NMR kinetic methods applicable to tRNA

Name	Quantity measured and range	Method	Remarks
Linewidth	Solvent exchange $50-200$ sec^{-1}	Observation of spectrum as a function of temperature	Usually qualitative
Real-time exchange	Solvent exchange $\gtrsim 10^{-2}$ sec^{-1}	Direct observation after solvent change H_2O to D_2O using microcolumn	High sensitivity essential
Saturation recovery	Solvent exchange $5-200$ sec^{-1}	Pinpoint saturation of specific resonance, then variable delay τ, then observation of saturated line	Pulsed observation required
Nuclear Overhauser effect (NOE)	Proximity < 4Å	Same as saturation recovery but $\tau \simeq 0$, and look elsewhere in spectrum	Qualitative, may get 3rd party effects

Linewidth The easiest way to obtain kinetic information is to study the linewidth versus temperature (13, 41). There are often two additive contributions to the width of an NMR line: a relatively temperature-independent part resulting from purely magnetic interactions with nearby nuclear spins (nitrogens and protons); and a chemical kinetic part that is proportional either to a chemical exchange rate with solvent, in the case of imino protons, or to the rate of a conformational change, in the case of carbon protons. Unfortunately, the linewidth method is often limited to rather qualitative studies, because as soon as an NMR line starts to broaden, it becomes difficult to observe in an overlapping spectrum.

Real-time exchange At much lower temperature, solvent exchange of some amino and imino protons can be observed by rapidly changing the solvent from H_2O to D_2O and observing the subsequent disappearance of resonances of exchangeable protons (54). Because this requires passing the tRNA through a Sephadex microcolumn, it is limited to rates of the order of 1 min^{-1}, and it is also quite laborious as the tRNA must be recycled back to H_2O solvent and reconcentrated for each measurement. Such as experiment is shown in Figure 5.

Saturation-recovery There is no way to use NMR for studies of tRNA rates between 1 min $^{-1}$ and 5 sec^{-1}, but the saturation-recovery method (55–57) can be used for rates greater than about 5 sec^{-1}; a relatively long (>0.1 sec), weak monochromatic pulse is applied at the frequency f_2 of an interesting resonance in the spectrum and saturates[1] the resonance. After a delay τ in the range 0 to 1 sec, the observation pulse is applied and the signal is measured. The experiment is repeated many times at a fixed delay τ, the signal is accumulated, and then Fourier transformed and recorded. If τ is small, the resonance at f_2 will be missing from the spectrum. The entire experiment is repeated for several τ values; for long τ the resonance at f_2 "forgets" that it was saturated and it reappears in the spectrum. The first-order recovery rate (exponential recovery versus τ) of the saturated resonance is extracted from such a series of runs (Figure 6). When this rate is highly temperature dependent, it is assumed to reflect solvent exchange kinetics; the protons resonating at f_2 recover from saturation by being replaced by solvent protons that were unaffected by the saturating pulse.

---

[1] Saturation, in NMR jargon, means the temporary obliteration of a resonance or resonances by on-resonance radio frequency power because of the destruction of the Boltzman population difference between up and down spin populations and resulting destruction of the nuclear spin magnetization which gives rise to the NMR signal.

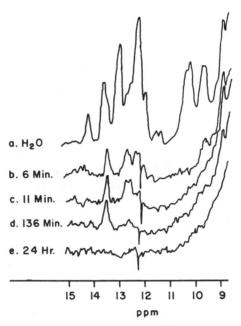

Figure 5 Real-time solvent exchange in yeast tRNA^{Phe} in 15 mM MgCl$_2$, 15°C. (*a*) Spectrum in H$_2$O buffer, at reduced gain; (*b*)–(*e*) Spectra obtained 6 min, 11 min, 136 min, and 24 hours, respectively, after starting to pass the sample through a Sephadex microcolumn equilibrated in D$_2$O. Each spectrum required 5 minutes, but acceptable spectra could now be obtained in one minute. Similar slow exchange is observable in unfractionated yeast tRNA and in the amino region of yeast tRNA^{Phe}. Adapted with permission from Reference 54.

Such measurements must be viewed cautiously because there is another contribution (56, 57) to the rate that was overlooked in an early paper (55) and that masks the chemical kinetics at low temperature; a typical proton in tRNA exchanges energy with its neighboring protons at a rate of 5–10 sec^{-1}, as a result of the magnetic dipolar interaction with these neighbors. This process is often called "spin diffusion" and is nearly temperature dependent. In macromolecules it is strongly analogous to Förster energy transfer between chromophores (29), and like such transfer it occurs at a rate that decreases as the inverse sixth power of the interspin distance.

NUCLEAR OVERHAUSER EFFECT Although this magnetic energy transfer limits the useful range of saturation recovery kinetic measurements to rates greater than about 10 sec^{-1}, it permits another very useful observation (56, 57), namely nuclear Overhauser effect (NOE); saturating one resonance with zero delay τ will partially saturate resonances of

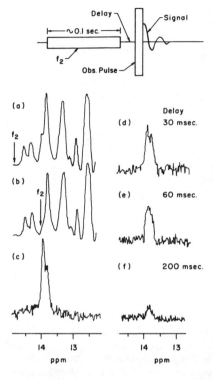

Figure 6 Saturation-recovery sequence and a run on a partially resolved resonance of the same sample as Figure 4. (a) Portion of a spectrum run with preirradiation at a point (indicated by the arrow) chosen not to saturate any resonance. (b) Spectrum obtained with preirradiation at 14 ppm. The intent was to saturate the resonance forming the downfield shoulder of the four-proton peak centered at 13.7 ppm. (c) Difference between (a) and (b). Both (a) and (b) are run with 5 ms delay between the preirradiation and observation pulses. The upfield shoulder in the difference spectrum is present because the large central peak was also partly saturated. (d)–(f) Similar difference spectra obtained with delays of 35, 65, and 205 ms. The downfield shoulder is seen to relax more rapidly than the central peak.

neighboring protons, and these can then be observed, even if they are in a hopelessly crowded region of the spectrum, as small "holes" in the spectrum. In practice this requires difference spectra (Figure 7). NOE spectra tell us that two protons having the observed resonances are physically close (usually within 3.5 Å) to each other in the solution structure. This information can sometimes be used to identify pairs of otherwise unidentifiable resonances (see below). Or, if one resonance of the NOE pair has been identified, it is often easy to identify the other, especially in nucleic acids where the distance between adjacent base pairs are so large that NOE's within single base pairs are dominant. Distances between protons can be estimated by NOE with an accuracy that probably exceeds 10% ($\pm$ .3 Å).

Decoupling and similar experiments have not been generally applied to tRNA because of technical difficulties in performing them in H_2O and because spin-spin couplings are obscured by the relatively large linewidth of the proton lines. However, decoupling has been used to locate dihydrouracil ring carbon proton resonances (94).

PHOSPHORUS AND CARBON   There have been only a few NMR experiments on tRNA using nuclei other than protons; these have used

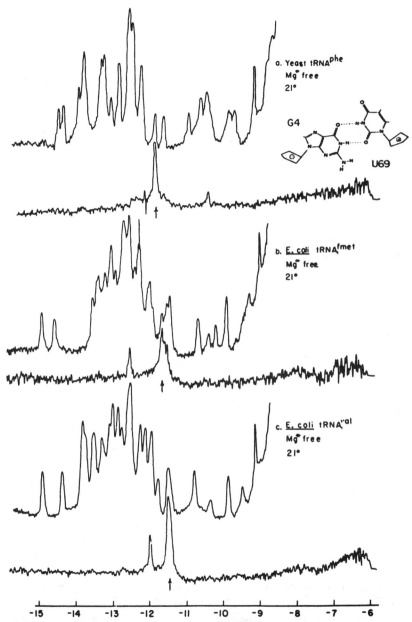

Figure 7 Nuclear Overhauser effect between the two ring N-H-carbonyl protons of GU wobble base pairs. (*a*) Yeast tRNA[Phe] spectrum and difference spectrum (below) obtained as described for Figure 6(*c*), with irradiation at the point indicated by the arrow. The NOE is at 10.4 ppm. (*b*) and (*c*) Similar NOE's for *E. coli* fMet and Val. Adapted with permission from Reference 57.

conventional FT NMR (78). It is relatively easy to obtain a ^{31}P spectrum, and it contains about ten resolved lines, split away from a large central peak that comes from standard W-C helical regions (37, 97). Natural abundance ^{13}C spectra have been obtained only from unfractionated samples (67). Samples enriched with ^{13}C have been produced biosynthetically using suitable auxotrophes (38, 119), and *E. coli* tRNA^{Val} with uridine replaced by 5-fluorouridine has been produced (46). Such samples may be useful for studies of tRNA in functional interaction with purified proteins of the biosynthetic machinery.

Resonance Assignments

Selected assignments of NMR lines are given in Table 3. The most important feature of Table 3 is the extent and variety of the resonances identified, both with respect to chemical nature and to location in the structure. We must make several qualifications concerning this table. Resonances have been listed very roughly in order of their reliability as judged by us. This is a matter of opinion and controversy in some cases, and this table may contain some incorrect assignments. We have left out assignments that seem casual or speculative or that we find unconvincing, but many of these may be correct. The references given are not necessarily those in which an assignment was first made, but rather are the references in which the assignment is fully discussed. Other references given may contain counterproposals for identification.

ASSIGNMENT STRATEGIES Methodology for identifying imino proton resonances has been reviewed elsewhere (except for the use of NOE) and will not be repeated here in detail (8, 64, 91, 93). Except in a few favorable cases involving chemical modification, no single observation can serve to identify a resonance with high probability of correctness, and the earlier literature abounds with "assignments" that were later questioned or shown to be incorrect. In the future, each assignment should be easier than in the past because one can build on past experience. The most detailed studies to date, by Reid, Hurd, McCollum & Ribiero and their collaborators (47, 49, 86, 87, 93), have concentrated on yeast tRNA^{Phe} and on other tRNA's that are expected to have closely analogous tertiary structure. Work on these species has been so extensive that the casual reader can get the impression that each downfield resonance in the spectrum is positively identified. This is by no means true, and there is still disagreement both about individual assignments and about predictive methods. Nevertheless, enough of the spectrum is understood for purposes of further problem-oriented work.

Table 3 Selected assignments of NMR lines in tRNA

Nucleus	Location	tRNA species	Chemical shift[a]	Assignment method[b]	References
^{1}H	S^{4}U8 – A14 Imino	All E. coli	14.8	Chem. modif.	17, 49, 127
	T54 Methyl	All	~1	Compar.	19, 61
	Y37 Ring CCH$_3$	Yeast Phe	2	Chem. modif.	19
	m^{7}G Imino	Yeast Phe[c]	13.4	Chem. modif.	49; 98[d]
	m^{7}G Imino	E. coli fMet	14.55	Chem. modif.	49
	m^{7}G C(8)	Yeast Phe[c]	9.1	Chem. modif.; NOE	49, 120
	m^{7}G Methyl	Yeast Phe[c]	3.9	Compar; NOE	19, 61, U0[e]
	GU Iminos	3 species	10–12	NOE	56, 57
	T54 C(6)	2 species	~7.2	NOE	U0[e]
	U8 – A14 Imino	Yeast Phe[c]	~14.4	Chem. modif.; Compar.	8, 49
	G15 – C48 Imino	Yeast Phe[c]	12.3–12.7	Magn. ion	47
	m^{5}C40,49 Methyl	Yeast Phe	~1.7	Compar.	19, 61
	DHU Ring carbon	Yeast Phe	2.7–4	Decoupl.; Compar.	19, 94
	AΨ31 Imino	Yeast Phe	13.1	Fragments; Buff. Varn.	70
	T54 – m^{1}A58 Imino	Yeast Phe[c]	~14.4	Fragments; Ring calc.	49; 8[d]
	U12 – A23 Imino	Yeast Phe	13.7	Magn. ion	47
	C13 – G22 Imino	Several	12.1	Magn. ion	47
	G19 – C56 Imino	Yeast Phe[c]	~13	Compar.; Buff. varn.; Ring calc.	55, 87
^{31}P	5′Terminal	Yeast Phe	~3.5	pH	37
	p36	Yeast Phe	3.5	Fragments; Chem. modif.	97
^{13}C	DHU C(5) (and others)	Yeast crude	30	Compar.	67
	ΨC(4)	Yeast (enriched)	163	Compar.	38
	T (and others)	E. coli (enriched)	11.9	Compar.	119

[a]Downfield in ppm from 2.2, dimethylsilapentane-5-sulfonate (protons), H$_3$PO$_4$ (phosphorous), or tetramethylsilane (carbon).
[b]Abbreviations used: Chem. modif. = chemical modification; Compar. = comparison between related species; Magn. ion = Magnetic ion shift or broadening; Buff. varn. = unusual sensitivity to temperature or to Mg^{++}; Fragments = fragment studies; Decoupl. = decoupling; Ring calc. = ring current calculation; pH = pH variation; NOE = nuclear Overhauser effect.
[c]And other species with similar local surroundings.
[d]References given after a semicolon give alternate assignments.
[e]U0 = J. Tropp and A. G. Redfield, unpublished observations.

Tertiary resonances The most important step in spectral dissection was to separate tertiary from secondary resonances (64, 87, 93).

It was assumed that tertiary resonances would be more sensitive to temperature and variations in magnesium concentration and that, as indicated by the X-ray structure, standard Watson-Crick helical regions of the molecule would have the same conformation in folded tRNA as in model compounds and fragments, at least for purposes of NMR prediction and with the possible exception of the strongly perturbed D-stem. Thus a subset of the lines in the downfield region were tentatively assigned to tertiary resonances (64, 87, 93) by their sensitivity to temperature and Mg^{++}, and the remaining insensitive peaks to secondary W-C imino protons. Several assignments of tertiary resonances made this way have been confirmed by other methods. The next step was to understand the positions of the latter set of resonances.

Ring current shifts The resonance shift of a secondary proton is assumed to be the sum of an intrinsic shift and a through-space shift, which depends on the relative position and distance of nearest and next nearest neighboring bases in the helix (64, 87, 91, 93). The intrinsic shift is, conceptually, that for the imino proton of an A-U or G-C base pair in a standard helix for which the diamagnetic ring currents of all neighbors are imagined to be turned off; and the through space shift is then the result of the classical magnetic fields produced by these diamagnetic ring currents. The ring currents are estimated from ring carbon proton shifts of uncomplexed monomers (36). The intrinsic shifts combine ring shifts from the donor and acceptor bases with the effect of hydrogen bonding. The through-space shifts can be calculated for yeast $tRNA^{Phe}$ from X-ray coordinates (8, 34, 91); but, according to Reid and coworkers (87, 93), it appears practical to use published tables, which give calculated shift values as a function of nearest and next nearest base pair neighbors for the expected elevenfold helix (2), and to make estimates for base pairs at the ends of helices. The intrinsic shifts have been varied to give the best overall fit to data for intact tRNA's and fragments thereof. Reid, Robillard, and their collaborators (87, 91) find that intrinsic shifts of about -14.35 and -13.4 ppm for AU and GC base pairs, respectively, fit the data best. The through-space shifts range from about $+0.3$ to $+2$ ppm. The positions of GU base pair resonances were predicted using a similar procedure (93), and once these were positively located by NOE (below), their observed positions in the spectrum could be rationalized in detail (34, 57).

Tertiary resonance positions can be predicted in the same way, but there is more uncertainty in the estimate of the intrinsic resonance position. Often it is estimated (49, 87), for modified bases, from a comparison of imino resonance positions with those for unmodified bases when both are dissolved in dry dimethylsulfoxide (these resonances are often too exchange-broadened to observe in water at neutral pH). Once a tertiary resonance has been identified in one species, then ring current shifts are useful as a method for predicting where an analogous resonance will occur in another tRNA, assuming a similar general structure.

Some of these conclusions are disputed by Bolton & Kearns (8), who favor more downfield, intrinsic shifts (-14.5 for AU, -13.6 for GC), which are obtained from model studies (34), and who feel that the tRNA structure is too distorted from the elevenfold RNA helix to use shift rules calculated from it. The most important assignment still under debate is that of the T-A base pair in the TψC loop (8, 49, 64, 87).

Thus it may be fair to say that theoretical estimates of resonance positions are well worth making, but that their main use at this point is to rationalize positions of resonances determined in other ways or to extend such determinations to other tRNA species and as a guide for analysis of NOE and other data. The predictions are hardly tested beyond their ability to predict spectra. It is not yet clear whether remaining disagreements and discrepancies of up to ± 0.3 ppm are a result of poor knowledge of structure or are due to possible variation in the "intrinsic" resonance positions resulting from local distortion of hydrogen bonds. Further experience gained from observations of methyl resonances and ring carbon resonances found by NOE may be illuminating.

Chemical modification Modifications used so far include excision of 7-methyl guanosine (49) and the Y base (18, 19) and modification or labeling of thiouracil (17, 49, 127). Such modifications could disrupt the structure over a widespread region of the molecule, but they do not appear to do so judging from the NMR spectrum. Only individual imino resonances are affected, which are then assigned to the modified bases.

Magnetic ions Another strategy is to look for changes in the spectrum when magnetic ions are added to various *E. coli* tRNA samples already stabilized by Mg^{++} (8, 47, 64, 93). The NMR spectrum shows that at low concentrations, perturbation is localized because only a few lines are affected. When Mn^{++} is added, one of the affected lines is the previously identified resonance of sU8-A14; this establishes the general area of one of the first binding states for Mn^{++}. This also corresponds to the general area of one of the Mg^{2+} sites observed in the X-ray crystallographic analysis (see above). Incidentally, in these experiments, moles of Mn added per mole of tRNA is less than one, yet every tRNA molecule has its NMR lines broadened, which means that Mn is being passed between molecules relatively rapidly, staying less than 10^{-3} sec on each molecule site.

Addition of Co^{++}, at higher concentration ($\gtrsim 250\ \mu M$) than used for the Mn^{++} experiment, produces shifts of several resonances (47). Again, the effects on the spectrum are selective, suggesting that there are only one or two Co^{++} sites and that sU8-A14 is close to one of these. By comparisons of data between several tRNA species and by using information about Co^{++} and Mn^{++} sites from X-ray studies, the tertiary resonance of G15-C48 (a reverse W-C pair) was assigned (47). Much more should be done along these lines, using nuclear spin relaxation-time measurements and careful titrations, to get geometrical information; to establish that there are only one or a few binding sites; and to locate them more positively.

Fragments Studies of homogeneous preparations of fragments of tRNA have proven useful as tests of ring current shift predictions. They have also been used for tertiary assignment, for the case of the T-A imino proton in fragments containing the TψC stem and loop (49), and for A31-ψ39 in the yeast Phe acceptor stem (70). It is assumed that these interactions are essentially the same as in intact tRNA; although this assumption may be challenged, there are observed resonances in the hairpin fragments which correlate well with candidates for the same resonances in intact tRNA.

Overhauser effect An important new method in tRNA NMR is the use of nuclear Overhauser effect (56, 57). NOE has been used to identify several unusual base pairs and will continue to be a very powerful tool in the future. For example, the single GU base pair resonances of three tRNA species were located by correlating their NOE patterns (two NOE-coupled resonances with similar chemical shifts in the upper field part of the imino region, shown in Figure 7) with the unique base pairing of the GU wobble base pairing (two carbonyl-to-ring hydrogen-bonded protons adjacent to each other). This technique is especially powerful when applied to methyl protons because their number is small, even in heavily methylated tRNA's, and all of them have slightly different NOE signatures. Thus, for example, the previously assigned m^7G methyl and C-8 proton resonances have been confirmed by observation of a methyl to C-8 proton NOE (J. Tropp and A. G. Redfield, unpublished).

A recent report (104) shows the aromatic region (6.5 to 8.5 ppm) of a tRNA spectrum that has been computer treated to bring out the narrower lines. It is stressed that the number of such lines is relatively small and tractable and that they probably come only from adenine C-2 protons and from single strand purine C-8 protons (which exchange slowly in D_2O). Observation and identification of many of these lines will be aided by use of NOE and the spin echo technique (9).

Applications of NMR

To some extent, interesting applications of NMR have been embedded in the extensive work devoted primarily to understanding spectra and to identification. Thus the existence of the standard W-C helical regions expected from sequence studies was confirmed in solution and prior to the completion of X-ray diffraction studies. Several important features of the X-ray structure have been shown to exist stably in solution at 20°C; these include the U8-A14 and m^7G tertiary interactions and the G-U wobble pairing in otherwise standard W-C helices. A fair, if perhaps not completely rigorous, generalization is that the local conformations observed by crystallographers are also dominant in solution. The dream of every NMR spectroscopist, to prove a blunder by the crystallographers, has not yet come true for tRNA, as it has for a few protein structures, although at least one structural revision has been proposed (48).

We now review the relatively few recent experiments directed toward other applications; these are in their infancy and only scratch the surface of what should be possible in the future. Most of these experiments are directed toward assaying changes, especially allosteric, in the structure that might occur during normal functions.

The NMR properties of tRNA show numerous minor shifts and other changes when buffer components such as Mg^{++}, other ions, salt, spermine, intercalators, temperature, and pH are varied (4, 6, 8, 37, 53, 58, 64, 87, 93, 117). Many of these have not been studied in much detail and are not yet well understood. Studies of the effect of aminoacylation have been reviewed and discussed previously; no well-established changes are found (8, 12). A recent study of yeast $tRNA^{Phe}$, which has been aminoacylated through a stable amide linkage at the 3' terminal ribose position, shows a change in structure as reported by the ribothymidine methyl resonance (19). This resonance appears in this charged species at 1.3 ppm, which is different from its position in uncharged tRNA at 20°C (1 ppm) but is the position found in the 47-76 fragment and also at intermediate temperatures in Mg^{++}-free uncharged tRNA.

Early NMR studies of tRNA-protein complexes were reviewed previously (8), and no recent studies have appeared. With the recent progress in spectral understanding and other methodology and in instrumentation, this should soon be a lively area of research.

CODON BINDING Geerdes et al (34, 35) have studied the spectra mainly of yeast $tRNA^{Phe}$ in the presence of various oligonucleotides. The methyl resonance of the ring carbon methyl group of the Y base shifts when oligonucleotides bind to the anticodon stem, providing a measure of the binding constant. Shifts are also observed for two phosphate resonances that have been identified with phosphates in the anticodon loop (97). Binding constants estimated in this way are consistent with previous non-NMR measurements. Unexpectedly, UUC, which binds less strongly than several tetra- and pentanucleotides complementary to the anticodon, appears to trigger a dimerization of the tRNA. This is reversed by addition of UUCA, which binds to the anticodon; therefore, the UUC-induced aggregation is associated with interaction of UUC with the anticodon.

At 35°C binding of polynucleotides produces only a small shift in the imino tRNA spectrum of a resonance thought to be from A31-ψ39 at the end of the acceptor stem. However, at 1°C, difference spectra (between tRNA with and without oligonucleotide) show definite peaks in the downfield region when the oligonucleotides UUCA, UUCG, UUCAG, and UUUU are added. These peaks are assigned to imino protons that provide the hydrogen bonds between the anticodon and the oligonucleotide (Figure 8). Previous non-NMR studies showed that UUCAG binds unusually tightly, suggesting that it forms a helix with CmUGmAA of the anticodon loop and that the anticodon loop undergoes a conformation change relative to the crystal structure conformation in order to permit stacking of these bases. The difference spectra

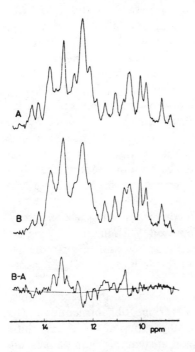

Figure 8 (*A*) 360 MH$_z$ NMR spectrum of the imino region of 1 mM yeast tRNA^{Phe} at 1°C. (*B*) The same, after adding 3.8 mM UUCAG. Under these conditions, 99% of the anticodon sites are thought to be occupied. (*B–A*) Difference between the two spectra. The negative peaks arise from tRNA internal hydrogen bonds whose resonances are shifted by UUCAG. Positive peaks at 13.3 and probably 11.6 ppm are attributed to the imino resonances of base pairs between the oligonucleotide and the tRNA anticodon loop. These sharp differences are real, because they are not observed at 35°C, even though the original spectra are sharp. Adapted with permission from Reference 34.

for UUCAG and UUCA were compared with ring current calculations and are in agreement with this idea. The GU base pair resonance that might be expected in the presence of UUCG was not observed, but it may be too unstable to be visible by NMR.

These workers also searched for evidence of a long-range interaction between the anticodon and the TψC loops. Such an interaction might be expected in view of the finding by Schwarz & Gassen (111) that $(pU)_8$, when complexed to yeast tRNA^{Phe}, induces binding of CGAA, presumably, to TψCG in the TψC loop. The sequence CGAA is found in ribosomal 5S RNA, and such an interaction could occur during protein synthesis. Geerdes & Hilbers could find no evidence for this (34), but their conditions were not precisely the same as those of Schwarz & Gassen because they used UUCA and UUUU instead of $(pU)_8$, and because the method of measurement is not identical.

This conclusion has been confirmed and extended by Davanloo et al (18, 19), who studied the methyl spectrum of yeast tRNA^{Phe} in association with *E. coli* tRNA^{Glu}; these species have mutually complementary anticodons and are known to form complexes that are stable, at ~ 0.5 mM tRNA, up to 70°C. The NMR spectra indicated no major structural shift in the TψC loops; in fact, temperature studies suggested that for yeast tRNA^{Phe} the overall structure is more stable in the complex. Thus there may be a long-range effect, but in the direction opposite to that suggested by Schwarz & Gassen, for this system at least; and the effect is on stability and not structure. The interaction of *E. coli* tRNA^{Glu} with an anticodon fragment excised from yeast tRNA^{Phe} was also investigated. Rather complex shifts were observed that could be explained as structural rearrangement of the anticodon loops. Removal of the Y base also produced interesting shifts in the anticodon loop, but a small decrease in stability of the TψC loop for the uncomplexed modified tRNA relative to the uncomplexed native tRNA.

LOCAL STABILITY AND FLEXIBILITY   A final area of application of NMR is directed toward understanding the relative stability of different parts of the tRNA structure. The finding of the influence of anticodon complexation or modification on stability of the TψC loop, just mentioned above, is an example. These studies use shifts of methyl resonances, observed as the temperature is raised, as markers for structural changes (19, 61, 62, 94). Such shifts can be correlated with optically observed melting equilibria and kinetics and can be used to identify optical transitions at specific locations on the molecule, as reviewed elsewhere (93). Davanloo & Sprinzl and their collaborators (20) have used this method to show that the high stability of tRNA^{fMet} from *Thermus thermophilus* is a result, at least in part, of the thiolation of

s^2T54 in that species. Yokayama et al have used a conformational study of modified 2-thio pyrimidines to reach a similar conclusion about the anticodon region (129).

Study of the methyl region has the advantage that the methyl resonance can be followed, in principle if not always in practice, through the transition so that detailed equilibrium, and limited kinetic, information can be inferred. It has the disadvantage that only a limited number of markers can be followed and that nothing observable happens until the transition occurs at rather high, nonphysiological temperatures.

Solvent exchange These shortcomings do not exist to such an extent for another class of experiments, in which the rate of imino proton exchange with solvent H_2O protons is measured. Earlier experiments have been reviewed by Hilbers (41). The interpretation of such rates is not necessarily simple (77), but the working hypothesis has been that solvent exchange either measures the opening rate of a helical or other structural feature or at least reflects the relative stability of the local structure. In some cases this hypothesis has been confirmed by comparison with optical T-jump studies (13). On the other hand, like any rate measurements, they do not yield rigorous equilibrium information.

The usefulness of this general approach was shown from early studies of several species by Crothers and others (4, 13). These investigators simply observed the spectrum as the temperature was raised. As mentioned above, a proton resonance broadens and soon becomes unobservable when the proton's exchange rate reaches a value of about 200 sec^{-1}. Groups of resonances are observed to disappear when their local structure has a lifetime of less than about 5 msec. Useful information is obtainable if there are separated thermal transitions, as is typical of zero or very low Mg^{++} conditions, or if there is some unusually stable feature in the structure (6, 42, 53, 92).

The advantage of such a study is that it is simple to perform and is potentially able to monitor almost the entire structure. An example is given in Figure 9 of a study by Johnston (53) on yeast tRNA^{Phe} in buffer containing a moderate Mg^{++} concentration. Several resonances persist to high temperature, and two of these have been tentatively identified with base pairs U8-A14 and U12-A23 (see below). This shows that part of the structure is especially stable, most probably the D-stem and the turn that occurs around phosphate 10, which is stabilized by a magnesium ion (see above).

Such studies suffer from the serious limitation that transitions are detectable only at relatively high temperatures where spectra may already be rather confused and overlapping. Johnston and coworkers have shown the usefulness of the other two techniques mentioned earlier

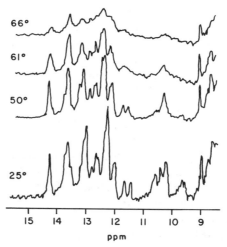

Figure 9 Spectra of tRNA containing 15 mM (total) MgCl$_2$, 7 mM EDTA, 0.1 mM NaCl, and 10 mM cacodylate buffer, pH 7, as a function of temperature. The peak which persists to 66° at 13.7 ppm is believed to be U12-A23 and appears to be the same one which persists for a long time in D$_2$O buffer in Figure 5. The most downfield peak at 66° is one of two coinciding resonances that are seen at the same place at lower temperature. It is tentatively assigned to U8-A14. Adapted with permission from Reference 53.

for rate studies, namely observation of real-time exchange after a change of solvent from H$_2$O to D$_2$O (54) and observation of saturation recovery at higher temperatures (53, 55–57). These techniques signal incipient kinetic instabilities at temperatures where the tRNA is likely to be in its native conformation, well below those temperatures where either optical studies or ordinary NMR spectra show transitions. As a method for correlating kinetics and assigning resonances to common structural regions, activation energies and absolute rates can be measured for the opening of each base pair (12). In principle, intrabase-pair proton transfer (51) might also be observable, but there is no indication that it occurs in tRNA.

Several conclusions could be drawn from the first investigations of this type (53, 55). In zero Mg^{++} for yeast RNA^{Phe} the tertiary structure melts first, followed by the acceptor stem. At moderate Mg^{++} concentration the bulk of the structure melts at about 60° but the D-stem and P-10 loop stabilize each other to higher temperatures, as described above (Figure 9). The assignment of the lowest field resonance in Figure 9 (top) to U8-A14 was accomplished with the aid of prior methods, combined with a detailed study of exchange rates as a function of very low amounts of Mg^{++}, and using the saturation recovery method (53). Real-time measurement at 15°C also showed that the D-stem was probably especially stable (54).

This work also shows that melting of individual helical sections of tRNA is not an all-or-none process, especially at low Mg^{++} and at least as far as NMR kinetics are concerned. Different base pairs assigned to the same helix show different exchange rates, though the rates are generally well within an order of magnitude and appear to have similar activation energies (53, 55). The same behavior is observed in short DNA double helices (41).

Reid & Hurd (86) have given a particularly elegant demonstration of this variation for *E. coli* tRNA^{Phe} in zero Mg^{++} above 50°C (Figure 10). The entire molecule is melted at this temperature, except for the acceptor stem which contains 6 G-C pairs and one A-U. They measured opening rates for several of the individual base pairs of the acceptor stem by saturation recovery and found at least fivefold variations in rates among them.

In concluding this section, it is of interest to compare the likely applications for NMR exchange studies to those for the tritium-labeling studies described earlier that monitor the slow H-8 exchange of adenine and guanine nucleotide units. The tritium method seems most likely to be useful for determining which parts of the molecule are "buried" and "free" in a stable structure and which parts are covered when tRNA is complexed with a protein. There is the great advantage that only small amounts of material are required (less than 1 mg) and that there is no problem in making assignments of exact labeling rates to specific bases in the structure. Thus, two of the drawbacks of NMR—relatively large amounts of sample are required and the difficulty in making assignments of rates to specific residues—are overcome in the H-8 labeling method.

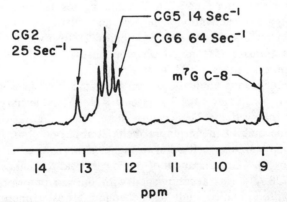

Figure 10 Downfield spectrum of *E. coli* tRNA^{Phe} at 58°C in 10 mM phosphate, 100 mM NaCl, pH 7.0. All imino protons except the acceptor stem have melted at this temperature. Three imino resonances are marked, together with exchange rates measured by the saturation-recovery method. Adapted with permission from Reference 86.

On the other hand, the slow ($10^{-3} - 10^{-4} \text{hr}^{-1}$ at 37°) H-8 exchange is useless for investigating questions on tRNA conformational dynamics, such as breathing rates of helical units or fluctuations in tertiary structure, which occur in times of seconds or less. Because the biological role of tRNA requires it to operate in a series of reactions that only take seconds, structural variations in this time range are of great interest, and here NMR exchange methods are well suited to tackle such problems. Moreover, NMR can also be used to study slower exchanges, such as H-8 exchange, but here the difficulty is in assigning specific resonances to specific residues. In short, for exchange studies the methods complement each other well.

CONCLUSIONS

Research on tRNA in solution continues at a vigorous pace. The major change that has occurred in the past few years is that investigations are now directed at specific molecular questions, such as conformational changes of, or the environment around, a particular residue, as opposed to the necessarily more general and phenomenological questions that were used in the earlier years. Using a battery of NMR methods and the C-8 labeling approach, it is now possible to examine specific residues with respect to their conformational mobility, exposure to the solvent, proximity to a bound protein, and so on. It is through the further expansion and refinement of these kind of studies that we will come to a more complete rational understanding of the behavior of tRNA in solution, particularly with respect to the subtle features that enable it to function in such a precise process as protein synthesis. In addition, the ongoing X-ray crystallographic studies will probably continue to help us understand specific structural aspects of tRNA in solution, as illustrated by recent crystallographic studies on metal ion-binding sites and their role in stabilizing the native structure (see above).

In the future, we expect more emphasis will be placed on understanding exactly what structural modulations take place as tRNA participates in various reactions and what parts of tRNA are used as recognition sites by proteins. NMR methods hold much promise for detecting subtle conformational shifts, such as those that may be crucial for aligning two tRNA molecules side-by-side on the ribosome as is required for peptide bond formation and peptidyl chain transfer. Much of the progress in this area will depend upon our ability to continue to resolve and unambigiously identify specific resonances in NMR spectra. Also, we expect to see more studies of protein-tRNA systems and of tRNA conformational alterations, with the C-8 labeling method. Hopefully, it will be possible to explain the unusual properties of some tRNA mutants, where single base changes can have a profound effect on biological function. Why, for example, should a mutation in the D stem of tRNA^{Trp} change its codon-anticodon reading properties (43)? And why should a $\psi \rightarrow U$ change in the anticodon stem-loop region of tRNA^{His} affect the way this tRNA participates in the regulation of specific gene expression (114)? It is these questions and others which we hope can be answered by further expansion and refinement of methods used for studying tRNA in solution.

ACKNOWLEDGMENT

We thank colleagues who kindly sent us preprints of their unpublished papers. Preparation of this manuscript was supported by United States Public Health Service Grants No. GM-15539 and GM-20168.

Literature Cited

1. Altman, S., ed. 1978. *Transfer RNA.* Cambridge, Mass: MIT Press
2. Arter, D. B., Schmidt, P. G. 1976. *Nucleic Acids Res.* 3:1437–46
3. Barnes, W. M. 1978. *Proc. Natl. Acad. Sci. USA* 75:4281–85
4. Bina-Stein, M., Crothers, D. M., Hilbers, C. W., Shulman, R. G. 1976. *Proc. Natl. Acad. Sci. USA* 70:2216–20
5. Bina-Stein, M., Stein, A. 1976. *Biochemistry* 15:3912–17
6. Bolton, P. H., Kearns, D. R. 1977. *Biochemistry* 16:5129–41
7. Bolton, P. H., Kearns, D. R. 1978. *Nucleic Acids Res.* 5:1315–24
8. Bolton, P. H., Kearns, D. R. 1978. In *Biological Magnetic Resonance,* ed. L. J. Berliner, J. Reuben, pp. 91–137. New York: Plenum
9. Campbell, I. D., Dobson, C. M., Williams, R. J. P., Wright, P. E. 1975. *FEBS Lett.* 57:96–99
10. Cohn, M., Danchin, A., Grunberg-Manago, M. 1969. *J. Mol. Biol.* 39:199–217
11. Crothers, D. M. 1979. See Ref. 102. In press
12. Crothers, D. M., Cole, P. E. 1978. See Ref. 1, pp. 197–247
13. Crothers, D. M., Cole, P. E., Hilbers, C. W., Shulman, R. G. 1974. *J. Mol. Biol.* 87:63–88
14. Cushley, R. J., Lipsky, S. R., Fox, J. J. 1968. *Tetrahedron Lett.* 52:5393–96
15. Danchin, A. 1972. *Biopolymers* 11:1317–33
16. Danchin, A., Guéron, M. 1970. *Eur. J. Biochem.* 16:532–36
17. Daniel, W. E. Jr., Cohn, M. 1975. *Proc. Natl. Acad. Sci. USA* 72:2582–86
18. Davanloo-Malherbe, P., Sprinzl, M., Cramer, F. 1978. In *Nuclear Magnetic Resonance in Molecular Biology,* ed. B. Pulman, pp. 125–36. Dordrecht: Reidel
19. Davanloo, P., Sprinzl, M., Cramer, F. 1979. *Biochemistry* 18: In press
20. Davanloo, P., Sprinzl, M., Watanabe, K., Albani, M., Kersten, H. 1979. *Nucleic Acids Res.* 6:1571–81
21. DiNocera, P. P., Blasi, F., DiLauro, R., Frunzio, R., Bruni, C. B. 1978. *Proc. Natl. Acad. Sci. USA* 75:4276–80
22. Dugas, H. 1977. *Acc. Chem. Res.* 10:47–54
23. Elvidge, J. A., Jones, J. R., O'Brien, C., Evans, E. A., Sheppard, H. C. 1973. *J. Chem. Soc. Perkin Trans.* 2:2138–41
24. Elvidge, J. A., Jones, J. R., O'Brien, C., Evans, E. A., Sheppard, H. C. 1974. *J. Chem. Soc. Perkin Trans.* 2:174–76
25. Englander, S. W. 1963. *Biochemistry* 2:798–807
26. Englander, S. W., Downer, N. W., Teitelbaum, H. 1972. *Ann. Rev. Biochem.* 41:903–24
27. Deleted in proof
28. Fink, R. M. 1965. *Arch. Biochem. Biophys.* 107:493–98
29. Förster, T. 1964. In *Modern Quantum Chemistry,* ed. D. Sinanoglu, 3:93–137. New York: Academic
30. Gamble, R. C. 1975. *I. Transfer RNA conformation in solution in-*

vestigated by isotope labeling. II. Relaxation kinetics of the order-disorder transition in phospholipid bilayers. PhD thesis. MIT, Cambridge, Mass. 220 pp.
31. Gamble, R. C., Schimmel, P. R. 1974. *Proc. Natl. Acad. Sci. USA* 71:1356–60
32. Gamble, R. C., Schoemaker, H. J. P., Jekowsky, E., Schimmel, P. R. 1976. *Biochemistry* 15:2791–99
33. Gardner, J. F. 1979. *Proc. Natl. Acad. Sci. USA* 76:1706–10
34. Geerdes, H. A. M. 1979. *NMR study of tRNA structure.* PhD thesis. The Catholic Univ., Nijmegen. 138 pp.
35. Geerdes, H. A. M., Van Boom, J. H., Hilbers, C. W. 1978. *FEBS Lett.* 88:27–32
36. Geissner-Prettre, C., Pullman, B. 1970. *J. Theor. Biol.* 27:87–95
37. Guéron, M., Shulman, R. G. 1975. *Proc. Natl. Acad. Sci. USA* 72:3482–85
38. Hamill, W. D. Jr., Grant, D. M., Horton, W. J., Lundquist, R., Dickman, S. 1978. *J. Am. Chem. Soc.* 98:1276–78
39. Hawkins, E. R., Chang, S. H. 1974. *Nucleic Acids Res.* 1:1531–38
40. Heller, S. R. 1968. *Biochem. Biophys. Res. Commun.* 32:998–1001
41. Hilbers, C. W. 1979. In *Magnetic Resonance Studies in Biology*, ed. R. G. Shulman. New York: Academic. In press
42. Hilbers, C. W., Robillard, G. T., Shulman, R. G., Blake, R. D., Webb, P. K., Fresco, R., Riesner, D. 1976. *Biochemistry* 15:1874–82
43. Hirsh, D. 1970. *Nature* 228:57–58
44. Holbrook, S. R., Sussman, J. L., Warrant, W., Church, G. M., Kim, S.-H. 1977. *Nucleic Acids Res.* 4:2811–20
45. Holley, R. W., Apgar, J., Everett, G. A., Madison, J. T., Marquise, M. Merrill, S. H., Penwick, J. R., Zamir, A. 1965. *Science* 147:1462–65
46. Horowitz, J., Ofengand, J., Daniel, W. E. Jr., Cohn, M. 1977. *J. Biol. Chem.* 252:4418–20
47. Hurd, R. E., Azhderian, E., Reid, R. E. 1979. *Biochemistry* 18:4012–17
48. Hurd, R. E., Reid, B. R. 1977. *Nucleic Acids Res.* 4:2747–55
49. Hurd, R. E., Reid, B. R. 1979. *Biochemistry* 18:4005–11
50. Iida, S., Wataya, Y., Kudo, I., Kai, K., Hayatsu, H. 1974. *FEBS Lett.* 39:263–66
51. Iwashi, H., Kyogoku, Y. 1978. *Nature* 271:277–78
52. Jack, A., Ladner, J. E., Rhodes, D., Brown, R. S., Klug, A. 1977. *J. Mol. Biol.* 111:315–28
53. Johnston, P. D. 1979. *Proton FT NMR studies of tRNA and dynamics.* PhD thesis. Brandeis Univ., Waltham, Mass.
54. Johnston, P. D., Figueroa, N., Redfield, A. G. 1979. *Proc. Natl. Acad. Sci. USA* 76:3130–34
55. Johnston, P. D., Redfield, A. G. 1977. *Nucleic Acids Res.* 4:3599–3615
56. Johnston, P. D., Redfield, A. G. 1978. *Nucleic Acids Res.* 5:3913–27
57. Johnston, P. D., Redfield, A. G. 1979. See Ref. 102, pp. 191–206
58. Jones, C. R., Bolton, P. H., Kearns, D. R. 1978. *Biochemistry* 17:601–7
59. Jones, C. R., Kearns, D. R. 1974. *Proc. Natl. Acad. Sci. USA* 71:4237–40
60. Kalman, T. I. 1971. *Biochemistry* 10:2567–73
61. Kan, L. S., T'so, P. O. P., Sprinzl, M., Haar, F. V. D., Cramer, F. 1977. *Biochemistry* 16:3143–54
62. Kastrup, R. V., Schmidt, P. G. 1978. *Nucleic Acids Res.* 5:257–59
63. Kayne, M. S., Cohn, M. 1974. *Biochemistry* 13:4159–65
64. Kearns, D. R., Bolton, P. H. 1978. In *Biomolecular Structure and Function*, ed. P. Agris, pp. 493–516. New York: Academic
65. Kim, S. H., Suddath, F. L., Quigley, G. J., McPherson, A., Sussman, J. L., Wang, A. H. J., Seeman, N. C., Rich, A. 1974. *Science* 185:435–39
66. Kim, S. H., Sussman, J. L., Suddath, F. L., Quigley, G. J., McPherson, A., Wang, A. H.-J., Seeman, N. C., Rich, A. 1974. *Proc. Natl. Acad. Sci. USA* 71:4970–74
67. Komoroski, R. A., Allerhand, A. 1974. *J. Am. Chem. Soc.* 13:369–72
68. Krakauer, H. 1971. *Biopolymers* 10:2459–90
69. Leroy, J.-L., Guéron, M., Thomas, G., Favre, A. 1977. *Eur. J. Biochem.* 74:567–74
70. Lightfoot, D. R., Wong, K. L., Kearns, D. R., Reid, B. R. 1973. *J. Mol. Biol.* 78:71–79
71. Litt, M. 1971. *Biochemistry* 10:2223–27
72. Litt, M., Greenspan, C. M. 1972. *Biochemistry* 11:1437–42
73. Livramento, J., Thomas, G. J. Jr. 1974. *J. Am. Chem. Soc.* 96:6529–31
74. Lynch, D. C., Schimmel, P. R. 1974. *Biochemistry* 13:1841–51

75. Lynch, D. C., Schimmel, P. R. 1974. *Biochemistry* 13:1852–61
76. Maeda, M., Saneyoshi, M., Kawazoe, Y. 1971. *Chem. Pharm. Bull.* 19:1641–46.
77. Mandal, C., Kallenbach, N. R., Englander, S. W. *Biochemistry* 18; In press
78. Patel, D. J. 1978. *Ann. Rev. Phys. Chem.* 29:337–62
79. Potts, R. O., Wang, C. C., Fritzinger, D. C., Ford, N. C. Jr., Fournier, M. J. 1979. See Ref. 102. In press
80. Privalov, P. L., Filimonov, V. V., Venkstern, T. V., Bayev, A. A. 1975. *J. Mol. Biol.* 97:279–88
81. Quigley, G. J., Teeter, M. M., Rich, A. 1978. *Proc. Natl. Acad. Sci. USA* 75:64–68
82. RajBhandary, U. L., Chang, S. H. 1968. *J. Biol. Chem.* 243:598–608
83. Redfield, A. G. 1978. *Methods Enzymol.* 49:253–70
84. Reeves, R. H., Cantor, C. R., Chambers, R. W. 1970. *Biochemistry* 9:3993–4001
85. Reid, B. R., Hurd, R. E. 1977. *Acc. Chem. Res.* 10:396–402
86. Reid, B. R., Hurd, R. E. 1979. See Ref. 102. In press
87. Reid, B. R., McCollum, L., Ribiero, N. S., Abbate, J., Hurd, R. E. 1979. *Biochemistry* 18:3996–4005
88. Rialdi, G., Levy, J., Biltonen, R. 1972. *Biochemistry* 11:2472–79
89. Rich, A., RajBhandary, U. L. 1976. *Ann. Rev. Biochem.* 45:805–60
90. Robertus, J. D., Lander, J. E., Finch, J. T., Rhodes, D., Brown, R. S., Clark, B. F. C., Klug, A. 1974. *Nature* 250:546–51
91. Robillard, G. T. 1977. In *NMR in Biology*, ed. R. P. Dwek, I. D. Campbell, R. E. Richards, R. J. P. Williams, pp. 201–30. London: Academic
92. Robillard, G. T., Hilbers, C. W., Reid, B. R., Gangloff, J., Dirheimer, G., Shulman, R. G. 1976. *Biochemistry* 15:1883–89
93. Robillard, G. T., Reid, B. R. 1979. In *Magnetic Resonance Studies in Biology*, ed. R. G. Shulman. New York: Academic. In press
94. Robillard, G. T., Tarr, C. E., Vosman, F., Reid, B. R. 1977. *Biochemistry* 16:5261–73
95. Romer, R., Hach, R. 1975. *Eur. J. Biochem.* 55:271–84
96. Rordorf, B. F., Kearns, D. R. 1976. *Biopolymers* 15:1491–1504
97. Salemink, P. J. M., Swarthof, T., Hilbers, C. W. 1979. *Biochemistry* 18: In press
98. Salemink, P. J. M., Yamane, T., Hilbers, C. W. 1977. *Nucleic Acids Res.* 4:3727–51
99. Sander, C., Ts'o, P. O. P. 1971. *J. Mol. Biol.* 55:1–21
100. Santi, D. V., Brewer, C. F. 1968. *J. Am. Chem. Soc.* 90:6236–38
100a. Scatchard, G., Coleman, J. S., Shen, A. L. 1957. *J. Am. Chem. Soc.* 79:12–20.
101. Schimmel, P. R. 1979. *CRC Critical Reviews in Biochemistry*. In press
102. Schimmel, P. R. 1979. In *Transfer RNA Part 1: Structure, Properties, and Recognition*, ed. P. Schimmel, D. Söll, J. Abelson. New York: Cold Spring Harbor Lab. In press
103. Schimmel, P. R., Schoemaker, H. J. P. 1979. *Methods Enzymol.* 59:332–50
104. Schimidt, P. G., Kastrup, R. V. 1978. See Ref. 64, pp. 517–25
105. Schoemaker, H. J. P. 1975. *Investigations of structure-function relationships of transfer RNA's and their complexes with aminoacyl transfer RNA synthetases*. PhD thesis. MIT, Cambridge, Mass. 303 pp.
106. Schoemaker, H. J. P., Gamble, R. C., Budzik, G. P., Schimmel, P. R. 1976. *Biochemistry* 15:2800–3
107. Schoemaker, H. J. P., Schimmel, P. R. 1976. *J. Biol. Chem.* 251:6823–30
108. Schoemaker, H. J. P., Schimmel, P. R. 1977. *Biochemistry* 16:5454–60
109. Schreier, A. A., Schimmel, P. R. 1974. *J. Mol. Biol.* 86:601–20
110. Schreier, A. A., Schimmel, P. R. 1975. *J. Mol. Biol.* 93:323–29
111. Schwartz, U., Gassen, H. G. 1977. *FEBS Lett.* 78:267–70
112. Shapiro, R., Servis, R. E., Welcher, M. 1970. *J. Am. Chem. Soc.* 92:422–24
113. Shulman, R. G., Hilbers, C. W., Söll, D., Yang, S. K. 1974. *J. Mol. Biol.* 90:609–11
114. Singer, C. E., Smith, G. R., Cortese, R., Ames, B. N. 1972. *Nature New Biol.* 228:72–74
115. Söll, D., Abelson, J., Schimmel, P. R., eds. 1979. In *Transfer RNA Part 2: Biological Aspects*, New York: Cold Spring Harbor Lab. In press
116. Stein, A., Crothers, D. M. 1976. *Biochemistry* 15:157–60
117. Steinmentz-Kayne, M., Benigno, R., Kallenbach, N. R. 1977. *Biochemistry* 16:2064–73
118. Tomasz, M., Olson, J., Mercado, C. M. 1972. *Biochemistry* 11:1235–41
119. Tompson, J. G., Hayashi, N., Paukstelis, J. V., Loeppky, R. N., Agris, P. F. 1979. *Biochemistry*

18:2079-85
120. Deleted in proof.
121. Vournakis, J. N., Scheraga, H. A. 1966. *Biochemistry* 5:2997-3005
122. Wataya, Y., Hayatsu, H., Kawazoe, Y. 1972. *J. Am. Chem. Soc.* 94:8927-28
123. Wechter, W. J. 1970. *Collect. Czech. Commun.* 35:2003-17
124. Willick, G. E., Kay, C. M. 1971. *Biochemistry* 10:2216-22
125. Willick, G., Oikawa, K., Kay, C. M. 1973. *Biochemistry* 12:899-904
126. Wolfson, J. M., Kearns, D. R. 1974. *J. Am. Chem. Soc.* 96:3653-54
127. Wong, K. L., Bolton, P. H., Kearns, D. R. 1979. *Biochim. Biophys. Acta* 383:446-51
128. Woodward, C. K., Hilton, B. D. 1979. *Ann. Rev. Biophys. Bioeng.* 8:99-127
129. Yokoyama, S., Yamaizumi, Z., Nishimura, S., Miyazawa, T. 1979. *Nucleic Acids Res.* 6:2611-26
130. Zurawski, G., Brown, K., Killingly, D., Yanofsky, C. 1978. *Proc. Natl. Acad. Sci. USA* 75:4271-75
131. Zurawski, G., Elseviers, D., Stauffer, G. V., Yanofsky, C. 1978. *Proc. Natl. Acad. Sci. USA* 75:5988-92

NERVE GROWTH FACTOR: MECHANISM OF ACTION

♦9146

Stanley Vinores and Gordon Guroff

Section on Intermediary Metabolism, Laboratory of Developmental Neurobiology, National Institute of Child Health and Human Development, National Institutes of Health, Bethesda, Maryland 20205

It is now more than thirty years since the discovery of the nerve growth factor (NGF), and although its chemistry has been almost completely elucidated, its mechanism of action remains one of the central problems in neurobiology. It is clear that nerve growth factor is required for the survival and development of certain sympathetic and sensory neurons. It is equally clear that nerve growth factor affects a wide variety of other cells as well. Indeed, the multiple cellular effects of nerve growth factor have complicated the study of its action. Recent experiments showing the presence of nerve growth factor receptors on the nucleus, on the plasma membrane, and almost certainly at the synaptic ending, as well, have made even its site of action unsure. In this review we try to present the current status of research on the nerve growth factor and to suggest at least the outlines of the possible mechanisms of action.

HISTORICAL

The original studies by Bueker in 1948 (30, 31) showed that fragments of certain tumors, specifically the mouse Sarcomas 180 or 37, implanted into chick embryos provoked a massive outgrowth of fibers from the sensory and sympathetic ganglia and an innervation of the tumor. Further investigation showed that the fiber outgrowth occurred even in ganglia which were widely separated from the site of implantation and which did not innervate the tumor. This finding suggested that the tumor was elaborating some material which acted on the ganglia, and the conclusion was confirmed by experiments of Levi-Montalcini &

[1]The US Government has the right to retain a nonexclusive, royalty-free license in and to any copyright covering this paper.

Hamburger (109, 110), in which the tumor was grafted onto the chorioallantoic membrane so that no direct contact between tumor and ganglia was possible. In these latter experiments also, the sympathetic and sensory nervous systems enlarged and elaborated a profuse outgrowth.

In order to purify the material an in vitro assay was developed. In this assay (111) sensory ganglia from 7- to 9-day-old chick embryos are explanted onto a plasma clot and a sample of the material to be assayed is included in the surrounding medium. The presence of nerve growth factor in the sample elicits a dense "halo" of neurite outgrowth. Using such an assay, Cohen attempted to fractionate extracts of the appropriate tumors and to isolate the factor responsible for the outgrowth. During the purification he used the lytic agent snake venom and found, unexpectedly, that the venom contained more activity than the tumor itself (41). In a systematic search of related tissues, he observed that the richest source of nerve growth factor was the submaxillary gland of the mature male mouse (42). This remains to the present day the tissue of choice for the preparation of the various forms of nerve growth factor.

The physiological significance of the nerve growth factor was established convincingly through studies of the nerve growth factor antibody. When purified nerve growth factor is injected into rabbits, the rabbits produce an antiserum. This antiserum, when administered to young rats or mice, causes a destruction or an atrophy of the sympathetic neurons (105). This deletion is irreversible and is the opposite of what happens to the animal upon treatment with nerve growth factor itself, namely, hypertrophy of the sympathetic neurons. The condition, known as immunosympathectomy, results presumably from a removal of the endogenous nerve growth factor, which is required if the sympathetic nervous system of the young animal is to survive.

The nerve growth factor was purified to homogeneity in two different forms. A high molecular weight form, or 7S, was prepared by Varon et al (187), and a low molecular weight form, or 2.5S, by Bocchini & Angeletti (25). The amino acid sequence of nerve growth factor was elucidated by Angeletti & Bradshaw (11), who have also explored the relationship of the structure of this molecule to that of other biologically active peptides. Recently, the nerve growth factor has been crystallized (204), paving the way for a complete description of its three-dimensional structure.

FORMS, SOURCES, AND STRUCTURE

The nerve growth factor is a small basic protein of molecular weight 13,259. The chain consists of 118 amino acids, has three intrachain disulfide bonds, and is basic, owing to the presence of a number of amide groups in the molecule. It is found as a noncovalently linked dimer in the salivary gland and can be isolated in milligram quantities from a few hundred grams of this tissue. The dimer is the active form and its activity is not diminished by chemical cross-linking (167).

The nerve growth factor can be isolated from salivary gland in two different forms. The 2.5S, or low molecular weight form (25), is the active dimer. The high molecular weight form of nerve growth factor, called the 7S (186, 187), contains, in addition to the dimer itself, two other types of subunits. The nerve growth factor here is called the β-subunit; the others are the α and the γ. The 7S nerve growth factor is composed of two α-subunits, one β-dimer, and two γ-subunits with molecular weights of 27,000, 26,500, and 25,500, respectively (74,166). The overall molecular weight of the complex is about 130,000. The β-dimer is identical to the 2.5S nerve growth factor, except that many of the chains of the 2.5S sustain minor proteolytic damage during the preparation (147). The functions of the α- and the γ-subunits are not known, but they may protect and store the nerve growth factor in the salivary gland. The γ-subunit is known to have arginine-specific esteropeptidase activity (72) and may participate in the processing of the nerve growth factor from a larger biosynthetic precursor. There are one or two molecules of Zn^{2+} present in the complex (144) and the Zn^{2+} is thought to participate in holding the structure together. In the absence of Zn^{2+} the subunits come apart (26), as they do at pH levels above 8 and below 5. An appreciation of this role of Zn^{2+} in the structure has clarified some conflicts in the recent literature regarding the structure of the 7S molecule in its biological milieu (36). Recently it has been shown that the 7S nerve growth factor itself is totally inactive and must come apart and release the β-subunit before any activity can be seen (168). These experiments, in which the 7S is cross-linked with dimethylsubarimidate (a treatment which does not influence the activity of the β-subunit itself), make the 7S totally inactive. Thus, the active subunit of the 7S is the β, the other subunits have no nerve growth factor activity of their own and do not influence the activity of the β, and the 7S must come apart and liberate the β before any activity can be expressed.

The primary sequence of the β-subunit has been determined by Angeletti et al (11) and confirmed independently by Mobley et al (126). The chain has an amino-terminal serine and a carboxyl-terminal arginine. The presence of this latter group supports the concept that the γ-subunit, an arginine-specific esteropeptidase, cleaves the active β-subunit from some larger precursor protein. The exact differences between the intact β-subunit and the 2.5S, modified during preparation, consist of an absence of the amino-terminal octapeptide and the carboxyl-terminal arginine (127, 128). Thus, in chains maximally modified during preparation, nine amino acids would be lost. Both modifications, the amino terminal and the carboxyl terminal, have been found to occur separately; therefore the 2.5S dimer, which is normally isolated, has a degree of molecular heterogeneity introduced by the preparative methods themselves. Although these modifications have some effect on the ability of the peptide to reassociate with the other subunits in the 7S complex, there is no discernable difference in the biological activities of the 2.5S form and the β dimer.

There are some important similarities between the primary structure of the β-subunit and that of other peptides. The initial observation was that the linear sequence of nerve growth factor shared substantial homology with that of proinsulin (59). When properly aligned, up to 50% of the chain of nerve growth factor and that of proinsulin are identical in some regions. Furthermore, the position of one of the disulfide bonds in the two molecules is identical.

The three-dimensional structures of the two molecules have been compared by indirect methods. These experiments, in which the reactive residues in the nerve growth factor molecule are compared with those known to be on the surface of the insulin structure, suggest that the two molecules share a three-dimensional similarity (60). Other studies using a mathematical analysis of secondary structures (12) indicate that the helical areas in the two molecules do not correspond. The recent availability of a crystalline preparation (204) for X-ray analysis should resolve the problem. In the meantime, the structures of several other peptides of interest have become available (28). Relaxin and the insulin-like growth factors resemble insulin in structure. These likenesses permit interesting speculations about the evolutionary relations among the growth factors and about similarities in their mechanisms of action (28). Whether or not the details of the structural analysis now underway will, in fact, bear out the initial expectations of similarity, the insight into the primary sequence relationship has produced a great deal of important research.

There have been several studies on the nerve growth factors from snake venom. A few of these proteins have been purified to homogeneity. Where complete structural information is available, the similarities between snake venom nerve growth factor and mouse nerve growth factor are quite extensive, but the proteins are not identical. The nerve growth factor from cobra venom is a dimer of molecular weight 28,000 (88). It has about 60% homology with the mouse nerve growth factor and probably a similar disulfide cross-linking. Its biological activity is lower, but of the same order of magnitude as that of the mouse form, and it cross-reacts to some extent with the mouse antibody. Other venoms have been inspected and their nerve growth factor components purified. In a few cases it has been reported that these materials are glycoproteins, but complete structures are not yet available.

There is less structural information about the nerve growth factor from other sources. The factor from the tumors Sarcoma 180 or 37 has not been purified. Biological and immunological methods have shown that nerve growth factor is contained in many of the tissues of the body. Elevated levels can be found in sympathetic ganglia, blood vessels, kidney, adrenal, vas deferens, and in all peripheral organs innervated by the sympathetic nervous system. Lesser quantities are present in heart, spleen, liver, placenta, thymus, and muscle (32, 78, 93, 162). As yet there is no information about the structure of the nerve growth factor or even the form in which it is found in these tissues. Nerve growth factor has been found in mammals, birds, reptiles, amphibians, and fish (15, 32, 40, 104, 201, 203). It is known that a number of cells in culture, e.g. 3T3 cells, L cells (138), fibroblasts (207), myoblasts (132), melanoma (161), glioma (118, 129), and neuroblastoma (131) will produce nerve growth factor. The product found in the L cell cultures (141) has been reported to have a molecular weight of 160,000, but none of these materials have been purified, let alone satisfactorily characterized, so it is not clear that they are new forms of nerve growth factor. Perhaps most useful is the recent discovery of nerve growth factor in human placenta (64). This promises a voluminous source of the human factor that may be useful also as a more accurate index of nerve growth factor levels in various clinical states. Structural information on the human protein is not yet available.

METHODS

It seems fair to say that methods used to assay nerve growth factor are difficult and relatively unreliable. Indeed, the slow progress in the field and the confusion in the literature, especially the clinical literature, are most certainly due to the time-consuming and semiquantitative nature of the methods available. In the original "halo" assay (111), dorsal root ganglia from embryonic chicks are placed in a semisolid medium, with graded doses of nerve growth factor, for 24 hrs; the outgrowth is scored on a scale of 1^+ to 4^+. An assay utilizing a similar outgrowth response in dissociated cells gives added sensitivity (68). Another sensitive assay uses the competitive reassociation of radio-labeled α-subunit with the 7S complex (162). A number of methods based on the use of nerve growth factor-antiserum have been proposed. A one-site assay (93), a two-site assay (78), and a complement fixation technique (10) have been developed. These all suffer, to a greater or lesser extent, from a lack of sensitivity, a variability in available reagents, and from the presence, even in normal serum, of proteins that bind nerve growth factor (87, 134). The purification of the nerve growth factor antibody to homogeneity, using affinity chromatography (171, 181), has allowed an improvement of the sensitivity and reliability of the existing immunological methods (176). Some of the quantitative inconsistencies now found in the literature may thus be sorted out.

TARGET CELLS

The classical targets of nerve growth factor are the sympathetic and certain of the sensory neurons. It has long been thought that the sympathetic neurons require nerve growth factor throughout their life. In contrast, the mediodorsal neurons of the dorsal root ganglia seem to need nerve growth factor only for a relatively short developmental period. Both these assumptions require some inspection. In the case of the sympathetic neurons, it is now known that they do not require nerve growth factor in the very early stages of embryonic development (44), but that they develop such a dependence between the 14th and the 18th prenatal day. Regarding the postnatal period, administration of anti-nerve growth factor antiserum to young rats deprives them permanently of about 80% of their sympathetic nervous system; administration of antiserum to adults causes only a temporary intoxication of the neurons (6, 23). Dorsal root neurons, thought to be insensitive to the presence of nerve growth factor after the embryonic period, still show a nerve growth factor dependence in postnatal life if dissociated and placed in culture (34). Thus, although it is clear that these two classes of neurons require nerve growth factor at certain periods in their life span, the exact time limits are still uncertain.

A number of other cells respond to nerve growth factor. There have been reports of changes in the adhesive properties of embryonic tectal cells (125), stimulation of mucopolysaccharide biosynthesis in chondrocytes (51), hypertrophy and increased numbers of microfilaments in glial cells in regenerating optic nerves (182), and increased outgrowth of processes in damaged noradrenergic neurons in the brain (24, 170)—all in response to nerve growth factor. In a most important paper, Aloe & Levi-Montalcini (2) have shown that adrenal medullary cells, at an early stage of development, respond to nerve growth factor in vivo and become sympathetic neurons. Comparable changes have also been produced in vitro (183). Thus, several normal cells respond to nerve growth factor and most of these are derived embryologically from the neural crest.

There is also a library of tumor cells that respond in one way or another to nerve growth factor. Fortunately, these lines exhibit a gradation of responses that might eventually be used to dissect the much more complicated actions of nerve growth factor on normal cells. There are a number of tumor lines that appear to have membrane nerve growth factor receptors, but for which no further response has been observed. These include certain human melanoma lines (52) and some clones of the murine neuroblastoma C1300 (149). The significance of such receptors on cells that have shown no response in terms of survival or growth rate is not known. Then there are at least four cell lines which exhibit some outgrowth of processes in response to nerve growth factor: the human neuroblastoma lines IMR-32 (152) and SH-SY5Y (J. R. Perez-Polo, K. Werrbach-Perez, E. Tiffany-Castiglioni, unpublished data), the anaplastic glioma clone F98 (192), and PC12, a clone of rat pheochromocytoma (73). Recently it has been shown that changes in the surface properties of F98 are caused by nerve growth factor. These changes, which can be measured in minutes, are initiated by concentrations of the order of 1 ng per ml and result in an increased adhesiveness of the cells to each other or to a plastic substrate (S. Vinores, G. Guroff, in preparation). One other clone of pheochromocytoma has been reported (66); it shows no outgrowth in response to nerve growth factor but does exhibit an induction of tyrosine hydroxylase, an enzyme also responsive to nerve growth factor in normal sympathetic neurons. This rapidly growing catalogue of responsive cells gives promise that the various actions of nerve growth factor can be studied independently and the mechanism of each understood. It may thus become clear whether nerve growth factor acts through a common mechanism on all responsive cells or whether the mechanisms involved are as varied as the cells themselves.

At the present time, the most interesting and the most thoroughly characterized clone is the PC12 (45, 70, 71, 73). An important property of this system is that it provides a unique in vitro tool by which nerve growth factor can be studied. That is, before the development of PC12, all in vitro work with the nerve growth factor was limited to systems containing sympathetic or sensory neurons. For all these systems the addition of nerve growth factor is necessary, or the cells will not survive. For this reason, every in vitro study on the action of nerve growth factor suffered from the criticism that the controls were dying and that any differences between nerve growth factor-treated and nerve growth factor-untreated cells were simply reflections of survival. The PC12 cells do quite well in the absence of nerve growth factor but in its presence undergo a startling series of alterations. The most obvious of these is the production, in a few days, of a dense outgrowth of neurites. These processes are long, vesiculated, and branched, and make functional synapses with co-cultured muscle cells (156). The cells develop excitable membranes (45), store and release catecholamines (71), and stop multiplying (73). They do not exhibit the specific induction of the catecholamine-synthesizing enzymes seen in normal sympathetic ganglia, but in most respects they respond by differentiating into sympathetic neurons. The differentiation is reversible in that within hours after the removal of nerve growth factor these cells retract their processes and revert to their original state. It is a most interesting model and one that has found extensive acceptance both as a tool for the study of nerve growth factor and as a model for neuronal development and differentiation.

Whether the central nervous system is a target for the action of nerve growth factor is still unclear. Reports of the presence of nerve growth factor receptors (57, 58, 178, 179) and of nerve growth factor itself (78, 93, 162, 198) in the brain suggest that nerve growth factor should have some role there. Studies on the action of exogenous nerve growth factor on repair of lesioned brains (24, 179) support such suggestions. But, as will be discussed later, hard evidence for such a role is still lacking.

BIOSYNTHESIS

Although a number of cells in culture have been shown to make nerve growth factor, definitive information on the mode of biosynthesis has been obtained only in two systems. The first of these studies, from the laboratory of Shooter, involves the incorporation of radioactive amino acids into nerve growth factor in explants of mouse salivary gland (17, 18, 36). When such explants are incubated under physiological conditions for several hours and then treated with anti-nerve growth factor antiserum, radioactive nerve growth factor can be precipitated and quantitated. In addition, a larger molecule, approximately 22,000 mol wt is also found in the immunoprecipitate. The kinetics of the labeling of this larger molecule indicate that it is precursor to the β-nerve growth factor. When the 22,000 mol wt precursor is treated with the γ-subunit, it is completely converted to β-nerve growth factor. Also using radioactive amino acids, Longo (117) has demonstrated the synthesis of nerve growth factor in C6 rat glioma cells. In this case a 24,000 mol wt protein with immunological cross-reactivity to β-nerve growth factor was also synthesized. The apparent precursor could be converted to β-nerve growth factor in vitro by treatment with the γ-subunit.

Based upon this data and on the data available on the biosynthesis of other secretory proteins, the following model has been suggested (18). The 22,000-24,000 mol wt species is one of perhaps several precursors of the β-subunit of nerve growth factor. It is processed to the active subunit by the action of the γ-subunit, an arginine esteropeptidase. Whether the γ-subunit is, indeed, the relevant or the only peptidase active in this regard is not known. Nevertheless, such a model is consistent with the known chemistry of the peptide and is similar to biosynthetic processes found for similar peptides.

SITES OF NERVE GROWTH FACTOR ACTION: THE CELLULAR RECEPTORS

There are three distinct kinds of cellular receptors for the nerve growth factor. Two of these have been described in detail and the third is known to be present through a series of retrograde transport experiments described below. Each receptor implies a different kind of mechanism by which the factor might act, and it is not yet known which receptor is the relevant one, or if they are all relevant. Thus, there are known to be receptors at the plasma membrane, on the nucleus, and at the synaptic ending as well.

The membrane receptors have been described by several groups (14, 56, 83). Basically, all these experiments involve the preparation of radioiodinated derivatives of nerve growth factor and the binding of them to intact cells or membrane fractions. In each case different procedures have been employed to label the nerve growth factor and somewhat different tissue preparations have been used. In all studies, specific, high affinity receptors for the nerve growth factor have been found on the target tissues.

Using the microsomal fraction from rabbit superior cervical ganglia, Banerjee et al (14) found saturable, reversible binding with a dissociation constant of the order of 0.2 nM. These sites were present only on the target tissues and the binding to them has been reported to be strictly dependent on the presence of Ca^{2+} (13). The receptors from this source have been solubilized by extraction with Triton X-100 and appear to exhibit binding of the same order of magnitude in solution as they do in the membrane.

Herrup & Shooter (83) measured the binding of iodinated nerve growth factor to dissociates of dorsal root ganglia and found receptors similar in character to those described in sympathetic ganglia by Banerjee et al. Again, these receptors had high affinity and were both saturable and specific. The binding of nerve growth factor to these receptors was not influenced by the presence of other similar proteins, such as cytochrome C, and the receptors, again, were present only on tissues known to be targets for nerve growth factor. Receptor binding of the molecule was shown to decline in parallel with biological activity as the molecule was oxidized, and, in a further work (84) the presence of receptors was correlated with the developmental stages at which nerve growth factor is thought to act on the sensory ganglia.

Using a solid-phase method for the iodination, Frazier et al (56) detected high-affinity, specific receptors on the plasma membranes of chick embryo dorsal root ganglia and sympathetic ganglia. The receptors reported in this work were not saturable even at moderately high levels of nerve growth factor, and the binding exhibited a complex kinetic picture. Furthermore, these receptors were found not only on the target tissues for nerve growth factor but were observed on many tissues including liver, heart, and brain (57). Nevertheless it is clear, even with the as yet unresolved differences in the receptor characteristics, that there are high-affinity, specific receptors on the membranes of the target neurons.

More recently it has been shown that there are also receptors on the nucleus. This work, perhaps keyed by the reports that receptors for insulin exist on the nuclei of insulin-sensitive cells (63) and by the similarities known to exist between the primary structures of the two proteins (59), was done first with chick dorsal root ganglia (4). The results showed that Triton solubilization of membrane receptors left a significant portion of the total binding to the cells unaffected. When the nature of the residual binding was further explored it was found to be distinct from binding to the membrane. The nuclear receptors are saturable, insoluble in Triton-X100, and exhibit linear Scatchard plots representative of a single classes of receptors. Of most interest is the observation that chromatin isolated from the nuclei bound nerve growth factor in the same manner as did the nuclei themselves.

Studies with the nerve growth factor-responsive clone PC12 have also revealed a nuclear locus for the binding of nerve growth factor. Yankner & Shooter (206) have incubated nuclear preparations from these cells with iodinated nerve growth factor and have found two classes of receptors, with dissociation constants of 0.08 nM and 9.0 nM, respectively. The binding to these receptors was saturable and specific but, in this case, was clearly not on the chromatin. The authors conclude that the binding sites are at the nuclear membrane.

Experiments initiated by Hendry & collaborators (82) and carried on by Thoenen, Stöekel and their colleagues (81, 142, 159, 172–175) implicate yet another site at which nerve growth factor interacts with the cell. This site is the synaptic ending of the neuron, and the experiments clearly imply the presence of specific nerve growth factor receptors on the presynaptic membrane. The experiments are straightforward and their interpretation seems unequivocal. If iodinated nerve growth factor is injected into the anterior chamber of the right eye of a mouse or a rat, more radioactivity can be found, after an appropriate period of time, in the right superior cervical ganglia of the animal than in the left. The protein must have moved from the eye, across the synaptic membranes of the neurons innervating the iris, up the axon in a retrograde direction, and into the neurons which terminate in the eye.

The rationale for these experiments seems quite clear. A possible mechanism for the targeting of neurons to their appropriate end-organs would be for the end-organs to elaborate some factor for which the neurons had an affinity and which could, in turn, encourage the neurons to continue to progress in the direction of the target tissue. In the case of sympathetic neurons that factor could be nerve growth factor. All that is necessary is that only the appropriate neurons have receptors on the synaptic ending and a mechanism for transporting the factor back to the anabolic center of the cell.

The properties of the retrograde transport are consistent with such a mechanism. It is specific; proteins of similar charge and similar size are not transported (81, 173, 175). Transport of nerve growth factor decreases in parallel with biological activity as the molecule is oxidized with N-bromosuccinimide (175). The nerve growth factor that appears in the ganglia is antigenically active and structurally intact (172). Most important, it retains and expresses its biological activity; the levels of tyrosine hydroxylase in the ganglia on the injected side are higher than those on the uninjected side (142). The nerve growth factor so taken up accumulates in a select fraction of the neurons of the ganglia—those innervating the iris (82). When the neurons are inspected by electron microscopy most of the nerve growth factor appears to be in the secondary lysosomes, the smooth vesicles, and the endoplasmic reticulum (59), although another study by a different group suggests that about 15% can be found in the nucleus (95). Transport occurs in other nerve growth factor-responsive ganglia; the injection of radioactive nerve growth factor into the forepaw of rats leads to its selective appearance in the dorsal root ganglia (174), and injection into chick embryo reveals a transport into dorsal root ganglia (29). These experiments strongly indicate that nerve growth factor receptors are present in the synaptic membranes of nerve growth factor–responsive neurons. These receptors have not been studied directly, however, because of the difficulty of obtaining synaptic particles from peripheral nerves.

A determination of the site or sites at which nerve growth factor interacts with the cell could point the way to most useful information on the mechanism of its action. Combination of a peptide with its membrane receptors frequently leads to increases in some "second messenger" such as cAMP, and a resultant intracellular chain of events. Internalization from a membrane receptor to a nuclear receptor could indicate that nerve growth factor is its own messenger and exerts a direct effect on nuclear events. Internalization via the synaptic receptor could suggest that the action of nerve growth factor is a trophic one and that the only nerve growth factor which is physiologically important is that produced by the end-organs.

There is some information on this last point; nerve growth factor given systemically can act directly on sympathetic ganglia without traversing the retrograde route. In the first place, some of the actions of nerve growth factor occur too rapidly to be due to its appearance by retrograde flow from the end-organs (119, 208). Secondly, as shown in a dramatic and surprising experiment by Levi-Montalcini & her collaborators (2, 3), ganglia from animals treated with 6-hydroxydopamine to destroy their sympathetic terminals respond even more vigorously to a treatment by nerve growth factor than do intact controls. So it is clear that neurons *can* respond via the nonretrograde route. But by which route and through which receptors the nerve growth factor works normally under physiological conditions is not known.

ACTIONS OF NERVE GROWTH FACTOR ON ITS TARGET CELLS

Classically, the actions of nerve growth factor on its various targets include hypertrophy, hyperplasia, neurite outgrowth, stimulation of transport systems, stimulation of anabolic reactions, and increased synthesis of transmitter-synthesizing enzymes (enzyme induction). More recent work in several laboratories has focused on the effects of nerve growth factor on the cell surface.

Hypertrophy of target cells has been well documented. In some of the earliest studies, it was reported that the mediodorsal neurons of the dorsal root ganglia in animals bearing S-180 Sarcoma were enlarged in volume (109). Similar changes have been reported when purified nerve growth factor was given to animals (104). Increases in neuronal size also have been seen in sympathetic ganglia. Neurons of fetal or adult superior cervical ganglia increase on the order of 30% in cell diameter when animals are treated with nerve growth factor (79). In the PC12 cells the average cellular diameter increases from between 6 and 14 μ upon treatment with nerve growth factor (45). Thus, the hypertrophic response has been seen in several nerve growth factor–responsive systems, and methods have been developed to compensate for its effect on cell counting (79). There is, of course, an enormous increase in cell volume due to the formation of, or the increase in, neurites. Recalculation of the data of Greene & Rein for PC12 (70) reveals an increase in cell volume of more than twofold. This kind of increase in neuronal volume is independent of any increases in the dimensions or volume of the cell body itself.

The original experiments also suggested that nerve growth factor elicited a hyperplastic response in appropriate ganglia. These data were based on the finding that ganglia from animals implanted with the appropriate tumors, or treated with nerve growth factor directly, had more neurons than those of controls (109) and that more mitotic figures were seen (80). Currently, however, it is thought that these results can be explained otherwise. The increase in mitotic figures may be in the nonneuronal compartment and may depend on the proliferation of support cells in response to the outgrowth of the neurons. The increased number of neurons appears to result from increased survival rather than increased cell division. That is, during the normal development of the ganglia in the first postnatal months there is a patterned degeneration in which a certain number of neurons die. The influences controlling this selection process are unknown, but it is known that to a certain extent this depends upon the availability and integrity of the synaptic connections the ganglia make with the end-organs. When nerve growth factor is given, fewer of these neurons die. In the PC12 system, nerve growth factor certainly does not cause hyperplasia. Indeed, the presence of nerve growth factor appears to inhibit cell division. Cell counts of appropriate cultures (73) indicate that cell division ceases as neurite outgrowth begins, and that it starts again a few days after the withdrawal of nerve growth factor. Measurements of thymidine incorporation into DNA are consistent with this finding (K. R. Huff, G. Guroff, in preparation), but direct chemical measurement of the DNA content of these cultures does not show an inhibition by nerve growth factor (65). In any case, there is no evidence that nerve growth factor stimulates cell division in this system and little current thought that it causes hyperplasia in any of its target neurons.

Neurite outgrowth is, of course, the most characteristic effect of nerve growth factor. It can be elicited in vitro and in vivo in sympathetic and sensory neurons (101, 104) and adrenal medulla (2, 183), and in vitro in PC12 (73), IMR-32 (150), SH-SY5Y (J. R. Perez-Polo, K. Werrbach-Perez, E. Tiffany-Castiglioni, unpublished data), F98 (192), and in primary neuroblastoma cultures (199). Indeed, neurite outgrowth has frequently been the only test of responsiveness to nerve growth factor, e.g. of central nervous system cells in culture, though this is probably not a sufficient test since some cells respond with metabolic or surface alterations or with specific enzyme inductions, but do not form neurites. The additional neurites that are formed in normal neurons following nerve growth factor are not distinguishable from those formed without treatment. The neurites that form in the PC12 cultures are neuronal-like and, after two weeks of treatment, are found on 80% of the cells. Fine, long, and highly branched, with numerous varicosities, these neurites resemble fibers produced by cultured primary sympathetic neurons.

There is no evidence in vivo that excess nerve growth factor inhibits neurite outgrowth, but in the in vitro outgrowth assay excess nerve growth factor in the medium causes an apparent inhibition of outgrowth. This is, in fact, due not to inhibition but to a reorientation and wrapping of the neurites around the ganglia itself (202). This dense intertwining of neurites is what gives the false impression of a lack of neurites. There is also no evidence for inhibition of neurite formation by high levels of nerve growth factor in PC12 or in any other cell which is responsive to low levels.

The direction of neuronal growth in vivo in animals treated systemically with nerve growth factor appears to proceed along normal channels. When the nerve growth factor is administered vectorially, the neurons grow toward it. The original implant experiments demonstrated that neurites grew toward the tumors, which were presumably elaborating nerve growth factor (31). Indeed, even if the source of nerve growth factor is positioned unnaturally, e.g. in the brain, the sympathetic ganglia will reach out to it (108). Comparable experiments have been done in vitro with pieces of tissue placed in culture dishes; it is reported that neurite outgrowth will proceed toward the target organs and not toward nontarget tissues, and furthermore, that this outgrowth is blocked by the addition of antiserum specific to nerve growth factor (38). No such experiments have been done with PC12. So, nerve growth factor can act in a vectorial sense, but it is clearly not necessary that it do so, because when ganglia are chemically axotomized by treating animals with 6-hydroxydopamine to destroy sympathetic terminals, nerve growth factor produces even greater outgrowth and that outgrowth is random (103).

Membrane alterations have been reported as a consequence of nerve growth factor addition to certain cells. In a rather anomalous report

Merrell et al (125) noted that nerve growth factor alters the cell-to-cell interactions among embryonic tectal cells. The levels of nerve growth factor necessary are rather high, and oxidized derivatives of nerve growth factor, which are inactive for dorsal root or sympathetic neurons, are active in this system. In spite of these surprising aspects, it is now clear that other targets of nerve growth factor also exhibit changes in cellular adhesiveness. Specifically, the PC12 show very rapid changes in their ability to bind to culture substrates or to each other (158), as do F98 (S. Vinores, G. Guroff, in preparation). These changes may be reflections of very rapid alterations in the configuration of the surface membrane detected in PC12 cells by scanning electron microscopy (43). Changes in adhesiveness and configuration occur within minutes, much faster than the changes seen in the tectal system, and much faster also than the recently reported changes in the glycosylation of membrane proteins in PC12 (124) and phosphatidyl inositol turnover in ganglia (98). But, taken together these studies indicate that nerve growth factor has both acute and long-term effects on the properties of the surface membranes of target cells.

Perhaps one reflection of these alterations is the changes seen in the transport systems serving the cell. Varon and his colleagues have studied the stimulation by nerve growth factor of the uptake of a number of small molecules in dorsal root ganglia (184). Among these have been nucleotides and amino acids, nonmetabolizable analogs of these materials, and 2-deoxyglucose, a model used to measure glucose transport. Many of these studies have used tissue preincubated in vitro for up to 6 hr in the absence of nerve growth factor. The addition of nerve growth factor to these tissues produces a rapid and total restoration of the normal transport rates of the cell (185). The rapidity of the changes seen places them among the fastest responses known to occur. The conditions of the restoration or stimulation, i.e. tissue deprived of factor for many hours, have led some to question the relevance of the experiments. But similar changes have now been seen in the transport of amino acids into PC12 cells (123). Coupled with the rapid alterations seen in membranes and membrane properties discussed above, these changes in transport properties make the membrane locus a likely primary site of nerve growth factor action. The recent demonstration that nerve growth factor causes rapid Na^+ extrusion from its target cells (165a) appears to put us one large step closer to the molecular site of nerve growth factor action on the membrane.

Nerve growth factor causes marked changes in the anabolic properties of the neurons. Early work demonstrated that the synthesis of RNA, protein, and lipid was increased (5, 8, 42, 115) and that the oxidation of glucose also increased (9, 116). Most of these alterations have been observed with in vitro preparations and are not evident until several hours after the addition of nerve growth factor, so again it is necessary to consider whether the controls are simply dying. The increased anabolic activities of the cells are probably not just due to increased uptake of precursors, although this question has been somewhat difficult to settle. The increases in these activities of the cell are, however, quite real and, although probably not primary or direct actions of nerve growth factor, are clearly an overall result of nerve growth factor action on the cell.

The nerve growth factor causes changes in the levels of specific enzymes in the cell. The first of these was observed by Thoenen & his colleagues (180). They showed that when neonatal rats were treated for 10 days with 10 μg of nerve growth factor per gram body weight, the specific activities of tyrosine hydroxylase and dopamine β-hydroxylase in the superior cervical ganglia rose sharply while the activity of the other enzyme in the biosynthetic pathway to norepinephrine, dopa decarboxylase, remained unchanged. Related enzymes such as monoamine oxidase, a degradative enzyme in norepinephrine metabolism, and dihydropteridine reductase, an enzyme serving the coupled cofactor reduction in tyrosine hydroxylation (137), were also unchanged, emphasizing the specificity of the induction. No such changes have been seen in the dorsal root ganglia; these cells do not contain tyrosine hydroxylase and, indeed, the transmitter of these ganglia is not known. Much smaller but nevertheless significant inductions of tyrosine hydroxylase have been observed in adrenal medulla under the influence of nerve growth factor. As mentioned before, no such induction occurs in PC12; the specific activity of tyrosine hydroxylase actually drops after exposure to nerve growth factor, perhaps as a result of the increase in cell volume and resultant increases in other cell constituents. But the induction of choline acetyltransferase in this clone has been reported (50) and the induction of tyrosine hydroxylase has been seen in another clone of pheochromocytoma (66).

The increase in the specific activity of tyrosine hydroxylase produced by nerve growth factor is due to an increase in the differential rate of synthesis of the enzyme. The normal level of synthesis in the ganglia is about 0.15% of the total soluble protein (121). A single administration of nerve growth factor to neonatal rats increases this rate threefold or more (120). Repeated administration of nerve growth factor can bring the level to about 0.5%; that is, the maximum level of the synthesis of tyrosine hydroxylase in the ganglia is about one half of one percent of all the soluble proteins made by these cells.

The administration of actinomycin D along with the nerve growth factor prevents the increase in the differential rate of the synthesis of tyrosine hydroxylase (120). This observation indicates that the increase in the synthesis is due to an increase in the synthesis of some RNA, presumably the messenger RNA for tyrosine hydroxylase. Other studies in which ganglia from adult animals were treated with nerve growth factor in culture have shown that some increase in the synthesis of tyrosine hydroxylase can also be produced in vitro (122). The increased synthesis under these conditions is less than that produced by in vivo administration of nerve growth factor to neonates and was found to be insensitive to inhibitors of RNA synthesis. The data have yet to be fully reconciled, but one interpretation is that nerve growth factor may have both a transcriptional and a translational effect on the synthesis of tyrosine hydroxylase.

One of the most interesting effects of nerve growth factor is its apparent ability to alter the response of its target cells to other factors. This response, observed in this laboratory using PC12 cells, involves the dual action on these cells of nerve growth factor and epidermal growth factor. Nerve growth factor, as mentioned above, causes a terminal differentiation and a cessation of cell division in PC12 cells. Epidermal growth factor in most of its responsive systems, is a powerful mitogen. Nerve growth factor in PC12 causes increased adherence to a substrate, an induction of ornithine decarboxylase, a lowering of thymidine incorporation into DNA, and an outgrowth of neurites. Epidermal growth factor in PC12 also causes an increased adherence to a substrate and an induction of ornithine decarboxylase, but it leads to an increase in thymidine incorporation into DNA and does not cause outgrowth of neurites. Specific, high-affinity receptors for epidermal growth factor can be found on these cells.

Since nerve growth factor stops cell division in PC12 and since epidermal growth factor in most responsive systems is a mitogen, it seemed of interest to investigate the possible interaction between them. Accordingly, cells were treated for three days with nerve growth factor and the response to epidermal growth factor was inspected (89). Although the level of ornithine decarboxylase in these cells was back to normal after its increase because of the nerve growth factor, the response to epidermal growth factor addition was much reduced. Concomitant measurements of epidermal growth factor binding showed that the specific receptor binding to these differentiated cells was reduced by approximately 80%.

The reduction in the ornithine decarboxylase response is not due to a generalized refractoriness of the system, since the ornithine decarboxylase activity responds equally in control and treated cells when cAMP derivatives are added. It is reasonable to assume that it is due to the reduction in receptor function. Whether or not it is due to such a reduction, the fact remains that the receptor binding is markedly reduced.

Whether this reduction in binding is caused by a reduction in the number of receptors or in the affinity of existing receptors for the ligand is not known. Preliminary data on the time course of the response would suggest the former, but more studies are needed. If a reduction in the number of receptors is indeed the cause, it suggests that nerve growth factor is acting transcriptionally to repress the synthesis, rather than at the membrane to alter the affinity. In either case, the data encourage the speculation that one of the effects of nerve growth factor is to alter the synthesis of the receptor for a mitogen, and that this action may be a mechanism or the mechanism by which nerve growth factor inhibits cell division in PC12. Further experiments are necessary to determine if nerve growth factor has a comparable action on normal neurons at some early stage in their development.

ABSENCE OF NERVE GROWTH FACTOR: IMMUNOSYMPATHECTOMY

The absence of nerve growth factor or the presence of sufficient anti-nerve growth factor antiserum to block the action of the available nerve growth factor results in a deterioration of dorsal root and sympathetic ganglia. The cytotoxic effects of anti-nerve growth factor on cultured sympathetic superior cervical ganglia can be observed within 2 to 6 hr by a cessation of all electrical activity in the neurons (100, 102, 148). Following this, several morphological changes occur, such as the folding and rupture of the nuclear membrane, the clumping of chromatin, the dilation of the perinuclear cisternae, alterations in the fine structure and the number of nucleoli, changes in the mitochondria, and eventual lysis (101, 102, 106, 154). By 12 to 17 hr cell death occurs (86, 148).

Treatment of developing rats or mice with anti-nerve growth factor antiserum results in an irreversible degeneration of the sympathetic nervous system (102, 105). This is characterized by a marked decrease in the dry weight of the sympathetic chain, a decrease in RNA and protein synthesis and the rate of glucose oxidation, and a decrease in synaptic transmission (102, 104). Similar treatment of adult rats or mice produces a similar but reversible deterioration. Immature animals that have been pretreated with nerve growth factor behave as adults in this respect, following antiserum treatment (102, 104).

Although passive immunization, i.e. administration of heterologous antibody, has been extensively explored, active immunization, i.e.

administration of mouse nerve growth factor, has not been heavily studied. Recent experiments (67) have shown that such immunization of rats produces atrophy of the sympathetic nervous system. Indeed, when such actively immunized females became pregnant and produce litters, the newborn are immunosympathectomized and appear to be missing a portion of their sensory nervous system as well. Exploration of the characteristics of these animals should be most rewarding.

ACTIONS OF NERVE GROWTH FACTOR ON THE BRAIN

The actions of nerve growth factor on the catecholaminergic neurons in the peripheral nervous system have prompted a number of experimental inquiries into the possible actions of nerve growth factor on the catecholaminergic portions of the central nervous system. The presence of receptors in the brain is unquestioned. Using embryonic chick brain as the experimental tool, Szutowicz et al (178, 179) have shown that the receptors for nerve growth factor in the brain are much like those found in the peripheral nerves. Subcellular fractionation studies revealed that a large proportion of the receptors occurred in the fraction containing the synaptosomal structures. There were, however, receptors also associated with heavier particles, possibly the nuclei.

There are persistent reports that nerve growth factor itself can be found in the brain (78, 93, 162) and that its level in the brain can be altered by specific hormones (198). A recent study shows that nerve growth factor is especially high in goldfish brain (15). However, the same methodological concerns which are involved with the measurement of low levels of nerve growth factor from any source lead to some scepticism about these reports. Certainly, the levels in brain are low, even compared to tissues which are not obviously target organs, such as liver.

Biochemical studies on the actions of nerve growth factor on the brain are quite scarce. Clearly, there are no actions comparable to those classical ones in the periphery. There are no reports of neurite outgrowth promoted by nerve growth factor in normal central nervous system cells in vivo, but in primary cultures of fetal rat cerebrum, established during late gestation, nerve growth factor was found to induce process formation and inhibit proliferation (194). Process formation could also be observed in primary cultures of medulla, trigeminal nerves, and pineal gland established in mid-gestation (S. Vinores, J. R. Perez-Polo, unpublished data), but not in cerebellum. There are no experiments which indicate that nerve growth factor or its antibody affects the normal development of the brain cells in vivo. Nerve growth factor has not been found to alter the transport of nutrients into or out of brain, nor has any effect on anabolism been reported. Finally, no effects on brain tyrosine hydroxylase levels have been seen, although there are some suggestions that the turnover of monoamines in the brain might be altered somewhat by nerve growth factor administration (113). The lack of effect of nerve growth factor antibody on brain tyrosine hydroxylase has been shown in several laboratories. In one very convincing experiment (134) purified nerve growth factor antibody, which was injected into the ventricles, lowered the levels of tyrosine hydroxylase in the peripheral nervous system by escaping from the brain or absorbing systemic nerve growth factor, but had no effect at all on tyrosine hydroxylase in the brain.

The only biochemical change noted has been an increase in the ornithine decarboxylase levels in the central nervous system. This change, within hours after the injection of nerve growth factor into the ventricles, was first observed by Roger, Schanberg, & Fellows (151) and explored in our laboratory (114, 133). The enzyme is of interest because it catalyzes the rate-limiting reaction in the synthesis of the polyamines. These are molecules which have been implicated in the control of growth and malignant change, and whose levels rise because of the action of a number of hormones on their target cells. The observation is of interest because a comparable rise in ornithine decarboxylase level has been shown to occur in sympathetic ganglia (119), in dorsal root ganglia, and in PC12 cells (69, 89) after the addition of nerve growth factor. In brain an exuberant increase in ornithine decarboxylase level occurs after the injection of nerve growth factor (114). This increase is fairly specific for nerve growth factor, although some other peptides will cause a similar increase (91). The increase is not localized to any particular portion of the brain, certainly not to the catecholamine-containing portions, and it is not even specific for the neurons (114).

The injection of nerve growth factor into the ventricles also causes an increase in the ornithine decarboxylase levels in several other tissues (133). These increases in liver, kidney, and adrenal are not caused by leakage of the nerve growth factor from the ventricles into the system, since systemic injection of the same quantity of nerve growth factor does not cause such changes. There is evidence from ablation experiments that these peripheral inductions are due to a neurohumoral response. Specifically, when the adrenal or the pituitary are removed, the administration of nerve growth factor to the animal has a reduced effect or is without effect entirely (133). The involvement of the hypothalamus-pituitary-adrenal axis is confirmed by the further finding that the level of cortical steroid in the blood rises rapidly after intraventricular administration of nerve growth factor. In confirmation of this, others have shown that the level of cAMP in the adrenal cortex of intact animals rises rapidly after systemic administration of nerve growth factor (140), but cAMP levels in hypophysectomized animals are unaffected. Thus, data from our laboratory and others suggest that nerve growth factor activates the hypothalamus or the pituitary in some as yet unknown way.

Administration of nerve growth factor to the brain has been reported to have functional consequences. Specific lesions have been made in the catecholaminergic tracts in the brains of rats and the regrowth of these transected tracts toward an implanted iris has been observed. Under normal conditions these transected tracts will sprout and innervate the implant. This regrowth can be seen histologically using the fluorescence of a derivative of the transmitter, norepinephrine, as an index. When nerve growth factor is administered, the regrowth is faster and more abundant (24,170). Conversely, when nerve growth factor antibody is administered, the regrowth is inhibited (22). Moreover, section of certain tracts leads to an altered behavior, which returns to normal over a period of time, presumably due to the normal repair of the lesion and regrowth of the tract. The administration of nerve growth factor to the brain of such damaged animals hastens their return to normal, possibly by hastening the repair of the lesion (16,76).

In summary, then, there are receptors for nerve growth factor in the brain and many reports of nerve growth factor itself in small amounts. Administration of nerve growth factor to the brain elicits at least one biochemical response, the induction of ornithine decarboxylase, and appears to affect both structure and functional recovery in brains containing experimental lesions. However, some evidence suggests that the biochemical action of nerve growth factor in the brain is due to an activation of the hypothalamic-pituitary-adrenal axis and that this action is not typical of nerve growth factor action on its other, more classical targets, nor is it competely specific for the nerve growth factor. It remains to be seen if the action of nerve growth factor on the morphology and function of the lesioned brain is also indirect and relatively nonspecific. As yet these studies are not at hand, but it can be suggested that peptides other than nerve growth factor may mimic the action of nerve growth factor on the morphology and the function of the lesioned brain and may do so indirectly by stimulation of the hypothalamus or the pituitary. Should this be the case, the important and intriguing question of the mechanism by which nerve growth factor activates the hypothalamus remains open.

CLINICAL INVOLVEMENTS: THE LEVELS OF NERVE GROWTH FACTOR IN NEUROLOGICAL DISEASE AND ITS EFFECT ON TUMORS

The effect of nerve growth factor on the peripheral nervous system has encouraged a search for the involvement of the peptide in various human disorders. Specifically, it has been of interest to look at various neurological diseases, especially those involving aberrations in the peripheral nervous system, for alterations in the circulating levels of nerve growth factor. Then, also, there have been some studies on the possible use of nerve growth factor or its antibody in the control of the several tumors arising from the nerve growth factor–responsive cells of the peripheral nervous system. The literature is sparse and contradictory, certainly in part because of the difficulties in the quantitative methodology mentioned above. At the moment, there is no generally accepted evidence for changes in nerve growth factor levels in any disease or for an effect of nerve growth factor or its antibody on any tumor. But there are some recent leads and the literature bears reviewing.

Most of the studies have consisted of an examination of the sera of patients for changes in nerve growth factor levels. Since in no case was the nerve growth factor isolated from these sera, the assays simply measured the "nerve growth factor-like" activity. The earliest studies were done with the dorsal root ganglia assay. More recent work has focused on the immunological approach, and in at least one study the sera have been inspected with several of the available nerve growth factor assays.

Using the dorsal root ganglia bioassay, Shenkein et al (160) showed that sera from children with disseminated neurofibromatosis (von Recklinghausen's disease) had higher levels of nerve growth factor activity than did controls. Neurofibromatosis is a disease in which tumors arise throughout the nervous system and appear to originate in support cells as well as in neurons. Using a radioimmunoassay, Siggers et al (163) could not confirm the earlier finding but did detect higher levels of nerve growth factor-like activity in several patients with bilateral acoustic neuroma, a tumor related in some way to the central form of neurofibromatosis. The most intriguing aspect of this latter report was the observation that some of the children genetically at risk for this tumor also had elevated levels, but that children beyond the age of onset who had no evidence of disease had normal levels. This would indicate that nerve growth factor levels could have predictive as well as diagnostic value in this disease. Fabricant et al (53) confirmed these results using a radioreceptor assay as well as a radioimmunoassay.

The most complete study has been done on patients with familial dysautonomia (Riley-Day syndrome). This is a hereditary condition characterized by a degeneration of the peripheral sensory neurons and a progressive loss of nerve fibers and sensory function. Several patients with this disorder were studied using three of the available methodologies (164). Results from bioassay and receptor radioassay were negative; no difference was found between patients and controls. Unexpectedly, radioimmunoassay revealed an increased level of nerve growth factor-like activity in patients. Although several possible molecular explanations were suggested by the authors, e.g. the presence of immunologically cross-reacting but nonfunctional precursors or products of nerve growth factor, the finding does little to explain the etiology of the disease.

Using bioassay and immunoassay procedures, an elevated level of a nerve growth factor-like material was also found in a patient with medullary carcinoma of the thyroid gland (20). The observation of high levels of nerve growth factor in the serum of a son showing clinical symptoms suggest that this may also be hereditary and that the assay may be of predictive value. The relationship of nerve growth factor to this tumor is not clear, but diffuse neural overgrowth, bilateral adrenal medullary hyperplasia, and multicentric pheochromocytoma, which are frequently associated with this disease, may play a role.

Fibroblasts are known to produce nerve growth factor, and a disorder involving rapid proliferation of fibroblasts, Paget's disease of bone, is also accompanied by elevated antinerve growth factor immunoreactive material in the serum. Following therapy, these levels markedly decrease (155). Other conditions in which increased levels of nerve growth factor or a nerve growth factor-like substance appear in humans include liposarcoma (196) and glioblastoma (155).

The previously described disorders involved elevated levels of nerve growth factor. Schizophrenia, on the other hand, has been reported to be associated with low levels of nerve growth factor measured by bioassay and radioimmunoassay (146). Although no supportive clinical data is available, decreased nerve growth factor levels are also characteristic of mice with muscular dystrophy (62, 130). Further investigation in these areas with refined methodology seems promising.

The neuroblastoma, a solid tumor arising in the peripheral nervous system, is the second most prevalent malignancy occurring in infants and children. Its suspected origin in the sympathetic nervous system from neural crest-derived cells led to an investigation of nerve growth factor involvement. Burdman & Goldstein (33) reported some years ago that children with neuroblastoma had increased levels of bioassayable nerve growth factor activity in their sera. Since then, however, two other, rather more detailed investigations (21, 197) have failed to reveal such an increase in the sera of other neuroblastoma patients.

The only recorded attempt to use nerve growth factor as a therapeutic agent in humans has been in children with advanced neuroblastoma. Kumar et al (97) administered 7S nerve growth factor for several days to three children with widespread disease. No clinical improvement was noted, but profuse sweating and headaches were reported.

Despite the confusion in the clinical literature, experiments with animals and with tumor cells in culture continue to provide tantalizing and encouraging information. As previously discussed, an increasing number of tumors are being found that will respond in one way or another to nerve growth factor in culture. All of these seem to be poorly differentiated tumors of neural crest origin. Especially in the case of pheochromocytoma and neuroblastoma, the response appears to be one of differentiation and cessation of growth, a response eminently useful clinically. Because neuroblastoma and pheochromocytoma represent malignant forms of the classic target cells of nerve growth factor, observing a response of these tissues in vitro may have been predictable; but nerve growth factor can also influence the induction or malignant behavior of other types of undifferentiated neural crest tumors.

Ethylnitrosourea (ENU) is a resorptive carcinogen which, when administered to rats transplacentally late in gestation or shortly after birth, induces central nervous system gliomas of various types and peripheral nervous system neurinomas in nearly 100% of treated rats (46–48, 92, 200). Nerve growth factor administered with ENU to fetal or neonatal rats substantially reduces the number of such neurogenic tumors, but this is accompanied by an increase in the number of nonneurogenic tumors (169, 191, 195). Antinerve growth factor antibody administered to fetuses prior to ENU treatment results in more intracerebral gliomas. Similar treatment of newborn rats results in more trigeminal nerve neurinomas (191). Ninety days after ENU treatment virtually all rats exhibit early neoplastic proliferation in the Schwann cells of trigeminal nerves (96, 177). Nerve growth factor can cause a reduction in the number of early neoplastic proliferations (193), which compares favorably with the reduced number of neurinomas that ultimately form in the nerves. These results suggest that increased systemic levels of nerve growth factor can reduce the neurotropic effect of ENU. This is supported by the observation that mice pretreated with antinerve growth factor antibody develop more neurogenic tumors following exposure to ENU than do controls (190).

Antinerve growth factor antiserum was also found to influence the induction of tumors by another carcinogen, benzo(a)pyrene. In this case, the number of tumors was reduced and the appearance of tumors that did occur was delayed (19).

Glia have not been thought of as nerve growth factor responsive cells, but since they originate from the same presursor cells as do neurons, there might be some point during their maturation at which they lose the capacity to respond. Except for the single study of Turner & Glaze (182), differentiated glial cells, normal or neoplastic, have not been found to respond to nerve growth factor, but undifferentiated glioma cells may still retain this ability. An example of such an undifferentiated cell is the anaplastic glioma F98. This tumor cell responds to nerve growth factor in vitro with a morphological change, an altered growth rate (192), and an increase in adhesiveness (S. Vinores, G. Guroff in preparation). Increased adhesiveness has been associated with decreased malignancy (49, 54) and the F98 does not appear to be an exception. The survival of animals to which such tumor cells had been given intracerebrally was increased because of a decreased tumor growth rate caused by treatment with nerve growth factor (192). Although the increase in the survival time was modest and the effect may be indirect, the results are understandable in terms of the biology of the cells and encourage further exploration with other regimens and other responsive tumors.

MECHANISMS OF ACTION

The molecular mechanism or mechanisms by which nerve growth factor affects its target cells are not known. There are several suggestions in the literature but none of the various theories has received wide experimental support. Basically they can be divided into three categories: membrane, structural, and transcriptional.

The membrane theories suggest that the site of action, or at least a site of action, of nerve growth factor on the cell is directly at the membrane. The evidence includes, of course, the presence of surface membrane receptors, but more directly, the observation of extremely rapid changes in the structure and morphology of the membranes of nerve growth factor–responsive cells, primarily PC12. Observations with scanning electron microscopy have shown that within minutes after the addition of nerve growth factor to such cells there are profound alterations in the surface of these cells (43). Studies on the ability of these cells to adhere to plastic or to one another show that there are equally rapid changes in the properties of the membrane (156). The latter observations have been accompanied by studies in which alterations in calcium fluxes and cAMP concentrations have been observed (157).

These experiments have led Schubert et al (157) to suggest that the combination of nerve growth factor with its receptor is followed by an increase in cAMP, a mobilization of calcium ions, and an increase in cell adhesiveness. The authors further suggest that this alteration in cellular adhesiveness could underlie the neurite outgrowth caused by nerve growth factor and also could participate in the trophic actions suggested for the factor in guiding normal neurons to their target organs. The general concept that nerve growth factor-induced alterations in cellular adhesiveness are fundamental to its action are consistent with earlier studies on normal neurons (188, 189). The finding that cellular adhesiveness can be increased by agents which do not induce neurite outgrowth (K. R. Huff, G. Guroff, in preparation) militates against such a postulate.

An alternate suggestion concerning membrane function has been advanced by Varon and his colleagues (184). Studies with dorsal root ganglia have shown that a very early action of nerve growth factor on the cells is an increase in the transport of a number of small molecules. These changes occur very rapidly, within minutes in nerve growth factor-depleted explants, and in the absence of corresponding increases in the synthesis of macromolecules. The fact that the changes are so rapid and that they do not depend upon transcriptional events, as well as the more recent observation of comparable increases in transport rates in PC12 (123), supports this theory. The consideration that most of the work has been done with explants cultured for several hours in the absence of nerve growth factor has cast some doubt on the physiological meaning of the studies. In any case, the theory holds that the combination of nerve growth factor with its receptor causes changes in the permeability of the membrane for small molecules (presumably an alteration in the active transport mechanisms) and because of this there is an increase in the amount of precursor available to the cells and an alteration in the anabolic processes of the cell. Since the transport systems for many small molecules are known to be linked to the sodium pump, the recent observation that nerve growth factor has a rapid effect on sodium extrusion in these cells (165a) may provide a common mechanism for the nerve growth factor-induced increase in all these transport systems.

The polymerization of microtubules from tubulin, the structural protein of the microtubules, is intimately connected with the formation of neurites. Thus, the suggestion has been advanced that the site of action of nerve growth factor is on the biosynthesis or the polymerization of tubulin (112). The postulate suggests that nerve growth factor binds to tubulin and thereby induces an increase in its biosynthesis or polymerization. Supporting this suggestion are the report that nerve growth factor binds to tubulin (37), several studies showing that stimulated neurons exhibit a much increased formation of microtubules (7, 107), and experiments indicating that nerve growth factor inhibits the action of vinblastine (39, 94). On the other hand, it is known that nerve growth factor will bind to many proteins, e.g. α- and γ-subunits, α_2-macroglobulin (153), bovine serum albumin (1), and even glass beads (139) or plastic (145). Also there is direct evidence that nerve growth factor has no specific effect on the synthesis of tubulin (175, 205). The possibility remains that nerve growth factor acts directly on the polymerization of tubulin subunits, but evidence for such an action and support for this site as a primary target of nerve growth factor in the cell is lacking.

There is substantial evidence that nerve growth factor has transcriptional effects on its target cells. Increases in the activities of a number of enzymes occur; included in these are ornithine decarboxylase, tyrosine hydroxylase, and dopamine β-hydroxylase in sympathetic ganglia, ornithine decarboxylase in dorsal root ganglia, and ornithine decarboxylase and choline acetyltransferase in PC12. The increases in ornithine decarboxylase and, based on evidence obtained in this laboratory, tyrosine hydroxylase are dependent upon transcriptional events. No information on the involvement or lack of involvement of transcription in the other inductions is yet available. Early experiments by Partlow & Larrabee (143) indicated that neurite outgrowth could proceed normally in the virtual absence of RNA synthesis, but more recent studies on PC12 (35) indicate that initial nerve growth factor–induced process formation in these cultures is RNA-synthesis dependent. So for at least some of the actions of nerve growth factor, nuclear transcriptional events are required.

These nuclear events could be initiated in either of two ways. Earlier studies on the immobilization of nerve growth factor by attachment to glass beads (55) showed that the material still had its characteristic effects on outgrowth, and these studies were interpreted to mean that the nerve growth factor could act without entering the cells. These studies have been reevaluated in the light of newer methodological information by the original authors (27) and by others and seem not to be as definitive as once supposed. Further studies in other laboratories (206) show that in PC12, at least, nerve growth factor can indeed be internalized. Retrograde transport experiments also show the uptake into the cell body of intact, active nerve growth factor (82, 142, 172). The presence of receptors for the nerve growth factor on the nucleus of responsive cells provides a means by which this internalized nerve growth factor might act to alter transcription, although there is still some dispute as to the exact nuclear location, e.g. chromatin (4) or membrane (206), of these receptors, and some dispute also as to whether nerve growth factor taken up by the retrograde route finds its way to the nucleus at all. Nevertheless these findings have led Bradshaw to postulate (27) the following comprehensive series of events. First is a complexation of the nerve growth factor with plasma receptors resulting in increased metabolite uptake. Second is a vesicularization of the hormone-receptor complex initiated by the polymerization of tubulin to microtubules and an internalization of the complex. Third is the transport of the vesicle and fusion of it with internal membrane structures, some to the lysosomes for degradation and some to the nucleus. Fourth is a transfer of the hormone to the nucleus, followed by binding of the nerve growth factor to its nuclear receptors, which then directly affects transcriptional events.

Our studies in the superior cervical ganglia from neonatal rats suggest yet another route. Nikodijevic et al (136) observed a moderate and transient increase in the cAMP content of ganglia in culture within minutes after the addition of nerve growth factor to the medium. Similar increases have now been seen in dorsal root ganglia (135, 165) and in PC12 cells (158) following the addition of nerve growth factor, although others have failed to observe such an increase using very similar techniques (61, 85, 99, 140). If cAMP is a "second messenger" for the actions of nerve growth factor, derivatives of cAMP should mimic the actions of nerve growth factor. This is the case. Cyclic AMP derivatives will maintain the levels of tyrosine hydroxylase in explanted ganglia (209). Derivatives of cAMP will induce ornithine decarboxylase in explanted ganglia (119). Derivatives of cAMP will promote outgrowth of neurites in explanted dorsal root ganglia (75), although these neurites are said to differ morphologically from those produced by the addition of nerve growth factor (152). So there is some evidence that nerve growth factor acts through cAMP as a second messenger. Other agents that raise cAMP levels also produce changes in protein kinase activity in the cell, but no such evidence exists for nerve growth factor at the present time.

Whether the effects of nerve growth factor are exerted directly on the nucleus after internalization or though a series of cytoplasmic messengers starting with cAMP, nuclear events ensue. Ornithine decarboxylase activity is elevated (69, 77, 119), RNA polymerase activity rises (90), and specific enzyme inductions occur (120). Furthermore, RNA-dependent neurite outgrowth is initiated. It is reasonable to expect that such changes in transcription would be preceded by changes in the chemistry of nuclear proteins, which apparently regulate transcription. Studies in this laboratory have shown one such change. Yu et al (208) have found that the addition of nerve growth factor to cultures of sympathetic ganglia or administration of nerve growth factor to the intact animal leads within one hour to an increase in phosphate incorporation into a specific nuclear protein. In more recent work we have found an increase in the phosphorylation of a second, possibly related protein. This increase is not due to the total synthesis of these proteins, since addition of cycloheximide does not alter the stimulation, and incorporation of leucine is not increased by nerve growth factor. The proteins are chromatin-bound and are present in the neurons of the ganglia rather than in the glia, since treatment of the animal with 6-hydroxydopamine, which selectively destroys the neurons, abolishes the stimulation of phosphate incorporation. Quantitative studies reveal that the stimulation of incorporation (or turnover) of phosphate is on the order of twofold. The identity of the proteins is not known, and the details of the sites of phosphorylation and the relevant phosphorylating enzymes still remain to be worked out. The phosphorylation does seem to be intimately involved with the action of nerve growth factor on the ganglia. Preliminary experiments in this laboratory have shown that 12-O-tetradecanoyl-phorbol-13-acetate (TPA), a tumor promoter and a molecule which has been shown (210) to interfere, at least temporarily, with the ability of the ganglia in culture to elaborate neurites, also prevents the increase in the phosphorylation of these nuclear proteins.

If all the relevant information is put together, the nerve growth factor would appear to influence the cell in several ways. It combines with a surface receptor and initiates conformational and functional changes in the membrane, including changes in cellular adhesion and alterations in membrane permeability. It raises cAMP levels moderately and transiently, and these may be a cause of or a consequence of the membrane changes and may serve intracellular second messenger function as well. Either through such a second messenger or through direct internalization, the factor affects the nucleus and acts to alter the phosphorylation of nuclear proteins. These phosphorylative alterations then pave the way for the transcriptional actions of nerve growth factor on the cell.

SUMMARY AND PROJECTIONS

Much is known about the nerve growth factor and much remains to be learned. The nerve growth factor protein has been sequenced and its three-dimensional structure should soon be known. Along with this information has come the insight that nerve growth factor is related to insulin structurally and, in turn, to the family of insulin-like growth factors. Its target cells include the sympathetic and sensory neurons, but it has a wide variety of effects on brain and other normal tissues and on a host of malignant cells. It is made in a number of tissues and, like so many other hormone-like peptides, originates in a larger peptide which undergoes posttranslational proteolytic processing. Receptors exist on the cell membrane of its target cells, on the nucleus as well, and at the synaptic ending. These latter receptors probably function to accept nerve growth factor from the end-organs and move it to the cell body via retrograde transport through the axon as a signaling mechanism. The actions of the nerve growth factor on these cells are varied, but generally include hypertrophy, neurite outgrowth, increased transport of nutrients, increased anabolic responses, and the induction of specific enzymes. In PC12, nerve growth factor also causes cessation of cell division, perhaps by repressing the synthesis of receptors for mitogens.

The molecular mechanism of action of nerve growth factor is not known but the outlines are appearing. Clearly, nerve growth factor has some actions on the cell membrane, mediated by combination with its surface receptor. Clearly, also it has transcriptional effects in the nucleus of the cell, mediated either by a series of second messengers or by direct combination of internalized nerve growth factor with its nuclear receptor.

We expect the molecular mechanism to become clear in the near future. The transcriptional effects, at least, seem amenable to solution through an understanding of the chemistry of the affected nuclear proteins and of the enzymes performing the alterations. We also expect that improved methodology and the wider availability of human nerve growth factor will lead to a reduction of the confusion in the clinical literature and to an increased ability to study a wide variety of clinical problems—from the occurrence of raised or lowered levels in disease states, to the clinical use of nerve growth factor and its antibody. Finally, it should become clear in the near future, with the emergence of more rapid ways of looking at the response of cells to various factors, whether nerve growth factor is one of a kind or simply the first of a class of factors that sustain and perhaps target the various cells in the nervous system.

Literature Cited

1. Almon, R. R., Varon, S. 1978. *J. Neurochem.* 30:1559–67
2. Aloe, L., Levi-Montalcini, R. 1979. *Proc. Natl. Acad. Sci. USA* 76:1246–50
3. Aloe, L., Mugnaini, E., Levi-Montalcini, R. 1975. *Arch. Ital. Biol.* 113:326–53
4. Andres, R. Y., Jeng, I., Bradshaw, R. A. 1977. *Proc. Natl. Acad. Sci. USA* 74:2785–89
5. Angeletti, P. U., Gandini-Attardi, D., Toschi, G., Salvi, M. L., Levi-Montalcini, R. 1965. *Biochim. Biophys. Acta* 95:111–18
6. Angeletti, P. U., Levi-Montalcini, R.,

Caramia, F. 1971. *Brain Res.* 27:343-55
7. Angeletti, P. U., Levi-Montalcini, R., Caramia, F. 1971. *J. Ultrastruct. Res.* 36:24-36
8. Angeletti, P. U., Liuzzi, A., Levi-Montalcini, R. 1964. *Biochim. Biophys. Acta* 84:778-81
9. Angeletti, P. U., Liuzzi, A., Levi-Montalcini, R., Gandini-Attardi, D. 1964. *Biochim. Biophys. Acta* 90:445-50
10. Angeletti, P., Vigneti, E. 1971. *Brain Res.* 33:601-4
11. Angeletti, R. H., Bradshaw, R. A. 1971. *Proc. Natl. Acad. Sci. USA* 68:2417-20
12. Argos, P. 1976. *Biochem. Biophys. Res. Commun.* 70:805-11
13. Banerjee, S. P., Cuatrecasas, P., Snyder, S. H. 1975. *J. Biol. Chem.* 250:1427-33
14. Banerjee, S. P., Snyder, S. H., Cuatrecasas, P., Greene, L. A. 1973. *Proc. Natl. Acad. Sci. USA* 70:2519-23
15. Benowitz, L. I., Greene, L. A. 1979. *Brain Res.* 162:164-68
16. Berger, B. D., Wise, C. D., Stein, L. 1973. *Science* 180:506-8
17. Berger, E. A., Shooter, E. M. 1977. *Proc. Natl. Acad. Sci. USA* 74:3647-51
18. Berger, E. A., Shooter, E. M. 1978. *J. Biol. Chem.* 253:804-10
19. Bhagat, B., Rana, M. W. 1971. *Proc. Soc. Exp. Biol. Med.* 138:983-84
20. Bigazzi, M., Revoltella, R., Casciano, S., Vigneti, E. 1977. *Clin. Endocrinol.* 6:105-12
21. Bill, A. H., Seibert, E. S., Beckwith, J. B., Hartmann, J. R. 1969. *J. Natl. Cancer Inst.* 43:1221-30
22. Bjerre, B., Björklund, A., Stenevi, U. 1973. *Brain Res.* 60:161-76
23. Bjerre, B., Wiklund, L., Edwards, D. C. 1975. *Brain Res.* 92:257-78
24. Björklund, A., Stenevi, U. 1972. *Science* 175:1251-53
25. Bocchini, V., Angeletti, P. U. 1969. *Proc. Natl. Acad. Sci. USA* 61:787-94
26. Bothwell, M. A., Shooter, E. M. 1977. *J. Biol. Chem.* 252:8532-36
27. Bradshaw, R. A. 1978. *Ann. Rev. Biochem.* 47:191-216
28. Bradshaw, R. A., Niall, H. D. 1978. *Trends Biochem. Sci.* 3:274-78
29. Brunso-Bechtold, J. K., Hamburger, V. 1979. *Proc. Natl. Acad. Sci. USA* 76:1494-96
30. Bueker, E. D. 1948. *Anat. Rec.* 100:735 (Abstr.)
31. Bueker, E. D. 1948. *Anat. Rec.* 102:369-90
32. Bueker, E. D., Schenkein, I., Bane, J. L. 1960. *Cancer Res.* 20:1220-28
33. Burdman, J. A., Goldstein, M. N. 1964. *J. Natl. Cancer Inst.* 33:123-33
34. Burnham, P., Raiborn, C., Varon, S. 1972. *Proc. Natl. Acad. Sci. USA* 69:3556-60
35. Burstein, D. E., Greene, L. A. 1978. *Proc. Natl. Acad. Sci. USA* 75:6059-63
36. Burton, L. E., Wilson, W. H., Shooter, E. M. 1978. *J. Biol. Chem.* 253:7807-12
37. Calissano, P., Cozzari, C. 1974. *Proc. Natl. Acad. Sci. USA* 71:2131-35
38. Chamley, J. H., Goller, I., Burnstock, G. 1973. *Dev. Biol.* 31:362-79
39. Chen, M. G. M., Chen, J. S., Calissano, P., Levi-Montalcini, R. 1977. *Proc. Natl. Acad. Sci. USA* 74:5559-63
40. Cohen, S. A. 1958. In *A Symposium on the Chemical Basis of Development*, ed. W. D. McElroy, B. Glass, pp. 665-67. Baltimore: Johns Hopkins Press
41. Cohen, S. 1959. *J. Biol. Chem.* 234:1129-37
42. Cohen, S. 1960. *Proc. Natl. Acad. Sci. USA* 46:302-11
43. Connolly, J. L., Greene, L. A., Viscarello, R. R. 1979. *Fed. Proc.* 38:1430 (Abstr.)
44. Coughlin, M. D., Boyer, D. M., Black, I. B. 1977. *Proc. Natl. Acad. Sci. USA* 74:3438-42
45. Dichter, M. A., Tischler, A. S., Greene, L. A. 1977. *Nature* 268:501-4
46. Druckrey, H., Ivankovic, S., Preussmann, R. 1966. *Nature* 210:1378-79
47. Druckrey, H., Ivankovic, S., Preussmann, R., Zülch, K. J., Mennel, H. D. 1972. In *Experimental Biology of Brain Tumors*, ed. W. M. Kirsch, E. G. Paoletti, P. Paoletti, pp. 85-147. Springfield, Ill: Thomas
48. Druckrey, H., Preussmann, R., Ivankovic, S., Schmähl, D. 1967. *Z. Krebsforsch.* 69:103-201
49. Easty, G. C., Easty, D. M. 1976. In *Scientific Foundations of Oncology*, ed. T. Symington, R. L. Carter, pp. 167-72. London: Heinemann Med. Books
50. Edgar, D. H., Thoenen, H. 1978. *Brain Res.* 154:186-90
51. Eisenbarth, G. S., Drezner, M. K., Lebowitz, H. E. 1975. *J. Pharmacol. Exp. Ther.* 192:630-34
52. Fabricant, R. N., DeLarco, J. E., Todaro, G. J. 1977. *Proc. Natl. Sci USA* 74:565-69

53. Fabricant, R. N., Todaro, G. J., Eldridge, R. 1979. *Lancet* 2:4–7
54. Fidler, I. J., Kripke, M. L. 1977. *Chemistry* 50:18–24
55. Frazier, W. A., Boyd, L. F., Bradshaw, R. A. 1973. *Proc. Natl. Acad. Sci. USA* 70:2931–35
56. Frazier, W. A., Boyd, L. F., Bradshaw, R. A. 1974. *J. Biol. Chem.* 249:5513–19
57. Frazier, W. A., Boyd, L. F., Pulliam, M. W., Szutowicz, A., Bradshaw, R. A. 1974. *J. Biol. Chem.* 249:5918–23
58. Frazier, W. A., Boyd, L. F., Szutowicz, A., Pulliam, M. W., Bradshaw, R. A. 1974. *Biochem. Biophys. Res. Commun.* 57:1096–103
59. Frazier, W. A., Angeletti, R. H., Bradshaw, R. A. 1972. *Science* 176:482–88
60. Frazier, W. A., Hogue-Angeletti, R. A., Sherman, R., Bradshaw, R. A. 1973. *Biochemistry* 12:3281–93
61. Frazier, W. A., Ohlendorf, C. E., Boyd, L. F., Aloe, L., Johnson, E. M. Jr., Ferrendelli, J. A., Bradshaw, R. A. 1973. *Proc. Natl. Acad. Sci. USA* 70:2448–52
62. Furukawa, S., Nishitani, H., Hayashi, K. 1977. *Biochem. Biophys. Res. Commun.* 76:1202–6
63. Goldfine, I. D., Smith, G. J. 1976. *Proc. Natl. Acad. Sci. USA* 73:1427–32
64. Goldstein, L. D., Reynolds, C. P., Perez-Polo, J. R. 1978. *Neurochem. Res.* 3:175–83
65. Goodman, R., Chandler, C., Herschmann, H. R. 1979. In *Cold Spring Harbor Cell Culture & Hormones*, ed. R. Roth, G. Sato. Cold Spring Harbor, NY: Cold Spring Harbor Lab. In press
66. Goodman, R., Herschmann, H. R. 1978. *Proc. Natl. Acad. Sci. USA* 75:4587–90
67. Gorin, P., Johnson, E. M. Jr. 1979. *Proc. Natl. Acad. Sci. USA*. In press
68. Greene, L. A. 1974. *Neurobiology* 4:286–92
69. Greene, L. A., McGuire, J. C. 1978. *Nature* 276:191–94
70. Greene, L. A., Rein, G. 1977. *Nature* 268:349–51
71. Greene, L. A., Rein, G. 1977. *Brain Res.* 129:247–63
72. Greene, L. A., Shooter, E. M., Varon, S. 1968. *Proc. Natl. Acad. Sci. USA* 60:1383–88
73. Greene, L. A., Tishler, A. S. 1976. *Proc. Natl. Acad. Sci. USA* 73:2424–28
74. Greene, L. A., Varon, S., Piltch, A., Shooter, E. M. 1971. *Neurobiology* 1:37–48
75. Haas, D. C., Hier, D. B., Arnason, B. G. W., Young, M. 1972. *Proc. Soc. Exp. Biol. Med.* 140:45–48
76. Hart, T., Chaimas, N., Moore, R. Y., Stein, D. G. 1978. *Brain Res. Bull.* 3:245–50
77. Hatanaka, H., Otten, U., Thoenen, H. 1978. *FEBS Lett.* 92:313–16
78. Hendry, I. A. 1972. *Biochem. J.* 128:1265–72
79. Hendry, I. A. 1976. *J. Neurocytol.* 5:337–49
80. Hendry, I. A., Campbell, J. 1976. *J. Neurocytol.* 5:351–60
81. Hendry, I. A., Stach, R., Herrup, K. 1974. *Brain Res.* 82:117–22
82. Hendry, I. A., Stöckel, K., Thoenen, H., Iversen, L. L. 1974. *Brain Res.* 68:103–21
83. Herrup, K., Shooter, E. M. 1973. *Proc. Natl. Acad. Sci. USA* 70:3884–88
84. Herrup, K., Shooter, E. M. 1975. *J. Cell Biol.* 67:118–25
85. Hier, D. B., Arnason, B. G. W., Young, M. 1973. *Science* 182:79–81
86. Hoffman, H. 1970. *J. Embryol. Exp. Morphol.* 23:273–87
87. Hogue-Angeletti, R. 1969. *Brain Res.* 12:234–37
88. Hogue-Angeletti, R. A., Frazier, W. A., Jacobs, J. W., Niall, H. D., Bradshaw, R. A. 1976. *Biochemistry* 15:26–34
89. Huff, K. R., Guroff, G. 1979. *Biochem. Biophys. Res. Commun.* 89:175–80
90. Huff, K. R., Lakshmanan, J., Guroff, G. 1978. *J. Neurochem.* 31:599–606
91. Ikeno, T., Guroff, G. 1979. *J. Neurochem.* In press
92. Ivankovic, S., Druckrey, H., Preussmann, R. 1966. *Naturwissenschaften* 53:410
93. Johnson, D. G., Gorden, P., Kopin, I. J. 1971. *J. Neurochem.* 18:2355–62
94. Johnson, E. M. Jr. 1978. *Brain Res.* 141:105–18
95. Johnson, E. M. Jr., Andres, R. Y., Bradshaw, R. A. 1978. *Brain Res.* 150:319–31
96. Koestner, A. 1973. *Proc. 10th Can. Cancer Conf.* 10:65–75
97. Kumar, S., Steward, J. K., Waghe, M., Pearson, D., Edwards, D. C., Fenton, E. L., Griffith, A. H. 1970. *J. Pediatr. Surg.* 5:18–22
98. Lakshmanan, J. 1978. *Biochem. Biophys. Res. Commun.* 82:767–75

99. Lakshmanan, J. 1978. *Brain Res.* 157:173-77
100. Larrabee, M. G. 1969. *Prog. Brain Res.* 31:95-110
101. Levi-Montalcini, R. 1966. *Harvey Lect.* 60:217-59
102. Levi-Montalcini, R. 1972. In *Immunosympathectomy*, ed. Steiner, G., Schonbaum, E., pp. 55-78. Netherlands: Elsevier
103. Levi-Montalcini, R., Aloe, L., Mugnaini, E., Oesch, F., Thoenen, H. 1975. *Proc. Natl. Acad. Sci. USA* 72:595-99
104. Levi-Montalcini, R., Angeletti, P. U. 1968. *Physiol. Rev.* 48:534-69
105. Levi-Montalcini, R., Booker, B. 1960. *Proc. Natl. Acad. Sci. USA* 46:384-91
106. Levi-Montalcini, R., Caramia, F., Angeletti, P. U. 1969. *Brain Res.* 12:54-73
107. Levi-Montalcini, R., Caramia, F., Luse, S. A., Angeletti, P. U. 1968. *Brain Res.* 8:347-62
108. Levi-Montalcini, R., Chen, M. G. M., Chen, J. S. 1974. *Abstr. 4th Ann. Meet. Soc. Neurosci. St. Louis*, p. 305
109. Levi-Montalcini, R., Hamburger, V. 1951. *J. Exp. Zool.* 116:321-62
110. Levi-Montalcini, R., Hamburger, V. 1953. *J. Exp. Zool.* 123:233-88
111. Levi-Montalcini, R., Meyer, H., Hamburger, V. 1954. *Cancer Res.* 14:49-57
112. Levi-Montalcini, R., Revoltella, R., Calissano, P. 1974. *Recent Prog. Horm. Res.* 30:635-99
113. Lewis, M. E., Brown, R. M., Brownstein, M. J., Hart, T., Stein, D. G. 1979. *Brain Res.* In press
114. Lewis, M. E., Lakshmanan, J., Nagaiah, K., MacDonnell, P. C., Guroff, G. 1978. *Proc. Natl. Acad. Sci. USA* 75:1021-23
115. Liuzzi, A., Angeletti, P. U., Levi-Montalcini, R. 1965. *J. Neurochem.* 12:705-8
116. Liuzzi, A., Pocchiari, F., Angeletti, P. U. 1968. *Brain Res.* 7:452-58
117. Longo, A. M. 1978. *Dev. Biol.* 65:260-70
118. Longo, A. M., Penhoet, E. E. 1974. *Proc. Natl. Acad. Sci. USA* 71:2347-49
119. MacDonnell, P. C., Nagaiah, K., Lakshmanan, J., Guroff, G. 1977. *Proc. Natl. Acad. Sci. USA* 74:4681-84
120. MacDonnell, P. C., Tolson, N., Guroff, G. 1977. *J. Biol. Chem.* 252:5859-63
121. MacDonnell, P., Tolson, N., Yu, M. W., Guroff, G. 1977. *J. Neurochem.* 28:843-49
122. Max, S. R., Rohrer, H., Otten, U., Thoenen, H. 1978. *J. Biol. Chem.* 253:8013-5
123. McGuire, J. C., Greene, L. A. 1979. *J. Biol. Chem.* 254:3362-67
124. McGuire, J. C., Greene, L. A., Furano, A. V. 1978. *Cell* 15:357-65
125. Merrell, R., Pulliam, M. W., Randono, L., Boyd, L. F., Bradshaw, R. A., Glaser, L. 1975. *Proc. Natl. Acad. Sci. USA* 72:4270-74
126. Mobley, W. C., Moore, J. B. Jr., Schenker, A., Shooter, E. M. 1974. *Mod. Probl. Pediatr.* 13:1-12
127. Mobley, W. C., Schenker, A., Shooter, E. M. 1976. *Biochemistry* 15:5543-52
128. Moore, J. B. Jr., Mobley, W. C., Shooter, E. M. 1974. *Biochemistry* 13:833-40
129. Murphy, R. A., Oger, J., Saide, J. D., Blanchard, M. H., Arnason, B. G. W., Hogan, C., Pantazis, N. J., Young, M. 1977. *J. Cell Biol.* 72:769-73
130. Murphy, R. A., Pantazis, N. J. 1978. *Abstr. 8th Ann. Meet. Soc. Neurosci. St. Louis* p. 604
131. Murphy, R. A., Pantazis, N. J., Arnason, B. G. W., Young, M. 1975. *Proc. Natl. Acad. Sci. USA* 72:1895-98
132. Murphy, R. A., Singer, R. H., Saide, J. D., Pantazis, N. J., Blanchard, M. H., Byron, K. S., Arnason, B. G. W., Young, M. 1977. *Proc. Natl. Acad. Sci. USA* 74:4496-4500
133. Nagaiah, K., Ikeno, T., Lakshmanan, J., MacDonnell, P., Guroff, G. 1978. *Proc. Natl. Acad. Sci. USA* 75:2512-15
134. Nagaiah, K., Lakshmanan, J., Montgomery, P., Yu, M. W., Guroff, G. 1978. *J. Neurochem.* 31:647-55
135. Narumi, S., Fujita, T. 1978. *Neuropharmacology* 17:73-76
136. Nikodijevic, B., Nikodijevic, O., Yu, M. W., Pollard, H., Guroff, G. 1975. *Proc. Natl. Acad. Sci. USA* 72:4769-71
137. Nikodijevic, B., Yu, M. W., Guroff, G. 1977. *J. Neurochem.* 28:851-52
138. Oger, J., Arnason, B. G. W., Pantazis, N., Lehrich, J., Young, M. 1974. *Proc. Natl. Acad. Sci. USA* 71:1554-58
139. Olender, E. J., Stach, R. W. 1977. *Biochem. Biophys. Res. Commun.* 79:561-68
140. Otten, U., Hatanaka, H., Thoenen, H. 1978. *Brain Res.* 140:385-89

141. Pantazis, N. J., Blanchard, M. H., Arnason, B. G. W., Young, M. 1977. *Proc. Natl. Acad. Sci. USA* 74:1492-96
142. Paravicini, U., Stoeckel, K., Thoenen, H. 1975. *Brain Res.* 84:279-91
143. Partlow, L. M., Larrabee, M. G. 1971. *J. Neurochem.* 18:2101-18
144. Pattison, S. E., Dunn, M. F. 1975. *Biochemistry* 14:2733-39
145. Pearce, F. L., Banthorpe, D. V., Cook, J. M., Vernon, C. A. 1973. *Eur. J. Biochem.* 32:569-75
146. Perez-Polo, J. R., Dy, P., Westlund, K., Hall, K., Livingston, K. 1978. *Birth Defects Orig. Artic. Ser.* 14:311-21
147. Perez-Polo, J. R., Shooter, E. M. 1975. *Neurobiology* 5:329-38
148. Pinkerton, H., Bhagat, B., Rana, M. W., Holtwick, S. 1971. *Cancer Res.* 31:1483-87
149. Revoltella, R., Bertolini, L., Pediconi, M., Vigneti, E. 1974. *J. Exp. Med.* 140:437-51
150. Reynolds, C. P., Perez-Polo, J. R. 1975. *Neurosci. Lett.* 1:91-97
151. Roger, L. J., Schanberg, S. M., Fellows, R. E. 1974. *Endocrinology* 95:904-11
152. Roisen, F. J., Murphy, R. A., Braden, W. G. 1972. *J. Neurobiol.* 3:347-52
153. Ronne, H., Anundi, H., Rask, L., Peterson, P. A. 1979. *Biochem. Biophys. Res. Commun.* 87:330-36
154. Sabatini, M. T., De Iraldi, A. P., De Robertis, E. 1965. *J. Exp. Neurol.* 12:370-83
155. Saide, J. D., Murphy, R. A., Canfield, R. E., Skinner, J., Robinson, D. R., Arnason, B. G. W., Young, M. 1975. *J. Cell Biol.* 67:A376 (Abstr.)
156. Schubert, D., Heinemann, S., Kidokoro, Y. 1977. *Proc. Natl. Acad. Sci. USA* 74:2579-83
157. Schubert, D., LaCorbiere, M., Whitlock, C., Stallcup, W. 1978. *Nature* 273:718-23
158. Schubert, D., Whitlock, C. 1977. *Proc. Natl. Acad. Sci. USA* 74:4055-58
159. Schwab, M., Thoenen, H. 1977. *Brain Res.* 122:459-74
160. Schenkein, I., Bueker, E. D., Helson, L., Axelrod, F., Dancis, J. 1974. *N. Engl. J. Med.* 290:613-14
161. Sherwin, S. A., Sliski, A. H., Todaro, G. J. 1979. *Proc. Natl. Acad. Sci. USA* 76:1288-92
162. Shine, H. D., Perez-Polo, J. R. 1976. *J. Neurochem.* 26:1-4
163. Siggers, D. C., Boyer, S. H., Eldridge, R. 1975. *N. Engl. J. Med.* 292:1134
164. Siggers, D. C., Rogers, J. G., Boyer, S. H., Margolet, L., Dorkin, H., Banerjee, S. P., Shooter, E. M. 1976. *N. Engl. J. Med.* 295:629-34
165. Skaper, S. D., Bottenstein, J. E., Varon, S. 1979. *J. Neurochem.* 32:1845-51
165a. Skaper, S. D., Varon, S. 1979. *Biochem. Biophys. Res. Commun.* 88:563-68
166. Smith, A. P., Varon, S., Shooter, E. M. 1968. *Biochemistry* 7:3259-68
167. Stach, R. W., Shooter, E. M. 1974. *J. Biol. Chem.* 249:6668-74
168. Stach, R. W., Shooter, E. M. 1979. *J. Neurochem.* In press
169. Stahn, R., Rose, S., Sanborn, S., West, G., Herschmann, H. 1975. *Brain Res.* 96:287-98
170. Stenevi, U., Bjerre, B., Björklund, A., Mobley, W. 1974. *Brain Res.* 69:217-34
171. Stoeckel, K., Gagnon, C., Guroff, G., Thoenen, H. 1976. *J. Neurochem.* 26:1207-11
172. Stoeckel, K., Guroff, G., Schwab, M., Thoenen, H. 1976. *Brain Res.* 109:271-84
173. Stoeckel, K., Paravicini, U., Thoenen, H. 1974. *Brain Res.* 76:413-21
174. Stoeckel, K., Schwab, M., Thoenen, H. 1975. *Brain Res.* 89:1-14
175. Stoeckel, K., Solomon, F., Paravicini, U., Thoenen, H. 1974. *Nature* 250:150-51
176. Suda, K., Barde, Y. A., Thoenen, H. 1978. *Proc. Natl. Acad. Sci. USA* 75:4042-46
177. Swenberg, J. A., Clendenon, N., Denlinger, R., Gordon, W. A. 1975. *J. Natl. Cancer Inst.* 55:147-52
178. Szutowicz, A., Frazier, W. A., Bradshaw, R. A. 1976. *J. Biol. Chem.* 251:1516-23
179. Szutowicz, A., Frazier, W. A., Bradshaw, R. A. 1976. *J. Biol. Chem.* 251:1524-28
180. Thoenen, H., Angeletti, P. U., Levi-Montalcini, R., Kettler, R. 1971. *Proc. Natl. Acad. Sci. USA* 68:1598-1602
181. Tomita, J. T., Varon, S. 1971. *Neurobiology* 1:176-90
182. Turner, J. E., Glaze, K. A. 1978. *Exp. Neurol.* 59:190-201
183. Unsicker, K., Krisch, B., Otten, U., Thoenen, H. 1978. *Proc. Natl. Acad. Sci. USA* 75:3498-3502

184. Varon, S. 1975. *Exp. Neurol.* 48(pt 2):75–92
185. Varon, S. Horii, Z. I. 1977. *Brain Res.* 124:121–35
186. Varon, S., Nomura, J., Perez-Polo, J. R., Shooter, E. M. 1972. In *Methods in Neurochemistry*, ed. R. Fried, pp. 203–29. New York: Dekker
187. Varon, S., Nomura, J., Shooter, E. M. 1967. *Biochemistry* 6:2202–9
188. Varon, S., Raiborn, C., Burnham, P. A. 1974. *Neurobiology* 4:231–52
189. Varon, S., Raiborn, C., Burnham, P. A. 1974. *J. Neurobiol.* 5:355–91
190. Vinores, S. A. 1976. *Diss. Abstr. Int. Sect. B* 37: 2159 (Abstr.)
191. Vinores, S. A. *Natl. Cancer Inst. Monogr.* In press
192. Vinores, S. A., Koestner, A. 1978. *J. Neuropathol. Exp. Neurol.* 37:704 (Abstr.)
193. Vinores, S. A., Koestner, A. 1979. *Fed. Proc.* 38:1269 (Abstr.)
194. Vinores, S., Perez-Polo, J. R. 1976. *Texas J. Sci. Spec. Publ.* 1:101–11
195. Vinores, S., Perez-Polo, J. R. 1976. *Texas J. Sci. Spec. Publ.* 1:113–24
196. Waddell, W. R., Bradshaw, R. A., Goldstein, M. N., Kirsch, W. M. 1972. *Lancet* 1:1365–67
197. Waghe, M., Kumar, S., Steward, J. K. 1970. *J. Pediatr. Surg.* 5:14–17
198. Walker, P., Weichsel, M. E. Jr., Fisher, D. A., Guo, S. M., Fisher, D. A. 1979. *Science* 204:427–29
199. Waris, T., Rechardt, L., Waris, P. 1973. *Experientia* 29:1128–29
200. Wechsler, W., Kleihues, P., Matsumoto, S., Zülch, K. J., Ivankovic, S., Preussmann, R., Druckrey, H. 1969. *Ann. NY Acad. Sci.* 159:360–408
201. Weis, J. S. 1968. *J. Embryol. Exp. Morphol.* 19:109–35
202. Weis, P. 1971. *J. Comp. Neurol.* 141:117–32
203. Winick, M., Greenberg, R. E. 1965. *Pediatrics* 35:221–28
204. Wlodawer, A., Hodgson, K. O., Shooter, E. M. 1975. *Proc. Natl. Acad. Sci. USA* 72:777–79
205. Yamada, K. M., Wessells, N. K. 1971. *Exp. Cell Res.* 66:346–52
206. Yankner, B. A., Shooter, E. M. 1979. *Proc. Natl. Acad. Sci. USA* 76:1269–73
207. Young, M., Oger, J., Blanchard, M. H., Asdourian, H., Amos, H., Arnason, B. G. W. 1975. *Science* 187:361–62
208. Yu, M. W., Hori, S., Tolson, N., Huff, K., Guroff, G. 1978. *Biochem. Biophys. Res. Commun.* 81:941–46
209. Yu, M. W., Nikodijevic, B., Lakshmanan, J., Rowe, V., MacDonnell, P., Guroff, G. 1977. *J. Neurochem.* 28:835–42

REFERENCE ADDED IN PROOF

210. Ishii, D. N. 1978. *Cancer Res.* 38:3886–93.

THE STRUCTURE OF PROTEINS INVOLVED IN ACTIVE MEMBRANE TRANSPORT[1]

♦9147

Ann S. Hobbs and R. Wayne Albers

Laboratory of Neurochemistry, National Institute of Neurological and Communicative Disorders and Stroke, National Institutes of Health, Bethesda, Maryland 20205

The scope of this topic is presently limited by the number of well-characterized active transport systems that are available for study. Primary active transport (226) may be defined as consisting of an enzyme-catalyzed reaction occurring on one side of a membrane that is linked to the vectorial transport of a substance not chemically altered by the process. This definition is useful because it permits one to compare the structural and mechanistic features of phenomenologically similar systems. The ATP-dependent transport systems fall easily within this category, as does the bacteriorhodopsin proton pump, and these are the topics of the present review. Among other membrane systems for which structural information is available, the pumping function of some is controversial, e.g. the electron-transporting oxidases (132, 227). These systems are not reviewed here, nor do we discuss systems that chemically transform their transport substrates in the process of accumulation (68, 105).

Reference to recent reviews of each of the four active transport systems is given in the appropriate section. The material discussed here is not intended to be comprehensive, but rather it is meant to summarize the available information about the structure of the proteins associated with each system and some of the strategies employed to obtain this information. In so doing, some interesting analogies and

---

[1]The US Government has the right to retain a nonexclusive, royalty-free license in and to any copyright covering this paper.

divergencies become apparent and may suggest further paths of investigation.

ATP-DEPENDENT SODIUM AND POTASSIUM TRANSPORT

The $(Na^+ + K^+)$-ATPase uses the hydrolysis of ATP as the source of energy for the active, coupled transport of Na^+ outward and K^+ inward across a wide variety of cell membranes. It requires the presence of Na^+, K^+, Mg^{2+}, and ATP, and it is thought to operate via a mechanism similar to the one shown in Figure 1 where E_1 and E_2 are different conformations of the unphosphorylated enzyme. $E_1 \sim P$ and $E_2 - P$ are different conformations of the phosphorylated enzyme, the former of higher energy and sensitive to ADP, which reverses the reaction, and the latter of lower energy and sensitive to K^+, which catalyzes its breakdown to E_2. It is still a matter of some controversy at exactly which step(s) the ion transport occurs, and whether the transports of Na^+ and K^+ are simultaneous or occur at different steps. However, there is strong evidence that phosphorylation of the enzyme is Na^+-dependent and dephosphorylation is K^+-dependent, and that K^+ inhibits the conversion of E_2 to E_1. Thus, the kinetic evidence is consistent with the transports of Na^+ and K^+ occurring at separate steps. This ATPase is very specifically inhibited by cardioactive steroids such as ouabain and strophanthidin, and these drugs have been used widely in attempts to elucidate its structure and function (32). The enzyme is thought to consist of two subunits, α and β; the smaller subunit, β, is a glycoprotein. That the $(Na^+ + K^+)$-ATPase is, in fact, the $Na^+ - K^+$ pump has been conclusively shown by the incorporation of the purified ATPase into lipid vesicles (76) and by demonstration of

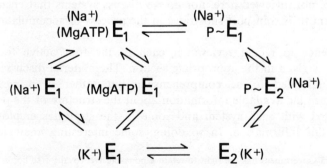

Figure 1 Proposed mechanism for the $(Na^+ + K^+)$-ATPase.

ATP-dependent and ouabain-sensitive Na^+ and K^+ fluxes in the reconstituted system. Several reviews of this enzyme are available for readers who are interested in more details of its structure, kinetics, and possible mechanism (4, 32, 56, 58, 224).

Ultrastructure

Negative staining of purified $(Na^+ + K^+)$-ATPase from *Squalus acanthias* by Hokin et al (78) revealed a rod and ring pattern with projecting subunits, the rods about 80 Å in diameter and the projections 35–55 Å. These authors suggested that the projections, which appeared to be rather hydrophilic based on their affinity for phosphotungstic acid, might be the glycoprotein subunits.

Freeze fracture studies on canine kidney (219), comparing a purified with a crude fraction, showed a class of particles of 95–120 Å, but "stalked" particles seen in the crude fraction were not seen in the purer ATPase. Freeze fracture studies by Vogel et al (222) on the pig kidney enzyme revealed globular particles of 80–120 Å diameter embedded in the membrane, and complementary negative staining experiments revealed long stalked particles, with a maximum diameter of 50 Å, extending into the cytoplasm. The "outer" surface, as interpreted by these authors, has a much finer granular structure. They conclude that the "knobs" and "core" comprise the α-subunit, with the "knob" perhaps containing the catalytic center, whereas the β-subunit projects from the core to the extracellular space and perhaps has a role in the orientation of the enzyme during its insertion into the membrane.

Deguchi et al observed similar 90–110 Å intramembranous particles in rabbit kidney, and smaller surface particles of 30–50 Å, but with a 3- to 4-fold greater frequency of the smaller particles (37). They proposed a model in which the large intramembranous particles were oligomers, made up of two or more of the units forming the surface particles.

Oligomeric Structure

The $(Na^+ + K^+)$-ATPase has two major subunits, α and β. The α-subunit has a molecular weight of about 95,000, and the β-subunit has one of about 55,000. Each of these differs somewhat in molecular weight from preparation to preparation as shown in Table 1. The differences shown in this table are probably real differences among the sources, as the weights also differ between different ATPases isolated and characterized in the same laboratory (156). The amino acid compositions of both large and small subunits are remarkably similar to each other and also are similar from preparation to preparation (77, 80). The stoichiometry of the subunits is a matter of some question. As shown in Table 2,

Table 1 Molecular weights of the α and β subunits of $(Na^+ + K^+)$-ATPase from selected sources

Source	Ref.	α	β
Electrophorus electricus	43	93,500	47,000
Squalus acanthias	78	97,000	55,000
Dog kidney	107	84,000	55,000
Beef brain	216	94,000	55,000
Carcharhinus obscurus	67	106,400	51,700
Anas platyrhynchos	80	94,000	60,000

Table 2 Stoichiometries suggested for the subunits of $(Na^+ + K^+)$-ATPase

$\alpha_4 \beta_2$	(67)	*C. obscurus*
$\alpha_2 \beta_2$	(48, 55, 89)	*S. acanthias*, dog kidney, rabbit kidney
$\alpha_2 \beta_3$	(48)	*S. acanthias*
$\alpha_2 \beta$	(80, 156)	*A. platyrhynchos, S. acanthias, E. electricus*

several different stoichiometries have been suggested. Estimates of molecular weight and subunit stoichiometry by gel electrophoresis are often subject to artifacts of staining and different dye affinities of the subunits. It seems to be fairly well agreed that each enzyme molecule has two α-subunits, suggested by chemical cross-linking studies (55) and by the fact that the whole enzyme has a molecular weight measured by radiation inactivation of about twice the sum of the individual weights of the subunits (32).

Infrared spectroscopy in the amide band region on the intact pig kidney enzyme (20) suggests that about 20% of the peptides are in highly ordered α-helical regions, 25% are in β-pleated sheets with antiparallel packing of chains, and these are contained in a hydrophobic core. The remainder of the protein appears to be unordered and accessible to water. These observations are in agreement with the known amino acid composition of the enzyme from several sources which include 50% hydrophobic residues (77).

Association of β, the glycoprotein subunit, with the $(Na^+ + K^+)$-ATPase is largely circumstantial, as no known function has been ascribed to it. The subunit has, however, copurified with the ATPase in stoichiometric association with the larger subunit (78), and antibodies against the glycoprotein subunit inhibit ATPase activity of the whole enzyme (85, 164). In addition, β can be cross linked to α using suberimidate (109) and ethyl acetimidate (202). β is far more resistant to proteolytic degradation than α (29), and the presence of Na^+ or K^+ enhances this resistance, suggesting that these ions may bind to the subunit. Part of the subunit, probably the carbohydrate portion, is on the surface of the membrane, since plant lectins are able to bind to the membrane-bound subunit (157). In addition to a wide variety of sugars, sialic acid has been reported to be a component in some preparations (32, 109). In addition to the major subunits, α and β, the $(Na^+ + K^+)$-ATPase has associated with it a small proteolipid subunit of about 12,000 mol wt. This has been observed in preparations of shark rectal gland (78), pig kidney (51), and eel electroplax enzymes (168). About half of the covalently bound amount of a photoaffinity derivative of ouabain is associated with this subunit, but the equivalent derivative of strophanthidin labels only the large subunit; this suggests that the sugar receptor for cardiac glycosides is located on the proteolipid, whereas the steroid receptor is located on the α-subunit (167). Forbush et al (51) have suggested that both the large subunit and the proteolipid are involved in the secondary (sugar) binding site.

Lipid Interactions

The native rabbit kidney enzyme has an estimated 267 phospholipid molecules per $(Na^+ + K^+)$-ATPase molecule, most of these consisting of phosphatidylcholine and phosphatidylethanolamine (40). Only about 90, however, are needed for proper functioning. Purified enzyme from *S. acanthias* or *Carcharhinus obscurus* has about 60 moles of phospholipid bound to each 300,000 g protein (48, 67).

The requirement of the $(Na^+ + K^+)$-ATPase for phospholipids has been known for a number of years, and certain phospholipids have been claimed as being essential for the functioning of the enzyme. For a partial summary, see Table 3 of Reference 99 and also the more detailed review of Dahl & Hokin (32). Many workers have considered negatively charged phospholipids as essential to activity, but contradictory conclusions have been drawn by different investigators, even when using the same enzyme source.

A recent study by de Pont et al (40) on the rabbit kidney enzyme led to their suggestion that negatively charged phospholipids are not absolutely required, although phosphatidylserine and phosphatidylinositol give the best reactivation of enzyme that has been inactivated by removal of the lipids. This conclusion is supported by the observation that ion transport is possible in vesicles consisting only of phosphatidylcholine, to which purified *S. acanthias* ATPase has been added (76). A similar study done on the rabbit kidney enzyme by Mandersloot et al (125) found that only negatively charged phospholipids were able to reactivate the ATPase after treatment with phospholipase A_2, but that phosphatidylcholine and phosphatidylethanolamine could reactivate if cholate were added simultaneously. They concluded that a negative charge on the lipid structure is essential for the functioning of the ATPase and that cholate is able to provide the necessary charge. This view may explain the results in the reconstituted vesicle system (76) in which a negatively charged phospholipid was not required.

Orientation of Binding Sites and Kinetics

Studies on the $Na^+ - K^+$ pump in intact systems provided some of the first evidence of numbers and affinities of monovalent cation-binding sites (41, 210). Post & Jolly first demonstrated in intact human erythrocytes that three Na^+ were transported outwardly for each two K^+ transported inwardly (158). Numerous studies since then have demonstrated the electrogenic nature of the pump (41, 210) and although there is controversy over the exact stoichiometry of Na^+/K^+ transported in many systems, it is clear that multiple binding sites for both Na^+ and K^+ must be accounted for in any structural model.

Recent ion-binding studies have reported 3 Na^+ sites per phosphorylation site [sheep kidney (95)] and 2 K^+ sites per ouabain-binding site [canine kidney (128)]. Ouabain blocked ion binding in both studies. These studies were the first direct measurements of ion binding to the ATPase, but numerous kinetic studies have revealed a high degree of cooperativity for both Na^+ and K^+ in stimulating their mutual transport and the full and partial ATPase reactions (4, 41, 56).

Ionophoric material has been isolated from the $(Na^+ + K^+)$-ATPase (16, 186, 188, 189), and will induce conductance of monovalent cations in lipid bilayers. The conductance is not specific for Na^+ or K^+, but Na^+ appears to be required for its incorporation into the bilayer (186). It has been suggested that the ionophore is on the β-subunit (188) and that histidine plays a role in the permeation mechanism (16), but, as yet, few specific characteristics of the ionophore have been determined. There is some kinetic evidence that the Na^+ and K^+ ionophores are separate (201).

Kinetic studies showing a negative cooperativity with respect to ATP suggest that both high and low affinity ATP sites exist (24, 52), but whether these sites exist independently of one another or are interconvertible is unknown. In contrast, in the absence of K^+, the enzyme shows hyperbolic kinetics and only a high affinity site (141). K^+ inhibits the enzyme at very low ATP concentrations (15, 141) and also when UTP is the substrate, suggesting that UTP binds only to the high affinity site (193). Recent studies with vanadate, a potent inhibitor of $(Na^+ + K^+)$-ATPase, suggest that this compound is a phosphate analog and binds at the low affinity nucleotide site (25).

Both the site for phosphorylation (11, 216) and the primary site for binding of cardiac steroids (32, 108) are located on the α-subunit. Since steroids bind to the outer surface (155) and phosphorylation occurs on the inner surface, α evidently spans the membrane. Some studies have measured a stoichiometry of phosphorylation sites to ouabain-binding sites of 2:1 (80, 89), whereas others have obtained a ratio of 1:1 (156) or 1:2 (108). From studies measuring ouabain binding and phosphorylation per large subunit, it appears that, at least in the dog kidney enzyme, less than half of the active sites can be phosphorylated simultaneously (108).

The validity of using these measurements for determining subunit stoichiometries has been questioned in a recent report comparing the binding of nucleotide, vanadate, and ouabain to both membrane-bound and detergent-treated mammalian kidney enzyme (62). In the detergent-treated enzyme the numbers of binding sites for the three were identical. In the membrane-bound preparation, however, the vanadate- or phosphate-supported ouabain binding was double the binding of nucleotide, vanadate, and nucleotide-supported ouabain binding. Although the system was not fully described with regard to the possible presence of vesicles, in order to explain the results obtained it was proposed that the membrane-bound enzyme consists of aggregations of eight interacting α-subunits.

Chemical and Physical Probes of Structure

Early studies showed that N-ethylmaleimide (NEM) and oligomycin inhibit the overall $(Na^+ + K^+)$-ATPase while stimulating a sodium ion-dependent ATP-ADP transphosphorylation reaction (49, 64). The ability of ATP and other ligands to alter the rate of inhibition by NEM suggests that sulfhydryl (SH) groups are present at the active site (194). Schoot et al (177) used NEM in conjunction with 5,5'-dithiobis-(2-nitrobenzoic acid), another sulfhydryl reagent, to identify two classes of essential SH groups on the rabbit kidney enzyme. There appear to be two different hydrophilic types—one located at the ATP-binding site, and one that is involved in the phosphatase activity. In addition, there are hydrophobic SH groups that appear to be unrelated to the active center. Fluorescence studies using fluorescein mercuric acetate and anthranoylouabain on *Electrophorus electricus* enzyme suggest that all the SH groups are far ($\sim$70 Å) from the ouabain site, probably on the intracellular portion of the enzyme (86).

The use of inhibitors and kinetic studies has implicated several amino acid residues in the functioning of the ATPase. De Pont et al (39) used butanedione to inhibit the enzyme and concluded that there is an essential arginine residue in the ATP-binding center of the rabbit kidney enzyme. Cantley et al (23) have obtained evidence for involvement of a tyrosine at the ATP site in eel enzyme. The kinetic studies of Karlish & Yates, using tryptophan fluorescence, led to the conclusion that a small number of tryptophan residues are involved in the conformational change from E_2 to E_1 (97).

Electron paramagnetic resonance (EPR) studies show a single, very tight Mn^{2+} binding site for each 250,000 daltons of enzyme protein, which presumably represents the binding site for Mg^{2+} (60). This divalent cation site is exposed to solvent, retains water ligands, and is close to the apparent phosphorylation site of the enzyme. Nuclear magnetic resonance (NMR) studies using $^{205}Tl^+$ and $^7Li^+$ suggest that the Na^+ transport site is close (4–5.4 Å) to the active site for ATP hydrolysis (58) and that a K^+ site is within 12.6 Å of this Na^+ site (59). Grisham & Mildvan (60) present an interesting but somewhat speculative model in which the phosphoryl group on the enzyme alternates as a Na^+ and K^+ transporter.

Energy transfer measurements between anthranoylouabain and trinitrophenyl-ATP indicate that the ouabain and ATP sites are about 64 Å apart, and measurements between anthranoylouabain and Co^{2+} suggest that the Mg^{2+} activation site is more than 40 Å from the ouabain site (134).

Primary Structure

Useful information, mainly about the α-subunit, has been gained by the use of proteolytic enzymes. In the presence of K^+, trypsin cleaves the subunit into two polypeptides of 58,000 and 48,000 mol wt; in Na^+ solutions, tryptic digestion results in a 78,000 dalton fragment (90). The fact that only one cleavage seems necessary to largely abolish ATPase activity, but two are needed to inhibit the K^+-dependent phosphatase activity suggests that the phosphatase site is remote from the ATP-binding site (91). A similar study led to the conclusion that the protein areas involved in the transformation from the ADP sensitive to the ADP insensitive form of the phosphoenzyme and those involved in the stimulation of dephosphorylation by K^+ are close or identical to one another, but are removed from the ATP-binding area (92). An intramembrane portion of the subunit of about 12,000 mol wt has been identified using 5-(^{125}I)-iodonaphthyl-1-azide and has been localized as being near the N-terminus of the subunit (96). Four separate segments have been identified by the use of trypsin and chymotrypsin in the presence of Na^+, K^+ or ouabain; the site for phosphorylation and that for carboxymethylation has been identified as being on the central

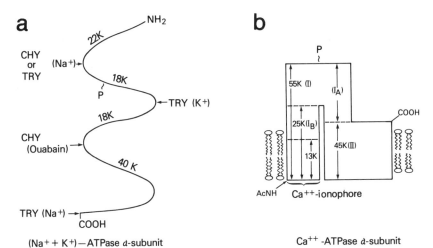

Figure 2 (*a*) Points of trypsin (TRY) and chymotrypsin (CHY) cleavage of the α-subunit of the (Na$^+$ + K$^+$)-ATPase. (*b*) Points of trypsin cleavage of the α-subunit of the Ca^{2+}ATPase, indicated by dashed lines. See text for details and references.

fragments; the phosphorylation site is closer to the N-terminus (26). This data is summarized in Figure 2*a*.

Little is known about the amino acid sequence of the (Na$^+$ + K$^+$)-ATPase. The N-terminal amino acid from the β-subunit of duck salt gland (80), dog kidney (109), eel electroplax (156), and shark salt gland (156) is alanine in each case. The N-terminal amino acid on the α-subunit seems to vary: Gly in the duck (80) and dog (109), Ala in the shark (156), and Ser in the eel (156). A short amino acid sequence from the N-terminal of α from the duck salt gland was determined to be Gly-Arg-Asn-Lys-Tyr-Glu-Thr-Thr-Ala-(?)-Ser-Glu- (80).

The amino acid sequence around the phosphorylation site in the guinea pig kidney enzyme has been tentatively characterized by somewhat indirect methods as -($^{Thr}_{Ser}$)-Asp(P)-Lys-, with a cysteine about four residues away from the Asp(P) on the N-terminal side (159). This is in agreement with the study of the enzyme from *Squalus acanthias* by Nishigaki et al (143), who characterized the phosphorylated intermediate as an aspartyl-β-phosphate residue.

Regulation

Numerous reports of hormones and other "serum factors" stimulating the Na$^+$ − K$^+$ pump or (Na$^+$ + K$^+$)-ATPase exist in the literature. Serum (170), as well as low doses of insulin (114, 170), epidermal growth factor (170), and prostaglandins (114, 170) stimulate Rb$^+$ uptake in mouse fibroblast cells. Lelievre et al (112) were able by washing to increase ouabain sensitivity of $(Na^+ + K^+)$-ATPase isolated from plasmocytoma cells grown in mice and to show that an EDTA extractable factor reversed this effect. Differential sensitivities to cardioactive steroids have been shown for two populations of the enzyme isolated from several brain tissues (203). These two forms, one from nonneuronal cells and the other from axolemma, showed slightly different molecular weights ($\Delta \cong 2,000$) for their α-subunits, as well as other chemical differences. It was suggested that the presence of two forms was related to different in vivo functions of the enzyme. Little is known in detail about these effects, but the implications concerning structure could be important for future work.

ATP-DEPENDENT CALCIUM TRANSPORT

There are several recent general reviews of Ca^{2+} transport (38, 65, 118, 119, 120, 121, 127, 207). ATP-dependent Ca^{2+} transport has been identified in several tissues other than striated muscle: for example, erythrocytes (173), brain (147, 215), and squid giant axon (42). The Ca^{2+} ATPase constitutes about 80 percent of the intrinsic membrane protein of skeletal sarcoplasmic reticulum (SR) (131).

The functional unit of the transport system probably consists of an oligomer of two or more large subunits (about 100,000 daltons), possibly a small proteolipid (about 12,000 daltons), and about 30 moles of essential phospholipid per large subunit. Recent evidence strongly suggests the association of a 55,000 dalton glycoprotein with the sarcoplasmic reticulum ATPase (Sr-ATPase) (121).

The mechanism of ATP hydrolysis involves a $(Ca^{2+} + Mg^{2+})$-dependent phosphorylation of the β-carboxyl of an aspartyl residue of the large subunit, followed by a Mg^{2+}-dependent hydrolysis of the phosphorylenzyme. Ca^{2+} transport is considered to result from a cycle of conformational transitions driven by the phosphorylation-dephosphorylation cycle as in the Na^+ system (Figure 1).

Ultrastructure

Negatively stained SR-vesicles are characterized by the presence of 35 Å projections from their outer surfaces (33, 63, 126, 171). Freeze-fracture replicas indicate the presence of 75–90 Å particles within the bilayer; the cytoplasmic leaflet exhibits a higher particle density than the lumenal leaflet (33, 66). In developing muscle, the rate of appearance of particles is well correlated with the rate of appearance of ATPase and Ca^{2+} transport activities (213). Both the particles and the projections are seen in vesicles reconstituted from the detergent-solubilized ATPase.

Tillack et al (213) have determined that the density of pumping sites (3,000 per μm^2) approximates the observed particle density in the SR of intact chicken muscle. Scales & Inesi (172) estimate a particle density of 5,700 per μm^2 for rabbit skeletal muscle. These authors were able to demonstrate the 35 Å projections by freeze-etching and to calculate their density to be 21,000 per μm^2. The density of ATPase large subunits was estimated to be 14,000 to 17,000 per μm^2, based upon the amount of phosphorylenzyme. Previous treatment with trypsin removes the evidence of projections in both negatively stained and freeze-etched vesicles (172). This treatment also produces a decrease in particle density within the cytoplasmic leaflet concurrently with an increase in the density within the lumenal leaflet. These observations suggest that the intramembranous particles may consist of three or four of the large subunits, whereas the projections might correspond to extensions of the individual protomers into the cytoplasm.

There are two reports of X-ray diffraction patterns of vesicles that have been oriented in a centrifugal field (45, 229). Both studies have concluded that the protein is asymmetrically distributed with respect to the lipid bilayer, but they disagree as to the sidedness of the distribution. Worthington & Liu (229) conclude that the average center-to-center particle separation is 73 Å, corresponding to a density of 20,000 per μm^2, the expected density of protomers. Dupont et al (45) estimate that 30 to 40 percent of the volume of the bilayer region is either water or protein.

Lipid Interactions

SR-vesicles contain about 80 molecules of phospholipid and 20 molecules of cholesterol per ATPase protomer. Several studies have found that the enzyme activity becomes unstable if the molar ratio of phospholipid to protomer is less than 30 (63, 74, 172). Warren et al have employed cholate extraction to progressively reduce the content of essential phospholipid or to progressively replace this lipid with cholesterol (225). They find that in the former case the inactivation is irreversible, whereas in the latter case, addition of phospholipid can restore the ATPase activity. They propose that the essential 30 moles of phospholipid are required to form an annulus about the protein, that the annulus is necessary for normal enzyme function, and that this amount of phospholipid interacts more strongly with the protein than with other lipids of the bilayer (74). They suggest that cholesterol replaces phospholipid in the maintenance of enzyme structure, but inhibits by virtue of the rigidity of a cholesterol annulus which might interfere with requisite conformational transitions of the pumping cycle.

Madden et al (124) find that a reversible inhibition of ATPase activity is produced by prolonged incubation of the intact vesicles with cholesterol-rich liposomes. This inhibition is proportional to the amount of cholesterol transferred into the vesicles. They conclude that cholesterol in the bulk phase of the bilayer can equilibrate with any annular phase that may be present.

Glycerol has been found to stabilize the ATPase activity when present during the process of phospholipid removal by deoxycholate (34, 35). By this means phospholipid content can be reduced as low as 1 mole of lipid per mole of large subunit. Although the ATPase is inactive in this state, it can be restored either by adding back phospholipid or by adding any of several non-ionic detergents. Dean & Tanford find that, despite the nearly complete absence of phospholipid, the detergent-activated ATPase displays the nonlinear Arrhenius plot characteristic of the native enzyme. This suggests that the nonlinear characteristic must be ascribed to something other than the phospholipid phase transition (35).

The V_{max} and temperature dependence of the detergent-activated enzyme reaction are similar to those of the native phospholipid-activated enzyme, but the substrate activation curves are quite different. The native enzyme produces a curve indicative of negative cooperativity, whereas the detergent-activated enzyme displays a simple hyperbolic response (35, but see also 208).

Oligomeric Structure

The electron micrographs of sarcoplasmic reticulum vesicles suggest that the enzyme may exist in the membrane as associations of three or four large subunits (87, 172). Similar oligomers appear as the initial products of solubilization by various detergents (113). Several attempts have been made to ascertain the state of association within the membrane by the use of cross-linking reagents (69, 116). A range of covalent associations have been found and interpreted as evidence for an oligomeric native ATPase. Hebdon et al have noted that the high density of subunits within the bilayer could react with these reagents to a large extent by random collisions lacking functional significance (69). They have reexamined some of these techniques and employed them at −10° to reduce bilayer fluidity and the frequency of random collisions. At this low temperature monomers were the predominant species found under all conditions of reaction with either dimethyl suberimidate or Cu^{2+} phenanthroline. Although random collisions may account for the aggregates produced at 25°, this does not rule out the existence of functionally significant association; the areas of subunit association may be inaccessible to the reagents or simply may not contain reactive amino acid residues.

Vanderkooi et al (218) have assessed the degree of protomer interaction by fluorescence energy transfer. Donor and acceptor fluorophors were reacted with different batches of purified ATPase. Simple mixing of the two preparations did not produce evidence of energy transfer. Fusion of the ATPase vesicles by addition of detergent brought about energy transfer. Although increasing the phospholipid to protein ratio, and thus presumably reducing the particle density, did not reduce energy transfer, the interaction could be diluted out by addition of unlabeled ATPase. They conclude that a dynamic equilibrium occurs between monomers and tetramers within the bilayer matrix.

Oligomeric association of the large subunits is clearly unnecessary for ATPase activity because the detergent-activated enzyme exists primarily as the monomer (35). The competency of the monomeric ATPase to effect Ca^{2+} transport is not known. Moreover, although the purified large subunit possesses intrinsic ATPase activity (117) and when reconstituted into liposomes can transport Ca^{2+} (103), there is recent evidence that a glycoprotein subunit may be necessary for normal transport. MacLennan and co-workers (121) have resolved the 55,000 dalton band seen on sodium dodecyl sulfate (SDS) gels of sarcoplasmic reticulum into 4 components. One of these is the previously defined high affinity Ca^{2+}-binding protein (149). The major component however is a glycoprotein that co-purifies with the 100,000 dalton subunit during membrane fractionation. It is a transmembrane protein that can be extracted into solution by low concentrations of deoxycholate, and thus it becomes separated from the large subunit in the more highly purified ATPase preparations. In such preparations it does not appear to modulate ATPase activity. However, addition of the glycoprotein to reconstituted SR-vesicles can activate Ca^{2+} transport several fold. MacLennan et al suggest that the glycoprotein may function as an anion channel in association with the pump.

Even highly purified SR-ATPase preparations contain, in addition to large subunit and phospholipid, a proteolipid of estimated 12,000 mol wt (123). This protein contains an unusually high amount of glutamic acid and one or two moles of covalently bound fatty acid. The relevance of this proteolipid to the transport function is unknown, although it has been reported to increase transport efficiency in reconstituted liposomes (162). It does not appear to be necessary for ATPase activity because the molar ratio of proteolipid to large subunit in fully active preparations may be as low as .1 (cited in 120). It also does not readily display ionophoric properties. In fact, addition of the SR proteolipid to phospholipid planar bilayers reduced their ion and water conductance (110).

Orientation Within the Membrane and Disposition of Binding Sites

From physiological considerations, one expects the Ca^{2+} pump to traverse the bilayer, either functionally or structurally, to bind Ca^{2+} with high affinity and interact with ATP from the cytoplasmic side, and to release Ca^{2+} within the lumen of the sarcoplasmic reticulum. A number of experimental observations confirm this conception. Miyamoto & Kasai (133) have obtained binding-site affinities and capacities for both inner and outer surfaces of the SR. By comparing their results with the known Ca^{2+}-binding properties of the isolated SR proteins, they conclude that a class of sites on the outer surface correspond to the Ca^{2+} pump high affinity sites with $K_d = 8.4 \times 10^{-7}$M. The binding capacity of these sites was $11.9 \pm .9$ nmoles per mg protein. On the basis of estimates that 80 percent of the membrane protein is large subunit (131), this corresponds to about 1.5 Ca^{2+} bound per subunit.

Nucleotide binding to intact SR vesicles has been measured by the flow-dialysis technique (209). Because membranes are generally impermeable to nucleotides, one may assume an external site is measured. In the presence of 1 mM EDTA, .35 to .45 ADP binding sites were estimated per large subunit. This was calculated from a Scatchard analysis assuming hyperbolic binding.

Yates & Duance (232) have measured nucleotide and Ca^{2+} binding to the purified ATPase. Measuring MgATP binding in the presence of EGTA, they estimate one nucleotide bound per large subunit but find a Hill coefficient of .82, i.e. either two binding affinities or negative cooperativity. They also find negative cooperativity for Ca^{2+} binding [in agreement with Meissner (130)] and estimate that there are 2 Ca^{2+} sites per large subunit.

Ikemoto (82) interpreted his earlier data on Ca^{2+} binding to detergent-solubilized ATPase as indicating one activating site per large subunit. This value was reduced in the presence of ATP. He suggested that ATP may induce the formation of dimers and thus negative cooperativity.

Kinetic Studies Relating to Structure

The ATPase reaction rate as a function of substrate concentration can be resolved into high ($K_m = 10^{-5}$ M) and low ($K_m = 10^{-3}$ M) affinity sites (83) in agreement with the flow dialysis studies. Verjovski-Almeida

& Inesi (220) have shown that Ca^{2+} transport activation at the high affinity site correlates with the kinetics of formation of the phosphorylenzyme, whereas occupation of the low affinity sites activates by a nonphosphorylating mechanism. Neet & Green (136) determined substrate-velocity curves for several different types of SR-ATPase preparations. They found negative cooperativity in all cases. Jørgensen and co-workers (88) solubilized the ATPase with deoxycholate in the presence of .4 M KCl and .3 M sucrose. Under these conditions the enzyme chromatographs as a monomer. In this form high affinity MgATP binding was retained, but the V_{max} was reduced to 35–50 percent of that of vesicular ATPase. Activation by ATP at the low affinity site, i.e. negative cooperativity, was not seen in this preparation.

The phospholipid-depleted ATPase, prepared in the presence of glycerol, exists largely in the monomeric form and can be reactivated by certain neutral detergents (77, 78). This type of monomeric ATPase was originally reported to exhibit only hyperbolic substrate activation. However, more recently Taylor & Hattan (208) have concluded that negative cooperativity is detectable in this preparation, although it is less marked than in the oligomeric forms. The high and low affinity nucleotide binding sites have different specificities. Thus pyrophosphate (44), α,β-methylene adenosine 5'-triphosphate (44), and β,γ-methylene adenosine 5'-triphosphate (190, 208) bind at the low affinity sites to activate ATP hydrolysis, but all are comparatively ineffective as either substrates or inhibitors at the high affinity sites.

Froehlich & Taylor (53) found that at 10 μM ATP, the level of phosphorylenzyme is a sigmoid function of Ca^{2+} concentration. This is consistent with the Ca^{2+}/ATP stoichiometry. Ikemoto (82) has shown that enzyme phosphorylation is accompanied by a decrease in Ca^{2+} affinity, whereas ATP binding is not.

The reversibility of the enzyme phosphorylation by ATP is not a function of the ATP concentration, but is sensitive to the concentration of Mg^{2+} and K^+ according to Shigekawa & Dougherty (191). The reaction of E-P with ADP is promoted by low Mg^{2+} and high K^+. At 0° the initially-formed E-P can react with ADP and, over a period of seconds, it is converted to an ADP-insensitive form (192). The rate of this conversion is a function of the ratio of Mg^{2+} to Ca^{2+} and the rate of hydrolysis of the ADP-insensitive form is accelerated by ATP and by both mono- and divalent cations.

Primary Structure of the SR-ATPase

Brief exposure of SR vesicles to trypsin results in cleavage of the large subunit into fragments I (55,000 daltons) and II (45,000 daltons) (165,

199, 212). Longer exposure to trypsin cleaves fragment I into 30,000 (I_A) and 20,000 (I_B) dalton fragments. The phosphorylation site is associated with fragments I and I_A (211). This is known to be the β-carboxyl of an aspartate residue (36). Ionophoric activity for divalent cations has been demonstrated for fragments I and I_B (2, 185, 187). The enzyme may be nicked to this extent and retain full ATPase activity. Ca^{2+} transport survives the first cleavage but not the second (180).

Allen & Green (5) have sequenced CNBr fragments of the large subunit to the extent of about 75 percent of the residues. The intact subunit, fragment I, and fragment I_B have N-acetyl methionine as N-termini so that the principal fragments may be ordered (Figure 2B) (101). Fragment I_B yields 4 CNBr fragments that have been further aligned. The 13,000 dalton fragment retains activity as a divalent cation ionophore.

After tryptic cleavage the major fragments are not appreciably dissociated by the neutral detergent $C_{12}E_8$ under conditions which preserve enzyme activity but dissociate most of the phospholipid (166). Solubilization of the "nicked" ATPase with deoxycholate or cholate produces only partial dissociation but complete inactivation. Circular dichroism (CD) spectral changes indicate that the dissociation is accompanied by substantial conformational changes. Following dissociation in sodium dodecylsulfate, the fragments remain soluble in aqueous buffer after removal of detergent. From the criterion of structural stability in guanidinium chloride as assessed by CD spectra, Rizzolo & Tanford (166) conclude that fragments IA, IB, and II each contain domains of stable structure. These are considered to be hydrophobic and thus constitute probable regions for association with the bilayer. Extensive proteolysis with the nonspecific protease, Nagarse, leaves about half of the intrinsic membrane protein associated with the membranes (231).

Regulation of Ca^{2+}-ATPases

Among the characterized primary transport systems, the Ca^{2+} pump has generated the largest literature pertaining to regulatory mechanisms. In the case of cardiac sarcoplasmic reticulum, ATPase activity and Ca^{2+} uptake are activated by a c-AMP-dependent protein kinase (205). This activation has been found to involve phosphorylation of a 22,000 dalton membrane protein (204), termed phospholamban. Tada et al have shown that phosphorylation of phospholamban stimulates ATP hydrolysis by increasing the rate of hydrolysis of the phosphorylenzyme (206). The mode of interaction between phospholamban and the large subunit has not been demonstrated.

Evidence for c-AMP regulation is less clear for skeletal muscle Ca^{2+} transport. A phosphorylatable protein with electrophoretic mobility similar to that of phospholamban has been noted in SR from slow skeletal muscle but not in fast-contracting muscles (100). However Schwartz et al (179) have found c-AMP-dependent stimulation of Ca^{2+} uptake in SR from cat fast muscle, although no associated protein phosphorylation was detected. Stimulation of rabbit muscle SR-ATPase by a Ca^{2+}-activated protein kinase and inhibition by a protein phosphatase has been reported (81).

The Ca^{2+}-ATPase of erythrocyte plasma membranes is activated by a soluble protein (17) that has subsequently been identified as calmodulin (57, 84). Niggli et al (142) have shown that the calmodulin-ATPase interaction is direct and can activate the detergent-solubilized enzyme.

Models of the Ca^{2+} Pump

Several models for integrating the structural data have been proposed. Although these are not mutually exclusive, they tend to emphasize different data. The existence of a phosphorylenzyme intermediate, the Ca^{2+} dependence of its formation, and other kinetic data have led to an emphasis on the operation of a cycle of conformational transitions. Ultrastructural studies, protease fragmentation data, and various physical studies indicate that the enzyme is firmly anchored to the lipid bilayer and that a major portion of the large subunit projects from the cytoplasmic surface.

A model involving rotation of the whole enzyme during transport has been proposed (214). Subsequent experiments in which macromolecular markers were bound to the ATPase without affecting transport (46) seem to rule out such large movements of the transport protein. Racker (161) has discussed a model that includes half-of-sites reactivity of a dimeric ATPase, with the transport resulting from conformationally gated channels. Shamoo & Abramson (2, 185) have outlined a topological model based on alignment of the tryptic fragments and their functional associations. This model assigns the primary lipophilic anchoring function to fragment II, which must then contain a channel to which fragment I_B acts as a gate in series between the channel and the catalytic center located in I_A.

This model is somewhat at variance with conclusions of other studies which (166, 212) indicate that lipophilic anchoring segments are distributed among all of the primary tryptic fragments. These latter observations suggest that the large subunit may insert into the bilayer at several points (Figure 2b).

ATP-DEPENDENT PROTON TRANSPORT

The F_1 ATPase and its associated proton channel, F_0, can function as an ATP synthetase when a proton gradient exists across the membrane with which it is associated, or they can function as a proton pump when ATP is present as a substrate. It is not clear, however, that these two functions are reversals of each other (154). The F_1F_0 complex is found widely throughout nature: in the inner membrane of mitochondria, the thylakoid membrane of plant chloroplasts, and the plasma membranes of mesophilic and thermophilic bacteria. Its structure when isolated from these very divergent sources appears to be remarkably similar. There are significant differences among the systems, however, and it is risky to generalize the findings from one system. For this reason we have specified as much as possible the particular system for which an observation was made. For more detailed expositions of this subject than are presented here, excellent reviews with emphasis on the chloroplast (8, 129), mitochondrial (19, 104, 154, 181) and bacterial (93) systems have appeared recently.

The F_1 ATPase

The molecular weight of the F_1 ATPase appears to be about 385,000 daltons when isolated from thermophilic bacteria (93) or mammalian mitochondria (27) and 325,000 daltons when isolated from chloroplasts (8). The enzyme is separated from the membrane and from F_0 by treatment with EDTA or concentrated salt solutions. It is widely agreed that it is composed of 5 major types of subunits: α, β, γ, δ, and ε. The molecular weights of these subunits obtained from different sources are shown in Table 3. The stoichiometry of the subunits is in some dispute. The two largest subunits, α and β, seem to occur in a 1:1 ratio in nearly all preparations. The number of smaller subunits varies somewhat, however. This has caused some workers to measure subunit ratios of $\alpha_3\beta_3\gamma\delta\varepsilon$ (27, 93, 104), whereas others calculate ratios of $\alpha_2\beta_2\gamma_{1-2}\delta_{1-2}\varepsilon_{1-2}$

Table 3 Molecular weights of subunits isolated from selected F_1 ATPase preparations

Preparation	α	β	γ	δ	ε
Rat liver mitochondria (27)	62,000	57,000	36,000	12,500	7,500
Thermophilic bacterium PS3 (233)	56,000	53,000	32,000	15,500	11,000
Spinach chloroplasts (see 8)	59,000	56,000	37,000	17,500	13,000

(7, 12). It should be pointed out that this is not simply a difference in the sources, since different methods applied to the same source have produced different conclusions (e.g. see 184). However, there are significant differences in the behavior and structure of F_1 as obtained from chloroplasts, mitochondria, and bacteria, and it may well turn out that the subunit ratios are not identical. In addition, it should be pointed out that the subunits have been named and classified by molecular weight, and they may not be functionally analogous from preparation to preparation, as is discussed below.

ULTRASTRUCTURE Electron microscopic studies indicate that F_1 from beef heart and rat liver mitochondria and from the thylakoid membranes of chloroplasts is a sphere of 90 Å, which appears in many cases to be joined to the membrane by a stalk-like structure (94, 148, 154). Small angle X-ray scattering and light-scattering experiments on F_1 from spinach chloroplasts lead Paradies et al (153) to the conclusion that this enzyme is an almost spherical particle with a diameter of 110 Å. However, similar studies of F_1 from *Escherichia coli* suggest that this molecule is an oblate ellipsoid of revolution, with half axes of $a = 43.5$ Å and $b = 60$ Å and a height of 68.5 Å (152).

Computer filtering of electron micrographs of two-dimensional crystals from a thermophilic bacterium, PS3, indicates that their shape is hexagonal, with pseudo 6-fold and 3-fold symmetry (223). The length of each side of the hexagon is about 38 Å. An area of lower electron density in the center of the hexagon may be a proton channel (223). Three patterns are seen in electron microscopy of crystals from beef heart mitochondria (197). From X-ray diffraction patterns, two of these have the plane group *pg*, and in the third, the periodicity across the crystal is doubled, and the plane group is *pgg*.

Only rat liver mitochondria F_1 ATPase has been crystallized sufficiently well that X-ray diffraction analysis appears possible (6). This enzyme forms large rhombohedral crystals with space group R32 and hexagonal cell dimensions $a_{hex} = 148$ Å and $c_{hex} = 368$ Å. The asymmetric unit is 190,000 daltons, about half the molecular weight, and the molecule probably has a 2-fold axis of symmetry corresponding to the 2-fold axis of symmetry of space group R32 (6).

It is difficult to make comparisons among these studies, as there may be several different crystalline forms. This is especially plausible in view of the evidence in the X-ray and light-scattering experiments (152, 153) that the shape of F_1 may be significantly different in the different sources.

Table 4 Suggested locations of subunits of F_1F_0 in bacterial (B) chloroplast (C) and mitochondrial (M) systems[1]

Head:	α, β, γ (B, C, M); ε (C, M?); δ (M?)
Stalk:	δ (B, C); ε (B); OSCP, F_6 (M)
Basepiece:	DCCD binding protein (B, C, M)

[1]See text for details and references.

ORGANIZATION OF THE SUBUNITS It is widely believed that the α, β, and γ subunits comprise the "head" of F_1, which is seen in electron micrographs, whereas δ (in chloroplasts and bacteria) and in some cases ε (bacteria) represent the "stalk" attaching the ATPase to F_0, embedded in the membrane. Some of this information comes from cross-linking studies, but much has been gleaned from reconstitution experiments. A partial summary of some of the proposed structural relationships is shown in Table 4.

Although, as mentioned above, the exact stoichiometry of the subunits is still in doubt and may in fact differ between sources, cross-linking studies have suggested which subunits may be adjacent, and the lack of certain combinations has suggested some spatial separation between the respective subunits. Baird & Hammes (7) have found that in F_1 from chloroplasts, α and β cross-link to each other and to all of the other subunits. They found the $\gamma\varepsilon_2$ combination but not ε_2, which suggests that the two ε's are not adjacent. δ did not cross link with either γ or ε. As pointed out by the authors, the lack of a given combination does not rule out the possibility that the subunits are adjacent, but simply do not have a close proximity of the proper reactive groups. They have developed a symmetrical model with stoichiometry $\alpha_2\beta_2\gamma\delta\varepsilon_2$, in which the α's and β's alternate, and γ and two ε's are found on one side of them while δ is on the opposite one. The authors emphasize that this stoichiometry is the minimal one.

This model differs somewhat from the model postulated from reconstitution experiments of F_1 from *E. coli* (198; and see Reference 8). In this model, δ and ε are adjacent as the "stalk". Since the role of the ε subunit can be different depending upon the source (see below), it is plausible that its location might also vary. Reconstitution experiments suggest that a $\gamma\delta\varepsilon$ combination is the "gate" of the proton channel in bacteria (233), so it seems likely that they are in close proximity to each other and to F_0.

STRUCTURE AND FUNCTION OF THE SUBUNITS Reconstitution experiments with various combinations of the 5 subunits show that no subunit is active alone, but that β is essential for ATPase activity in a preparation from thermophilic bacteria, and that a variety of combinations containing β are active (235). The $\alpha\beta\gamma$ combination seems to comprise most of the ATPase activity. This, along with evidence that aurovertin (221) and ATP analogs (104) bind only to β, suggests that the active site for ATP hydrolysis is on the β-subunit (93).

A variety of "tight" and "loose" nucleotide binding sites exist on all types of F_1 studied so far (8, 19, 154, 195). Their role in ATPase activity is uncertain, however. In both the mitochondrial (61) and chloroplast (139) systems a high degree of cooperativity is seen with respect to ATP when ADP is present. Unlike the Ca^{2+} and $(Na^+ + K^+)$-ATPases, phosphorylated intermediates have not been isolated (154). There is some belief that nucleotide binding provides the energy necessary to carry out the functions of F_1 (3, 98). Fluorescence resonance energy transfer studies have provided some information on the relative distances of some of the sites identified in the chloroplast system (9, 79). The fact that critical sites seem to be far from each other suggests that conformational changes must play a very important role in the mechanism.

The δ-subunit appears to be required for binding of the F_1 ATPase to the membrane in chloroplast and bacterial preparations (138, 196, 233). Isolated from chloroplasts, it has a molecular weight of about 21,000, and its physical and X-ray scattering characteristics suggest that it is a prolate ellipsoid of revolution (175), with gross dimensions $2a = 25$ Å, $2b = 28$ Å, and $2c = 90$ Å. This subunit may represent the "stalk" that appears in electron micrographs (196). The oligomycin sensitivity-conferring protein (OSCP) of mammalian mitochondria appears to fulfill at least part of the role that δ plays in the chloroplast and bacterial systems (122). This protein has an apparent molecular weight of 18,000 when purified from beef heart (122). A second soluble polypeptide from mitochondria, F_6, binds OSCP and facilitates the binding of solubilized F_1 (163).

The ϵ-subunit of F_1 from chloroplasts is thought to be analogous to the endogenous inhibitor isolated from mitochondria by Pullman & Monroy (160) and to play a role as a regulator of F_1 (140). In mitochondria, however, the ϵ-subunit and the inhibitor are not identical (22). The behavior of the two inhibitors is quite different. Their inhibitory actions do not cross-react with their respective ATPases, and the amino acid compositions are dissimilar (140). The ϵ-subunit is more insoluble in aqueous solution and more difficult to detach than the mitochondrial inhibitor, and it has been proposed that differences in this subunit account for the fact that chloroplast ATP synthetase does not reverse under ordinary conditions, whereas mitochondrial does (140). Heat activation involves the removal of the ε-subunit in *Alcaligenes faecalis*, and deactivation may involve the covering of negative charges (3).

Schmidt & Paradies (174) have used X-ray and inelastic light scattering to show that the ε-subunit isolated from spinach chloroplasts is a nearly spherical particle in solution with a diameter of 32 Å. They conclude that it is a prolate elipsoid of revolution, with half axes $a = b = 12.7$ Å and $c = 25.4$ Å. The molecular weight of their subunit, estimated by sedimentation and diffusion coefficients, is 11,900, which is in fairly good agreement with the estimate from SDS gel electrophoresis (Table 3).

Additional subunits A number of additional factors involved in the F_1F_0 complex have been isolated. OSCP, isolated from mitochondrial systems, is generally defined as a component of F_0 (154). An endogenous inhibitory factor of about 10,000 daltons was isolated by Pullman & Monroy from bovine heart mitochondria (160). In rat liver and yeast mitochondria, similar factors have molecular weights of 12,300 (30) and 7,000 (47). The amino acid composition of this inhibitor has been determined for at least four species (47). F_6 (see above) may be one of the subunits of F_1 (30). F_{C1} and F_{C2}, two factors involved in oligomycin and dicyclohexylcarbodiimide (DCCD) sensitivity and isolated by Knowles et al (102) from mitochondria, are probably identical to OSCP and F_6.

The Proton Channel

F_0 is a collection of extremely hydrophobic polypeptides consisting of at least three different subunits; it remains embedded in the membrane under extraction procedures that detach the F_1 ATPase. One of these subunits is a polypeptide of about 8,000 daltons which probably exists as a hexamer (182) and which binds DCCD, a potent inhibitor of the membrane-bound, but not the solubilized F_1 ATPase. This subunit is sufficient to support proton transport when it is isolated from chloroplasts and reconstituted into phospholipid vesicles (31, 137, 236); however, the transport, when reconstituted from a thermophilic bacterial system, is not substantial unless all three of the F_0 components are present (93). Cooperativity with respect to H^+ has been observed for proton transport (176). The amino acid sequence has been determined for the DCCD binding protein of F_0 from *Neurospora crassa*, *Saccharomyces cerevisiae*, and *E. coli* by Sebald & Wachter (183), and is

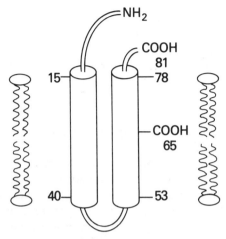

Figure 3 Probable orientation of DCCD-binding protein. Residues 15-40 and 53-78 are hydrophobic regions probably forming helical segments across the bilayer. Residue 65 is labeled with ^{14}C-DCCD (183).

quite similar in the three preparations. Although the hydrophilic sequences in the N- and C-termini appear to be different in these species, which are widely separated in evolution, all three species have a short hydrophilic amino acid sequence in the center of the protein, surrounded by two very hydrophobic segments of about 25 residues each. Close to the center of the hydrophobic chain that is located nearest the C-terminus is an acidic side chain which reacts with DCCD. This same hydrophobic chain is also connected with oligomycin sensitivity. Sebald & Wachter (183) postulate that the two hydrophobic segments span the lipid bilayer and that the polar regions are in contact with the aqueous phase and the other subunits of the F_1F_0 complex, as shown in Figure 3. Even with a hexameric form of the DCCD-binding protein embedded in the membrane, there do not appear to be enough hydrophilic groups to construct a proton channel consisting of a chain of hydrogen bonds (cf 135, 228); thus, a model such as that of Boyer (18), which involves migration of a charged amino acid side chain, seems more consistent with the current evidence (183).

BACTERIORHODOPSIN

Several recent comprehensive reviews of this subject have appeared (71, 72, 111, 178, 200). Bacteriorhodopsin functions in the plasma membranes of *Halobacterium halobium* as a proton pump driven by the photoisomerization of its chromophor, retinal (146). The retinal is covalently linked as a Schiff's base to a lysine residue (145). When the bacterium is maintained under conditions of low O_2 the bacteriorhodopsin accumulates as patches of about .5 m diameter (14). These purple patches can be separated from other components of osmotically lysed cells by density gradient centrifugation.

Ultrastructure

The purple membrane is made up of about 75 percent protein and 25 percent lipid (145). Evidence from a neutron diffraction study (cited in Reference 72) indicates that water is present only as a thin layer on the surface of the lipid bilayer. About 30 percent of the lipid is glycolipid and over 50 percent is phosphatidylglycerophosphate (106). Most of the glycolipid occurs on the outer surface. The protein can be seen in freeze-fracture electron micrographs as extensive regular arrays of 120 Å particles, each of which consist of 3–4 of the unit cells discussed below (50).

From the X-ray diffraction patterns of aqueous suspensions, the unit cell dimension was found to be 63 Å (14). Measurements of X-ray diffraction of oriented and dried purple membranes demonstrated the occupation of each unit cell by three polypeptide chains (13, 70). Additional features of the pattern have been attributed to α-helical structures oriented nearly perpendicular to the membrane plane. Unwin & Henderson (73, 217) have combined data from electron diffraction patterns and electron micrographs to produce a three-dimensional structure of the purple membrane to about 7 Å resolution. The bacteriorhodopsin molecule contains seven helical segments about 40 Å long and 10 Å apart, extending nearly across the 45 Å membrane and almost perpendicular to the membrane plane.

The helical segments of each molecule are disposed in two roughly parallel arcs; the outer has four and the inner, three segments (Figure 4a). From the density difference profiles of the X-ray intensities of wet and dry membranes, it is concluded that intervening spaces contain phospholipids. Three molecules constitute a unit cell, arranged so that the three inner arcs approximate a circle. The segments of the outer arcs are tilted somewhat more from the membrane perpendicular than are those of the inner circle.

The helices show some degree of supercoiling from the electron diffraction data (71). This is also suggested by the infrared amide-I absorption which is shifted upfield from the normal range for α-helical polypeptides (169). Because of the rigid lattice arrangement of the bacteriorhodopsin molecules within the membrane, chromophore-chromophore interactions can occur. Heyn and co-workers (75) have de-

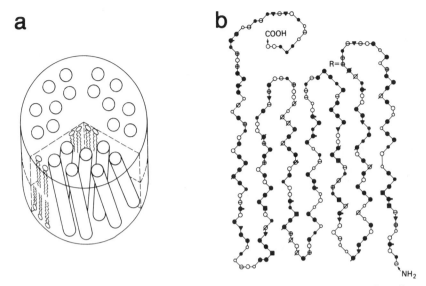

Figure 4 (*a*) Schematic diagram of the unit cell of bacteriorhodopsin and the orientation of helices of one molecule within the unit cell. The unit cell is integrated into the lipid bilayer. Adapted from Reference 72. (*b*) Primary structure of bacteriorhodopsin. (Adapted from Reference 150). The polar amino acids are represented by the following open symbols: ○ = Gly; ○ = Ser, Thr, Asn, Gln; ⊖ = Asp, Glu; ⊕ = Lys, Arg; ⌀ = Tyr. The less polar amino acids are represented by closed symbols: • = Ala; • = Val, Leu, Met; • = Ile; ⌀ = Phe; ■ = Try; ▼ = Pro. R = retinal. Helix-favoring residues are set obliquely with respect to the preceding residue, whereas helix-breaking residues are set orthogonally to the preceding residue.

tected exciton bands in the CD spectrum which are associated with the 570 nm absorption. These bands disappear under conditions that dissociate the trimers without affecting secondary structure.

Primary Structure

Oesterhelt & Schumann (144) demonstrated that the retinal Schiff's base of bacteriorhodpsin can be reduced to a stable secondary amine. Bridgen & Walker obtained the sequence of a thermolysin fragment containing the retinylamine (21). From the total amino acid analysis, the molecular weight was estimated at about 25,000, corresponding to seven lysines per molecule. Because only five lysine-containing peptides are produced by thermolysin treatment, the protein was shown to contain a single peptide.

Partial sequences have been published from several laboratories (1, 21, 54). The complete sequence has now been described by Ovchinnikov and co-workers (150).

The bacteriorhodopsin molecule within the intact purple membrane can be hydrolyzed by papain at five points (1, 150), presumably near the membrane surfaces. From this data certain segments of the primary sequence can be identified as serving to connect adjacent helices. Gerber et al (54) have shown that carboxypeptidase A removes amino acids from the protein both in intact membranes and in "inside-out" vesicles. Thus the C-termini are normally exposed at the cytoplasmic surface, and because of the odd number of helical segments the N-termini must be within the periplasmic space. Ovchinnikov et al have shown in the right-side-out vesicles that the 71–72 peptide bond is susceptible to chymotrypsin, suggesting that this region connects the second and third helices. Similar reasoning has led to a tentative two-dimensional map of the disposition of the primary sequence within the membrane (150). This map places most of the charged residues near the membrane surfaces; in particular, Lys-41, bearing the retinal, is near the first bend on the cytoplasmic side (Figure 4b).

Structure of the Active Site

Packer & Konishi have discussed experiments suggesting that folding of the helical segments within the plane of the membrane may bring several aromatic residues near the chromophore (151).

Light absorption is known to initiate a cycle of interactions that have been spectroscopically defined and that are coupled to proton pumping (146). In its ground state, the retinylidene exists in a protonated form that is accessible to the aqueous environment (115). This form becomes deprotonated during the pumping cycle. The rate of deprotonation appears to be controlled by an interaction with an adjacent residue that has a pK near 10.2. This pK is consistent with the participation of the adjacent Lys 40. Lewis and co-workers (115) propose a model in which the structural transitions subsequent to light absorption act to lower the pK of the Schiff's base complexed with Lys-40; this promotes transfer of a proton to an adjacent residue with a high pK (Arg?). The retinylidene would be subsequently reprotonated by a third residue (Tyr?).

Such a chain of proton transfers might continue across the membrane via closely spaced, hydrogen-bonding residues, according to a proposal of Nagle & Morowitz (135). No single helical segment of the bacteriorhodopsin map could form such a chain, but appropriate folding within the membrane plane might permit residues from two or three segments to interdigitate and to form a proton-conducting path through the chromophoric site. The interactions among adjacent helical segments must in fact be rather strong because the "nicked" bacteriorhodopsin remains functional (1).

RELATION OF STRUCTURE TO FUNCTION IN ACTIVE TRANSPORT SYSTEMS

The ATP-dependent Ca^{2+} and Na^+ transport systems appear similar with respect to macromolecular organization, active site structure, and kinetic mechanism. Although this is true in broad outline, there may be major differences. For example, the Ca^{2+}-ATPase activity is retained in a monomeric detergent-solubilized 100,000 dalton subunit, whereas the $(Na^+ + K^+)$-ATPase has not been separated in active form from its associated glycoprotein and appears to be dimeric in its detergent-solubilized form. Sequencing of the $(Na^+ + K^+)$-ATPase should provide more conclusive information about structural homologies in the near future. At least in the case of the Ca^{2+}-ATPase, the catalytic and ionophoric domains appear to be separate, and coupling of these functions must involve conformational interactions. Although this concept suffers from lack of concreteness, there are data that demonstrate substantial structural changes in the proteins consequent to cation (90) and nucleotide (97) binding. Large scale reorientation of functional domains in response to substrate binding may be a common feature of ATP-dependent kinases, as is suggested by X-ray data (10). It seems reasonable to expect that similar properties may be associated with ATP-dependent transport systems.

The F_1F_0-proton transport system is readily dissociable into ATPase and ionophoric domains consisting of separate subunits. If the proton transport proves to reside exclusively within the F_0 complex, one must again invoke conformational coupling between the catalytic and ionophoric domains. To the extent that a mechanism can be inferred from the primary structure of the DCCD-binding protein, the single carboxyl embedded in a hydrophobic region (Figure 3) cannot constitute a channel but perhaps can transport by a saltatory mechanism which, in turn, may energize ATP synthesis. In contrast, the bacteriorhodopsin structure, with proper folding of the helices within the bilayer plane (Figure 4), may offer the possibility of forming the continuous channel of ionizable residues required for the Morowitz mechanism. At the molecular level, cation transport mechanisms probably require more complex structures than those for protons because of the larger size of the ions and because there must be discrimination against smaller as well as larger cations. A variety of conceptual models have been advanced (28, 214, 230), but few have been analyzed in detail. There is little likelihood that transport mechanisms involve major rotations or inversions of subunits (46). The evidence, particularly in the cases of F_1F_0 and Ca^{2+} ATPases, suggests that there may be gated channels in active transport systems not unlike those that have been described for the voltage-dependent cation channels of nerve and muscle. In such a model, the gate would have some features of Maxwell's demon, in that it must recognize the cation and the direction of its movement. The energy for this informational process could derive from the enzyme reaction. If the gates are opened and closed at the proper times, diffusional events may be sufficient to load and discharge the channels.

Models with a somewhat more propulsive aspect have been advanced by Chandler et al (28) and by Yager (230). These involve progressive segmental transitions of peptide helices between two helical forms of differing radii. Ion transport could occur either within the core of a single helix or within an interstitial channel formed by two or more helices. The role of the enzyme reaction here would be to propagate the transition along the length of the helical segments spanning the bilayer and thus to do vectorial work.

If in general the catalytic and ionophoric functions of transport systems are localized in different structural domains, the concept of conformational coupling is implicit. A high degree of cooperativity among different subunits must be required to effectively channel the energy between these domains. Although the Ca^{2+}-transport ATPase can hydrolyze ATP as a monomer (35, 88), in none of the cases reviewed here is there evidence for transport by a monomeric protein. Bacteriorhodopsin displays coupling at least among elements of a trimer (75). The kinetics of proton conductance (176) and nucleotide substrate binding (61, 139) are highly cooperative in the case of F_1F_0-ATPases. If the oligomeric complex of F_1 is joined to an oligomeric F_0 complex by monomeric γ- and δ-subunits, do the latter react dynamically and equivalently or statically and nonequivalently with the elements of the oligomers during the transport and synthetase cycles?

Negative cooperativity is a prominent feature of substrate activation of the cation transport ATPases and there is also evidence for half-of-sites phosphorylation in these cases (52, 53, 90). The ultrastructural and recent ligand-binding studies (62) suggest even more complex subunit interactions. A particularly fundamental and unresolved problem is the structural basis of the observed 3:2 Na^+ to K^+ transport stoichiometry.

Clearly these structural studies have contributed much information about transport mechanisms. Conversely, kinetic studies have provided clues about structure. However, neither alone provides a satisfactory understanding of transport functions.

ACKNOWLEDGMENTS

We thank members of the laboratory, in particular L. Amende, S. Chock and N. Krishnan, for helpful discussion. We also thank the many authors who aided in the preparation of this review by sending us reprints and preprints of their work.

Literature Cited

1. Abdulaev, N. G., Feigina, M. Yu., Kiselev, A. V., Ovchinnikov, Yu. A., Drachev, L. A., Kaulen, A. D., Khitrina, L. V., Skulachev, V. P. 1978. *FEBS Lett.* 90:190–94
2. Abramson, J. J., Shamoo, A. E. 1978. *J. Membr. Biol.* 44:233–57
3. Adolfsen, R., Moudrianakis, E. N. 1976. *Biochemistry* 15:4163–70
4. Albers, R. W. 1976. In *The Enzymes of Biological Membranes*, ed. A. Martonosi, pp. 283–301. New York: Plenum
5. Allen, G., Green, N. M. 1978. *Biochem. J.* 173:393–402
6. Amzel, L. M., Pedersen, P. L. 1978. *J. Biol. Chem.* 253:2067–69
7. Baird, B. A., Hammes, G. G. 1976. *J. Biol. Chem.* 251:6953–62
8. Baird, B. A., Hammes, G. G. 1979. *Biochim. Biophys. Acta* 549:31–53
9. Baird, B. A., Pick, U., Hammes, G. G. 1979. *J. Biol. Chem.* 254:3818–25
10. Banks, R. D., Blake, C. C. F., Evans, P. R., Haser, R., Rice, D. W., Hardy, G. W., Merrett, M., Phillips, A. W. 1979. *Nature* 279:773–77
11. Bastide, F., Meissner, G., Fleischer, S., Post, R. L. 1973. *J. Biol. Chem.* 248:8385–91
12. Binder, A., Jagendorf, A., Ngo, E. 1978. *J. Biol. Chem.* 253:3094–3100
13. Blaurock, A. E. 1975. *J. Mol. Biol.* 93:139–58
14. Blaurock, A. E., Stoeckenius, W. 1971. *Nature New Biol.* 233:152–55
15. Blostein, R., Chu, L. 1977. *J. Biol. Chem.* 252:3035–43
16. Blumenthal, R., Shamoo, A. E. 1974. *J. Membr. Biol.* 19:141–62
17. Bond, G. H., Clough, D. L. 1973. *Biochim. Biophys. Acta* 323:592–99
18. Boyer, P. D. 1975. *FEBS Lett.* 58:1–6
19. Boyer, P. D., Chance, B., Ernster, L., Mitchell, P., Racker, E., Slater, E. C. 1977. *Ann. Rev. Biochem.* 46:955–1026
20. Brazkinkov, E. V., Chetverin, A. B., Chirgadze, Y. N. 1978. *FEBS Lett.* 93:125–28
21. Bridgen, I., Walker, I. D. 1976. *Biochemistry* 15:792–98
22. Brooks, J. C., Senior, A. E. 1970. *Arch. Biochem. Biophys.* 147:467–70
23. Cantley, L. C., Gelles, J., Josephson, L. 1978. *Biochemistry* 17:418–25
24. Cantley, L. C., Josephson, L. 1976. *Biochemistry* 15:5280–87
25. Cantley, L. C., Josephson, L., Gelles, J., Cantley, L. G. 1979. In *Symp. 2nd Int. Conf. Prop. Funct. ($Na^+ + K^+$)-ATPase, Sønderborg, Denmark.* New York: Academic. In press
26. Castro, J., Farley, R. A. 1979. *J. Biol. Chem.* 254:2221–28
27. Catterall, W. A., Pedersen, P. L. 1971. *J. Biol. Chem.* 246:4987–97
28. Chandler, H. D., Woolf, C. J., Hepburn, H. R. 1978. *Biochem. J.* 169:559–65
29. Churchill, L., Hokin, L. E. 1976. *Biochim. Biophys. Acta* 434:258–64
30. Cintrón, N. M., Pedersen, P. L. 1979. *J. Biol. Chem.* 254:3439–43
31. Criddle, R. S., Packer, L., Shieh, P. 1977. *Proc. Natl. Acad. Sci. USA* 74:4306–10
32. Dahl, J. L., Hokin, L. E. 1974. *Ann. Rev. Biochem.* 43:327–56
33. Deamer, D. W., Baskin, R. J. 1969. *J. Cell Biol.* 42:296–307
34. Dean, W. L., Tanford, C. 1977. *J. Biol. Chem.* 252:3551–53
35. Dean, W. L., Tanford, C. 1978. *Biochemistry* 17:1683–90
36. Degani, C., Boyer, P. D. 1973. *J. Biol. Chem.* 248:8222–26
37. Deguchi, N., Jorgensen, P. L., Maunsbach, A. B. 1977. *J. Cell Biol.* 75:619–34
38. de Meis, L., Vianna, A. L. 1979. *Ann. Rev. Biochem.* 48:275–92
39. de Pont, J. J. H. H. M., Schoot, B. M., van Prooijen-van Eeden, A., Bonting, S. L. 1978. *Biochim. Biophys. Acta* 482:213–27
40. de Pont, J. J. H. H. M., van Prooijen-van Eeden, A., Bonting, S. L. 1978. *Biochim. Biophys. Acta* 508:464–77
41. De Weer, P. 1975. In *MTP International Review of Science Physiology Series: Neurophysiology*, ed. C. C. Hunt, 3:231–78. Baltimore, Md: Baltimore Press/Univ. Park Press
42. Di Polo, R. 1977. *J. Gen. Physiol.* 69:795–813
43. Dixon, J. F., Hokin, L. E. 1974. *Arch. Biochem. Biophys.* 163:749–58
44. Dupont, Y. 1977. *Eur. J. Biochem.* 72:185–90
45. Dupont, Y., Harrison, S. C., Hasselbach, W. 1973. *Nature* 244:555–58
46. Dutton, A., Rees, E. D., Singer, S. J. 1976. *Proc. Natl. Acad. Sci. USA* 73:1532–36

47. Ebner, E., Maier, K. L. 1977. *J. Biol. Chem.* 252:671–76
48. Esmann, M., Skou, J. C., Christiansen, C. 1979. *Biochim. Biophys. Acta* 567:410–20
49. Fahn, S., Hurley, M. R., Koval, G. J., Albers, R. W. 1966. *J. Biol. Chem.* 241:1890–95
50. Fisher, K. A., Stoeckenius, W. 1977. *Science* 197:72–74
51. Forbush, B. III, Kaplan, J. H., Hoffman, J. F. 1978. *Biochemistry* 17:3667–76
52. Froehlich, J. P., Albers, R. W., Koval, G. J., Goebel, R., Berman, M. 1975. *J. Biol. Chem.* 251:2186–88
53. Froehlich, J. P., Taylor, C. W. 1975. *J. Biol. Chem.* 250:2013–21
54. Gerber, G. E., Gray, C. P., Wildenauer, D., Khorana, H. G. 1977. *Proc. Natl. Acad. Sci. USA* 74:5426–30
55. Giotta, G. J. 1976. *J. Biol. Chem.* 251:1247–52
56. Glynn, I. M., Karlish, S. J. D. 1975. *Ann. Rev. Physiol.* 37:13–55
57. Gopinath, R. M., Vincenzi, F. F. 1977. *Biochem. Biophys. Res. Commun.* 77:1203–9
58. Grisham, G. C. 1979. In *Advances in Inorganic Biochemistry*, ed. G. L. Eichhorn, Luigi G. Marzilli, 1:193–218. New York: Elsevier-North Holland
59. Grisham, C. M., Hutton, W. C. 1978. *Biochem. Biophys. Res. Commun.* 81:1406–11
60. Grisham, C. M., Mildvan, A. S. 1974. *J. Biol. Chem.* 249:3187–97
61. Hammes, G. G., Hilborn, D. A. 1971. *Biochim. Biophys. Acta* 233:580–90
62. Hanson, O., Jensen, J., Norby, J. G., Ottolenghi, P. 1979. *Nature* 280:410–12
63. Hardwicke, P. M. D., Green, N. M. 1974. *Eur. J. Biochem.* 42:183–93
64. Hart, W. M. Jr., Titus, E. O. 1973. *J. Biol. Chem.* 248:4674–81
65. Hasselbach, W. 1978. *Biochim. Biophys. Acta* 515:23–54
66. Hasselbach, W., Elfrim, L. G. 1967. *J. Ultrastruct. Res.* 17:598–622
67. Hastings, D. F., Reynolds, J. A. 1979. *J. Biol. Chem.* 18:817–21
68. Hayes, J. B. 1978. In *Bacterial Transport*, ed. B. P. Rosen, 4:43–102. New York: Dekker
69. Hebdon, G. M., Cunningham, L. W., Green, N. M. 1979. *Biochem. J.* 179:135–39
70. Henderson, R. 1975. *J. Mol. Biol.* 93:123–38
71. Henderson, R. 1977. *Ann. Rev. Biophys. Bioeng.* 6:87–109
72. Henderson, R. 1979. In *Membrane Transduction Mechanisms*, ed. R. A. Cone, J. E. Dowling, pp. 3–15. New York: Raven
73. Henderson, R., Unwin, P. N. T. 1975. *Nature* 257:28–32
74. Hesketh, T. R., Smith, G. A., Houslay, M. D., McGill, K. A., Birdsall, N. J. M., Metcalfe, J. C., Warren, G. B. 1976. *Biochemistry* 15:4145–51
75. Heyn, M. P., Bauer, P. J., Dencher, N. A. 1977. In *FEBS Symp. No. 42 Biochem. Membr. Trans.* ed. G. Semenza, E. Carafoli, pp. 96–104. Berlin/New York: Springer
76. Hilden, S., Hokin, L. E. 1975. *J. Biol. Chem.* 250:6296–6303
77. Hokin, L. E. 1974. *Ann. NY Acad. Sci.* 242:12–23
78. Hokin, L. E., Dahl, J. L., Deupree, J. D., Dixon, J. F., Hackney, J. F., Perdue, J. F. 1973. *J. Biol. Chem.* 248:2593–2605
79. Holowka, D. A., Hammes, G. G. 1977. *Biochemistry* 16:5538–45
80. Hopkins, B. E., Wagner, H. J., Smith, W. T. 1976. *J. Biol. Chem.* 251:4365–71
81. Hörl, W. H., Heilmeyer, L. M. G. 1978. *Biochemistry* 17:766–72
82. Ikemoto, N. 1976. *J. Biol. Chem.* 251:7275–77
83. Inesi, G., Goodman, J. J., Watanabe, S. 1967. *J. Biol. Chem.* 242:4637–43
84. Jarrett, H. W., Penniston, J. T. 1977. *Biochem. Biophys. Res. Commun.* 77:1210–16
85. Jean, D. -H., Albers, R. W., Koval, G. J. 1975. *J. Biol. Chem.* 250:1035–40
86. Jesaitis, A. J., Fortes, P. A. G. 1979. *J. Biol. Chem.* In press
87. Jilka, R. L., Martonosi, A. N., Tillack, T. W. 1975. *J. Biol. Chem.* 250:7511–24
88. Jørgensen, K. E., Lind, K. E., Roigaard-Petersen, H., Moller, J. D. 1978. *Biochem. J.* 169:489–98
89. Jørgensen, P. L. 1974. *Biochim. Biophys. Acta* 356:53–67
90. Jørgensen, P. L. 1975. *Biochim. Biophys. Acta* 401:399–415
91. Jørgensen, P. L. 1977. *Biochim. Biophys. Acta* 466:97–108
92. Jørgensen, P. L., Klodos, I. 1975. *Biochim. Biophys. Acta* 507:8–16
93. Kagawa, Y. 1978. *Biochim. Biophys. Acta* 505:45–93
94. Kagawa, Y., Racker, E. 1966. *J. Biol. Chem.* 241:2475–82

95. Kaniike, K., Lindenmayer, G. E., Wallick, E. T., Lane, L. K., Schwartz, A. 1976. *J. Biol. Chem.* 251:4794-95
96. Karlish, S. J. D., Jørgensen, P. L., Gitler, C. 1977. *Nature* 269:715-17
97. Karlish, S. J. D., Yates, D. W. 1978. *Biochim. Biophys. Acta* 527:115-30
98. Kayalar, C., Rosing, J., Boyer, P. D. 1977. *J. Biol. Chem.* 252:2486-91
99. Kimelberg, H. K. 1976. *Mol. Cell. Biochem.* 10:171-90
100. Kirschberger, M. A., Tada, M. 1976. *J. Biol. Chem.* 251:725-29
101. Klip, A., MacLennan, D. H. 1978. *Front. Biol. Energ.* 2:1137-47
102. Knowles, A. F., Guillory, R. J., Racker, E. 1971. *J. Biol. Chem.* 246:2672-79
103. Knowles, A. F., Racker, E. 1975. *J. Biol. Chem.* 250:3538-44
104. Kozlov, I. A., Skulachev, V. P. 1977. *Biochim. Biophys. Acta* 463:29-89
105. Kundig, W. 1976. See Ref. 4, pp. 31-74
106. Kushwaha, S. C., Kates, M., Martin, W. G. 1975. *Can. J. Biochem.* 53:280-92
107. Kyte, J. 1971. *J. Biol. Chem.* 246:4157-65
108. Kyte, J. 1972. *J. Biol. Chem.* 247:7634-41
109. Kyte, J. 1972. *J. Biol. Chem.* 247:7642-49
110. Laggner, P., Graham, D. E. 1976. *Biochim. Biophys. Acta* 433:311-1
111. Lanyi, J. K. 1978. In *Light Transducing Membranes*, ed. D. W. Deamer, pp. 157-65. New York: Academic
112. Lelievre, L., Charlemagne, D., Paraf, A. 1976. *Biochem. Biophys. Res. Commun.* 72:1526-33
113. le Maire, M., Moller, J. V., Tanford, C. 1976. *Biochemistry* 15:2336-43
114. Lever, J. E., Clingan, D., Jimenez de Asua, L. 1976. *Biochem. Biophys. Res. Commun.* 71:136-43
115. Lewis, A., Marcus, M. A., Ehrenberg, B., Crespi, H. 1978. *Proc. Natl. Acad. Sci. USA* 75:4642-46
116. Louis, C. F., Sanders, M. J., Holroyd, S. A. 1977. *Biochim. Biophys. Acta* 493:78-92
117. MacLennan, D. H. 1970. *J. Biol. Chem.* 245:4508-18
118. MacLennan, D. H., Holland, P. C. 1975. *Ann. Rev. Biophys. Bioeng.* 4:377-404.
119. MacLennan, D. H., Holland, P. C. 1976. In *The Enzymes of Biological Membranes*, ed. A. Martonosi, 3:221-59. New York: Plenum
120. MacLennan, D. H., Klip, A. 1979. See Ref. 72, pp. 61-75
121. MacLennan, D. H., Klip, A., Reithmeier, R., Michalak, M., Campbell, K. 1979. In *Membrane Bioenergetics*, ed. C. P. Lee, G. Schatz, L. Ernster. Reading, Mass: Addison-Wesley. In press
122. MacLennan, D. H., Tzagoloff, A. 1968. *Biochemistry* 7:1603-10
123. MacLennan, D. H., Yip, C. C., Iles, G. H., Seaman, P. 1972. *Cold Spring Harbor Symp. Quant. Biol.* 37:469-78
124. Madden, T. D., Chapman, D., Quinn, P. J. 1979. *Nature* 279:539-41
125. Mandersloot, J. G., Roelofsen, B., de Gier, J. 1978. *Biochim. Biophys. Acta* 508:478-85
126. Martonosi, A. 1975. *Biochim. Biophys. Acta* 415:311-33
127. Martonosi, A., Lagwinska, E., Oliver, M. 1974. *Ann. NY Acad. Sci.* 227:549-67
128. Matsui, H., Hayashi, Y., Homareda, H., Kimimura, M. 1977. *Biochem. Biophys. Res. Commun.* 75:373-80
129. McCarty, R. E. 1978. In *Current Topics in Bioenergetics*, ed. D. R. Sanadi, L. P. Vernon, 7:245-78. New York: Academic
130. Meissner, G. 1973. *Biochim. Biophys. Acta* 298:906-26
131. Meissner, G., Fleischer, S. 1971. *Biochim. Biophys. Acta* 241:356-77
132. Mitchell, P., Moyle, M. 1979. In *Developments in Biochemistry: Cytochrome Oxidase*, ed. T. King, 5:361. New York: Elsevier-North Holland
133. Miyamoto, H., Kasai, M. 1979. *J. Biochem.* 85:765-73
134. Moczydlowski, E. G., Jesaitis, A. J., Fortes, P. A. G. 1979. See Ref. 25. In press
135. Nagle, J. F., Morowitz, H. J. 1978. *Proc. Natl. Acad. Sci. USA* 75:298-302
136. Neet, K. E., Green, N. M. 1977. *Arch. Biochem. Biophys.* 178:588-97
137. Nelson, N., Eytan, E., Notsani, B. E., Sigrist, H., Sigrist-Nelson, K., Gitler, C. 1977. *Proc. Natl. Acad. Sci. USA* 24:2375-78
138. Nelson, N., Karny, O. 1976. *FEBS Lett.* 70:249-53
139. Nelson, N., Nelson, H., Racker, E. 1972. *J. Biol. Chem.* 247:6506-10
140. Nelson, N., Nelson, H., Racker, E. 1972. *J. Biol. Chem.* 247:7657-62
141. Neufeld, H. H., Levy, H. M. 1969. *J. Biol. Chem.* 244:6493-97
142. Niggli, V., Ronner, P., Carafoli, E., Penniston, J. 1979. *Arch. Biochem.*

143. Nishigaki, I., Chen, F. T., Hokin, L. E. 1973. *J. Biol. Chem.* 249:4911–16
144. Oesterhelt, D., Schumann, L. 1974. *FEBS Lett.* 44:262–65
145. Oesterhelt, D., Stoeckenius, W. 1971. *Nature New Biol.* 233:149–52
146. Oesterhelt, D., Stoeckenius, W. 1973. *Proc. Natl. Acad. Sci. USA* 70:2853–57
147. Ohtsuka, O., Ohtsuki, I., Ebashi, S. 1965. *J. Biochem. Tokyo* 58:188–90
148. Oleszko, S., Moudrianakis, E. N. 1974. *J. Cell Biol.* 63:936–48
149. Ostwald, T. J., MacLennan, D. H. 1974. *J. Biol. Chem.* 249:974–79
150. Ovchinnikov, Yu. A., Abdulaev, N. G., Feigina, M. Yu., Kiselev, A. V., Lobanov, N. A. 1979. *FEBS Lett.* 100:219–24
151. Packer, L., Konishi, T. 1978. In *Energetics and Structure of Halophilic Microorganisms*, ed. S. R. Caplan, M. Ginsberg, pp. 143–59. New York: Elsevier-North Holland Biomedical
152. Paradies, H. H., Schmidt, U. D. 1979. *J. Biol. Chem.* 254:5257–63
153. Paradies, H. H., Zimmermann, J., Schmidt, U. D. 1978. *J. Biol. Chem.* 253:8972–79
154. Pedersen, P. L., Amzel, L. M., Soper, J. W., Cintrón, N., Hullihen, J. 1978. In *Energy Conservation in Biological Membranes*, ed. G. Schafer, M. Klingenberg, pp. 159–94. New York: Springer
155. Perrone, J. R., Blostein, R. 1973. *Biochim. Biophys. Acta* 291:680–89
156. Perrone, J. R., Hackney, J. F., Dixon, J. F., Hokin, L. E. *J. Biol. Chem.* 250:4178–84
157. Perrone, J. R., Hokin, L. E. 1978. *Biochim. Biophys. Acta* 525:446–54
158. Post, R. L., Jolly, P. C. 1957. *Biochim. Biophys. Acta* 25:118–28
159. Post, R. L., Kume, S. 1973. *J. Biol. Chem.* 248:6993–7000
160. Pullman, M. E., Monroy, G. C. 1963. *J. Biol. Chem.* 238:3762–69
161. Racker, E. 1977. In Ca^{++}-*binding Proteins and* Ca^{++}-*Function*, ed. R. H. Wassermann, R. A. Corradino, E. Carafoli, R. H. Kretsinger, D. H. MacLennan, F. L. Siegel, pp. 155–63. New York: North Holland
162. Racker, E., Eytan, E. 1975. *J. Biol. Chem.* 250:7533–34
163. Racker, E., Horstman, L. L., Kling, D., Fessenden-Raden, J. M. 1969. *J. Biol. Chem.* 244:6668–74
164. Rhee, H. M., Hokin, L. E. 1975. *Biochem. Biophys. Res. Commun.* 63:1139–45
165. Rizzolo, L. J., le Maire, M., Reynolds, J. A., Tanford, C. 1976. *Biochemistry* 15:3433–37
166. Rizzolo, L. J., Tanford, C. 1978. *Biochemistry* 17:4049–55
167. Rogers, T. B., Lazdunski, M. 1979. *Biochemistry* 18:135–40
168. Rogers, T. B., Lazdunski, M. 1979. *FEBS Lett.* 98:373–76
169. Rothschild, K. J., Clark, N. A. 1979. *Science* 204:311–12
170. Rozengurt, E., Heppel, L. A. 1975. *Proc. Natl. Acad. Sci. USA* 72:4492–95
171. Sarzala, M. G., Pilarska, M., Zubrzycka, E., Michalak, M. 1975. *Eur. J. Biochem.* 57:25–34
172. Scales, D., Inesi, G. 1976. *Biophys. J.* 16:735–51
173. Schatzmann, H. J. 1973. *J. Physiol.* 235:551–69
174. Schmidt, U. D., Paradies, H. H. 1977. *Biochem. Biophys. Res. Commun.* 78:383–92
175. Schmidt, U. D., Paradies, H. H. 1977. *Biochem. Biophys. Res. Commun.* 78:1043–52
176. Schonfeld, M., Neumann, J. 1977. *FEBS Lett.* 73:51–54
177. Schoot, B. M., de Pont, J. J. H. H. M., Bonting, S. L. 1978. *Biochim. Biophys. Acta* 522:602–13
178. Schreckenbach, T. 1979. *Top. Photosynth.* 3:189–209
179. Schwartz, A., Entman, M. L., Kaniike, K., Lane, L. K., Van Winkle, W. B., Bornet, E. P. 1976. *Biochim. Biophys. Acta* 426:57–72
180. Scott, T. L., Shamoo, A. E. 1977. *Biophys. J.* 17:A185
181. Sebald, W. 1977. *Biochim. Biophys. Acta* 463:1–27
182. Sebald, W., Graf, T., Lukins, H. H. 1979. *Eur. J. Biochem.* 93:587–99
183. Sebald, W., Wachter, E. 1978. See Ref. 154, pp. 228–36
184. Senior, A. E. 1975 *Biochemistry* 14:660–64
185. Shamoo, A. E., Abramson, J. J. 1977. See Ref. 161, pp. 173–80
186. Shamoo, A. E., Albers, R. W. 1973. *Proc. Natl. Acad. Sci. USA* 70:1191–94
187. Shamoo, A. E., Goldstein, D. A. 1977. *Biochim. Biophys. Acta* 472:13–53
188. Shamoo, A. E., Myers, M. 1974. *J. Membr. Biol.* 19:163–78
189. Shamoo, A. E., Myers, M., Blumenthal, R., Albers, R. W. 1974. *J. Membr. Biol.* 19:129–40
190. Shigekawa, M., Akowitz, A., Katz, A. 1978. *Biochim. Biophys. Acta* 526:591–96

191. Shigekawa, M., Dougherty, J. P. 1978. *J. Biol. Chem.* 253:1451–57
192. Shigekawa, M., Dougherty, J. P. 1978. *J. Biol. Chem.* 253:1458–64
193. Siegel, G. J., Goodwin, B. 1972. *J. Biol. Chem.* 3630–37
194. Skou, J. C. 1974. *Biochim. Biophys. Acta* 339:234–45
195. Slater, E. C., Kemp, A., van der Kraan, I., Muller, J. L. M., Roveri, D. A., Verschoor, G. J., Wagenvoord, R. J., Wielders, J. P. M. 1979. *FEBS Lett.* 103:7–11
196. Smith, J. B., Sternweis, P. C. 1975. *Biochem. Biophys. Res. Commun.* 62:764–71
197. Spitsberg, V., Haworth, R. 1977. *Biochim. Biophys. Acta* 492:237–40
198. Sternweis, P. C. 1978. *J. Biol. Chem.* 253:3123–28
199. Stewart, P. S., MacLennan, D. H., Shamoo, A. E. 1976. *J. Biol. Chem.* 251:712–19
200. Stoeckenius, W., Lozier, R. H., Bogomolni, R. A. 1979. *Biochim. Biophys. Acta* 505:215–78
201. Swann, A. C., Albers, R. W. 1975. *Biochim. Biophys. Acta* 382:437–56
202. Sweadner, K. J. 1977. *Biochem. Biophys. Res. Commun.* 78:962–69
203. Sweadner, K. J. 1979. *J. Biol. Chem.* 254:6060–67
204. Tada, M., Kirchberger, M. A., Katz, A. M. 1975. *J. Biol. Chem.* 250:2640–47
205. Tada, M., Kirchberger, M. A., Repke, D. I., Katz, A. M. 1974. *J. Biol. Chem.* 249:6174–80
206. Tada, M., Ohmori, F., Yamada, M., Abe, H. 1979. *J. Biol. Chem.* 254:319–26
207. Tada, M., Yamamoto, T., Tonomura, Y. 1978. *Physiol. Rev.* 58:1–79
208. Taylor, J. S., Hattan, D. 1979. *J. Biol. Chem.* 254:4402–7
209. Tenu, J. P., Ghelis, C., Yon, J., Chevallier, J. 1976. *Biochimie* 58:513–19
210. Thomas, R. C. 1972. *Physiol. Rev.* 52:563–94
211. Thorley-Lawson, D. A., Green, N. M. 1973. *Eur. J. Biochem.* 40:403–13
212. Thorley-Lawson, D. A., Green, N. M. 1975. *Eur. J. Biochem.* 59:193–200
213. Tillack, T. W., Boland, R., Martonosi, A. 1974. *J. Biol. Chem.* 249:624–33
214. Tonomura, Y., Morales, M. F. 1974. *Proc. Natl. Acad. Sci. USA* 71:3687–91
215. Trotta, E. E., de Meis, L. 1975. *Biochim. Biophys. Acta* 394:239–47
216. Uesugi, S., Dulak, N. C., Dixon, J. F., Hokin, L. E. 1971. *J. Biol. Chem.* 246:531–43
217. Unwin, P. N. T., Henderson, R. 1975. *J. Mol. Biol.* 94:425–40
218. Vanderkooi, J. M., Ierokomas, A., Nakamura, H., Martonosi, A. 1977. *Biochemistry* 16:1262–67
219. Van Winkle, W. B., Lane, L. K., Schwartz, A. 1976. *Exp. Cell Res.* 100:291–96
220. Verjovski-Almeida, S., Inesi, G. 1979. *J. Biol. Chem.* 254:18–21
221. Verschoor, G. J., van der Sluis, P. R., Slater, E. C. 1977. *Biochim. Biophys. Acta* 462:438–49
222. Vogel, F., Meyer, H. W., Grosse, R., Repke, K. R. H. 1977. *Biochim. Biophys. Acta* 470:497–502
223. Wakabayashi, T., Kubota, M., Yoshida, M., Kagawa, Y. 1977. *J. Mol. Biol.* 117:515–19
224. Wallick, E. T., Lane, L. K., Schwartz, A. 1979. *Ann. Rev. Physiol.* 41:397–411
225. Warren, G. B., Houslay, M. D., Metcalfe, J. C., Birdsall, N. J. M. 1975. *Nature* 255:684–87
226. Wilbrandt, W., Rosenberg, T. 1961. *Pharmacol. Rev.* 13:109–83
227. Wikström, M., Saori, H., Pentilla, T., Saroste, M. 1978. In *FEBS Symp.* 45:88–91
228. Williams, R. J. P. 1978. *FEBS Lett.* 85:9–19
229. Worthington, D. R., Liu, S. C. 1973. *Arch. Biochem. Biophys.* 157:573–79
230. Yager, P. 1977. *J. Theor. Biol.* 66:1–11
231. Yamanaka, N., Deamer, D. W. 1976. *Biochim. Biophys. Acta* 426:132–47
232. Yates, D. W., Duance, V. C. 1976. *Biochem. J.* 159:719–28
233. Yoshida, M., Okamoto, H., Sone, N., Hirata, H., Kagawa, Y. 1977. *Proc. Natl. Acad. Sci. USA* 74:936–40
234. Yoshida, M., Sone, N., Hirata, H., Kagawa, Y. 1975. *J. Biol. Chem.* 250:7910–16
235. Yoshida, M., Sone, N., Hirata, H., Kagawa, Y. 1977. *J. Biol. Chem.* 252:3480–85
236. Younis, H., Winget, G. D., Racker, E. 1977. *J. Biol. Chem.* 252:1814–18

MAGNETIC CIRCULAR DICHROISM OF BIOLOGICAL MOLECULES

♦9148

John Clark Sutherland

Biology Department, Brookhaven National Laboratory, Upton, New York 11973

Barton Holmquist

Biophysics Research Laboratory, Department of Biological Chemistry, Harvard Medical School; Division of Medical Biology, Peter Bent Brigham Hospital, Boston, Massachusetts 02115

INTRODUCTION

Magnetic circular dichroism (MCD) has now been developed to the point where it should be one of the several tools at the disposal of all scientists who use visible, ultraviolet, and near infrared spectroscopies to probe the structure and function of biological molecules. As in other spectral techniques, the diversity of the method is large, with applications extending from analytical analysis (it provides perhaps the most direct method for tryptophan in proteins) to the theoretical description of Zeeman splitting in the porphyrins. This review describes MCD and several closely related experiments, summarizes the theory of MCD from the viewpoint of practical applications to biochemical problems, surveys available instrumentation, and reviews the literature on the MCD of major classes of biological molecules. More specialized reviews describe the use of MCD in particular areas in the biological realm (24, 48, 67, 73, 78, 135, 153, 164, 189).

TERMS, UNITS, AND TYPES OF EXPERIMENTS

Transmission-Detected MCD

Magnetic circular dichroism is the difference in absorption of left and right circularly polarized light induced by an external magnetic field. The field must be parallel, or have a component that is parallel, to the direction of propagation of the light. Since MCD is a difference in absorption, it is described using the same terms and units used in absorption spectroscopy. In the biochemical literature, the two most frequently used quantities are the absorbance (A), an extrinsic parameter, and the molar absorbance (ϵ), the corresponding intrinsic parameter. Suppose that a collimated beam of unpolarized monochromatic radiation of wavelength λ and intensity I_o is incident on a parallel slab of homogeneous, isotropic material of thickness d, containing absorbing centers of concentration c (moles/liter). If the intensity of the beam emerging from the other side of the slab is I, ignoring refractive losses we can write

$$I(\lambda) = I_o(\lambda) 10^{-A}. \qquad 1.$$

This familiar equation defines the absorbance, A, which is a function of wavelength, the concentration of absorbers, and the distance traveled through the absorbing medium. Beer's law relates the extrinsic and intrinsic parameters and separates the wavelength, concentration and path length dependence:

$$A = \epsilon c d. \qquad 2.$$

Assuming that the incident beam is left or right circularly polarized, one can write equations analogous to Equation 1, thus defining the absorbancies for left and right circularly polarized light, A_L and A_R, respectively. The net circular dichroism (CD) is given by $\Delta A = A_L - A_R$. Assuming that the intensity incident on the sample is the same for the two polarizations, it is easy to show that

$$\Delta A \text{ is proportional to } (I_L - I_R)/(I_L + I_R) \qquad 3.$$

where I_L and I_R are the transmitted intensities of the left and right circularly polarized incident radiation. The proportionality holds so long as $(I_L - I_R) \ll (I_L + I_R)$.

The net or observed CD is the sum of inherent and induced components. The inherent CD, which we shall refer to as natural CD, is due to molecular asymmetry. Thus we can write ΔA as a sum of inherent and induced components:

$$\Delta A = \Delta A_{CD} + H \Delta A_{MCD}{}^{1} \qquad 4.$$

where H is the intensity of the applied field (measured in Teslas, T). It is implicit in Equation 4 that the contributions to A of the inherent and induced components are independent and that the induced component

---

[1]Other external perturbations capable of inducing circular dichroism will not be discussed.

is a linear function of the strength of the field; good theoretical and experimental evidence supports both assumptions under the conditions encountered in almost all biological work.

The extrinsic parameter, ΔA_{MCD}, can be converted to the corresponding intrinsic parameter by Beer's law, i.e. $\Delta A_{MCD} = \Delta \varepsilon_{MCD} \, cd$. Measurement of ΔA_{MCD} and $\Delta \varepsilon_{MCD}$ requires subtraction of the natural CD spectrum from the observed CD and other baseline corrections discussed in the review by Holmquist & Vallee (78).

The units of ΔA_{MCD} and $\Delta \varepsilon_{MCD}$ are $(T)^{-1}$ and $(M \cdot cm \cdot T)^{-1}$ respectively. A variety of other systems of units have been used; Holmquist & Vallee (78) list most of these and give conversion factors to transform to the system just described.

In MCD one observes the radiation that has traversed a sample. Related effects can be observed by monitoring radiation scattered by or reemitted as luminescence from a specimen. All such measurements are inherently less sensitive than MCD, but in some situations they may provide otherwise unobtainable information. In luminescence measurements it is critical to distinguish between fluorescence-detected MCD and the completely different phenomenon, magnetic circular emission.

Fluorescence-Detected Magnetic Circular Dichroism (FDMCD)

The apparatus for FDMCD is similar to that for MCD except that the detector is positioned to intercept fluorescence emitted from the sample rather than the transmitted beam (171). By monitoring the difference in fluorescence intensities for the two incident circular polarizations, the total fluorescence, the absorption, and the (transmission-detected) MCD of the sample, it is possible to calculate the MCD of the moieties that give rise to fluorescence at a particular wavelength and *independently* the MCD of the components that do not give rise to fluorescence. Thus FDMCD may prove useful in analyzing complex-multichromophore systems. The only systems studied to date are tryptophan and cytochrome *c* (171). If the emitting species does not undergo complete rotational relaxation before emission, artifacts due to linear polarization may distort the measured spectrum (185).

Magnetic Circular Emission (MCE)

Here, one measures the difference in intensity of left and right circularly polarized emitted radiation. Thus MCE provides information about the configuration of a molecule in its excited state. Forbidden transitions such as the triplet-to-singlet transitions involved in phosphorescence can be studied with MCE. Most work on biological molecules has involved porphyrins, either at cryogenic temperatures (106, 140, 141) or room temperature (167). Whenever rotational relaxation is slow compared to excited state lifetime, as is the case at low temperatures, care must be taken to avoid effects due to photoselection (68).

Magnetically Induced Circular Differential Scattering

A magnetic field applied parallel to the Poynting's vector of a beam of circularly polarized radiation results in differential Rayleigh and Raman scattering for right and left polarizations (10). The incident beam, generated by a laser, is phase modulated before reaching the sample, and a monochromator is placed between the sample and the detector. Results on ferrocytochrome c demonstrate that the method is sensitive to both the symmetry of Raman active vibrational modes (8) and, as with unpolarized Raman scattering, to the wavelength of the incident radiation (9). This type of measurement holds promise of complementing the information available from unpolarized Raman experiments, much as MCD complements absorption spectroscopy. Since Raman scattering is an inherently weak process and the differential effects are typically more than three orders of magnitude smaller, the experimental problems are formidable.

THEORY

First we shall summarize the theory of MCD for well-resolved spectral transitions (30); then we shall discuss the case of broad overlapping spectral bands.

MCD of Discrete Line Spectra

There are three distinct mechanisms that can give rise to MCD; the resulting spectra are characterized as A, B, and C type, respectively. To illustrate these concepts, we use a hypothetical molecule, with the electronic ground and excited states shown in Figure 1. The ground state, labeled S_0, and the second and third excited states, S_2 and S_3, are orbitally nondegenerate, whereas the lowest excited state, S_1, is assumed to be doubly degenerate with components S_{1+} and S_{1-}. Since it contains an orbitally degenerate state, the molecule must, according to group theory, possess at least threefold symmetry. This degree of symmetry implies that in the absence of an external perturbation, such as a magnetic field, transitions from S_0 to S_1, S_2, and S_3 will absorb left and right circularly polarized light with precisely the same probability; no natural CD is observed. The designation of the states as S_0, S_1 etc. indicates that all are singlet states with no net electron spin. The

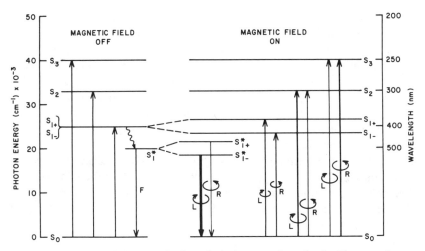

Figure 1 Energy level diagram of a hypothetical symmetric molecule. The ground state (S_0) and the second and third excited states (S_2 and S_3) are nondegenerate. The lowest excited state (S_1) is doubly degenerate. All are presumed to be singlet states, i.e. the net spin is zero. Solvent relaxation lowers the energy of S_1 relative to S_0; the relaxed excited state is designated S_1^*. A magnetic field lifts the degeneracy of S_1; the two resolved states, S_{1-} and S_{1+}, absorb only right and left circularly polarized light respectively. The same magnetic field causes S_2 and S_3 to absorb one circular polarization slightly more strongly than the other. Fluorescence emission is enhanced from S_{1-}^* because of the Boltzmann distribution, which determines the relative populations of S_{1+}^* and S_{1-}^*; the emission of left circularly polarized fluorescence is thus enhanced. This effect increases as the temperature is lowered.

complications resulting from nonzero spin (Kramers degeneracy) are discussed later.

We shall assume that after absorbing a photon, the molecule undergoes rapid vibrational relaxation so that fluorescence is from the first excited state S_1^* to the ground state, S_0. We designate the emitting state S_1^* to indicate that it is shifted to lower energy, compared to S_1, as a result of interactions with solvent. We also assume that the lifetime of the excited state is much longer than the time required to achieve thermal equilibrium with the medium and random orientation, with respect to a coordinate system fixed in the laboratory. Both vibrational and solvent relaxation presuppose a condensed medium, but for the moment we shall ignore other effects of the surrounding medium on the MCD.

"A" TYPE OR ZEEMAN MCD   One effect of an external magnetic field is to lift the degeneracy of the two components of S_1. This can also be described as a Zeeman splitting of the degenerate components. One component, call it S_{1-}, is shifted to a slightly lower energy while S_{1+} is shifted to higher energy by the same amount. Furthermore, the transition $S_0 \to S_{1-}$ absorbs only one sense of circularly polarized light while $S_0 \to S_{1+}$ absorbs only the opposite sense. For our illustration we shall assume that $S_0 \to S_{1-}$ absorbs only right circularly polarized light (RCPL) and $S_0 \to S_{1+}$ absorbs only left circularly polarized light (LCPL).

The magnitude of the splitting is proportional to the strength of the magnetic field. For most molecules of biological interest, the total splitting between S_{1-} and S_{1+} is small compared to the width of the absorption band. Thus the MCD is the difference between the absorption of two components of equal magnitude which are slightly displaced from one another in energy. The resulting MCD resembles the derivative of the corresponding absorption band. For discrete spectral bands, the derivative-like shape of the MCD is diagnostic for an "A" type spectrum (Figure 2).

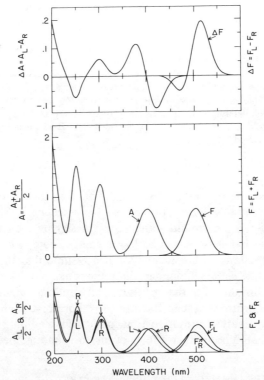

Figure 2 The circularly polarized absorption spectrum and circular dichroism and corresponding fluorescence spectra observed in the presence of a magnetic field for the molecule with energy levels shown in Figure 1. An "A" type MCD spectrum is observed at 400 nm while "B" type MCD bands are found at 250 and 300 nm. The MCE is the sum of "A" and "C" components.

In the bottom panel of Figure 2 we show the absorption spectra of the two circularly polarized components (A_R and A_L respectively) for the molecule characterized by the energy level scheme shown in Figure 1. The center panel of Figure 2 is the absorption spectrum that would be observed using unpolarized light $A = (A_L + A_R)/2$ and the top is the MCD ($\Delta A = A_L - A_R$). In this illustration, the effects of the magnetic field are greatly exaggerated compared to what is observed in real molecules where ΔA_{MCD} is typically one thousandth of A or less.

"B" TYPE SPECTRA Besides splitting the degenerate components of S_1, an external magnetic field also perturbs the various electronic states in such a way as to favor the absorption of either RCPL or LCPL in the various transitions. These effects can be described using nondegenerate perturbation theory which represents the effect of the field as a "mixing" of the states. In our example, we shall assume that the transition dipoles for absorption from $S_0 \rightarrow S_2$ and $S_0 \rightarrow S_3$ are mutually perpendicular and that the magnetic field mixes S_2 and S_3. Thus in the presence of the field, transition $S_0 \rightarrow S_2$ acquires some of the character of $S_0 \rightarrow S_3$ and vice versa. The perturbed transitions can no longer be considered linear dipoles, because they contian the field-induced component perpendicular to their original orientation. Rather they are elliptically polarized and thus will absorb RCPL slightly more effectively than LCPL or vice versa. There are, in fact, some rules which indicate that for two transitions perturbed by the magnetic field, the excess RCPL absorption induced in one will be balanced by excess absorption of LCPL by the other. In this situation, the MCD will have the same shape as the corresponding absorption band. The "B" type MCD spectra resulting from the mixing of S_2 and S_3 are shown in the top panel of Figure 2, at 300 and 250 nm respectively. In this illustration the MCD for $S_0 \rightarrow S_2$ is shown as positive and that for $S_0 \rightarrow S_3$ as negative. Note that the field can perturb any of the states of the molecule, including ground states such as S_0 and components of degenerate pairs such as S_{1+} and S_{1-} as well as nondegenerative excited states like S_2 and S_3. The magnitude of the "B"type spectrum is inversely proportional to the energy separation of the two states being mixed by the field. The integrated intensities of the MCD of the two bands are of equal magnitude but opposite sign. If one of the transitions were much weaker than the other, the existence of the weaker component might be more readily detected in MCD than by absorption spectroscopy. An example of this effect has been reported in the ultraviolet MCD of the ethidium cation (81).

"C" TYPE SPECTRA The ground state of the molecule whose energy level diagram is shown in Figure 1 is nondegenerate. Since S_0 is the initial state for all normal absorption events, "C" type spectra will not be observed in absorption MCD for this molecule, although they can be observed in other molecules with degenerate ground states. The situation is different for fluorescence. Since S_1^* is the initial state and is degenerate, a "C" type spectrum will be observed in the magnetic circularly polarized emission spectrum (MCE).

Just as in absorption, the transitions between S_0 and the two components of S_1^* are circularly polarized. The transition $S_{1-}^* \to S_0$ will emit only LCPL and $S_{1+}^* \to S_0$ will emit only RCPL—in both cases the opposite of what they absorbed. Yet there is an important difference from the case of absorption. Since the initial state is degenerate, the populations of S_{1-}^* and S_{1+}^* are determined by a Boltzmann distribution. Thus more molecules will emit from state S_{1-}^* than from S_{1+}^*. This situation is illustrated in Figure 2 for an "intermediate" temperature where the effects of Zeeman splitting and the Boltzmann distribution are comparable. At "high" temperatures, functionally defined as kT much greater than the Zeeman splitting, the MCE has the same derivative shape as the corresponding absorption event (albeit reversed in sign). At low temperatures, i.e. kT much less than the Zeeman splitting, the MCE is due solely to the transition $S_{1-}^* \to S_0$ and will thus have the shape of the total emission and not a derivative shape. The intermediate temperature MCE shown in Figure 2 can be decomposed into the sum of one component with the shape of the derivative of the emission spectrum, and another with the shape of the total emission. These are the "A" and "C" components respectively. So long as kT is larger than the Zeeman splitting, the magnitude of the "C" component is proportional to $1/kT$, whereas near absolute zero it will asymptotically approach the magnitude of the corresponding absorption or emission spectrum. Although "B" and "C" type spectra have similar shapes, they can be distinguished by their temperature dependence.

SUMMARY OF SPECTRAL TYPES In the "discrete line" approach to MCD theory, the spectrum can be expanded as a sum of three "terms" according to the equation

$$\Delta \varepsilon_{\text{MCD}}/\nu = -1.002 \times 10^2 \left(Af' + [B + C/kT] f^\circ \right) \qquad 5.$$

where ν is photon energy expressed in wave numbers (cm^{-1}), and f° and f' are the shape of the absorption band and its first derivative normalized such that $\int f^\circ(\nu) d\nu = 1$. The corresponding expression for the unpolarized absorption spectrum is $\varepsilon/\nu = 1.089 \times 10^{-2} Df^\circ$. The coefficients A, B, C, and D are related to molecular parameters by sums of matrix elements, expressions for which may be found elsewhere (153).

Usually the relative strengths of the three components increase in the sequence $B < A < C$ and are roughly proportional to the reciprocal of the energy separation of the interacting states, the half-width of the absorption spectrum, and the reciprocal of absolute temperature, respectively. The contributions of the three types are additive. Thus a band exhibiting an "A" type spectrum may also have smaller contributions of "B" type. Similarly, a band dominated by a "C" type spectrum may also have contributions from "A" and "B" terms.

Overlapping Lines and Continuum Band Spectra

From the discussion of discrete line spectra it might seem to follow that the shape of the MCD spectrum could be used to identify "A" type spectra and hence lead to conclusions about the symmetry of the absorbing or emitting species. Such, however, is not the case. With the aid of examples taken from the experimental literature we shall explain the modifications to the theory required to deal with real molecules that interact with the medium about them; the theory of MCD has also been formally modified to encompass the effects we shall discuss (152).

ZERO FIELD SPLITTINGS OF DEGENERATE STATES The theory that leads to Equation 5 was based on the assumption that all states are either exactly degenerate or, if not degenerate, well resolved. Thus the effects of an applied field can be deduced with the aid of degenerate perturbation theory or nondegenerate perturbation theory. In all real molecules, however, degeneracy will always be removed by the Jahn-Teller effect, although the magnitude of the Jahn-Teller splitting may be very small. The degenerate levels of molecules in condensed media will be further split by interactions with their environment. These interactions can be static (e.g. a molecule in a vitrified medium) or dynamic (e.g. a molecule in solution). For porphyrin molecules in solution (165) or heme moieties in a protein (170) the zero field splittings have been shown to be on the order of 100 cm^{-1}.

ACCIDENTAL DEGENERACY A derivative-like MCD spectrum may also be observed for a molecule of low symmetry which happens to have two closely spaced excited states that are not resolved in the absorption spectrum. Formally, the MCD is composed of two "B" terms of opposite sign; functionally, the spectrum often cannot be distinguished from the "A" type spectrum of a slightly perturbed symmetric molecule. We refer to this class of spectra as "pseudo A type". Adenine and related purine derivatives exhibit pseudo "A" type MCD spectra in their lowest energy electronic absorption bands (194).

"C" TYPE SPECTRA UNRESOLVED IN ABSORPTION   Multiple "C" type spectra, sometimes unresolved in absorption, can have either the same sign or opposite signs. In the latter case the MCD has the derivative-like shape characteristic of an "A" or pseudo "A" type spectrum. Reducing the temperature causes both the peak and the trough to increase in absolute magnitude (23). Thus, unlike closely spaced "B" spectra, multiple "C" spectra can be distinguished from "A" spectra on the basis of their temperature dependence. Nor is high symmetry a precondition for observing temperature dependent MCD; a low energy "excited" state could accidentally be located within kT of the ground state.

IMPLICATIONS FOR INTERPRETATION OF SPECTRA   A derivative-like MCD spectrum can be due to degenerate or nearly degenerate excited states of a three- or higher-fold symmetric molecule or closely spaced "B" spectra of accidentally degenerate states of a molecule of less than threefold symmetry *or* closely spaced "C" type bands of either type of molecule. The dispersion (i.e. shape) of the MCD at a single temperature and in the absence of other information about the molecule being investigated thus provides no information on spectral type or on the symmetry of the molecule. The appearance of a derivative-like MCD *does* immediately imply the existence of more than one transition, which may not be resolved in absorption. More generally, multiple components are indicated whenever the MCD does not have the same shape as the corresponding absorption band. This is one of the important practical applications of MCD spectroscopy.

Extraction of Molecular Parameters

Two methods exist for determining A, B, C and D from experimental spectra. In *dispersion analysis* one fits the shape and magnitude of the experimental spectrum (154, 157), whereas in *moment analysis* certain integrals are performed over the absorption and MCD bands (154, 156, 162). Among biologically relevant molecules, most work has been done on porphyrin-like molecules (157, 163, 170), and the two methods give generally similar results. Both methods also have limitations.

Dispersion analyses invariably postulate that the magnetic field has the same effect, be it shift in wavelength or modification in absorption strength, on all portions of the band. For "B" type spectra this hypothesis has been verified; the MCD of the Q_{oo} band of coproporphyrin free base has the same shape as the corresponding absorption spectrum (164). For "A" type spectra the notion that the magnetic field shifts all portions of the band by an equal energy increment [rigid shift model (154)] remains unproven. Most workers who have used dispersion theory to extract values of "A" from experimental data have presumed some analytical shape for the absorption band, usually Gaussian or Lorentzian. It is possible, however, to write expressions for MCD while assuming an arbitrary shape for the absorption envelope of the individual components *and* allowing for unresolved components and the random orientation of molecules with respect to the magnetic field (170). The difficulty with this approach is that, in the presence of zero field splittings, the shape of the absorption envelope of the individual components is not an observable quantity.

Moment analysis avoids the issue of the rigid shift hypothesis and the shape of the absorption bands. The usefulness of this approach is limited, however, by the requirement that the absorption band under analysis be completely isolated from other bands so that limits of integration are determined unambiguously. This situation obtains for a relatively small number of biologically interesting molecules in the visible but almost never in the ultraviolet. In contrast, dispersion analysis can be truncated on the sides of an absorption band to avoid interference from other transitions.

What Is It Good For?

We list five ways in which MCD is useful in elucidating the structure and function of biological materials. Typical examples are mentioned for each class; some of these examples are discussed in greater detail in subsequent sections.

MCD can be extremely useful in detecting transitions unresolved in the unpolarized absorption spectrum. In this role it is complementary to fluorescence polarization, linear dichroism, and natural CD. In contrast to fluorescence polarization, MCD can be used to detect multiple transitions of molecules that are nonfluorescent or that cannot conveniently be obtained in a high viscosity environment. No physical orientation of the individual chromophores is required, as is the case with linear dichroism, and the molecule need not lack a rotation-reflection axis, as is required in natural CD. An example of this application is the detection of multiple transitions in the longest wavelength absorption band of the antibiotic netropsin (169).

MCD can detect the presence and determine the concentration of certain "marker" molecules, chromophores characterized by high MCD anisotropy even in the presence of absorption from other chromophores and high turbidity. Examples include the nondestructive determination of tryptophan concentration in proteins (15, 72, 77) and the measurement of small concentrations of reduced cytochromes in highly colored extracts from photosynthetic plants (21) and turbid preparations of submitochondrial particles from beef heart (3).

MCD can differentiate between structures of different symmetry classes when the general chemical composition of the system is known and reference spectra of appropriate models are available. Both the shape of the spectrum and the magnitude are important in such work. If concentrations are unknown, MCD/absorption ratios can be used. Examples include determination of the symmetry of cobalt ions in various proteins (187), the axis ligation of cytochrome P450 (67), and the cluster type of iron-sulfur proteins (155, 159, 160).

MCD is a sensitive means of determining molecular parameters and can serve as a stringent test of theoretical predictions. Examples of this class of application are determination of the excited state angular momenta of porphyrins (156, 157) and elucidation of the excited states of the iron-sulfur proteins (131, 180).

Finally, "C" type MCD can reflect the Kramers degeneracy of paramagnetic species. In this role, MCD complements electron paramagnetic resonance spectroscopy (EPR). Although lacking some of the analytical capabilities achievable with EPR, MCD is considerably less sensitive to paramagnetic contaminants, since only species with an appropriate absorption spectrum are monitored. The chief applications of this aspect of MCD have involved heme proteins (73, 78, 188, 189) and nonheme-iron proteins (180). One hitherto unexploited possibility, as far as biological materials are concerned, is the unambiguous assignment of EPR signals by means of a simultaneous EPR:MCD experiment (82).

APPARATUS FOR MCD SPECTROSCOPY

The apparatus required for measuring MCD consists of a dichrograph to measure circular dichroism and a magnet to apply a field to the sample. It is desirable to have a spectrophotometer to measure the absorption spectrum of the sample and a computer capable of collecting and processing spectral data from both the spectrophotometer and the dichrograph. A cryostat is also desirable since the standard method of identifying "C" type spectra is measurement of MCD as a function of temperature.

Dichrometers

Commercially available dichrometers operate from about 180 nm in the ultraviolet to 800 or 1000 nm in the near infrared. The performance of the latest instruments is enhanced significantly by the use of photoelastic modulators (PEM) (17, 17a, 20, 32, 84, 92, 110, 183), which replaced electrooptical modulators (i.e. Pockells' cells). The present generation of commercial instruments can mount superconducting magnets; one is available with an electromagnet (12). The performance of commercial dichrometers deteriorates below 250 nm because of the decreased output of the xenon arc and in the near infrared because of the inherently low dispersion of quartz-prism monochromators.

One rationale for building a dichrometer is the improved performance in the visible and near infrared, achievable with a grating monochromator (173). Assembled instruments can also be designed to use a greater variety of magnets. Several "homebuilt" instruments measure both CD/MCD and the absorption spectrum of the sample (37, 95, 183). One spectrometer was designed to measure both CD/MCD and emission spectra (166). This combination is attractive since both CD and fluorescence are single-beam, photon-limited experiments and thus require intense light sources and both can use a phase-sensitive or "lock-in" amplifier as the major component for analogue-signal processing. The single instrument is also easier to interface to a computer (168) and can be used for additional experiments such as FDMCD (171) and MCE (167). MCD spectrometers operating in the near infrared (to 5μ) (123) and the vacuum ultraviolet (to 125 nm) have been described (62, 134).

Spurious signals caused by the polarization sensitivity of the detection system in the infrared have been suppressed by a second PEM placed after the sample; the second modulator scrambles the polarization of the transmitted beam (33). Improvements in sensitivity have been achieved by mechanically chopping the incident beam and using two lock-in amplifiers in series to detect the CD signal (115).

The light sources used for CD measurements must be intense and provide continuous coverage over a broad spectral region and should possess good short term stability. MCD spectroscopy is one area in which lasers have had little impact except in MCE (31, 142), magnetic circular polarized Raman scattering (89), and situations that do not require scanning of wavelength (66). Synchrotron radiation, in contrast, promises to be an excellent source of radiation for CD experiments from the vacuum ultraviolet to the infrared (168).

Magnets

Superconducting magnets are easily accommodated by present commercial dichrographs and provide high fields, typically 4 to 6 T. The disadvantages of superconducting magnets are high initial cost, high cost of the liquid helium required for operation, and greater complexity of operation.

Electromagnets, in contrast, are initially less expensive and much simpler and less expensive to operate. Their chief disadvantage is lower field strength, typically 1 to 2 T.

Permanent magnets are relatively inexpensive, easy to use, and can be fitted to most commercial spectrometers. Unfortunately, they usually provide low fields (usually less than one Tesla) and restrict the path length of the sample. A sophisticated design capable of generating a field of 1.3 T and automatic field reversal was recently reported (117).

There is no simple answer to the question of whether superconducting magnets are better than permanent magnets and electromagnets or vice versa. Because of their simplicity of use, electromagnets are welcomed in biochemical laboratories. On the other hand, the sensitivity of an MCD spectrometer is directly proportional to the strength of the magnetic field. The high anisotropies of *some* biological materials (heme proteins, porphyrins, and tryptophan for example) are conducive to the use of electromagnets or even permanent magnets. When the MCD anisotropy is low and the CD anisotropy is high, a superconducting magnet may be necessary. A quantitative analysis of the significance of CD and MCD anisotropies in the selection of a magnet has appeared elsewhere (164). The type of magnet chosen also determines one's options for measuring the MCD of a sample at cryogenic temperatures (see below).

Spectrophotometers

Absorption spectra greatly facilitate the interpretation of the corresponding MCD and are required for the determination of anisotropies and for analysis by the method of moments. Ideally, every published MCD spectrum should be accompanied by the corresponding absorption spectrum. Usually the absorption spectrum is measured with a separate double-beam spectrophotometer. Low temperature measurements may be an exception since unambiguous assignment and quantitation of "C" type MCD spectra require that an increase be observed in the MCD, which is not due to an increase in the magnitude or a change in the shape of the absorption spectrum. Thus the MCD and absorption should be measured at the same temperature and under exactly the same conditions. Simultaneous measurement of MCD and absorption is also important when, as in dispersion analysis, a very careful comparison must be made of the MCD and the absorption spectrum. The use of separate spectrometers with different wavelength calibrations and spectral-band widths might introduce significant artifacts. Instruments that permit measurement of both CD and absorption spectra have been described (37, 95, 183).

Computers

MCD spectroscopy benefits from digital data collection and processing more than do most UV-visible-near infrared experiments because of the need to separate the MCD spectrum from the natural CD present in most biological molecules. Digital data collection and processing can also improve spectrometer sensitivity via signal averaging and can greatly facilitate analysis of spectra by dispersion theory or the method of moments. Schemes for obtaining data in digital form range from simple devices that digitize spectra previously recorded on chart paper (78) to instruments that are operated under control of a computer (166, 173).

Cryostats

The definitive test for "C" type MCD spectra is the increase of the MCD at cryogenic temperatures compared to room temperature (30). Superconducting magnets are themselves cryogenic devices and it is possible to design them such that the cryostat for the sample is an integral part of the magnet's cryogenic system. In the simplest and least satisfactory design the sample is mounted in position before the magnet is cooled. Thus only one sample can be studied per run. A much better, but more expensive, configuration permits changing of the sample while the magnet is at operating temperature. Schemes exist for controlling sample temperature at any point between 4.2 and 300 K (22). It is practically impossible to upgrade a room-temperature bore magnet to obtain cryogenic sample temperatures by tapping the magnet's own cryogenic system. Adding an auxiliary cryostat requires careful engineering because of limited access to the sample area (131).

Electromagnets and permanent magnets provide good access to the sample from the plane perpendicular to the light path and can be fitted with cryostats for operation down to liquid nitrogen or liquid helium temperature (7). Since the cryostat and the magnet are separate units, they need not be purchased at the same time.

PROTEINS

Peptide Bonds

The theory of the optical properties of the isolated amide bond predicts four transitions between 150 and 240 nm: $n \to \pi^*$ at 220–235 nm, $\pi_1 \to \pi^*$ around 180–190 nm, $\pi_2 \to \pi^*$ near 160 nm, and $n \to \sigma^*$ (a Rydberg transition on the oxygen atom) at around 150 nm (79, 126). MCD appears to offer an excellent means of confirming these assignments;

certainly it should be more informative than absorption spectroscopy. For example, since the two $\pi \to \pi^*$ transitions are nonparallel, they would be expected to give rise to coupled, "B" type MCD. Considering this promise, experimental studies of the amide chromophore and polypeptides have been few in number and restricted to wavelengths greater than 200 nm as a result of the deterioration in the performance of conventional dichrometers in the far ultraviolet.

Barth et al (15) reported the MCD of poly-L-lysine down to 205 nm. The weak negative band they found is presumably the long wavelength edge of the $\pi_1 \to \pi^*$ transition. This band is split by conformation-dependent exciton interactions in polypeptides, giving rise to strong, conformation-dependent CD and absorption spectra. The spectra published by Barth et al indicate that the MCD for poly-L-lysine in the random coil configuration is about 20% more negative at 210 nm than the corresponding α-helical configuration. This difference was somewhat less than the experimental uncertainty and Barth et al thus concluded that the MCD intensity of the peptide band is insensitive to peptide conformation. This conclusion may not apply to other wavelengths. The absorption spectra of Rosenheck & Doty (132) show that whereas the absorption of the poly-L-lysine in a random coil exceeds that of the α-helical form by only about 10% at 210 nm, at 200 nm they differ by nearly a factor of 2. Thus MCD could be at least as sensitive to conformation as is the absorption spectrum. Barth et al (15) also reported a very weak band at about 232 nm in MCD, which they assign to the $n \to \pi^*$ peptide transition.

Gabriel et al (56–59) reported the MCD of several amino acids down to about 200 nm. The spectrum for alanine showed a weak negative signal, whereas cysteine, histidine, tyrosine, and phenylalanine showed more complicated spectra presumably reflecting the influence of nonpeptide chromophores.

The potential for MCD in elucidating the excited states of the peptide bond has certainly not been exhausted. Significant advances await only improved instrumentation for the far and vacuum ultraviolet.

Aromatic Amino Acids

Although the MCD of amino acids such as cysteine, histidine, phenylalanine, tyrosine, and tryptophan shows evidence of nonpeptide contributions below 230 nm, only the latter three have significant absorption at longer wavelengths. Phenylalanine has an extremely weak negative MCD centered near 267 nm; the somewhat stronger MCD of tyrosine is also negative, with a maximum absolute value near 275 nm at neutral pH. In contrast, tryptophan has a very large MCD with a positive maximum value between 290 and 295 nm and a negative maximum between 265 and 275 nm. Since the contributions of phenylalanine and tyrosine are negligible above 290 nm, the positive MCD of tryptophan near 292 nm can be used to determine the concentration of tryptophan in solution and in proteins (14, 15, 72, 77, 107), using the MCD molar extinction coefficient of 2.35 M^{-1} cm^{-1} T^{-1} (78).

Ionization of tyrosine at higher pH shifts its absorption and MCD to longer wavelengths. The maximum negative value of the MCD of tyrosinate is near 295 nm. Thus the MCD of a denatured protein measured first at pH 7 and then at pH 13 provides the ratio of tryptophan to tyrosine content (11). Since tyrosine is more easily quantitated in chemical analysis of amino acid composition than is tryptophan, this procedure permits an accurate determination of the number of tryptophan residues in a protein even if the concentration of protein in the sample is otherwise unknown.

Without question, determination of the tryptophan content of proteins is one of the most important biological applications of MCD to date. For a more complete discussion see the review of Holmquist & Vallee (78).

The indole chromophore appears in a large number of biologically important compounds in addition to tryptophan, e.g. indole alkaloids [Barth et al (13)] and serotonin [Sprinkel et al (151)]. Barth et al (13) and Miles & Eyring (108) believe that for $\lambda > 250$ nm the MCD of tryptophan is due largely to the mixing of the B_{1u} and B_{2u} states by the magnetic field. Sprinkel et al (151) offer a more complicated interpretation in which $\sigma \rightarrow \pi^*$ transitions located near 200 nm influence the spectrum at longer wavelengths. Barth et al (131) point out that the signs of the two long wavelength bands are reversed by saturation of the 2–3 double bond. Miles & Eyring (108) correlate the signs observed in MCD with the presence of electron-donating or -accepting groups.

Heme Proteins

The most extensive application of MCD in biological systems has been in the study of heme proteins; this has led to a more complete understanding of the electronic structure of the porphyrin-protein system and its relation to biological function. These proteins produce probably the most easily measured MCD of any biological material. Indeed, the earliest application of magnetic optical activity by Shashoua (139) dealt with redox states of cytochrome *c* and was the forerunner to the expanded activity in this area in the last ten years. The MCD spectra of heme proteins are sensitive to the redox and spin states of the metal and to the nature of its axial ligands, as well as to the proximity of a second heme group and to the nature of the protein environment. Thus, in complementing other spectral techniques, MCD of heme proteins provides both corroborative and unique information on the electronic properties of heme-containing proteins.

Recent reviews have appeared concerning the magnetic optical activity of heme proteins (67, 73, 78, 188) and of porphyrins (164), the determination of spin states of heme proteins by MCD (189), and the theory of the MCD approach of heme protein spectral analysis (52, 67). Recent MCD investigations of heme proteins include studies on cytochromes c (18, 22, 102, 127, 130, 190, 197), b_5 (39), b_2 (130), and P-450 (39–42, 45, 49, 120, 144, 147, 150, 193), chloroperoxidase (43, 44), catalase (29, 88, 175), hemoglobins (52, 94, 99, 102, 118, 128, 136–138, 158, 184, 198), heme-hemopexin (112), horseradish peroxidase (93, 119, 136, 161), myoglobin (87, 89, 121, 136, 143, 145, 150), and nitrite reductase (122, 192).

SPIN-STATE ANALYSIS   Much of the application of MCD in heme protein studies centers on the capacity to determine heme spin states in a straightforward manner using dilute solutions. The earliest studies on hemoglobin derivatives suggested a relationship to the spin state that later was extended and extensively documented by Vickery et al (190, 191) and by Thomson and his collaborators (150, 178). Application to various proteins has been recently reviewed (189).

Studies of metmyoglobin in various thermal spin-state mixtures induced by added ligands (CN^-, imidazole, azide, OH^-, SCN^-, H_2O, and F^-) (Figure 3) illustrate this sensitivity. The Soret bands are all similar in shape, but their intensities in the low-spin form of the protein are much greater than in high-spin forms. When the Soret band intensity is plotted against spin-state equilibria obtained from the literature, based on absorption and magnetic susceptibility data, a good correlation is observed (Figure 4). The mixed-spin complexes found largely in the high-spin state, such as F^- and H_2O, exhibit low intensity Soret MCD bands, whereas the CN^- derivative, which is largely low-spin, produces a large signal. Cytochromes b_5 and c and the hemopexins, low-spin proteins when in their oxidized states, were found to have a large Soret MCD as predicted by the results from myoglobin and hemoglobin. Obviously, MCD offers a means to readily determine spin equilibria and investigate the effects of temperature, ligands, etc. (189). Recent examples where MCD has proven useful for spin-state determinations include catalase (29), cytochrome c oxidase (26, 28, 178, 179), myoglobin (150), cytochrome P-450 (45), and horseradish peroxidase (94, 119).

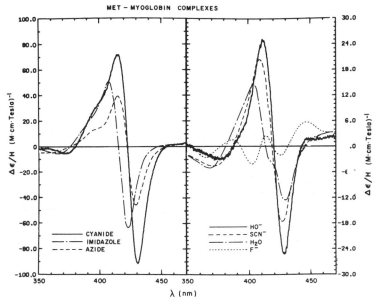

Figure 3 MCD spectra of the Soret band of ferrimyoglobin complexes: cyanide, imidazole, azide, OH^-, SCN^-, H_2O, F^-. (Reprinted with permission from Vickery et al (191), copyright by the American Chemical Society).

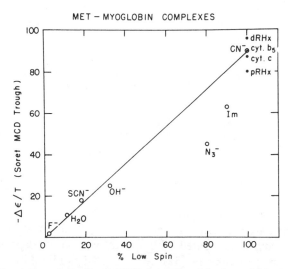

Figure 4 Correlation of the Soret MCD intensity of ferrimyoglobin complexes with low-spin content. dRHx refers to deuteroheme hemopexin and pRHx to protoheme hemopexin. (Reprinted with permission from Vickery et al (191), copyright by the American Chemical Society).

The near infrared MCD of high, low, and intermediate-spin derivatives of methemoglobin (158), cytochrome c' (127), and various myoglobin derivatives (121) have been reported. In the studies of hemoglobin and cytochrome c', MCD was found to be very sensitive both to the spin state and for the monitoring of spin-state populations, thus providing a new method for assignment of near infrared bands and assessment of other perturbations of heme electronic structure. Spin-state analysis in the near infrared has advantages over UV-visible methods, owing to less complexity and spectral overlap of low- and high-spin bands (158).

CYTOCHROME P-450   The origin of the anomalous spectral properties of cytochrome P-450 is believed to be associated with the presence of an axial thiolate ligand from a cysteine residue occupying the fifth axial position. The ferrous carbonyl form exhibits a Soret maximum at an unusually long wavelength, 450 nm, indicating that the state of the heme is considerably different from other heme proteins. MCD studies of cytochromes P-450 and P-420 in the free and membrane bound states have been reviewed in detail (67). Additional evidence obtained by MCD for the presence of an axial thiolate ligand from cysteine in P-450 has appeared, complementing earlier studies. Thus three forms of oxidized rabbit liver cytochrome P-450 (which differ by the agent employed for their induced synthesis) were examined and the first MCD spectra of the oxidized, reduced, and reduced + CO states of cytochrome P-420 were presented and integrated with previous P-450 studies (45). Strong support for the orbital mixing hypothesis of Hanson et al (64) has been obtained from studies comparing the spectra of hyperporphyrin complexes, e.g. (octaethylporphyrinate)bismuth(III) nitrate with P-450 (44). Although chloroperoxidase differs in activity from P-450, many of its physical properties mimic the cytochrome P-450 proteins, and comparison of its MCD spectrum with models and various P-450 forms provided strong evidence for thiolate ligation in this protein as well (43).

Models for cytochrome P-450 in the form of synthetic porphyrin complexes with various axial ligands have been prepared; those with mercaptan ligands were found to have MCD properties much in common with the parent protein. Thus, iron protoporphyrin IX diethyl ester and the dimethyl ester treated with sodium methyl mercaptide (38, 39) or with p-(nitro)-thiophenol (42) as an axial ligand exhibit properties remarkably similar to various forms of P-450, further substantiating the presence of an axial mercaptide ion. Additionally, metmyoglobin complexes with alkyl mercaptans and with cysteine exhibit MCD spectra (143) analogous to oxidized cytochrome P-450.

The ethyl isocyanide complexes of ferri- and ferroheme complexes have been also studied as potential models for cytochrome P-450 (144). Although the ferroheme complex exhibits a 455 nm absorption maximum, much like P-450, its MCD shows only a weak "B" term rather than the strong Faraday "A" term of P-450, consistent, at least, with the absence of a thiol ligand. In response to the possibility that histidine may be the transaxial ligand with oxy-P-450$_{cam'}$, Dawson & Cramer (40) measured and observed substantial differences between the MCD of oxy-P-450$_{cam}$ and oxymyoglobin (the latter known to have axial histidine ligands), providing evidence that the thiolate ligand is also present in oxy-P-450$_{cam}$.

CYTOCHROME c OXIDASE   MCD in combination with EPR and magnetic susceptibility has been employed to characterize the electronic state of the hemes in cytochrome c oxidase (4, 124, 125).

The protein is oligomeric and contains two spectroscopically distinguishable heme groups, designated a and $a_{3'}$, differences probably endowed on a single heme a by its site of occupation in the protein. It also contains two functional copper ions. The chemical reactivity of the heme is a function of its spin state, which, in turn, is determined by the coordination of the iron by the pyrrole nitrogens, the protein milieu, and the binding of exogenous ligands. Spin-state studies by MCD have proven valuable in describing the overall chemical state of the resting enzyme and its alterations in reductive titrations (5, 6, 178, 179). High- and low-spin complexes of isolated heme a have been prepared as models and their MCD compared with that of cytochrome a and a_3. The bis-imidazole and bis-pyridine low-spin ferrous complexes of heme a are analogous to those calculated for the cytochrome a component of cytochrome c oxidase (27). The MCD of the high-spin heme a and its CO adduct are analogous to the cytochrome a_3 component. By difference MCD measurements, the spectrum of the holoenzyme has been interpreted as the sum of two MCD curves (Figure 5): the reduced a_3 component exhibits an intense asymmetric band much like deoxymyoglobin, which is high-spin, and the reduced a component exhibits a weaker symmetric band, which is diamagnetic, low-spin (6). In the oxidized state only a_3 has significant MCD intensity. From analysis of such spin-state determinations by MCD during reductive titrations, the heme-heme interaction in this protein manifests not as an optically detectable perturbation but as a cooperative redox shift: a change of oxidation state of one of the hemes lowers the redox potential of the second. This is observed primarily when cytochrome a_3 is stabilized by ligand binding, e.g. formate (26), which lowers the redox potential of the cytochrome a form. In studies at 100° K where protein mobility is near

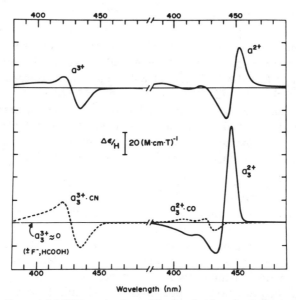

Figure 5 Computed MCD spectra for oxidized and reduced cytochrome a (*above*) and for oxidized and reduced cytochrome a_3 (*below*). Also shown are computed MCD spectra for low-spin cytochrome a_3 in the oxidized (as the CN^- complex) and the reduced (as the CO complex) states (*broken curves*). (Reprinted with permission from Babcock et al (6)).

zero, spin-state changes of cytochrome a_3 and a were detected with the CO complex (28), further indicating that the heme-heme interaction is not conformational in nature but arises from an electronic state change.

In the case of cytochrome oxidase from *Pseudomonas aeruginosa*, which also contains two hemes (heme c and heme d_1), MCD studies (192) indicate that all forms (oxidized, reduced, and the CO complex) of the intact protein also exhibit spectra that can be accounted for by the sum of the individual heme c and d_1 contributions, with no evidence for a spectrally active "heme-heme" interaction, as had been reported earlier (122).

HEMOPEXIN The heme complex of hemopexin (112), a serum glycoprotein that binds heme iron-protoporphyrin IX, presumably for transport, exhibits MCD spectra very similar to cytochrome b_5 and other bisimidazole-coordinated heme derivatives, e.g. imidazole myoglobin (191), in contrast to the methionyl-imidazole cytochrome c or the P-450 low-spin forms. They are low spin in both the reduced and oxidized states. Spectra of deuteroheme-hemopexin and cobalt- and nickel-deuteroporphyrin IX-hemopexin spectra were also recorded (112).

PROTEIN ENVIRONMENT EFFECTS   Numerous attempts have been made to employ MCD as a monitor of secondary effects on the heme electronic state, that is, alterations of spectral features responsive and unique to a particular protein environment, some of which have been reviewed (73, 130). In general, to the extent that structural or conformational alterations do not alter the basic energy levels of the hemochromophore, MCD is not sensitive to its protein environment, though minor perturbations can be expected. A recent assessment of the influence of the protein environment on the heme electronic state by Sharonov et al (136), extending their earlier studies, was made by comparing the MCD of the ferrous forms of myoglobin and hemoglobin tetramers, dimers, and isolated chains, and of horseradish peroxidase. The protoheme group of all of these systems is in the pentacoordinated high-spin ferrous state, such that difference in their MCD spectra reflects distortions induced by the protein. In contrast to the Soret MCD which are similar in all cases, the ratio of the "A" term of the Q_{oo} transition to that of the Q_V bands in the visible near 560 nm provided an assessment of structural alterations of the heme environment. Sharonov singled out deoxyhemoglobin and ferroperoxidase as exhibiting a significant difference in this ratio, presumably from constraints on the heme group imposed by quaternary and/or tertiary protein structure.

MICROPARTICULATE HEME PROTEINS   MCD has decided advantages over other forms of optical spectrometry when applied to suspensions, since turbidity and light scattering problems are an order of magnitude smaller. This advantage has been dramatized in studies of microparticulate systems of heme proteins, where active proteins free of phospholipids or membranes cannot be obtained. Clear identification of cytochrome c and c_1 was made in MCD studies of submitochondrial particles of beef heart (3). Similarly, microsomal preparations containing induced cytochrome P-450 exhibited MCD spectra and could be entirely accounted for by cytochromes P-450 and b_5 (49, 193). A distinction between cytochrome P-450 and P-448 is even possible in microsomal fractions of 3-methylcholanthrene-induced rats, by monitoring crossover points of the Soret bands (41), further illustrating the resolving capacity and sensitivity of MCD.

Iron-Sulfur Proteins

MCD has been valuable in resolving the broad, unresolved absorption spectra of iron-sulfur proteins (16, 131, 155, 159, 160, 172, 180, 186). Eaton & Lovenberg (51) proposed detailed assignments of the sulfur-to-iron charge transfer bands of rubredoxin (a one-iron-per-cluster protein), based in part on room temperature MCD. Rivoal et al (131) refined these assignments using the MCD of oxidized rubredoxin measured at very low temperatures. Thomson et al (180) measured the MCD of several two-iron-per-cluster proteins at very low temperatures and interpreted the spectra using rubredoxin as a model. The low temperature MCD spectra of Thomson et al (180) also demonstrate that the ground state of the reduced two-iron ferredoxins are Kramers degenerate, whereas the oxidized protein has a nondegenerate ground state as a result of antiferromagnetic coupling. Muraoka et al (114) reported the MCD of several inorganic iron-sulfur complexes that might be models for the complexes found in proteins.

Stephens and his co-workers (159, 160) reported the room temperature MCD of several two-iron- and four-iron-per-cluster proteins in the visible (sulfur-to-iron charge transfer transitions) and near infrared (iron d→d transitions). The qualitative and quantitative differences between the CD and MCD spectra of the two-iron and four-iron type proteins lead to the proposal that CD/MCD spectroscopy can be used to assign cluster type. Stephens et al (155) have applied this approach to two classes of iron-sulfur proteins involved in nitrogen fixation. The "Fe" proteins from *Azotobacter vinelandii* and *Klebsiella pneumoniae* were shown to contain a single four-iron cluster. The "Mo-Fe" proteins from the same organisms resemble four-iron clusters in the visible and near ultraviolet. Unlike previously characterized four-iron-per-cluster proteins, they do not exhibit MCD at wavelengths greater than 1.4μ, suggesting some as yet undefined difference in structure.

The utility of MCD as a probe of iron-sulfur proteins will be enhanced when it becomes possible to make measurements as a function of temperature in the infrared as well as the visible and ultraviolet regions. MCD-ESR (electron spin resonance) saturation experiments (82) would also aid in unambiguous assignments of EPR spectra.

Cobalt, Nickel, and Copper Proteins

Of the transition metals, Co(II) has received most attention, primarily because of the relatively strong MCD signals it offers and because of its probe capacities when replacing Zn^{2+} in the zinc metalloenzymes. Such substitutions usually result in a more active enzyme. Since it is a paramagnetic ion, details of Co(II) spectra reveal features of the metal coordination and its functional participation in catalysis (187).

The MCD spectra of a number of Co(II)-substituted proteins and Co(II) complex ions of various geometries have been measured, the comparison of which permits deductions of the overall coordination geometry of the Co(II)-substituted proteins.

High-spin Co(II) can accept various geometries and coordination numbers. Octahedral and tetrahedral are most numerous but 5-coordi-

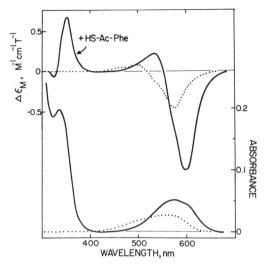

Figure 6 Effect of mercaptan inhibitor-mercaptoacetyl-*D*-phenylalanine on the MCD (*above*) and absorption (*below*) spectra of carboxypeptidase A. Inhibitor (0.9 equivalents) was added to the cobalt enzyme (······) to produce the spectra of the inhibited enzyme (——). MCD was measured at a field of 4 Tesla and enzyme was 1.9×10^{-4}M. (From Holmquist & Vallee, *Proc. Natl. Acad. Sci. USA*, in press).

nate complexes of various geometries are known. The MCD spectra of a considerable number of Co(II) complexes have been examined (35, 85, 90), with most attention on tetrahedral geometries. These studies have revealed that the MCD spectra fall into 3 general categories, in accord with the coordination number of the metal (Figure 6). Apparently, the overall geometry of Co(II) complexes is significant in determining the shape of the MCD signal; the sensitivity to distortions in the coordination sphere is much less than in either CD or absorption. This has allowed qualitative comparisons to be made of the overall coordination geometry of the metal sites of proteins.

Cobalt-substituted proteins that have been examined include carboxypeptidase A (76) and P (19), procarboxypeptidase A (76), thermolysin (76, 113) and *Bacillus cereus* neutral protease (71), carbonic anhydrase from various sources (35, 76), concanavalin A (111, 129), rabbit muscle pyruvate kinase (97), azurin, stellacyanin and plastocyanin (149), superoxide dimutase (133), alkaline phosphatase (1, 2, 176), rubredoxin (105), and alcohol dehydrogenase (174). In most instances, the MCD spectra suggest tetrahedral Co(II), as confirmed in a number of proteins by X-ray crystallography. Examples of octahedral coordination have been indicated in concanavalin A and pyruvate kinase and 5-coordinate-like geometries are exhibited by carbonic anhydrase and in certain forms of alkaline phosphatase.

Charge transfer bands between sulfur and cobalt(II) exhibit intense MCD bands near 340 nm that have now been observed both in proteins where a cysteine sulfur serves as a natural metal ligand and in protein-inhibitor complexes where mercaptans are components of metal-directed inhibitors of proteolytic enzymes. Azurin, stellacyanin, and plastocyanin, copper blue proteins in which Co(II) has been substituted for the copper atom, exhibit intense S→Co(II) charge transfer bands near 340 nm (149), as does cobalt-substituted horse liver alcohol dehydrogenase (174), α-lactamase I (A. Galdes, personal communication), and yeast alcohol dehydrogenase (174). In both plastocyanin and horse liver alcohol dehydrogenase, cysteine sulfurs serve as metal ligands, as shown by X-ray crystallography, a factor considerably aiding S→Co(II) band assignment.

Excellent inhibitors of zinc proteases have been designed on the basis of an active site directing an ancillary metal ligand to form inhibitory complexes of the type E:M:I where inhibitor (I) possesses affinity for substitute sites on the enzyme (E) and for the metal (M). Various ligands including hydroxamic acids, phosphoramidates, and mercaptans have been employed where these are attached to amino acids that mimic substrates. By utilizing mercaptan-derived inhibitors with the cobalt-

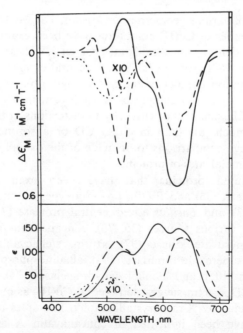

Figure 7 Magnetic circular dichroism (MCD) spectra (*above*) and absorption spectra (*below*) of Co(II) model complexes: $Co(H_2O)_6^{2+}$ (······) $Co(Me_6tren)Br_2$(-----) and $Co(OH)_4^{2+}$(———). From Reference 76.

substituted proteases carboxypeptidase A and thermolysin, proof that the inhibitor binds directly to the metal was obtained by MCD and absorption spectroscopy (187). For example, N-(2-mercaptoacetyl)-D-phenylalanine (Ki = 200 nM), the inhibitor of carboxypeptidase A, binds stoichiometrically to the cobalt-substituted enzyme producing the spectral alterations shown in Figure 7. The geometry of the cobalt atom remains tetrahedral, but the interaction results in a new spectral band, clearly evident in MCD, and an absorbance that is due to S→Co(II) charge transfer.

The S→Co(II) transition is also observed with analogous inhibitors with thermolysin. However, here the energy of the transition is such that in absorption only a shoulder is observed, but in MCD resolution of a discrete band is achieved, further illustrating a chief advantage of MCD —to reveal or resolve transitions that are either hidden or only weakly observable in other forms of spectroscopy.

Although numerous copper proteins are known, little has been reported on the MCD of this metal. Superoxide dismutase (133, 196) and hemocyanin (55, 113) exhibit very broad and extremely weak negative bands near 600 nm, much like simple complexes of Cu(II) (133). Theoretically Cu(II) with the d^9 configuration is expected to exhibit up to 4 d→d transitions, which MCD appears not to resolve, although a suggestion of band structure was observed in the studies of superoxide dismutase (196).

The MCD spectra of the d→d transitions of a number of Ni(II) complexes have been measured (36, 65, 98, 103), but extension to Ni(II)-containing proteins has yet to be made. Studies of d→d as well as charge transfer bands in Ni(II) analogous to those observed with Co(II) could facilitate assignment of the metal-binding site of urease, a native nickel protein, and of nickel-substituted zinc enzymes (e.g. carboxypeptidase A) and copper proteins (e.g. azurin, stellacyanin, and plastocyanin) (177).

Lanthanides

The lanthanide ions, primarily Tb^{3+}, Nd^{3+}, Sm^{3+}, and Eu^{3+}, have recently received considerable attention as spectroscopic probes of metal binding sites of proteins and nucleic acids (54, 74, 96, 116). This has been primarily through the recognition that they will nearly universally substitute for calcium (25, 54). The fluorescence (Tb^{3+}, Eu^{3+}, Sm^{3+}) and MCD (Ho^{3+}, Sm^{3+}, Nd^{3+}, Pr^{3+}, Er^{3+}, Tb^{3+}) of these metals allows measurements at concentrations commensurate with biological studies; in contrast, their absorption and circular dichroism, which are extremely weak, seem to preclude the direct application of these techniques.

The lanthanide ion Nd^{3+}, which provides the largest signal in MCD, (91), has been substituted into thermolysin (72) and parvalbumin (50), and their MCD spectra have been compared to those of a limited number of model complexes. In complexes with different donor atoms to the metal, the complex MCD bands are considerably broadened and weakened compared to the aquo ion (75, 148), thus indicating a sensitivity of the metal to its environment as monitored by MCD. In thermolysin, which binds a single Nd^{3+} with very high affinity, the prominent "A" term structure of the aquo ion at 790 nm is split into two positive and one negative extrema; while bound to paravalbumin, Nd^{3+} displays a unique spectrum of weak intensity.

Eu^{2+}, prepared by electrolytic reduction, has been shown to bind to the calcium-binding site (S2) of concanavalin A (80). In contrast to the Laporte forbidden f→f transitions of the tripositive lanthanides, the Eu^{2+} ion exhibits intense f→d transitions that are much more observable in MCD than the f→f bands. The MCD of Eu^{2+} is highly sensitive to its environment, and alterations in the MCD spectrum in the 370 and 415 nm region with time were interpreted to reflect the evolution of a strong Eu^{2+}-binding site resulting from a slow conformational change induced by binding the metal. With the gathering interest in the lanthanides as probes of the metal environment and the increased sensitivity of modern instrumentation, extension of the MCD properties of lanthanide ions can be expected.

NUCLEIC ACIDS

Purines

The ability of MCD to resolve multiple transitions contained in a single absorption envelope is nowhere better demonstrated than by the MCD of adenine, inosine, hypoxanthine, and their nucleosides (194). Although only a single absorption band is observed in the 230–300 nm region, the MCD is of the pseudo "A" type, with a negative maximum on the long-wavelength side and a positive maximum on the short-wavelength side of the absorption peak. The MCD spectra of guanine, xanthine, and their derivatives are qualitatively similar, but in these chromophores the two components are partially resolved in the absorption spectrum (194). MCD thus provided definitive confirmation that the near UV absorption band of purines is a superposition of nonparallel transitions, presumably related to the $A_{1g} \rightarrow B_{2u}$ and $A_{1g} \rightarrow B_{1u}$ transitions of benzene (see Reference 34 and references cited therein).

The MCD spectra of a series of 7-ribofuranosyl purines are similar to the corresponding chromophores studied above (109), except that the

MCD of 7-ribosylpurine has the opposite sign from all the compounds substituted with amino or carboxyl groups at the 6 position. The MCD of purine was not reported in Reference 194, but a series of N-methyl derivatives of purine were all shown to have inverted type (long wavelength positive maximum) pseudo "A" spectra (182). The inverted pseudo "A" MCD of the purine chromophore, while confirming the composite composition of the near UV absorption band, reveals subtle differences due to substituent effects not revealed by other methods. MCD is also more sensitive to the position of methyl substitution than is absorption in both methylpurines (181), methyladenines (101), and 8-azapurines (195). Molecular orbital calculations have been successful in correctly predicting the sign of the pseudo "A" MCD spectrum in the 240–300 nm region (195). MCD has also been used in conjunction with natural CD to probe solvent effects on the adenine transitions (47) and to study the interactions of the adenine ring of NAD with tryptophan residues located proximal to the NAD-binding sites of dehydrogenases (83). MCD was used in conjunction with CD and absorption spectroscopy to study solvent effects on guanosine and several of its derivatives (46). The MCD of several adenosine-cyclonucleosides has been reported (53).

Pyrimidines

The MCD spectra of the biologically important pyrimidines (cytosine, uracil and thymine) are substantially different from those of the purines. The longest wavelength absorption consists only of the $A_{1g} \rightarrow B_{2u}$ transition (34) and thus a "B" type, negative MCD is observed (194). A positive "B" type component is observed at shorter wavelengths, presumably corresponding to the $A_{1g} \rightarrow B_{1u}$ transition. Since the excited states mixed by the magnetic field are further apart in energy, the magnitude of the MCD is less by a factor of 3 or 4 than for the purines.

Guschlbauer & Holy (63) used MCD to demonstrate that changes in the natural CD of thymine derivatives substituted at the 1 and 3 positions with saturated ribose moieties were not due to changes in the wavelengths or orientation of the transition dipoles of the thymine ring.

Polynucleotides, DNA and Viruses

Maestre et al (104) reported the MCD of several di-, tri-, oligo-, and polynucleotides, and DNA from phage T2 and T5 as well as intact and degraded T2 and T5 phage. Conformational sensitivity of MCD was demonstrated in most of these systems. Absorption spectra were not presented; it is thus difficult to determine whether MCD is a more sensitive probe of the conformation of these materials than unpolarized absorption spectroscopy. MCD is usually considered to be less sensitive to conformational perturbations than is natural circular dichroism.

NON-HEME PORPHYRINS

The $\pi \rightarrow \pi^*$ absorption bands of porphyrins are some of the best understood spectra of all biological molecules. MCD provides excellent experimental support for the theoretical interpretation developed by Gouterman. For reviews of the theory of porphyrin spectra and the MCD of the $\pi \rightarrow \pi^*$ bands of porphyrins see references (61) and (60, 164) respectively. Attention should also be called to recent theoretical work on the MCD of porphyrins (86) and to studies of the MCD of the closely related phthalocyanine molecules (69, 70, 100, 162, 163). The MCD of porphyrin derivatives has recently been reported in the ultraviolet (146).

ACKNOWLEDGMENTS

Preparation of this review was supported by the US Department of Energy, a Research Career Development Award (CA00465) from the National Cancer Institute, US NIH to JCS and grant in aid GM-15003 from the National Institutes of Health, Department of Health, Education and Welfare. We thank Kathleen Griffin for assistance in proofreading the manuscript.

Literature Cited

1. Anderson, R. A., Kennedy, F. S., Vallee, B. L. 1976. *Biochemistry* 15:3710–14
2. Anderson, R. A., Vallee, B. L. 1977. *Biochemistry* 16:4388–92
3. Arutyunyan, A. M., Konstantinov, A. A., Sharonov, Y. A. 1974. *FEBS Lett.* 46:317–20
4. Babcock, G. T., Van Steelandt, J., Palmer, G., Vickery, L. E., Salmeen, I. 1977. In *Cytochrome Oxidase*, ed. T. E. King, B. Chance, Y. Orii, pp. 105–15. Amsterdam:Elsevier
5. Babcock, G. T., Vickery, L. E., Palmer, G. 1976. *J. Biol. Chem.* 251:7907–19
6. Babcock, G. T., Vickery, L. E., Palmer, G. 1978. *J. Biol. Chem.* 253:2400–11.
7. Badoz, J., Billardon, M., Boccara, A. C., Brita, B. 1969. *Symp. Faraday Soc.* 3:27–39
8. Barron, L. D. 1975. *Nature* 257:372–74
9. Barron, L. D. 1977. *Chem. Phys. Lett.* 46:579–81
10. Barron, L. D., Buckingham, A. D. 1972. *Mol. Phys.* 23:145–50
11. Barth, G., Bunnenberg, E., Djerassi, C. 1972. *Anal. Biochem.* 48:471–79
12. Barth, G., Dawson, J. H., Dolinger, P. M., Linder, R. E., Bunnenberg, E., Djerassi, C. 1975. *Anal. Biochem.* 65:100–8
13. Barth, G., Linder, R. E., Bunnenberg, E., Djerassi, C. 1972. *Helv. Chim. Acta* 55:2168–78
14. Barth, G., Records, R., Bunnenberg, E., Djerassi, C., Voelter, W. 1971. *J. Am. Chem. Soc.* 93:2545–47
15. Barth, G., Voelter, W., Bunnenberg, E., Djerassi, C. 1972. *J. Am. Chem. Soc.* 94:1293–98
16. Bason, R., Zubieta, J. A. 1973. *Angew. Chem.* 12:390
17. Billardon, M., Badoz, J. 1966. *C. R. Acad. Sci. Ser. B* 262:1672–75
17a. Billardon, M., Badoz, J. 1966. *C. R. Acad. Sci. Ser. B* 263:26–29
18. Bolard, J., Garnier, A. 1972. *Biochim. Biophys. Acta* 263:535–49
19. Breddam, K., Bazzone, T., Holmquist, B., Vallee, B. L. 1979. *Biochemistry* 18:1563–70

20. Breeze, R. H., Ke, B. 1972. *Anal. Biochem.* 50:281–303
21. Breton, J., Hilaire, M. 1972. *C. R. Acad. Sci. Ser. D* 274:678–81
22. Briat, B. 1973. In *Fundamental Aspects of Recent Development in Optical Rotary Dispersion and Circular Dichroism*, ed. F. Ciardelli, P. Salvadori, pp. 375–401. London: Heydon
23. Briat, B., Berger, D., Leliboux, M. 1972. *J. Chem. Phys.* 57:5606–7
24. Briat, B., Billardon, M., Badoz, J. 1977. *Physica* 89B:27–37
25. Brittain, H. G., Richardson, F. S., Martin, R. B. 1976. *J. Am. Chem. Soc.* 98:8255–60
26. Brittain, T., Greenwood, C., Johnson, J. P. 1977. *Biochem. J.* 167:531–34
27. Brittain, T., Greenwood, C., Springall, J. P., Thomson, A. J. 1978. *Biochem. J.* 173:411–17
28. Brittain, T., Springall, J., Greenwood, C., Thomson, A. J. 1976. *Biochem. J.* 159:811–13
29. Browett, W. R., Stillman, M. J. 1979. *Biochim. Biophys. Acta.* In press
30. Buckingham, A. D., Stephens, P. J. 1966. *Ann. Rev. Phys. Chem.* 17:399–432
31. Cavenett, B. C., Sowersby, G. 1975. *J. Phys. E* 8:365–68
32. Cheng, J. C., Nafie, L. A., Allen, S. D., Braunstein, A. I. 1976. *Appl. Opt.* 15:1960–65
33. Cheng, J. C., Nafie, L. A., Stephens, P. J. 1975. *J. Opt. Soc. Am.* 65:1031–35
34. Clark, L. B., Tinor, I. 1965. *J. Am. Chem. Soc.* 87:11 15
35. Coleman, J ..., Coleman, R. V. 1972. *J. Bic. Chem.* 247:4718–828
36. Collingwood, J. C., Day, P., Denning, R. G. 1973. *J. Chem. Soc. Faraday Trans. 2* 69:591–607
37. Collingwood, J. C., Day, P., Denning, R. G., Quested, P. N., Snellgrove, T. R. 1974. *J. Phys. E* 7:991–96
38. Collman, J. P., Sorrell, T. N. 1975. *J. Am. Chem. Soc.* 97:4133–34
39. Collman, J. P., Sorrell, T. N., Dawson, J. H., Trudell, J. R., Bunnenberg, E., Djerassi, C. 1976. *Proc. Natl. Acad. Sci. USA* 73:6–10
40. Dawson, J. H., Cramer, S. P. 1978. *FEBS Lett.* 88:127–30
41. Dawson, J. H., Dolinger, P. M., Trudell, J. R., Barth, G., Linder, R. E., Bunnenberg, E., Djerassi, C. 1974. *Proc. Natl. Acad. Sci. USA* 71:4594–97
42. Dawson, J. H., Holm, R. H., Trudell, J. R., Barth, G., Linder, R. E., Bunnenberg, E., Djerassi, C., Tang, S. C. 1976. *J. Am. Chem. Soc.* 98:3707–9
43. Dawson, J. H., Trudell, J. R., Barth, G., Linder, R. E., Bunnenberg, E., Djerassi, C., Chiang, R., Hager, L. P. 1976. *J. Am. Chem. Soc.* 98:3709–10
44. Dawson, J. H., Trudell, J. R., Barth, G., Linder, R. E., Bunnenberg, E., Djerassi, C., Gouterman, M., Connell, C. R., Sayer, P. 1977. *J. Am. Chem. Soc.* 99:641–42
45. Dawson, J. H., Trudell, J. R., Linder, R. E., Barth, G., Bunnenberg, E., Djerassi, C. 1978. *Biochemistry* 17:33–41
46. Delabar, J. -M., Guschlbauer, W. 1973. *J. Am. Chem. Soc.* 95:5729–35
47. Delabar, J. -M., Guschlbauer, W., Schneider, C., Thiery, J. 1972. *Biochimie* 54:1041–48
48. Djerassi, C., Bunnenberg, E., Elder, D. L. 1971. *Pure. Appl. Chem.* 25:57–90
49. Dolinger, P. M., Kielczewski, M., Trudell, J. R., Barth, G., Linder, R. E., Bunnenberg, E., Djerassi, C. 1974. *Proc. Natl. Acad. Sci. USA* 71:399–403
50. Donato, H. Jr., Martin, R. B. 1974. *Biochemistry* 13:4575–79
51. Eaton, W. A., Lovenberg, W. 1973. In *Iron-Sulfur Proteins*, ed. W. Lovenberg, pp. 131–62. New York: Academic
52. Eaton, W. A., Hanson, L. K., Stephens, P. J., Sutherland, J. C., Dunn, J. B. R. 1978. *J. Am. Chem. Soc.* 100:4991–5003
53. Elder, D. L., Bunnenberg, E., Djerassi, C., Ikehara, M., Voelter, W. 1970. *Tetrahedron Lett.* 10:727–30
54. Ellis, K. J. 1975. *Inorganic Perspectives in Biology & Medicine* 1:101–35. Amsterdam:Elsevier/North Holland Biomed. Press
55. Gabriel, M., Godbillon, G., Larcher, D., Rinnert, H., Thirion, C. 1972. *Experientia* 28:1019–20
56. Gabriel, M., Larcher, D., Rinnert, H., Thirion, C. 1973. *FEBS Lett.* 35:148–50
57. Gabriel, M., Larcher, D., Rinnert, H., Thirion, C., Grange, J. 1972. *C. R. Acad. Sci. Ser. B* 275:935–37
58. Gabriel, M., Larcher, D., Rinnert, H., Thirion, C., Grange, J. 1972. *C. R. Acad. Sci. Ser. B* 275:861–63
59. Gabriel, M., Larcher, D., Rinnert, H., Thirion, C., Grange, J. 1973. *C. R. Acad. Sci. Ser. B* 276:39–41
60. Gale, R., McCaffery, A. J., Rowe, M. D. 1972. *J. Chem. Soc. Dalton Trans.* 5:596–604

61. Gouterman, M. 1978. In *The Porphyrins*, ed. D. Dolphin, pp. 1–165. New York:Academic
62. Gross, K. P., Schnepp, O. 1977. *Rev. Sci. Instrum.* 48:362–63
63. Guschlbauer, W., Holy, A. 1972. *FEBS Lett.* 20:145–47
64. Hanson, L. K., Eaton, W. A., Sligar, S. C., Gunsalus, I. C., Gouterman, M., Connell, C. R. 1976. *J. Am. Chem. Soc.* 98:2672–74
65. Harding, M. J., Mason, S. F., Robbins, D. J., Thomson, A. J. 1971. *J. Chem. Soc. A* 19:3058–62
66. Harms, H., Feldtkeller, E. 1973. *Rev. Sci. Instrum.* 44:742–43
67. Hatano, M., Nozawa, T. 1978. *Adv. Biophys.* 11:95–149
68. Hipps, K. W. 1978. *J. Phys. Chem.* 82:602–5
69. Hollebone, B. R., Stillman, M. J. 1974. *Chem. Phys. Lett.* 29:284–86
70. Hollebone, B. R., Stillman, M. J. 1978. *J. Chem. Soc. Faraday Trans. 2* 74:2107–27
71. Holmquist, B. 1977. *Biochemistry* 16:4591–94
72. Holmquist, B. 1975. In *Protein Nutritional Quality of Foods and Feed, Part 1*, ed. M. Friedman, pp. 463–78. New York:Dekker
73. Holmquist, B. 1978. In *The Porphyrins*, ed. D. Dolphin, 3:249–70. New York:Academic
74. Holmquist, B. 1979. In *Methods for Investigating Metal Ion Environment in Proteins*. Amsterdam:Elsevier, In press
75. Holmquist, B., Horrocks, W. D. 1975. *Fed. Proc.* 34:594 (Abstr.)
76. Holmquist, B., Kaden, T. A., Vallee, B. L. 1975. *Biochemistry* 14:1454–61
77. Holmquist, B., Vallee, B. L. 1973. *Biochemistry* 12:4409–17
78. Holmquist, B., Vallee, B. L. 1978. *Methods Enzymol.* 69:P. G, pp. 149–79
79. Holzwarth, G., Doty, P. 1965. *J. Am. Chem. Soc.* 87:218–28
80. Homer, R. B., Mortimer, B. D. 1978. *FEBS Lett.* 87:69–72
81. Hudson, B., Jacobs, R. 1975. *Biopolymers* 14:1309–12.
82. Izen, E. H., Modine, F. A. 1972. *Rev. Sci. Instrum.* 43:1563–67
83. Jallon, J. H., Risler, Y., Schneider, C., Thiery, J. M. 1973. *FEBS Lett.* 31:251–55
84. Jasperson, J. N., Schnatterly, S. E. 1969. *Rev. Sci. Instrum.* 40:761–67
85. Kaden, T. A., Holmquist, B., Vallee, B. L. 1974. *Inorg. Chem.* 13:2585–90
86. Kaito, A., Nozawa, T., Yamamoto, T., Hatano, M. 1977. *Chem. Phys. Lett.* 52:154–60
87. Kajiyoshi, M., Anan, F. K. 1975. *J. Biochem. Tokyo* 78:1087–95
88. Kajiyoshi, M., Anan, F. K. 1977. *J. Biochem. Tokyo* 81:1319–25
89. Kajiyoshi, M., Anan, F. K. 1977. *J. Biochem. Tokyo* 81:1327–33
90. Kato, H., Akimoto, K. 1974. *J. Am. Chem. Soc.* 96:1351–57
91. Kato, Y., Nagai, T., Nakaya, T. 1976. *Chem. Phys. Lett.* 39:183–87
92. Kemp, J. C. 1969. *J. Opt. Soc. Am.* 59:950–54
93. Kobayashi, H., Shimizu, M., Fujita, I. 1970. *Bull. Chem. Soc. Jpn.* 43:2335–41
94. Kobayashi, N., Nozawa, T., Hatano, M. 1977. *Biochem. Biophys. Acta* 493:340–51
95. Koning, R. E., Vliek, R. M. E., Zandstra, P. J. 1975. *J. Phys. E.* 8:710–11
96. Koreneva, L. G., Zolin, V. F., Serbinova, T. A. 1975. *Biofizika* 20:767–71
97. Kwan, C. -Y., Erhard, K., Davis, R. C. 1975. *J. Biol. Chem.* 250:5951–59
98. Larcher, D., Gabriel, M. 1975. *J. Inorg. Nucl. Chem.* 37:2117–19
99. Linder, R. E., Records, R., Barth, G., Bunnenberg, E., Djerassi, C., Hedlund, B. E., Rosenberg, A., Benson, E. S., Seamans, L., Moscowitz, A. 1978. *Anal. Biochem.* 90:474–80
100. Linder, R. E., Rowlands, J. R. 1971. *Mol. Phys.* 21:417–37
101. Linder, R. E., Weiler-Feilchenfeld, H., Barth, G., Bunnenberg, E., Djerassi, C. 1974. *Theor. Chim. Acta* 36:135–43
102. Livshitz, M. A., Arutyunyan, A. M., Sharonov, Y. A. 1976. *J. Chem. Phys.* 64:1276–80
103. Looney, Q., Douglas, B. E. 1970. *Inorg. Chem.* 9:1955–57
104. Maestre, M. F., Gray, D. M., Cook, R. B. 1971. *Biopolymers* 10:2537–53
105. May, S. W., Kyo, J. -Y. 1978. *Biochemistry* 17:3333–38
106. McCaffery, A. J., Gale, R., Shatwell, R. A., Sichel, K. 1976. In *Excited States of Biological Molecules*, ed. J. Birks, pp. 327–32. New York:Wiley
107. McFarland, T. M., Coleman, J. E. 1972. *Eur. J. Biochem.* 29:521–27.
108. Miles, D. W., Eyring, H. 1973. *Proc. Natl. Acad. Sci. USA* 70:3754–58
109. Miles, D. W., Inskeep, W. H., Townsend, L. B., Eyring, H. 1972. *Biopolymers* 11:1181–1207
110. Mollenauer, L. F., Downie, D., Engstrom, H., Grant, W. B. 1969.

111. Morel, C., Gabriel, M., Larcher, D., Grange, J. 1976. *C. R. Acad. Sci. Ser. B* 283:61–64
112. Morgan, W. T., Vickery, L. E. 1978. *J. Biol. Chem.* 253:2940–45
113. Mori, W., Yamauchi, O., Nakao, Y., Nakahara, A. 1975. *Biochem. Biophys. Res. Commun.* 66:725–38
114. Muraoka, T., Nozawa, T., Hatano, M. 1978. *Bioinorg. Chem.* 8:45–59
115. Nafie, L. A., Keiderling, T. A., Stephens, P. J. 1976. *J. Am. Chem. Soc.* 98:2715–23
116. Nieboer, E. 1975. *Struct. Bonding Berlin* 22:1–46
117. Norden, B., Hakansson, R., Danielsson, S. 1977 *Chem. Scr.* 11:52–56
118. Nozawa, T., Hatano, M., Yamamoto, H., Kwan, T. 1976. *Bioinorg. Chem.* 5:267–73.
119. Nozawa, T., Kobayashi, N., Hatano, M. 1976. *Biochim. Biophys. Acta* 427:652–62.
120. Nozawa, T., Shimizu, T., Hatano, M., Shimada, H., Iziuka, T., Ishimura, Y. 1978. *Biochim. Biophys. Acta* 534:285–94
121. Nozawa, T., Yamamoto, T., Hatano, M. 1976. *Biochem. Biophys. Acta* 427:28–37
122. Orii, Y., Manabe, M., Yoneda, M. 1977. *Biochem. Biophys. Res. Commun.* 76:983–88
123. Osborne, G. A., Cheng, J. C., Stephens, P. J. 1973. *Rev. Sci. Instrum.* 44:10–15
124. Palmer, G., Antalis, T., Babcock, G. T., Garcia-Iniquez, L., Tweedle, M., Wilson, L. J., Vickery, L. E. 1978. In *Mechanisms of Oxidizing Enzymes*, ed. T. Singer, R. Ordarza, pp. 221–38. Amsterdam:Elsevier
125. Palmer, G., Babcock, G. T., Vickery, L. E. 1976. *Proc. Natl. Acad. Sci. USA* 73:2206–10
126. Peterson, D. L., Simpson, W. T. 1957. *J. Am. Chem. Soc.* 79:2375–82
127. Rawlings, J., Stephens, P. J., Nafie, L. A., Kamen, D. M. 1977. *Biochemistry* 16:1725–29
128. Rein, H., Ruckpaul, K., Haberditzl, W. 1973. *Chem. Phys. Lett.* 20:71–76
129. Richardson, C. E., Behnke, W. D. 1976. *J. Mol. Biol.* 102:441–51
130. Risler, J. -L., Groudinsky, O. 1973. *Eur. J. Biochem.* 35:201–5
131. Rivoal, J. C., Briat, B., Cammack, C., Hall, D. O., Rao, K. K., Douglas, I. N., Thomson, A. J. 1977. *Biochim. Biophys. Acta* 493:122–31
132. Rosenheck, K., Doty, P. 1961. *Proc. Natl. Acad. Sci. USA* 47:1775

Appl. Opt. 8:661–65

133. Rotilio, G., Calabrese, L., Coleman, J. E. 1973. *J. Biol. Chem.* 248:3855–59
134. Schnepp, O., Allen, S., Pearson, E. F. 1970. *Rev. Sci. Instrum.* 41:1136–41
135. Sharonov, Y. A. 1976. *Mol. Biol. USSR* 8:70–161
136. Sharonov, Y. A., Mineyev, A. P., Livshitz, M. A., Sharonova, N. A., Zhurkin, V. B., Lysov, Y. P. 1978. *Biophys. Struct. Mech.* 4:139–58
137. Sharonov, Y. A., Sharonova, N. A., Atanasov, B. P. 1976. *Biochim. Biophys. Acta* 434:440–51
138. Sharonova, N. A., Sharonov, Y. A., Volkenstein, M. V. 1972. *Biochim. Biophys. Acta* 271:65–76
139. Shashoua, V. E. 1964. *Nature* 203:972–73
140. Shatwell, R. A., Gale, R., McCaffery, A. J., Sichel, K. 1975. *J. Am. Chem. Soc.* 97:7015–23
141. Shatwell, R. A., McCaffery, A. J. 1973. *J. Chem. Soc. Chem. Commun.* 546–47
142. Shatwell, R. A., McCaffery, A. J. 1974. *J. Phys. E* 7:297–99
143. Shimizu, T., Nozawa, T., Hatano, M. 1976. *Biochim. Biophys. Acta* 434:126–36
144. Shimizu, T., Nozawa, T., Hatano, M. 1976. *Bioinorg. Chem.* 6:1–9
145. Shimizu, T., Nozawa, T., Hatano, M. 1976. *Bioinorg. Chem.* 6:119–31
146. Shimizu, T., Nozawa, T., Hatano, M. 1976. *Bioinorg. Chem.* 6:77–82
147. Shimizu, T., Nozawa, T., Hatano, M., Imai, Y., Sato, R. 1975. *Biochemistry* 14:4172–78
148. Sipe, J. P., Martin, R. B. 1974. *J. Inorg. Nucl. Chem.* 36:2122–24
149. Solomon., E. I., Rawlings, J., McMillan, D. R., Stephens, P. J., Gray, H. B. 1976. *J. Am. Chem. Soc.* 98:8046–48
150. Springall, J., Stillman, M. J., Thomson, A. J. 1976. *Biochim. Biophys. Acta* 453:494–501
151. Sprinkel, F. M., Shillady, D. D., Strickland, R. W. 1975. *J. Am. Chem. Soc.* 97:6653–57.
152. Stephens, P. J. 1970. *J. Chem. Phys.* 52:3489–3516
153. Stephens, P. J. 1974. *Ann. Rev. Phys. Chem.* 25:201–32
154. Stephens, P. J. 1976. *Adv. Chem. Phys.* 35:197–264
155. Stephens, P. J., McKenna, C. E., Smith, B. E., Nguyen, H. T., McKenna, M. -C., Thomson, A. J., Devlin, F., Jones, J. B. 1979. *Proc. Natl. Acad. Sci. USA* 79:2585–89

156. Stephens, P. J., Mowery, R. L., Schatz, P. N. 1971. *J. Chem. Phys.* 55:224–31
157. Stephens, P. J., Suetaka, W., Schatz, P. N. 1966. *J. Chem. Phys.* 44:4592–4602
158. Stephens, P. J., Sutherland, J. C., Cheng, J. C., Eaton, W. A. 1976. See Ref. 106, pp. 434–42
159. Stephens, P. J., Thomson, A. J., Dunn, J. B. R., Keiderling, T. A., Rawlings, J., Rao, K. K., Hall, D. O. 1978. *Biochemistry* 17:4770–78
160. Stephens, P. J., Thomson, A. J., Keiderling, T. A., Rawlings, J., Rao, K. K., Hall, D. O. 1978. *Proc. Natl. Acad. Sci. USA* 75:5273–75
161. Stillman, M. J., Hollebone, B. R., Stillman, J. S. 1976. *Biochem. Biophys. Res. Commun.* 72:554–59
162. Stillman, M. J., Thomson, A. J. 1974. *J. Chem. Soc. Faraday Trans. 2* 70:790–804
163. Stillman, M. J., Thomson, A. J. 1974. *J. Chem. Soc. Faraday Trans. 2* 70:805–14
164. Sutherland, J. C. 1978. See Ref. 61, 3:225–48
165. Sutherland, J. C., Axelrod, D., Klein, M. P. 1971. *J. Chem. Phys.* 54:2888–98.
166. Sutherland, J. C., Cimino, G. D., Lowe, J. T. 1976. *Rev. Sci. Instrum.* 47:358–60
167. Sutherland, J. C., Cimino, G. D., Lowe, J. T. 1976. See Ref. 106, pp. 28–33
168. Sutherland, J. C., Desmond, E. J., Takacs, P. Z. 1979. *Nucl. Instrum. Methods* (In press)
169. Sutherland, J. C., Duval, J. F., Griffin, K. P. 1978. *Biochemistry* 17:5088–91
170. Sutherland, J. C., Klein, M. P. 1972. *J. Chem. Phys.* 57:76–86
171. Sutherland, J. C., Low, H. 1976. *Proc. Natl. Acad. Sci. USA* 72:276–80
172. Sutherland, J. C., Salmeen, I., Sun, A. S. K., Klein, M. P. 1972. *Biochim. Biophys. Acta* 263:550–54
173. Sutherland, J. C., Vickery, L. E., Klein, M. P. 1974. *Rev. Sci. Instrum.* 45:1089–94
174. Sytkowski, A. J., Vallee, B. L. 1978. *Biochemistry* 17:2850–57
175. Takeda, A., Samejima, T. 1977. *J. Biochem. Tokyo* 82:1025–33
176. Taylor, T. S., Lau, C. Y., Meredith, A. L., Coleman, J. E. 1973. *J. Biol. Chem.* 248:6126–6220
177. Tennent, D. L., McMillan, D. R. 1979. *J. Am. Chem. Soc.* 101:2307–11
178. Thomson, A. J., Brittain, T., Greenwood, C., Springall, J. 1976. *FEBS Lett.* 67:94–98
179. Thomson, A. J., Brittain, T., Greenwood, C., Springall, J. P. 1977. *Biochem. J.* 165:327–36
180. Thomson, A. J., Cammack, R., Hall, D. O., Rao, K. K., Briat, B., Rivoal, J. C., Badoz, J. 1977. *Biochim. Biophys. Acta* 493:132–41
181. Townsend, L. B., Long, R. A., McGraw, J. P., Miles, D. W., Robins, R. K., Eyring, H. 1974. *J. Org. Chem.* 39:2023–27
182. Townsend, L. B., Miles, D. W., Manning, S. J., Eyring, H. 1973. *J. Heterocycl. Chem.* 10:419–21
183. Treu, J. I., Callender, A. B., Schnatterly, S. E. 1973. *Rev. Sci. Instrum.* 44:793–97
184. Treu, J. I., Hopfield, J. J. 1975. *J. Chem. Phys.* 63:613–23
185. Turner, D. H. 1978. *Methods Enzymol.* 49:119–214
186. Ulmer, D. D., Holmquist, B., Vallee, B. L. 1973. *Biochem. Biophys. Res. Commun.* 51:1054–61
187. Vallee, B. L., Holmquist, B. 1979. See Ref. 74. In press
188. Vickery, L. E. 1975. In *Studies of Metalloproteins with Paramagnetic Probes*, ed. A. Tasaki, M. Kotani, pp. 27–50. Osaka:Osaka Univ. Press
189. Vickery, L. E. 1978. *Methods Enzymol.* 54:284–302.
190. Vickery, L. E., Nozawa, T., Sauer, K. 1976. *J. Am. Chem. Soc.* 98:351–57
191. Vickery, L. E., Nozawa, T., Sauer, K. 1976. *J. Am. Chem. Soc.* 98:343–50
192. Vickery, L. E., Palmer, G., Wharton, D. C. 1978. *Biochem. Biophys. Res. Commun.* 80:458–63
193. Vickery, L. E., Salmon, A., Sauer, K. 1975. *Biochim. Biophys. Acta* 386:87–98
194. Voelter, W., Records, R., Bunnenberg, E., Djerassi, C. 1968. *J. Am. Chem. Soc.* 90:6163–70
195. Weiler-Feilchenfeld, H., Linder, R. E., Barth, G., Bunnenberg, E., Djerassi, C. 1977. *Theor. Chim. Acta* 46:79–88
196. Weser, V., Bunnenberg, E., Cammack, R., Djerassi, C., Flohe, L., Thomas, G., Voelter, W. 1971. *Biochim. Biophys. Acta* 243:203–13
197. Whanger, P. D., Pedersen, N. D., Weswig, P. H. 1973. *Biochem. Biophys. Res. Commun.* 53:1031–35
198. Yoshida, S., Iizuka, T., Nozawa, T., Hatano, M. 1975. *Biochim. Biophys. Acta* 405:122–35

RADIOIMMUNOASSAY ♦9149

Rosalyn S. Yalow[1]

Solomon A. Berson Research Laboratory, Veterans Administration Medical Center, 130 West Kingsbridge Road, Bronx, New York 10468

Introduction

Radioimmunoassay (RIA) is a technique used in thousands of laboratories throughout the world to measure the concentrations in blood, tissues, and other biologic fluids of peptidal and nonpeptidal hormones, drugs, enzymes, viruses, bacterial antigens, and other organic substances of biologic interest. Depending upon the particular needs, the method can be expeditious, highly sensitive, and remarkably specific. Because of its widespread applicability, most current users employ commercially supplied reagents and/or kits containing all needed reagents and specific directions for use and therefore are not greatly concerned with the fundamental theoretical and experimental bases for RIA. This review will place its emphasis on certain aspects of the physics and mathematics of RIA.

RIA is simple in principle (Figure 1). The concentration of the unknown, unlabeled antigen is obtained by comparing its inhibitory effect on the binding of radioactively labeled antigen to a specific antibody with the inhibitory effect of known standards. RIA is an in vitro test, i.e. the ingredients, which are labeled antigen, specific antibody, and standards or unknowns, are incubated together in test tubes. After an appropriate reaction time, the antibody-bound and free fractions of radioactive antigen are separated, the radioactivity in each fraction is determined, and a calibration curve is drawn from the data on standards. The concentration of the unknown sample is then determined from the calibration curve (Figure 2).

The concept of RIA developed out of an earlier study (13) in which it was demonstrated that virtually all patients treated with commercial preparations of insulin develop insulin-binding antibodies. To determine the concentration of antibody in the plasma, labeled insulin was used as

---

[1]Supported by the Medical Research Program of the Veterans Administration.

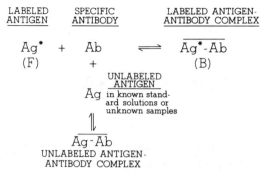

Figure 1 Competing reactions that form the basis of radioimmunoassay (RIA).

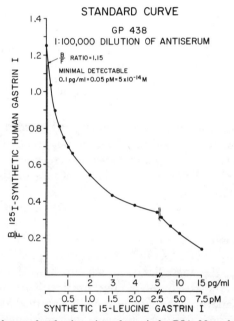

Figure 2 Standard curve for the detection of gastrin by RIA. Note that as little as 0.1 pg gastrin/ml incubation mixture (5×10^{-14}M) is readily detectable. (Reproduced from R. S. Yalow, 1978, in *Les Prix Nobel En 1977*, pp. 237–41, Nobel Foundation, Stockholm, Sweden.)

a tracer with increasing amounts of unlabeled insulin in order to determine the maximum amount of insulin that could be bound to antibody. It was soon appreciated, from the observation that the fraction of labeled insulin bound to antibody decreased with increase in the concentration of unlabeled insulin, that an immunoassay method for the determination of insulin concentrations in body fluids was possible (6). RIA was first applied to the measurement of exogenously administered insulin in the plasma of rabbits (7). However, several years were to pass before sufficient sensitivity was finally achieved (47, 48) to permit measurement of the concentration of endogenous insulin in man.

Sensitivity of Equilibrium RIA

What factors determine the sensitivity of a RIA procedure? In the mathematical analysis which follows, it is assumed that labeled and unlabeled antigen behave identically in the immune reaction and that the reaction follows the law of mass action (8, 12, 49). This condition simplifies the mathematical analysis but is not necessary for the validation of an RIA, which requires only that the standards and unknowns behave identically in the immune system and does not require identity of labeled and unlabeled antigen.

Consider then the simplest system, i.e. the biomolecular reaction between an antigen containing a single reactive site [Ag] and a single order of homogeneous antibody-combining sites [Ab], resulting in formation of an antigen-antibody complex [AgAb]. From the mass-action law, we derive Equation 1:

$$\underset{F}{[Ag]} + [Ab] \underset{k'}{\overset{k}{\rightleftharpoons}} \underset{B}{[\overline{AgAb}]}. \qquad 1.$$

F designates the concentration of unbound or free antigen and B represents the concentration of bound antigen. Then, after equilibration,

$$B/F = K([Ab_0] - B) \qquad 2.$$

where the equilibrium constant for the reaction $K = k/k'$ and $[Ab_0]$ is the total molar concentration of antibody-binding sites.

It is evident from Equation 2 that when B is much less than $[Ab_0]$, B/F decreases only slightly for large changes in B; thus if B increases 10-fold from 0.001 $[Ab_0]$ to 0.01 $[Ab_0]$ the change in B/F is less than 1%. For a sensitive assay, therefore, $[Ab_0]$ must be reduced by dilution so that $[Ab_0]$ is not much larger than B. H, the minimal detectable antigen concentration, must also be greater, but not much greater than B. To meet both these conditions it follows that $[Ab_0] \cong H$. It is, of course, inadvisable in a RIA to employ an amount of labeled tracer antigen whose immunochemical concentration is large compared to the concentration of unlabeled antigen in the unknown; for example, if the tracer concentration is 10 times the antigen concentration, a random 5% error in the tracer produces a 50% error in the observed hormone concentration. Thus, if we wish to start with $B/F = 1$ in the absence of added unlabeled antigen ("trace" conditions), then from Equation 2

$$1 \lesssim K[Ab_0] \lesssim KH$$

and, therefore,

$K \gtrsim 1/H$.

Thus, assay sensitivity is limited ultimately by K, the equilibrium constant that characterizes the reaction of antigen with the predominating antibodies. In a more sophisticated analysis (49), it has been shown for a given antiserum that maximum sensitivity is achieved when the concentration of labeled antigen (Ag*) is negligibly small and when the antiserum is diluted to the extent that the initial $B/F = 0.5$. In this analysis, sensitivity is defined in terms of the rate of change of B/F with change in hormone concentration. It is assumed that random experimental errors such as those associated with pipetting, detection of radioactivity in bound and free fractions, etc., are negligibly small. It is evident that these errors should be minimized if the maximal theoretical sensitivity is to be achieved.

In practice, for the standard curve to be useful over a wider range of antigen concentrations, an initial $B/F \sim 1$ is generally employed even though this results in a slight theoretical loss in sensitivity. To visualize the role of K and the concentration of labeled antigen in determining assay sensitivity, theoretical standard curves have been constructed using Equation 2. Curves are shown for three homogeneous antisera with equilibrium constants $K = 10^{10}$, 10^{11}, and 10^{12} liter/mole, using as tracers two different concentrations of labeled antigen, 10^{-11}M and 10^{-12}M (Figure 3). It is evident that with increase in equilibrium constant there is a striking decrease in the minimal hormone concentration detectable. When $K = 10^{10}$ the addition of 2×10^{-11}M unlabeled antigen results in a decrease in B/F ratio of about 10%. The same decrease in B/F ratio is effected by the addition of about 0.2×10^{-11}M unlabeled antigen when $K = 10^{11}$ and by 0.2×10^{-11}M when $K = 10^{12}$. However, in the latter case this sensitivity is achieved only with a tracer concentration of 10^{-12}M. If a tracer of 10^{-11}M is used, the concentration of unlabeled antigen required to produce the same decrease in B/F is threefold higher. It is also evident from these curves that there is little advantage in employing a tracer that is significantly less than $1/K$. Thus when $K = 10^{10}$, the curves obtained when tracer is 10^{-12}M or 10^{-11}M are virtually indistinguishable. When $K = 10^{11}$ there is some slight improvement in sensitivity when the smaller tracer is employed. However it should be noted that, using the same labeled antigen preparation, it would be necessary to prolong the counting time more than threefold in order to achieve the same statistical accuracy in detection of the radioactivity if the lower concentration of tracer were used. This would generally militate against employing the smaller tracer to achieve so minimal an improvement in sensitivity.

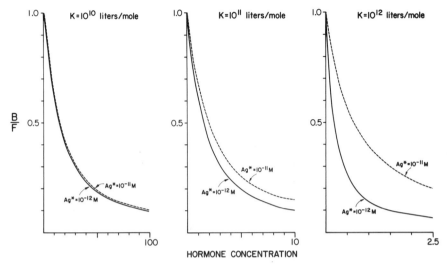

Figure 3 Theoretical curves of B/F as a function of unlabeled hormone in homogeneous antisera with equilibrium constants as shown. The antisera are diluted to yield a B/F ratio of 1.0 in the presence of two different concentrations of labeled antigen ($Ag^* = 10^{-11}M$ or $Ag^* = 10^{-12}M$). Note from the scale of hormone concentrations that the sensitivity depends on the equilibrium constant for the reaction with antibody. The most sensitive assay is obtained with $K = 10^{12}$ liters/mole (*right*) and the least sensitive with $K = 10^{10}$ liters/mole (*left*). Note also that there is little gain in sensitivity if Ag^* is reduced below $1/K$. When $K = 10^{12}$ liters/mole, use of a tracer of $10^{-12}M$ enhances the sensitivity (*right*); when $K = 10^{11}$ liters/mole the sensitivity is improved only slightly when tracer is decreased to $10^{-12}M$ (*middle*); the standard curves are virtually identical when $K = 10^{10}$ liters/mole and tracers are $10^{-11}M$ or $10^{-12}M$ (*left*).

The conclusions reached from this analysis, based on the assumptions of homogeneous antibodies and identity of labeled and unlabeled hormone, are that assay sensitivity is limited by the K of the reaction of antigen with antibody and that this sensitivity can be achieved only if the tracer antigen is $\cong 1/K$. There is no advantage in employing a smaller tracer. If labeled and unlabeled antigen are not identical, the sensitivity of the assay will generally be reduced but its validity will not be affected. Random errors due to pipetting, detection of radioactivity, etc. further limit the sensitivity of the assay.

In general, antisera are not homogeneous. They usually contain antibodies with more than one order of antibody binding sites; each site has a different K for the reaction with antigen. The theoretical standard curve (Figure 4) generated by assuming that identical amounts of high energy ($K = 10^{12}$) and low energy ($K = 10^8$) antibodies are mixed reveals that the initial slope of this curve resembles that obtained with the high

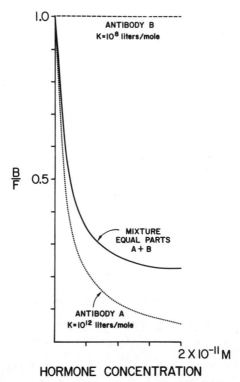

Figure 4 Theoretical curves of B/F as a function of unlabeled hormone employing a tracer of 10^{-13}M with a high energy antibody, A, ($K=10^{12}$ liters/mole); with a low energy antibody, B, ($K=10^{8}$ liters/mole); and with an antiserum made up of equal parts of A and B. In all cases the antisera were diluted sufficiently so that the initial B/F ratio was 1.0. Note that the initial slope of the standard curve with (A + B) was not much different from that with A alone. The curves diverge as the high energy antibody binding sites approach saturation. The sensitivity (initial slope of the standard curve) reflects the reaction with the higher energy-binding sites.

energy antibodies. A low concentration of antigen (10^{-13}M) was employed to avoid occupation of the higher energy sites by a large tracer. At higher concentrations of unlabeled antigen the lower energy sites become more prominent and the curves of the high energy antibodies and the mixture tend to diverge. Thus the ultimate sensitivity of a heterogeneous antiserum is limited by those of the antibody-binding sites having the highest K.

This theoretical analysis provides the basis for understanding the essential features of a sensitive RIA. However, there is no proven method for producing an antiserum with a predetermined K value. Antisera from different animals immunized on the same regimen vary greatly in antibody titer (the dilution required to bind the same percentage of labeled antigen) and in sensitivity. A practical way to select for the antiserum that provides the most sensitive assay is the following: the dilution of each antiserum that yields a $B/F \sim 1$ with a fixed amount of labeled antigen is first determined by testing over a range of antiserum dilutions. Then the amount of labeled antigen is reduced to the minimum amount consistent with a tolerable counting error or to the level of a true trace, i.e. such that further reduction of labeled antigen no longer results in an increase in B/F ratio. Then each antiserum is tested with various concentrations of unlabeled antigen to determine which standard curve has the sharpest initial slope.

High sensitivity assays are particularly necessary for measurement of the peptide hormones that are present in plasma in the unstimulated states at concentrations ranging from 1 to 100×10^{-12}M. Since there are substances in plasma that can interfere nonspecifically in the immune reaction, it is usually advisable to employ dilutions of plasma of 1/10 or greater in the assay; thus for some of the hormones assay sensitivities of 10^{-13}M are required if the measurement of plasma hormone is to be practical. For nonpeptidal hormones or drugs that are present in much higher concentrations the requirements for assay sensitivity are much less stringent.

Labeled Antigen

It is evident that the long half-lives of ^{14}C or ^{3}H limit the attainable specific activity unless two or more radioactive atoms are incorporated into a molecule. However, as described below, labeling of a molecule with more than one radioactive atom per molecule does limit its stability. Thus these radioactive isotopes are not useful as tracers in assays for the peptide hormones. Although ^{14}C is less commonly used in RIA, ^{3}H has been employed, particularly for the RIA of drugs (see Reference 15 for review) and steroid hormones (see Reference 2 for review). When tracers of high specific activity are required, as for the assay of peptide hormones, labeling with radioiodine has long been the method of choice. Radioiodine can substitute onto tyrosyl or histidyl residues of most peptides or can be incorporated using other chemical techniques into other substances of biologic interest. The two radionuclides that have been used most extensively are ^{131}I ($T_{1/2} = 8$ days) and ^{125}I ($T_{1/2} = 60$ days). In principle, if ^{131}I were available without carrier, its use would confer a more than sevenfold advantage in specific activity. However, direct determination of the isotopic abundance of these two radionuclides revealed that the theoretical advantage is not realized in practice (10, 51). The isotopic abundance of commercially available ^{125}I at the time of receipt in the laboratory is generally greater than 90%; that of ^{131}I, whether produced as a fission product or by irradiation in a reactor of nonenriched telluruim oxide targets, generally does not exceed 15%, and this of course decreases as the ^{131}I decays (10, 51). Moreover, the sensitivity in the usual well-type scintillation counter for the detection of the lower energy photons (<35 KeV) from ^{125}I is greater than for the higher energy γ-rays (364 KeV) of ^{131}I. These physical factors, taken together, have been responsible for the use of ^{125}I as the radionuclide of choice when a maximal counting rate is required with a minimal chemical amount of tracer.

One could lower the chemical amount of substance used for tracer and maintain the same counting rate by incorporating more than one ^{125}I atom per molecule. This is generally not desirable for at least two reasons. First, overiodination not infrequently results in an alteration in the three-dimensional configuration of the molecule (11). This is particularly important if the radioiodine is incorporated in or near the site of reaction of antigen with antibody. A second reason relates to a phenomenon that we have called "decay catastrophe" (11, 12, 50). When a radioactive atom undergoes decay, the residual molecule to which it is attached generally dissociates. The reason for this molecular disintegration has not been well studied. It may be due in part to the "hole" left in the molecular configuration by the substitution of an atom of a different element; it may, as in the case of ^{125}I, be due to the intensity of the cascade radiation associated with the decay process, and the recoil momentum and energy imparted to the molecule by internal conversion electrons. It is hypothesized that when the initial labeled molecule contains two or more radioactive atoms, decay of the first atom disintegrates the molecule into labeled fragments or unlabeled fragments and free radioiodide. Thus, with time, the residual radioactivity is no longer associated only with intact unaltered molecules. These changes would limit the shelf life of the labeled antigen during storage. They might also result in increased damage to labeled antigen during the incubation period required for reaction with antibody.

Decay catastrophe is distinctly different from the chemical alterations arising from what are considered the secondary effects of radiation, i.e. those induced by oxidizing or reducing radicals that are produced secondary to absorption of ionizing radiation by the water molecules of solutions containing radioactive materials.

The decreased stability of molecules labeled with two ^{125}I atoms is evident from studies (26) with relatively simple molecules, the thyronines. In members of this family of amino acids, there are one to four positions, as shown in Figure 5, that may contain iodine. The

SUBSTRATES AND PRODUCTS OF IODINATION OF THYRONINES

$$R-\!\!\!\left\langle\begin{array}{c}3\\ \\5\end{array}\right\rangle\!\!\!-O-\!\!\!\left\langle\begin{array}{c}3'\\ \\5'\end{array}\right\rangle\!\!\!-OH$$

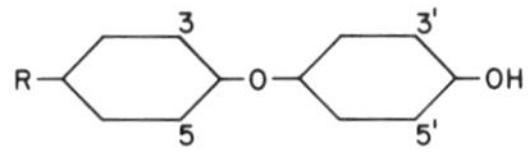

SUBSTRATES	RADIOIODOTHYRONINE PRODUCTS		
	EXCHANGE	+1 ^{125}I	+2 ^{125}I
$3T_1$		$3,3'T_2$	rT_3
$3,3'T_2$	$3,3'T_2$	rT_3	
$3,5 T_2$		T_3	T_4
$rT_3 (3,3',5'T_3)$	rT_3		
$T_3 (3,5,3'T_3)$	T_3	T_4	
T_4	T_4		

Figure 5 Substrates and products of iodination of thyronines. Each ring is unsaturated and R = $-CH_2-CH(NH_2)-COOH$. T is thyronine and the subscript denotes the number of iodine atoms in the compound. Note that rT_3 has two iodine atoms in phenolic ring; T_3 has one.

various radioiodothyronines can be produced by substitution or addition in the phenolic ring (3' or 5' positions). The theoretical specific activities of the products produced by addition reactions are given in Table 1. The stability of several of the radioiodothyronines dissolved in an organic solvent is shown in Figure 6. T_3 or $3,3'T_2$ have only one iodine and therefore only one ^{125}I atom in the phenolic ring. Over a three-month period, the radioactivity in these preparations is found only in intact, unaltered molecules. In fact, the specific activity of each of these thyronines remains unchanged as the ^{125}I undergoes decay. This is because decay catastrophe reduces the chemical amount of the thyronine at the same rate as the loss of ^{125}I. The other amino acids, rT_3 and T_4, contain two iodine atoms in the outer ring and appear to be less stable chemically than are T_3 or $3,3'T_2$ even when produced by an addition reaction (Figure 6). Perhaps this is due in part to their inherent chemical instability. However, there is some exchange with the existing ^{127}I as well as addition of one ^{125}I atom during the iodination procedure, and therefore some fraction of the molecules contain two ^{125}I atoms. The radioiodothyronines, rT_3 and T_4, produced by a double addition reaction are evidently always much less stable than the more lightly labeled preparations.

The hypothesis concerning decay catastrophe has also been verified for small peptides labeled with ^{125}I. Catt & Baukal (16) in a very elegant study demonstrated that, following almost one-half a year of storage, the specific activity of the octapeptide monoiodoangiotensin II declined only about 10% and there was virtually no increase in free iodide or labeled fragments. Thus the immunologic activity of the residual mole-

Table 1 Theoretical maximum specific activity (mCi/μg) for radioiodothyronines

	Number of ^{125}I atoms/molecule	
n	One	Two
T$_4$	3.0	6.0
rT$_3$	3.5	7.0
T$_3$	3.5	—
3,3'-T$_2$	4.4	—

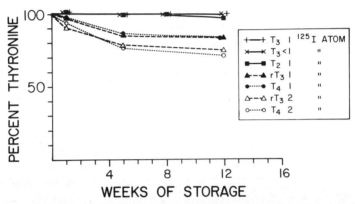

Figure 6 Stability of radioiodothyronines stored in an organ c solvent as a function of time after purification. The most stable preparations were those with a single iodine atom in the phenolic ring. There were no differences in the stabilities of the T$_3$'s prepared by an addition reaction or by an exchange reaction. The least stable preparations were those with two ^{125}I atoms in the phenolic ring. (Reproduced from data in Reference 26.)

cule is destroyed at essentially the same rate as the radioactive decay of the ^{125}I.

It is somewhat more difficult to verify the hypothesis that loss of immunoreactivity parallels decrease of ^{125}I radioactivity in the case of peptides of larger molecular size. This is due in part to the fact that employing the usual techniques of iodination and purification results in labeled peptides that are not pure monoiodopeptides. Let us consider the case of insulin, a 6000 dalton peptide consisting of two chains coupled through two disulfide bonds. It is known that each chain, A or B, contains two tyrosyl residues and that those in the B chain cannot be iodinated easily unless the peptide is at least partially denatured. In addition, there appears to be an almost equal probability for iodination in each of the four positions on the two tyrosines in the A chain. Calculation of the distribution of radioiodine in the insulin molecule as a function of the average amount of iodine substitution was performed using a Monte Carlo simulation (40). The calculation was made initially by considering that there were 250 molecules, each containing four equally probable sites for iodine incorporation. The percentages of molecules with zero, one, two, three, or four iodine atoms were then calculated on the assumption that a total of 50, 100, 150, 200, 250 iodine atoms were distributed among the 1,000 sites completely on the basis of a random process. To smooth out statistical uncertainties the calculation was repeated for 25,000 molecules and the same relative number of iodine atoms. The calculated distribution frequencies were in agreement to better than 0.1%. It is evident from Figure 7 that when insulin is iodinated at an *average* of one iodine atom per molecule almost 60% of the radioactivity is attached to molecules containing two or more iodine atoms. These theoretical considerations were shown to be consistent with experimental determinations of iodine content of radioiodoinsulins prepared by the chloramine T method (40). The distribution of radioactivity among the iodoinsulins appears to be independent of the iodination technique, since our observed distribution is virtually the same as that obtained by Massaglia et al (31) who employed an electrolytic method and by DeZoeten & DeBruin (17) who used KIO_3 as the oxidizing agent. It appears, therefore, that in aqueous solution, iodination up to an average of one iodine atom per molecule proceeds on a

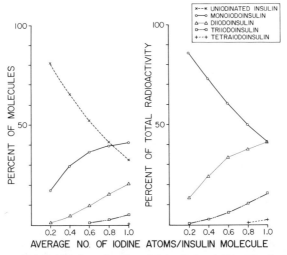

Figure 7 Theoretical distribution of radioactivity calculated from Monte Carlo simulation based on assumption of exclusive A-chain iodination and random distribution of ^{125}I among the four reactive sites on the two A-chain tyrosines: (*left*) percentage of insulin molecules with zero to four iodine atoms as a function of the average number of iodine atoms/molecule; (*right*) percentage of the total radioactivity in each form as a function of the average number of iodine atoms/molecule. (Reproduced from Reference 40.)

random basis, irrespective of the iodination method employed, and that a statistical method, the Monte Carlo, can be used to predict the distribution of iodine atoms incorporated in iodoinsulin.

In order to have most of the radioiodine associated with mono-iodoinsulin, it is necessary to iodinate to a very low average number of iodine atoms per molecule. But under these circumstances the labeled preparation has a low specific activity since it contains a large fraction of unlabeled molecules. As the ^{125}I undergoes decay the specific activity of the preparation decreases at a rate dependent on the initial fraction of unlabeled molecules.

Pure monoiodopeptides are generally produced, as was described for insulin (see above), by iodination at a moderate specific activity and then by purification of the labeled peptides using some fractionation system that depends on molecular charge, such as starch gel (40) or ion-exchange chromatography (16, 27), to separate the monoiodopeptides from the unlabeled or more highly labeled species. This separation is not as readily effected with peptides much larger in molecular weight than insulin. Thus, although it is evident that highly iodinated preparations of large peptides are less stable than lightly iodinated preparations, there has been no experimental verification that for larger peptides there is total loss of immunologic properties following radioactive decay of the self-contained ^{125}I.

Non-Equilibrium RIA

The preceding discussion assumed that the incubation period for RIA was sufficiently long so that equilibrium was reached, i.e. the rate of formation of antigen-antibody complexes was equal to their rate of dissociation. The rate of association depends on the molar concentrations of the reactants. Since antibody concentration is kept constant, the association rate increases with, and is limited by, the concentration of antigen. In the terminal, flatter portion of the standard curve, corresponding to higher antigen concentrations, equilibrium is achieved more rapidly than in the initial portion of the curve. Thus an assay would not be as sensitive if the reaction is stopped before equilibration, since the initial slope of the standard curve would be less sharp.

As first described by Hales & Randle (23), a two-stage assay in which mixtures containing unlabeled hormone and a homogeneous antibody are preincubated for an appropriate interval (following which the labeled tracer is added and the incubation continued for another interval) theoretically can result in an improvement in sensitivity when compared to an equilibrium method employing the same antiserum. Generally the first incubation is carried to equilibrium, and the second is stopped short of equilibrium in order to effect the improvement in sensitivity. Usually the antiserum will be more concentrated in the two-stage than in the one-stage assay in order to have the same percentage binding of the tracer. When a heterogeneous antiserum is employed, the use of a higher concentration of antibody may bring into prominence lower energy antibody-binding sites and the theoretical advantage may not be achieved. In practice, it is advisable to test the system for optimal conditions to determine whether there is a significant improvement in sensitivity. The two-stage assay has the further advantage that the labeled tracer is exposed to the incubation mixtures for a shorter period of time, thereby reducing damage which the tracer might be subject to during incubation.

The two-stage immunoassay can result in a considerable improvement in sensitivity only if the reverse velocity constant for dissociation of antigen-antibody complexes is small compared to the association constant, i.e. for a virtually irreversible reaction. In practice, the soluble complexes readily dissociate, so that the improvement in sensitivity is usually not sufficient to offset the inconvenience associated with later addition of tracer. A theoretical analysis and experimental data comparing equilibrium assays with two stage assays have been described by Rodbard et al (37).

Immunoradiometric Assay

The immunoradiometric assay (IMRA), first described by Miles & Hales in 1968 (32, 33), employs labeled antibodies for the assay of soluble antigens. The method involves (*a*) preparation of an immunoadsorbent, (*b*) purification and iodination of antibodies on the immunoadsorbent, and (*c*) acid elution of the labeled antibody (^{125}I-Ab) from the immunoadsorbent. In the assay an excess of the ^{125}I-Ab is incubated with unknown antigen or standards. At the end of the incubation, unreacted ^{125}I-Ab is precipitated by addition of excess antigen bound to an immunoadsorbent.

The labeled antibody technique was subsequently modified to a two-stage IMRA (3, 29, 44). In this method unlabeled antibody is coupled to a solid phase; multivalent antigen in standards or unknowns is added and then incubated for the first reaction. The tubes are washed, and purified ^{125}I-Ab is added for a second reaction. At the end of the second reaction the tubes are again washed to remove unreacted ^{125}I-Ab and then are counted. Two-stage IMRA is unlikely to be of any value in assay of small peptides or haptens because of the requirement that each antigen bind two antibody molecules, one unlabeled and the other labeled. The method has been used most extensively for assay of hepatitis B antigen (29). This assay is routinely used in blood banks for detection of the virus that is implicated in transfusion hepatitis. As generally employed, there is little need for high sensitivity or for precisely determined dose-response curves.

At present, radioiodine-labeled antibody assays have not been widely used for quantitative assays. They certainly have the disadvantage of the need to prepare highly purified antibody and to assure that iodination of the antibody does not alter its stability or its immunochemical behavior. The several reaction steps followed by washes are cumbersome, and they are complicated by the resultant dissociation of antigen-antibody complexes. To date, there have been no theoretical or experimental data comparing optimized IMRA with optimized RIA employing the same antiserum.

Specificity of RIA

RIA is based on the chemical reaction of antigen with antibody. This reaction can be remarkably specific, as described below. Nonetheless, there are a number of factors that interfere in the chemical reaction in a nonspecific fashion: the reaction is pH dependent and can be inhibited by various salts, proteins, and other substances such as bacteriostatic or anticoagulant agents, etc. To eliminate these nonspecific factors it is generally advisable to assure that standards and unknowns are prepared in an identical environment and are treated identically during the incubation and separation procedures.

The unique specificity of the immune reaction is evidenced by the observation that some human antisera against insulin can recognize structural differences between pig and dog insulins (9) whose amino acid sequences have been reported to be identical (41). What can account for the apparent differences in configuration? It is now known that insulin is synthesized as a single chain 9000 dalton peptide, proinsulin, in which the A and B chains are joined by a connecting peptide (C-peptide) (42). Although the A and B chains of dog and pig insulins are identical, the sequences of their C-peptides are strikingly different (36). Therefore at the time of synthesis, conformational differences between the two prohormones are not surprising. If the subsequent removal of the C-peptide leaves the conformational folding of the remaining molecule unaltered, then the 6000 dalton pig and dog insulins, in spite of identity of their two-chain primary structures, remain immunochemically distinguishable.

Antibody specificity permits distinction between cortisol and corticosterone (14) and between digitoxin and digoxin (35)—steroids that are identical except for a single OH group. Depending on the relative concentration of various steroids in body fluids and the specificity of the antisera employed, it is often possible to assay steroids directly without prior chromatographic purification as would be required for their assay by other chemical methods.

Nonetheless RIA of peptide hormones has been complicated by an appreciation that many, if not all, of these are found in more than one form in plasma and in glandular and other tissue extracts (see Reference 46 for review). These different forms may or may not have biologic activity. They may include, in addition to the usual biologically active form, precursor molecules such as proinsulin (42) or hormonal fragments. Depending on the specificity of the individual antiserum, these different forms with different biologic potencies may have equal, slightly different, or very different immunologic potencies. The problem of heterogeneity of hormonal forms has complicated interpretation of measurements by immunoassay and by bioassay as well. However, it has generated new insights concerning the biosynthesis of the peptide hormones.

Competitive Assays

The principle of radioimmunoassay has been extended to nonimmune systems in which the binding agent or specific reactor is other than an antibody and to systems in which the labeling agent is other than a radioisotope (Figure 8). The term *competitive radioassay* may be used to describe those assays in which the tracer is radioisotope-labeled but in which the specific reactor is not an antibody. This system was first applied to the measurement of vitamin B_{12} in a liver receptor system using ^{60}Co-vitamin B_{12} as tracer and intrinsic factor as the binding substance (24, 25). The method was rendered more sensitive and applied to the measurement of serum B_{12} by Rothenberg (39) and Barakat &

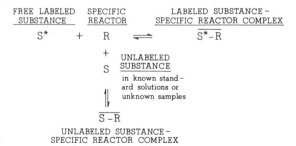

Figure 8 Competitive assay principle. Labels which have been employed include: bacteriophage, fluorescent group, stable free radical, or enzymes as well as radioisotopes. Specific reactors include: serum-binding globulins, tissue receptor sites, intrinsic factor, or enzymes as well as antibodies.

Ekins (5). Subsequently, competitive radioassays for thyroxine (19) and cortisol (34) using serum-binding globulins as specific reactors were described. However, at present RIA is more generally used for assay of thyroidal and steroidal hormones. There is now an extensive literature on the use of tissue receptor sites as specific reactors for assaying a variety of peptide and steroid hormones (see Reference 38 for early review). Tissue receptor assays are reported to be advantageous because they are assumed to reflect biologic potency. Nonetheless, their sensitivities for detection of peptide hormones are generally too low to be useful for routine assays. Receptor assays for steroids have proven to be more satisfactory since these hormones are present in plasma in much higher concentrations than the peptide hormones.

Considerable interest has been expressed in using other than radioactive labels for antigen or antibody. This is due in part to the mistaken notion that handling of any amount of radioactivity is unsafe and that the detection of radiation requires highly sophisticated and expensive equipment. It should be appreciated that the basic cost of radiation detectors is quite low; the expensive equipment is necessary only for the automation and computerization that is considered desirable to minimize technician costs when thousands of samples are analyzed. In the United States institutions already licensed for the use of radioisotopes for other purposes may easily be licensed for RIA applications. Physicians, clinical laboratories, or hospitals not licensed for other purposes may apply to the Nuclear Regulatory Commission for a registration certificate under a general license which permits them to receive RIA kits and to maintain, on hand, amounts of radioactivity sufficient for thousands of assays. Information concerning these regulations is found in the United States Nuclear Regulatory Commission Rules and Regulations, Title 10, Chapter 1, Part 31.

One nonradioactive label which has been used is bacteriophage, the so-called "viroimmunoassay." In this method bacteriophage is chemically modified by attaching to it various small molecules (21, 30) or proteins (22). The modified phages that survive the coupling process can be neutralized with antibodies to the attached moieties. This neutralization can be inhibited by the "free" moiety in standards or unknowns. This method has been reported to be useful for measurement of peptide hormones (20), prostaglandins (18), and steroids (4, 18). It is of interest that even those using the method do not believe it will be suitable for direct measurement of substances in plasma because of problems with nonspecific inhibition (4). Although this method of labeling was first applied to immunoassay almost a decade ago, it has still not been widely used.

Other methods of labeling that have found some application are those involving a fluorescent group (1) or a stable free radical (28). However, by far the most widely used label is one of a variety of enzymes (see Reference 45 for extensive review). These assays resemble RIA's in that either antigen [enzyme-immunoassay (EIA)] or antibody (immunoenzymometric assays) may be labeled, and one or two stage assays may be used. These assays have been used most extensively for measurement of drugs, serum proteins, thyroxine, etc. which are likely to be present in high concentration and which do not require the high sensitivity possible with RIA. They have also been reported to be quite useful in screening for a variety of parasitic diseases, particularly under circumstances in which precision and sensitivity are not necessarily required (43).

Problems which can be anticipated using enzyme-labeling relate to possible changes in the configuration of the enzyme or the labeled substance because of the necessary conjugation procedure. These changes can result in loss of enzyme activity or decrease in antigen-antibody reactivity, inevitably decreasing the sensitivity of the assays. Since enzymes are not small compared to the size of the reactants in the immune system, steric hindrance in the immune reaction or in the later reaction between enzyme and its substrate is quite possible. Generally, iodination of proteins or peptides or attachment of an iodotyrosyl peptide to other substances is simpler than is the preparation of enzyme-labeled substances, especially since there has been no general agreement as to which enzyme is best for substances as small as steroids or drugs or as large as viruses. Furthermore, the presence in plasma and tissue fluids of nonspecific inhibitors or activators of the enzyme system which are unrelated to the immune system being analyzed may introduce artifacts into the assay. At present it appears that although enzyme-linked substances may be useful in some immunoassay procedures, they are not likely to replace radioactive labeling as the method of choice for a competitive assay requiring high sensitivity. In addition their usefulness may be further limited because of problems with nonspecific interference in the enzyme reaction by substances in plasma or tissue fluids, even in assays which are less demanding with respect to sensitivity.

Conclusions

The general principles involved in the development of RIA procedures were well established in the first decade of its use. At present the majority of workers employ RIA in routine diagnostic procedures and are dependent on the use of kits or bulk reagents. Their concerns are now largely directed to automation and computerization and problems of quality control, such as intraassay and interassay reproducibility. In general commercial suppliers provide the automated equipment, computer software, and reference samples needed to effect quality control.

For the first decade following the development of RIA primary emphasis was given to its importance in endocrinology. The appreciation that the gastrointestinal tract is a rich endocrine organ followed the application of RIA to gastroenterology. More recently, pharmacology and neuroendocrinology have been enriched by RIA methodology. It is quite likely that the next field to feel the impact of these procedures is that of infectious diseases. These diseases remain a major public health problem throughout the world and simple inexpensive methods of identifying carriers of disease would facilitate their eradication. RIA can provide those methods. There remains a continuing need for development of new RIA methodology, and familiarity with basic principles can serve as a guide in such investigations.

Literature Cited

1. Aalberse, R. C. 1971. *Clin. Chim. Acta* 48: 109–12
2. Abraham, G. E. 1974. *Proc. Symp. Radioimmunoassay Relat. Proc. Med. Istanbul, Turkey, 1974*, IAEA-SM-177/207, 2:3–29.
3. Addison, G. M., Hales, C. N. 1971. In *Radioimmunoassay Methods*, ed. K. E. Kirkham, W. M. Hunter, pp. 481–87. London: Churchill Livingstone
4. Andrieu, J. M., Mamas, S., Dray, F. 1974. See Ref. 2, IAEA-SM-177/67:47–55
5. Barakat, R. S., Ekins, R. P. 1961. *Lancet* 2:25–26
6. Berson, S. A., Yalow, R. S. 1957. *J. Clin. Invest.* 36:873 (Abstr.)
7. Berson, S. A., Yalow, R. S. 1958. *Adv. Biol. Med. Phys.* 6:349–430
8. Berson, S. A., Yalow, R. S. 1959. *J. Clin Invest.* 38:1966–2016
9. Berson, S. A., Yalow, R. S. 1966. *Am. J. Med.* 40:676–90
10. Berson, S. A., Yalow, R. S. 1966. *Science* 152:205–7
11. Berson, S. A., Yalow, R. S. 1969. *Proc. Int. Diabetes Fed. Congr., 6th, Stockholm, Sweden, 1967*, Excerpta Medica Found., pp. 50-67.
12. Berson, S. A., Yalow, R. S. 1973. *Methods In Investigative and Diagnostic Endocrinology, Part I—General Methodology*, ed. S. A. Berson, R. S. Yalow, pp. 84–120, Amsterdam: North-Holland
13. Berson, S. A., Yalow, R. S., Bauman, A., Rothschild, M. A., Newerly, K. 1956. *J. Clin. Invest.* 35:170–90
14. Boiti, C., Yalow, R. S. 1978. *Endocr. Res. Commun.* 5:21–33
15. Butler, V. P. Jr., 1973. *Metabolism* 22:1145–53
16. Catt, K. J., Baukal, A. 1973. *Biochim. Biophys. Acta* 313:221–25
17. DeZoeten, L. W., DeBruin, O. A. 1961. *Rec. Trav. Chim. Pays-Bas* 80:907–16
18. Dray, F., Andrieu, J. M., Charbonnel, B. 1972. *Techniques Radioimmunologiques*, pp. 285–88. Paris: Cent. Rech. Inserm Hosp. St. Antoine
19. Ekins, R. P. 1960. *Clin. Chim. Acta* 5:453–59
20. Givol, D. 1969. *Proc. Symp. "In Vitro" Proced. Radioisot. Clin. Med., Vienna, Austria*, IAEA-SM-124/72:515–24
21. Haimovich, J., Sela, M. J. 1966. *Immunology* 97:338–43
22. Haimovich, J., Sela, M. J. 1969. *Science* 164:1279–80
23. Hales, C. N., Randle, P. J. 1963. *Biochem. J.* 88:137–46
24. Herbert, V. 1959. *Am. J. Clin. Nutr.* 7:433–43
25. Herbert, V., Castro, Z., Wasserman, L. R. 1960. *Proc. Soc. Exp. Biol. Med.* 104:160–4
26. Kochupillai, N., Yalow, R. S. 1978. *Endocrinology* 102:128–35
27. Lefkowitz, R. J., Roth, J., Pastan, I. 1972. *Ann. NY Acad. Sci.* 185:195–209
28. Leute, R., Ullman, E. F., Goldstein, A. 1972. *J. Am. Med. Assoc.* 221:1231–34

29. Ling, C. M., Overby, L. R. 1972. *J. Immunol.* 109:834–41
30. Mäkelä, O. 1966. *Immunology* 10:81–86
31. Massaglia, A., Rosa, U., Rialdi, G., Rossi, C. A. 1969. *Biochem. J.* 115:11–18
32. Miles, L. E. M., Hales, C. N. 1968. *Biochem. J.* 108:611–18
33. Miles, L. E. M., Hales, C. N. 1968. *Nature* 219:186–89
34. Murphy, B. E. P. 1964. *Nature* 201:679–82.
35. Oliver, G. C. Jr., Parker, B. M., Brasfield, D. L., Parker, C. W. 1968. *J. Clin. Invest.* 47:1035–42
36. Peterson, J. D., Nehrlich, S., Oyer, P. E., Steiner, D. F. 1972. *J. Biol. Chem.* 247:4866–71
37. Rodbard, D., Ruder, H. J., Vaitukaitis, J., Jacobs, H. S. 1971. *J. Clin. Endocrinol. Metab.* 33:343–55
38. Roth, J. 1973. *Metabolism* 22:1059–73
39. Rothenberg, S. P. 1961. *Proc. Soc. Exp. Biol. Med.* 108:45–48
40. Schneider, B. S., Straus, E., Yalow, R. S. 1976. *Diabetes* 25:260–67
41. Smith, L. F. 1966. *Am. J. Med.* 40:651–772
42. Steiner, D. F., Clark, J. L., Noland, C., Rubenstein, A. H., Margoliash, E., Aten, B., Oyer, P. E. 1969. *Recent Prog. Horm. Res.* 25:207–82
43. Voller, A., Bartlett, A., Bidwell, D. E. 1976. *Trans. R. Soc. Trop. Med. Hyg.* 70:98–106
44. Wide, L. 1971. *Radioimmunoassay Methods*, ed. K. E. Kirkham, W. M. Hunter, pp. 405–12. London/Edinburgh: Churchill Livingstone
45. Wisdom, B. G. 1976. *Clin. Chem. NY* 22:1243–55
46. Yalow, R. S. 1974. *Recent Prog. Horm. Res.* 38:597–633
47. Yalow, R. S., Berson, S. A. 1959. *Nature* 184:1468–69
48. Yalow, R. S., Berson, S. A. 1960. *J. Clin. Invest.* 39:1157–75
49. Yalow, R. S., Berson, S. A. 1968. *"In Vitro Studies" AEC Symp. Med.* 11:7–41
50. Yalow, R. S., Berson, S. A. 1968. *Proc. Int. Symp. Protein Polypeptide Horm., Pt I, Liege, Belgium, 1968, ICS No. 161*, ed. M. Margoulies, pp. 36–44. Amsterdam: Excerpta Medica Found.
51. Yalow, R. S., Berson, S. A. 1970. *Proc. Symp. "In Vitro" Proc. Radioisot. Clin. Med. Res., Vienna, Austria, 1969*, IAEA-SM-124/106:455–79

SPECIAL TECHNIQUES FOR THE AUTOMATIC COMPUTER RECONSTRUCTION OF NEURONAL STRUCTURES

♦9150

I. Sobel, C. Levinthal, and E. R. Macagno

Facility for Computer Graphics and Image Processing, Department of Biological Sciences, Columbia University, New York, NY 10027

INTRODUCTION

Reason to Automate

In spite of impressive gains in recent years, the field of neuroscience remains in its infancy in terms of our understanding of neural mechanisms of behavior and nervous system development. One serious problem is the difficulty of obtaining precise, specific anatomical descriptions of neurons and their interconnections down to the synaptic level. We assume the reader has read our previous review (12) in which we described the rapidly evolving use of techniques for Computer-Aided Reconstruction by Tracing Of serial Sections (CARTOS). Here we focus on the use of the computer in such systems to extract topographic and/or topologic information directly from the raw data—whether it be serial sections on film photographed with a light or electron microscope, or "optically sectioned" whole mounts seen in a light microscope.

In most CARTOS systems—hereafter called "manual systems"—an operator using some sort of coordinate input device is required to trace manually anatomical data such as cell-section boundary contours and/or centers of fibers. This is a very slow and tedious task, particularly in view of the fact that a single neuron commonly has numerous branches which appear in many sections. It may take many hours for an experienced operator to trace such a cell. Moreover the tracing is fatiguing and cannot be done for long, uninterrupted stretches of time. In view of the large numbers of cells in even the simplest invertebrate, not to mention vertebrate, nervous system we believe it imperative to speed up this data-inputting process.

How fast is fast enough to be useful? Surely the faster the better. Speed is directly related to the cost of the computer system, but there is a minimum speed of operation for a given cost below which the machine does not justify its expense. The following argument should make this fact clear: At the present time any device which purports to automate the inputting of image information into a computer will not eliminate completely the need for a human operator to guide its search and monitor its operation. Such a device can be thought of as a manual CARTOS system with additional hardware and programs for image acquisition and feature extraction. It will thus cost more than a basic manual CARTOS system. To be cost effective the automated system and its operator must be able to process more data per unit time than that number of manual systems and operators which could be obtained for the same price.

Having presented our criterion for cost effectiveness, we will not say much more about it. An adequate evaluation of existing systems is a separate study in itself. Cost effectiveness is difficult to evaluate because the costs and performance figures in general have not been published. Moreover, costs themselves are hard to evaluate for research systems such as these where one well-placed group may have easy and free access to expensive instrumentation and/or highly skilled software and hardware designers, whereas another group may have to pay heavily for these services. Performance also must be carefully defined for various modes of operation of the different systems and various types of data specimens. The cost effectiveness idea, however, is well worth keeping in mind, lest automation costs grow out of proportion to the benefits to be derived therefrom.

Automated nerve tracing is a new field. Until recently the necessary computer hardware has been very expensive. Moreover the necessary programming effort is beyond the reach of most neuroscientists. For these reasons there are currently only a handful of groups working on this problem. Since the field is so young and rapidly changing we have chosen to organize this article around the fundamental issues of the task, with references to how particular groups have solved the various problems. Fortunately the price of hardware is coming down and the performance is going up so that we can expect to see rapid development in the near future.

IMAGE INPUT DEVICES

We are thus concerned with the automatic or semiautomatic extraction of anatomical information from images. These images may be on film, as serial section photomicrographs; they may be taken from a light microscope objective, as in the technique of optical sectioning (12); or they may be acquired directly in electronic form from a scanning transmission electron microscope (STEM). In any case, the computer must be provided with direct access to the information in the images. This will be in the form of a two-dimensional array of quantities. Each quantity is usually an integer representing a digitized quality of the input material at a particular location. It may alternatively represent optical density, transmittance, or fluorescence of a light microscope slide, or electron density of an electron microscope image. Often the raw measurements are of light intensity at a particular photodetector. Transmittance is computed as the ratio of light intensity through the film or specimen, to the light intensity entering it. Optical density is defined as the negative logarithm of the transmittance: $d = -\log_{10}(T)$. If the measurements of light through the specimen and light entering the specimen are made simultaneously, this technique has the benefit of cancelling noise introduced by fluctuations in the intensity of the light source.

For completeness we also mention the possibility that a typical sample measurement may be more than one quantity. A common example is to measure optical density in more than one spectral band in stained tissue visualized in a light microscope. Thus, "true-color" imagery is a special case where the three primary visible bands red, green, and blue are used. However, for microanatomical purposes the number and characteristics of the spectral bands are chosen to best enhance the features of interest in a particular study. Such vector-valued information can be displayed in "pseudo-color" by mapping the measured values into the true-color red, green and blue space. There are many such mappings to chose from for any particular data space. Presumably the one which best enhances objects of interest will be chosen.

Input devices are characterized by spatial resolution, photometric (or grey-scale) resolution, and speed. These quantities are usually traded against each other in the design of an instrument. Many readers of this article are no doubt familiar with commercially available drum or flat-bed mechanical-scanning microdensitometers that give high spatial and photometric accuracy, albeit at low speed. Machines of this sort are often unacceptable for the task at hand because they are too slow.

Faster devices are commercially available off-the-shelf for imagery in the form of a standard television-raster signal. These are usually sold with a video "frame-buffer" memory for refreshing a TV monitor as a display device. Some of these are supplied as options by "raster-display" manufacturers (e.g. IIS, DeAnza, Grinnell, Comtal). Others are primarily sold as image acquisition devices, with the raster-display as the added feature (e.g. Quantex, SDS, CVI). The constraint of standard TV-raster format limits the size of an image array, at best, to about

700×500 samples or "pixels." Such an array can be digitized in a minimum time of 33 msec, the standard video frame time. At these rates and full raster resolution, inherent sensor and amplifier noise limit photometric accuracy to four or five bits of grey-scale at best. Higher photometric accuracy is achievable at the expense of either spatial resolution or speed. One of the early systems (Quantimet/Imanco) achieved six bits of grey-scale and an 800×600 array by using a slow-scan video format. The whole array was read out of the plumbicon camera in 100 msec. To monitor this low frame-rate video image without flicker and without the use of a then expensive digital refresh memory, they used a video monitor with a very slow phosphor screen.

Images that contain more spatial information than can be had in standard video-frame format must either be handled by special purpose devices or be broken up into TV-frame sized "windows" that can be presented to a TV camera by a computer-controlled window-seeking mechanism (8, 10).

There are two other off-the-shelf commercially available input devices. The first is an "image dissector" camera (EMR/Schlumberger, DICOMED, ITT, SDS/PDP8-EYE) also called a "vidisector." These devices were originally used as TV cameras. They have the disadvantage that for TV-quality photometric accuracy they are slower than modern vidicon-type cameras. However, as computer input devices they have the advantage of program-controlled scan-formatting as opposed to a standard TV-raster format. This allows the programmer to input only those pixels actually needed for a given calculation and to trade off speed for photometric accuracy as needed. They also have larger array sizes—typically in the range of 1000 to 2000 independently resolvable spots per axis. If only small parts of an image need be digitized, their slower speed may be acceptable for CARTOS (e.g. see 4).

The second device is a "flying spot film scanner." At least one company (CELCO) offers these off-the-shelf. A flying spot scanner has a program-controlled high resolution cathode ray tube (CRT) as a light source. It is enclosed in a light-tight box with the film transparency to be scanned. The face of the CRT is imaged on the film plane. Light from a CRT-generated spot is collected both before and after the film, and transmittance and/or density is computed as described above. This device can be used for output to "write" film as well as to "read" it. It also has a large array size similar to that of the vidisector. Array sizes in both of these are limited, for a given spot size, by the accuracy of their deflection electronics. Moreover, an increase in spatial resolution is accompanied by a decrease in speed. That is, higher spatial resolution demands a smaller sampling spot, which in turn reduces light flux—or electron flux, in the vidisector case—to the detector(s). This in turn requires more light collection time at a spot to achieve a given photometric signal-to-noise ratio.

Llinas et al (11) have recently constructed an electron beam flying spot scanner by interfacing their JEOL 100C STEM to a minicomputer. Their computer can position the electron beam under program control and subsequently read the output of the electron detector. They can thus obtain electron microscope images without the use of photographic film.

The CRT-type flying spot film scanner has the disadvantage that the film cannot be viewed by an operator while it is being scanned. At the Columbia University Computer Graphics Facility (Sobel, manuscript in preparation) we use a laser scanner specially built to our specifications by Altman Associates of Stamford, Conn. This unique device uses a Helium-Neon laser source for densitometry and an ordinary projector lamp to allow for simultaneous operator viewing of the film as it is scanned. A part of the scanning laser beam is visible to the operator overlaid on the projected image. It can be programmed to act as an operator-controlled pointing device or "cursor." The addressable input array is determined by the laser spot size of 20 μm and is 1200×1800 pixels for a 24×36 mm film frame. Currently, only transmittance is measured and converted to a 10-bit number (1024 grey-levels), although we intend to add a logarithmic amplifier and optionally digitize density for better resolution of high density images. The deflection system can be programmed to fetch from this array any 32×32 pixel window within 60 msec, a time which in the future we hope to reduce to 20 msec.

Several biological image-processing laboratories have built their own light microscope scanners. An early version (17, 20) used galvanometer mirror deflectors to scan a microscope image across a pinhole, behind which was a photomultiplier tube. Later models have used a linear solid state array to replace the single detector and one of the galvanometer deflectors. The arrays can be purchased in lengths of from 128 to 1024 elements (Reticon, Fairchild, IPI). These deflection systems have better scanning accuracy than standard vidicon type deflections systems. They are generally slower. The linear array versions trade photometric accuracy for increased speed. The mirror deflector scan axes generally allow for a program-controlled scan. In addition the longer linear arrays allow for sizes of 2D arrays that are larger than the standard TV-raster. We are aware of one manufacturer (Eikonix) who recently began offering a complete computer image input system built around a Reticon 1024 element linear array with linear stepping drive for the second scan-axis, and 8-bit photometrically corrected (see below) sensor output. If the slow axis is capable of random access positioning in a short enough time, the device may be suitable for automatic CARTOS work.

A new generation of solid-state 2D photodetector arrays is now becoming available at nearly full TV-frame sizes (RCA, Fairchild, GE). These are being incorporated into TV cameras and are eliminating the poor scanning accuracy of most vidicon-type cameras. Presently their main drawback (also shared by the linear arrays above) is element-to-element static variation in sensitivity and dark-current. Unless these variations are measured and compensated for, they limit the grey-scale accuracy to about 4-bits. Such compensation can be effected in the computer by storing two or more frames of reference data, e.g. a dark field and a bright-light field. These can then be used to determine a two or more parameter transform for each element which normalizes its output.

There are also several commercial image analysis systems on the market. These usually include an image input device, an interactive display, and a small computer with software for extracting object boundaries and shape and density statistics from single images. A partial list in the form Manufacturer/System-Name is: Imanco/Quantimet; Bausch & Lomb/Omnicon; Leitz/TAS; Joyce-Loebl/Magiscan; Micro-Measurements/Optomax. Rather than list all their characteristics here, we refer the reader to the manufacturers' specifications.

In summary there are a wide range of image input devices that differ in speed, spatial resolution, photometric resolution, cost, and availability. We have given an exemplary rather than exhaustive survey of them. The choice of which to use will of necessity be guided by both the specific automation problem and the size of one's pocketbook.

MAJOR DETERMINANTS OF SYSTEM DESIGN

Type of Information Desired

In recording anatomical detail we are essentially looking for shape and/or location of cellular features. We may be recording topographic information as the shape of the surface of a neuron. Alternatively we may be recording only the topologic information defining its branching pattern, or there may be a requirement which is partway between the two; for example, the locations of the longitudinal axes of all branches, i.e. the "skeleton." We may also be recording the location of synapses, cell nucleii, and/or other organelles whose shape is assumed a priori. Which of these attributes is to be automatically traced is of prime importance in determining the form and complexity of an automatic tracing system. At one extreme, if the data is of sufficiently high contrast in aligned micrographs on film, the neuron branches are typically many sections long, and neuron topology or a rough skeleton is required, the following strategy suffices: The operator runs the projector in ciné-mode (i.e. fast) and dynamically tracks the center of a given fiber with a cursor whose position is read by the computer. In fact if mainly topology, e.g. branches and branch lengths and not spatial arrangement, is required only the branch and termination points need be marked. Note that this strategy requires no image input to the computer and only requires changes to existing manual tracing programs. It is thus potentially very cheap and simple for these special types of specimens and probably more cost-effective than any foreseeable scheme using image processing to extract the same information. At the other extreme, if very detailed surface topography of neurons is required for aligned micrographs on film and if the data is sufficiently high contrast, a system that uses a computer and associated image input device for automatic contour extraction will almost certainly prove more cost-effective than any manual system. We say this because at

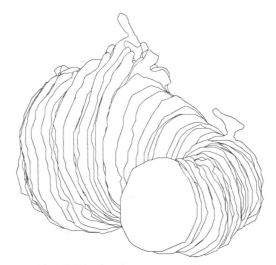

Figure 1 Display of contours extracted by program-controlled laser scanner at Columbia University, Department of Biological Sciences. Giant lateral fiber from ventral nerve cord of giant prawn *Macrobrachium rosenbergii*. Biological investigations, care, feeding, sectioning, staining, microscopy, and aligned movie by D. Friedlander. Contours are automatically extracted in smoothed chain-coded form (programs by I. Sobel). They are then converted to (x,y,z) point coordinate lists with minimal loss of detail (programs by N. Kropf, I. Sobel, S. Somech). These point lists are then displayed and interactively processed by scaling and/or translation and/or rotation to produce a desirable view (programs by N. Kropf). The transformed point lists are then output to programs (by C. Tountas) that remove hidden lines and display the result as shown.

Columbia we have programmed a computer to trace cell boundaries in high contrast photomicrographs. The computer traces in detail (limited only by the resolution of our laser scanner) at much higher speeds than a human operator working at the same level of detail (Figure 1). On the basis of our preliminary experiments with this prototype, which uses part of a time-shared minicomputer, it appears that a stand-alone system which is quite cost-effective can be built for a reasonable price.

We note in concluding this section that the skeleton of a neuron is not a physical object but a convenient shape descriptor. Any automatic system must do some amount of shape or at least boundary point extraction to compute it. The distinction between skeleton and surface tracing strongly affects the design of manual systems and the decision of whether to choose the manual or automatic approach. However once the decision to use the automatic approach is made, one is committed to extract surface information first, even if it is only to be used for defining the skeleton.

Sources of Data

Under this heading we treat the effects of the type of sectioning (optically or mechanically cut), the type of staining (cell-specific or block), and the type of microscopy (electron or light). The emphasis is on special problems for automatic CARTOS.

It is interesting to compare the properties of data generated with optical sectioning in a light microscope with those of cut sections, whether photographed in a light or electron microscope. An obvious advantage of the optical sectioning technique over cut sections is that successive images within a block are in registration and need no further alignment. At worst, alignment is only needed between successive blocks. Cut sections also introduce section-to-section variations in thickness which, at the least, cause frame-to-frame changes in film average density and contrast. These variations must be removed either in the photographic process or in the computer. Reddy et al (15) did this by automatically varying the lens aperture of their image-dissector to maximize the dynamic range of the digitized image. At Columbia we are currently testing an algorithm which normalizes such frame-to-frame changes by mapping densities, such that the resultant density histograms from successive frames have the same mean and variance. It does this only if the shapes of successive mapped histograms are sufficiently similar—the underlying assumption being that there are no radical changes in the optical quality of the tissue. For an initial section, or if the similarity condition is not satisfied, we use a heuristic which maps the central 98% of the histogram area into the full 8-bit density range.

We find that clipping 2% of the extreme values of the raw data is a good technique for minimizing the effects of artifacts such as dust and scratches in the film. These tend to stretch the density range of the raw data and thus reduce the number of quantization levels available for the grey-level regions of interest.

Another fact which is both a limitation and an advantage for optical sectioning is that it works best with cells that are stained at high contrast and are sparse in the tissue. This is because any single optical section image is degraded by out-of-focus images of all the dense material in the block. Thus the more stain in the block, the lower the contrast of the optical section image. Shantz (see below) has developed a 3D linear filter that removes much of the out-of-focus contamination.

It is not surprising then that optical sectioning is primarily used for reconstructing in 3D from slides containing few Golgi-stained or dye-injected neurons. In these situations the staining technique also solves another problem—namely that of separating, or optically dissecting, the neurons to be traced from the rest of the tissue. This fact is quite important in that it guarantees high contrast object boundaries which are quite amenable to automatic detection and tracking. Unless exceptionally large cells are being studied, block staining techniques are not suited for optical sectioning and raise the problem of effectively separating structures of interest from their surround. At the same time they present the opportunity of recovering spatial relationships and interactions among many if not all the cells in a tissue specimen. At least two automatic systems (8, 15) take advantage of the optical dissection implicit in cell-specific stains to eliminate the need to a priori identify cell branches and then follow each through the tissue as a separate entity. They first extract boundaries of all suitably stained areas within an image. After all such boundaries have been extracted from all sections (and the sections aligned if necessary), they automatically assemble contours in successive frames into branches (see section on 2D tracing systems, below). This can yield a significant speed-up in operation if the assembly algorithms are fast, since all the necessary data can be extracted in a single pass through the sections. The procedure cannot be used with block-stained specimens, since the presence of stain at a point does not guarantee that it belongs to a cell of interest.

Finally we note that local tissue distortions are introduced in the electron microscope case (12) which require special alignment techniques before multi-frame automated following can succeed. J. K. Stevens, T. L. Davis, N. S. Friedman, and P. Sterling (manuscript in preparation) have developed a method for compensating for these distortions. It uses a film transport mounted on an x,y translation stage (rotation soon to be added) and a video frame–buffer. The operator views the area of interest at high magnification in an aligned serial section movie and alternately displays the area in the current film frame, as seen via a 1000-line TV camera, and the area shown in the previous frame, as stored in the frame–buffer. The stage is then interactively moved so as to get best alignment of the local areas, and the computer is signaled to record the resultant motions. Subsequently, any time that processing or viewing a particular area is desired, the stored stage movements are used to maintain the alignment of each frame. Thus there is no need for rephotographing the sections for every small area. The goal of Stevens et al (private communication) is to build a fast enough stage so that prerecorded alignment motions can be accomplished during projector frame-changing at ciné playback speeds. This would obviate the need for any aligned movie.

SYSTEMS AND ALGORITHMS

We classify automatic CARTOS systems into two major categories. Those of the first category treat their data as a 3D array of densities (4, 17) and do surface-following within a volume. We call this 3D tracing. Those of the second category (8, 10, 11, 13, 15 and our Columbia system) work within a 2D array of densities from a single serial section and do not carry any detailed density information other than that for the current section. They introduce the third dimension not at the pixel, but at the contour, level. We call this 2D tracing. 3D tracing allows surface-following along arbitrary trajectories, whereas the 2D approach allows only contouring within section planes. One intriguing possibility with the 3D approach, especially for skeleton extraction and detection of branching, is to extract contours in planes normal to the axis of a fiber instead of the section planes.

A fact which strongly mediates against 3D tracing is that image input devices are much faster at retrieving nearby density information from within the current section plane than from a different one. These devices are constructed to access a rectangular pixel array (be it 32×32 or 800×600) either at electronic communication speeds or at very fast mechanical motion speeds (i.e. μsec to msec). On the other hand, changing a section plane—whether by moving the focal plane of a light microscope objective or by changing frames on a film transport—always entails mechanical motion that is much slower (msec to sec). Another serious problem with 3D tracing is the difficulty of efficiently segmenting and accessing the data when dealing with many sections and large image arrays.

3D Tracing Systems

We describe two systems (4, 17) of the 3D tracing type. Both used the optical sectioning technique with Golgi-stained specimens.

Shantz (17) did not automate shape extraction, but he developed a promising filter technique for removing out-of-focus artifacts from digitized optical section density data. His raw data was Golgi-stained frog retinas in whole mounts. Each depth level was scanned into an array of 512×512 pixels. There were 15 to 30 optical sections per cell studied. Slightly less than 8-bits per pixel of grey-level (226 levels) was stored. Shape features were interactively extracted from the cleaned-up density data. The microscope scanner, alluded to earlier, has x and y galvanometer mirror deflectors to sweep the image across a small hole, behind which is a photomultiplier tube. The filter was an improvement on an earlier one demonstrated by Weinstein & Castleman (20). The basic idea is to clean up a given optical section image plane by subtracting a fraction of the defocused images of the planes immediately above and below it. Weinstein & Castleman used the fraction .8 of the average of the two defocused images. Shantz used the maximum of the two densities. Shantz also carefully measured the defocused point-spread function for his microscope and stain, whereas the earlier work only used a rough approximation.

There has been work on extracting shape automatically from such 3D density arrays in several related fields. Liu & Herman (9) have a program that extracts surfaces bounding an object from density arrays generated by Computer Assisted Tomography (CAT) scanners. Their criteria for acceptable surface boundary points are based upon density, density gradient, connectivity, and global a priori knowledge such as curvature-bound constraints. Greer (7) has developed programs for extracting skeletons of protein molecules from their X-ray crystallographic electron density maps. His algorithms have many protein-specific features in them, but could probably be used in modified form to effectively extract neuronal skeletons.

Coleman et al (4) have a highly automated system for tracking dendrites in Golgi-stained tissues. Their system has been extensively field-tested over the last few years in the process of gathering a data base of neuron morphology. A vidisector is used for image input. It is attached to a light microscope with computer-controlled x, y, and z axes. Skeletons are extracted with minimal boundary following. The system is initialized by an operator pointing—in x, y with a joystick and in z with a focus potentiometer—at the soma of a neuron to be tracked. The operator then points to the start of all primary dendrites exiting from the soma. In addition he supplies for each primary dendrite a direction of initial travel in x, y. The system then enters its automatic mode in which it samples densities on a semicircular arc about its current position and centered upon its current tracking direction. It looks for a density profile indicating the cross section of a dark object. Having found one it varies the focus (z coordinate) up and down until it maximizes the density of the profile. It then updates its position and direction and iterates. Double profiles indicate branching. These are marked, the operator is alerted for verification, and they are enqueued for later following. A fiber termination is presumed whenever the density profile contains no points darker than a preset threshold. The operator is alerted to verify the end or reorient the follower. If the end was real the operator specifies whether it is a cut end or not, and the program returns to the most recently unfinished branch region.

2D Tracing Systems

The rest of the systems we discuss all do 2D tracing. That is, they all extract contours from a given section before going on to the next section. There is an enormous literature on boundary detection and extraction under the heading "Image Processing and Pattern Recognition." We will not attempt to review it here. For an excellent introduction to the subject we recommend Rosenfeld & Kak (16, especially Ch. 6, 8) and Duda & Hart (5, especially Ch. 7–9). In addition, there are both national and international joint conferences annually under the same title.

In general, an outline of an object is a very context-dependent concept. Parts of such an outline are very often missing from an image due to poor contrast or are obscured by artifacts. Any successful contour extraction technique must therefore rely on both local and global image data and a priori knowledge of the class of objects being sought. In the systems under discussion local information is provided by thresholding grey-levels and/or grey-level gradients. Grey-level thresholding has the advantage that it always yields closed contours (or their intersection with the film frame). However, it will only generate meaningful contours if the object of interest has boundary points that are consistently darker (or lighter) than the surround. Grey-level gradient thresholding constrains boundary point locations to a rapidly changing grey-level. In addition to detecting boundary point candidates it can also provide a boundary (i.e. edge) direction. It however does not yield continuous connected sets of boundary points as the former technique does, and gap-filling is a serious problem for any such boundary extraction algorithms. A good compromise between these approaches is to enhance the grey-level of visual edges in a picture by high-pass spatial prefiltering. This is equivalent to adding some fraction of the smoothed gradient magnitude to the original image before applying grey-level thresholding. Unfortunately it has not been used much to date because it is computationally expensive and thus slows down most systems. However fast hardware to do this is becoming affordable both in processing video raster-displays (e.g. DeAnza, Grinnell, Ikonas, ADI, IIS, Comtal), and general purpose array processors (e.g. FPS, CSPI).

Some global information is introduced by density normalization and frame-to-frame matching of contours. Some a priori information is also introduced in smoothing contour boundaries. However, much global and a priori information is still left to the operator to introduce via designation of thresholds and search locations and editing of extracted contours.

Automatic contour extraction is accomplished in (8, 13) and our Columbia system by thresholding the grey-level data and extracting and smoothing contours from the resultant binary images. In (13), our system, and the "vector-branching-optical-sectioning" mode in (8), only one interactively selected fiber or stack of contours is followed at a time. Thus assembly of contours into discrete objects is implicit in the order of extraction. On the other hand in the system of Reddy et al (15) all contours of interest in a single section are extracted before going on to the next one. Specimens were cut sections of dye-injected cells and a gradient-directed boundary-following algorithm was used with thresholding to constrain the acceptable boundary points to areas having densities below the threshold. Contour assembly was done automatically by identifying the successor to a contour as one that overlapped it or was sufficiently close to it. If a contour had no such neighbors in either adjacent section it was eliminated.

The system of Hillman, Llinas et al (8, 10, 11) uses a Quantimet 720 for image input from light microscopes and film, and a STEM/computer interface for direct acquisition of EM images (see section on image input devices, above). The 720 has special purpose hardware for grey-level thresholding and displaying the resultant binary (i.e. black and white) image on its slow-scan TV monitor. A direct-view storage tube display (Tektronix 611) interfaces between the 720 and a PDP15 minicomputer. The 611 stores and displays the binary TV image. At the same time, programs on the PDP15 can access the binary pixel value at any x, y location without TV frame latency delays. These programs extract the boundary contours between black and white areas, with operator interaction to accept, delete, or edit specific contours. Presumably the system will run fastest with optically sectioned Golgi-stained or dye-injected specimens. If the cell to be traced is wholly contained within the TV camera's field of view, the operator need only interact with the system to change the z-coordinate of the microscope stage and in principle can extract all contours in all optical sections via the 720-611 interface and later assemble them. For cut sections or ends of thick optically sectioned blocks, contour assembly is done after interactive alignment of adjacent cut sections (pages) and marking of cut ends of fibers in cases where the pages contain more than one optical section plane. In the case of thin pages that contain only one optical section plane, all contours are marked as having at least one and usually two cut ends. An algorithm that connects closest cut ends in the aligned adjacent pages is then applied.

Weinstein & Castleman (20) identified and displayed boundary points by both threshold density and thresholded density gradient methods. They did not however extract the boundaries from their 3D array format. They just generated stereo pair displays by computing the projections of the whole 3D array onto a pair of image planes.

In the case of cut serial sections, all of the 2D systems other than ours work with unaligned section images, require operator intervention to change sections, and do the necessary alignment interactively at tracing time. Of these, the system of Reddy et al (15) was most automated in that it computed a best in-plane translation and rotation to align successive outline contours of a ganglion structure that enclosed the cells of interest. The computed alignment was then presented to an operator for verification and/or correction.

Our system accepts prealigned film strips mounted on a computer-controlled film transport/projector and makes use of the 3D continuity of the structure being analyzed. We have taken advantage of this format to predict the location, area, and the shape measure, $S = (\text{perimeter}^2)/\text{area}$, of the next contour in sequence from the previous three contours. The program automatically changes frames and extracts a contour from the thresholded, normalized grey-level data at the predicted location. If the area and shape measure S are within preset tolerances of their predicted values, it accepts the contour and continues automatically. Otherwise the program alerts the operator for verification, or change of start point or threshold, or manual tracing of data. We routinely compute the centers of gravity of all contours and can use this data for skeleton information.

Computing Skeletons From Contour Data

Finding the skeleton of an object defined by a set of contours is very straightforward when the object is fiber-like and the section planes are normal to the fiber axis. An algorithm as simple as connecting the centers of gravity of successive contours yields good results, at least for near convex cross-section data. Problems arise for objects such as cell bodies which do not generally have an obvious axis. There are also problems when the section planes are nearly parallel to the fiber axis. Most CARTOS systems patch up these cases interactively (for example see Reference 6). Soroka (18, 19) dealt with these issues. He proposed and tested an elaborate set of programs to fit stacks of section contours piecewise with volumetric primitives, called generalized cylinders, which are defined by straight-line axes and linearly tapered elliptical cross sections. The algorithm always yields an axis, although it is not unique to an object and depends strongly on the direction of the section planes. We would prefer an algorithm which rapidly determines when the data is nondegenerate, as defined above, and finds the fiber axis, and which in the other cases returns an indication of degeneracy with or without proposed axis. An operator could then fill in these cases which would probably consist of cell bodies, branch junctions, and pieces of fiber turned parallel to the section planes. That is, let the computer handle the easy cases and the operator the hard ones.

Related Work

There are related areas of research where similar automation has been done. We have already mentioned some relevant CAT scanner and X-ray crystallographic work. In addition there is a wealth of literature on tracking moving objects in sequential images (1–3). There are even two complete commercial systems (DBA video target tracker; PDS film motion analyser) for this purpose. We know of no attempts to adapt these to automated CARTOS. Perhaps this is due to their high cost and the need for significant software modification. No discussion of this sort is complete without mention of cellular logic operations and their implementation in very high speed parallel image processors. These operations transform 2D arrays. The value of an element in the output array is a programmable function of the element values in the local neighborhood or "cell" of the corresponding element in the input array. Neighborhoods for image processing are usually defined to contain an element and either four or eight neighbor elements in the Cartesian grid case, or an element and its six neighbors in the hexagonal grid case. An excellent review of the capabilities of cellular logic machines with some applications in medical image processing is given in (14).

SUMMARY

In this article we have restricted our discussion to automation of the data input phase of CARTOS. This is a relatively new field of research and development. It is still small in terms of numbers of active researchers but is certain to expand rapidly as the necessary computer tools become cheaper and more widely available. The technology is changing so fast that any description of it is likely to be outdated before it is published. For this reason we have tried to emphasize the fundamental problems posed by the task and the various approaches toward their solution. We look forward to rapid progress in the next few years.

ACKNOWLEDGMENT

The work at Columbia University referred to in this review was supported by Grant RR00442 from the National Institutes of Health, Division of Research Resources, Biotechnology Branch.

Literature Cited

1. Badler, N. I., Aggarwal, J. K., eds. 1979. *Abstr. Workshop on Time-Varying Imagery*, Philadelphia: IEEE Comput. Soc. & Comput. Sci. Dep., Univ. Pa. 153 pp.
2. Booth, J. R., Childers, D. G. 1979. *IEEE Trans. Biomed. Eng.* 26:185–92
3. Clayton, P. D., Harris, L. D., Rumel, S. R., Warner, H. R. 1974. *Comput. Biomed. Res.*, 7:369–94
4. Coleman, P. D., Garvey, C. F., Young, J. N., Simon, W. 1977. In *Computer Analysis of Neuronal Structures*, ed. R. D. Lindsay, pp. 91–109. New York: Plenum. 210 pp.
5. Duda, R. O., Hart, P. E. 1973. *Pattern Classification and Scene Analysis*. New York: Wiley. 482 pp.
6. Glasser, S., Miller, J., Xuong, N. G., Selverston, A. 1977. See Ref. 4, pp. 21–58
7. Greer, J. 1974. *J. Mol. Biol.* 82:279–301
8. Hillman, D. E., Llinas, R., Chujo, M. 1977. See Ref. 4, pp. 73–90
9. Liu, H. K., Herman, G. T. 1977. *Comput. Graph. Image Process.* 6:123–34
10. Llinas, R., Hillman, D. E. 1975. In *Golgi Centennial Symposium Proceedings*, ed. M. Santini, pp. 71–80. New York: Raven. 668 pp.
11. Llinas, R., Spitzer, R., Hillman, D. E., Chujo, M. 1979. In *Scanning Electron Microscopy, 1979* 2:367–74. O'Hare, Ill: SEM Inc., AMF
12. Macagno, E. R., Levinthal, C., Sobel, I. 1979. *Ann. Rev. Biophys. Bioeng.* 8:323–51
13. Mazziotta, J., Hamilton, B. L. 1977. *Comput. Biol. Med.* 7:265–79
14. Preston, K., Duff, M. J. B., Levialdi, S., Norgren, P. E., Toriwaki, J. I. 1979. *Proc. IEEE* 67:826–56
15. Reddy, D. R., Davis, W. J., Ohlander, R. B., Bihary, D. J. 1973. In *Intracellular Staining in Neurobiology*, ed. S. D. Kater, C. Nicholson, pp. 227–53. Berlin/New York: Springer. 332 pp.
16. Rosenfeld, A., Kak, A. C. 1976. *Digital Picture Processing*. New York: Academic. 456 pp.
17. Shantz, M. J. 1976. In *Computer Technology in Neuroscience*, ed. P. B. Brown, pp. 113–29. Washington DC: Hemisphere. 650 pp.
18. Soroka, B. I. 1979. In *Proc. NSF Wkshp. Representation of 3D Objects*, pp. P1–31. Philadelphia: Comput. Sci. Dep., Univ. Pa.
19. Soroka, B. I. 1979. *Understanding objects from slices: Extracting generalized cylinder descriptions from serial sections*. PhD thesis. Univ. Pa. Available as TR-79-1 from Univ. Kansas Comput. Sci. Dep., Lawrence, Kans. 411 pp.
20. Weinstein, M., Castleman, K. R. 1971. *Proc. Soc. Photo-Opt. Instrum. Eng. Biomed.* 26:131–37

BIOPHYSICAL APPLICATIONS ♦9151
OF NMR TO PHOSPHORYL
TRANSFER ENZYMES AND
METAL NUCLEI
OF METALLOPROTEINS

Joseph J. Villafranca and Frank M. Raushel[1]

Department of Chemistry, The Pennsylvania State University, University Park, Pennsylvania 16802

INTRODUCTION

In the past few years exciting new applications of nuclear magnetic resonance (NMR) spectroscopy have been made in the study of biological macromolecules. This review is intended to highlight two areas of current NMR research that have mainly been concerned with elucidating the mechanism of action of enzymes. The first area deals with the newly discovered effect of the isotopic shift of ^{18}O on the ^{31}P-NMR spectrum of phosphorus-containing compounds (15); the second area deals with the study of metal nuclei in metalloenzymes.

Until recently mass spectrometry was the method used to study ^{18}O-^{16}O exchange reactions in enzymes that carry out various phosphoryl transfer reactions. The method has the advantage of requiring only small amounts of material for analysis (nanomoles), but volatile phosphate esters must be prepared, which is very time consuming, and continuous observation of the enzymatic process is not possible. The NMR method permits the continuous study of an enzymatic reaction and affords kinetic and isotopic distribution data simultaneously. However, micromoles of material are required. Selected examples of a wide

---

[1]Partial support for some of the studies reported herein was provided by the National Science Foundation PCM-7807854 and the Public Health Service, GM-23529, AM-21785, AM-05966. JJV is the recipient of an Established Investigatorship Award of the American Heart Association.

variety of enzymatic reactions are covered in this review to stimulate further research in the area.

The study of metal nuclei by NMR is not new but the application to biological systems is. Most metal nuclei have low gyromagnetic ratios, some have low natural abundance, and many are quadrupolar (spin > 1/2). The advent of fourier transform technology to NMR spectroscopy can overcome these disadvantages, and this review deals with some theoretical considerations in applying NMR studies to metalloenzymes. Future applications of NMR to such systems are foreseen and some of the more recent work with metal nuclei binding to macromolecules is reviewed.

NMR STUDIES OF PHOSPHORYL TRANSFER ENZYMES

^{18}O-^{31}P-NMR Shift

Recently, Cohn & Hu (15) and others (46, 48) have demonstrated that there is a small but easily measured chemical shift when ^{18}O is substituted for ^{16}O. The chemical shift is ~0.021 ppm upfield for each ^{18}O

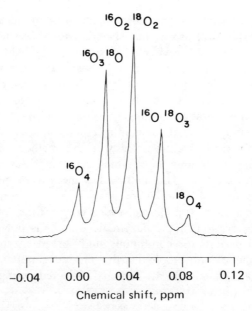

Figure 1 ^{31}P NMR of a randomized sample of [$^{16}O^{18}O$]P_i [From Cohn & Hu (15)].

Table 1 Effect of ^{18}O on the ^{31}P chemical shifts of ATP (16)

P of ATP	Oxygen position	$\Delta\delta$(ppm) per ^{18}O atom
α	α-β bridge	0.017
β	α-β bridge	0.017
	nonbridge	0.028
	β-γ bridge	0.016
γ	β-γ bridge	0.022
	nonbridge	0.022

substituted for ^{16}O and is additive. Thus, the chemical shift for $HP^{16}O_4^-$ is 0.084 ppm downfield from $HP^{18}O_4^-$. Shown in Figure 1 is the ^{31}P-NMR spectrum at 145.7 MHz of inorganic phosphate containing 44% ^{18}O. Each of the five possible species can clearly be distinguished and each species appears in the expected ratio for a random distribution of ^{18}O and ^{16}O. In phosphate-containing species such as ATP where the phosphorous (α, β, and γ) and oxygen (bridge and nonbridge) atoms are in different environments the effect of ^{18}O substitution on the ^{31}P chemical shift is dependent on the oxygen environment. The shift is largest for those oxygens with most double bond character (16). Shown in Table I is the effect of ^{18}O on the ^{31}P chemical shifts of ATP (16). The 235 MHz spectrum of the β-P of ATP labeled with 63% ^{18}O is shown in Figure 2. Each of the six possible ^{18}O-^{16}O species can be clearly resolved.

As pointed out by Cohn & Hu (15) this phenomenon can be used in two ways: (a) for labeling phosphate groups in much the same way as ^{32}P is used, and (b) for determining oxygen-phosphate exchange reactions and the site of bond cleavage in phosphate esters such as ATP.

The advantages of this spectroscopic method over the currently used methods are many. Since the method is nondestructive the labeled compounds can be recovered and reused if necessary. The need for using radioactive materials is eliminated. In the case of exchange reactions the kinetics of the reaction can be followed continuously without increasing the amount of material needed to generate a large number of data points. There is also no need for a complicated workup of reaction mixtures to isolate the compound of interest. The main disadvantage appears to be the larger amounts of material needed owing to the insensitivity of the NMR method. The amount of material

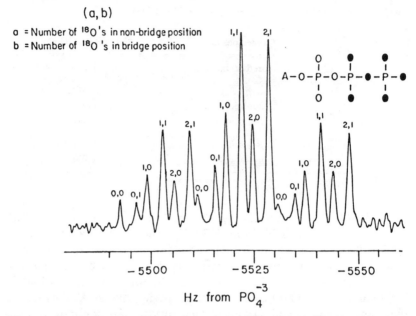

Figure 2 ^{31}P NMR spectrum of the βP of ~63% ^{18}[0]ATP-β, γ^{18}O$_6$ recorded at 235 MHz by correlation spectroscopy [from Cohn & Hu (16)].

required is about 10 μmoles, which in most cases is not excessive. If oxygen-phosphorus exchange is to be monitored, there is also the need for a high field spectrometer in order that sufficient resolution be obtained to detect changes in each individual species.

This technique has already found widespread use with a number of different enzymes. The salient results are summarized below.

Uses in Individual Enzyme Systems

POLYNUCLEOTIDE PHOSPHORYLASE   Cohn & Hu (15) found that upon the incubation of [^{18}O]P$_i$ and ADP with polynucleotide phosphorylase from *Micrococcus luteus* an exchange of the β-P of ADP with P$_i$ occurred. It was also shown that all four oxygen atoms of P$_i$ were incorporated into the ADP as follows:

1.

The site of bond cleavage is thus between the α-P and the α-β bridge oxygen.

INORGANIC PYROPHOSPHATASE Inorganic pyrophosphatase catalyzes the hydrolysis of pyrophosphate to two moles of inorganic phosphate. Cohn & Hu (15) found that upon incubation of inorganic phosphate (93.4% ^{18}O) and pyrophosphatase the enzyme catalyzed the exchange of ^{18}O out of P_i and into H_2O and that the reaction could be followed at 25 MHz, although the resolution was not good enough to get precise rate constants. This exchange reaction is consistent with the reversible formation of enzyme-bound pyrophosphate and H_2O from two molecules of P_i. No detectable formation of pyrophosphate occurs in solution.

MYOSIN This enzyme catalyzes the hydrolysis of ATP to ADP and P_i. It has also been shown that myosin will catalyze the slow exchange of oxygens between water and P_i in the presence of MgADP (21). Webb et al (70) have examined this later reaction in more detail by following the ^{31}P NMR spectrum of [^{18}O]P_i at 145.7 MHz. They found that all molecules of [^{18}O]P_i that bound to the enzyme underwent complete exchange with H_2O and were released as [$^{16}O_4$]P_i, with no intermediate formation of [^{18}O^{16}O]P_i in the bulk solution. This is interpreted as occurring by a rapid equilibration of enzyme-MgADP-P_i and enzyme-ATP, with very slow release of P_i from the enzyme, in comparison to the interconversion of the central complexes.

ALKALINE PHOSPHATASE This enzyme catalyzes the hydrolysis of a large variety of phosphate esters and is known to involve a phosphoenzyme intermediate. Eargle et al (25) have shown that Zn^{2+} alkaline phosphatase will catalyze an H_2O-P_i exchange with only one ^{18}O exchanged from [^{18}O]P_i per enzyme encounter. Bock & Cohn (8) have confirmed this observation using ^{31}P NMR and have also shown with the Co^{2+}-substituted enzyme that the exchange pattern is different. On the average more than one ^{18}O is exchanged from [^{18}O]P_i per enzymatic encounter. From a computer analysis of the data it was determined that the ratio of the rates for the formation of the covalent phosphoenzyme from the $E \cdot P_i$ Michaelis complex and the rate for the dissociation of the P_i from $E \cdot P_i$ was 3.0 ± 0.5.

ACID PHOSPHATASE Human prostatic acid phosphatase is another phosphatase that involves a phosphoenzyme intermediate, and Van Etten & Risley (69) have shown by ^{31}P NMR at 40.5 MHz that the enzyme will catalyze an H_2O-P_i exchange. The rate constants for this process are such that only about 1 atom of ^{18}O is exchanged per encounter.

GLUTAMINE SYNTHETASE Glutamine synthetase catalyzes the formation of glutamine from ATP, glutamate, and ammonia. The other products are ADP and P_i. The reaction mechanism is thought to proceed through a γ-glutamyl phosphate intermediate. Balakrishnan et al have shown using ^{31}P NMR that incubation of ADP, glutamine, and $[^{18}O]P_i$ with glutamine synthetase resulted in the loss of ^{18}O from the P_i (6). Analysis of the data showed that only one ^{18}O was lost per encounter and that the rate constant for exchange was 5–7 times faster than net turnover of products. This was also demonstrated by Stokes & Boyer (66) using mass spectrometry.

This exchange reaction is accommodated by assuming that Reactions 2 and 3 are happening on the enzyme surface and that the γ-carboxyl of glutamate (II) is free to rotate, thus allowing, upon reversal of Equation 3, the formation of γ-glutamyl phosphate (I) with an ^{16}O in the bridge position. Reaction of I with NH_3 would then produce P_i with only 3 atoms of ^{18}O instead of the original 4. Since a random distribution of ^{18}O is maintained at all times, only one ^{18}O is lost per encounter with the enzyme and thus P_i must be dissociating from the enzyme faster than glutamine.

The recently introduced positional isotope exchange technique of Midelfort & Rose (50) is also amenable for use by ^{18}O chemical shift technique. Briefly, this technique follows the exchange of label from one part of a substrate to another due to rotational equivalence of some intermediate. The method was first applied to glutamine synthetase in the reaction:

$$\text{Ado-O-}\overset{\overset{O}{\|}}{\underset{\underset{O^-}{|}}{P}}\text{-O-}\overset{\overset{\bullet}{\|}}{\underset{\underset{O^-}{|}}{P}}\text{-O}^- + \gamma\text{-Glutamyl-P} \rightleftharpoons \text{Ado-O-}\overset{\overset{O}{\|}}{\underset{\underset{O^-}{|}}{P}}\text{-O-}\overset{\overset{\bullet}{\|}}{\underset{\underset{O^-}{|}}{P}}\text{-O-}\overset{\overset{O}{\|}}{\underset{\underset{O^-}{|}}{P}}\text{-O}^- + \text{Glu}$$

5.

(This is a simplified scheme presented for clarity. In fact all four oxygens of the γ-P were labeled with ^{18}O, but the conclusion is the same for the positional isotope exchange). ATP is synthesized with ^{18}O in the β-γ bridge position. If the enzyme catalyzed the formation of γ-glutamyl-P from ATP and glutamate in the absence of NH_3, then ADP would be formed; because of rotational equivalence it would allow, upon the reversal of the above reactions, the formation of ATP with ^{18}O in the nonbridge position of the β-P. To determine if any isotope exchange occurred they degraded the ATP using acetyl CoA synthetase and glycerokinase and then derivatized the resulting phosphate with diazomethane. The trimethylphosphate derivative was then analyzed for ^{18}O content by mass spectrometry. This reaction is easily detected by the ^{18}O chemical shift technique because the γ-P of ATP will have lost its ^{18}O and thus will give rise to a new resonance for the γ-P of ATP. There will also be a shift in the resonance of the β-P because of different shifts caused by a bridge or nonbridge ^{18}O. This experiment has not, as yet, been repeated using NMR to detect the ^{18}O exchange.

CARBAMYL PHOSPHATE SYNTHETASE Carbamyl phosphate synthetase catalyzes the synthesis of carbamyl-P from HCO_3^-, glutamine, and 2 moles of ATP. The enzyme also catalyzes the HCO_3^--dependent hydrolysis of ATP. Raushel & Villafranca (unpublished observations) followed the exchange of ^{18}O from the bridge to the nonbridge position of [γ-^{18}O]ATP after incubation with enzyme and bicarbonate. The exchange rate was 0.5 times the rate of ADP formation. Shown in Figure 3 is the resonance for the γ-P of ATP before reaction with carbamyl phosphate synthetase and after the chemical reaction had proceeded to ~50% completion. The increase in the [$^{16}O^{18}O_3$] peak shows that there is loss of one ^{18}O from the γ-P. Under somewhat different reaction conditions and using a mass spectral analysis of the [^{18}O] ATP, Wimmer et al (73) have obtained an exchange rate that is 1.4–1.7 times the chemical rate. The experiments described above using ^{31}P-NMR were done in the presence of 10 mM L-ornithine, a positive allosteric effector of the enzyme which tightens the binding of ATP to the enzyme but does not affect the V_{max}. The experiments of Wimmer et al were done without any ornithine. Our experimental findings differ from those of Wimmer et al in that the ratio of the exchange rate to the chemical rate is ~0.5 in the presence or absence of ornithine. This is inconsistent with a slower rate for dissociation (k_{off}) of ATP from the enzyme relative to

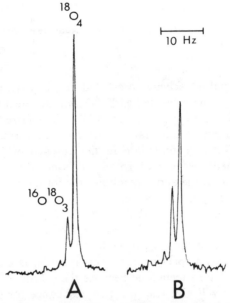

Figure 3 ^{31}P NMR spectra of the γ-P of [γ-^{18}O]ATP at 81 MHz. Only one half of the doublet is shown. (*A*) Before incubation with carbamyl phosphate synthetase and HCO_3^-. (*B*) After incubation with carbamyl phosphate synthetase. The reaction was stopped after 50% completion and the spectrum recorded.

the k_{cat} for the ATPase reaction. The rate of formation of intermediates in the reaction was studied using rapid reaction techniques by Raushel & Villafranca (56) and the data agree with the ^{31}P-NMR studies presented above.

The positional isotope exchange has also been measured with ^{31}P NMR in the reverse reaction of carbamyl phosphate synthetase:

$$\text{Carbamyl-P} + \text{ADP} \rightarrow \text{ATP} + \text{NH}_3 + \text{HCO}_3^-. \qquad 6.$$

Carbamyl phosphate was synthesized with ^{18}O in all oxygens except the carbonyl oxygen of carbon. The exchange of the bridge oxygen into the carbonyl oxygen was followed by NMR and was evidence for the following series of reactions

$$\underset{}{NH_2-\overset{O}{\overset{\|}{C}}-\bullet-\overset{\overset{\bullet}{\|}}{\underset{\bullet^-}{P}}-\bullet^- + ADP} \rightleftharpoons NH_2-\overset{O}{\overset{\|}{C}}-\bullet^- + ATP \qquad 7.$$

$$\underset{}{NH_2-\overset{\bullet}{\overset{\|}{C}}-O-\overset{\overset{\bullet}{\|}}{\underset{\bullet^-}{P}}-\bullet^- + ADP} \rightleftharpoons NH_2-\overset{\bullet}{\overset{\|}{C}}-O^- + ATP \qquad 8.$$

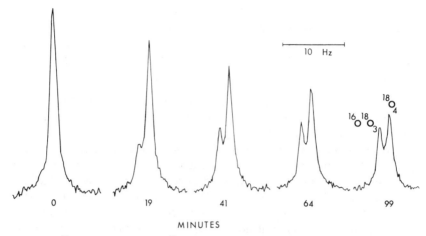

Figure 4 ^{31}P NMR spectra of [^{18}O] carbamyl phosphate at various times after the addition of carbamyl phosphate synthetase and MgADP. Initially all oxygens were labeled with ^{18}O except the carbonyl oxygen.

Shown in Figure 4 is a series of spectra taken at different times showing the exchange of ^{18}O from the bridge to nonbridge position of carbamyl phosphate. The exchange rate was 3.9 times the chemical rate.

PYRUVATE KINASE Using ATP that was labeled in the α-β bridge and β nonbridge positions (III)

$$\text{III} \qquad \text{IV}$$

Lowe & Sproat (46) have shown using ^{31}P NMR at 36.4 MHz that pyruvate kinase catalyzes the exchange of ^{18}O from a β nonbridge position of ATP to a β-γ bridge position (IV) in the absence of any obvious acceptor molecule. They have argued that this is evidence for the formation of a metaphosphate intermediate. However, the possibility of transient transfer of the phosphoryl group to a nucleophile on the enzyme or to compounds such as bicarbonate, acetate (buffer), or even H$_2$O have yet to be ruled out.

Tsai (67) has developed a technique using ^{17}O substituted phosphate. Since ^{17}O has a spin quantum number of 5/2 and a large electric quadrupole moment, it causes the ^{31}P NMR signal of phosphate to broaden substantially when ^{17}O has been substituted for ^{16}O. Tsai (67) has used this method to determine whether the attack of acetate on adenosine-5'-(1-thiotriphosphate) proceeded with inversion or retention of configuration at the α-P in the acetyl coenzyme A synthetase reaction. The reaction catalyzed by this enzyme is:

Acetate + coenzyme A + ATP→acetyl CoA + PP$_i$ + AMP.

One of the oxygens of acetate is incorporated into AMP. This enzyme was found to use the B isomer of adenosine-5'-(1-thiotriphosphate) (ATPαs), which has the R configuration at the α-P (9, 12).

To determine the stereochemical course of the reaction, Tsai used [^{17}O] acetate and then determined whether the ^{17}O was incorporated into the pro-S or pro-R position of adenosine 5'-thiophosphate (AMPS). The isolated AMPS was reacted with myokinase and pyruvate kinase which are known to make ATP(αS), with the S configuration at the α-P (64). The pro-S oxygen will now be in the nonbridge position and the pro-R oxygen in the bridge position. Thus incorporation of ^{17}O in the pro-S position will decrease the intensity of only the α-P, and incorporation of ^{17}O into the pro-R configuration will decrease the intensity of both the α and β phosphate. Tsai found that both the α and β intensities were reduced 20% using 20% labeled ^{17}O acetate, thus showing that the reaction proceeded with inversion of configuration. This is shown in the following scheme:

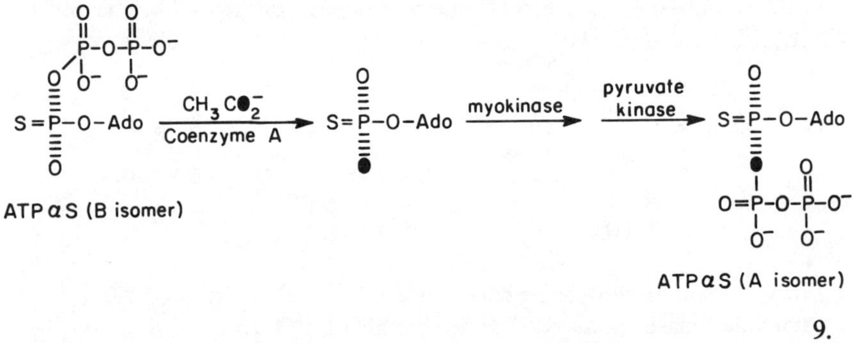

9.

Midelfort & Sarton-Miller have also reported that acetyl coenzyme A synthetase proceeds with inversion of configuration (51).

The same experiments could have been performed using ^{18}O. The only advantage of using ^{17}O over ^{18}O appears to be that the effect caused by ^{17}O, since only the intensities are reduced, is easily detected by low field spectrometer, whereas a high field spectrometer is needed to get significant resolution to separate the ^{18}O and ^{16}O ^{31}P resonances.

Presumably the ^{17}O method could also be applied to problems analogous to those solved by the ^{18}O shift method, but since the ^{17}O method cannot distinguish cases where more than one oxygen will become labeled it does not appear to be potentially as versatile as the ^{18}O method. This is especially true if the phosphate-water exchange were to be measured because the ^{17}O method cannot distinguish the number of ^{17}O atoms bonded to a particular phosphate atom. Therefore it could not be determined if all of the oxygens were exchanged with solvent per enzyme encounter or if they were exchanged one at a time.

Stereochemistry of MgATP Complexes

Since the α and β phosphorous atoms of ATP are prochiral they are potentially asymmetric centers once they are complexed to divalent cations. There will be two diastereomers if complexation by Mg^{2+} is on the γ and β phosphates and 4 diastereomers if complexation is to all three phosphates. Since Mg^{2+} exchanges ligands very rapidly it is not possible to isolate individual isomers. However, Cr^{3+} and Co^{3+} exchange ligands much more slowly and thus should allow preparation of individual isomers. The Cr^{3+} and Co^{3+} derivatives of ATP have been synthesized in Cleland's laboratory (19, 22).

Since Cr(III) is paramagnetic it is not possible to analyze the products of the synthesis by NMR. However, Co^{3+} is diamagnetic and the β, γ derivative, Co(III) $(NH_3)_4$ATP, distinctly shows that two isomers are present as determined by ^{31}P NMR (20). Cornelius & Cleland were also able to show that yeast hexokinase used only one of the isomers in the following reaction (20):

D-glucose + $Co(NH_3)_4$ATP → CoADP · glucose-6-P.

Both the ADP and the glucose-6-P are complexed with the cobalt. The unreacted isomer was degraded and crystallized and its absolute stereochemistry determined by X-ray crystallography (49). The active isomer (substituting Mg^{2+} for Co^{3+}) had the following structure:

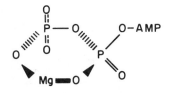

Mg ATP

This isomer has been labeled the Λ isomer and the other possible structure, the Δ isomer (20).

Ekstein & Goody have also prepared derivatives of ATP where sulfur has replaced oxygen in one of the nonbridge positions of the α and β phosphoryls (26). There are two diastereomers for each labeled phosphate. Jaffe & Cohn (36) showed that only the B isomer (nomenclature of Ekstein & Goody (26)) of ATPβS is a substrate for yeast hexokinase. Assuming that only oxygen is a ligand for Mg^{2+} they therefore proposed that the structure of the B isomer had the following configuration:

MgATPβS(B)

Thus the B isomer of ATPβS has an R configuration at the β phosphoryl. They also showed that the two diastereomers of ATPβS and ATPαS could be distinguished by their NMR spectra (37). The substitution of S for O results in a 40–50 ppm downfield chemical shift of phosphorus resonance for the phosphorus atom directly bonded to the sulfur atom. In ADPαS the α-P resonances of the A diastereomer are centered 0.4 ppm downfield from those of the B diastereomer (37). The β-P chemical shift of the A diastereomer of ATPβS is 0.14 ppm upfield from that of the B diastereomer (37). The absolute configuration of the ATPαS diastereomers has recently been completed and thus it is now possible to define the geometry of the Mg-ATP complexes that are active with kinases. The A isomer of ATPαS has the S configuration at α-P (9, 12).

With yeast hexokinase the preferred isomer of ATPβS was reversed when Cd^{2+} was used at the activating divalent cation (36). Jaffe & Cohn reasoned that this is because Cd^{2+} will bind preferentially to sulfur. However, the active compound still has the same geometric arrangement.

CdATPβS(A)

The reversal of specificity when Cd^{2+} is used as the activating divalent cation is a necessary requirement for establishing whether the metal ion is actually ligated to a particular phosphoryl group. It is also possible for side chains of the protein to determine the specificity by hydrogen bonding to one of the oxygens. This appears to be the case in hexokinase with ATPαS. The A isomer is used by hexokinase 20 times faster than the β isomer with either Mg^{2+} or Cd^{2+} and thus the metal ion is probably not coordinated to the α-P during the reaction of hexokinase.

Distance Determinations

In addition to being useful in the elucidation of the stereochemistry of metal ligation to ATP complexes at enzyme active sites, CrATP and CoATP are useful reagents for distance determinations of bound substrates using NMR. CrATP is paramagnetic and thus can be used as a spin-labeled center for distance measurements to other substrate nuclei. CoATP is diamagnetic and can serve as a replacement for MgATP in distance measurements from other paramagnetic centers on the enzyme. This also determines the geometrical conformation of the bound nucleotide. The theory for using paramagnetic molecules to affect the relaxation rates of bound substrates in order to determine intermolecular distances has been adequately reviewed (24).

PYRUVATE KINASE   Gupta et al (29) have used CrATP as a paramagnetic probe for the enzyme pyruvate kinase. Using the effect of enzyme bound CrATP on the longitudinal relaxation times (T_1) of pyruvate they were able to determine the distance from the Cr to the methyl protons and to the C-1 and C-2 carbon atoms of pyruvate. The correlation time for the interaction was estimated from the frequency dependence of $1/T_1$ of the water protons. The distances were 6.1 ± 0.4, 6.1 ± 0.3, and 7.9 ± 0.5 Å to the C-1, C-2 carbon atoms, and methyl protons respectively. The results clearly put the pyruvate atoms in the second sphere of the Cr^{3+} and thus establish a close proximity of the phosphate donor substrate (ATP) to the phosphate acceptor substrate on a kinase.

Gupta & Benovic (31) also measured the effect of CrADP on the longitudinal relaxation times of the protons and phosphorus atoms of phosphoenolpyruvate in the enzyme·PEP·CrADP complexes. Using a correlation time of 0.35 nsec they obtained a Cr(III) to phosphorus distance of 5.9 ± 0.4 Å and Cr(III) to proton distances of 8.5 ± 0.6 and 9.6 ± 1.3 Å. The results indicated van der Waals contact between a phosphoryl oxygen of phosphoenolpyruvate (PEP) and the hydration sphere of the nucleotide-bound metal.

Gupta (30) has recently introduced a novel method for determining the distance between two paramagnetic species bound to a protein. This technique has been applied to pyruvate kinase where it is known that two divalent cations are needed (29). One site is a free divalent cation site and the other is a metal-nucleotide site (29). This technique is based on the assumption that two paramagnetic metal ions with similar electron resonance frequencies may affect the relaxation properties of each other via concerted spin flips. Gupta has used this technique to measure the distance between Mn^{2+} bound at the free divalent cation site and CrATP at the metal nucleotide site. CoATP was used as the diamagnetic control. In this case where one of the spins, such as Cr(III) in CrATP, has a relaxation time (τ_s^{Cr}) much shorter than that of the other spin such as Mn^{2+} (τ_s^{Mn}), and the rotational correlation time of the entire molecule, τ_R, is long compared to τ_s^{Cr}, the cross-relaxation effect shortens the effective spin lattice time of the slowly relaxing spin according to the following equation:

$$\left(\frac{1}{\tau_s^{Mn}}\right)_{Cr} - \left(\frac{1}{\tau_s^{Mn}}\right)_{Co} = \frac{2}{15} \frac{S^{Cr}(S^{Cr}+1)\gamma_{Cr}^2\gamma_{Mn}^2 h^2}{r^6} \cdot \tau_s^{Cr} \qquad 10.$$

S^{Cr} and τ_s^{Cr} are the net unpaired spin and the electron spin relaxation time of Cr(III). γ_{Cr} and γ_{Mn} are the electronic gyromagnetic ratios for the Cr^{3+} and Mn^{2+}, and r is the Mn^{2+} to Cr^{3+} distance. $\left(\frac{1}{\tau_s^{Mn}}\right)_{Co}$ is the electron spin relaxation time in the presence of the diamagnetic control, CoATP.

The relaxation times (τ_s^{Mn}) are estimated from the enhancement by Mn^{2+} on the longitudinal relaxation rate of water protons (ε^{Mn}) in the appropriate enzyme complex.

At 24.3 MHz with CrATP and CoATP, the enhancements, ε^{Mn}, by Mn^{2+} are 6 ± 1 and 12 ± 2, which calculates to a distance of 4.8–5.6 Å, depending on the number of assumed water molecules and using $\tau_s^{Cr} = (2.3\pm0.5)\times10^{-10}$ sec. From data also at 8 MHz the calculated Cr^{3+} to Mn^{2+} distance is 5.2 ± 0.9 Å, which indicates van der Waals contact between the hydration sphere of Cr^{3+} and Mn^{2+} in the pyruvate kinase $\cdot Mn^{2+} \cdot$ CrATP complex.

This technique should also be useful in other kinase reactions that require two divalent cations such as glutamine synthetase, carbamyl phosphate synthetase, and PEP carboxy kinase. The chief difficulty with this technique is knowing the exact number of fast exchanging water molecules in the enzyme complexes. However, because of the sixth-root dependence on the distance, this does not introduce unnecessarily high errors. Another important assumption is that the replacement of CrATP with CoATP has no other effect than replacing a paramagnetic complex with a diamagnetic one.

GLUTAMINE SYNTHETASE   The distance between two metal ions bound to a protein has also been measured using electron paramagnetic resonance (EPR). This experiment is based on the theory of dipolar electronic relaxation developed by Leigh (40). The theory predicts that if the EPR spectrum of enzyme-bound Mn^{2+} can be observed, the intensity of that signal will diminish upon addition of a second paramagnetic species, such as CrATP, without any change in lineshape. The magnitude of the diminution depends on the distance between the two metal ions. Balakrishnan & Villafranca (37) measured the effect of CrATP on the EPR spectrum of enzyme-bound Mn^{2+} in glutamine synthetase and determined the distance to be 7.1 ± 0.5 Å. The distance between the metal ions of glutamine synthetase was also estimated using the method of Gupta. A distance of 6 ± 1 Å was obtained (S. Ransom and J. J. Villafranca, unpublished results).

BOVINE HEART PROTEIN KINASE   Granot et al (27) have used $Co(NH_3)$ ATP to measure the conformation of ATP bound to protein kinase. They had previously determined that in the presence of ATP, protein kinase will bind two atoms of Mn^{2+}. One Mn^{2+} is at the ATP site and one is at an unknown site. In the presence of CoATP only one additional Mn^{2+} was bound, indicating the Co was occupying one site usually held by the other bound Mn^{2+} (24). Using bound Mn^{2+} as the paramagnetic center, the effect of Mn^{2+} on the longitudinal and transverse relaxation times of the proton and phosphorus atoms of CoATP was determined. The results indicated that the CoATP was bound very closely to the Mn^{2+}. The Mn^{2+} was 3.0 Å from each of the phosphorus atoms and 6.4–8.1 Å to the protons on the adenine ring and ribose. From the data gathered it was not possible to determine whether the adenine ring was in an *anti* or *syn* conformation.

On the basis of these data they have suggested the following structure

$$E \frac{}{} ATP\ Mg^{2+}$$
$$\diagdown \diagup$$
$$Mg^{2+}$$

which indicates that the extra metal is binding to both the enzyme and MgATP.

PRPP SYNTHETASE   PRPP synthetase catalyzes the synthesis of phosphoribosyl pyrophosphate (PRPP) from ATP and ribose-5-phosphate. This is an unusual reaction for ATP because of nucleophilic attack at the β-phosphorus. Li et al (41) had previously shown that the Δ isomer of β-γ CoATP is a good substrate for the enzyme. Li et al (42) have used tridentate CrATP to establish the substrate topography on the enzyme. From the paramagnetic effect of enzyme-bound CrATP on the longitudinal relaxation rates of the anomeric proton and phosphorus atoms of ribose 5-phosphate, they found that both isomers of ribose 5-phosphate bind near the CrATP and that H-1 of the β-isomer is 1.0 Å closer to bound CrATP than is the H-1 of the α-anomer. Therefore the 1-OH group of the α isomer is 1.6 Å closer than the 1-OH group of the β isomer. Assuming that the polyphosphate chain is between the Cr^{3+} and the ribose 5-phosphate, the β-P of ATP is 3.8±0.8 Å from the 1-OH of α-ribose 5-phosphate. This distance approximates the sum of the van der Waals radii and allows no intervening atoms. This is consistent with a direct nucleophilic attack and an associative mechanism as previously deduced from the inversion of configuration of the β-P of $Co(NH_3)ATP$ (41).

NMR STUDIES OF METAL NUCLEI

Background on Nuclei With Spin $>1/2$

The direct study of metal ions (other than Na^+) in biological systems by NMR techniques is a field in its adolescence. Table 2 presents the properties of a number of metal nuclei that either have been shown to have a direct biological function or can serve as substitutes for naturally

Table 2 Properties of nuclei[a]

Nucleus	Spin	Quadrupole moment	Frequency, MHz	Natural abundance	Relative sensitivity[b]
^{1}H	1/2	—	200.1	99.99	1.00
^{6}Li	1	0.00046	29.45	7.42	6.31×10^{-4}
^{7}Li	3/2	−0.042	77.77	92.58	2.71×10^{-1}
^{13}C	1/2	—	50.32	1.11	1.75×10^{-4}
^{23}Na	3/2	0.11	52.93	100	9.25×10^{-2}
^{25}Mg	5/2	0.22	12.25	10.13	2.70×10^{-4}
^{39}K	3/2	0.055	9.34	93.08	4.73×10^{-4}
^{43}Ca	7/2	0.06[c]	13.46	0.145	9.28×10^{-6}
^{63}Cu	3/2	−0.15	53.04	69.09	6.43×10^{-2}
^{67}Zn	5/2	0.18	12.52	4.11	1.17×10^{-4}
^{113}Cd	1/2	—	44.39	12.26	1.34×10^{-3}
^{133}Cs	7/2	−0.003	26.25	100	4.74×10^{-2}
^{139}La	7/2	0.21	28.27	99.91	5.91×10^{-2}

[a]From the *Handbook of Chemistry and Physics* (39).
[b]The sensitivity of a nucleus (from Reference 39) multiplied by the natural abundance and normalized to ^{1}H.
[c]Calculated from the T_1 data in Reference 43.

occurring metal ions. The Table includes NMR frequencies at constant field for a spectrometer operating at 200 MHz for ^{1}H. A number of companies offer spectrometers operating at this field strength (47 kG) that have multinuclear capabilities; thus many scientists will be able to study some or all of the nuclei listed in Table 2 in the near future. This section of the review explores practical and theoretical aspects of high resolution fourier transform NMR studies of metal ions and reviews some of the current literature.

One of the most important considerations in the design of NMR experiments is the sample (nuclei) concentration, as this directly determines the signal-to-noise (S/N) ratio that one can hope to achieve. All of our applications deal with metal ions binding to proteins. The solubility of a protein in a buffer at a certain pH, temperature, salt concentration, etc., is usually a limiting factor, since the practical range for protein concentrations in NMR experiments is 10^{-4}–10^{-2}M. Certain enzymes have been studied extensively by high resolution ^{1}H and ^{13}C NMR, in part because of their unusually high solubility and low molecular weight (e.g. lysozyme and ribonuclease). Solubility of a protein governs the S/N problem for a metal ion bound to a protein, since only 1–2 metal ions are usually bound per monomer. However studies of metal ion binding to proteins can be carried out by using excess metal ion and titrating with protein (for fast exchange conditions), thus permitting information on the bound metal ion to be evaluated from the data. However, direct observations of the bound metal ion is always desirable, since changes in chemical environment can be directly followed by changes in chemical shifts.

In Table 2 the relative natural abundance sensitivities of metal nuclei at constant field are compared to ^{1}H and ^{13}C. The comparison of sensitivities for various nuclei is presented in this way so that one can judge how practical a particular experiment would be (time-wise) to achieve the same S/N for the same "concentration" of material. This assumes that all other conditions are equal e.g. relaxation rates, nuclear Overhauser effect (NOE), instrumental conditions, quadrupole effects, chemical environment, exchange, etc., which is never the case. However the comparison provides an idealized starting point. Isotopic enrichment of ^{6}Li, ^{25}Mg, ^{43}Ca, ^{67}Zn, and ^{113}Cd will substantially improve the desired S/N conditions, and enriched samples of these nuclei are available commercially or from government laboratories.

Of the metal nuclei listed only ^{113}Cd has a nuclear spin of one half (other metal nuclei with a spin of one half, e.g. Pt, Hg, Sn, Pb, are not discussed). All the other nuclei have spin greater than one half, which means that they possess a nuclear quadrupole moment. These values are also listed in Table 1.

One example will suffice. For a sample of 0.05 M lithium formate in water, a single scan in a fourier transform spectrometer is all that is needed to observe the ^{1}H of formate and the ^{7}Li$^+$. Several thousand spectra would have to be averaged to achieve a spectrum with good S/N for ^{13}C and ^{6}Li$^+$ however. Incidentally, the time required for 1000 scans for the ^{13}C spectrum is roughly 4 hr since the T_1 is about 5 sec and the time between pulses should be $\sim 3 \times T_1$. However, the T_1 for aqueous samples of ^{6}Li$^+$ is ~ 170 sec at 30°C, and 1000 scans would take ~ 6 days (!) under the same conditions. Isotopic enrichment would seem to be a must for biological experiments with ^{6}Li.

Since most of the metal nuclei in Table 2 have quadrupole moments, a discussion of quadrupolar relaxation and its important aspects with regard to biological applications are presented before a discussion of each nucleus. An excellent compilation of physical properties of the alkali (44) and alkaline earth (45) ions has recently appeared, and the purpose of the current review is to stress applications to biomacromolecules.

Hertz has presented a detailed mathematical model for quadrupolar relaxation (33, 34). The magnetic relaxation of nuclei with spin greater than one half is caused by the interaction of their electric quadrupole moment with an electrical gradient at the nuclear site, which is the result of the local charge distribution. This charge distribution can be due to ions or dipoles or both in the first or outer hydration shells of metal ions. For solvated ions Hertz has presented a "fully random distribution model" formulated as follows

$$\frac{1}{T_{1(o)}} = \frac{1}{T_{2(o)}} = \frac{24\pi^3}{5} \cdot \frac{2I+3}{I^2(2I-1)} \left(\frac{PeQ(1+\gamma^\infty)}{h} \right)^2 \frac{M^2 \cdot C \cdot \tau}{r^5} \qquad 11.$$

The longitudinal ($1/T_1$) and transverse ($1/T_2$) relaxation rates are equal for infinite dilution, and definitions of the pertinent quantities for our discussion are: I, spin quantum number; eQ, nuclear quadrupole moment; and $(1+\gamma^\infty)$, the Sternheimer antishielding factor (47). A common misconception about nuclei with quadrupole moments is that they have excessively broad NMR spectra that will make high resolution experiments unobtainable. This fallacy can be overcome by examining the data for some ions. T_1 values in parenthesis for some aqueous ions at ~ 25°C are ^{7}Li$^+$ (25 sec), ^{6}Li$^+$ (170 sec), ^{23}Na$^+$ (0.07 sec), ^{133}Cs$^+$ (15 sec), ^{43}Ca^{+2} (1.3 sec), ^{25}Mg^{2+} (0.14 sec), ^{139}La^{3+} (0.003 sec) (44, 45, 60, 71, 72). The corresponding line widths (not considering instrumental broadening) are ^{7}Li$^+$ (0.01 Hz), ^{6}Li$^+$ (0.002 Hz), ^{23}Na$^+$ (4.5 Hz), ^{133}Cs$^+$ (0.02 Hz), ^{43}Ca^{2+} (0.25 Hz), ^{25}Mg^{2+} (2.3 Hz), ^{139}La^{3+} (106 Hz). Thus, with the exception ^{139}La^{3+}, narrow lines are expected in this limited selection and are actually observed for metal ions with I values from 1 to 7/2. The major structural difference between metal ions and compounds involving $^{14}N (I=1, Q=0.071)$ and $^{17}O (I=5/2, Q=0.004)$ is that metal ions have "ionic" bonding, whereas compounds of ^{14}N and ^{17}O are covalently bonded to other atoms. The paramagnetic effect of the p electrons involved in bonding produces a large asymmetric electric field gradient at the nucleus, making quadrupole relaxation quite efficient and hence the lines very broad (100–3000 Hz). Comparison of the spin quantum number, the Sternheimer antishielding factor, and the quadrupole moments will point out why narrow lines are observed for $^{133}Cs^+$ and broad lines for $^{139}La^{3+}$ in dilute ionic solutions. For both nuclei $I=7/2$, so this factor is not important. The Sternheimer antishielding factor is larger for Cs $[(1+\gamma^\infty) \simeq 104)]$ compared to La $[(1+\gamma^\infty) \simeq 66]$, and since this enters Equation 11 as a squared term this factor is ~2.5 times larger for Cs. However, the ratio of the quadrupole moments squared differs by about 5000, which is the dominant factor in the narrow lines widths for Cs^+ compared to La^{3+}.

For consideration of NMR experiments of metal ions with quadrupole moments bound to macromolecules the following expression holds:

$$1/T_{1,obs} = P_f(1/T_{1f}) + P_b(1/T_{1b}). \qquad 12.$$

This equation is valid for rapid chemical equilibrium between free (f) and bound (b) sites where P is the fraction of nuclei in each environment. An analogous expression obtains for T_2. Titration experiments of protein into metal ion solutions are usually performed for weakly binding ions, and both binding constants and $1/T_{1b}$ (or $1/T_{2b}$ or both) can be obtained. One can expect the symmetries of the free and bound states to be different, and hence their quadrupole coupling constants, since the free ion is hydrated and the bound ion is ligated by amino acid residues and perhaps also by water molecules. Ideally such titration experiments would reach an end point for the fully bound metal and this "bound" species could then be examined in detail. For the case of enzymes, which are of most interest to the authors, substrate and inhibitor binding at the metal ion site could then be explored to look for chemical shift differences in tightly bound metal ion-enzyme complexes. Examination of Table 2 again reveals that this may be practical, in the high resolution sense, for the quadrupole nuclei $^{43}Ca^{2+}$, $^{133}Cs^+$, $^{7}Li^+$, and $^{6}Li^+$. But only Ca^{2+} is known to bind very tightly to proteins and such experiments would have to be performed with enriched $^{43}Ca^{2+}$.

In addition to changes in the quadrupole coupling constant when a metal ion binds to a macromolecule, there is a drastic change in the correlation time, τ_c. This arises because in the hydrated ion τ_c is 10^{-11}–10^{-12} sec, and when the ion is bound to a protein τ_c is the tumbling time for the macromolecule (10^{-9}–10^{-8} sec for small proteins of $<60,000$ mol wt). Bull has treated this problem for the $I=3/2, 5/2$, and $7/2$ cases (10, 11). For these nuclei and nonextreme narrowing conditions (when $\omega\tau_c \sim 1.5$ as would be the case for many macromolecules), a single T_1 and T_2 value is still obtained since the magnetization decay is approximately exponential. For fast exchange conditions the difference in relaxation rates in the presence $(1/T_{i,\text{obs}})$ and absence of $(1/T_i)$ of macromolecules are

$$1/T_1 = 1/T_{1,\text{obs}} - P_f(1/T_{1f})$$

$$= \frac{3(2I+3)\tau_c}{40I^2(2I-1)} \cdot K^2 \cdot P_B \left(\frac{0.8}{1+4\omega^2\tau_c^2} + \frac{0.2}{1+\omega^2\tau_c^2} \right) \quad 13.$$

$$1/T_2 = 1/T_{2,\text{obs}} - P_f(1/T_{2f})$$

$$= \frac{3}{40} \frac{(2I+3)\tau_c}{I^2(2I-1)} \cdot K^2 \cdot P_B \left(0.3 + \frac{0.5}{1+\omega^2\tau_c^2} + \frac{0.2}{1+4\omega^2\tau_c^2} \right) \quad 14.$$

where K^2 is the quadrupole coupling constant. When $\omega^2\tau_c^2 \gtrsim 1$, τ_c can be obtained from a frequency dependence of T_1 or T_2 or from the T_1/T_2 ratio. The above equations are both valid for $I=3/2$; but for $I=5/2$ and $7/2$, $1/T_1$ is approximated by a single exponential for the case of two-site exchange, whereas $1/T_2$ is the sum of three and four exponential for $5/2$ and $7/2$ respectively. This could in practice produce one relatively narrow line and one very broad line or a non-Lorentzian line shape. This has been recently observed (S. Forsén, personal communication).

For $^{43}Ca^{2+}$ bound to a protein of $\sim 40,000$ mol wt with $\tau_c \sim 1.5 \times 10^{-8}$ sec, one can expect a line of the bound ion of ≥ 250 Hz. If all of the $^{43}Ca^{2+}$ is bound, one could observe this resonance with modern fourier transform spectrometers. However, if the relaxation is also modulated by chemical exchange (between several protein conformers, each having a different chemical environment), then the spectrum may be broadened beyond detection. This will be shown for some $^{113}Cd^{2+}$-proteins in a later section.

^7Li NMR Studies

There have been relatively few studies of this nucleus with biological macromolecules. ^7Li is the most abundant naturally occurring isotope; its narrow line width in aqueous solution and high sensitivity make it an ideal NMR probe. Li^+ has been shown to substitute for Na^+ or K^+ with many enzymes that require a monovalent cation for catalytic activity. Perhaps the best studied enzymes that are activated by monovalent cations are pyruvate kinase, the Na^+-K^+ ATPases, and the Ca^{2+}-ATPases.

Li$^+$ provides a low level of activation of pyruvate kinase and has a weak binding constant of ~11 mM. Two groups have studied the interaction of ^{7}Li$^+$ with this enzyme (3, 35). Hutton et al (35) conducted studies of the $1/T_1$ relaxation rates of ^{7}Li$^+$ interacting with enzyme in the presence of Mn^{2+} and the substrate phosphoenolpyruvate (PEP). They calculated distances between Mn^{2+} and ^{7}Li$^+$ based on paramagnetic enhancement of the $1/T_1$ rate of ^{7}Li$^+$ by enzyme-bound Mn^{2+}. Their distances (E-Mn^{2+}-Li$^+$, $r=11.0$ Å; E-Mn^{2+}-Li$^+$-PEP, $r=5.8$ Å) were longer than the corresponding Mn^{2+} to ^{205}Tl$^+$ distances (8.2 Å and 4.9 Å respectively) obtained by Kayne & Reuben (38, 59) who conducted ^{205}Tl$^+$-NMR ($I=1/2$) studies with this enzyme. Some of the discrepancies were clarified by Ash et al (3) who carried out ^{7}Li$^+$-NMR experiments with this same enzyme at 5 and 30°C. Previous EPR studies (58) showed that the pyruvate kinase-Mn^{2+}-Li$^+$-PEP complex exists as an equilibrium between two different enzyme conformers. Low temperature favors a complex characterized by a Mn^{2+}-Li$^+$ distance of 9 Å, whereas at high temperature this distance is 4.7 Å in agreement with Kayne & Reuben.

Recent ^{7}Li$^+$ data with Cr^{3+}-ATP as the paramagnetic probe show that the Li-Cr distance is 4.9 Å in the absence of Mg^{2+} (there is another divalent metal ion site in addition to the metal-nucleotide site) but changes to 6.0 Å with Mg^{2+} (68). The overall conclusion from these data is that in the active enzyme complex the divalent-monovalent cation distance is the same, whether the monovalent cation gives low or high activation; thus other factors than the spatial relationship between enzyme-bound cations must be responsible for the differences in enzyme activity. One can now see how powerful NMR studies can be for answering fundamental structural questions, especially when two nuclei (^{7}Li$^+$ and ^{205}Tl$^+$) can be studied.

Several other enzymes are in the preliminary stages of investigation by ^{7}Li$^+$-NMR. Very interesting studies on Na$^+$-K$^+$ATPase from kidney by Grisham & Hutton (28) suggest that Li$^+$ will occupy only the K$^+$ site at 10 mM Na$^+$. Using this observation, these authors studied the effect of Mn^{2+} and Cr^{3+}-ATP on the $1/T_1$ relaxation rates of ^{7}Li$^+$. A Mn^{2+}-Li$^+$ distance of 7.2 Å was computed in the presence of Na$^+$ but the Li$^+$-Cr^{3+} data were not analyzed. More recent studies with Ca^{2+}-ATPase, with Gd^{3+} as the paramagnetic probe, show two Gd^{3+} binding sites that are 7.0 and 9.1 Å from the Li$^+$ (K$^+$) binding site (65). A picture is thus emerging from these data showing that the activating monovalent cation site(s) are close to the active site for ATP hydrolysis.

Carbamoyl-phosphate synthetase is activated by divalent and monovalent cations in addition to having two ATP sites (see earlier section) (55, 57). In preliminary studies from our laboratory a Li$^+$-Mn^{2+} distance of 7.9 Å was calculated between the two cation activator sites

(F. M. Raushel and J. J. Villafranca, unpublished observations). Further studies are underway to establish the spatial relationship between these sites and the catalytic site.

In future work $^7Li^+$ relaxation rate studies should be conducted at two magnetic field strengths (for diamagnetic complexes) so that both τ_c and K^2 can be evaluated for the bound $^7Li^+$. In this way the change in quadrupole coupling constant can be used to evaluate certain aspects of the bonding at the monovalent cation binding site in macromolecules. This parameter should be sensitive to change in local environment when various other substrates or inhibitors bind; these data may then be correlated with the paramagnetic effects already observed.

If only one field strength is available but the spectrometer has the capability for observations of $^6Li^+$ and $^7Li^+$, then the T_1 ratio for these two nuclei will provide the ratio of the K^2 values; this could provide a parameter for studying conformational changes at the monovalent cation site of macromolecules, especially since τ_c is expected to stay the same for both bound $^6Li^+$ and $^7Li^+$.

Another nucleus that should receive more attention for biological studies is $^{133}Cs^+$. As can be seen in Table 2, the quadrupole moment is very small, the natural abundance is 100%, T_1 values are long, and the sensitivity quite reasonable. In practice, only a few scans are required per τ value for a T_1 experiment at a reasonable concentration of ions (10–50 mM) for biochemical studies (F. M. Raushel and J. J. Villafranca, unpublished observations). The crystal ionic radius of 1.67 Å suggests that it would be a good replacement for K^+ (1.33 Å), considering that Tl^+ (1.47 Å) has been used as a replacement for K^+ in pyruvate kinase.

Studies could of course be conducted with $^{23}Na^+$ and $^{39}K^+$ but would be more difficult because of excess broadening in the bound state.

^{43}Ca, ^{25}Mg, and ^{139}La Studies

Only a few studies on biological molecules have been reported with these nuclei. Robertson et al (62) reported $^{43}Ca^{2+}$ and $^{25}Mg^{2+}$ binding to γ-carboxy-glutamate containing peptides as models of proteins involved in the blood coagulation processes. This newly discovered amino acid is the proposed site for binding the Ca^{2+} and Mg^{2+} ions in the proteins of the blood-clotting process.

As expected the $^{25}Mg^{2+}$ line width broadened from a few Hz in free solution to ~160 Hz in the peptide complex. This is due to the drastic change in τ_c upon complexation and also perhaps to an increase in K^2.

For this study 98.25 atom % enriched $^{25}Mg^{2+}$ was used, and the authors report that from 1000-30,000 scans were acquired per spectrum in a titration experiment to achieve a S/N of 5-10 to 1.

When $^{43}Ca^{2+}$ (79.98 atom %) binding was studied with the peptide, the authors report a shift upon binding to the peptide with little broadening (~2 Hz). The line widths were sharp (as expected for the small Q value) and the reported shift was about 14 Hz for full complexation. No discussion is given of whether fast exchange prevails under these experimental conditions, however.

Lindman's laboratory has presented preliminary NMR data on $^{25}Mg^{2+}$ (13), $^{43}Ca^{2+}$ (54), and $^{113}Cd^{2+}$ (13, 23) binding to parvalbumin, a Ca^{2+}-binding protein of 11,500 mol wt. When parvalbumin was added to a 0.1 M solution of $^{43}Ca^{2+}$ (61.63 atom %), no significant variation of the line width (2-3 Hz) was observed up to 58°C. This is attributed to slow exchange of Ca^{2+} on the NMR time scale ($k_{off} \leq 10^3$ sec^{-1}). Broadening of the $^{43}Ca^{2+}$ signal was observed from 65 to 95°C (up to ~20 Hz) and the protein is known to undergo a disorganization of tertiary structure at these elevated temperatures. Rapid exchange conditions are thus present and the authors argue that exchange of metal ion from the native Ca^{2+}-binding sites is being monitored under these conditions. An increase in the $^{43}Ca^{2+}$ line width was also observed at 54°C when the pH was varied from 7 to 12. This phenomenon was reversible, but none of these studies allows a distinction between the two Ca^{2+}-binding sites.

$^{113}Cd^{2+}$ NMR studies of parvalbumin have been conducted by this group. These studies show two distinct $^{113}Cd^{2+}$ resonances, with parvalbumin at about -94 and -98 ppm. These signals of equal intensity are nearly invariant when the pH is changed from 7 to 9. An interesting competition experiment was performed in which Gd^{3+} was titrated into a solution of $^{113}Cd_2^{2+}$-parvalbumin. The two signals selectively and progressively disappear without broadening, which is consistent with the stepwise displacement of the two $^{113}Cd^{2+}$ ions.

Experiments with $^{25}Mg^{2+}$ binding to parvalbumin were not as definitive as those with $^{43}Ca^{2+}$ and $^{113}Cd^{2+}$. When Ca_2^{2+}-parvalbumin is titrated into a solution of $^{25}Mg^{2+}$ (0.1 M, 97.9 atom %), broadening of the 25 Mg^{2+} signal is observed. It is unclear whether Mg^{2+} is displacing the Ca^{2+}. Since the broadening implies that fast exchange conditions exist, it is possible that $^{25}Mg^{2+}$ is binding to ancillary metal ion sites on the protein. Further experimentation seems necessary. Overall, one can see from these experiments by Lindman's group that the use of multiple NMR nuclei in the study of metal ion binding to proteins is indeed a powerful technique.

Reuben has reported binding of $^{139}La^{3+}$ to bovine serum albumin (60). In all cases the line widths were broad and the T_1 values very short (61), as was expected for the Q value (Table 2) and large Sternheimer antishielding factor (47). Nonetheless, when $^{139}La^{3+}$ replacement for Ca^{2+} or Mg^{2+} is a desirable experiment, one may choose to study this nucleus to provide additional information on the metal ion binding site. The $^{139}La^{3+}$ binding study to bovine serum albumin was conducted at two magnetic field strengths and the τ_c value obtained, 2.25×10^{-8} sec, was in good agreement with the tumbling time of the macromolecule.

^{63}Cu and ^{67}Zn

No NMR studies of these nuclei have been reported for biomacromolecules. $^{67}Zn^{2+}$ should possess many of the NMR properties of $^{25}Mg^{2+}$ but would have to be obtained in enriched form for studies to be practical. The line widths of bound nuclei would be ⩾ 350 Hz but may be observable for some small proteins. ^{63}Cu could be studied in the diamagnetic states as Cu^+ or Cu^{3+} (with tetragonal distortions). Many small molecule complexes have been synthesized and could be studied with respect to their chemical shift differences (expected to be large due to the difference in charge); NMR could be used to evaluate whether some Cu^{2+}-containing proteins change oxidation state in their catalytic reactions (32) to Cu^+ or Cu^{3+}.

^{113}Cd NMR Studies

Since ^{113}Cd has a nuclear spin of one half it does not have a quadrupole moment. Recently many studies of $^{113}Cd^{2+}$ binding to macromolecules have appeared, mainly from the laboratories of Coleman at Yale (18) and Ellis at the University of South Carolina (4). $^{113}Cd^{2+}$ compounds have a shift of over 800 ppm which is attributable to distortions of the electron cloud of the filled orbitals and the paramagnetic contribution of these electrons to the "shielding" at the nucleus. Coleman's laboratory has observed a chemical shift range of ~200 ppm for $^{113}Cd^{2+}$ bound to enzymes where the protein ligands are nitrogen and/or oxygen.

In Figure 5A there is the spectrum of $^{113}Cd_2^{2+}$-alkaline phosphatase (14, 17) (2 Cd^{2+} bound to the dimer of 86,000 mol wt from *Escherichia coli*.) One line is observed (with a line width of ~50 Hz) at 170 ppm referenced to cadmium perchlorate. This is interpreted as both Cd^{2+} ions being in the same chemical environment in the protein dimer. Addition of one equivalent of phosphate to the protein gives the spectrum shown in Figure 5B. The $^{113}Cd^{2+}$-NMR spectrum now consists of two lines—one at 142 ppm and one at 55 ppm. The conclusion

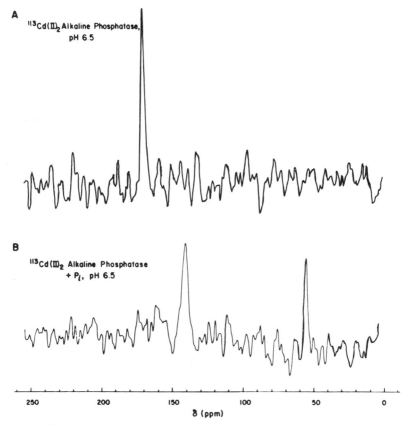

Figure 5 $^{113}Cd^{2+}$ NMR spectra of Cd_2^{2+}-alkaline phosphatase (*A*) and Cd_2^{2+}-phosphoryl alkaline phosphatase (*B*) (from Reference 14, with permission).

drawn from this study is that phosphorylation of one of the two active sites by P_i has altered the chemical environment of both subunits and thus of both metal ion sites. Since neither $^{113}Cd^{2+}$ resonance is at the same resonance position as in the sample with no P_i added, phosphorylation destroys the twofold symmetry of the subunit, and this is interpreted in terms of the negative cooperativity exhibited in catalysis by this enzyme.

An even more dramatic result (for a spectroscopist) is the observation of ^{31}P-^{113}Cd coupling in the ^{31}P spectrum of alkaline phosphatase (52). A doublet with ~30 Hz coupling constant is observed for the ^{31}P resonance assigned to the binary $E \cdot P_i$ complex (Figure 6). This coupling is found in the enzyme-Cd_4^{2+} complex (2 active site metal ions and two structural) and is proof that the metal ion at the active site is ligated to the P_i in an inner sphere complex. The overall catalytic scheme with alkaline phosphatase and P_i is

$$E + P_i \rightleftharpoons E \cdot P \rightleftharpoons E-P + H_2O$$

where $E \cdot P_i$ is a non-covalent complex and $E-P$ is covalently attached phosphate. The ^{31}P-^{113}Cd coupling is only seen for the ^{31}P resonance corresponding to $E \cdot P_i$ and these data suggest that when the covalent complex forms, the $^{113}Cd^{2+}$ is no longer binding to the phosphoryl group.

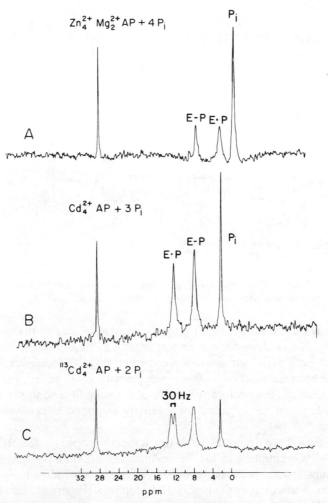

Figure 6 ^{31}P NMR spectrum of $^{113}Cd_4^{2+}$-alkaline phosphatase; conditions in Reference 52. The 30 Hz coupling constant in *C* is from $^{113}Cd^{2+}$ (enriched).

$^{113}Cd^{2+}$-NMR has also been applied to the study of carbonic anhydrase by Coleman's laboratory (18, 63). These data point out some of the unique advantages and disadvantages in studying metal ion-NMR spectra of protein-bound metal ions. One consequence of the large range of chemical shifts of $^{113}Cd^{2+}$ bound to macromolecules is that if the protein is fluctuating between various conformers at rates between 10^2-10^4 sec^{-1} and the chemical shift of the $^{113}Cd^{2+}$ in these

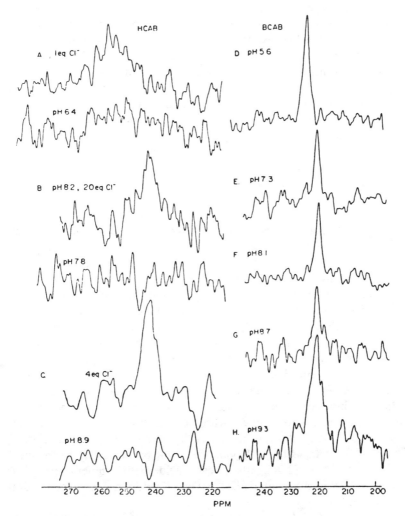

Figure 7 $^{113}Cd^{2+}$ NMR spectra of human carbonic anhydrase (HCAB) and bovine carbonic anhydrase (BCAB) (from Reference 18, with permission).

different states is 50–200 ppm, this corresponds to the slow or intermediate exchange condition encountered in NMR experiments. The result is severe broadening of the $^{113}Cd^{2+}$ resonance, which in practice leads to a lack of observation of the bound spectrum. Coleman has treated this problem and his calculations show that the populations of the two (or three) different protein conformers need not be equal in order to give broadening of peaks that are beyond detection in conventional spectrometers (17). This consideration along with the NOE and possibly long T_1's of $^{113}Cd^{2+}$ make quantitation of various observed $^{113}Cd^{2+}$ resonances problematical. Figure 7 shows the spectrum of human carbonic anhydrase B (HCAB) and bovine carbonic anhydrase B (BCAB) at various pH values and with and without Cl^- (63). The differences in line width and chemical shift are interpreted in terms of varying degrees of chemical exchange broadening between different conformational states of the enzyme, each with a different chemical shift. The near complete lack of spectrum for Cl^--free HCAB at all pH values is taken as an intermediate (10^3 sec^{-1}) chemical exchange process of a monodentate ligand (24) giving two discrete species. A broad resonance appears when the temperature is lowered, as is expected in going from an intermediate to slow exchange process. In BCAB this slow exchange process either is absent or is faster than the difference in chemical shift of various enzyme species in solution. Cl^- alters the exchange rate such that spectra are observed, but the line widths are still dependent on pH. This example shows how difficult it may be to observe the NMR spectrum of bound metal ions and that one should carry out many experiments before attempting to correlate chemical environments of the metal ion with catalytic functions. This field is fresh and new and many exciting results can be anticipated.

Ellis and coworkers (4, 53) have been studying $^{113}Cd^{2+}$ binding to concanavalin A and have obtained data showing a large difference in chemical shift between the S1 (46 ppm) and S2 (−125 ppm) metal ion sites of this protein. The original report assigned the resonance at 68 ppm to another protein site but this was later shown to be due to a cadmium chloride complex in solution. Competitive experiments with Zn^{2+} show that it can compete with Cd^{2+} at both the S1 and S2 sites whereas Ca^{2+} competes for binding at the S2 site. The chemical shift of $^{113}Cd^{2+}$ at S2 changes to −118 ppm when Zn^{2+} occupies S1. The chemical shift of $^{113}Cd^{2+}$ at S2 suggests a very shielded environment relative to $Cd(H_2O)_6^{2+}$; this site contains six oxygen ligands in a near octahedral environment.

Other studies of $^{113}Cd^{2+}$ substitution for Zn^{2+} in superoxide dismutase (1, 5) show that the $^{113}Cd^{2+}$ resonance can be observed when the copper at the Cd-Cu site (~6 Å apart) is reduced to Cu^+, but the $^{113}Cd^{2+}$ resonance cannot be observed when the copper is oxidized to Cu^{2+}. This suggests a paramagnetic broadening of the $^{113}Cd^{2+}$ by Cu^{2+} as expected for two ions at this distance. The $^{113}Cd^{2+}$ resonance in these samples has a line width of 27 Hz and a T_1 of 1.2 sec.

In conclusion, several metal ion nuclei have been studied with proteins. As spectrometers become more sensitive and more available to biophysicists and biochemists, we can expect to see studies of many other metal ion nuclei in very interesting biochemical systems. The purpose of this section of the review was to point out the properties of metal ion nuclei that either have not been previously studied with proteins or have been studied only in a few systems.

Literature Cited

1. Armitage, I. M., Schoot-Uiterkamp, A. J. M., Chelbowski, J. F., Coleman, J. E. 1978. *J. Magn. Reson.* 29:375–92
2. Armstrong, R. N., Konda, H., Granot, J., Kaiser, E. T., Mildvan, A. S. 1979. *Biochemistry* 18:1230–38
3. Ash, D. E., Kayne, F. J., Reed, G. H. 1978. *Arch. Biochem. Biophys.* 190:571–77
4. Bailey, D. B., Ellis, P. D., Cardin, A. D., Behnke, W. D. 1978. *J. Am. Chem. Soc.* 100:5236
5. Bailey, D. B., Ellis, P. D., Fee, J. A. 1980. *Biochemistry* 19:591–96
6. Balakrishnan, M. S., Sharp, T. R., Villafranca, J. J. 1978. *Biochem. Biophys. Res. Commun.* 85:991–98
7. Balakrishnan, M. S., Villafranca, J. J. 1978. *Biochemistry* 17:3531–38
8. Bock, J. L., Cohn, M. 1978. *J. Biol. Chem.* 253:4082–85
9. Bryant, F., Benkovic, S. J. 1979. *Biochemistry* 18:2825–28
10. Bull, T. E. 1972. *J. Magn. Reson.* 8:344–53
11. Bull, T. E., Forsen, S., Turner, D. L. 1979. *J. Chem. Phys.* 70:3106–11
12. Burgers, P. M. J., Eckstein, F. 1978. *Proc. Natl. Acad. Sci. USA* 75:4798–800
13. Cave, A., Parello, J., Drakenberg, T., Thulin, E., Lindman, B. 1979. *FEBS Lett.* 100:148–52
14. Chelbowski, J. F., Armitage, I. M., Coleman, J. E. 1977. *J. Biol. Chem.* 252:7053–61
15. Cohn, M., Hu, A. 1978. *Proc. Natl. Acad. Sci. USA* 75:200–3
16. Cohn, M., Hu, A. 1980. *J. Am. Chem. Soc.* 102:913–16
17. Coleman, J. E., Armitage, I. M., Chelbowski, J. F., Otvos, J. D., Schoot-Uiterkamp, A. J. M. 1979. In *Biophysical Applications of Magnetic Resonance*, ed. R. G. Shulman, pp. 345–96. New York: Academic. 450 pp.
18. Coleman, J. E. 1980. *Symp. Biophys. Physiol. Carbon Dioxide*, pp. 133–50. Berlin:Springer.
19. Cornelius, R. D., Hart, P. A., Cleland, W. W. 1977. *Inorg. Chem.* 16:2799–2805
20. Cornelius, R. D., Cleland, W. W. 1978. *Biochemistry* 17:3279–86
21. Dempsey, M. E., Boyer, P. D., Benson, E. S. 1963. *J. Biol. Chem.* 238:2708–15
22. DePamphilis, M. L., Cleland, W. W. 1973. *Biochemistry* 12:3714–24
23. Drakenberg, T., Lindman, B., Cave, A., Parello, J. 1978. *FEBS Lett.* 92:346–50
24. Dwek, R. A. 1973. *NMR in Biochemistry*. Oxford: Clarendon. 395 pp.
25. Eargle, D. H., Licko, V., Kenyon, G. L. 1977. *Anal. Biochem.* 81:186–95
26. Eckstein, F., Goody, R. S. 1976. *Biochemistry* 15:1685–91
27. Granot, J., Kondo, H., Armstrong, R. N., Mildvan, A. S., Kaiser, E. T. 1979. *Biochemistry* 18:2339–45
28. Grisham, C. M., Hutton, W. C. 1978. *Biochem. Biophys. Res. Commun.* 81:1406–11
29. Gupta, R. K., Fung, C. H., Mildvan, A. S. 1976. *J. Biol. Chem.* 251:2421–30
30. Gupta, R. K. 1977. *J. Biol. Chem.* 252:5183–85
31. Gupta, R. K., Benovic, J. L. 1978. *J. Biol. Chem.* 253:8878–86
32. Hamilton, G. A., Libby, D. B., Hartzell, C. R. 1973. *Biochem. Biophys.*

Res. Commun. 55:333–40
33. Hertz, H. G. 1973. *Ber. Bunsenges. Phys. Chem.* 77:531–40
34. Hertz, H. G. 1973. *Ber. Bunsenges. Phys. Chem.* 77:688–97
35. Hutton, W. C., Stephens, E. M., Grisham, C. M. 1977. *Arch. Biochem. Biophys.* 184:166–71
36. Jaffe, E. K., Cohn, M. 1978. *J. Biol. Chem.* 253:4823–25
37. Jaffe, E. K., Cohn, M. 1978. *Biochemistry* 17:652–57
38. Kayne, F. J., Reuben, J. 1970. *J. Am. Chem. Soc.* 92:220–22
39. Lee, K., Anderson, W. A. 1972. In *Handbook of Chemistry and Physics*, ed. R. C. Weast 52:E57–61. Cleveland, Ohio: Chem. Rubber Co.
40. Leigh, J. S. 1970. *J. Chem. Phys.* 52:2608–12
41. Li, T. M., Mildvan, A. S., Switzer, R. L. 1978. *J. Biol. Chem.* 253:3918–23
42. Li, T. M., Switzer, R. L., Mildvan, A. S. 1979. *Arch. Biochem. Biophys.* 193:1–13
43. Lindman, B., Forsén, S., Lilja, H. 1977. *Chem. Scr.* 11:91–92
44. Lindman, B., Forsén, S. 1978. In *NMR and the Periodic Table*, ed. R. K. Harris, B. E. Mann, pp. 129–81. London/New York/San Fransisco: Academic. 459 pp.
45. Lindman, B., Forsén, S. 1978. See Ref. 44, pp. 183–94
46. Lowe, G., Sproat, B. S. 1978. *J. Chem. Soc. Perkin Trans.* 1:1622–30
47. Lucken, E. A. C. 1969. *Nuclear Quadrupole Coupling Constants.* New York: Academic. 450 pp.
48. Lutz, O., Nolle, A., Staschewski, D. 1978. *Z. Naturforsch. Teil A* 33:380–82
49. Merritt, E. A., Sundaralingam, M., Cornelius, R. D., Cleland, W. W. 1978. *Biochemistry* 17:3274–78
50. Midelfort, C. F., Rose, I. A. 1976. *J. Biol. Chem.* 251:5881–87
51. Midelfort, C. F., Sarton-Miller, I. 1978. *J. Biol. Chem.* 253:7127–29
52. Otvos, J. D., Alger, J. R., Coleman, J. E., Armitage, I. M. 1979. *J. Biol. Chem.* 259:1778–80
53. Palmer, A. R., Ellis, P. D., Behnke, W. D., Bailey, D. B., Cardin, A. D. 1979. Submitted for publication
54. Parello, J., Lilja, H., Cave, A., Lindman, B. 1978. *FEBS Lett.* 87:191–95
55. Raushel, F. M., Anderson, P. M., Villafranca, J. J. 1978. *Biochemistry* 17:5587–91
56. Raushel, F. M., Villafranca, J. J. 1979. *Biochemistry* 18:3424–29
57. Raushel, F. M., Rawding, C., Anderson, P. M., Villafranca, J. J. 1979. *Biochemistry* 18: 5562–66
58. Reed, G. H., Cohn, M. 1973. *J. Biol. Chem.* 248:6436–42
59. Reuben, J., Kayne, F. J. 1971. *J. Biol. Chem.* 246:6227–34
60. Reuben, J. 1975. *J. Am. Chem. Soc.* 97:3823–24
61. Reuben, J. 1975. *J. Phys. Chem.* 79:2145–57
62. Robertson, P. Jr., Hishy, R. G., Koehler, K. A. 1978. *J. Biol. Chem.* 253:5880–83
63. Schoot-Uiterkamp, A. J. M., Armitage, I. M., Coleman, J. E. 1980. *J. Biol Chem.* 260: In press
64. Shue, K. F. R., Frey, P. A. 1977. *J. Biol. Chem.* 252:445–48
65. Stephens, E. M., Grisham, C. M. 1979. *Biochemistry* 18:4876–85
66. Stokes, B. O., Boyer, P. D. 1976. *J. Biol. Chem.* 251:5558–64
67. Tsai, M. D. 1979. *Biochemistry* 18:1468–72
68. Van Divender, J. M. 1979. *Pyruvate kinase: A structural study using ^{7}Li and water proton nuclear magnetic resonance.* MS Thesis, Univ. Virginia, Va. 76 pp.
69. Van Etten, R. L., Risley, J. M. 1978. *Proc. Natl. Acad. Sci. USA* 75:4784–87
70. Webb, M. R., McDonald, G. G., Trentham, D. R. 1978. *J. Biol. Chem.* 253:2908–11
71. Wehrli, F. W. 1976. *J. Magn. Reson.* 23:527–32
72. Wehrli, F. W. 1977. *J. Magn. Reson.* 25:575–80
73. Wimmer, M. J., Powers, S. G., Meister, A., Rose, I. A. 1979. *J. Biol. Chem.* 254:1854–59

MACHINE-ASSISTED PATTERN CLASSIFICATION IN MEDICINE AND BIOLOGY

♦9152

Ching-Chung Li[1]

Department of Electrical Engineering, University of Pittsburgh, Pittsburgh, Pennsylvania 15261

King-Sun Fu[1]

School of Electrical Engineering, Purdue University, West Lafayette, Indiana 47907

INTRODUCTION

The use of computer methods in classifying biomedical information covers a very broad spectrum. The artificial intelligence approach has been applied to computer-based medical consultations in which quantitative analyses and experts' medical knowledge are incorporated. Several systems have been developed, such as INTERNIST, MYCIN, PUFF, DIG, CASNET/glaucoma, HELP, CONSULT, etc., which provide advice to physicians for medical decision-making with the ultimate goal of helping to formulate a plan of patient treatment in each case (86, 87, 91, 104, 124, 127). Although these systems involve computer-aided interpretation of the patient's disease state, they are not within the scope of this review. The subject of interest here is restricted to the computer-aided pattern classification of a set of physiological measurements or medical images. Computerized methods are desirable for providing quantitative information, consistency and objectivity in classification, and speed. They may also enable the early detection of abnormalities which physicians alone would find difficult to recognize.

---

[1] Work was supported by National Science Foundation grants ENG 75-14852 A02 to C. C. Li and ENG 78-16970 to K. S. Fu.

During the past decade, a tremendous amount of work has been done in the area of biomedical pattern classification. This includes machine-assisted classifications of discrete measurements (e.g. pulmonary function test data), physiological signals (e.g. EEG, EMG, ECG, blood pressure wave, and hormonal signals), and medical images (e.g. radiographs, ultrasound images, microscopic images of cells and tissues, and computerized tomography (CT) images). These are briefly reviewed in the following sections.

Classification of Physiological Measurements

Pattern analyses of patients' pulmonary function test data have been recently reported (42, 66, 117). Based upon interrelated spirometric data including forced vital capacity, forced expiratory volume in one second, and forced expiratory flow rates at various percentages of vital capacity, etc., the Karhunen-Loeve expansion was used to evaluate the relative discriminatory powers of these pulmonary function indices. A Bayesian classifier and a linear discriminant function classifier were trained to classify measured data into categories of normal, obstructive lung disease, and restrictive lung disease. The very short time series of circadian patterns of adrenocortical hormones in human blood plasma, sampled every half hour and analyzed by radioimmunoassay, have been studied in both normal subjects and patients with Cushing's syndrome (120, 129). The differentiation in their circadian patterns can help confirm the diagnosis of Cushing's disease.

Waveforms of EEG, EMG, ECG, and blood pressure, which can be easily measured, have been extensively studied with regard to noise suppression, feature extraction, and classification strategy. The results of these investigations have been reported in hundreds of references. Recent reviews for EEG analysis have been included in References (10, 22, 36, 103, 123) where power spectral analysis and transient analysis were discussed in connection with automatic recognition of neurological disorders (epilepsy and schizophrenia), detection of brain tumors, real-time detection of brain events, and assessment of sleep stages. Pattern recognition of multiple EMG's for a description of human gait was reported in Reference (8). ECG analysis has been reviewed in References (59, 90, 121, 122) where identification of P, QRS, and ST waves and analysis of intrinsic components of multiple lead ECG's were discussed in relation to cardiac arrhythmia, myocardial infarction, etc. Applications of the newly developed syntactic approach to pattern recognition of events in EEG, of QRS waves in ECG, and of structural variations in carotid pulse waves have been described in References (9, 35, 51, 109).

Classification of Medical Images

In hospitals and laboratories around the world, millions of medical images taken each year demand a tremendous amount of time for examination by medical specialists. New imaging systems are also being developed and constantly improved to enhance the accuracy of diagnosis. Together, these have provided the impetus for the development of automated image analysis and classification in medicine. Some major efforts in this area prior to 1976 were reviewed in a book edited by Preston & Onoe (95); an additional review on automated analysis in cytology by Preston (94) and another on radiographic image processing by Harlow et al (48) provided two excellent supplements. In the latter, feasibility studies on computer classification of rheumatic heart disease and congenital heart disease from chest radiographs were well summarized.

Since that time, much progress has appeared in this field. We review in separate sections the recent advances in computer analysis of chest images and blood cell images. New research has been carried out on automated classification of lung tissue (115), tissues of liver and kidney (96), and cervical smears (13, 68, 111), regarding segmentation of components, feature extraction, and classification schemes for identification of abnormalities. Improvement in the automatic detection of microcalcifications in mammograms has been reported (106). Computerized measurement of contrast indices has demonstrated their effectiveness in detecting small lesions in liver scintillograms (27). Video densitometric studies of pulmonary perfusion measurements are in progress (12); computer analysis of properly aligned lung perfusion image and ventilation image has shown promise in classifying pulmonary embolism, parenchymal disease, and normal lung (83).

Early detection and quantification of atherosclerosis has been a subject of research for a number of years. Methods of computer grading of arterial wall irregularity have recently been developed, including measurements of vessel edge roughness, local width, and local contrast density change (21). The result of these computer measurements from femoral angiograms shows a fairly high correlation with cholesterol content of the arterial specimens. A pilot study was carried out to apply this quantitative method to some coronary segments selected interactively from cineangiograms (102). Automatic determination of the left ventricular contour and volume, as well as analysis of its shape and contraction patterns from cineangiograms, has been investigated in many laboratories (50, 101); and the classification of normal and abnormal left ventricular wall motion has been studied (38).

Advances in computerized tomography (X-ray transmission, radionuclide emission, nuclear magnetic resonance imaging, and ultrasound) have added a new dimension to disease classification through so-called noninvasive vivisection (32, 130). The latter provides an effective means for detecting small tumors in brain or other organs and for examination of coronary diseases. The development of methodology for three-dimensional image reconstruction, object segmentation, surface determination, and display in four dimensions (x,y,z,t) is an integral part of the automatic classification problem of three-dimensional medical images.

The report of a Dahlem workshop in 1979 on biomedical pattern recognition and image processing discusses many of the above-mentioned topics (32). The scope of this paper, however, is limited to the description of recent progress in the automated classification of nucleated blood cells and of lung images in chest radiographs.

PATTERN RECOGNITION METHODOLOGY

An automatic pattern classification system for medical images generally consists of four components as shown in Figure 1. The input medical images are two-dimensional pictures (for example, chest radiographs or microscopic images of blood smears) that must be digitized into $n \times m$ array of pixels (picture elements), each of which may have one of N optical intensity levels (or gray levels). The digitizer may be a TV scanner, a flying spot scanner, or a drum scanner that possesses the desired spatial resolution and optical density resolution. However, the digitization step is not needed if medical images are generated directly in digital form, for example, images obtained from CT and computerized electronic radiography. The original data of a digitized image are usually preprocessed to enhance the image quality. This may include histogram equalization, noise suppression (low-pass filtering), and edge enhancement (high-emphasis filtering). The preprocessed image is then segmented into regions of interest and the boundary of the individual object is obtained (for example, lung boundary in a chest radiograph, or cytoplasm and nucleus boundaries in a blood cell image). Measurements are then made on the object or regions of interest, including object size and shape, optical intensity parameters, and texture parameters in the regions. These parameters characterize essential properties of the objects or regions under consideration and may be used for their classification. Very often, many measurements are made initially; some of the measured parameters may contain redundant information and, hence, feature selection is performed to consider only a subset of the measured features to be used for classification. In other instances,

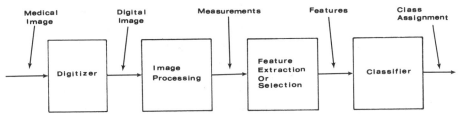

Figure 1 Block diagram of a medical image pattern recognition system.

feature extraction is performed by applying a certain mathematical transformation to the measured parameters and choosing a subset of effective features after the transformation. The selected features form pattern descriptors and are the input to the classification component which makes the decision on class assignment of the object or image under consideration.

The above description refers to the statistical pattern recognition system commonly in use (33). Another type is the syntactic pattern recognition system (30) whose application to biomedical pattern classification problems has only recently begun. It will not be included in the following review because of the space limitation.

Preprocessing

Depending upon the image quality and processing requirement, one or more steps of preprocessing may be desirable before image segmentation takes place. The original digitized image is usually at a high spatial resolution. For initial segmentation, a coarse outline of an object of interest may be obtained at a lower spatial resolution in order to save computation effort. Image consolidation can be performed to provide such an option. For example, a digitized chest radiograph of 1024×1024 array of pixels may be consolidated into 256×256 array of pixels by replacing every nonoverlapping 4×4 square subarray of pixels in the original image by one pixel with the mean gray level in that subarray.

Histogram equalization is another preprocessing step commonly performed for normalizing the image contrast and for facilitating the computation of statistical texture parameters. Let a graph $h(z)$ represent the histogram of a given image, which is defined as the relative frequency with which any gray level z occurs in the image. Many raw images show a gray level histogram highly skewed toward the darker levels, such that details in the darker region of the image are often not visible. Histogram equalization by equal probability quantizing is a process to rescale the image so that the new histogram becomes uniform (44). The number of new quantizing levels is usually reduced from N to

K, where $K = N/2^r$ and r is a positive integer. This often simplifies the computation of texture parameters since it greatly reduces the effort in computing the joint probability of a pair of gray levels (z_i, z_j) occurring at two pixels separated by a distance d; this joint probability is needed in computing texture measures by the spatial gray level dependence method. Equal probability quantizing follows a monotonic transformation of image intensity; hence, the texture measures of the image are invariant to this transformation. In fact, it has been shown that digital image contrast is normalized and a "near optimal" way is provided to reduce the number of gray levels in an image and yet still retain an accurate representation of the original image (17). Therefore, the differences in image contrast of different radiographs caused, for example, by slight differences in exposure time, development temperature, and development time may be minimized. Equal probability quantizing has been used in texture measures of both radiologic images and nucleated blood cell images.

For low contrast images, enhancement is desirable before any further processing (92). High-emphasis filtering is used to enhance high spatial frequency components and to depress low frequency components so that edges of objects which have predominantly high frequency components will be more readily detectable by a gradient algorithm. Median filtering can be used for noise suppression in images (52). Zonal notch filtering may be used to preserve edges between regions of widely differing average gray levels (100).

For detection of edge elements (pixels), a gradient operation followed by a threshold operation may be used. A digital approximation of the gradient operation can be given by convolving the image array with a set of eight 3×3 compass gradient masks and by taking the maximum convolution (92). These masks represent eight possible quantized gradient directions separated successively by 45°. The direction of the mask giving the maximum convolution is the quantized gradient direction at the center pixel of the 3×3 array. Edge elements are detected by thresholding the result of the gradient operation. Different types of linear or nonlinear edge enhancement can be performed by choosing appropriate operators before thresholding. Edge element extraction may also be approached as an edge-fitting problem. With a Hueckel operator one can attempt to find the best approximation of the image data in a circular neighborhood by an ideal step edge in a certain direction and of a certain contrast; if the approximation is sufficiently close, an edge is obtained (53). Persoon's operator is defined in a window of 5×5 pixels where gray levels in the two columns to the left and to the right of the central one are approximated by linear functions; both edge measure and edge direction are evaluated by maximizing a modified derivative over eight quantized directions (89).

Image Segmentation

Image segmentation is the process of dividing a given image into different regions, each of which has certain specific properties, so that objects of interest may be located. In constructing an algorithm to extract the object, prior knowledge about the object and image should be utilized. Most of the available segmentation techniques fall into the following three categories: characteristic feature thresholding or clustering, edge detection, and region growing. Thresholding or clustering techniques are based on the detection of similar properties within individual regions. Edge detection techniques are based on the detection of discontinuity or any sharp change in gray level between two neighboring regions. Region-growing techniques are based on a set of growth rules applied to atomic regions initially obtained according to certain similar properties of neighboring pixels; procedures of region merging, region dividing, or a combination of both are used to extract the regions of interest (88). The first two categories of segmentation techniques have been widely used in biomedical image-processing problems.

Characteristic feature (e.g. gray level) thresholding is a frequently used technique for segmentation of simple images. The global histogram of an image, if bimodal, is smoothed first and a gray level threshold is chosen at the valley between two peaks. Pixels with gray levels higher than the threshold are considered to belong to one group, and those lower than the threshold to belong to another group. Often it may be difficult to accurately locate the minimum point; thus, various techniques have been developed to sharpen the valley between two modes. If the background is not quite uniform, it is necessary to adapt the threshold to the local mean gray level; this may be achieved by subdividing the image into small zones and determining the local threshold for each zone. Techniques for local thresholding and adaptive thresholding are reviewed in Reference (128).

Characteristic feature clustering is another segmentation technique in practical use. Typically, two or more characteristic features (for example, histograms of multispectral images, and simple texture measures defined on a small window centered at each pixel) are used, and each class of regions is assumed to form a distinct cluster in the space of these characteristic features (14, 45). An appropriate clustering method is used to group the points in the characteristic feature space. These clusters are then mapped back to the original spatial domain to produce a segmentation of the image. A clustering procedure on the bivariate "color-density" histogram has been used in the segmentation of nucleated blood cell images and is discussed later (78).

Edge detection is another commonly used segmentation technique by which outlines of objects within an image are obtained. A boundary or edge usually follows a closed contour or some structured arc segment that can be determined from the edge elements extracted during preprocessing. Portions of the extracted edge elements may be generated by noise in the image and hence are not true edge elements. On the other hand, some true edge pixels may not be found because the corresponding gradient magnitudes or edge measures are too weak to be detected. Boundary formation (edge elements combination) consists of eliminating false edge elements, merging the edge elements into streaks, linking streaks into boundaries, and, finally, eliminating false boundaries. This can be achieved by heuristic search (72), dynamic programming (75), relaxation (98), or curve fitting (18, 23) techniques to optimize a certain edge evaluation function representing the heuristic information about the boundary. Search techniques and a combination of a search technique and a curve-fitting technique have been used in the automatic determination of lung and tumor boundaries in chest radiographs (5).

Feature Measurement

Various measurements can be made on the extracted objects or regions in an image. The measured quantities may be grouped into geometric parameters, optical density (or gray level) parameters, texture parameters, etc. The geometric parameters of an object include its size and shape. Size parameters are usually area (A), perimeter (p), average length, and average width of an object. Shape parameters can be defined in a variety of ways: for example, shape factor $p^2/4\pi A$, elongation index, circularity measure, and specularity measure. The chain code of an object boundary may be conveniently used to define certain shape parameters. When an object boundary is traced in a clockwise (or counter clockwise) direction, the direction code of each line segment connecting two adjacent boundary pixels is given by one of the octal numbers representing eight quantized directions. A high peak of the histogram of boundary chain code indicates straightness of boundary segments. Variations of the smoothed boundary chain divide it into several segments, each of which corresponds to a concave or convex region; several shape parameters can be derived from the number and length of those concave segments and have been used in the classification of nucleated blood cell images and of candidate nodules in lung images. Fourier descriptors of the boundary may also be utilized.

The optical density or gray level parameters include the integrated optical density of an object, average optical density, standard deviation, and coefficient of variation of the optical density in an object. Several indices have been defined for gray level contrast of a small object from the background. Each object is completely surrounded by its outside edge pixels. The outside edge pixels are adjacent to object member pixels but are not member pixels themselves. The boundary pixels are a part of the object member pixels. The isolation contrast integral is defined as the summation of differences between the minimum gray level of all the outside edge pixels and the gray level of each member pixel (99). This feature has been used in the recognition of pneumoconiosis opacities. The area contrast is defined as the difference between the average gray level of the outside edge pixels and the average gray level inside the object. The contrast depth is the difference between the average gray level of the outside edge pixels and the minimum gray level of the object member pixels. The edge contrast is defined as follows (27). Consider a 3×3 neighborhood for a boundary pixel; some elements in the neighborhood are outside edge pixels and others are object member pixels. Let us compute the difference between the average gray level of the outside edge pixels in this neighborhood and the average gray level of the object boundary pixels in the neighborhood; this difference is normalized by the sum of these two average gray levels to give a contrast measure for this boundary pixel. The edge contrast of the object is the average of such contrast measures over all its boundary pixels.

Within each segmented region, a first-order histogram of gray levels and a histogram of gradient directions may provide some discriminatory information. Let $h(z)$ be the first-order histogram or probability distribution, with z varying in discrete levels from 0 to $N-1$. Four parameters are commonly measured for each histogram. They are mean $m = \Sigma z h(z)$, variance $\sigma^2 = \Sigma(z-m)^2 h(z)$, skewness $\gamma = [\Sigma(z-m)^3 h(z)]/\sigma^3$, and kurtosis $K = [\Sigma(z-m)^4 h(z)]/\sigma^4$, where skewness measures the departure from symmetry and kurtosis measures the tendency of the distribution to either cluster about the mean or spread out towards the sides.

The image texture is related to two-dimensional arrays of optical density variations. Coarseness or fineness is related to the spatial repetition period of the local structure and, hence, texture is a neighborhood property of an image point. Various models for texture have been proposed, and both statistical and syntactic texture analyses have been developed (46, 58, 71). For statistical texture analyses of medical images, the spatial gray level dependence method (44) and the gray level run length method (34) have been applied (16, 24, 60, 80, 93, 118). The spatial gray level dependence method has been found to be the most effective in classification of radiographic images (17). It is based on the joint probability function $p(i,j,d,\theta)$ of the pairs of gray levels (i,j) that occur at pairs of pixels separated by an intersampling distance d radial units at an angle θ with respect to the horizontal axis. This second-order discrete probability function can be described by a gray level co-occurrence matrix for each pair of (d,θ). It is obtained by normalizing the two-dimensional histogram of pairs of pixels. It will tend to be skewed toward the diagonal probability matrix if an image region contains coarse texture, and it will be more uniform if the region contains fine texture. In order to obtain statistical confidence in estimating the joint probability function, the two-dimensional histogram must contain a reasonably large average occupancy level. One way to achieve this is to reduce the number of quantization levels of grayness, for example, from $N=256$ levels to $K=16$ levels. This may be done by equal probability quantizing as discussed previously; such processing will, however, result in a loss of accuracy in the measurement of low gray level texture. Eight texture measures based on gray level co-occurrence matrix have been defined as given below:

$$T_1(d,\theta) = \sum_{i=0}^{K-1} \sum_{j=0}^{K-1} [p(i,j,d,\theta)]^2 \quad \text{(energy)};$$

$$T_2(d,\theta) = \sum_{i=0}^{K-1} \sum_{j=0}^{K-1} (i-j)^2 p(i,j,d,\theta) \quad \text{(inertia or contrast)};$$

$$T_3(d,\theta) = \frac{1}{\sigma_1 \sigma_2} \sum_{i=0}^{K-1} \sum_{j=0}^{K-1} (i-\mu_1)(j-\mu_2) p(i,j,d,\theta) \quad \text{(correlation)}$$

where

$$\mu_1 = \sum_{i=0}^{K-1} i \left[\sum_{j=0}^{K-1} p(i,j,d,\theta) \right]$$

$$\mu_2 = \sum_{j=0}^{K-1} j \left[\sum_{i=0}^{K-1} p(i,j,d,\theta) \right]$$

$$\sigma_1^2 = \sum_{i=0}^{K-1} (i-\mu_1)^2 \left[\sum_{j=0}^{K-1} p(i,j,d,\theta) \right]$$

$$\sigma_2^2 = \sum_{j=0}^{K-1} (j-\mu_2)^2 \left[\sum_{i=0}^{K-1} p(i,j,d,\theta) \right];$$

$$T_4(d,\theta) = -\sum_{i=0}^{K-1} \sum_{j=0}^{K-1} p(i,j,d,\theta) \{\log p(i,j,d,\theta)\} \quad \text{(entropy)};$$

$$T_5(d,\theta) = \sum_{i=0}^{K-1} \sum_{j=0}^{K-1} \frac{1}{1+(i-j)^2} p(i,j,d,\theta) \quad \text{(inverse difference moment or local inhomogeneity)};$$

$$T_6(d,\theta) = \sum_{l=0}^{K-1} l \cdot p_{ad}(l,d,\theta) \quad \text{(mean of absolute difference)}$$

where

$$p_{ad}(l,d,\theta) = \sum_{i=0}^{K-1} \sum_{\substack{j=i\pm l \\ j>0}} p(i,j,d,\theta) = \text{probability density function of absolute difference } |i-j|;$$

$$T_7(d,\theta) = - \sum_{l=0}^{K-1} p_{ad}(l,d,\theta)\{\log p_{ad}(l,d,\theta)\}. \quad \text{(absolute difference entropy)};$$

and

$$T_8(d,\theta) = \sum_{l=0}^{K-1} [p_{ad}(l,d,\theta)]^2 \quad \text{(angular second moment of absolute difference)}.$$

Usually, four distances ($d = 1, 2, 4, 8$) and four directions ($\theta = 0°, 45°, 90°, 135°$) are considered. The distance of diagonal elements is approximately equal to d. Hence, if the first pixel in the pair is at the location $(0,0)$, the second pixel along $\theta = 45°$ is located at $(1,1)$ for $d=1$, at $(2,2)$ for $d=2$, at $(3,3)$ for $d=4$, and at $(6,6)$ for $d=8$. Similarly, the second pixel along $\theta = 135°$ is located at $(-1,1)$, $(-2,2)$, $(-3,3)$, and $(-6,6)$ respectively for $d=1, 2, 4,$ and 8. If a texture measure of interest is angular invariant, the mean texture measure averaged over the four directions can be used as a texture feature. These textural features have been used in classifications of lung images for pulmonary diseases (60, 118).

The gray level run length method of texture analysis considers histograms of run length of individual gray levels along a certain direction. A run length is defined as the number of members of a set of consecutive, colinear pixels having the same gray level. Coarse texture tends to create long runs, whereas fine texture gives short runs. Let $P(i,j,\theta)$ denote the number of times that the image contains a run of length j, in the θ direction, consisting of pixels having gray level i. A gray level run length matrix, having $P(i,j,\theta)$ as its (i,j)th element, is obtained for each θ, where $\theta = 0°, 45°, 90°,$ and $135°$. Let M be the maximum run length that may occur. Five texture measures based on gray level run length have been used:

$$T_{r1}(\theta) = \left\{ \sum_{i=0}^{K-1} \sum_{j=1}^{M} \frac{P(i,j,\theta)}{j^2} \right\} \bigg/ \left\{ \sum_{i=0}^{K-1} \sum_{j=1}^{M} P(i,j,\theta) \right\} \quad \text{(short run emphasis)};$$

$$T_{r2}(\theta) = \left\{ \sum_{i=0}^{K-1} \sum_{j=1}^{M} j^2 P(i,j,\theta) \right\} \bigg/ \left\{ \sum_{i=0}^{K-1} \sum_{j=1}^{M} P(i,j,\theta) \right\} \quad \text{(long run emphasis)};$$

$$T_{r3}(\theta) = \left\{ \sum_{i=0}^{K-1} \left[\sum_{j=1}^{M} P(i,j,\theta) \right]^2 \right\} \bigg/ \left\{ \sum_{i=0}^{K-1} \sum_{j=1}^{M} P(i,j,\theta) \right\} \quad \text{(gray level nonuniformity)};$$

$$T_{r4}(\theta) = \left\{ \sum_{j=1}^{M} \left[\sum_{i=0}^{K-1} P(i,j,\theta)^2 \right] \right\} \bigg/ \left\{ \sum_{i=0}^{K-1} \sum_{j=1}^{M} P(i,j,\theta) \right\} \quad \text{(run length nonuniformity)};$$

and

$$T_{r5}(\theta) = \left\{ \sum_{i=0}^{K-1} \sum_{j=1}^{M} P(i,j,\theta) \right\} \bigg/ A \qquad \text{(run percentage)}$$

where $A =$ area of the region.

Again, each of these texture measures can be averaged over four directions if it is angular invariant. The texture features derived from both gray level co-occurrence matrices and gray level run length matrices have been used in the classification of nucleated blood cell images (80).

Feature Selection and Feature Extraction

In pattern analysis during the design stage of a classifier, a large number of features are often measured, some of which may be correlated and may contain redundant information. In actual operation of a pattern classifier, it is desirable to use only those features which are most effective for discrimination of pattern classes. Suppose that there are L measures $(x_1, x_2, \ldots, x_L)$ from training samples of known classes. Feature selection means that a subset of $(x_1, \ldots, x_L)$ is selected according to a certain evaluation criterion. On the other hand, feature extraction is a process in which m features $(y_1, y_2, \ldots, y_m)$ are generated from $(x_1, \ldots, x_L)$ by some sort of transformation $h_i(\cdot)$, $(i = 1, 2, \ldots, m; m < L)$, i.e. $y_i = h_i(x_1, \ldots, x_L)$ for the purpose of reducing the dimensionality of feature space and achieving effective classifications. Many different methods, of which a few are mentioned below, have been developed for feature extractions (28, 33, 114, 132). Clustering transformation and feature ordering can be used for maximizing the interset distance between pattern classes while minimizing the intraset distance. Entropy minimization can be used to derive a linear transformation useful for reducing the feature dimensions if the feature distribution for each class is known to be Gaussian and all distributions have equal covariance matrices. If this assumption cannot be justified, the discrete version of the Karhunen-Loéve expansion can be used for dimensionality reduction; this method minimizes the mean square error in the approximation of all available measurements by an orthonormal expansion in terms of the eigenvectors of the covariance matrix of the measurement vector. Divergence is defined as a distance measure between two pattern populations; if they are Gaussian distributed, divergence maximization can also be used for feature dimensionality reduction.

The Karhunen-Loéve expansion is probably the best known and most useful method of feature extraction, especially when the number of classes is large (132). Let x be a measurement vector with L components

$(x_1, x_2, \ldots, x_L)$, $\mathbf{y}$ be a feature vector with m components $(y_1, y_2, \ldots, y_m)$, and the set of $h_i(\cdot)$ for $(i=1, 2, \ldots, m)$ be represented by a linear transformation matrix $\mathbf{A}$ such that $\mathbf{y} = \mathbf{Ax}$. It is assumed that $E\{\mathbf{x}\} = \mathbf{0}$ or $E\{\mathbf{y}\} = \mathbf{0}$ where $E\{\cdot\}$ denotes the expectation operation. Let $\mathbf{R} = E\{\mathbf{xx'}\}$ be the covariance matrix of $\mathbf{x}$, where $\mathbf{x'}$ denotes the transpose of $\mathbf{x}$. Compute eigenvalues $\lambda_i's$, $(i=1, 2, \ldots, n)$, and the corresponding normalized eigenvectors of $\mathbf{R}$. The normalized eigenvectors $\mathbf{u}_i's$, $(i=1, 2, \ldots, n)$, are a set of orthonormal base vectors with $\lambda_1 \geq \lambda_2 \ldots \geq \lambda_n$. The transformation matrix $\mathbf{A}$ is formed by the m eigenvectors corresponding to the m largest eigenvalues, i.e. $\mathbf{u}_j'$ becomes the jth row of $\mathbf{A}$. Hence, the extracted feature vectors are obtained by $\mathbf{y} = \mathbf{Ax}$. These m feature components are uncorrelated since the covariance matrix of $\mathbf{y}$ is diagonal. The Karhunen-Loéve expansion optimizes the mean square approximation error $E\{\|\mathbf{x} - \mathbf{A'y}\|^2\}$. However, it gives only a suboptimal result if the condition $E\{\mathbf{y}\} = \mathbf{0}$ or $E\{\mathbf{x}\} = \mathbf{0}$ is not satisfied. Another comment on this method is that it does not take discrimination of classes into consideration. We have mentioned this method in the introductory section in connection with the pattern recognition of pulmonary function test data.

As to the problem of selection of a feature subset from all the available measurements, a choice should be made on the feature subset evaluation criterion and feature subset search procedure. The probability of error should be the most pertinent evaluation criterion; however, computation cost is often a major consideration. Two simple criteria, which are frequently used, are mentioned below. One criterion function is given by $J_1 = |\mathbf{W}|/|\mathbf{T}|$, where $\mathbf{W}$ is the within-class scatter matrix that is equal to the summation of the individual class scatter matrices, $\mathbf{T} = \mathbf{W} + \mathbf{B}$ is the total scatter matrix, and $\mathbf{B}$ is the between-class scatter matrix. Another criterion function is given by $J_2 = (\mathbf{m}_i - \mathbf{m}_j)'[(\Sigma_i + \Sigma_j)/2]^{-1}(\mathbf{m}_i - \mathbf{m}_j)$ for two-class problems where $\mathbf{m}_i$ and Σ_i are the mean vector and covariance matrix of class i, respectively. This is a scaled component of the Bhattacharyya distance for normal distributions due to the difference in mean of the two classes. For M-class problems, J_2 should be summed over all possible distinct pairs of classes. A feature subset may be selected to either minimize J_1 or maximize J_2. A number of feature subset search procedures have been developed (15, 20, 29, 77, 82, 108). The sequential forward selection procedure (29, 77) has been used in both the chest radiographic image classification and the nucleated blood cell image classification. This procedure can be briefly described as follows. First, the best feature is selected, for which the chosen criterion function J_i is optimized. This best feature, together with the remaining features, is then examined to find the best two features for which the criterion function is optimized. This process is repeated until the value of the criterion function decreases or a predetermined fixed number of features are selected. This method has the advantages that its computation cost is moderate when a small number of features are selected from a large set and that it takes into account the correlation among the features.

Classification Strategies

A classifier receives a feature vector that consists of m feature components of a pattern or object under consideration and makes a decision on its class assignment. Let $\mathbf{x}$ be a feature vector, and $G_k(\mathbf{x})$ be a decision function (or discriminant function) ($k=1,2,\ldots,M$) where M is the number of classes. If $G_i(\mathbf{x}) > G_k(\mathbf{x})$ for all $k \neq i$, $\mathbf{x}$ is assigned to class ω_i. The discriminant function $G_i(\mathbf{x})$ may be either a linear or nonlinear function of $\mathbf{x}$. In the context of statistical pattern classification, a Bayesian decision rule that minimizes the probability of error is the best strategy one can design (28, 33, 132). If the a priori probability $p(\omega_i)$ of class ω_i and the conditional probability density function $p(\mathbf{x}/\omega_i)$ of $\mathbf{x}$ are known for all ω_i, and if the zero-one loss function is used, i.e. zero loss for correct classification and unity loss for misclassification, the decision rule of the Bayesian classifier is to maximize $G_i(\mathbf{x}) = p(\mathbf{x}/\omega_i)p(\omega_i)$ over ω_i. If $p(\mathbf{x}/\omega_i)$ is a Gaussian distribution with the mean feature vector $\mathbf{m}_i$ and the covariance matrix Σ_i for every class ω_i, the maximization of $G_i(\mathbf{x})$ is equivalent to the decision rule of minimizing $(\mathbf{x} - \mathbf{m}_i)'\Sigma_i^{-1}(\mathbf{x} - \mathbf{m}_i) + \ln|\Sigma_i| - 2\ln p(\omega_i)$, which is a quadratic decision function. Furthermore, if all covariance matrices are equal, $\Sigma_i = \Sigma$ for all i, then the decision function is reduced to a linear function of $\mathbf{x}$. In short, to construct a Bayesian classifier, $p(\mathbf{x}/\omega_i)$ and $p(\omega_i)$ must be known.

If $p(\omega_i)$ is not known, as is true in many practical situations, a maximum likelihood classifier may be used where the decision rule is to minimize $(\mathbf{x} - \mathbf{m}_i)'\Sigma_i^{-1}(\mathbf{x} - \mathbf{m}_i) + \ln|\Sigma_i|$. However, if $p(\mathbf{x}/\omega_i)$ is unknown or its estimations are not reliable because of the insufficient number of available training samples, nonparametric classifiers may be used without any assumption about the knowledge of probability functions. Two of the nonparametric classifiers are mentioned below. One is the minimum distance classifier that utilizes a normalized distance metric $d_i^2(\mathbf{x}, \mathbf{m}_i) = (\mathbf{x} - \mathbf{m}_i)'\Sigma_i^{-1}(\mathbf{x} - \mathbf{m}_i)$ where $\mathbf{m}_i$ and Σ_i are the sample mean and sample covariance matrix respectively of training samples of class ω_i. If $d_i^2(\mathbf{x}, \mathbf{m}_i) + B_{ij} < d_j^2(\mathbf{x}, \mathbf{m}_j)$ for all $j \neq i$, then $\mathbf{x}$ is classified as ω_i. This is a quadratic classifier where B_{ij} is a constant chosen such that the number of misclassifications in the training samples is minimized. This type of classifier has been used in both lung image classification and blood cell image classification. It may be interpreted as a Bayesian classifier if B_{ij} is equal to $\ln\{|\Sigma_i|/|\Sigma_j|\} - 2\ln\{p(\omega_i)/p(\omega_j)\}$. Another nonparametric classifier is k-nearest-neighbor classifier with a certain distance metric. The decision rule is to assign x to the class to which belong the majority of the k samples that are closest to x.

Sequential decision is another procedure that has been applied to pattern classification (28). For two-class problems, Wald's sequential probability ratio test can be used. At each stage of the sequential process, take one additional feature into consideration. At the kth stage when the kth feature x_k is taken, the classifier computes the sequential probability ratio

$$\lambda_k = p(x_1, \ldots, x_k/\omega_2)/p(x_1, \ldots, x_k/\omega_1).$$

If $\lambda_k \geqslant A$, x is classified into class ω_2; if $\lambda_k \leqslant B$, x is classified into class ω_1 where A and B are two stopping boundaries related to error probabilities. The computation cost for λ_k's may be excessive if the feature components are not statistically independent.

A structure called tree classifier has been used in the automated classification of nucleated blood cells (81). Its graph representation consists of a root node, with which is associated the entire set of classes, a number of nonterminal nodes, and a number of terminal nodes, each of which corresponds to a terminal decision on the class assignment of a pattern (see Figure 5). A nonterminal node has an ascendant node and two or more descendant nodes; it is associated with a subset of classes following the decision made at its ascendant node, and a decision is made whose outcomes are respectively associated with its immediate descendant nodes. The root node has no ascendant node, and a terminal node has no descendant node. From a node to each of its descendant nodes, there is an edge which represents the information flow of the decision made at that node. If there are exactly two immediate descendant nodes for each nonterminal node in a tree classifier, it is called a binary tree classifier. A tree classifier classifies a pattern by starting the pattern at the root node and traversing a path of the tree where each nonterminal node encountered involves a decision until the pattern ends at a terminal node, whose label is the class assigned to the pattern. Decision rules of differing complexity may be used at different nodes, and only those features which are pertinent to a specific classification are used at individual nodes. Hence, a tree classifier may provide improved overall classification accuracy. However, it is difficult to design an optimum tree classifier because there are too many possible tree structures and too many possible combinations of feature subsets for an optimum search.

Classifier Training and Testing

Among the methods for using available samples of known classes in classifier training and testing, the following three are commonly used: (a) resubstitution, (b) leave-one-out, and (c) rotation (33, 69, 116). In the resubstitution method, the classifier is both trained and tested on the whole set of available samples; the probability of error obtained is an optimum estimate. The rotation method consists of the following steps. At the ith training and testing stage, take a small subset S_i of P patterns out of the whole set of N available samples and train the classifier on the remaining $(N-P)$ samples. This trained classifier is then tested on the removed subset S_i of P samples to obtain a proportion of classification errors denoted by P_{ei}. P is a small number chosen such that N/P is a positive integer. Repeat the same process for $i = 1, 2, \ldots, N/P$ such that all the subsets S_i's are disjoint and the union of all S_i's is the whole set of available samples. The resulting estimate of the probability of error is computed as $P_e = [\Sigma_{i=1}^{N/P} P_{ei}] P/N$. Usually P/N is taken as $1/10$; this is called the 10% jackknife testing method. If $P = 1$, this special case of rotation method becomes the leave-one-out method.

The resubstitution method is optimistically biased and gives an overly optimistic estimate of performance. The leave-one-out method is more pessimistically biased and gives an error estimate that has a larger variance than the resubstitution method; it also requires excessive computation in most cases. The combination of the above two methods has been used in evaluating the nucleated blood cell classification system (81). The difference between the two error probabilities gives some indication about the sufficiency of the size of the available samples versus the number of features used; however, the ten-to-one ratio of the number of training samples per class to the number of features used should be followed initially in the feature selection process (25). The 10% jackknife testing method is a compromise that requires less computation than the leave-one-out method; it has been used in the evaluation of textural classification of lung images (118).

COMPUTER ANALYSIS OF CHEST RADIOGRAPHS FOR CLASSIFICATION OF LUNG DISEASES

Segmentations of Lung Images, Ribs, and Pulmonary Arteries

The computer analysis of chest radiographs presents a very complicated pattern recognition problem: the highly structured picture pattern contains two-dimensional projections of several organs which overlap to some degree and whose boundaries may not be well defined. To detect lung diseases from chest radiographs one must first preprocess the digitized radiograph in order to reduce the irrelevant data and locate the lung images within which the search for pulmonary lesions is restricted. It is also desirable to determine dorsal rib contours to facilitate the texture analysis within the respective interrib and rib spaces, to provide a frame of reference for the location and description of lesions detected in the lung field, and to facilitate the removal of false positives in the tumor detection. The extraction of the image of major pulmonary arteries in each lung region provides a means, if needed, for excluding this subimage from further processing, thus reducing the possible error in recognition of lesions.

The first attempt to use a top-down descriptive approach to extract the lung boundary in a chest X-ray image was at the University of Missouri at Columbia (47, 110). Using an embedding metric, Fu & Chien of Purdue University developed an algorithm for lung boundary location (16, 31). Both local and global information are used to first determine the coarse lung boundary and five key points on each lung—the lung apex, the intersection point of clavicle and right boundary of the lung, the intersecting point of clavicle and left boundary of the lung, cardiac-diaphragm intercept, and costophrenic angle vertex. The detailed lung boundary is determined as a succession of five fourth-order polynomial curves between every pair of successive key points. This algorithm operates on consolidated images of the size of 256×256 with 8 bits of gray level information per pixel.

A method for coarse lung boundary detection, which is computationally less expensive, was implemented by Hall et al in the automatic classification of pneumoconiosis (40, 41). The chest radiograph is scanned at low spatial resolution (or consolidated) to obtain 64×64 array of pixels. A global threshold gray level is selected from the image histogram $h(z)$; the thresholding operation creates the first estimate of the lung region which is refined with local operations. From the horizontal and vertical signatures of the image, a rectangular frame for each lung is determined. Three anatomic constraint points are located by local searches on each lung image: they are the cardiac-diaphragm intercept, the costophrenic angle vertex, and the lung apex. A logic operation is also incorporated to reduce false recognition. Three endpoint constrained, fourth-order polynomials are fitted by the least squares to represent three segments along each lung boundary. They can be easily interpolated upon the same chest image at a higher (512×512 pixels) spatial resolution. This software has been implemented at the NIOSH (National Institute of Occupational Safety and Health) Laboratory at Morgantown, West Virginia.

The basic concepts of Fu and Hall, as discussed above, have been incorporated by Toriwaki et al in Japan in their software system called

AISCR-V3—a modification of AISCR-V2 which was previously developed for lung boundary determination, rib boundary and vessel shadow recognition, and abnormal lesion detection in chest radiographs (49, 112). Five key points on each lung boundary are located. Linear filtering, thinning, and border following operations are used to recognize the borders of thorax, heart, and diaphragm; lung apex borders and clavicles are determined by quadratic approximations. Rib edges are fitted by polynomials of the second degree after a template matching. Recognition of the major and small vessel shadows is achieved through filtering, extraction of connected components, and classification according to some criteria on location and size parameters of the connected components (113).

Ballard & Sklansky constructed an algorithm for finding lung boundary, which is a two-stage process consisting of a plan generator and a plan follower (5). The chest image of size 256×256 is consolidated to 64×64 array of pixels. High-emphasis filtering and a gradient operation enhance the edges in the image and give at each pixel both gradient magnitude and the directional element, which is defined as an arrow of unit length pointing $+90°$ to the gradient direction. The plan generator traces a coarse approximation to the lung boundary at the low spatial resolution by following a heuristic search procedure. The plan follower has a structure similar to the plan generator but operates at the higher spatial resolution (256×256 array) to give a refined lung boundary.

Persoon & Fu (31) constructed an algorithm for automatic extraction of rib boundaries. Edge pixels are first detected by using a local edge detector (89). A filter thins out the edges pixels, eliminating those with weaker edge measure and favoring the continuity of lines; then "joining" and "filling" operations follow to trace out the longest convex contour as the first approximation to the major portion of a dorsal rib contour. The convex contour is fit by a second-degree curve to extend the computed rib contour close to the lung boundary. An algorithm for automatic determination of both dorsal and ventral rib contours was developed by Wechsler & Sklansky (125). It consists of high-emphasis filtering, local edge detection (using a window of size 5×5 and a combination of both gradient and Laplacian operations), a global boundary detector, and a rib linker. Elliptic curves are obtained as rough approximations to dorsal rib contours; these computed second-degree curves are used as guides or "plans" in a heuristic search for refined rib contours. Each pair of corresponding dorsal-ventral rib contours can be linked by a fourth-degree curve. Wechsler & Fu evaluated the above-mentioned two algorithms for rib boundary detection (126). Both were implemented on a minicomputer DEC PDP 11/45 system. The recognition accuracy of the two methods seems to be comparable. Ballard proposed a computational model for rib cage anatomy seen in a chest radiograph, and developed an algorithm for model-directed detection of ribs (6).

Kulick et al implemented a method for dorsal rib contour detection, where hysteresis smoothing was applied to gray level profiles in the vertical direction for enhancing rib edges (62). A similar, simple algorithm for dorsal rib extraction was also developed by Shiek (see Reference 64). Fong et al developed an algorithm for extracting boundaries of major descending pulmonary arteries (26). It involves first locating the center line of a major descending artery, spatial filtering and edge detection for artery boundaries, and then tracing, linking, and tree-searching operations for the recognition of the major artery and some of its prominent branches (26).

Classification of Pulmonary Infiltrations

One way of providing quantitative descriptions of lung images in chest radiographs is to perform statistical texture analysis of the images. From the anatomical point of view, the lung consists of alveoli (air spaces), the interstitium between, pulmonary vessels, and bronchi. The lung image on a chest radiograph represents the projection of these components plus ribs, etc. The texture analysis of pulmonary vascular patterns has been reported in References (24, 76). Computer analysis of texture patterns of a special kind of interstitial infiltrate disease of the lung, the coal workers' simple pneumoconiosis, has been studied for the past eight years; this is described later. In general, there are two types of pulmonary infiltration: alveolar and interstitial. Tully et al have conducted a study on computerized texture discrimination among images of alveolar infiltration, interstitial infiltration, and normal lung (118).

A series of $14'' \times 17''$ chest radiographs of 129 subjects were selected as the data base. Of those studied, 48 had normal lungs, 39 had pure interstitial infiltration, and 42 had pure aveolar infiltration. The qualitative descriptions of these chest radiographs are as follows. The alveolar infiltration pattern is a fluffy texture with air bronchograms, the interstitial infiltration pattern is a finely linear, nodular and/or cystic texture, and the normal lung pattern is the absence of the above two patterns, with tapering vessels toward the periphery. A square region of 3.2 cm $\times$ 3.2 cm in the middle of the right lung, about an inch or so up from the diaphragm, was chosen from each radiograph for digitization. It lies mostly in the interrib space and away from hila so as to minimize the biasing caused by ribs and hilar structures. The digitization over this square region provided 128×128 array of pixels with 256 possible gray levels. At this fine spatial resolution, each pixel represents an image area of 0.25×0.25 mm^2. The equal probability quantization was applied to each digitized image for histogram modification. This was done in order to normalize the contrast variations of radiographs and to reduce the number of gray levels from 256 to 16, thereby facilitating the estimation of the joint probability functions. Five texture measures based on the spatial gray level dependence method were computed. They were energy $T_1(d,\theta)$, contrast $T_2(d,\theta)$, correlation $T_3(d,\theta)$, entropy $T_4(d,\theta)$, and local homogeneity $T_5(d,\theta)$, as defined previously. Four values of θ ($\theta = 0°, 45°, 90°, 135°$) and two values of $d(d=1,2)$ were considered in the feature selection. The forward sequential search procedure was used in selecting 4 best features among 40 for each pairwise classification of this three-class problem. Bayes classifier with zero-one loss function was used in each pairwise classification, i.e. normal-versus-interstitial, normal-versus-alveolar, and alveolar-versus-interstitial. The final class assignment to a pattern was made by the joint decision of these three classifiers.

For the normal/interstitial classifier, the texture features selected were $T_1(1,45°)$, $T_1(1,0°)$, $T_1(1,90°)$ and $T_4(1,135°)$. For the normal/alveolar classifier, the features were $T_1(2,135°)$, $T_1(1,45°)$, $T_3(2,90°)$ and $T_5(1,0°)$. For the interstitial/alveolar classifier, the texture features were $T_1(2,135°)$, $T_1(2,90°)$, $T_1(2,45°)$ and $T_3(1,0°)$. The classifiers were trained and tested on 129 images by the 10% jackknife method. Among the 39 images with interstitial infiltration, 34 were correctly classified, 1 was misclassified as normal, and 4 were misclassified as alveolar infiltration. Of the 42 images with alveolar infiltration, 39 were correctly classified and 3 were misclassified as normal. Of the 48 normal images, 44 were correctly classified and 4 were misclassified as alveolar infiltration. The overall classification accuracy was 90.6% with 3.1% false negatives. This feasibility study has demonstrated the potential of using quantitative texture measures for automated classification of pulmonary infiltrations.

Classification of Coal Workers' Pneumoconiosis by Texture Analyses

Coal workers' simple pneumoconiosis (CWP) is a lung disease caused by long-term inhalation of coal dust and local tissue reaction to accumulated dust particles. The radiological symptom of simple pneumoconiosis is the appearance of small opacities in coal workers' chest X-rays. Based on an internationally agreed upon ILO U/C classification system (55), in simple pneumoconiosis there are three types of small rounded opacities, p, q, and r, according to the approximate diameter of the predominant nodules, and also three types of irregular opacities, s, t, and u, according to their fine linear or coarse irregular structure. These opacities may be observed anywhere in a lung. According to the profusion of small opacities, four categories 0–3 have been established to indicate the severity of the disease, where category 0 means normal and category 3 means very numerous small opacities. The scale is further divided by using two digits, x/y, where x is the assigned category but y category was seriously considered, $(x,y=0,1,2,3)$, thus providing a full scale of 12 classes.

The automated classification of pneumoconiosis by texture analysis of coal workers' chest X-rays has been studied by several groups. The first report on this study was made by Hall et al (39). After a coarse lung boundary was determined, the rectangle enclosing each lung was divided into thirds to correspond to the three lung zones used in the ILO U/C classification. Each X-ray was digitized into 512×512 pixels with 10-bit gray levels. By the stepwise discriminant analysis procedure, they selected 62 histogram moments and spatial moments measured from the zone image, the horizontal gradient image, and the intercostal sub-image within the zone (40). Based on the assumption of multivariate normal distributions of features, a linear, maximum likelihood, pairwise classification scheme was adopted. A set of 38 sample X-ray films were used for training and testing. For classification into four categories, an overall accuracy rate of about 60% was obtained; for normal-abnormal classification, an accuracy rate of 87% was reported. Kruger et al studied a classification system by utilizing statistical texture features in the manually extracted interrib spaces based on the second-order spatial gray level dependence method (60, 119). Five textural measurements $\{T_i(d,\theta); i=2,3,4,5,6\}$ were made. Mean $m_i(d) = E\{T_i(d,\theta)\}$, variance $v_i(\theta) = \text{Var}\{T_i(d,\theta)\}$, and range $r_i(d) = [\text{Max } T_i(d,\theta) - \text{Min } T_i(d,\theta)]$, averaged over four directions, were computed as features for each texture measure at a given spacing d to eliminate the directional bias. Divergence maximization was used in selecting six features $[r_4(3), m_5(1), v_4(3), m_6(3), v_6(3), m_6(1)]$ for the normal-abnormal classification and seven features $[r_4(3), m_6(1), r_6(7), v_4(3), m_6(3), v_6(3), m_3(3)]$ for the four-category classification. The selected features were assumed to be multivariate Gaussian, and a linear maximum likelihood classifier was used for each interrib space. The data base utilized consisted of $4'' \times 5''$ zonal reproductions from 141 chest radiographs. The spatial resolution of the digitized zonal images was 9.8 pixels/mm. Each zonal film consisted of three to four interrib spaces; the classification of zonal film was based on that of interrib spaces by following a certain majority rule. 95 digitized zonal images were tested. Classification accuracy was 96.8% for normal-abnormal classification and 66.3% for the four-category classification, as shown in Table 1. The Fourier domain measure of

Table 1 Test results of machine classifications of 4-category pneumoconiosis

Pneumoconiosis category	Computer classification				Total X-rays	Percentage correct
	0	1	2	3		
Kruger's result [a,b]						
0	30	1	0	0	31	96.8%
1	1	5	7	2	15	33.3%
2	0	3	15	1	19	78.9%
3	1	3	13	13	30	43.3%
Total X-rays	32	12	35	16	95	
Overall classification accuracy						66.3%
Jagoe's result [c]						
0	8	15	0	0	23	34.7%
1	8	29	4	0	41	70.7%
2	0	14	16	2	32	50.0%
3	0	0	0	0	0	—
Total X-rays	16	58	20	2	96	
Overall classification accuracy						55.2%

[a] Based on interrib images in zonal films (adapted from Table 3 in Reference 56).
[b] False negative rate = 22%. False negatives are defined as those films that were classified by the computer into a category lower than their respective true categories. These are shown on the lower left of the diagonal.
[c] False negative rate = 23% (adapted from Reference 57).

visual texture was also studied by using a coherent optical system, and a similar performance on classification experiments was obtained. It was demonstrated in this feasibility study that the classification accuracy of the automated system was comparable to that of the human readers, which ranges from 48% to 64%.

The combined efforts of Hall, Kruger & Turner designed a prototype system which has been installed at NIOSH laboratory (41, 61). At first each chest radiograph is digitized at low resolution by an image dissector camera into 64×64 array of pixels for film quality examination and for determining coarse boundaries of the right and left lung regions as described previously. Centroids of the lung boundary within each of six lung zones are computed. At this point, the film may be transported under computer control into the path of an optical subsystem. Or, the film may follow the path in the digital subsystem to be scanned with a high resolution digitizer (digitized into 512×512 pixels with 256 gray levels) for digital processing and feature extraction, which includes histogram moments, spatial moments, and features computed from spatial gray level dependence matrix. This option is presently being used. Each of the six lung zones is classified separately by a linear Gaussian classifier, with an average of eight digitally extracted texture features per zone; and the results are integrated into an overall film classification. Altogether, 130 chest X-rays (57 normal, 73 pneumoconiosis) were used as training samples and 339 X-rays (107 normal, 232 pneumoconiosis) as testing samples. Classification accuracy during training was 92% for normal-abnormal classification. In the testing experiment, classification accuracy was 74% (41). Continued research on improvement of this system has been conducted by Hankinson. Equal probability quantization has been incorporated for image normalization before the computation of texture features (43). This may lead to an improvement in the accuracy of machine classification of pneumoconiosis.

A hybrid optical-digital system was also investigated by Stark & Lee (107). With features extracted from the optical Fourier transform of the chest X-ray, a k-nearest-neighbor classifier was used. The normal-abnormal classification rate was about 91% based on a training set of 64 sample X-ray films. Ledley et al introduced a set of specially defined texture measures and used 64 features for classifying coal miners' chest X-rays into one of four profusion categories (63). Based on a training set of 40 X-ray films (right lung only), classification accuracy of 82% was reported; the test result on another set of 28 films was reported to be satisfactory.

Jagoe & Paton of the British Medical Research Council earlier introduced a diversity measure in digitized chest X-ray images that have a spatial resolution of 1.2×1.2 mm^2/pixel with 256 possible gray levels (56). For each small square of 2×2 pixels, the rate of increase in optical density is computed in each of eight quantized directions; the greatest value yields the gradient for that small square. The diversity in a 4×4 square grid is defined as the number of different gradients taken by the nine small squares in that grid. The mean diversity over all relevant grids in both lung fields is chosen as a single feature for their classifier, which gave about 80% classification accuracy for 36 training X-ray films, with 9 films in each category. Recently, Jagoe reported a study on the application of gradient pattern coding to pneumoconiosis classification (57). In the square grid mentioned above, each combination of the 3×3 gradient directions forms a gradient pattern. Consider two subsets of five gradient elements: one is an axially adjacent set and the other is a diagonally adjacent set. There are 4144 different gradient patterns in each set, after the rotational and reflectional symmetries are considered. By looking for gradient patterns that contribute most towards the visual texture of pneumoconiosis in the 36 training radiographs, 77 patterns in the two sets were identified. Classification of four pneumoconiosis categories may be performed by examining the frequency counts of these 77 gradient patterns. This was tested on an independent set of 96 chest radiographs; the result is shown in Table 1. The overall classification accuracy was 55%, which is comparable to the result obtained by Kruger's study on interrib images in zonal films and is also comparable to radiologists' performance. The false negative rate was 23%, again comparable to the radiologists' performance ranging from 7 to 44%.

Detection of Small Opacities of Pneumoconiosis

During the early phase of simple pneumoconiosis, opacities are often small, non-numerous, and have weak contrast in chest images. Computer-aided recognition of individual small opacities will provide a significant help for the early detection of pneumoconiosis. The small rounded opacities, types p, q, and r are defined as follows (55). Type p rounded opacities are up to about 1.5 mm in diameter, type q opacities range from 1.5 to 3 mm in diameter, and type r opacities from 3 to 10 mm in diameter. When a chest radiograph is digitized into 1024×1024 array of pixels with 256 possible gray levels, the spatial resolution of the digitized image is about 0.345×0.345 mm^2 per pixel. Thus, a small type p rounded opacity of 1 mm in diameter consists of about 3×3 pixels. A type q rounded opacity ranging from 1.5 to 3 mm in diameter may have 4×4 to 7×7 pixels. Small irregular opacities of types s, t, and u are, however, only qualitatively defined. Type s are fine, irregular linear opacities, type t are medium irregular opacities, and type u are coarse irregular opacities. Figures 2(A) and (B) are two PA chest radiographs showing simple pneumoconiosis. There are only rounded opacities in Figure 2(A); 36 of them have been identified by at least two radiologists and their distributions in the upper, middle, and lower zones of right and left lungs are 10, 12, 3, and 5, 3, 3 respectively. Only irregular opacities of type t exist in Figure 2(B); 18 of them have been identified by radiologists, with distributions of 5, 10, and 5 in the right middle, right lower, and left lower lung zones respectively.

Li & Savol have developed an adaptive object-growing algorithm for detection of small rounded opacities in lung images (65, 99). The algorithm operates by starting from a "seed" pixel (whose gray level is less than or equal to that of its eight immediate neighbors) and by adding pixels in a geometrically flexible manner to maximize a figure of merit called the isolation contrast integral (ICI) (defined in the section on feature measurement). All the outside edge pixels surrounding the seed pixel are examined and the one with the minimum gray level is taken as a test point for an additional member pixel. The isolation

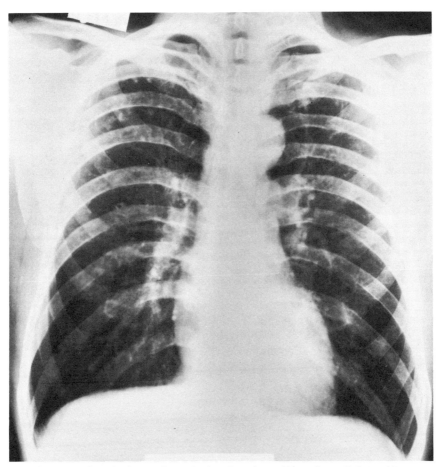

Figure 2A Sample chest radiograph of coal worker: with small rounded opacities.

contrast integral of this enlarged set of member pixels is computed. If it is nondecreasing as compared to that of the previous set, this test point is accepted as an additional member pixel. Then the outside edge pixels surrounding the enlarged set of member pixels are examined, and the "growing" process continues. The growth of an object terminates when its isolation contrast integral reaches a maximum. At this stage the object is a "best" defined object in the area of search. The growth process may be subject to the effect of noise on an intermediate outside edge pixel whose gray level becomes slightly lower than that of the test pixel, and may thus be terminated prematurely. To reduce this error, an averaging process is incorporated before the actual termination. For an

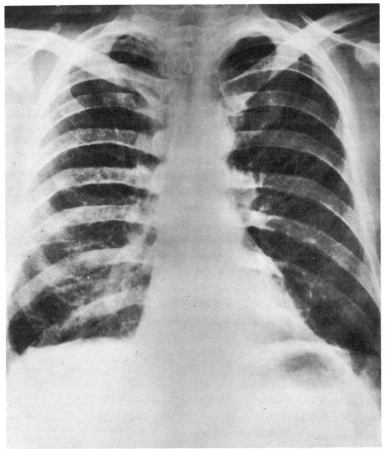

Figure 2B Sample chest radiograph of coal worker: with small irregular opacities.

outside edge pixel that has a minimum observed gray level and that tends to stop the "growing" process, an estimated gray level is obtained by averaging the gray level of the pixel itself and all those of its immediate neighboring pixels which were not included in the set of object member pixels previously determined. The isolation contrast integral is then computed with this estimated gray level being used for the outside edge pixel under consideration. If it is greater than or equal to the isolation contrast integral obtained in the previous step, the growth process will continue. This averaging process is allowed to take place only once in order to avoid excessive growth that may, in turn, lead to erroneous results. Figure 3 illustrates a candidate opacity detected by the "growing" algorithm. For each grown object, measurements of area, perimeter, and isolation contrast integral are readily

```
 43  47  50  55  59  59  60  61  65
 40  45  53  61  65  64  66  70  74
 44  52  62  68  77  82  89  90  97
 68  73  81  89  98 109 115 122 131
 92  92  96  99 110 113 120 131 132
100 101 107 109 116 121 122 128 135
106 106 110 116 120 131 135 136 135
111 108 109 112 124 128 134 138 141        S **
110 109 109 105 105 115 124 133 141     *L*@@*
115 108 100  95  93 102 113 129 138    *@@@@@*
122 110  96  89  86  95 101 119 135    *@@C@@*
127 113  99  93  91  95  99 120 138    *@@@@@*
130 119 112 103 100 101 110 126 145     *@@@*
138 131 126 121 116 117 122 136 158        ***
143 138 145 137 133 142 141 154 164
153 150 153 156 156 158 167 164 168
156 158 162 165 165 173 178 176 176
167 164 169 174 179 177 182 179 178
174 177 173 174 179 181 177 175 177
176 174 175 180 174 175 176 179 183
178 175 178 175 178 172 178 177 180

            (A)                          (B)
```

Figure 3 A rounded opacity in a digitized chest radiograph as detected by the adaptive object growing algorithm: (A) gray values in the region under consideration, (B) the algorithm-generated result of a small rounded opacity whose gray values are shown in (A). C = the "seed" pixel, @ = a member pixel of the object, L = the last member pixel obtained by the algorithm, * = an outside edge pixel, and S = the minimum outside pixel that terminates the growing process. Point C in (B) corresponds to point 86 in (A).

available. A "grown" object is classified as a small, rounded opacity of a specific type according to the parameters p/A and A; a simple nonlinear decision boundary has been trained. Experimental studies indicate that an overall recognition accuracy of this algorithm was about 67%. For example, for the chest radiograph shown in Figure 2(A), application of the algorithm resulted in recognition of 25 of the 36 rounded opacities, among which 4 are of type r, 19 of type q, and 2 of type p. However, it also gave a very large number of false positives. The method has also been extended to classify short linear opacities or segments of longer ones by the use of indices p^2/A and the ratio of object length to width. It is assumed that a small irregular opacity may have multiple seed pixels aligned linearly; the individually "grown" segments from these seed pixels can be merged for the recognition of a full, linear opacity. For example, the machine correctly recognized 9 of the 18 irregular opacities in the chest radiograph shown in Figure 2(B); it misclassified 3 (segments) of these as rounded opacities and rejected 6, of which 2 appeared crossing over rib contours (67). Work is in progress to improve the recognition accuracy for both the rounded and irregular small opacities.

Detection of Lung Tumors

It has been reported that radiologists routinely miss about 30% of the lung tumors which are supposed to be visible in chest radiographs. They may also recognize as many or more false positives than false negatives in their examinations (105). For early detection of lung cancer, therefore, it is very important to have an effective diagnostic method which is aided by computer analysis of chest radiographs. Ballard & Sklansky have developed a hierarchical recognition system involving a ladder-structured decision tree (5). It is capable of detecting nodular tumors as small as 1 cm in diameter. For each chest radiograph digitized into 256 gray levels, three spatial resolutions 1024×1024, 256×256, and 64×64, were used. A lung boundary detection procedure, reviewed earlier, is first applied to the coarse-resolution image, and then a refined lung boundary is obtained at the intermediate resolution. Enhancement filtering and gradient operation are performed in the intermediate-resolution and fine-resolution images. A "circularity finder" algorithm using Hough accumulator array technique is applied to measure the circularity of the gradient field at each pixel within the lung region in the intermediate-resolution image. A circularity measure is defined as the number of edge pixels (on a circumference at radius r about a search pixel) whose gradient direction points toward this pixel. For a given location and radius, if the circularity value is high and, furthermore, if the radius is larger than 2 cm, this region is recognized as a large tumor; if the radius is below that value, it is considered a "candidate nodule site." A dynamic programming procedure is used to search for a detailed, closed boundary of a possible nodule at each candidate nodule site in the fine-resolution image. The best curves of length equal to two pixels, emanating from every edge pixel, are chosen as primitives in the dynamic programming algorithm. The algorithm repeats this generation of primitives in a sequence of stages, concatenating the primitives to form longer and longer curves in an optimal manner; the criterion for stopping is closure. When such a boundary is found, the resultant object is accepted as a candidate nodule. A two-stage binary tree classifier, with the k-nearest-neighbor decision rule, is used to classify each candidate nodule. All together, eight features are measured for each candidate nodule; they include one size parameter (radius), one location parameter (distance from hilar region), three measures related to contrast depth, edge contrast, and root mean square value of the Fourier transform, and three shape parameters (number and size of major concavities, and straightness of boundary segments). At the first node of the tree classifier, five features are used to label a candidate nodule as either a nodule or a ghost. At the second node, six features are used to further classify the nodule as tumor or nontumor. An experimental study of six radiographs showed that 9 out of a total of 16 tumors were detected by the computer method. One radiograph was correctly diagnosed as tumor-free.

The most recent work of Sklansky et al has shown the capability of detecting smaller lung tumors with improved accuracy (105). Preprocessing of the lung image at the intermediate resolution consists of the following four steps: (a) filtering by a normalized zonal notch filter, (b) local edge detection by Sobel gradient operator, (c) suppression of false edges and edge equalization by adaptive thresholding in a 9×9 window, and (d) suppression of straight line edge segments that are not portions of tumor boundaries. The last step is accomplished by first detecting peaks on the projections of gradient elements in a 20×20 window along a set of orientations separated by 7° intervals. It is followed by weighted subtraction of the detected line segment from the gradient field and then by thresholding for noise suppression. The result of the preprocessing is the line-suppressed gradient field in the lung region. The same "circularity finder" algorithm as described before is used to find circular regions that are suspected nodule sites. An annular-ring-shaped potential well is generated in the fine-resolution image and is used as a plan for guiding a minimum-cost tree search for the nodule boundary. The search space is partitioned into four quadrants. By starting from pixels on the vertical line inside the annular region, the search follows the minimum-cost paths in a counter-clockwise direction until each path reaches the end point on the other side of the quadrant. The cost function at a pixel P on a path depends on the accumulated modulus of the line-suppressed gradient field, the value of the plan's potential well, and the distance of P from the end point. Such end points are used as the starting points for the search in the next quadrant. After completing a revolution, the search process continues for one more quadrant so that the closed boundary of a candidate nodule can be found. Features such as number and length of straight segments are readily available from the computed boundary. Using these features a classifier has been designed to assign the detected object as tumor or nontumor. Experiments were performed on five radiographs showing 14 small nodules between 5 mm and 2.5 cm in diameter. This new method detected 13 of them and gave 35 false positives. It indicates good progress toward the early detection of lung tumors.

AUTOMATED CLASSIFICATION OF NUCLEATED BLOOD CELLS

White Blood Cell Differential Counters

Most of the research on automation of blood smear analysis during the past fifteen years (1, 2, 7, 11, 131, 54, 84, 97) has focused on the machine classification of white blood cells into six normal types, i.e. lymphocyte, eosinophil, basophil, monocyte, and neutrophil, which are further classified into band neutrophil (young immature) and segmented neutrophil (mature). This has led to the development of four machines, LARC, HEMATRAK, Coulter diff3, and ADC-500, which have become commercially available (96). Corning Electronics developed LARC, which is based on the research work of Bacus (2, 73); it can classify white cells into the normal six types plus a class called others. In an evaluation test on 4800 cells (19, 73), it showed a mean classification accuracy of 88.3%; in a more recent evaluation of LARC with 101 routine specimens containing 12,456 cells, 67% of band neutrophils (bands) and 5% of segmented neutrophils (segs) were misclassified (3). Miller of Geometric Data Corporation developed HEMATRAK (74) and reported an accuracy rate of 96.6% for the classification of 18,194 cells into the normal five classes (without subdivisions of neutrophils) plus a suspect class; another evaluation performed by Niitani on 39 smears showed 61% of bands and 1.4% of segs were misclassified (85). No evaluation result on Perkin-Elmer's Coulter diff3 has been published; however, an analysis using diff3 was described by Preston (96). Green reported on Abbott's ADC-500 differential counter which uses a hierarchical classifier and can classify white cells into 14 types (37). Using a set of 2894 testing cells, he obtained a mean classification accuracy of 80.5%; 19% of bands and 13% of segs were misclassified.

The percentages of band neutrophils and segmented neutrophils in the differential white blood cell count represent a significant sign of the body's reaction to an inflammatory process. Therefore, the misclassification of bands into segs or vice versa should be kept very small. The difference between a band neutrophil and a segmented neutrophil is that the nucleus of the latter has lobes connected by a filament. However, the cell image is a two-dimensional projection of a three-dimensional structure, and the nuclear lobes may overlap or clump together so that the filament becomes hidden. Hence, neutrophil classification into bands and segs is a rather difficult problem; improvements in machine classification have been attempted by Liu (70) and Mui et al (78, 79). The latter developed a scene segmentation technique that improves nucleus extraction accuracy for neutrophils; this is described later.

Classification of Nucleated Blood Cells into 17 Classes

In addition to the six normal cell types, it is desirable to automate more fully the differentiation of young and abnormal cell types among the nucleated blood cells. By nucleated blood cells are meant red and white blood cells that have nuclei. A specific subset of these nucleated blood cells contains the following 17 cell types: myeloblast (MYB), promyelocyte (PRO), myelocyte (MYC), metamyelocyte (MET), band neutrophil (BAN), segmented neutrophil (SEG), basophil (BAS), eosinophil (EOS), lymphocyte (LYM), atypical lymphocyte (LYA), immunocyte (LYI), monocyte (MON), nucleated red cell A (NRA), nucleated red cell B (NRB), nucleated red cell C (NRC), plasmacyte (PLA), and lymphoblast (LYB). The nucleated red blood cells are included because they are located by the algorithm for searching cells with nuclei and because they are clinically important.

The team of Brenner, Neurath, & Gelsema made the first attempt to classify nucleated blood cells into 17 types (11, 84). Images of nucleated blood cells were taken with Kodak Plus-X film through two color filters [Kodak Wratten Number 44 (blue-green) and Wratten Number 22 (yellow-orange)]. The film negatives were scanned using a flying spot scanner system with a digitization resolution of 0.1μ as referred to the microscope slides. A threshold value chosen from the yellow image histogram was used to delineate the cell from the background. For each of the blue and yellow image histograms, a nucleus threshold was independently selected and the one that gave the smallest nucleus area was retained. A single-stage linear classifier with 20 features was used; there were 7 geometric, 5 optical density, 4 color, and 4 texture features. A set of 1296 cells, consisting of 17 classes of normal and abnormal cells, were divided equally into training and testing sets. Experimental results showed that the mean classification accuracy is 67.3% for the normal and abnormal 17 classes and is 90.2% for the normal 6 classes. For neutrophil classification, 6 out of 38 bands (16%) and 8 out of 33 segs (24%) were misclassified. As mentioned before, Green (37) discussed the use of an ADC-500 machine to classify white blood cells into 14 types (not considering 3 types of nucleated red cells); however, no detailed information was given regarding training and testing methods or data acquisition. The most recent work on automated classification of nucleated blood cells uses a binary tree classifier (80, 81). This work is discussed below.

Segmentation of Cell Nucleus and Cytoplasm, and Feature Measurement

For each nucleated blood cell, two images are taken: one is a blue-filtered image and the other is a yellow-filtered image; both are digitized to 64 levels of optical density. When a smoothed monochromatic optical density histogram is examined, the three peaks along the direction of increasing density values correspond to the pixels of background, of white cell cytoplasm plus possibly red cells, and of white cell nucleus.

When the two-dimensional histogram of both blue and yellow images is examined, image pixels tend to cluster around four modes but are distributed more or less along the 45° line in the two-dimensional space spanned by blue density versus yellow density. This bivariate "blue-yellow" distribution may be normalized by finding two eigenvectors of the covariance matrix of the blue- and yellow-filtered optical densities. Since the principal eigenvector normally lies close to the 45° line in the original bivariate space and the variances in both directions are relatively constant, this normalization may be approximated by a 45° rotational transform plus appropriate scaling and shifting that can be summarized by: $C_{ij} = K_1(b_{ij} - y_{ij}) + K_3$, $D_{ij} = K_2(b_{ij} - y_{ij}) + K_4$, where b_{ij} and y_{ij} are respectively the blue-filtered and yellow-filtered density values at pixel (i,j), and C_{ij} and D_{ij} represent the "color" information and "density" information of pixel (i,j) respectively. Scaling factors K_1 and K_2 are introduced so that the dynamic ranges of C_{ij} and D_{ij} are not more than 64; otherwise they may be set to one. The constants K_3 and K_4 are added to shift the resulting transformed images into suitable density ranges. This is the so-called whitening transformation of two-color cell images (4). This transformation results in a "color image," a "density image," and a "color-density" histogram defined in two nearly uncorrelated axes (D-axis along the original 45° direction, and C-axis orthogonal to D-axis). In the transformed color image, the lighter gray levels represent blueness and the darker gray levels represent redness. Hence, the red cell pixels tend to be separated from the cytoplasm pixels in the bivariate color-density histogram, thereby providing four color–density contrast modes to facilitate image segmentation.

Mui et al developed an improved segmentation technique based on clustering of the color density histogram and incorporation of a priori spatial information (78). The procedure may be divided into four steps. The first step is to make an initial segmentation by two threshold values corresponding to two valleys in the smoothed histogram of the density image. The boundaries between background and cytoplasm and also between background and red cells are searched for points of maximal concavity, where the angle subtended by the boundary at each of such points is less than a certain predetermined value (e.g. 2.35 radians) when measured on the outside of the cells. If there is no point of maximal concavity, there is no other cell touching the cell under consideration and, hence, it is assumed that there are only three clusters in the bivariate color-density histogram. Otherwise, there are red cells touching the white cell and, hence, there are four clusters in the color-density histogram. The second step is to estimate the initial cluster centers in the color-density histogram. The center of mass and the area of the cell nucleus are first estimated from pixels whose density values are greater than that of the second valley in the density histogram. A square whose area is four times the estimated nucleus area is centered in the mass of the nucleus as shown in Figure 4(A). All pixels within this square region, whose density values are between those of two valleys, are used to calculate averages of density and color values in order to estimate a cluster center for cytoplasm pixels. The cluster center for red cells pixels, if present, is estimated next. A square window of 13×13 pixels is placed around each point of maximal concavity and also around the end points of short boundary curves of red cell cores, if any, as shown in Figure 4(B). The averages of density and color values of the six "reddest" pixels from each window are calculated as the initial cluster center of red cell pixels. The cluster centers for background and nucleus are estimated by choosing the density values corresponding to the background and nucleus peaks in the density histogram and then searching for the highest peak under each of these two density values in the color-density histogram. The third step is a clustering operation in the bivariate color-density histogram. A ceiling-lowering clustering algorithm, which incorporates a minimum distance classifier, is used with respect to a group of class means. The results of clustering are mapped back to the original blue-filtered and yellow-filtered cell images as shown in Figure 4(C). The fourth step is to clean the segmented images obtained, thus far. This is accomplished by removing the unconnected cytoplasm pixels and taking out the red cells and by contracting the cell boundary twice (each contraction is the stripping of one layer of boundary pixels) and then expanding it twice (each expansion is the attachment of one layer of boundary pixels). The latter operation removes all unwanted cytoplasm pixel fringes bordering the red cells. The boundaries of all the cytoplasm and nucleus are thus obtained; a final segmented cell image is shown in Figure 4(D).

During the course of research, a total of 133 features were measured from the cytoplasm and nucleus of both blue and yellow images. There were 15 geometric features, 19 density features, 25 color features, and 74 statistical texture features. The latter contained 64 texture feature $T_i(d)$'s based on the gray level co-occurrence matrix, and 10 texture feature T_{ri}'s based on the gray level run length matrix; one half of them were measured from the blue-filtered image and the other half from the yellow-filtered image. Before texture measurements were taken, each image was normalized by equal probability quantizing to 16 gray levels. These texture measures in blood cells are assumed to be rotationally invariant; hence, each texture feature was obtained by averaging four values of θ. The density features are defined by the optical densities in

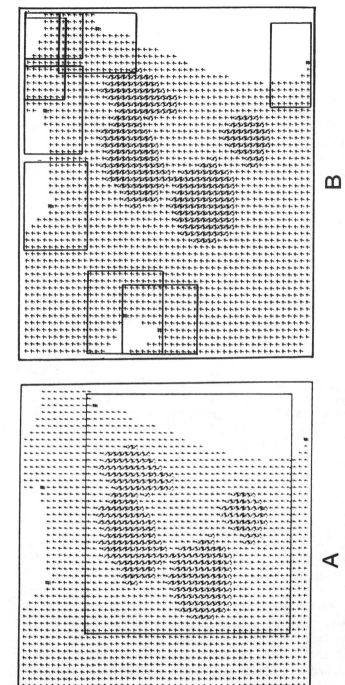

Figure 4A–B Segmentation of a nucleated blood cell. (*A*) The "density" image after the initial segmentation. # denotes a point of maximal concavity. The square window is used for estimation of an initial cluster center for cytoplasm pixels in the "color-density" histogram. (*B*) The same "density" image with small windows around points of maximal concavity for estimation of an initial cluster center of red cell pixels. Two small windows are also placed around end points of a short boundary curve of a red cell core in the upper right corner.

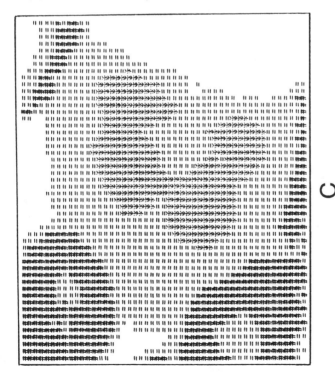

Figure 4C–D Segmentation of a nucleated blood cell. (*C*) Scene segmentation after clustering results are mapped back to an original cell image. (*D*) Final segmented cell image showing boundaries of cytoplasm and nucleus.

either cytoplasm or nucleus in terms of the whitened color-density histogram, which is also the basis of computation for quadrant moments of the cell. A geometric feature that is effective in separating seg from band is termed the ratio of two shrinks, g_{15}, which is defined as the minimum number of shrinks required to break a nucleus into two or more fragments divided by the minimum number of shrinks required to completely reduce the nucleus to zero area. Here, each shrink means the stripping of one layer of boundary pixels. For the tree classifier, discussed below, only 12 geometric features (g_i), 14 density features (d_i), 20 color features (c_i), and 28 texture features (t_i) have been selected. For detailed definitions of these features, see Reference (80).

A Binary Tree Classifier

The binary tree classifier designed by Mui & Fu (81) for classifying blood cells into 17 types is shown in Figure 5. A quadratic discriminant function with not more than 10 features is used at each nonterminal node. Nonterminal nodes are designated by a letter of the alphabet. Each cell type is designated by a class number from 1 to 17 as shown in Table 3. The numbers within parentheses that appear at each nonterminal node in Figure 5 are the classes to be separated, at that node, into

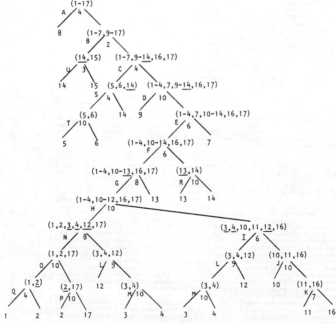

Figure 5 The binary tree classifier for the 17-class nucleated blood cell classification.

Table 2 List of the features and the bias used at each nonterminal node of the binary tree classifier in Figure 5

Nonterminal node	Features used in classification at individual node	Bias B_{12} in decision rule
A	c_1, d_{17}, c_{12}, d_7	0.0
B	d_{18}, d_9	0.0
C	g_3, c_1, c_6, c_{25}	0.0
D	$g_1, c_6, d_{17}, t_{13}, d_{12}, d_{13}, t_{15}, t_{59}, c_{13}, t_{66}$	0.0
E	$c_1, c_{14}, d_5, d_{13}, g_7, t_{51}$	-1.8
F	$g_3, d_{18}, c_{15}, t_{72}, g_{12}, d_{17}$	0.0
G	$g_3, d_{11}, t_{47}, c_{22}, t_{36}, t_7, t_{59}, t_{73}$	-1.8
H	$t_{51}, t_{45}, t_{23}, d_{16}, t_{63}, g_3, c_{24}, c_{13}, t_{36}, c_5$	0.0
I	$c_1, t_{40}, g_1, d_5, d_{19}, c_5$	0.0
J	$d_4, g_1, c_5, t_{11}, t_{51}, t_{25}, t_{71}, t_9, c_{24}, c_{16}$	-1.2
K	$c_4, t_3, c_{17}, c_8, g_7, t_{29}, t_{32}$	3.6
L	$c_6, d_{17}, t_9, c_7, d_4, c_4, c_{19}, c_{16}, g_2$	0.0
M	$c_1, g_{10}, g_{12}, d_{17}, c_{15}, t_{59}, d_7, c_{21}, g_3, t_{34}$	-2.0
N	$c_{14}, t_{21}, c_1, d_{16}, c_{13}, d_4, c_{11}, t_{58}$	1.8
O	$d_9, c_{14}, g_{14}, t_{37}, d_6, t_{57}, d_5, c_{16}, g_4, d_1$	0.0
P	$g_1, c_8, t_{72}, t_{71}, c_{21}, d_5, c_{14}, d_6, c_9, g_5$	0.0
Q	g_6, c_6, d_{12}, c_8	0.0
R	$d_9, g_{12}, d_{18}, t_{51}, d_6, t_{66}, g_3, c_{23}, g_9, t_{72}$	6.8
S	g_6, c_9, c_8, t_{43}	0.0
T	$g_{15}, c_7, c_{23}, c_{16}, d_{15}, c_5, g_7, t_{43}, t_{19}, c_{17}$	-3.4
U	g_6, d_4, t_7	2.0

two groups associated with the two descendant nodes, respectively. The underlined class numbers are the classes repeated, and the single class number without parentheses appears at each terminal node. The quadratic discriminant functions used at each node are the distance functions $d_i^2(\mathbf{x}, \mathbf{m}_i) = (\mathbf{x} - \mathbf{m}_i)' \Sigma_i^{-1} (\mathbf{x} - \mathbf{m}_i)$, $(i = 1, 2)$, where $\mathbf{m}_i$ and Σ_i are the mean and covariance matrix of class group ω_i respectively. The decision rule at each node is that $\mathbf{x}$ is classified into class group ω_1 if $d_1^2(\mathbf{x}, \mathbf{m}_1) + B_{12} < d_2^2(\mathbf{x}, \mathbf{m}_2)$; otherwise, $\mathbf{x}$ is classified into class group ω_2. The number below the vertex of each nonterminal node in Figure 5 indicates the number of features used in the discriminant functions at that node. The list of features and the bias value B_{12} used at each node is given in Table 2, where the features are listed in the order they were chosen by the sequential forward selection procedure.

This binary tree classifier has been designed with a data base of 1294 cells supplied by Tufts New England Medical Center (11). For the 17-class problem, the average number of samples per class was 75. The maximum number of features used at each nonterminal node was limited to 10 because of the number of features versus sample size consideration. The tree skeleton or hierarchical ordering of the class labels was interactively designed as described below (81). First, a canonical expansion on a chosen set of features, for example, 20 features used by Brenner et al (11), was performed. The within-class scatter matrix W and the total scatter matrix T were computed from the sample feature vectors, and the generalized eigenvalue problem: $Tv_i = \lambda_i W v_i$, $v_i' W v_j = \delta_{ij}$, (δ_{ij} is the Kronecker delta), was solved for the eigenvalues λ_i and the corresponding eigenvectors v_i. Consider the eigenvectors v_1 and v_2 corresponding to the two largest eigenvalues, λ_1 and λ_2, (hence, to the two largest values of the generalized variance ratio of total to within-class variances). Let all sample feature vectors be projected onto the two-dimensional subspace spanned by v_1 and v_2: $y_1 = v_1'x$ and $y_2 = v_2'x$; and let each projection be labeled by its class symbol. These projections were examined to see which class forms a cluster that is separable from the rest of the classes. For example, Class 8 (EOS) was identified as one group ω_1, and the remaining classes (1-7, 9-17) as the other group ω_2. The best features used for this two-category classification were selected, however, among the 133 measured features, by the sequential forward selection procedure according to the feature subset evaluation criterion J_2 which was defined previously. Hence, four features $(c_1, d_{17}, c_{12}, d_7)$ were selected at node A. Next, the group of classes (1-7, 9-17) was separated into two subgroups. From the display of the two-dimensional projections previously discussed, it was observed that some elements of Class 14 (NRB) were mingled inside the cluster of Class 15 (NRC). Classes 14, 15 were then considered as one group to be separated from the rest of the classes, excluding Class 8, which had already been taken care of. However, Class 14 was also included in the second group (1-7, 9-14, 16, 17) to facilitate the training of a two-category classifier at node B, with the goal of separating Class 15 at one of its descendent nodes and recognizing Class 14 at two different nodes. During this construction process, the canonical expansion on another feature subset could be examined to determine the class groupings at some intermediate nodes. For example, at node G, the canonical expansion of the best 20 features from 74 texture features was considered. The projections in the corresponding two-dimensional subspace were examined. It was observed that Class 13 (NRA) forms one group, Classes 10, 11, and 16 form another group, Classes 1, 2, and 17 form the third group, and Classes 3, 4, and 12 form the fourth group. Therefore, it was determined that Class 13 was to be separated from Classes 1-4, 10-12, 16, 17 at node G; and seven features were selected from the 74 measured features. This process of binary tree construction was repeated until all nonterminal nodes had been examined.

Table 3 Classification of nucleated blood cells into 17 classes using the binary tree classifier and the leave-one-out method of training and testing[a]

Hematologist classification	Class No.	Computed classification																	Accuracy (%)
		1	2	3	4	5	6	7	8	9	10	11	12	13	14	15	16	17	
Myeloblast (76)	1	50	10	0	0	0	0	1	0	0	2	0	6	0	0	0	1	6	66
Promyelocyte (42)	2	10	22	4	0	0	0	0	0	0	2	0	2	0	0	0	0	2	52
Myelocyte (47)	3	1	1	29	8	0	0	2	0	0	0	1	4	0	0	0	1	0	62
Metamyelocyte (61)	4	0	1	15	31	0	3	3	0	0	0	0	7	1	0	0	0	0	51
Band neutrophil (97)	5	0	0	1	0	86	5	0	0	0	0	0	0	3	2	0	0	0	89
Segmented neutrophil (124)	6	0	0	2	0	9	110	2	0	0	0	0	0	0	1	0	0	0	89
Basophil (100)	7	0	0	1	2	0	3	88	0	1	0	0	0	0	1	0	0	4	88
Eosinophil (77)	8	0	0	0	0	0	0	0	76	0	0	0	0	1	0	0	0	0	99
Lymphocyte (81)	9	0	0	0	0	0	0	2	0	64	9	0	2	0	3	0	0	1	79
Atypical lymphocyte (75)	10	2	2	1	0	0	0	3	0	8	49	3	4	0	0	0	1	2	65
Immunocyte (64)	11	1	0	0	0	0	0	2	0	0	6	39	1	1	0	0	12	2	61
Monocyte (84)	12	2	1	4	2	0	0	1	0	0	7	1	64	2	0	0	0	1	76
NRBC-A (29)	13	0	0	1	1	0	0	0	0	0	3	2	1	14	5	0	0	1	48
NRBC-B (110)	14	0	0	0	0	0	1	0	0	2	3	2	0	3	88	10	0	0	80
NRBC-C (106)	15	0	0	0	0	0	0	0	0	0	0	0	0	0	8	98	0	0	93
Plasmacyte (48)	16	1	0	0	0	0	0	2	0	0	2	7	1	0	1	0	34	0	71
Lymphoblast (73)	17	6	5	0	0	0	0	2	0	1	3	1	2	0	0	0	0	53	73
*	(%)	69	52	50	71	91	90	81	100	84	57	70	68	56	81	91	69	74	77 #

[a]NRBC = Nucleated red blood cell; * = percentage of cells classified for a class is actually from that class; # = overall accuracy. The number in parentheses is the number of cells in that class (from Reference 80).

The binary tree classifier was trained and tested on all of the 1294 available sample cells. The confusion matrix obtained by the leave-one-out method of training and testing is shown in Table 3. The best classification result is 99% for eosinophil. This is followed by 93% for nucleated red blood cell C and 89% for both band and segmented neutrophils. The worst classification result is 48% for nucleated red blood cell A. The overall classification accuracy is 77%. The result obtained by the resubstitution method of training and testing gives an overall accuracy of 83%, which can be regarded as the upper bound of the classification accuracy. These two testing results of the binary tree classifier are compared with the testing results of the single-stage Fisher linear classifier (11) using the same data base. The comparison demonstrates that this binary tree classifier is more accurate than the single-stage classifier by about 6 to 12%. The improvements in classification accuracy are especially good for myeloblast, band and segmented neutrophils, basophil, and nucleated red blood cells B and C.

CONCLUSION

In the diverse area of machine classification in medicine and biology, we have chosen to discuss the classification problems in lung images and blood cell images to illustrate the recent progress and the future promise. Pattern classification of the lung image presents two types of problems: One is to distinguish between normal vascular pattern and diffuse patterns of pulmonary infiltrations; the other is to recognize individual small lesions. It has been demonstrated that detailed quantitative analysis of chest radiographs makes it feasible to detect certain classes of low contrast small nodules or opacities which are sometimes missed by radiologists in their initial examinations. The false positive rate is rather high at the present stage of development. However, the potential exists for early detection of lung diseases. It has also been shown that statistical texture measure by the spatial gray level dependence method is effective in the classification of normal and pulmonary infiltration patterns, with an accuracy comparable to visual examination by radiologists. Various degrees of success have been reported on computer extraction of rib and vessel structures from chest radiographs. It has yet to be seen how these results can be efficiently incorporated to improve classification accuracy.

There has been significant progress in automating the classification of blood cells into 17 types of normal and abnormal cells; performance has been demonstrated comparable to that of hematologists. In particular, the classification accuracies of band and segmented neutrophils are surprisingly good. This improvement has been achieved through the new segmentation technique, selection of effective features, and classifier design. Some confusion, for example, between myeloblast and lymphoblast and between lymphocyte and atypical lymphocyte is also common to hematologists. But it can be expected that more quantitative analysis will further improve the machine performance.

These two endeavors in computer-assisted classification of medical images have followed the same approach of image processing and statistical pattern recognition. Innovative and systematic studies have led to the current level of success. It is apparent that computer-assisted classification methods can provide more quantitative information with higher reliability and consistency in performance than manual classification. At the present time, a great deal of processing time is required to perform machine classification tasks; this is especially true in the computer classification of radiographic images. However, the problem of speed can be solved by using new hardware designs and by developing new software techniques, once the methodology for meeting the desirable accuracy has been successfully developed. Hopefully, machine classification will be more accurate, consistent, and faster than human performance. Here lies a major area for further research in methodology appropriate for biomedical applications.

Literature Cited

1. Anbalagan, S. 1976. *A sequential classifier design for automatic leukocyte recognition.* PhD thesis. State Univ. New York, Stony Brook, NY. 278 pp.
2. Bacus, J. W., Gose, E. E. 1972. *IEEE Trans. Syst. Man Cybern.* SMC2: 513–26
3. Bacus, J. W., Aggarwal, R. K., Belanger, M. G. 1975. *IEEE EASCON '75 Record, Washington, DC*, pp. 73 A–F
4. Bacus, J. W. 1976. *Pattern Recognition*, 8:53–60
5. Ballard, D. H., Sklansky, J. 1976. *IEEE Trans. Comput.* C25:503–13
6. Ballard, D. H. 1978. *Proc. Int. Jt. Conf. Pattern Recognition, 4th, Kyoto*, pp. 907–10
7. Bartels, P. H., Chen, Y. P., Durie, B. G. M., Olson, G. B., Vaught, L., Salmon, S. E. 1978. *Acta Cytol.* 22:530–37
8. Bekey, G. A., Chang, C. W., Perry, J., Hoffer, M. M. 1977. *Proc. IEEE* 65:674–81
9. Belforte, G., DeMori, R., Ferraris, F. 1979. *IEEE Trans. Biomed. Eng.* BME 26:125–36
10. Bodenstein, G., Praetorius, H. M. 1977. *Proc. IEEE* 65:642–52
11. Brenner, J. F., Gelsema, E. S., Necheles, T. F., Neurath, P. W., Selles, W. D., Vastola, E. 1974. *J. Histochem. Cytochem.* 22:697–706
12. Bürsch, J. H., Heintzen, P. H. 1978. *Ann. Radiol.* 21:349–53
13. Cambier, J. L., Wheeless, L. L. 1978. *Acta Cytol.* 22:523–29
14. Carlton, S. G., Mitchell, O. R. 1977. *Proc. IEEE Conf. Pattern Recognition Image Process.*, Troy, NY, pp. 387–91
15. Chang, C. Y. 1973. *IEEE Trans. Syst. Man Cybern.* SMC3:166–71
16. Chien, Y. P., Fu, K. S. 1974. *IEEE Trans. Syst. Man Cybern.* SMC4:145–56
17. Conners, R. W., Harlow, C. A. 1978. *Comput. Graph. Image Process.* 8:447–63
18. Cooper, D. B., Elliott, H. 1978. *Proc. IEEE Conf. Pattern Recognition Image Process.*, Chicago, pp. 25–31
19. Cotter, D. A. 1973. *Am. J. Med. Technol.* 39:383–91
20. Cover, T. M. 1973. *IEEE Trans. Syst. Man Cybern.* SMC-4:116–17
21. Crawford, D. W., Brooks, S. H., Selzer, R. H., Barndt, R. Jr., Beckenbach, E. S., Blankenhorn, D. H. 1977. *J. Lab. Clin. Med.*

89:378-92
22. Dolce, G., Kunkel, H., eds. 1975. *CEAN Computerized EEG-Analysis.* Stuttgart: Fisher. 466 pp.
23. Duda, R. O., Hart, P. E. 1972. *Comm. ACM* 15:11-15
24. Dwyer, S. J. III. 1976. See Ref. 95, pp. 271-90
25. Foley, D. H. 1972. *IEEE Trans. Inf. Theory* IT18:618-26
26. Fong, C. P., Li, C. C., Savol, A. M., Shiek, C. F., Preston, K. Jr., Hoy, R. J., Sze, T. W., Sashin, D. 1978. See Ref. 6, pp. 919-24
27. Fritz, S. L. 1979. *Proc. Am. Coll. Radiol./IEEE Comput. Soc. Conf. Comput. Appl. Radiol. Comput.-Aided Anal. Radiol. Images, 6th, Newport Beach, Calif.*, pp. 129-31
28. Fu, K. S. 1968. *Sequential Methods in Pattern Recognition and Machine Learning.* New York: Academic. 227 pp.
29. Fu, K. S., Min, P. J., Li, T. J. 1970. *IEEE Trans. Syst. Man Cybern.* SMC6:33-39
30. Fu, K. S. 1974. *Syntactic Methods in Pattern Recognition*, New York: Academic. 295 pp.
31. Fu, K. S., Chien, Y. P., Persoon, E. 1975. See Ref. 3, pp. 72A-P
32. Fu, K. S., Pavlidis, T., eds. 1979. *Biomedical Pattern Recognition and Image Processing.* Berlin: Dahlem Konferenzen. 443 pp.
33. Fukunaga, K. 1972. *Introduction to Statistical Pattern Recognition.* New York: Academic. 369 pp.
34. Galloway, M. M. 1975. *Comput. Graph. Image Process.* 4:172-79
35. Giese, D. A., Bourne, J. R., Ward, J. W. 1979. *IEEE Trans. Syst. Man Cybern.* SMC9:429-35
36. Givens, A. S., Yeager, C. L., Diamond, S. L., Spire, J. P., Zeitlin, G. M., Gevins, A. H. 1975. *Proc. IEEE* 63:1382-99
37. Green, J. R. 1978. See Ref. 18, pp. 492-98
38. Hachimura, K., Kuwahara, M., Kinoshita, M. 1978. See Ref. 6, pp. 911-13
39. Hall, E. L., Crawford, W. O. Jr., Preston, K. Jr., Roberts, F. E. 1973. *Proc. Int. Jt. Conf. Pattern Recognition, 1st, Washington DC*, pp. 77-87
40. Hall, E. L., Crawford, W. O. Jr., Roberts, F. E. 1975. *IEEE Trans. Biomed. Eng.* 22:518-27
41. Hall, E. L., Kruger, R. P., Turner, A. F. 1975. *Proc. IEEE Conf. Comput. Graph. Pattern Recognition Data Struct.*, Los Angeles, pp. 75-83
42. Hankinson, J. L. 1977. *Proc. IEEE Int. Conf. Cybern. Soc. Washington, DC*, pp. 419-21
43. Hankinson, J. L., Stewart, J. E. 1979. *Proc. IEEE Conf. Pattern Recognition Image Process., Chicago*, pp. 353-55
44. Haralick, R. M., Shanmugam, K., Dinstein, I. 1973. *IEEE Trans. Syst. Man Cybern.* SMC3:610-21
45. Haralick, R. M., Dinstein, I. 1975. *IEEE Trans. Circuits Syst.* CAS22:440-50
46. Haralick, R. M. 1979. *Proc. IEEE* 67:786-804
47. Harlow, C. A., Eisenbeis, S. A. 1973. *IEEE Trans. Comput.* C22:678-89
48. Harlow, C. A., Dwyer, S. J. III, Lodwick, G. 1976. In *Digital Picture Analysis*, ed. A. Rosenfeld, pp. 65-150. Berlin: Springer
49. Hasegawa, J. I., Toriwaki, J. I., Fukumura, T., Takagi, Y. 1978. See Ref. 6, pp. 925-27
50. Heintzen, P. H., Bürsch, J. H., eds. 1978. *Roentgen Video Techniques for Dynamic Studies of Structure and Function of the Heart and Circulation.* Stuttgart: Thieme. 338 pp.
51. Horowitz, S. L. 1975. *Comm. ACM* 18:281-85
52. Huang, T. S., Yang, G. J., Tang, G. Y. 1979. *IEEE Trans. Acoust. Speech Signal Process.* ASSP27:13-18
53. Hueckel, M. H. 1971. *J. Assoc. Comput. Mach.* 18:113-25
54. Ingram, K., Preston, K. Jr. 1970. *Sci. Am.* 223:77-82
55. Jacobson, G., Lainhart, W. S., eds. 1972. *Med. Radiogr. Photogr.* 48:66-103
56. Jagoe, J. R., Paton, K. A. 1975. *Br. J. Ind. Med.* 32:267-72
57. Jagoe, J. R. 1979. *Comput. Biomed. Res.* 12:1-15
58. Julesz, B., Gilbert, E. N., Shepp, L. A., Frisch, H. L. 1973. *Perception* 2:391-405
59. Kinius, P., Fozzard, H. A. 1979. In *Computer Techniques in Cardiology*, ed. L. D. Cady Jr., pp. 97-122. New York: Dekker
60. Kruger, R. P., Thompson, W. B., Turner, A. F. 1974. *IEEE Trans. Syst. Man Cybern.* SMC4:40-49
61. Kruger, R. P., Hall, E. L., Turner, A. F. 1977. *Appl. Opt.* 16:2637-46
62. Kulick, J. H., Challis, T. W., Brace, C., Christodoulakis, S., Merritt, I., Neelands, P. 1976. *Proc. Int. Jt. Conf. Pattern Recognition., 3rd, Coronado, Calif.* pp. 233-37
63. Ledley, R. S., Huang, H. K., Rotolo,

L. S. 1975. *Comput. Biol. Med.* 5:53–67
64. Li, C. C., Savol, A. M., Fong, C. P., Shiek, C. F., Sze, T. W., Preston, K. Jr., Sashin, D., Hoy, R. 1977. *Proc. IEEE Conf. Cybern. Soc.*, Washington, DC, pp. 422–28
65. Li, C. C., Savol, A. M., Fong, C. P. 1978. See Ref. 18, pp. 390–95
66. Li, C. C., Strintzis, M. G., Stanis, J. W., Newman, J., Kunkel, G. A., Wadhwani, B. J. 1979. *Proc. IEEE Eng. Med. Biol. Soc. Ann. Conf., 1st*, Denver, pp. 110–13
67. Li, C. C., Tai, T. H., Todhunter, J. S., Ayres, M., Preston, K. Jr., Hoy, R. J., Kunkel, G. A. 1979. *Proc. IEEE Int. Conf. Cybern. Soc.*, Denver, pp. 144–46
68. Lin, Y. K., Fu, K. S. 1979. See Ref. 43, pp. 479–83
69. Lissack, T., Fu, K. S. 1976. *IEEE Trans. Inf. Theory* IT22:34–45
70. Liu, H. C. 1976. *Shape description and characterization of continuous change.* PhD thesis. State Univ. New York, Stony Brook, NY. 244 pp.
71. Lu, S. Y., Fu, K. S. 1978. *Comput. Graph. Image Process.* 7:303–30
72. Martelli, A. 1976. *Comm. ACM* 19:73–83
73. Melga, G. K. 1973. *Acta Cytol.* 17:3–14
74. Miller, M. N. 1974. *Design and clinical results of Hematrak—an automated differential counter.* Presented at Int. Jt. Conf. Pattern Recognition, 2nd., Copenhagen
75. Montanari, U. 1971. *Comm. ACM* 14:335–45
76. Moore, G. E. 1978. *A systematic approach to computer analysis of pulmonary vascular patterns.* PhD thesis. Univ. Missouri, Columbia, Mo. 273 pp.
77. Mucciardi, A. N., Gose, E. E. 1971. *IEEE Trans. Comput.* C20:1025–31
78. Mui, J. K., Bacus, J. W., Fu, K. S. 1976. *Proc. IEEE Symp. Comput.-Aided Diagn. Med. Images*, Coronado, Calif. pp. 99–106
79. Mui, J. K., Fu, K. S., Bacus, J. W. 1977. *J. Histochem. Cytochem.* 7:633–40
80. Mui, J. K., Fu, K. S. 1978. *TR-EE 78-47*, School Electr. Eng., Purdue Univ., West Lafayette, Ind.
81. Mui, J. K., Fu, K. S. 1979. *Automated classification of nucleated blood cells using a binary tree classifier.* Presented at Int. Symp. Pattern Recognition Cell Images, Univ. Chicago, Ill.
82. Narendra, P. M., Fukunaga, K. 1977. *IEEE Trans. Comput.* C26:917–22
83. Nejdl, J. F., Gose, E. E., Kaplan, E. 1978. See Ref. 6, pp. 914–18
84. Neurath, P. W., Brenner, J. F., Selles, W. D., Gelsema, E. S., Powell, B. W., Gallus, G., Vastola, E. 1974. In *International Computing Symposium*, ed. A. Gunther, B. Levart, H. Lipps, pp. 399–405. Amsterdam: North-Holland
85. Niitani, K. 1976. *Evaluation of Hematrak Automatic Counter.* Presented at Int. Congr. Hematol., 16th, Kyoto
86. Patrick, E. A. 1979. *Decision Analysis in Medicine, Methods and Applications.* West Palm Beach, Fla: CRC. 352 pp.
87. Pauker, S. G., Gorry, G. A., Kassirer, J. P., Schwartz, W. B. 1976. *Am. J. Med.* 60:981–96
88. Pavlidis, T. 1978. *Structured Pattern Recognition.* Berlin: Springer. 302 pp.
89. Persoon, E. 1976. *Comput. Graph. Image Process.* 5:425–47
90. Pipberger, H. V., Dunn, R. A., Berson, A. S. 1975. *Ann. Rev. Biophys. Bioeng.* 4:15–42
91. Pople, H. E., Myer, J. D., Miller, R. A. 1975. *Proc. Int. Jt. Conf. Artif. Intell., 4th*, Tbilisi, USSR, pp. 848–55
92. Pratt, W. K. 1978. *Digital Image Processing.* New York: Wiley. 750 pp.
93. Pressman, H. J. 1976. *J. Histochem. Cytochem.* 24:138–44
94. Preston, K. Jr. 1976. See Ref. 48, pp. 216–49
95. Preston, K. Jr., Onoe, M., eds. 1976. *Digital Processing of Biomedical Images.* New York: Plenum. 442 pp.
96. Preston, K. Jr. 1979. *Proc. IEEE* 67:826–56
97. Prewitt, J. M. S. 1978. *Technical Report*, Dep. Comput. Sci., Uppsala Univ., Dep. Clin. Cytol., Uppsala Univ. Hosp., Sweden
98. Rosenfeld, A. 1977. See Ref. 14, pp. 14–18
99. Savol, A. M., Li, C. C., Hoy, R. J. 1978. *Proc. BIOSIGMA 78*, Paris 2:245–50
100. Schwartz, A. A., Soha, J. M. 1977. *Appl. Opt.* 16:1779–81
101. Schwartz, M. D. 1979. See Ref. 59, pp. 243–59
102. Selzer, R. H., Blankenhorn, D. H., Crawford, D. W., Brooks, S. H., Barndt, R. Jr. 1978. See Ref. 99. 2:267–75

103. Shipton, H. W. 1975. *Ann. Rev. Biophys. Bioeng.* 4:1–13
104. Shortliffe, E. H. 1976. *Computer-based Medical Consultations: MYCIN.* New York: Am. Elsevier. 264 pp.
105. Sklansky, J., Sankar, P., Katz, M., Towfig, F., Hassner, D., Cohen, A., Root, W. 1979. See Ref. 43, pp. 484–87
106. Spiesberger, W. 1979. *IEEE Trans. Biomed. Eng.* 26:213–19
107. Stark, H., Lee, D. 1976. *IEEE Trans. Syst. Man Cybern.* SMC6:788–93
108. Stearns, S. D. 1976. See Ref. 62, pp. 71–75
109. Stockman, G., Kanal, L., Kyle, M. C. 1976. *Comm. ACM* 19:688–95
110. Sutton, R. N., Hall, E. L. 1972. *IEEE Trans. Comput.* C21:667–76
111. Taylor, J., Bartels, P. H., Bibbo, M., Wied, G. L. 1978. *Acta Cytol.* 22:261–67
112. Toriwaki, J. I., Suenaga, Y., Negoro, T., Fukumura, T. 1973. *Comput. Graph. Image Process.* 2:252–71
113. Toriwaki, J. I., Hasegawa, J. I., Fukumura, T. 1976. See Ref. 78, pp. 1–8
114. Tou, J. T., Gonzalez, R. C. 1974. *Pattern Recognition Principles.* Reading, Mass: Addison-Wesley. 377 pp.
115. Tou, J. T., Liu, H. H. 1979. See Ref. 43, pp. 453–59
116. Toussaint, G. T. 1974. *IEEE Trans. Inf. Theory* IT20:472–79
117. Tsai, M. J., Pimmel, R. L., Donohue, J. F. 1979. *IEEE Trans. Biomed. Eng.* 26:293–98
118. Tully, R. J., Conners, R. W., Harlow, C. A., Lodwick, G. S. 1978. *Invest. Radiol.* 13:298–305
119. Turner, A. F., Kruger, R. P., Thompson, W. B. 1976. *Invest. Radiol.* 11:256–66
120. Vagnucci, A. H. 1979. *Am. J. Physiol.* 236:R268–81
121. Van Bemmel, J. H. 1977. In *MEDINFO-77*, ed. D. B. Shires, H. Wolf, pp. 819–27. Amsterdam: North-Holland
122. Van Bemmel, J. H., Willems, J. L., eds. 1977. *Trends in Computer-Processed Electrocardiograms.* Amsterdam: North-Holland. 437 pp.
123. Vidal, J. J. 1977. *Proc. IEEE* 65:633–41
124. Warner, H. R. 1972. *Comput. Biomed. Res.* 5:65–74
125. Wechsler, H., Sklansky, J. 1977. *Pattern Recognition*, 9:21–30
126. Wechsler, H., Fu, K. S. 1978. *Comput. Graph. Image Process.* 7:375–90
127. Weiss, S. G., Kulikowski, A., Safir, A. 1978. *Comput. Biol. Med.* 8:25–40
128. Weszka, J. S., Rosenfeld, A. 1978. *IEEE Trans. Syst. Man Cybern.* SMC8:622–29
129. Wong, A. K. C., Vagnucci, A. H., Liu, T. S. 1976. *IEEE Trans. Syst. Man Cybern.* SMC6:33–45
130. Wood, E. H., Robb, R. A., Ritman, E. L. 1978. In *Coronary Heart Disease*, ed. M. Kaltenbach, P. Litchlen, R. Balcon, W. D. Bussman, pp. 106–15. Stuttgart: Thieme
131. Young, I. T. 1972. *IEEE Trans. Biomed. Eng.* B19:291–98
132. Young, T. Y., Calvert, T. W. 1974. *Classification, Estimation and Pattern Recognition.* New York: Am. Elsevier. 366 pp.

CERTAIN SLOW SYNAPTIC RESPONSES: THEIR PROPERTIES AND POSSIBLE UNDERLYING MECHANISMS

♦9153

JacSue Kehoe and Alain Marty

Laboratoire de Neurobiologie, Ecole Normale Supérieure, 46, rue d'Ulm, Paris, France 75005

INTRODUCTION

A recent series of elaborate kinetic studies (see 137 for review) has taught us much about the mechanisms underlying rapid potential changes elicited by chemical transmitters in muscle fibers and in neurons. Most rapid synaptic events are assumed to result from the interaction of the transmitter substances with a specialized membrane receptor which, in turn, directly activates a well-defined, otherwise inoperative permeation process. These synaptic events have been well accounted for by a two-state model incorporating only one rate-limiting reaction.

In contrast, many other transmitter-induced events described over the last 15 years are difficult to explain in these relatively simple terms because of the peculiarity of the permeability changes being controlled, the especially long latencies often observed, the slow time courses involved, and/or the seeming necessity to implicate a second messenger intervening between the receptor activation and the change in membrane permeability.

These rather disparate synaptic events have been grouped together under the heading of slow synaptic potentials and are often illustrated by the slow inhibitory postsynaptic potential (IPSP) and slow excitatory postsynaptic potential (EPSP) seen in neurons of the sympathetic ganglion, or by the potential changes elicited in smooth muscle by catecholamines or muscarinic agents. Because recent reviews have described the state of knowledge concerning these classical examples of slow synaptic potentials (17, 103), we have selected other examples which we hope will illustrate the variety of mechanisms proposed for explaining slow synaptic events.

We describe, first of all, a series of slow potassium-dependent inhibitory potentials which, like classical EPSPs and excitatory junctional potentials (EJPs), seem to result from the activation, direct or indirect, of a "new" conductance channel, i.e. one that does not contribute to the resting current-voltage (I/V) relationship. Different two-state and multistep kinetic models that have been proposed to account for certain temporal characteristics of these inhibitory potentials are discussed.

We then turn to another series of synaptic events which, in contrast to the K-dependent inhibitory responses, seem to be due to the increase or decrease by synaptic transmitters of voltage-dependent conductances that are present even when the transmitter receptors are not activated. To explain these "modulatory" effects, many authors have suggested that the agonist-receptor binding does not act directly to cause a change in membrane conductance, but rather controls cyclic nucleotide levels which, in turn, determine the amplitude of the voltage-dependent membrane permeability. The experimental support for this second-messenger hypothesis is discussed for the different synaptic events presented.

This review is not intended to provide an exhaustive coverage, even of these selected points. Rather, it is designed to describe the type of experiments that have been performed for analyzing each of the synaptic events described, and to evaluate the adequacy of the information that has been obtained for telling us something about the mechanisms underlying these selected synaptic events.

SLOW SYNAPTIC RESPONSES MEDIATED BY AN INCREASE IN PERMEABILITY TO POTASSIUM IONS

K-dependent inhibitory potentials, all having a relatively slow time course, have been described in many preparations. We consider four preparations where such responses occur and where a special effort has been made to understand their kinetic peculiarities.

Description of Four Experimental Preparations Showing K-Dependent Responses

HEART MUSCLE Application of acetylcholine (ACh) to heart muscle activates a specific K conductance, as shown by electrophysiological (52, 60, 170, 173) and tracer flux (71, 79) studies.

PARASYMPATHETIC NEURONS OF THE HEART   Parasympathetic neurons of the heart of the mudpuppy present both a fast, nicotinic, excitatory and a slow, muscarinic, inhibitory response upon application of ACh. Postsynaptic effects with the same kinetic and pharmacological properties are obtained when the vagus nerve is stimulated. Hartzell et al (72) first described the slow inhibitory response and provided firm electrophysiological evidence that it is due to a conductance increase selective for K ions. They underlined the similarity between the muscarinic responses obtained in muscle cells and in parasympathetic neurons in the same preparation.

SALIVARY GLANDS OF MAMMALS AND INSECTS   Secretory cells of mammalian salivary glands possess α-adrenergic, muscarinic, and peptidic receptors that mediate slow, K-dependent electrical responses (see 62, 145, 152). In the salivary gland of the cockroach, acinar cells have receptors for catecholamines, and dopamine may be the neurotransmitter at the corresponding neuroglandular synapse (18, 64). The hyperpolarizing response to dopamine application or to salivary nerve stimulation is due to an increase in permeability to K ions (63).

MOLLUSCAN NEURONS   During the past twenty years a large variety of synaptic responses has been described in the nervous system of molluscs. In these preparations, it is not uncommon that a given neurotransmitter gives rise to several types of electrical responses associated with different receptor properties (6, 57, 98 for reviews).

Specific increases in K conductance have been demonstrated upon application to molluscan neurons of ACh (56, 92, 94, 95), dopamine (5, 13, 14), 5-hydroxytryptamine (5HT) (58, 59), histamine (68, 120, 187, 188), γ-aminobutyric acid (GABA) (192), glutamate (113, 169), octopamine and phenylethanolamine (33, 119, 159). From electrophysiological and/or biochemical evidence, it appears that these K-dependent responses serve a synaptic role at least for ACh (93–95), dopamine (13, 14), 5HT (59), and histamine (120, 187, 188) (see also 6, 98 for reviews). Recently, neuronal responses to a small peptide (37) and to a 6000 dalton protein (19, 26) have been reported where the substances involved seem to have the role of a neurohormone rather than of a neurotransmitter (see also 20). In certain cells, responses to these peptides are, at least in part, due to a K-conductance increase.

It has been stressed (6, 97, 168) that the different K-dependent responses share specific kinetic properties which stand in sharp contrast to those of the excitatory or Cl-mediated inhibitory responses obtained on the same cells with the same compounds. All K-dependent responses are characterized by a slow time course and a delayed onset, apparent both with ionophoretic application and when the neurotransmitter is released by presynaptic stimulation. In addition, they present a high sensitivity to cooling (see following section).

Failure of Diffusion Models to Account for Latency and Kinetics of K-Dependent Responses

TESTS OF DIFFUSION MODELS WITH IONOPHORESIS EXPERIMENTS Del Castillo & Katz (41) showed that the inhibitory response obtained by vagal stimulation in the sinus venosus of the frog exhibits a marked delay (greater than 100 ms). Ionophoretic application of ACh to frog heart muscle cells also evokes a response with a pronounced latency (150). A similarly delayed response was obtained when applying ionophoretically bethanecol (a muscarinic agonist) on parasympathetic ganglion cells (72), dopamine on cockroach salivary gland cells (15, 16), or ACh on *Aplysia* neurons (92).

These results show that the origin of the delay is not presynaptic, and they strongly suggest that it is not due to diffusion of the neurotransmitter to the receptors. However, several authors have recently considered the possibility that diffusion of the neurotransmitter could be hindered by several factors: (*a*) the receptors could be located at the bottom of narrow invaginations of the cell membrane; (*b*) a space of restricted diffusion speed (such as a basement membrane) could be interposed between the ionophoretic pipette and the receptors; or, (*c*) diffusion of the neurotransmitter could be affected by its binding to the cell membrane (75). None of these models can account quantitatively for the data (15, 16, 75, 150). In particular, the models predict larger variations of the delay and time to peak of the response with ionophoretic dose or micropipette positioning than those which are observed. Furthermore, morphological evidence in favor of the structures proposed in (*a*) and (*b*) above is lacking.

TEMPERATURE SENSITIVITY The K-dependent responses in molluscan neurons to at least some of the neurotransmitters [ACh (94, 97); dopamine (5); histamine (187); 5HT, Paupardin-Tritsch, personal communication] share a remarkably high sensitivity to cooling. These responses become much slower (an effect particularly noticeable during the initial rise of the response) and smaller as the temperature is reduced, and they are often virtually abolished below 8°C. In contrast, excitatory or Cl-dependent inhibitory responses to the same transmitters are much less affected by cooling.

A similarly high temperature sensitivity has been described for the ACh response in heart muscle (148) and in the principal cells of the parasympathetic ganglion of the heart (72), as well as for the response to dopamine in the salivary glands of the cockroach (16). In these preparations, although the rising phase of the response is very temperature sensitive, the total area under the response (as defined by $\int_0^\infty (\Delta v) dt$, where Δv is the agonist-induced hyperpolarization, and t the lag after the ionophoretic pulse) is not much affected by temperature (e.g. see Figures 8 and 9 in 16). This seems to contrast with the results obtained in molluscs, where cooling eventually abolishes the response. However, the difference may merely reflect the range of temperatures explored, which was usually confined to between 5 and 20°C in molluscs and between 15 and 35°C in vertebrate or insect preparations.

The high temperature sensitivity of the rising phase of K-dependent responses is yet another argument for rejecting diffusion as the origin of the latency (e.g. see 16, 72, 75).

Evidence in Favor of a Channel Type Permeation Mechanism

In the preparations where appropriate tests have been made, the evidence strongly favors a channel rather than a carrier system in the permeation mechanisms of the K-dependent response.

In *Aplysia* neurons, the instantaneous current associated with the K-selective response to ACh shows changes with membrane potential and with outer K concentration which are consistent with diffusion models through ionic channels (65, 162). Such changes would be very difficult to reconcile with a carrier mechanism. Further, cesium ions have a voltage-dependent blocking action that is also very suggestive of a channel mechanism (R. Kado, personal communication). In the rabbit sinoatrial node, recent experiments have demonstrated an increase in noise variance in the presence of bath-applied ACh (139, 141). The size of the elementary conductance obtained from noise analysis (4 pS) is too large to be accounted for by a carrier mechanism. In addition, the shape of the noise power spectrum associated with the ACh-induced current (see below) is similar to that found with channels rather than with carriers.

Evidence for Long Channel Open Times

The mean open time of synaptic channels may be determined from the analysis of synaptic current decay (114), noise power spectrum of agonist-induced current (2, 89), voltage-jump induced relaxations, (1, 135) and direct recordings of the opening of single channels (136). All of these methods were first developed at the frog end-plate where the results may be explained by a two-state model (42, 115, and 137 for review) which assumes a fast binding reaction to one or several agonist molecules followed by a slower, rate-limiting conversion between "bound closed" and "bound open" states. It is tempting at first to adopt the same type of kinetic model in the case of slow synaptic responses, with the simple change that the channel open time would be much longer than at the frog end-plate. Certain channels studied in artificial bilayers have very long open times (tens of seconds; see 137). In line with this type of interpretation, three of the methods mentioned above were recently applied to slow K-dependent responses to ACh and all revealed channels with very long open times.

SYNAPTIC CURRENT DECAY   Hartzell et al (72) have analyzed the time course of decay of muscarinic responses in voltage-clamped frog parasympathetic neurons. They found that this decay may be fitted with an exponential function with a time constant of about 2 s. This time constant was found to increase with cell hyperpolarization, thus suggesting that neither hydrolysis of ACh nor diffusion away from the synaptic area was the rate-limiting step governing the decay phase.

Similar results have been obtained at a cholinergic synapse in *Aplysia* where the K-mediated synaptic currents were found to decay with a time constant of about 1 s (116). Here again, the time course of decay was found to increase with hyperpolarization.

By analogy with the more thoroughly studied situation found at the neuromuscular junctions of vertebrates, these results suggest that the mean open time τ of ACh sensitive channels in parasympathetic neurons and in *Aplysia* neurons is very long. In addition, they suggest that τ increases with cell hyperpolarization. This is again parallel to what is seen at the frog neuromuscular junction, as well as when studying several other types of "fast" ACh sensitive cationic channels (137).

NOISE POWER SPECTRA   Noma & Trautwein (139) have been able to detect an increase in the variance of the membrane noise in the presence of ACh in rabbit sinoatrial node. The Fourier analysis of the extra noise associated with the ACh-induced current (which yields the "noise power spectrum") demonstrated that the main part of the variance increase occurred in the very low frequency range (141). This indicates that the corresponding channels must have a very long duration. The noise power spectrum could be fitted with a single Lorentzian function, as is expected from the two-state model. The value obtained for τ was 160 ms. However, the preparation used is subject to several technical limitations for noise analysis, so that the more quantitative conclusions (i.e.

the fact that the noise power spectrum is Lorentzian, and the value of τ) should be considered as tentative.

VOLTAGE JUMPS The voltage jump method has been applied to K-dependent responses in *Aplysia* neurons (116) and in rabbit heart muscle cells (139, 140). In both preparations, exponential relaxations were obtained, with time constants of about 2 s (in *Aplysia*) and 0.1 s (in rabbit heart). These time constants were found to increase with hyperpolarization. These results indicate again that the mean channel open time, τ, is quite large, and that it increases as the cell potential is made more negative.

Two-State vs Multistep Models

TWO-STATE MODEL According to the two-state model that has been used at the vertebrate end-plate, the activation of a channel may be written

$$nA + R \underset{\text{fast}}{\rightleftharpoons} RA_n \underset{\text{rate-limiting}}{\rightleftharpoons} (RA_n)^* \qquad 1.$$

Here, n agonist molecules (A) bind to a receptor (R) in a comparatively rapid set of reactions. This is followed by a rate-limiting equilibrium between the closed (RA_n) complex and the activated $(RA_n)^*$ complex, corresponding to the open channel (42, 115). Because only one reaction is rate limiting, this model is kinetically equivalent to a single equilibrium between two states of the receptor channel complex, closed and open.

The two-state model (see Equation 1, above) has been used by Noma & Trautwein (139, 140) and Noma et al (141) to account for their noise and relaxation results on heart cells. In support of this interpretation, they found that, as in the frog end-plate (see 137), the steady-state conductance and the time constant of the relaxation displayed similar variations with membrane potential. In addition, the time constants found in noise and relaxation experiments were similar. (They should be equal according to the two-state model).

In contrast to these findings, the time constant of relaxations obtained in *Aplysia* neurons has a weaker dependence on voltage than that of the steady state number of open channels. In addition, this time constant is larger than the time constant of decay of synaptic currents. Although this last result may also reflect differences between synaptic and extrasynaptic receptors, the results obtained in *Aplysia* are not readily explained by the two-state model and have been interpreted using a more complex kinetic scheme (116).

A stronger argument against the applicability of the two-state model to slow K-dependent responses comes from the sigmoid rise of these responses. Following a short ionophoretic application of bethanecol, the current obtained in heart parasympathetic neurons or muscle cells rises as kt^3, where t is the time following the application (72). A kt^3 rise was also found by Pott (148) in heart muscle cells, and by Ger et al (55) in molluscan neurons. The results of Blackman et al (16) in cockroach salivary glands were also compatible with a kt^3 or kt^4 rise if an additional delay was introduced to account for the diffusion time of dopamine. It is rather remarkable that these results do not appear to depend on temperature (55, 148).

If the two-state model (see Equation 1, above) held true for slow synaptic responses, a sudden and sustained agonist application would lead to an immediate exponential rise of current. This is not compatible with the observations made on the rising phase of ionophoretic responses (above). Similar observations made by many authors have led them to point out that the sigmoid rise of slow synaptic responses requires that several steps rather than one rate-limiting reaction govern the opening of the channels (e.g. 15, 16, 55, 66, 72, 75, 116, 148).

Thus, in our opinion (but see 140) the kinetic differences between "fast" and "slow" synaptic responses must extend beyond a difference in the mean channel open times.

MULTISTEP MODELS   To interpret their results on the shape of the initial part of the response and on its temperature sensitivity, Hartzell et al (72) proposed that several (at least 3) reactions occur in series between the binding of the cholinergic agonist and the opening of the channel, and that none of these reactions is rate limiting. In the simplest case this sequence may be written

$$A \underset{k_{-1}}{\overset{k_1}{\rightleftharpoons}} B \underset{k_{-2}}{\overset{k_2}{\rightleftharpoons}} C \underset{k_{-3}}{\overset{k_3}{\rightleftharpoons}} D$$

where A represents the agonist-receptor complex and D corresponds to the open channel. The steps between A and D may simply reflect successive conformation changes of the agonist-receptor complex, but other mechanisms may also contribute to one or several of these steps. Two examples of such possible intermediate reactions are now considered.

Cuatrecasas (39) suggested that some hormone responses depend on separate receptor and adenylate cyclase units. In this model, both units "float" freely in the membrane in the absence of hormone. Following hormone-receptor interaction, the two units associate, leading to the activation of the cyclase. This theory has gained much experimental

SLOW SYNAPTIC RESPONSES   445

support in the last five years (e.g. 85, 171). It appears that a third proteic unit, containing a GTP-binding site, is probably involved in the receptor cyclase interaction (see 111 for review). A similar mechanism could operate in the case of slow K responses if the opening of the channels depends on intracellular cyclic AMP (cAMP). Even if cyclic nucleotides are not involved in these responses, one can envisage that receptor and channel are separate proteins and that the opening of a channel results from its collision with an activated receptor molecule.

A second example of a multistep model has been recently proposed to account for relaxation and synaptic current decay properties of the K-dependent response to ACh in *Aplysia* neurons (116). This model is similar to the Hodgkin-Huxley model of K currents in axons and explains the sigmoid rise of the current, during a sustained application of the agonist, by the necessity to activate simultaneously several independent subunits. Because the current is occasionally still increasing several hundred milliseconds after the end of the ACh application, it was further assumed that the equilibrium between the activated subunits and the active complex (corresponding to an open channel) is not immediate. These hypotheses may be written:

$$A + S \underset{v}{\overset{u}{\rightleftharpoons}} AS \quad \text{(activation of a subunit S by one agonist molecule A)}$$

2.

$$\alpha(AS) \underset{s}{\overset{r}{\rightleftharpoons}} (A_\alpha S_\alpha)^* \quad \text{(opening step per se)}$$

3.

Equations 2 and 3 are equivalent to the following multistep model (see 4)

$$S_\alpha \underset{v}{\overset{\alpha u}{\rightleftharpoons}} S_\alpha A \underset{2v}{\overset{(\alpha-1)u}{\rightleftharpoons}} S_\alpha A_2 \underset{3v}{\overset{(\alpha-2)u}{\rightleftharpoons}} \ldots \underset{\alpha v}{\overset{u}{\rightleftharpoons}} S_\alpha A_\alpha \underset{s}{\overset{r}{\rightleftharpoons}} (A_\alpha S_\alpha)^*$$

$[(A_\alpha S_\alpha)^*$ being the open channel].

The value $\alpha = 3$ accounts for the experimental results.

Possible Involvement of Intermediate Compounds

Many authors have considered the possibility that the opening of the channel would require the accumulation of an intracellular compound. This may be considered as a multistep model, with the additional feature that the intermediate compound (e.g. Ca or cAMP) is likely to be buffered by a set of cellular reactions (see 8 for an example of such a scheme).

INTRACELLULAR CALCIUM AS AN INTERMEDIATE COMPOUND   Intracellular Ca concentrations as low as $10^{-6}M$ are known to increase the K permeability of molluscan (122) and mammalian (49) neurons (124 for review). A first step in slow K-dependent responses could be provided by an increase in intracellular Ca concentration. Activation of a membrane receptor could directly open a calcium selective pathway (pore or carrier), leading to a movement of Ca from the extracellular solution into the cell where it would act to increase K permeability. Such an increase in K permeability could also be triggered by a release of Ca from intracellular stores. In rat parotid slices, Selinger et al (163) found that extracellular Ca is necessary for stimulating K release by epinephrine. Furthermore, Selinger et al (164) showed that introduction of Ca via the ionophore A23187 produces massive K release. From these observations, they suggested that epinephrine opens a Ca-selective pathway and that the resulting increase in intracellular Ca provokes the K release. Putney (reviewed in 152) distinguished an early part of the K release that is Ca independent and a later part that requires the presence of extracellular Ca. According to Putney (151) the first part of the response would be due to the release of a finite pool of intracellular Ca.

Since electrophysiological studies often reveal a depolarization followed by a slow K-mediated hyperpolarization in salivary glands upon stimulation by ACh (145), it is tempting to ascribe Ca entry to the former part of the response. However, whereas Douglas & Poisner (44) showed that the secretion of saliva is Ca dependent, Peterson & Pedersen (146) showed that the electrical response to epinephrine is not suppressed in a Ca-free EGTA-containing solution. It is therefore possible that the Ca-dependent K release in mammalian salivary glands is linked to exocytosis, rather than to an electrical response due to the opening of membrane channels.

The Ca accumulation hypothesis has also been proposed for a number of other preparations (see review, 152). In smooth muscle of guinea pig taenia coli, α-adrenergic agonists cause an increase in K permeability (28, 86) which is dependent on extracellular Ca (29). Likewise, in insect salivary glands the K-dependent response to repeated iono-phoretic applications of dopamine disappears in calcium-free solutions (62).

In some of the preparations considered here, however, the present evidence fails to support the hypothesis that Ca accumulation is an intermediate step in slow synaptic responses. In *Aplysia*, Kehoe (92) showed that the slow ACh response persists in a Ca-free EGTA-containing solution. Even more clear-cut evidence against the calcium hypothesis in this preparation is provided by the experiment illustrated in Figure 1. EGTA was injected into an *Aplysia* neuron so that the increase in K permeability following a train of action potentials was suppressed. This K permeability increase is mainly related to Ca entry during the action potential (123). The fact that the after-burst hyperpolarization was

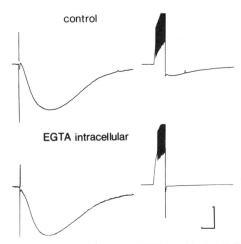

Figure 1 Failure of an intracellular injection of EGTA to block a K-dependent response to ionophoretically applied ACh in a cell in which a potassium conductance increase known to result from an influx of calcium was completely eliminated. *Left*: Responses to ionophoretic application of ACh in control (*upper record*) and after intracellular injection of EGTA (*lower record*). Holding potential: −60 mV. Calibration: 5 mV, 10 s. *Right*: The after-discharge hyperpolarization (*upper record*) is abolished (*lower record*) following EGTA injection. Calibration: 20 mV, 10 s. Holding potential: −50 mV.

suppressed shows that EGTA was effective in preventing the "Meech effect." However, the ACh-induced K response, elicited on the soma of the same cell, was unaffected. These results strongly suggest that the ACh response is not linked to an increase in intracellular calcium concentration.

CYCLIC NUCLEOTIDES AS INTERMEDIATE COMPOUNDS It has been suggested that the primary effect of a number of slow responses to neurotransmitters could be a variation in the intracellular level of cyclic nucleotides (see 67). In the case of K-dependent responses described in this section, evidence in favor of this hypothesis is either lacking or negative. For example, although administration of ACh increases the level of cyclic GMP (cGMP) in rabbit heart (54) and 8-bromo cGMP has a negative inotropic effect (133, 174), application of this cGMP derivative does not modify the efflux of labeled K from rat auricles (134; see also 50).

MODIFICATION BY TRANSMITTERS OF VOLTAGE-DEPENDENT CONDUCTANCES

The slow inhibitory potentials described above resemble classical IPSPs and EPSPs in that they involve a conductance that can be opened only following an interaction between the transmitter and a specialized membrane receptor (for possible exception, see discussion in 52; but see also 140). Many other slow synaptic effects, however, may involve a qualitatively different mechanism; that is, they seem to result from the modification, by the transmitter, of a conductance that can be activated by changes in the membrane potential, even in the absence of the agonist.

The best studied examples of such synaptic events are provided by certain effects of adrenergic and cholinergic compounds on heart muscle. These rather well-described events are treated below in detail; then we discuss other, more recently discovered and not as well analyzed examples taken from neuro-neuronal synapses.

Transmitter-Induced Modifications in Voltage-Dependent Calcium Currents

HEART MUSCLE

Increase in slow inward current by adrenergic compounds It has been repeatedly observed that catecholamines, acting via β-adrenergic receptors, increase the plateau of the action potential in a variety of cardiac muscle fibers (see 154, 179 for reviews).

Even prior to the use of voltage clamp techniques, the adrenaline sensitive plateau phase of the cardiac action potential was attributed to the presence of a time- and voltage-dependent increase in membrane permeability, primarily to calcium ions (see 38, 154, 156, 172 for reviews). Agents known to block calcium permeability were shown to reduce markedly the plateau phase (e.g. 70), which was furthermore shown to change in height as a function of external calcium concentration (e.g. 138, 153). The existence of an adrenaline-sensitive calcium conductance was also revealed: in fibers where the regenerative sodium conductance was (*a*) blocked by tetrodotoxin (TTX) (e.g. 32, 160), (*b*) eliminated by use of Na-free solutions (157), or (*c*) inactivated by high external potassium concentrations (e.g. 31, 32, 48, 143), it was found that adrenaline elicits (or lowers the threshold for eliciting) a slow action potential which is sensitive to changes in external calcium concentration and is blocked by manganese ions (see 154 for review).

Voltage clamp experiments performed in recent years have confirmed that a slow, predominantly calcium-dependent inward current is indeed elicited by depolarization of cardiac muscle fibers (see 154, 156, 172 for reviews), that this current is increased in the presence of adrenaline (24, 155, 182), and that it can account for the plateau phase of the normal action potential as well as for the slow potentials elicited in the presence of adrenaline in fibers where the regenerative sodium conductance has been blocked or inactivated (11, 24, 154, 182).

Recently, Reuter & Scholz (158) analyzed in detail the effect of adrenaline on mammalian cardiac muscle and showed that the adrenaline-induced increase in the slow inward current (I_{si}) does not reflect a change in the rate constants describing its activation, but rather is due exclusively to an increase in the limiting conductance of this calcium-dependent current. Furthermore, two sets of data demonstrated that the selectivity of the channel is unchanged by adrenaline. First, measurements of the inversion potential of the inward current (taken from instantaneous current voltage relations of I_{si}) were unchanged by adrenaline. Second, the effect of changes in calcium concentration on the tail currents (measured at holding potential after full activation of the conductance during a preceding clamp step) was the same in the presence and absence of adrenaline. Reuter & Scholz (158) proposed that the effect of adrenaline on the slow inward current could best be explained by assuming that it causes an increase in the number of voltage-dependent calcium channels (see also 166; see 181 for an alternative interpretation).

Decrease in slow inward current induced by acetylcholine In addition to inducing an increase in potassium permeability (see above), vagal stimulation (76, 80) and application of exogenous acetylcholine (3, 30, 51, 53, 60, 76, 82, 83, 170, 173, 184) have been shown to reduce the duration of the cardiac action potential. A series of studies of growing precision demonstrates that ACh exerts this inhibitory action by reducing the slow inward current which is carried mainly by calcium ions. Prokopczuk and coworkers (149) first proposed such a mechanism after observing that calcium-dependent action potentials, activated by adrenaline in canine and rabbit atria rendered quiescent by high potassium solutions, could be blocked by ACh. Since the effect of ACh was additive with the effect of the calcium antagonist manganese ions, they suggested that ACh acts by inhibiting the slow inward, predominantly calcium (see 154, 172 for reviews) current.

This interpretation of the action of ACh was reinforced by observations that TTX-resistant action potentials in frog atrial fibers are also markedly reduced by ACh (60, 61). Finally, a series of voltage clamp studies (51, 60, 61, 82–84, 142, 170) have shown that ACh selectively reduces the slow inward calcium current of normal action potentials as well as the calcium current underlying slow potentials elicited by adrenaline in fibers where the regenerative sodium current has been blocked or inactivated.

Thus, ACh acts to diminish the voltage-dependent conductance that adrenaline enhances. As was shown to be the case for adrenaline, the

ACh-induced reduction in inward current is due to a modification of the limiting calcium conductance, with no effect on either the activation or inactivation parameters of this current (83).

Certain authors (3, 78, 80, 170, 173) proposed that the action potential shortening could be explained as an indirect effect of ACh due to the increase in potassium permeability induced in these fibers by cholinergic stimulation (see K-dependent responses above). However, further voltage clamp studies in the frog atrium have shown (52, 60) that the ACh-induced potassium current, because of its marked inward rectification, is reduced to a very low level in the voltage range of the action potential plateau. Consequently, the alteration in action potential duration cannot be attributed to an ACh-induced potassium conductance (see, however, 170 for mammalian fibers). Furthermore, in frog atria it has been demonstrated that there is a differential dose-dependence for the increase in K conductance and the shortening of the action potential, such that the latter cannot be explained by the former at low agonist concentrations (53, 60). In mammalian atria, in contrast, the ACh concentrations required for affecting the slow inward current are higher than those needed for causing membrane hyperpolarization (170), thus leaving open the possibility that in these fibers the increase in potassium permeability could account, at least in part, for the action potential shortening.

Relation of transmitter-induced changes in slow inward current to contraction amplitude: inotropic effects The amplitude of the cardiac muscle contraction is clearly related to the amount of calcium entering the fiber during the action potential; however, the immediacy of the effect of the calcium influx seems to depend upon the preparation studied. In mammalian fibers, the amplitude of the contraction usually seems to be determined primarily by the amount of calcium in the intracellular stores. A given action potential liberates previously sequestered calcium (provoking the contraction) while simultaneously replenishing the stores it is emptying (see 11, 130, 190, and 179 for discussion). In contrast, the amplitude of a given twitch in frog cardiac fibers is a direct reflection of the calcium entering during the action potential that triggers that particular twitch. Increases in calcium current are thus immediately reflected by changes in twitch tension, without having to first cause an increase in intracellular stores.

In either case, the means by which contraction amplitude is affected by adrenaline and ACh (which increase and decrease, respectively, the calcium current of the spike) becomes evident. Recent studies have confirmed that there is indeed a very high correlation between drug-induced changes in the action potential (132) or in the underlying slow inward current (138a) and contraction amplitude.

MOLLUSCAN AND VERTEBRATE GANGLIONIC NEURONS

5HT-induced increase in voltage-dependent calcium conductance in molluscan neurons Effects strikingly similar to those obtained with adrenaline and acetylcholine on heart muscle have recently been observed in neurons in response to other transmitter substances. The most thorough demonstration of this phenomenon at a neuro-neuronal synapse is offered by the work of Klein & Kandel (99) in *Aplysia* central ganglia. These authors showed that serotonin (5HT), like the transmitter liberated when one of the connectives leading into the ganglion is stimulated, causes an increase in the height and duration of the calcium-dependent phase of the somatic action potential of identified sensory neurons.

The presence of a calcium component in the action potential was demonstrated by (*a*) the persistence of a regenerative potential in sodium-free seawater; (*b*) the elimination of this "Na-free" spike by the calcium channel blocking agent, cobalt (15 mM); and (*c*) the finding that in tetraethylammonium (TEA)-containing seawater, the height of the action potential plateau shows a Nernstian relationship to changes in external calcium concentration.

Since the effects of 5HT and of the natural transmitter were found to be greatest when both the sodium and potassium voltage-dependent conductances were eliminated (by replacing sodium with sucrose and by adding 100 mM TEA), it was suggested that 5HT and the transmitter act primarily, if not exclusively, by enhancing the voltage-dependent calcium current. The possibility was not excluded, however, that the increase in calcium current is simply the indirect effect of a diminution in, or slowing down of, a remaining potassium current (e.g. the calcium-dependent potassium conductance; see 124).

Both the natural transmitter and 5HT usually caused a depolarization of the sensory neurons that was enhanced by lowering the membrane potential. When hyperpolarizing constant current pulses were passed across the membrane, the voltage displacement caused by these pulses increased during the action of the transmitter, indicating a decrease in slope conductance. This decrease in slope conductance would be expected whether the transmitter increased an inward current or decreased an outward current. Full voltage-current relations taken in the presence and absence of 5HT might help discriminate between these two possibilities, since one would expect the curves to cross at E_K if the depolarization represented a decrease in K permeability. On the other hand, a failure of the curves to cross would reinforce the hypothesis proposed by the authors: that the transmitter acts to increase an inward current. The use of voltage clamp techniques coupled with further pharmacological isolation of the various underlying currents should provide a clearer understanding of this phenomenon.

Transmitter-induced decreases in voltage-dependent calcium conductance in molluscan and vertebrate ganglionic neurons Whereas Klein & Kandel (99) observed neuro-neuronal synaptic actions that resemble the effects of adrenaline on the cardiac action potential (i.e. an increase in the voltage-dependent calcium conductance), other investigators observed transmitter effects on neurons that resemble those of ACh on the action potential of heart muscle. Mudge et al (131) have shown that a pressure injection of enkephalin or one of its analogues on the soma of cultured dorsal root ganglion cells causes a shortening of the action potential. Since the alteration of the action potential still occurred in cells poisoned with TTX and since the shortening was found to be most pronounced in cells in which barium was used (to enhance the current carried through the calcium channels and to block the delayed potassium conductance), these authors concluded that enkephalin acts by reducing the contribution of the voltage-dependent calcium currents to the action potential.

On the same neurons, Dunlap & Fischbach (45) observed similar effects when applying the putative neurotransmitters GABA, noradrenaline, and serotonin. None of these transmitter substances change the resting potential. The degree of shortening elicited by these transmitters was found to be dose-dependent and, in the case of noradrenaline, could be blocked by α-receptor antagonists.

Similar effects, mediated by α-adrenoceptors, have been demonstrated in bullfrog sympathetic ganglion cells (128) and in postganglionic neurons of the superior cervical ganglion of the rat (77). In both investigations it was shown that noradrenaline reduced the rate of rise, the plateau phase, and the afterhyperpolarization (resulting from the calcium influx during the spike) in calcium-dependent action potentials in these cells.

A similar, synaptically activated phenomenon was observed under voltage clamp conditions in an identified cell (L10) of *Aplysia californica* in which the regenerative sodium conductance channel and the delayed potassium channel were pharmacologically blocked with TTX and TEA, respectively. Shapiro et al (see 88) passed .4 sec depolarizing voltage steps in L10, bringing the cell from -35 mV to $+4$mV and back. These pulses elicited an initial calcium current that was followed by a potassium-dependent outward current (assumed to result directly from the influx of calcium ions; see 124).

The initial inward current measured in the soma of L10 in response to these voltage steps was shown to be reduced by activation of certain synaptic inputs triggered by nerve trunk stimulation. This effect is reminiscent of the inhibition by ACh of the calcium currents measured by Giles & Noble (60) in heart muscle. However, in the case of the investigation on L10, further experiments in which the secondary potassium currents are eliminated will permit a more serious analysis of the effects of synaptic stimulation since, in the preliminary records that have been published (88), the potassium current seems to increase concomitantly with the decrease in the inward calcium current upon nerve stimulation. Such a relationship is difficult to understand if the potassium current is indeed activated, as the authors state, by the entrance of calcium. On the other hand, an increase in the K current could be the cause of an apparent decrease in inward current.

Relation of transmitter-induced changes in calcium conductance to liberation of transmitter from the "modulated" cell Many of the investigators studying the effects of transmitters on neuronal action potential duration used their results to explain other observations they had made concerning the action of the same substance on transmitter release. For example, Klein & Kandel (99) showed that connective stimulation or 5HT application increased the release of transmitter from the neurons whose action potential was enlarged by similar stimulation. Furthermore, when experiments were performed in TEA-containing seawater, the increase in liberation was shown to be coincident with an increase in the duration of the somatic action potential. Conversely, the neuron L10, in which the calcium current was diminished by connective stimulation (whether directly or indirectly, see above), released less transmitter (as measured by postsynaptic potential amplitude) following this stimulation; the degree of release was correlated with the lowered calcium current in the soma of the cell. Similarly, in dorsal root ganglion cells, the substances shown to decrease the action potential duration also were shown to reduce the liberation of substance P from the same neurons (131; see also 121).

In view of the data obtained on somatic spikes, these facilitatory and inhibitory effects on transmitter release have been attributed by the various authors to synaptically activated (or transmitter-activated) alterations in the voltage-dependent calcium current in the region of the terminals. A similar hypothesis based, however, on less direct evidence was proposed earlier by Shimahara & Tauc (165) to account for similar modifications in transmitter release at central synapses in *Aplysia*.

Modification by Transmitters and Hormones of Pacemaker Currents

HEART MUSCLE   Three transmitter effects on heart muscle have already been discussed: cholinergic K-dependent inhibition; adrenergic increase in slow, voltage-dependent calcium conductance; and cholinergic diminution in this same conductance. These slow synaptic events were shown to have counterparts at neuro neuronal synapses. An additional type of transmitter effect for which an equivalent has not yet been demonstrated in neuronal transmitter effects is the acceleration by adrenaline of the cardiac rhythm.

Adrenergic activation is known to accelerate the cardiac rhythm, but the mechanism underlying the acceleration seems to depend upon the type of cardiac cell being stimulated. In Purkinje fibers, adrenaline has been shown (73, 74, 177, 178) to cause this acceleration by shifting the activation curve of a voltage-dependent potassium current (iK_2) in the positive direction. This shift then permits the inward currents, which bring the fibers to spike threshold, to dominate at potentials where they are normally overridden by K currents (118). Adrenaline has another effect on the fibers affecting the same current: it apparently increases the potassium gradient by stimulation of the Na-K pump, thereby causing a 5–15 mV increase in the reversal potential of the K current (35).

In atrial fibers (25), as well as at the sino atrial node (22), adrenaline acts to increase the amplitude of another potassium current (i_x) (also present and sensitive to adrenaline in Purkinje fibers, see 180). This effect, of course, tends to slow, rather than accelerate, the cardiac rhythm. It is presumably by increasing the slow inward current (see above) and by decreasing a recently described K current (i_f) (see 23) that adrenaline manages to cause an acceleration in these fibers.

MOLLUSCAN NEURONS   Certain molluscan neurons fire in bursts of action potentials that are elicited by an endogenous pacemaker potential attributed generally to a self-feeding inward and outward current cycle. Presumably due to the voltage-dependent inward limb of this pacemaker potential, there is a distinctive negative slope conductance region in the current voltage (I/V) curves generated in these voltage-clamped neurons (see 69, 125 for reviews).

Recently, the *inhibitory* effects of transmitters on certain bursting neurons [effects of dopamine (189) and nerve stimulation on R15 in *Aplysia* and F1 in *Helix* (109) as well as the effects of cholinergic stimulation on left quadrant cells of *Aplysia* abdominal ganglion (189) and of egg-laying hormone and bag cell stimulation on these same neurons (26)] have been accounted for by assuming that the transmitters act by turning off this voltage-dependent, TTX-insensitive inward current. This explanation has been offered for four reasons: 1. These slow inhibitory potentials often disappear with increasingly negative membrane potentials and fail to invert in a manner expected for an increase in membrane conductance to Cl or K ions. 2. During the action of the transmitter, the negative slope region in the current voltage curve is eliminated, rendering the I/V relationship more or less linear. 3. Both the negative slope conductance and the inhibitory responses disappear when the sodium in the seawater is replaced by sucrose. 4. Changes in external potassium concentration often do not affect the transmitter-induced change in the I/V relationship or do not affect it in a way expected for a simple increase in potassium conductance.

All of these findings are compatible with the hypothesis that the transmitters act to turn off this voltage-dependent inward current. However, certain reservations must be expressed before concluding that this is the only possible interpretation.

First, the failure of many of these responses to invert with increasing membrane potentials may simply be related to a failure to clamp the very large low impedance cells in which these responses have been found (97). Although such an explanation cannot account for the failure of high external potassium to affect the transmitter-induced currents in the more depolarized zone (see 189), it might account for the erratic behavior as a function of membrane potential of these responses for which a potassium-dependent component has clearly been demonstrated (5, 26, 109).

Second, a concomitant elimination of the effect of the transmitter and of the negative slope conductance does not necessarily support the hypothesis that the transmitter acts by shutting off the inward current. Even a pure potassium-dependent potential (elicited in the soma of the left quadrant cells or in the medial cells of the pleural ganglion by ACh) is markedly reduced in sodium-free solutions (J. S. Kehoe, unpublished observations).

Finally, if one assumes that the membrane potential read in the soma of the clamped cell reflects the potentials in the axon where the synaptic events occur, it is difficult to understand how a synaptically activated hyperpolarization that can often persist at values beyond -100 mV (see, for example, 96, 97) is due to the turning off of a "voltage-dependent" pacemaker current that presumably is activated only at potentials less negative than -70 mV.

Certain of these noninverting IPSPs were previously explained by invoking a synaptic activation of the electrogenic Na/K pump. Although the experiments reported for demonstrating this involvement

(147) were inadequate (see 96–97a), more recent investigations on isolated molluscan neurons provide rather clear evidence that certain neurotransmitters can activate the Na/K pump (90).

It is highly possible that many of these potentials that have resisted clear analysis do in fact represent a combination of either two or even three of these elements: an increase in K conductance, the activation of the Na/K pump, and the inhibition of pacemaker inward currents. The use of isolated, internally perfused neurons might help in the clear demonstration and eventual separation of these three possible contributors to these inhibitory synaptic potentials (see below).

Other experiments have suggested that these same pacemaker potentials can be *accelerated* by peptides, one of which is liberated by stimulation of the bag cells in the abdominal ganglion (7, 19, and 104 for review). The precise ionic nature of the change in membrane characteristics underlying the change in firing frequency remains to be defined. However, it has been shown that the transmitter-induced acceleration in rhythm is often associated with an accentuation of the negative slope conductance found in the depolarizing zone of the I/V curve obtained with bursting neurons.

A similar accentuation of a voltage-dependent TTX-insensitive current has been proposed to explain atypical potential changes elicited in molluscan neurons by 5HT (144).

Use of a Receptor–Second Messenger Model for Explaining Transmitter-Mediated Changes in Voltage-Dependent Conductances

Many authors contend that the slow, potassium-dependent synaptic potentials described in the first section of this review can be explained by using relatively simple "receptor-ionophore" models that are similar in many respects to those used to describe more rapid synaptic events. Such models do not involve the activation, by the agonist-receptor interaction, of a second molecule which in turn would control (directly or indirectly) the change in membrane permeability.

In contrast, to explain the transmitter-induced changes in voltage-dependent conductances, a second messenger type of model has often been invoked. Such a hypothesis has most frequently been employed to explain the increase by adrenergic compounds of the voltage-dependent calcium conductance in heart muscle and, to a lesser extent, to account for the antagonistic effects of ACh on the same conductance. We give examples of the types of electrophysiological experiments that have been performed to evaluate these hypotheses (in particular, relating adrenergic increases in calcium conductance to increases in cAMP) and then note the similar tests that attempt to relate cyclic nucleotide levels to transmitter-induced changes in voltage-dependent conductances at neuro-neuronal synapses.

Similar hypotheses have been proposed for explaining slow synaptic potentials in sympathetic ganglia and in smooth muscle. However, since critical reviews dealing with the adequacy of these hypotheses for explaining slow potentials observed in these preparations have recently been published (17, 91, 103, 112), these systems are not discussed here.

CYCLIC NUCLEOTIDES AND TRANSMITTER EFFECTS ON CALCIUM CURRENTS IN HEART MUSCLE Many authors have proposed that the increase in calcium current caused by activation of β-receptors in cardiac fibers is mediated by changes in cAMP levels. Beta-receptor activation in heart muscle has indeed been shown to stimulate the production of cAMP via a receptor-associated adenylate cyclase (167). These original experiments stimulated a series of other investigations designed to show that the cAMP thus produced is the agent responsible for the change in membrane permeability, which is then expressed as a change in muscle contraction (for review, see 179).

First, this second messenger hypothesis predicted that the increase in calcium conductance caused by β-adrenergic agonists should be amplified when enzymatic destruction of cAMP is blocked by phosphodiesterase inhibitors. Such an effect has been demonstrated using the methylxanthine compound theophylline in Purkinje fibers (180), in frog atrial fibers (36), and in mammalian ventricular muscle (12). Similar results were obtained in Purkinje fibers using a phosphodiesterase inhibitor (Ro 7 2956) that is chemically distinct from the methylxanthines and that may avoid the complicating increase in free intracellular calcium caused by theophylline (see 179; see also effects of papavarine on papillary muscle, 47).

Another corollary of this second messenger hypothesis is that exogeneously applied cAMP or its more permeant derivatives should be able to imitate the action of β-adrenergic agonists on the action potential plateau (or on the underlying slow inward current). Such was shown to be the case in mammalian Purkinje fibers when either monobutyryl cAMP (180) or dibutyryl cAMP (155) was added to the bathing medium (see also 126). Furthermore, dibutyryl cAMP was shown to mimic isoproterenol in activating presumably calcium-dependent action potentials in hearts depolarized by potassium rich solutions (175, 185) or pretreated with TTX (185).

Doubts have been expressed about the means by which these agents exert their influence on the action potential, since the possibilities remain that they can act (*a*) via other membrane receptors, (*b*) intracellularly, but directly, rather than by causing an increase in cAMP, or (c) by simply inhibiting the phosphodiesterase (see 179 for review). However, the limits of this approach have been bypassed by intracellular injection of cAMP itself, which was shown to cause the same effect on the action potential of Purkinje cells (176) or of frog atrial fibers (84) as that caused by β-receptor activation or externally applied cAMP derivatives (see also 191 for effects of cAMP and dibutyryl cAMP on sinoatrial node).

The cAMP hypothesis receives further support from experiments using a GTP analogue, 5' guanylimidodiphosphate [GPP(NH)P], which has been shown to activate irreversibly the adenylate cyclase in broken cell preparations (40, 107). In TTX-treated ventricular fibers, it was shown that this compound, like β-adrenergic agonists, induces slow calcium-dependent action potentials and that these potentials persist after the GTP analogue has been washed out (87).

Beta-adrenergic agonists have been shown to cause an increase in cAMP, and the dose dependence of the cAMP increase parallels that of the increase in twitch tension (43; see also 138a). Since twitch tension is known to reflect, under certain circumstances at least, alterations in the voltage-dependent, slow inward current (11, 130, 132, 138a, 155, 190), these data support, albeit indirectly, a role for cAMP in the adrenergic effects on the action potential plateau.

Finally, if cAMP is truly the agent controlling the change in membrane permeability, the adrenergically stimulated increase in cAMP must antedate the change in action potential form. An increase in cAMP has been shown to occur prior to an increase in twitch tension (43, 161, 183) or prior to the appearance of excitability (in response to adrenaline) in TTX or potassium-inactivated fibers (185). Again, although these measurements are indirect, these findings strongly suggest that an increase in cAMP precedes the increase in voltage-dependent calcium conductance.

These indications do not suffice to prove the hypothesis that cAMP mediates the β-adrenergic effect on the plateau of the cardiac action potential, nor do they tell us much about how cAMP might mediate this effect. However, they certainly are all compatible with such a hypothesis.

In contrast, recent findings suggest that the role of cAMP as the mediator of this permeability change needs to be seriously reconsidered. Experiments have been performed showing that highly localized β-receptor activation causes no detectable elevation in cAMP levels (183) but does cause an increase in the plateau of the action potential, not only in those fibers in direct contact with the adrenergic agonist but in fibers throughout the whole muscle (9). These experiments were accomplished by using, in the studies of the action potential form, about 15 covalently bound isoproterenol beads of 150–200 μm in diameter. These beads, placed on one end of the preparation, caused a change in the action potential shape, even on the opposite end of the muscle. Such localized application of isoproterenol was also capable of lowering the threshold for the calcium-dependent slow potential in TTX-treated muscles or in muscles bathed in high potassium solutions. Once more, the lowered threshold was manifest across the entire muscle, not only in the fibers in the region of the beads.

It is true that an increase in cAMP restricted to the fibers in direct contact with isoproterenol would not have been detected by the methods used and that such a localized increase in cAMP might have occurred and been involved in the initiation of the generalized effect on calcium permeability. But the data do not permit one to assume that cAMP was responsible for the propagation of the effect from the fibers in direct contact with the isoproterenol to the rest of the muscle. Seemingly contradictory evidence has, however, been obtained by Tsien & Weingart (181a) who exposed a "cut end" of ventricular muscle to either ^{3}H cAMP or to "cold" cAMP and then, in a test region that had not been exposed to cAMP, measured over time the increase in radioactivity or in the amplitude of contraction. These authors found that the time dependence of the increase in twitch tension (maximal at 50 min following exposure of the cut end) agreed with their estimates (based upon their measurements of radioactivity) for diffusion of the cAMP into the test region of the muscle. The discrepancies in the conclusions drawn from these two types of experiments illustrate that the role of cAMP in the control of these membrane permeability changes still requires further clarification.

Even if it becomes an indisputable conclusion that the β-adrenergic effects on the slow inward current in cardiac fibers are mediated by cAMP, such a mechanism must not automatically be assumed to be involved in all other transmitter-induced increases in voltage-dependent calcium conductance. Even in cardiac muscle itself, it appears that pure α-adrenergic stimulation, which seems to cause no measurable increase in cAMP (21, 46), can induce calcium-dependent action potentials in rabbit papillary muscle in which the fast Na current has been inactivated by high external potassium (129). These slow action potentials, elicited by phenylephrine in the presence of a β-adrenoreceptor blocking drug (bufetolol), could be blocked by phentolamine, an α-adrenoreceptor antagonist. Further analyses under voltage clamp are needed before concluding that this α-adrenergic control of calcium permeability truly resembles the selective β-adrenergic modification of the limiting calcium conductance (158); these data do suggest, however, that transmitters can induce changes in voltage-dependent calcium conductance without an associated change in cAMP levels (see also 138a).

A few parallel experiments have been performed in an attempt to relate the inhibitory effect of ACh on the voltage-dependent calcium current in heart muscle to an increase in cGMP. In particular, cGMP injected into bullfrog atrial fibers caused a decrease in the overshoot and in the plateau of the action potential while having no effect on the resting membrane potential (84). Similarly, in rat atrial muscle, application of the permeant GMP derivative 8-bromo cGMP was found to cause a shortening of the action potential similar to that obtained with ACh. The change in action potential form was correlated with a reduced uptake (in beating preparations) of radioactive calcium, whereas no change could be detected in the uptake of radioactive potassium (134). Likewise, the upstroke velocity of the calcium-dependent slow action potential elicited in mammalian atrial fibers in high potassium solutions was markedly decreased by 8-bromo cGMP (100) as it was by ACh. These effects could be partially counteracted by activating the β-adrenergic receptors with isoproterenol. Using the more indirect measure of contractile force, it was found that the inotropic effects of ACh could be mimicked by 8-bromo cGMP (100, 133). Furthermore, it was found that ACh caused an increase in cGMP (117, 186), although the effects on cGMP and on twitch tension were not always correlated.

These data are once more consistent with, though inadequate for proving, the hypothesis that there is a causal relationship between the ACh-induced increase in cGMP and the reduction in the slow inward current that occurs in the presence of this transmitter.

CAMP AND TRANSMITTER-INDUCED INCREASES IN CALCIUM CURRENTS IN NEURONS The increase in action potential duration caused by synaptic stimulation in certain molluscan neurons was shown to be accentuated by the phosphodiesterase inhibitors 3-isobutyl-1-methylxanthine (IBMX) and by Ro 20-1724 (99) (see also 165). Furthermore, the action potential prolongation could be imitated either by extracellular application of a permeant cAMP derivative or by intracellular application of cAMP (99) (see also 27, 102, 165). The effects of phosphodiesterase inhibitors and cAMP and its derivatives were seen only in cells in which 5HT caused an increase in the action potential duration. Finally, incubation of the ganglion in 5HT was shown to stimulate an increase in cAMP levels in the ganglion (34). All of these findings are consistent with the hypothesis that cAMP can mediate such a change (directly or indirectly), but they still do not demonstrate conclusively that it is in fact cAMP that must intervene for these transmitter-induced changes to occur.

Levitan and collaborators (108, 109) have found that physiological agents (vertebrate and invertebrate peptides) that increase the duration of the burst in R15 and F1 (*Aplysia* and snail neurons, respectively) increase cAMP and, less consistently, cGMP in these preparations (110). Furthermore, they have shown that these same compounds activate adenylate cyclase activity in a crude membrane preparation (110) and that the physiological effect on the burst and the adenylate cyclase activation are blocked by another factor extracted from *Helix* and *Aplysia* (81). Phosphodiesterase inhibitors imitate the effects of these peptides, as do bath applied, 8-substituted derivatives of cAMP [which are unaffected by, but tend to inhibit, phosphodiesterase (127)].

In contrast, in the same cells they find that an intracellular injection of the adenylate cyclase activator GPP(NH)P, like the physiological compound liberated by nerve stimulation, causes a complete elimination of the burst. This GTP derivative is assumed to cause a selective increase in cAMP levels. The initial effect of an intracellular injection of 8-substituted cAMP derivatives likewise causes an elimination of the burst, contrary to the effect it has when applied extracellularly.

The authors propose that the net effect on the burst pattern caused by the different physiological agents will depend on the relative elevation it causes in cAMP and cGMP levels, with a pure increase in cAMP causing an elimination of the burst, and a mixed increase (cAMP and cGMP) causing a facilitation of the burst. It is clear that if such opposing effects can be obtained as a function of delicate balancing between cAMP and cGMP levels, the interpretation of data obtained when using phosphodiesterase inhibitors or compounds with side effects on phosphodiesterase (179) becomes very difficult.

Many of the hypotheses proposed for explaining the slow synaptic events discussed in this review have as yet to be unequivocally proven. Recently developed techniques permitting the isolation and internal perfusion of heart muscle fibers (106) and molluscan neurons (101, 105) (among other cells) should permit us to obtain a clearer understanding of some of these phenomena.

For example, a more critical series of tests of the cyclic nucleotide hypothesis should be possible—with the ability to manipulate in a quantitative way the intracellular levels of cAMP and/or cGMP and with the possibility of attacking the internal surface of the membranes with pharmacological agents previously used only for membrane fragment preparations. Likewise, the study of transmitter-induced changes in calcium conductance will clearly benefit from the ability to use cells containing no intracellular potassium and/or containing a known quantity of calcium chelating agents, etc. Furthermore, the great improvement in the voltage clamp that is obtained in neurons from which the axon has been removed and closed off, or in individual (as opposed to coupled) heart cells, combined with the control over intracellular ion concentrations, should permit a separation and a more serious analysis of the many possible membrane changes contributing to the transmitter effects on pacemaker and bursting cells.

ACKNOWLEDGMENT

We would like to thank Dr. R. Tsien for his careful reading of the manuscript and for his helpful suggestions.

Literature Cited

1. Adams, P. R. 1975. *Br. J. Pharmacol.* 53:308–10
2. Anderson, C. R., Stevens, C. F. 1973. *J. Physiol. London* 235:655–91
3. Antoni, H., Rotmann, M. 1968. *Pfluegers Arch.* 300:67–86
4. Armstrong, C. M. 1975. In *Membranes: A Series of Advances*, ed. G. Eisenman, 3:325–58. New York: Dekker
5. Ascher, P. 1972. *J. Physiol. London* 225:173–209
6. Ascher, P., Kehoe, J. S. 1975. In *Handbook of Psychopharmacology*, ed. L. L. Iversen, S. D. Iversen, S. Snyder. 6:265–310 New York: Plenum
7. Barker, J. L., Smith, T. J. Jr. 1977. *Neurosci. Symp.* 2:340–73
8. Baylor, D. A., Hodgkin, A. L., Lamb, T. D. 1974. *J. Physiol. London* 242:759–91
9. Becker, E., Ingebretsen, W. R. Jr., Mayer, S. E. 1977. *Circ. Res.* 41:653–60
10. Beeler, G. W., Reuter, H. 1970. *J. Physiol. London* 207:191–209
11. Beeler, G. W. Jr., Reuter, H. 1970. *J. Physiol. London* 207:211–29
12. Beresewicz, A., Reuter, H. 1977. *Naunyn Schmiedeberg's Arch. Pharmacol.* 301:99–107
13. Berry, M. S., Cottrell, G. A. 1973. *Nature New Biol.* 242:250–53
14. Berry, M. S., Cottrell, G. A. 1975. *J. Physiol. London* 244:589–612
15. Blackman, J. G., Ginsborg, B. L., House, C. R. 1979. *J. Physiol. London* 287:67–80
16. Blackman, J. G., Ginsborg, B. L., House, C. R. 1979. *J. Physiol. London* 287:81–92
17. Bolton, T. D. 1979. *Physiol. Rev.* 59:606–718
18. Bowser-Riley, F., House, C. R. 1976. *J. Exp. Biol.* 64:665–76
19. Branton, W. D., Arch, S., Smock, R., Mayeri, E. 1978. *Proc. Natl. Acad. Sci. USA.* 75:5732–36
20. Branton, W. D., Mayeri, E., Brownell, P., Simon, S. B. 1978. *Nature* 274:70–72
21. Brodde, O. E., Motomura, S., Endoh, M., Schumann, H. J. 1978. *J. Mol. Cell. Cardiol.* 10:207–19
22. Brown, H. F., DiFrancesco, D., Noble, S. J. 1979. *J. Physiol. London* 290:31–32P
23. Brown, H. F., DiFrancesco, D., Noble, S. J. 1979. *Nature* 280:235–36
24. Brown, H. F., McNaughton, P. A., Noble, D., Noble, S. J. 1975. *Philos. Trans. R. Soc. London. Ser. B* 270:527–37
25. Brown, H. F., Noble, S. J. 1974. *J. Physiol. London* 204:717–36
26. Brownell, P., Mayeri, E. 1979. *Science* 204: 417–20
27. Brunelli, M., Castellucci, V., Kandel, E. R. 1976. *Science* 194:1178–80
28. Bulbring, E., Tomita, T. 1969. *Proc. R. Soc. London Ser. B* 172:89–102
29. Bulbring, E., Tomita, T. 1977. *Proc. R. Soc. London Ser. B* 197:271–84

30. Burgen, A. S. W., Terroux, K. G. 1953. *J. Physiol.* 120:449–64
31. Carmeliet, E. 1967. *Arch. Int. Physiol. Biochim.* 75:542–43
32. Carmeliet, E., Vereecke, J. 1969. *Arch. Gesamte Physiol.* 313:300–15
33. Carpenter, D. O., Graubatz, G. L. 1974. *Nature* 252:483–85
34. Cedar, H., Schwartz, J. H. 1972. *J. Gen. Physiol.*, 60:570–87
35. Cohen, I., Eisner, D. A., Noble, D. 1978. *J. Physiol. London* 280:155–68
36. Coraboeuf, E. 1971. In *Actualites Pharmacologiques.* 24:153–82. Paris: Masson
37. Cottrell, G. A. 1978. *J. Physiol. London* 284:130–31P
38. Cranefield, P. F. 1975. *The Conduction of the Cardiac Impulse.* New York: Futura
39. Cuatrecasas, P. 1974. *Ann. Rev. Biochem.* 43:169–214
40. Cuatrecasas, P., Bennett, V., Jacobs, S. 1975. *J. Membr. Biol.* 23:249–79
41. del Castillo, J., Katz, B. 1955. *Nature* 175:1035
42. del Castillo, J., Katz, B. 1957. *Proc. R. Soc. London. Ser. B* 146:369–81
43. Dobson, J. G., Ross, J., Mayer, S. E. 1976. *Circ. Res.* 39:388–95
44. Douglas, W. W., Poisner, A. M. 1963. *J. Physiol. London* 165:528–41
45. Dunlap, K., Fischbach, G. D. 1978. *Nature* 276:837–39
46. Endoh, M., Brodde, O. E., Schumann, H. J. 1976. *Naunyn-Schmiedeberg's Arch. Pharmacol.* 295:109–15
47. Endoh, M., Schumann, H. J. 1975. *Eur. J. Pharmacol.* 30:213–20
48. Engstfeld, G., Antoni, H., Fleckenstein, U. A. 1961. *Pfluegers. Arch. Gesamte Physiol.* 273:145–63
49. Feltz, A., Krnjevic, K., Lisiewicz, A. 1972. *Nature New Biol.* 237:179–81
50. Fleming, B., Giles, W., Lederer, J. 1979. *J. Physiol. London* 292:69–70P
51. Garnier, D., Goupil, N., Nargeot, J., Ojeda, C., Rougier, O. 1976. *J. Physiol. Paris* 72:8A
52. Garnier, D., Nargeot, J., Ojeda, C., Rougier, O. 1978. *J. Physiol. London* 274:381–96
53. Garnier, D., Nargeot, J., Ojeda, C., Rougier, O. 1978. *J. Physiol. London* 276:27–28P
54. George, W. J., Polson, J. B., O'Toole, A. G., Goldberg, N. D. 1970. *Proc. Natl. Acad. Sci. USA* 66:398–403
55. Ger, B. A., Katchman, A. N., Zeimal, E. V. 1979. *Brain Res.* 171:355–59
56. Ger, B. A., Zeimal, E. V. 1976. *Nature* 259:681–84
57. Gerschenfeld, H. M. 1973. *Physiol. Rev.* 53:1–119
58. Gerschenfeld, H. M., Paupardin-Tritsch, D. 1974. *J. Physiol. London* 243:427–56
59. Gerschenfeld, H. M., Paupardin-Tritsch, D. 1974. *J. Physiol. London* 243:457–81
60. Giles, W., Noble, S. J. 1976. *J. Physiol. London* 261:103–23
61. Giles, W., Tsien, R. 1975. *J. Physiol. London* 246:64–66P
62. Ginsborg, B. L., House, C. R. *Ann. Rev. Biophys. Bioeng.* 9: In press
63. Ginsborg, B. L., House, C. R., Silinsky, E. M. 1974. *J. Physiol. London* 236:723–31
64. Ginsborg, B. L., House, C. R., Silinsky, E. M. 1976. *J. Physiol. London* 262:489–500
65. Ginsborg, B. L., Kado, R. T. 1975. *J. Physiol. London* 245:713–25
66. Glitsch, H. G., Pott, L. 1978. *J. Physiol. London* 279:655–68
67. Greengard, P. 1976. *Nature* 260:101–8
68. Gruol, D. L., Weinreich, D. 1979. *Brain Res.* 162:281–301
69. Gulrajani, R. M., Roberge, F. A. 1978. *Fed. Proc.* 37:2146–52
70. Hagiwara, S., Nakajima, S. 1966. *J. Gen. Physiol.* 49:793–806
71. Harris, E. J., Hutter, O. F. 1956. *J. Physiol. London* 133:58–59P
72. Hartzell, H. C., Kuffler, S. W., Stickgold, R., Yoshikami, D. 1977. *J. Physiol. London* 271:817–46
73. Hauswirth, O., Noble, D., Tsien, R. W. 1968. *Science* 162:916–17
74. Hauswirth, O., Wehner, H. D., Ziskoven, R. 1976. *Nature* 263:155–56
75. Hill-Smith, I., Purves, R. D. 1978. *J. Physiol. London* 279:31–54
76. Hoffman, B. F., Suckling, E. F. 1953. *Am. J. Physiol.* 173:312–20
77. Horn, J. P., McAfee, D. A. 1979. *Science* 204:1233–35
78. Hutter, O. F. 1957. *Br. Med. Bull.* 13:176–80
79. Hutter, O. F. 1961. In *Nervous Inhibition*, ed. F. Florey, pp. 114–23. New York: Pergamon
80. Hutter, O. F., Trautwein, W. 1956. *J. Gen. Physiol.* 39:715–33
81. Ifshin, M. S., Gainer, H., Barker, J. L. 1975. *Nature* 254:72–74
82. Ikemoto, Y., Goto, M. 1975. *Proc. Jpn. Acad.* 51:501–5
83. Ikemoto, Y., Goto, M. 1977. *J. Mol. Cell. Cardiol.* 9:313–26

84. Ikemoto, Y., Goto, M. 1978. In *Recent Advances in Studies on Cardiac Structure and Metabolism*, ed. T. Kobayashi, T. Sano, N. S. Dhalla, 11:57–61. Baltimore: Univ. Park Press
85. Jacobs, S., Cuatrecasas, P. 1976. *Biochim. Biophys. Acta* 433:482–95
86. Jenkinson, D. H., Morton, I. K. M. 1967. *J. Physiol. London* 188:373–86
87. Josephson, I., Sperelakis, N. 1978. *J. Mol. Cell. Cardiol.* 10:1157–66
88. Kandel, E. R. 1979. *Harvey Lect.* 73:19–92.
89. Katz, B., Miledi, R. 1972. *J. Physiol. London* 224:665–99
90. Kazachenko, V. N., Musienko, V. S., Gakhova, E. N., Veprintsev, B. N. 1979. *Comp. Biochem. Physiol.* 63C:67–72
91. Kebabian, J. W. 1977. *Adv. Cyclic Nucleotide Res.* 8:421–508
92. Kehoe, J. S. 1967. *Nature* 215: 1503–5
93. Kehoe, J. S. 1969. *Nature* 221: 866–68
94. Kehoe, J. S. 1972. *J. Physiol. London* 225:85–114
95. Kehoe, J. S. 1972. *J. Physiol. London* 225:115–46
96. Kehoe, J. S. 1972. *J. Physiol. London* 225:147–72
97. Kehoe, J. S. 1973. In *Drug Receptors*, ed. H. P. Rang, pp. 63–86. London: MacMillan
97a. Kehoe, J. S., Ascher, P. 1970. *Nature* 225:820–23
98. Kehoe, J. S., Marder, E. 1976. *Ann. Rev. Pharmacol. Toxicol.* 16:245–68
99. Klein, M., Kandel, E. R. 1978. *Proc. Natl. Acad. Sci. USA* 75:3512–16
100. Kohlhardt, M., Haap, K. 1978. *J. Mol. Cell. Cardiol.* 10:573–86
101. Kostyuk, P. G., Krishtal, O. A. 1977. *J. Physiol. London* 270:545–68
102. Krnjevic, K., van Meter, W. G. 1976. *Can. J. Physiol. Pharmacol.* 54:416–21
103. Kuba, K., Koketsu, K. 1978. *Prog. Neurobiol. Oxford* 11:77–169
104. Kupfermann, I. 1979. *Ann. Rev. Neurosci.* 2:447–65
105. Lee, K. S., Akaike, N., Brown, A. M. 1978. *J. Gen. Physiol.* 71:489–507
106. Lee, K. S., Weeks, T. A., Kao, R. L., Akaike, N., Brown, A. M. 1979. *Nature* 278:269–71
107. Lefkowitz, R. J. 1974. *J. Biol. Chem.* 249:6119–24
108. Levitan, I. B. 1979. In *The Neurosciences: Fourth Study Program*, ed. F. O. Schmidt, pp. 1043–55. Cambridge, Mass: MIT Press
109. Levitan, I. B., Harmar, A. J., Adams, W. B. 1979. *J. Exp. Biol.* 81:131–52
110. Levitan, I. B., Treistman, S. N. 1977. *Brain Res.* 136:307–17
111. Levitzki, A., Helmreich, E. J. M. 1979. *FEBS Lett.* 101:213–19
112. Libet, B. 1979. *Life Sci.* 24:1043–58
113. Lowagie, C., Gerschenfeld, H. M. 1973. *J. Physiol. Paris.* 67:350–51A
114. Magleby, K. L., Stevens, C. F. 1972. *J. Physiol. London* 223:151–71
115. Magleby, K. L., Stevens, C. F. 1972. *J. Physiol. London* 223:173–97
116. Marty, A., Ascher, P. 1978. *Nature* 274:494–97
117. McAfee, D. A., Whiting, G. J., Siegel, B. 1978. *J. Mol. Cell. Cardiol.* 10:705–16
118. McAllister, R. E., Noble, D., Tsien, R. W. 1975. *J. Physiol. London* 251:1–59
119. McCaman, M. W., McCaman, R. E. 1978. *Brain Res.* 141:347–52
120. McCaman, R. E., McKenna, D. G. 1978. *Brain Res.* 141:165–71
121. McDonald, R. L., Nelson, P. G. 1978. *Science* 199:1449–50
122. Meech, R. W. 1974. *J. Physiol. London* 237:259–78
123. Meech, R. W. 1974. *Comp. Biochem. Physiol.* 48A:387–95
124. Meech, R. W. 1978. *Ann. Rev. Biophys. Bioeng.* 7:1–18
125. Meech, R. W. 1979. *J. Exp. Biol.* 81:93–112
126. Meinertz, T., Nawrath, H., Scholz, H. 1973. *Naunyn Schmiedeberg's Arch. Pharmacol.* 279:327–38
127. Meyer, R. B. Jr., Miller, J. P. 1974. *Life Sci.* 14:1019–40
128. Minota, S., Koketsu, K. 1977. *Jpn. J. Physiol.* 27:353–66
129. Miura, Y., Inui, J., Imamura, H. 1978. *Naunyn Schmiedeberg's Arch. Pharmacol.* 301:201–5
130. Morad, M., Goldman, Y. 1973. *Prog. Biophys. Mol. Biol.*, 27:257–313
131. Mudge, A. W., Leeman, S. E., Fischbach, G. D. 1979. *Proc. Natl. Acad. Sci. USA* 76:526–30
132. Nathan, D., Beeler, G. W. 1975. *J. Mol. Cell. Cardiol.* 7:1–15
133. Nawrath, H. 1976. *Nature* 262:509–11
134. Nawrath, H. 1977. *Nature* 267:72–74
135. Neher, E., Sakmann, B. 1975. *Proc. Natl. Acad. Sci. USA* 72:2140–44
136. Neher, E., Sakmann, B. 1976. *Nature* 260:799–801
137. Neher, E., Stevens, C. F. 1977. *Ann. Rev. Biophys. Bioeng.* 6:345–81
138. Niedergerke, R., Orkand, R. K. 1966. *J. Physiol. London* 184:291–311
138a. Niedergerke, R., Page, S. 1977. *Proc. R. Soc. London.* 197:333–62

139. Noma, A., Trautwein, W. 1978. *J. Physiol. London* 284:97–98P
140. Noma, A., Trautwein, W. 1978. *Pfluegers Arch.* 377:193–200
141. Noma, A., Peper, K., Trautwein, W. 1979. *Pfluegers Arch.* 38:255–62
142. Ochi, R., Hino, N. 1978. *Proc. Jpn. Acad.* 54:474–77
143. Pappano, A. J. 1970. *Circ. Res.* 27:319–90
144. Pellmar, T. C., Wilson, W. A. 1977. *Nature* 269:76–78
145. Petersen, O. H. 1976. *Physiol. Rev.* 56:535–77
146. Petersen, O. H., Pedersen, G. L. 1974. *J. Membr. Biol.* 16:353–62
147. Pinsker, H., Kandel, E. R. 1969. *Science* 163:931–35
148. Pott, L. 1979. *Pfluegers Arch.* 380:71–77
149. Prokopczuk, A., Lewartowski, B., Czarnecka, M. 1973. *Pfluegers Arch.* 339:305–16
150. Purves, R. D. 1976. *Nature* 251:149–51
151. Putney, J. W. 1977. *J. Physiol. London* 268:139–49
152. Putney, J. W. 1979. *Pharmacol. Rev.* 30:209–45
153. Reuter, H. 1967. *J. Physiol. London* 192:479–92
154. Reuter, H. 1973. *Prog. Biophys. Mol. Biol.* 26:1–43
155. Reuter, H. 1974. *J. Physiol. London* 242:429–51
156. Reuter, H. 1979. *Ann. Rev. Physiol.* 41:413–24
157. Reuter, H., Scholz, H. 1968. *Pfluegers Arch. Ges. Physiol.* 300:87–107
158. Reuter, H., Scholz, H. 1977. *J. Physiol. London* 264:17–47
159. Saavedra, J. M., Ribas, J., Swann, J., Carpenter, D. O. 1977. *Science* 195:1004–6
160. Scholz, H., Reuter, H. 1968. *Naunyn Schmiedeberg's Arch. Pharmacol.* 260:196–7
161. Schumann, H. J., Endoh, M., Brodde, O. E. 1975. *Naunyn Schmiedeberg's Arch. Pharmacol.* 209:291–302
162. Schwartz, T. L., Kado, R. T. 1977. *Biophys. J.* 18:323–49
163. Selinger, Z., Batzri, S., Eimerl, S., Schramm, M. 1973. *J. Biol. Chem.* 248:369–72
164. Selinger, Z., Eimerl, S., Schramm, M. 1974. *Proc. Natl. Acad. Sci. USA* 71:128–31
165. Shimahara, T., Tauc, L. 1977. *Brain Res.* 127:168–72
166. Sperelakis, N., Schneider, J. A. 1976. *Am. J. Cardiol.* 37:1079–85
167. Sutherland, E. W., Robison, G. A., Butcher, R. W. 1968. *Circulation* 37:279–306
168. Swann, J. W., Carpenter, D. O. 1975. *Nature* 258:751–54
169. Szczepaniak, A. C., Cottrell, G. A. 1973. *Nature New Biol.* 241:62–64
170. Ten Eick, R., Nawrath, H., McDonald, T. F. Trautwein, W. 1976. *Pfluegers Arch.* 361:207–13
171. Tolkovsky, A. M., Levitzki, A. 1978. *Biochemistry* 17:3795–3817
172. Trautwein, W. 1973. *Physiol. Rev.* 53:793–835
173. Trautwein, W., Dudel, J. 1958. *Pfluegers Arch. Gesamte Physiol.* 266:324–34
174. Trautwein, W., Trube, G. 1976. *Pfluegers Arch.* 366:293–95
175. Tritthart, H., Volkmann, R., Weiss, R., Fleckenstein, A. 1973. *Naunyn Schmiedeberg's Arch. Pharmacol.* 280:239–52
176. Tsien, R. W. 1973. *Nature New Biol.* 245:120–22
177. Tsien, R. W. 1974. *J. Gen. Physiol.* 64:293–319
178. Tsien, R. W. 1974. *J. Gen. Physiol.* 64:320–42
179. Tsien, R. W. 1977. *Adv. Cyclic Nucleotide Res.* 8:363–420
180. Tsien, R. W., Giles, W., Greengard, P. 1972. *Nature New Biol.* 240:181–83
181. Tsien, R. W., Siegelbaum, S. 1978. In *Physiology of Membrane Disorders*, ed. T. E. Andreoli, J. F. Hoffman, D. D. Fanestil, pp. 517–38. New York: Plenum
181a. Tsien, R. W., Weingart, R. 1976. *J. Physiol. London* 260:117–41
182. Vassort, G., Rougier, O., Garnier, D., Sauviat, M. P., Coraboeuf, E., Gargouil, Y. M. 1969. *Pfluegers Arch.* 309:70–81
183. Venter, J. C., Ross, J., Kaplan, N. O. 1975. *Proc. Natl. Acad. Sci. USA* 72:824–28
184. Ware, F., Graham, C. D. 1967. *Am. J. Physiol.* 212:451–55
185. Watanabe, A. M., Besch, H. R. 1974. *Circ. Res.* 35:316–24
186. Watanabe, A. M., Besch, H. R. 1975. *Circ. Res.* 37:309–17
187. Weinreich, D. 1977. *Nature* 267:854–56
188. Weinreich, D., Weiner, C., McCaman, R. 1975. *Brain Res.* 84:341–45
189. Wilson, W. A., Wachtel, H. 1978. *Science* 202:772–75
190. Wood, E. H., Heppner, R. L., Weidmann, S. 1969. *Circ. Res.* 24:409–45
191. Yamasaki, Y., Fujiwara, M., Roda, N. 1974. *J. Pharmacol. Exp. Ther.* 190:15–20
192. Yarowski, P. J., Carpenter, D. O. 1978. *Brain Res.* 144:75–94

COMPARATIVE PROPERTIES AND METHODS OF PREPARATION OF LIPID VESICLES (LIPOSOMES)[1]

♦9154

Francis Szoka, Jr.[2]

Department of Experimental Pathology, Roswell Park Memorial Institute, Buffalo, New York 14263

Demetrios Papahadjopoulos

Cancer Research Institute and Department of Pharmacology, University of California, San Francisco, California 94143

INTRODUCTION

In the fifteen years following the demonstration by Bangham et al (8) that aqueous dispersions of phospholipids form closed structures that are relatively impermeable to entrapped ions, there has been a great expansion in the use of lipid vesicles as a model membrane system and more recently as a drug delivery system. The main features of lipid

---

[1] Abbreviations used in this article are as follows: AraC = 1-β-d arabinofuranosyl cytosine, Chol = cholesterol, DNA = deoxyribonucleic acid, DMPA = dimyristoyl phosphatidic acid, DMPC = dimyristoyl phosphatidylcholine, DMPE = dimyristoyl phosphatidylethanolamine, DOPC = dioleoyl phosphatidylcholine, DOPE = dioleoyl phosphatidylethanolamine, DPPA = dipalmitoyl phosphatidic acid, DPPC = dipalmitoyl phosphatidylcholine, DPPG = dipalmitoyl phosphatidylglycerol, DPPS = dipalmitoyl phosphatidylserine, DSPC = distearoyl phosphatidylcholine, EPC = egg phosphatidylcholine, EDTA = ethylene diamine tetracetic acid, HDL = high density lipoprotein, HPLC = high performance liquid chromatography, LUV = large unilamellar vesicle, MLV = multilamellar vesicle, NTA = nitrilotriacetic acid, NMR = nuclear magnetic resonance, PA = phosphatidic acid, PC = phosphatidylcholine, PE = phosphatidylethanolamine, PG = phosphatidylglycerol, PS = phosphatidylserine, REV = reverse-phase evaporation vesicle, RNA = ribonucleic acid, SUV = small unilamellar vesicle, Tc = transition temperature.

[2] Present address: Department of Physiology, Tufts University School of Medicine, Boston, Massachusetts 02111.

vesicles that have made them a valuable investigative tool are the following: 1. their characteristic morphology, where a relatively impermeable lipid bilayer completely encloses an aqueous space, and 2. their ability to encapsulate various solutes present in the aqueous phase during their formation. A number of review articles have described the general properties of liposomes and their use as model membranes (5, 7, 156) or the reconstitution of membrane proteins into vesicles (172). Numerous recent reviews have discussed the use of liposomes as carriers of drugs and macromolecules in vivo and their interaction with eucaryotic cells in vitro (44, 48, 61, 100, 150, 168, 169, 201). A recent symposium has covered the area of liposomes in considerable breadth (153), although the methodological aspects of liposome preparation have not as yet been adequately reviewed. A number of new methods for the preparation and characterization of liposomes have appeared since the publication of the excellent review addressing these topics by Bangham et al (7), and it now seems timely to compare the various methodologies. In this review we point out the salient points in vesicle preparation and the relative advantages of each type of vesicle. Our aim is to acquaint the reader with the pitfalls in liposome preparations, the lipid compositions that can be used, methods of preparation, size distributions of the resulting vesicles, efficiencies of encapsulation of the aqueous space, and methods to characterize the resulting vesicles.

Preparation and Precautions in Handling of Lipids

The sine qua non of preparing well-defined liposomes is to make certain of the purity of the lipids. Commercial sources of lipids vary widely in their purity and it is essential for the investigator to ascertain the purity of these products. Even small amounts of free fatty acids (96) or lysophosphatides (101, 192, 198) can have significant effects on the surface charge and permeability properties of liposomes. Detection of oxidation products from unsaturated lipids can be monitored by the ultraviolet absorption ratio at 232/215 nm in absolute ethanol (7) or by the formation of thiobarbituric acid reaction compounds (60, 67). To quantitate phospholipids, the phosphate content is assayed following digestion by 70% perchloric acid or by the wet digestion method of Morrison (136). Phosphate can be quantitated at the μmole level by the Fiske & Subbarow method (49) or at the nmole level by one of the more sensitive procedures that employ malachite green (74, 164). All phospholipids should be examined before use by thin layer chromatography (7) in at least two solvent systems. Compounds that run close together should be spotted at low loading levels. This is particularly important for the detection of 1, 3 diacyl phosphatidylcholines which can arise by acyl migration during the synthesis of defined PC and can run slightly in front of the 1, 2 diacyl derivatives (110). When minor contaminants are being sought, heavy loadings (250–500μg) should be used.

Phosphorus-containing lipids can be detected by spraying with the molybdenum blue reagent (40), and organic compounds can be visualized by spraying with 50% sulfuric acid, followed by heating at 110°C. Phospholipids containing a free amino group can be detected by spraying with 0.25% ninhydrin in acetone and heating at 110°C for 5 min.

The isolation of phosphatidylcholine from egg yolks (7), phosphatidylserine from bovine brain (157), phosphatidic acid (157) and sphingomyelin from bovine brain (191), and the preparation of defined phosphatidylcholines (177) and defined phosphatidylglycerols (30, 155) are all procedures that give good yields of purified phospholipids. Recent improvements in the synthesis of defined chain phospholipids are reported to reduce acyl chain migration during synthesis (110) and to give higher yields (66).

The most important new development in the preparation of pure lipids is the use of high performance liquid chromatography (HPLC) to quickly and conveniently isolate pure components from mixtures (66, 94, 143). The use of HPLC with a n-hexane-isopropyl solvent system permits the resolution of most lipid components in a total lipid extract. The individual components are detected directly by ultraviolet absorption at 206 nm (204). In addition to the separation of phospholipids of different classes, separation of phospholipids of the same class can be achieved. The advantages of HPLC have been exploited by Porter and his colleagues (167) who have developed a procedure that permits the resolution of individual phosphatidylcholines from complex mixtures, e.g. DMPC form DPPC. The use of these HPLC techniques for the analysis of the purity of phospholipid samples and their adaptation to preparative HPLC (131) should facilitate the production of large quantities of pure phospholipids.

Once the phospholipids have been purified they can be conveniently stored under an inert gas (nitrogen or argon) in organic solvents at −20°C. Even more rigorous precautions to protect radiolabeled lipids from degradation should be employed, particularly if the label is on the acyl chains. Degradation in this case can result in labeled fatty acids that will remain with the liposome during preparation but may then rapidly exchange when placed in contact with biological material.

Lipids and Liposome Composition

Liposomes can be prepared from a variety of lipids and lipid mixtures (for a compilation of some of the compositions used to date see

Reference 61), with phospholipids the most commonly used. However, vesicles can also be prepared from single chain amphiphiles (55, 56, 68, 107), lysophosphatides in the presence of equimolar cholesterol (101), and dicetyl phosphate (137). Some of the phospholipids that have been the focus of much of the work with lipid vesicles are compiled in Table I. For convenience we refer to the individual phospholipids by the abbreviations given in this table. When referring to lipid vesicles we use the nomenclature that was informally agreed upon at the New York Academy of Sciences meeting on "Liposomes and Their Uses in Biology and Medicine" (Reference 153, p. 367). In this nomenclature all types of lipid bilayers surrounding an aqueous space are considered to be in the general category of liposome. The different types of liposomes are then referred to by a three letter acronym, e.g. multilamellar vesicles (MLV). Unilamellar vesicles are substituted into two classes on the basis of size; vesicles under 1000 Å are usually considered small unilamellar vesicles (SUV), whereas those with a greater diameter are large unilamellar vesicles (LUV). The lipid composition with the ratio of the lipids enclosed in parenthesis follows the acronym, e.g. SUV (DPPC/Chol, 1:1). When selecting the liposome composition, one should keep in mind that phospholipids form smectic mesophases that undergo a characteristic gel-liquid crystalline phase transition. Vesicles composed of phospholipids that are at temperatures below this transition temperature (Tc) are considered "solid"; when above Tc they are considered fluid, (19, 114, 133, 165). This transition temperature is a function of acyl chain length; in phospholipids composed of the same acyl chain in both positions, Tc increases about 14–17° with every 2-methylene unit increase in chain length. The presence of unsaturated acyl chains, branched chains, or those carrying bulky side groups, e.g. cyclopropane rings, produces a considerable decrease of the transition temperature. In mixed acyl chain phospholipids composed of saturated fatty acids that differ by two methylene units, the isomer that has the longer chain in the sn-1 position of glycerol has the lower transition temperature of the pair (97). The head group of the acidic phospholipid, such as phosphatidylserine, phosphatidylglycerol, or phosphatidic acid, can also have a considerable influence on the transition temperature (Table 1). The transition temperature of these phospholipids can be modulated by interactions with divalent cations or by H^+ ion titration of the head group. Calcium and magnesium ions in the physiological range (0.1–10mM) can increase the transition temperature of PA, PS, and PG (86, 152, 154, 160, 204a). The effect of pH on the transition temperatures of these phospholipids has also been well characterized. The addition of a hydrogen ion increases the temperature by 17°C in DPPG (46, 207) and by 6–8° in DMPA and DPPA (86, 152, 199). It should be

Table 1 Some properties of phospholipids used in liposomes

Lipid	Abbreviation	Charge[a]	Tc(°C)[b]	Ref.
Egg phosphatidylcholine	EPC	0	−15 to −7	109
Dilauryloylphosphatidylcholine (C12:0)	DLPC	0	−1.8	122
Dimyristoylphosphatidylcholine (C14:0)	DMPC	0	23	109
Dipalmitoylphosphatidylcholine (C16:0)	DPPC	0	41	109
Distearoylphosphatidylcholine (C18:0)	DSPC	0	55	109, 122
1-Myristoyl-2-palmitoylphosphatidylcholine (C14:0, 16:0)	MPPC	0	27	97
1-Palmitoyl-2-myristoyl phosphatidylcholine (C16:0, 14:0)	PMPC	0	35	97
1-Palmitoyl-2-stearoyl phosphatidylcholine (C16:0, 18:0)	PSPC	0	44[c]	97
1-Stearoyl-2-palmitoyl phosphatidylcholine (C18:0, 16:0)	SPPC	0	47[c]	97
Dioleoylphosphatidylcholine (C18:1)	DOPC	0	−22	109
Dilauryloylphosphatidylglycerol	DLPG	−1	4	46
Dimyristoylphosphatidylglycerol	DMPG	−1	23	158
Dipalmitoylphosphatidylglycerol	DPPG	−1	41	86
Distearoylphosphatidylglycerol	DSPG	−1	55	153
Dioleoylphosphatidylglycerol	DOPG	−1	−18	46
Dimyristoyl phosphatidic acid	DMPA	−1[d]	51	86
Dimyristoyl phosphatidic acid	DMPA	−2[e]	45	153
Dipalmitoyl phosphatidic acid	DPPA	−1[f]	67	153
Dipalmitoyl phosphatidic acid	DPPA	−2[g]	58	153
Dimyristoyl phosphatidylethanolamine	DMPE	−[h]	50	153
Dipalmitoyl phosphatidylethanolamine	DPPE	—	60	203
Dimyristoyl phosphatidylserine	DMPS	—	38	203
Dipalmitoyl phosphatidylserine	DPPS	—	51	120
Brain phosphatidylserine	PS	—	6–8°	86
Brain sphingomyelin	BSP	0	32	191
Dipalmitoyl sphingomyelin	DPSP	0	41	18
Distearoyl sphingomyelin	DSSP	0	57	42

[a]Charge at pH 7.0 unless otherwise indicated.
[b]Values determined by differential scanning calorimetry and rounded to the nearest degree.
[c]Due to acyl group migration during synthesis of the phospholipid the measurements were made on a mixture of the two derivatives; hence the pure PSPC would probably have a slightly lower Tc, and SPPC a slightly higher Tc (97).
[d]pH 6.0.
[e]pH 9.0.
[f]pH 6.5.
[g]pH 9.1.
[h]Phosphatidylethanolamine is partially titrated at pH 7.0 (7).

noted that most studies examining the transition temperatures of pure phospholipid components or the presence of lateral phase separations between mixtures of phospholipids have used MLV. When the Tc is measured in SUV, it is broader and appears at 4–5°C lower than in the MLV. This is attributed to the packing constraints on the acyl chains imposed by the high radius of curvature of the sonicated vesicles. In forming liposomes it is necessary to hydrate the dried lipid at a temperature *higher* than the Tc of the lipid. This will be obvious to those attempting to hydrate dried phospholipids below their Tc, since it is very difficult to remove the lipids from the sides of the vessel. When preparing mixtures of phospholipids it is necessary to mix them thoroughly in the organic phase before drying and to hydrate them at temperatures above the Tc of the higher melting component. The ability to prepare vesicles with defined transition temperatures is an important parameter for drug delivery, since the permeability of vesicles to entrapped compounds is relatively low below the Tc (see below). Furthermore, an anomalous increase in permeability of ions (15, 32a, 155) and in penetration of various molecules into the bilayer of the vesicle in the vicinity of the Tc (87, and references therein) has recently been exploited to achieve a measure of drug targeting by the use of hyperthermia (209, 217).

In many cases liposomes have been prepared from mixtures of different phospholipids. Saturated phospholipids of the same head group, but differing by only two methylene units in the acyl chains, will exhibit complete miscibility in all proportions. However, nonideal mixing will occur with evidence of lateral phase separation of the phospholipids when the chain lengths differ by four or more methylene groups (120a, 166, 190). Similar results can be observed with mixtures of saturated phosphatidylglycerols and phosphatidylcholines, both in the presence and absence of calcium (46). In the latter system, calcium does not induce a phase separation when the PG chain lengths are equal to, or two carbons longer than, the PC, but does so when the PG chain lengths are two carbons shorter (46). Such calcium-induced lateral phase separation has been observed initially in PS/PC mixtures (144, 158) and later with PA/PC mixtures (54, 186, 125) and DPPS/DPPC mixtures (120). Mixtures of phosphatidylcholines and phosphatidylethanolamines of equal chain lengths also exhibit lateral phase separations in the gel phase in MLV (20, 114, 190) which appear to be accentuated in SUV (115). Phosphatidylethanolamine appears to be unique among the phospholipids in not forming closed bilayer vesicles at pH 7.0 (115, 157), but rather a hexagonal phase structure (171). In fact, vesicles cannot be formed from PE (at neutral pH) if the mole fraction of PE is greater than 70% (115, 117, 157).

In addition to lateral phase separations occurring in lipid mixtures in liposomes, the radius of curvature of SUV can lead to an asymmetric distribution of the different components between the inner and outer monolayers (13, 117, 134). Such an asymmetric arrangement can be expected on theoretical grounds (84, 85). In vesicles composed of mixtures of ^{13}C enriched egg PC with DPPC, dilinoleyl PC or diarachidomyl PC, the more unsaturated species were preferentially found in the outer surface (219). This is thought to be due to the inability of the unsaturated acyl chains to pack as closely as saturated chains; thus, the head group of the unsaturated chains occupies a larger area than the saturated phospholipids and hence prefers the outer monolayer. In mixtures of PC and sphingomyelin it is the sphingomyelin that has a preference for the outer monolayer (12). When PC is mixed with a low ratio of negatively charged phospholipids (PG, PS, and phosphatidylinositol) the negatively charged phospholipids prefer the inside, particularly at low pH (12, 205). At high ratios of PG to PC, the PG component prefers the outside monolayer (134). Egg phosphatidylethanolamine when mixed with PC prefers the outer monolayer at low concentrations (10 mol%) but the inner monolayer at high concentrations (118). However, in DMPE/DMPC mixtures, the PE prefers the outer monolayer until a 1:1 or higher PE/PC ratio is achieved and it appears to be proportionally distributed across the bilayer (115).

In cholesterol-phosphatidylcholine mixtures the composition of the two monolayers is found to be similar up to 30% mole ratio of cholesterol (34, 35, 80), but at higher mole ratios of cholesterol it is found to preferentially reside in the inner monolayer. Thus, when using mixtures of lipids to form SUV, it is possible that the composition of the outer monolayer may not be the same as the composition of the initial mixture. The resulting distribution depends upon the packing requirements of the individual head groups, the acyl side chain composition, the charge of the phospholipids, and the pH.

At present there is no method that forms fully asymmetric vesicles, although by the use of phospholipid exchange proteins and a suitable source of phospholipid donors, vesicles with an asymmetric distribution of phospholipid can be prepared (149, 179, 181). Due to the long half-time for the exchange of phospholipid between the inner and outer monolayers, these vesicles are suitable for studies relating asymmetric lipid distribution to membrane function.

Cholesterol is a prominent component of eucaryotic biological membranes, and liposomes containing cholesterol have been important in relating the physicochemical properties of cholesterol-phospholipid interactions to the possible physiologic role of cholesterol in cell membranes. Recent reviews (88, 145, 156, 165) have discussed in detail the effects of cholesterol on the properties of phospholipid membranes, such as the elimination of the Tc at a 33% mole ratio of cholesterol to phospholipids, the condensation of the area of phospholipid molecules in monolayers, the inhibition of motion within the outer segment of the phospholipid acyl chains in bilayers, the increase in width of bilayers composed of short chain phospholipids, and the increased perpendicular orientation of the acyl chains. Of considerable importance for the use of liposomes as drug carriers is the ability of cholesterol to decrease the permeability of phospholipid bilayers to ions and small polar molecules and to reduce the ability of a number of proteins to penetrate and increase disorder within the bilayer (151, 156).

There is considerable disagreement as to the exact mode of interaction of cholesterol with the phospholipids in bilayers—in particular, whether the 3-β-hydroxyl on the steroid ring interacts with the carbonyls of the phospholipids. In liposomes, the 3-α-hydroxysterols are less effective in reducing permeability in liposomes than the corresponding 3-β-hydroxysterols (37); thus the orientation of the sterol hydroxyl group at the hydrophilic-hydrophobic interface must be an important factor in the ability of cholesterol to reduce the permeability of liposomes to ions and hydrophilic molecules. Cholesterol is able to reduce the permeability of MLV liposomes composed of either diester, diether, or alkyl analogues of PC of hydrophilic nonelectrolytes at temperatures above their respective Tc (24). This suggests that carbonyl oxygens of the ester groups in either position 1 or 2 on the glycerol backbone are not essential for interactions with cholesterol (24). Furthermore, the motions of the phosphocholine group are essentially unaffected by the incorporation of cholesterol in PC bilayers when examined by a variety of nuclear magnetic resonance (NMR) techniques (26, 98, 193, 218). However, there are some indications from NMR studies that a hydroxyl group is interacting with the carbonyl oxygens of PC, although it is not clear whether this is from a water molecule or from the 3-β-OH of cholesterol (220). Neutron and X-ray diffraction studies have been interpreted to place the hydroxyl group in the vicinity of the carbonyl groups (51, 214). In support of such an interaction, cholesterol at 30 mol% was observed to have no significant effect on the efflux of glucose, Na^+, or chloride from small unilamellar liposomes prepared from an unsaturated diether PC, as opposed to the results from SUV liposomes prepared from the corresponding diester PC (187) where cholesterol decreased the permeability to these compounds.

These results must be contrasted with those observed with MLV preparations of diether PC (24). However, in the former study, the use of unsaturated phospholipids and small unilamellar vesicles, which have a high radius of curvature, may be the source of this apparent contradiction. In addition to the 3-β-hydroxyl group which is important for the interaction of cholesterol with the bilayer, the complete cholesterol side chain is required for maximum ordering (25). Sidechains shorter or longer (by 3 carbons) than those normally found in cholesterol cause significantly less ordering of spin labels in EPC liposomes (25). The maximal mole ratio of cholesterol that can be inserted into phospholipid bilayers is 2:1 for EPC and sphingomyelin (121, 175). However, SUV composed of cholesterol/phospholipid at such a high ratio appear to be metastable, forming larger structures upon standing (121). Vesicles formed from a 1:1 mole ratio of cholesterol/phospholipid form stable structures (121 and references therein).

In mixtures that do not phase separate, there is no clear evidence for preferential interaction of cholesterol with different chain length PC or PC-sphingomyelin. In those mixtures where phase separation occurs, cholesterol interacts preferentially with the lower melting component (18, 36). In mixed systems of phosphatidylcholine and phosphatidylethanolamine there is a preferential interaction of cholesterol with the phosphatidylcholine component. The extent of this interaction depends upon the chain length, degree of unsaturation, and the transition temperature of individual components in the mixtures (27, 28, 203). In equimolar mixtures of PE and PC that phase separate, cholesterol preferentially abolishes the transition of the PC component, which suggests that in the liquid crystalline state cholesterol is preferentially associated with the PC component in the mixture (28, 203). Since PE dispersions exhibit complex phase behavior with evidence for both hexagonal and lamellar phases (171), the preferential interactions of cholesterol with the PC component of PC/PE mixtures that form a stable lamellar phase (bilayer structure) result in perturbations of the bilayer structure (27). Such perturbations have been reported for equimolar mixtures of DPPC/DOPE, where the addition of cholesterol stabilizes the bilayer structure, and for equimolar mixtures of DOPC/DOPE, where the addition of cholesterol destabilizes the bilayer structure by enhancing the formation of a "hexagonal" phase (28). Thus, when working with ternary mixtures of PC, PE, and cholesterol for the formation of liposomes, the possibility exists that stable bilayer structures may not be formed.

Early studies showed that the inclusion of cholesterol at a 1:1 mole ratio with saturated PC up to C16 increases the width of the bilayer, but for the distearoyl PC (C18) cholesterol reduces the bilayer width (132). In terms of SUV structure, the inclusion of cholesterol in the bilayers at up to 20 mol% has no effect on SUV size (50, 90), but as the mole ratio increases above 30 there is a sharp and progressive increase in the diameter of the vesicles. This is observed in SUV composed of both EPC (90) and DMPC (50). SUV containing 50 mol% cholesterol exhibit about a 30% increase in diameter.

Thus, it is obvious from the above studies that, depending on the lipid compositions used for the formation of liposomes, wide variations may occur in the specific composition of the external surface of the bilayer. Lateral phase separations, transbilayer asymmetries, and in certain cases, nonbilayer structures may arise from different mixtures of lipids commonly used for the preparations of liposomes.

Methods for Characterizing the Size Distribution of Lipid Vesicles

For many physical and biological studies, the average size and size distribution of the liposomes used is an important parameter for the final interpretation of results. Most of the biophysical techniques for measuring the size of vesicles require a homogeneous population of vesicles with a well-defined shape. Thus, light scattering techniques, analytical ultracentrifugation, and NMR spectroscopy are most appropriately applied to small unilamellar vesicles. For studies using liposomes for drug delivery the MLV and LUV preparations seem most appropriate because of their higher aqueous space-to-lipid ratio. These preparations are often quite heterogeneous in size, and the size distributions can only be determined by microscopic techniques.

Light microscopy has been utilized to examine the gross size distribution of large vesicle preparations such as MLV (7), the large unilamellar vesicles produced by the Reeves & Dowben (173) method, or the large vesicles produced from single chain amphiphiles (68). The inclusion of a fluorescent hydrophobic probe in the bilayer (138) permits the examination of the liposomes under a fluorescent microscope and makes it possible to obtain an estimate of at least the upper end of the size distribution. The resolution of the light microscope limits this technique for obtaining the complete size distribution of the preparation but, in conjunction with a knowledge of the volume of aqueous space captured per mole of lipid, it can yield a rough approximation as to whether large vesicles comprise an appreciable fraction of the preparation. By also using negative stain electron microscopy one can then obtain an estimate of the lower end of the size distribution. For large vesicles ($>5\mu$m), negative stain electron microscopy is not suitable for determination of the size distribution, since vesicle distortion during preparation of the specimen makes it difficult to obtain an estimate of the diameter of the original particle. The advantages and disadvantages of negative stain electron microscopy have been discussed by Bangham et al (7). In general, negative stain using molybdate or phosphotungstate has proven to be satisfactory for the analysis of SUV yielding values that agree to within 10 or 20% of values determined by freeze fracture, hydrodynamic studies, and NMR (13, 78, 208). The analysis of the size distribution of MLV by negative stain electron microscopy is a formidable problem since they consist of a heterogeneous distribution of sizes and shapes. Often large elliptical or cylindrical shaped structures are observed. To deal with this problem, Olson and colleagues (146) took the mean of the major and minor axis of such structures and assumed it represented a sphere of equivalent volume when calculating its contribution to the size distribution of the MLV preparation. Furthermore, since the observed structures appeared collapsed, they assumed that the observed structure was a disc of two bilayers that had arisen from a spherical particle and they multiplied the observed particle diameter by a factor of 0.71 which would convert the surface area of a disc to that of the precursor sphere from which it had arisen. Comparing similar preparations by freeze fracture electron microscopy, they were able to show a close correspondence for the mean diameter and size distribution when measured by the two methods. The discrepancy between the two methods in estimating the vesicle size was the greatest for the distributions that contained the largest diameter particles and the greatest heterogeneity (146). It was reported in the same study that sequential extrusion of MLV through polycarbonate membranes of defined pore size reduces the mean size of the MLV and size distribution and also reduces the discrepancy between the values obtained by the two techniques (146).

A technical difficulty in obtaining good negative stains of liposomes is the spreading of the vesicles on the carbon-coated grid. Treating the grid with a 0.1 mg/ml solution of bacitracin (63) usually permits a satisfactory spreading of liposomes on the support film. The coating of support films with silica by the evaporation of silicon monoxide has recently been reported to give excellent spreading of liposomes for negative staining (111) and should be a useful method for examining liposome preparations by negative stain electron microscopy. Freeze etch and freeze fracture electron microscopic techniques have been extensively used to study vesicle size and structure. Evaluation of SUV yields estimates within 10% of those obtained by hydrodynamic measurements, light scattering, and NMR techniques, (13, 78, 208). The freeze etch technique is particularly suitable for the measurement of vesicles of small diameter because the effects of random cleavage that can occur through and around the vesicle, not necessarily through the midplane, can be compensated for by the etch step. Cleavages that occur above the midplane of the vesicle and that would have revealed a smaller diameter particle by freeze fracture are enlarged by the etching process to approach the maximum diameter of the vesicle. Thus, vesicles that are clearly exposed by the etch process can be measured.

For populations of large size vesicles, freeze fracture can yield a representative morphological view of the liposomes; it has been useful for examining the morphological changes that occur in the bilayer surface as the phospholipids go through the gel-liquid crystalline transition (120, 120a, 158a, 203, 204a) or a lamellar to a hexagonal transition (202a, 204a) or following the incorporation of membrane proteins into reconstituted liposome (202). However, the freeze fracture technique has a serious drawback for estimating the size distribution and mean vesicle size of a heterogeneous population of vesicles. The fracture plane passes through vesicles that are randomly positioned in the frozen section, resulting in nonmidplane fractures. Thus, the observed profile diameter depends on the distance of the vesicle center from the plane of the fracture. A homogeneous population of vesicles will therefore yield a distribution of profile sizes, with the largest being equal to the true radius of the vesicle. Mathematical methods have been described to correct for this nonequational fracture (211) and have been applied to freeze fracture data to obtain an estimation of the size of SUV (65).

It is important to point out that there is no reason to suspect that a unimodal distribution of vesicle sizes will always be observed in a preparation of liposomes (111). This may be particularly true when an MLV preparation is sonicated for a short period of time (62, 92, 180). It has been suggested by one report (12) that SUV may be formed immediately from MLV in a discrete process, whereas another report suggests they are formed by the gradual and sequential reduction of the larger vesicles. (71). These differences could possibly be a consequence of differing sonication intensity used by the two groups, and they point out the need for measuring the size distribution of MLV that have been exposed to a short period of sonication.

Techniques to obtain the size distribution of small homogeneous vesicle populations include analytical ultracentrifugation (78), NMR (13), and quasi-elastic laser light scattering (188). The techniques for the hydrodynamic analysis of SUV have been extensively worked out using SUV (EPC) by Huang and colleagues (78, 79, 124). The most fruitful application of these techniques to the investigation of vesicle size required a homogeneous population of vesicles. To achieve this goal, gel chromatography on large pore agarose gels was introduced by Huang (78) to separate SUV from residual MLV in the preparation. However, high speed centrifugation can also be used to remove MLV (10, 208). An accurate estimate of the sedimentation coefficient depends on the measurement of the partial specific volume of the particle being studied. One method, introduced by Huang & Charlton, to accurately calculate this value for phospholipid vesicles is sedimentation analysis using mixtures of hydrogen and deuterium oxide (79). The application of hydrodynamic measurements to determine the size, shape, surface area, inner and outer monolayer areas and molecular weights of SUV has recently been reviewed by Mason & Huang (124). Hydrodynamic measurements have also been employed to analyze the influence of cholesterol and negative surface charge density on the size of SUV (90) and on the dimensional changes of SUV (DMPC) upon going through the phase transition (208). Among all groups there is a general agreement that SUV (EPC) or SUV (DMPC) have a particle weight of approximately 2×10^6 daltons and are spherical in shape.

The use of nuclear magnetic resonance to determine the size of vesicle preparations depends upon the use of paramagnetic ions to shift or broaden the NMR signal from the choline trimethylammonium protons or the phosphate moiety of the phospholipid. This methodological approach has recently been reviewed by Bergelson (13). Essentially, the paramagnetic species is added to a dispersion of vesicles and due to its low permeability through the bilayer, only the external headgroups are exposed. The intensity ratio of the perturbed signal to the unperturbed is independent of the shift reagent employed and depends on the ratio of phospholipids in the outer and inner monolayers. Thus, the ratio of the phospholipids in the two monolayers can be determined either from the peak areas of the two resonances when a shift reagent is used or from the percentage reduction in peak area when a broadening reagent is employed. Use of this technique for size determination rests on the assumptions that the vesicles are unilamellar, spherical, and of a defined bilayer thickness. ^{3}H and ^{31}P NMR are limited to small unilamellar vesicles since they do not yield high resolution spectra on larger vesicles which have a restricted motion (13). Furthermore, when the inner to outer ratio is unity, the technique can merely set a lower limit on the size of the vesicle suspension. Thus, even ^{1}C NMR of the N-methycholine group, which experiences only a limited resonance broadening in structures with restricted motion and can yield information on the transbilayer distribution of the choline groups in large structures, will give no further information on the size of the larger vesicles (57). It can be utilized, however, to determine whether a population of LUV is strictly unilamellar (in which case the ratio of inner to outer signal would be unity) or whether it contains an appreciable population of MLV (in which case the ratio would be less than unity).

The use of classical light scattering techniques to determine vesicle size has been discussed in some detail in (7), and methods have been proposed to estimate the average size of lipid vesicles dispersed in water (22). If a log-normal distribution of vesicle sizes is assumed, this technique can be used to obtain a size and heterogeneity estimate for polydispersed vesicles preparations (22). For monodispersed collections of small and intermediate sized unilamellar vesicles, dynamic light scattering (quasi-elastic light scattering) can be employed to obtain an estimate of the vesicle size (4, 188). This procedure measures the hydrodynamic diameter of the diffusing vesicles from the time-dependent fluctuations of scattered light. The limitations of this technique for obtaining a size estimate of vesicles have been discussed by Selser and colleagues (188).

Preparation of Liposomes

MULTILAMELLAR VESICLES (MLV) Multilamellar vesicles were the first liposome preparation to be described in detail (8) and are the most widely used, although in many respects they are still poorly characterized. The preparation of MLV is exceptionally simple and, as pointed out by Bangham (7), it can even be claimed that they form spontaneously. The lipids are deposited from organic solvents in a thin film on the wall of a round bottom flask by rotary evaporation under reduced pressure. If mixtures of lipids are used they must be adequately mixed in organic solvents prior to deposition. The aqueous buffer is added and the lipids are hydrated at a temperature above the Tc of the lipid or above the Tc of the highest melting component in the mixture. A thin film of lipid is desirable to facilitate the efficient hydration of the bilayer. The hydration time is important for obtaining maximal encapsulation. A similar lipid concentration can encapsulate 50% more of the aqueous phase per mole of lipid when hydrated for 20 hr with gentle shaking, compared to a hydration period of 2 hr, despite the fact that the two preparations exhibit a roughly similar size distribution (146). A similar MLV preparation hydrated for only 30 min with vigorous shaking yields a suspension that not only has a lower capture of aqueous space per mole of phospholipid but also a smaller mean diameter (146). Thus the hydration time and method of resuspension of the lipids can result in decidedly different preparation of MLV, despite an identical lipid concentration and composition and volume of suspending aqueous phase.

The measurement of the surface area of various MLV preparations has been described by Bangham and colleagues (6) using titration of the exposed headgroups by uranyl ion; this procedure has been elaborated in a review (7) on the preparation of liposomes. However, it would be technically difficult for most laboratories to perform such a titration, and alternate procedures such as the reduction of spin-labeled lipids in the outer monolayer (186) or the chemical modification of phosphatidylethanolamine (118) can be conveniently employed. Schwartz & McConnell (186) using the spin-label reduction method have observed that liposomes prepared under specified conditions from DMPC, DPPC, or mixtures of DPPC-cholesterol have the same proportion of external lipid. Short hydration times and gentle vortexing result in 8% of the spin label exposed to the reductant, whereas vigorous vortexing or long hydration times with gentle shaking result in 5.5% of the spin label exposed on the surface. By assuming that the spin-labeled lipid is randomly distributed between the lamellae of the MLV and that the diameter of the liposomes is much greater than the thickness of the bilayers, they calculated the average number of lamellae from the percentage of the exposed spin label. Vigorous vortexing or longer gentle shaking favors a statistically greater number of lamellae, and under these conditions the authors conclude all three compositions have an average of 10 lamellae. However, the function relating percentage of exposed spin label to the number of lamellae has apparently been derived by assuming successive lamellae have the same surface area. This does not take into account the size of the aqueous core, the interlamellar spacing, or the fact that each successive lamella going towards the core has a smaller surface area. Accounting for all these factors would tend to considerably increase the calculated number of lamellae in the MLV.

Besides alterations in liposome size or the average number of lamellae per liposome, the intralamellar volume and hence the entrapped volume of the MLV can also be increased (7). This is accomplished by including charged lipids in the bilayer, usually at 10–20 mol%. Vigorous vortexing, brief sonication, or extrusion through polycarbonate membranes (146) can be employed to obtain a smaller or/and more uniform preparation of MLV. The size reduction brought about by a brief sonication depends upon the time and power level of the sonication and in general is difficult to standardize. On the other hand, the extrusion technique is easy to employ and gives a reproducible size distribution of MLV (146). Gregoriadis (61) has compiled both a list of compounds encapsulated and the lipid compositions used for the preparations. This attests to the versatility of the MLV preparation. The aqueous space per mole of phospholipid can range between 1 and 4 liters per mole and under appropriate conditions up to 15% of protease inhibitors (47) or 26% of polynucleotides (106) appear to be entrapped in MLV. Recent studies have demonstrated that high molecular weight DNA can become associated and entrapped in MLV (75) and even intact chromosomes can become liposome-associated when co-deposited from organic solvent with lipids prior to hydration (138). It was found that the liposome-associated chromosomes increased the rate of gene transfer substantially over existing methods (138).

In summary, MLV are suitable for the encapsulation of a variety of substances and can be made with a wide variety of lipid compositions. Of all the types of liposome preparations, they are the simplest to prepare and, in conjunction with recent methods for obtaining a more homogeneous size distribution of liposomes, could be the choice for many experiments on drug delivery. The major drawback with MLV is the relatively low encapsulation capacity in terms of liters of aqueous space per mole lipid, which is several fold less than that of large unilamellar vesicles of comparable size.

SMALL UNILAMELLAR VESICLES The now classical procedure of sonicating a dispersion of phospholipids to form optically clear suspensions with particle weight of approximately 2×10^6 daltons was initially introduced by Saunders (182) and Abramson and colleagues (1). Subsequently, it was demonstrated by Papahadjopoulos and colleagues (157, 163) that the structures formed by sonication were microvesicles of about 500 Å that enclosed an aqueous space. This preparation was rigorously characterized by Huang (78) who introduced the use of gel filtration chromatography to separate the residual larger multilamellar vesicles from the smaller unilamellar vesicles and who studied in detail their hydrodynamic properties. This careful characterization of the SUV preparation conclusively proved that the structures observed by electron microscopy were consistent with the hydrodynamic measurements. The preparation of SUV has been covered in some detail by Bangham et al (7). To briefly recapitulate, the usual MLV preparation is subsequently sonicated either with a bath type sonicator (91, 163) or a probe sonicator under an inert atmosphere (usually nitrogen or argon) (1, 70, 78, 182). Although sonication with a probe usually can deliver a higher power density, hence faster breakdown of the MLV to structures with a minimal radius, it has the disadvantage of contaminating the preparation with metal from the tip of the probe and if one is not careful, it can lead to degradation of the phospholipid (70). Moreover, this technique can generate aerosols from solutions containing radioactive traces, carcinogenic chemicals, or infectious agents that have been added to the preparations. These can be a serious biohazard. Bath type sonicators, on the other hand, avoid these difficulties but require more care in the positioning of the closed tube that contains the lipid dispersion within the sonicating bath and require more attention to the time of sonication in order to obtain vesicles of minimal diameter. In addition, bath sonication has the advantage of being a closed system in which the temperature of the preparation can be carefully regulated. It is equally essential in preparing SUV from defined phospholipids that the sonication be performed at a temperature above the Tc of the highest melting lipid in the mixture and that the vesicles be allowed to anneal the Tc for at least 30 min in order to form a stable preparation (112). Sonication below the Tc of the lipids produces structures that contain defects within the bilayer. These allow the rapid permeation of ions and when raised above the Tc permit vesicle-vesicle fusion. This property can be used to form large unilamellar vesicles that can encapsulate an appreciable fraction of the aqueous phase (see below; 112).

SUV with diameters of between 215 and about 500 Å are formed by sonication, depending on the phospholipid composition, time of sonication, and the molar fraction of cholesterol in the lipid mixture (50, 78, 90). Due to the high radius of curvature in these vesicles there is a higher percentage of phospholipid in the outer monolayer (approximately 60–70%) compared to the inner monolayer. In mixtures of lipids this can lead to an asymmetric distribution of the components in the two bilayers. Thus when preparing SUV for studies relating lipid composition to some functional properties, the possibility for such asymmetric distributions must be taken into account. The SUV are drastically limited in terms of the encapsulation of aqueous space per mole lipid; values range from 0.2 to 1.5 liters per mole lipid, depending on lipid composition of the vesicle. The inclusion of cholesterol and charged lipids increases the captured volume. The small aqueous space also limits the size of macromolecules that can become encapsulated; thus, molecules above 40,000 daltons have restricted encapsulation (2). Sonicated vesicles can be prepared from a wide variety of phospholipids including such simple phosphate diesters as dihexadecyl phosphate (137), and most likely they can be prepared from any other amphipathic molecule with a sufficiently long nonpolar side chain and a hydrophilic headgroup (84).

The most important attribute of SUV is the fact that they form a small homogeneous population of vesicles that can be separated from contaminating MLV by simple techniques (10). Their principal drawbacks are the low encapsulation efficiency of the aqueous space, usually in the range of 0.1 to 1.0% depending on the lipid concentration, the low ratio of captured volume per mole of lipid, and the possibility for the asymmetric distribution of various lipids between the inner and outer monolayer.

DETERGENT REMOVAL   An essentially different method for the preparation of phospholipid vesicles depends upon the removal of detergents from detergent-phospholipid mixtures, which results in the formation of unilamellar vesicles. This method has been widely used for the reconstitution of membranes and has been reviewed by Razin (172); more recent examples of membrane reconstitution have been reviewed by Korenbrot (103).

The detergent dialysis method for phospholipid vesicle preparation was initially introduced by Kagawa & Racker (95). These authors removed cholate or deoxycholate from lipid-protein mixtures to form lipid vesicles that incorporated protein. More recent methods remove the detergent from the phospholipids by centrifugation (206), gel filtration (17, 41), or by a fast controlled dialysis (135). The treatment of EPC with sodium cholate in a 1:2 molar ratio followed by gel filtration to remove the bile salts results in the formation of a homogeneous population of unilamellar vesicles witth a mean diameter of 300 Å (17). This technique has been modified by increasing the molar ratio of EPC to deoxycholate to 2:1 (41). Either the detergent is added to preformed SUV or the dried phospholipids are sonicated in the presence of the detergent. In either method the mixture is then passed through two consecutive columns of Sephadex G-25 that remove the detergent. These vesicles have an average diameter of 1000 Å and are unilamellar. They are capable of entrapping proteins and low molecular weight compounds such as glucose; at 50 µmol lipid/ml, 12% of the aqueous phase can be entrapped in the vesicles. However, the trapping of sodium was not achieved, indicating that sodium leakage may precede detergent removal. These vesicles are stable when stored under nitrogen at 4°C for at least one month and are reported to contain less than one deoxycholate molecule per 1000 phospholipids. The method is applicable to a wide range of phospholipids and mixtures of phospholipids but may not be suitable for the formation of vesicles from acidic phospholipids alone. The essential step in the formation of the 1000 Å vesicles is to start with a phospholipid:detergent ratio of 2:1. Either deoxycholate or cholate is a suitable detergent for use in this method. This technique should prove very useful for the reconstitution of membrane proteins in uniform unilamellar vesicles of intermediate size.

Another modification of the cholate removal technique is one where the rate of efflux of detergent from the mixture is controlled (135). This procedure employs a phospholipid:detergent ratio of 0.625 and removes the detergent in a flow-through dialysis cell. The procedure forms a homogeneous population of vesicles with a mean diameter of 570 Å, although the method of formation essentially precludes the efficient entrapment of small water soluble compounds.

Another promising method for membrane reconstitution studies, recently described by Gerritsen and his colleagues (58), is vesicle formation from Triton-phospholipids mixtures. This method is based on the ability of Bio-beads SM-2 to absorb Triton X-100 selectively and rapidly (76). The dried lipid is dispersed in a 0.5% Triton X-100 solution and the dispersion is treated with 0.3 g wet Bio-beads SM-2/ml of dispersion for 2 hr at 4°C. The beads are removed by filtration through glass wool and the vesicles are sedimented at $150,000 \times g$ for 90 min and washed once with buffer (58). The mean size and size distribution of the resulting vesicles, the effect of lipid composition, temperature, and detergent:phospholipid ratio on the formation of the vesicles, as well as the amount of Triton remaining in the vesicles, are all parameters that require a more thorough examination. However, the technique appears applicable to a large number of membrane proteins that have been solubilized by Triton X-100 and seems to have circumvented the problem of protein aggregation when Triton is removed by slower dialysis techniques (72).

ETHANOL INJECTION An alternative method for the preparation of small unilamellar vesicles that avoids both sonication and detergents is the ethanol injection technique described by Batzri & Korn (11). Lipids dissolved in ethanol are rapidly injected into a buffer solution where they spontaneously form small unilamellar vesicles. This procedure is simple, rapid, and gentle. The major drawback of this technique is that one obtains a relatively dilute preparation of liposomes which decreases the encapsulation efficiency of the aqueous phase. The vesicles can be concentrated by ultrafiltration under N_2 pressure (10 lbs/in^2) in various commercially available ultrafiltration apparatuses. Preparations formed in this manner have internal volumes similar to sonicated preparations, i.e. about 0.5 liters per mole lipid (11). A recent modification of this procedure permits the formation of vesicles of variable diameter by altering the concentration of lipids in the ethanolic solution (104). At low concentrations of lipid (3 mM) 300 Å diameter vesicles are formed, whereas at high concentrations (36 mM) 1100 Å diameter vesicles are formed. The size and homogeneity of the resulting vesicles can be modulated to a certain degree by including ethanol in the aqueous buffer, by the injection rate of ethanolic solution into the aqueous phase, and by the rate of stirring of the aqueous phase (104). The ability of this method to encapsulate macromolecules or to incorporate membrane proteins has not as yet been investigated. Furthermore, like the original preparation, the modified ethanol injection procedure will have a low encapsulation efficiency of the aqueous phase owing to the dilute suspension of liposomes formed in the process.

FRENCH PRESS EXTRUSION Dispersions of multilamellar vesicles can be reduced in size by extrusion at high pressures through a French press.

This method is originally ascribed to Hamilton & Goerke (150), although no details were reported. A similar method has been documented in greater detail by Barenholz and colleagues (9). Dispersions of lipids are placed in the French press and extruded at 20,000 lbs/in^2 at 4°C. Dispersions extruded only once consist of a heterogeneous collection of vesicles, including MLV, with approximately 60% of the vesicles occurring in the 250–500 Å size range. Multiple extrusion of the preparation resulted in a progressive decrease in the size heterogeneity, and after four extrusions 94% of the vesicles ranged in size from 315–525 Å in diameter. The resulting vesicles are somewhat larger than sonicated unilamellar vesicles and the technique should be applicable to a wide variety of lipid compositions (9). It appears to be a simple, reproducible, and nondestructive technique that makes it possible to prepare large volumes of vesicles at high lipid concentrations with a minimal dilution. Thus the encapsulation of the aqueous phase should be relatively efficient, and this preparation should be a valuable tool to compliment other methods for forming small unilamellar vesicles.

ETHER INFUSION Techniques based upon the addition of phospholipids in organic solvents other than ethanol and subsequent removal of the organic solvents by evaporation have been attempted in the past (23, 163, 176) but until recently have resulted in the formation of multilamellar vesicles. The ether infusion technique introduced by Deamer & Bangham (32) avoids many of the difficulties of these previous methods and has been carefully characterized in several publications (31, 32, 184). The phospholipids are dissolved in diethylether or diethylether-methanol mixtures and are injected into an aqueous solution of the material to be encapsulted either at 55–65°C or under reduced pressure at 30°C. Upon removal of the solvent by evaporation, large unilamellar vesicles are formed (32). When these are initially filtered through a 1.2 μm millipore filter, a preparation of large unilamellar vesicles with a mean size of between 1500–2500 Å (31) is obtained. Unlike the ethanol injection technique, the lipid concentration in the organic solvent does not appear to affect the size of the resulting liposomes. It is usually 2 μmol/ml and typically 2 ml of this solution is infused into a 4 ml aqueous phase at a rate of 0.2 ml/min at 50–60°C (31).

In a modification of this procedure, petroleum ether replaced diethylether, and the volume of the lipid solution (2 μmol/ml) infused was increased to 30 ml (184). A detailed description of the apparatus used for the infusion has also been published (184). Both proteins (31) and RNA (147) have been encapsulated using this method. However, the attempted encapsulation of protease inhibitors substantially reduced the entrapped volume of the vesicles (184).

The infusion technique is applicable to a number of lipid mixtures and produces large unilamellar liposomes, although the use of organic solvents and high temperatures may denature macromolecules or inactivate heat soluble compounds. With this method, uncharged liposomes tend to form aggregates, the size distribution of the liposomes is rather heterogeneous, and the efficiency of encapsulation is relatively low, although the captured volume per mole of lipid is high, 8–17 μmol (31).

REVERSE PHASE EVAPORATION Large unilamellar vesicles can also be formed from water-in-oil emulsions of phospholipids and buffer in an excess organic phase, followed by removal of the organic phase under reduced pressure (194). Vesicles formed by this technique, which are referred to as reverse-phase evaporation vesicles (REV), have a high aqueous space-to-lipid ratio and encapsulate a high percentage of the initial aqueous phase. The phospholipids are dissolved first in organic solvents such as diethylether, isopropylether, or mixtures of organic solvents such as isopropylether and chloroform (1:1). The aqueous material is added directly to this phospholipid-solvent mixture. The ratio of aqueous phase to organic solvent is 1:3 for diethylether and 1:6 for isopropylether-chloroform. The preparation is then sonicated for a brief period, forming a homogeneous emulsion. The organic solvents are removed under reduced pressure, resulting in the formation of a viscous gel-like intermediate phase, which spontaneously forms a liposome dispersion when residual solvent is removed by continued rotary evaporation under reduced pressure. To facilitate the formation of vesicles it is best to match the amount of the mixture to the size of the vessel used for the preparation. For small preparations (0.2 ml aqueous phase) a 12×100 mm screw cap can be used; larger preparations (3 ml aqueous phase) are prepared in a vessel with a 50 ml capacity. If the ratio of the surface area/volume of the preparation during rotary evaporation is too large the organic solvent may be removed too rapidly and an appreciable fraction of the water may evaporate from the preparation. This can lead to the formation of MLV; in extreme cases the preparation will be dried onto the sides of the vessel. A typical preparation contains 60 μmol of lipid in 3 ml of diethylether or 6 ml of 1:1 mixture of isopropylether:chloroform with 1 ml of aqueous buffer. It is sonicated for 5 min under nitrogen (bath-type sonicator) at temperatures between 5 and 30°C, depending on the material to be encapsulated. The organic solvent is removed by rotary evaporation under a partial vacuum provided by a water aspirator. For material that is sensitive to sonication such as DNA, shorter (1–2 min) sonication can be employed. The technique is applicable to a large number of phospholipids and solvents, either alone or in mixtures. Maximal encapsulation of the aqueous phase (65%) is attained when the ionic strength of the aqueous buffer is low (0.01 m NaCl); encapsulation decreases to 20% as the ionic strength is increased to 0.5 m NaCl (194). This method can encapsulate large macromolecules, such as ferritin, and 25S RNA as efficiently as low molecular weight compounds such as sucrose. The size range of the resulting vesicles is sensitive to the percentage of cholesterol included in the lipid mixtures. Vesicles prepared from 1:1 mole ratio of cholesterol/phospholipid form a heterogeneous sized dispersion of vesicles with a mean diameter based upon entrapped volume of 0.5 μm and a size range from 0.16 to 0.9 μm. Vesicles prepared from similar phospholipid mixtures lacking cholesterol have a mean size of 0.16 μm and a size range from 0.08 to 0.24 μm (F. Szoka, F. Olson, D. Papahadjopoulos, unpublished observation). It is necessary to redistill solvents used for the preparation of REV from sodium bisulfite immediately prior to use in order to reduce the presence of peroxides which might result in degradation of either the phospholipids or the material to be encapsulated. In addition it is prudent to remove traces of the solvents following the evaporation step, either by dialysis or by column chromatography shortly after preparation, since the solvents can generate peroxides resulting in appreciable formation of lysophosphatides upon storage of the vesicles at 4°C (F. Szoka, unpublished observation). The REV preparation has the advantage of forming large unilamellar and oligolamellar vesicles with aqueous space-to-lipid volumes comparable to those obtained with the ether infusion technique; in addition, high efficiency encapsulation of the aqueous phase (20–68%) is attained. The REV preparation is unaffected by the presence of water-soluble protein and vesicles can be prepared from a wide variety of phospholipids. Preparations with as little as 0.2 ml of aqueous phase or even larger volumes (at least 10 ml or more) of aqueous phase can be easily handled. The principal disadvantage of this method is the exposure of the material to be encapsulated to organic solvents and to short periods of sonication—conditions that may possibly result in denaturation of sensitive proteins or breakage of DNA. The size distributions obtained for most phospholipid mixtures are quite uniform, although preparations which contain cholesterol have a heterogeneous size distribution. This can be reduced by extrusion through polycarbonate membranes.

This method has also been used for the reconstitution of membrane proteins previously extracted into organic solvents (29). Rhodopsin has been efficiently incorporated by this technique into large unilamellar vesicles (75% were greater than 0.5 μm and 40% greater than 1 μm, with more than 90% of the vesicles being unilamellar). The spectral properties of the rhodopsin in these vesicles are similar to those found in photoreceptors, and bleached rhodopsin could be chemically regenerated (29). Vesicles were formed either from diethylether with a 3:1 ratio of organic phase to aqueous phase as specified in the original technique (194) or from pentane with a 20:1 ratio of organic to aqueous phase. More than 80% of the rhodopsin from the original preparations was recovered in the vesicles. The important parameters in the successful reconstitution of membrane proteins are the selection of the apolar solvent and the volume ratio of the aqueous to nonaqueous phases (29). The large size and encapsulation ability of the vesicles should allow them to become excellent systems for the reconstitution of transport and other membrane proteins.

It appears feasible that the REV method could be used with purified membrane proteins solubilized in Triton X-100 to obtain large unilamellar vesicles, followed by removal of the Triton with Biobeads SM-2 as previously described (58). This should allow the reconstitution of many membrane proteins into large unilamellar vesicles.

CALCIUM-INDUCED FUSION A unique method to prepare large unilamellar vesicles from acidic phospholipids has been described by Papahadjopoulos and colleagues (159, 160, 161, 162). This procedure is based on the observation that calcium addition to appropriate SUV induces fusion and results in the formation of large cylindrical, folded, multilamellar structures in a spiral configuration (cochleate cylinders). Addition of ethylenediaminetetraacetic acid (EDTA) to these preparations produces large closed, spherical, unilamellar vesicles. A variety of acidic phospholipids can be used in this preparation including PA, PG, PS, and cardiolipin—pure or with cholesterol (160, 161, 162, 202a). However, the most commonly used phospholipid is bovine brain PS. The preparation of LUV starts with preformed SUV from PS that are then mixed with a calcium-containing solution either by dialysis or by the direct addition of calcium chloride (159). In both cases the SUV fuse into the cochleate structures, forming a flocculant white precipitate that can be removed from the dialysis bag and collected by centrifugation. If PA is used the calcium concentration to form cochleates is the same as PS ($\frac{1}{2}$ mole ratio to PS plus enough for 2 mM solution); for cardiolipin it is 4 mM and for PG it is 10 mM. Once formed, the cochleates can be stored in buffer under nitrogen either at 4°C or at $-20°C$ for extended periods. In order to form LUV and to encapsulate the aqueous space, the cochleate-containing pellet is resuspended in a minimal volume of the material to be encapsulated, and a solution of 0.1 M tetrasodium EDTA is added until the suspension clears. The pH of the solution is adjusted to 7.4 by adding sodium hydroxide. The addition of EDTA causes the milky white suspension to become opalescent as the cochleate cylinders reform into LUV. The LUV can be pelleted by centrifugation for 20 min at 48,000×g and nonencapsulated material can be removed by washing. Alternatively, they can be placed on 5–20% Ficoll gradients and centrifuged at high speeds (212). The LUV will form a diffuse band in the 5% Ficoll region of the gradient. In addition to pure acidic phospholipids, mixtures with up to equimolar cholesterol can be used in this procedure to form the cochleates and then can be reformed into LUV. Proteins of at least the size of ferritin (159), picornaviruses such as poliovirus (195, 212), picornavirus RNA (213), messenger RNA (38), and DNA (123) have been successfully encapsulated in these vesicles. The efficiency of the encapsulation process for such macromolecules under optimal conditions is in the range of 10%. For small compounds such as sucrose, up to 15% of the aqueous phase can be encapsulated when 20 μmol of PS cochleates are incubated in 1 ml of buffer. The captured volume for this type of LUV is on the order of 7 liters/mole phospholipid, and the vesicles have a heterogeneous size distribution of between 0.2 to 1.0 μm, with the majority of the vesicles having diameters in the lower size range.

A modification of this procedure that employs both acidic phospholipids SUV and positively charges vesicles has recently been described (215). In this technique SUV (PS 1 μmol) are mixed with SUV composed of stearylamine/PC/Chol:1/7/2 (1.7 μmol) to form a hybrid vesicle aggregate. Calcium chloride is added to this mixture, forming a precipitate that is collected by centrifugation and mixed with a small volume of an RNA-containing solution. The LUV are subsequently formed by the addition of EDTA. The inclusion of stearylamine in the preparation is thought to allow the formation of an electrostatic interaction between the negatively charged nucleic acid and the aggregated lipid complex. The entrapment of 7–11% of messenger RNA and 5–7% of ribosomal RNA is attained by this method (215). This percentage of capture is similar to that obtained by the original procedure.

The major advantage of the calcium-induced fusion method in the formation of LUV is that macromolecules can be encapsulated under exceptionally gentle conditions. The resulting vesicles are largely unilamellar, although of a heterogeneous size range. The principal disadvantage is that their formation is restricted to acidic phospholipids or to mixtures containing a preponderance of acidic phospholipids.

MISCELLANEOUS METHODS Thin films of phospholipids hydrated in solutions of very low ionic strength form extremely large unilamellar and oligolamellar vesicles. This technique was introduced by Reeves & Dowben (173) and was utilized to determine osmotic water permeability (174). The vesicles are prepared by drying the phospholipid over a large surface area and rehydrating with distilled water or 0.25 M sucrose solution. Vesicles of up to 300 μm in diameter can be formed which will include 20 μm latex beads (3). The vesicles are stable for several weeks in nitrogen-containing solutions and can easily be impaled by glass microelectrodes. They have been used for the reconstitution of calcium-magnesium ATPase (139) and for the measurement of the lateral diffusion of fluorescent phospholipid analogues (using fluorescence recovery after photobleaching) (43). The principal advantage of these vesicles is their large size, which permits measurements on single bilayer vesicles. However, the rigorous conditions required for their preparation, the fragility of such large bilayer structures, and their large size, which precludes their use in drug delivery experiments, suggests they will be most useful in very specific experimental situations.

Liposomes can also be prepared by a sonication of MLV below the Tc of the phospholipids. Under these conditions the resulting bilayer vesicles incorporate structural defects. These structural defects can be eliminated by incubating the vesicles at or slightly above the Tc, which results in fusion of the small vesicles into larger, stable vesicles (12). This method is applicable to phospholipids with a high transition temperature such as DMPC, DPPC, and DSPC but has not as yet been applied to vesicles formed from charged phospholipids. Mixtures of lipids such as DPPC/DSPC and cholesterol/DPPC:1/4 can undergo this annealing process, although vesicles composed of a equimolar ratio of cholesterol/phospholipid cannot. Vesicles of DSPC formed by this annealing technique are heterogeneous in size; some are as large as 1 μm. The capture efficiency of the aqueous space of a 20 μmol/ml DSPC preparation is between 15–20%; thus vesicles are formed with an average aqueous space of approximately 10 liters/mole phospholipid (F. Szoka, unpublished observation). This process is simple, does not require exposure of the compound being encapsulated to sonication, and results in a stable preparation of large vesicles. The principal disadvantages are that the phospholipid composition which can be used appears severely limited and that elevated temperatures must be employed for some of the phospholipids.

A method has been described that starts with the formation of "pre-vesicles" in an organic phase, in which a small amount of aqueous phase is surrounded by a phospholipid monolayer (200). This mixture is then centrifuged through an interface between an organic solvent and aqueous phase. As the pre-vesicles are forced through the interface by the centrifugal force, the pre-vesicles pick up a second monolayer of phospholipid, thus forming a bilayer vesicle (200). To form the pre-vesicles, 8 μmol of EPC is dissolved in 20 ml of benzene or di-N-butylether, and 0.05 ml of a 1M CsCl solution is added. This mixture is sonicated for 15 min at 35°C. Then 2 ml of this mixture is layered over 0.5 ml of organic solvent containing 0.8 μmol of EPC, which is layered over 3 ml of 0.7 M CsCl. The centrifugation is carried out for 3 hr at $100,000 \times g$ and the vesicles are driven into the aqueous phase. The authors of this report point out that these vesicles are relatively unstable and are especially sensitive to changes in osmotic pressure (200). This technique was believed to be capable of forming vesicles with an asymmetric phospholipid distribution in the bilayer (200); if this proves to be the case, it will provide an additional tool for studies of such asymmetric phospholipid distribution in model systems. However, the technique is limited to relatively small volumes of aqueous phase and when the density difference between the encapsulated solution is less than that used above, the majority of the phospholipid is trapped at the interface and does not form liposomes (F. Szoka and D. Papahadjopoulos, unpublished observation).

Lipid vesicles have also been produced by a combination of techniques described above (126). Phospholipids and detergents (Span-80) are used to stabilize water-in-hexane emulsions. The hexane is removed by reduced pressure and the resulting suspension is mixed with a 5–10 mM detergent solution for 3 min. This entire mixture is dialyzed for 20 hr to remove excess detergent. The resulting lipid vesicles contain a very high proportion of detergent but have been reported to retain entrapped glucose for prolonged periods (126).

Besides phospholipids and mixtures of phospholipids and detergents, lipid vesicles can be prepared from single chain amphiphiles (55, 56, 68, 107). Liposomes can be made from dilute aqueous dispersions $C(8)$–$C(18)$ of single-chain amphiphiles which must be in an uncharged state. The resulting vesicles are heterogeneous in size, and the size depends upon the lipid composition and the conditions of formation. In some cases the diameter can vary from 1 to 100 μm (68). In particular, liposomes prepared from dodecanol/sodium dodecylsulfate (7/3) form stable bilayer structures that retain entrapped compounds for a number of weeks. Similarly, large vesicles could be formed from octadecylsulfate or octadecylphosphate. The exact conditions for their preparation, permeability properties, and the size range of vesicles formed from various single chain amphiphiles, either alone or in mixtures, have been described by Hargreaves & Deamer (68). For certain applications, liposomes prepared from such single chain amphiphiles will certainly prove useful; with many of the mixtures, however, rapid transitions from the lamellar state to other structural forms can occur when the temperature, pH, or ionic strength is altered, suggesting that the use of such vesicles will be limited to specific experimental systems.

Methods for Obtaining Defined Size Distributions of Liposomes

Obtaining defined size distributions of liposomes has not received a great deal of attention in the past, since many biophysical investigations examining such phenomena as lipid phase transitions or lateral phase separations could be accomplished with either heterogeneous size dispersions of MLV or with the SUV preparations containing small vesicles of relatively homogeneous size distribution.

Huang has introduced the use of sepharose chromatography to separate MLV from SUV to obtain a homogeneous population of SUV (78). It was observed that total recovery of lipid from the column was not always achieved (78, 108). This problem has also been studied by Sharma and colleagues (189) who found complete recovery of neutral and anionic SUV but noticed that cationic liposomes usually showed lower recoveries. MLV showed appreciable losses on column chromatography that were attributed to physical blocking of the gel pores. Large unilamellar vesicles prepared by ether-infusion also gave poor recoveries. With both MLV and LUV preparations, neutral and cationic liposomes gave appreciably lower recoveries than anionic liposomes (189).

Light scattering, radiotracers, or phosphate analysis can be employed to follow the elution of liposomes from the gel filtration column. The inclusion in the lipid mixture of a fluorescent phospholipid analogue such as nitrobenzo-2-oxa-1,3-diazole phosphatidylethanolamine permits fluorescence to be followed either with a hand-held UV lamp or a fluorescence detector. With this method, vesicles that fail to enter the column are easily seen on the top of the gel. Alternatively, a sensitive fluorescence assay for the lipid-containing fraction can be performed by mixing an aliquot of the isolated fractions from the column with a small amount of diphenylhexatriene in tetrahydrofuran (119).

In addition to column chromatography, centrifugation can be used to separate MLV from SUV. Ultracentrifugation at $100,000 \times g$ for 10 min is sufficient to remove MLV from SUV preparations (208). However, in order to obtain a homogeneous-sized preparation of SUV, the method of choice used by Barenholz and colleagues (10) is $159,000 \times g$ centrifugation for about 3 hr. Applicable to a wide variety of lipid compositions, it is the most thoroughly documented technique for obtaining a homogeneous population of SUV. The centrifugation techniques have the advantage over column chromatography of not diluting the sample and not exposing the vesicles to a potentially reactive matrix that can alter the size of the vesicles or lead to vesicle adsorption on the column.

The methods described above only produce homogeneous populations of SUV; until recently there were no techniques to obtain more homogeneous populations of larger vesicles. Filtration of vesicle preparations through millipore filters has been attempted in the past and although this procedure can remove larger vesicles it also results in an appreciable adsorption of small vesicles to the membrane (185). This problem has been overcome by the sequential extrusion of liposomes through defined pore size polycarbonate membranes (146). The polycarbonate membranes have straight-through pores of defined size and the polycarbonate does not interact with vesicles containing charged groups. This simple technique was also used to form a more homogeneous preparation of REV when it was first observed that a large percentage of the lipid was recovered following extrusion (194). The extrusion technique can reduce a heterogeneous preparation of MLV or REV to a more homogeneous suspension of vesicles whose mean size approaches that of the pores through which they have been extruded. MLV with a mean diameter of 0.26 μm, (based upon negative stain electron microscopy) can be obtained following extrusion through a 0.2 μm pore size polycarbonate membrane; 75% of the encapsulated volume resides in vesicles between 0.17 and 0.37 μm. Compared to the SUV this is still a heterogeneous collection of vesicles, but when compared to the initial MLV preparation, which has a 1.32 μm mean diameter with 75% of the encapsulated volume occurring in a 0.57–2.14 μm range, it represents a considerable reduction both in mean size and in heterogeneity of the preparation (146). In the case of drug delivery where variations in size distribution could prove useful for specific tissue uptake, this technique should prove valuable for obtaining a more homogeneous preparation of MLV or LUV.

Encapsulation of Water Soluble Compounds into Liposomes

The spontaneous rearrangement of anhydrous phospholipids in the presence of water into a hydrated bilayer structure is coupled with the entrapment of a portion of the aqueous space within a continuous closed bilayer. This very property suggests that the encapsulation of water soluble compounds into liposomes should be a trivial process; however, there are several pitfalls and limitations to the encapsulation of compounds into the aqueous space.

The successful encapsulation of water soluble compounds can be measured by two parameters: (a) the efficiency of encapsulation, which is the percentage of the compound initially added to the preparation that becomes entrapped in the aqueous space; and (b) the volume (liters) of aqueous space encapsulated per mole of phospholipid. For water soluble compounds of relatively small molecular weight, the former will depend upon the relative amounts of lipid and water present in the preparation, whereas the latter will be a function of liposome size and number of lamellae. When comparing the encapsulation efficiency of various liposome preparations, it is obviously necessary to do so at the same lipid concentration. The maximum encapsulation volume of a suspension of lipid is the volume of the corresponding single unilamellar bilayer sphere that can be formed from the available surface area of the lipid molecules. For 1 μmol suspension of EPC with a headgroup surface area of 70 Å^2, a total of 4200 cm^2 will be available to form a single bilayer sphere. At first approximation the outer surface area of the sphere is 2100 cm^2, which corresponds to a sphere with a diameter of 25.9 cm. Thus, 1 μmol of EPC can form a single unilamellar spherical bilayer of about the size of a basketball, with an internal volume of 9100 cm^3. Reducing the diameter of the vesicles by a factor of 10 while maintaining the same lipid concentration will also reduce the captured volume per μmol of lipid by approximately 10. This is because the reduced volume per vesicle is partially compensated for by the greater number of smaller vesicles. This rule of thumb holds true until the volume occupied by the phospholipids in the vesicle bilayer is an appreciable fraction of the total volume of the vesicle. To obtain a value for the encapsulated aqueous space, hydrophilic nonelectrolytes with low permeability across the bilayer (e.g. sucrose) or ionic compounds that do not bind to the head groups of the phospholipid can be employed. With such markers the nonencapsulated component can be removed by dialysis, column chromatography, or centrifugation and repeated washing of the vesicles. Column chromatography has the disadvantage of diluting the liposome suspension. This can be circumvented by removing the excess fluid from small sephadex columns by centrifugation, applying the liposome sample to the column, and centrifuging again. The liposomes and encapsulated material are forced through the column while the free solute is retained (52).

Depending on the vesicle type (SUV, LUV, etc.), large macromolecules such as proteins or nucleic acids may exhibit a size-dependent efficiency of encapsulation. For example, the encapsulation efficiency of proteins in SUV has been demonstrated to be dependent upon the molecular weight of the protein (2). In small unilamellar vesicles prepared by ethanol injection, proteins of less than 40,000 daltons can be entrapped with the same efficiency as sucrose, but larger proteins show a progressive decease in trapping efficiency (2). In hand-shaken MLV and LUV prepared by ether-injection, proteins up to at least 120,000 daltons exhibit the same trapping efficiency as sucrose (2).

Deoxyribonucleic acid has also been shown to exhibit size-dependent encapsulation into negatively charged LUV. DNA fragments up to

1×10^6 daltons exhibit encapsulation similar to smaller molecular weight fragments. Fragments larger than this size exhibit a progressive decrease in encapsulation efficiency (123).

To confirm that macromolecules which can interact with the phospholipids through electrostatic interactions (75, 210) or by inserting into the bilayer are indeed encapsulated, digestion of the macromolecules can be attempted with proteolytic enzymes for proteins (2), RNAse for RNA (106, 194, 195), or DNAse for DNA (75, 123). When enzymes are encapsulated into liposomes it is advisable to demonstrate latency by the absence of activity when the intact liposome suspension is exposed to substrate and by a corresponding increase in activity when the liposome bilayer is ruptured by detergent. Such latency of enzyme activity has been used to demonstrate the encapsulation of enzymes (194, 210).

In addition to the passive entrapment of water soluble compounds into liposomes, a number of techniques have been devised to actively concentrate specific water-soluble compounds and ions into the aqueous space of vesicles. Catecholamines can be concentrated by a pH gradient across the bilayer of the liposome. The aqueous space of the vesicle is acidic with respect to the external buffer and concentrations of 12- to 23-fold were achieved with various catecholamines (142). Concentration of indium was also accomplished by using 8-hydroxyquinoline as a mobile carrier and a chelating agent, nitrilotriacetic acid (NTA), as the encapsulated reservoir in the vesicle. The transport of indium across the bilayer occurs rapidly under these conditions and a large fraction (68%) of the added indium can be concentrated into the vesicle (81). It has been shown that the encapsulated indium-NTA complex remains in the vesicles and the addition of serum proteins does not result in release of the complex, although the disruption of the vesicles by the addition of detergent brings about an immediate release of the indium-NTA complex (81). A similar procedure that uses the ionophore A23187 for loading indium into NTA-containing liposomes has been described (127). Conditions were established that permit the loading of greater than 90% of the added indium into SUV (127). The ionophore A23187 can also be used as a mobile carrier for ferrous ions across the bilayer (221). Furthermore, a number of ion-carrying ionophores have been described (170), which when coupled with suitable chelation agents entrapped in liposomes should allow an extension of this approach to a large number of different metals. This will greatly increase the encapsulation efficiency for valuable radiolabeled tracers.

Incorporation of Lipid Soluble Compounds into Vesicles

In addition to encapsulation within the aqueous space of vesicles, lipophilic compounds can intercalate into the bilayer. The mode of interaction of such compounds with the bilayer and the resulting alterations in vesicle properties such as permeability, size, and stability of the bilayer will depend both on the amount and structure of the added lipophilic compound. Amphipathic compounds such as detergents (e.g. Triton and lysophospholipids) can intercalate into the bilayer below their critical micelle concentration (CMC), increasing the permeability to entrapped compounds (82, 83, 101), whereas concentrations above the CMC can lead to disruption of the bilayer. The detailed effect of Triton on the vesicles depends on the liposome composition, and cholesterol at an equimolar ratio stabilizes the bilayer against disruption by Triton (101). Such compositional effects on Triton disruption have also been observed in MLV with varying ratios of sphingomyelin to PC (73). At high ratios of sphingomyelin to PC the vesicles are solubilized with less Triton. The process of Triton and lysolecithin disruption of vesicles is complex and depends upon the vesicle composition, the detergent to phospholipid ratio, the detergent concentration, and the phase transition of the phospholipids (69, 73, 101). A more complete description of detergent action on bilayers can be found in a review on detergent action (72).

The effect of the incorporation of nonamphipathic lipophilic molecules into vesicles is an area that is not as well studied, particularly in relation to the use of these molecules as a drug delivery system in which it is desirable to have large amounts of the drug associated with the bilayer. The ability of the bilayer to form a "solvent" for such molecules is limited by the dimensions and molecular structure of the bilayer. Compounds such as chlorophyll (113) or Gramicidin S (216) exhibit a complex interaction with the bilayer that depends upon the acyl chain composition. At a 0.1 mole ratio of Gramicidin S to lipid, the Gramicidin S is almost totally incorporated into liposomes composed of DMPC while only 45% is incorporated into liposomes composed of DPPC. At an equimolar ratio of cholesterol to DMPC, gramicidin S is laterally excluded from multibilayers, an effect that was not observed in EPC/cholesterol bilayers. A similar lateral exclusion of chlorophyll from MLV composed of DMPC/cholesterol has also been described (113). At high Gramicidin S to DPPC ratios (2.0), the extended bilayer structure is lost, with an apparent breakdown into smaller lipid-peptide aggregates (148).

In order to maximize the association of lipophilic compounds with liposomes the most common practice is to coevaporate the lipid and the compound or drug of interest from organic solvents (93) onto a round bottom vessel. This mixture is then rehydrated in buffer and the liposome-associated drug is separated from the nonassociated drug. Under these conditions, if the preparation is rehydrated with an insufficient aqueous phase to solubilize the added lipophilic compound, at least two complexes can be formed: the lipid-drug complex and an insoluble drug precipitate. These complexes can be quite difficult to separate. Centrifugation will often bring the complexes down together, whereas attempts to dialyze out the unencapsulated compound will be frustrated by its low solubility in water. The limiting concentration of cortisol 21-palmitate in liposomes has been found to be about 13 mol%. Steroid in excess of this amount forms a discrete phase that is difficult to separate from the liposome-associated compound (45). Cholesterol esters also have been shown to have a limited solubility of about 5 mol% in the bilayer (18). This is probably related to the absence of a free 3-β-hydroxyl group and the necessity to accommodate both a long acyl chain and the sterol ring structure in the bilayer. The proposed horseshoe configuration places both the acyl chain and the cholesterol moiety parallel to the acyl chains of the phospholipid and the ester group near the hydrophilic interface (64).

The affinity for, and the location in, the bilayer of a number of derivatives of adriamycin has been shown to be highly dependent upon modifications of the compound (59). The parent compound adriamycin has the lowest affinity for the bilayer and is thought to reside close to the hydrophilic interface (59). These studies reinforce the observation that the bilayer is not an isotropic solvent for lipophilic compounds and they certainly point out the limitations of liposomes as carriers of lipophilic drugs and the possibility of nonliposomal lipid-drug complexes.

Permeability Properties of Liposomes

The permeability properties of liposomes and methods for measuring permeabilities of various compounds through the bilayer have been reviewed elsewhere (7, 156). To briefly summarize, permeability through the bilayer generally increases with decreasing acyl chain length and degree of unsaturation of the acyl chains. In liposomes composed of defined phospholipids the permeability is relatively low below Tc, exhibits an anomalous increase and decrease at the vicinity of the Tc, and increases further at temperatures above Tc (32a, 86, 87, 155). When the lipid composition of the liposome includes at least 33 mol% cholesterol, the permeability is decreased and the anomalous permeability increase in the vicinity of the transition temperature is eliminated (151, 155). The permeability of various molecules through liposomes occurs in the following order: water $>$ $>$ small nonelectrolytes $>$ $>$ anions $>$ cations. The permeability of various organic molecules is determined by the type of compound, size, and charge. For small compounds, permeability increases with increasing lipid solubility, whereas multivalent ions and macromolecules are essentially impermeable through intact liposomes. Exogenous proteins may have a pronounced effect on the permeability barrier of lipid vesicles (for reviews, see 88, 99, 156, 165). The extent of such protein-mediated increase in permeability depends upon the charge and hydrophobicity of the protein and the surface charge of the liposome. Cholesterol markedly reduces the ability of exogenous proteins to mediate such increases in permeability (99, 151). Thus, when employing liposomes as a drug carrier system, it is important to confirm that the permeability properties that are measured in the absence of biological fluids represent the actual permeability of the compound in their presence. To minimize this protein mediated leakage, a 1:1 mole ratio of cholesterol/phospholipid has been used to form liposomes for in vivo studies (128, 129, 131).

For certain hydrophobic compounds, however, cholesterol has been reported to increase the permeability through unilamellar vesicles containing membrane proteins (108). At cholesterol/phospholipid ratios of between 20–30 mol% the ability of nigericin, A23187, or NH_3 SCN to collapse a proton gradient in a bacteriorhodopsin vesicle is increased. It is not clear if this is due to an inhibition of the protein pump or an enhancement of ionophore translocation. A similar enhancement of the translocation of an ion by valinomycin or nigericin was observed with protein-free SUV, but not in MLV of a similar composition (33, 108). The reason for the difference between the effect of cholesterol on the valinomycin translocation of $Rb+$ in MLV and SUV is not apparent, since the inclusion of cholesterol into liposomes of varying types has usually resulted in a qualitatively similar effect on permeabilities.

Other compounds such as the tocopherols have also been demonstrated to decrease the permeability of liposomes, composed of various lecithins, to a number of hydrophilic compounds (39, 53). These studies used between 5 and 20 mol% of tocopherol but did not define the upper limit of tocopherol/phospholipid ratio that can be used to form stable liposomes. Analogues of tocopherol that either have shorter isoprenoid side chains or lack them completely were less effective in reducing permeability, whereas at least in DPPC liposomes, phytol had an effect similar to tocopherol (53).

In most cases, fluxes through the bilayer can be measured in either direction and it is merely experimental convenience that dictates the method and direction chosen to study the permeability (7). For certain studies such as reconstituted transport proteins, it would be convenient to be able to measure the influx of the transported compound into the vesicle. A limitation to doing so is often the size of the vesicle (a small internal volume will rapidly come to equilibrium with the external fluid), which will complicate the determination of initial rates (77). One way to alleviate this problem is to remove the transported compound by a chemical alteration, following its translocation into the vesicle. A recent report described the use of encapsulated D-amino acid oxidase inside MLV liposomes for the polarographic measurement of D-amino acid uptake (140). This method yields values for the permeabilities of amino acids that are consistent with those from efflux experiments (102). Such a technique used in conjunction with a reconstituted transport protein may facilitate studies on various membrane transport systems.

Stability and Storage of Liposomes

Studies on the stability and the effective storage of liposomes preparations are areas of research that will assume increasing importance for the development of liposome preparations as a drug delivery system. Stability encompasses a number of parameters and the following aspects should be examined: (a) the chemical stability of the lipids; (b) the maintenance of the vesicle size, including an examination for aggregation; (c) the maintenance of the structure of the vesicles; (d) retention of entrapped contents; and (e) the influence of biological fluids on the integrity and permeability properties of the liposomes. Since, for most biochemical and biophysical experiments, small preparations are made which are used immediately, this subject had received only minimal attention. SUV composed of DPPC have been observed to retain their size for at least 48 hr when stored both above or below the Tc (116). Moreover, it has been reported that SUV can maintain both their structure and the asymmetric distribution of phosphatidylethanolamine for at least 12 days when stored under nitrogen (178). Vesicles prepared by extrusion through a french press maintained a similar size and bilayer integrity for at least two days following their preparation (9). Large unilamellar vesicles formed by detergent dialysis were reported to be stable for several weeks in terms of vesicle size, and it was suggested that the overall stability is limited by the chemical stability of the unsaturated lipid components (41). Liposomes formed from single-chain amphiphiles, particularly sodium dodecyl sulfate and dodecanol, are reported to be stable for at least seven weeks when incubated in the presence of azide (68). Mayhew, Papahadjopoulos, and colleagues (unpublished observations) have examined the stability of several liposome types composed of a variety of lipids for the long-term retention of 1-β-D-arabinofuranosylcytosine (AraC). They found that liposomes stored under sterile conditions in phosphate buffered saline under nitrogen can retain their entrapped AraC for extended periods. The relative retention is a function of liposome type (MLV > REV > SUV), temperature ($4° > 25° > 37°$), and lipid composition (saturated phospholipids > saturated phospholipids with equimolar cholesterol >

unsaturated phospholipids with equimolar cholesterol > unsaturated phospholipids). They also observed that the larger the mean vesicle size, the longer it took to lose 10% of the entrapped AraC. At 4°C all of the liposomes and lipid compositions examined (except those composed of PG/PC:1/4) retained greater than 50% of the encapsulated drug after one year. E. G. Mayhew and colleagues (unpublished observation) have also lyophilized and reconstituted REV composed of either PG/PC:1/4/5 or DPPC/chol:1/1. Following such a procedure, REV composed of PG/PC/chol rehydrated satisfactorily, retaining 60–70% of the encapsulated Ara-C, and had a similar size distribution when examined by electron microscopy. However, REV composed of DPPC/chol retained only 30–40% of the entrapped drug and formed irregularly shaped vesicles when examined by electron microscopy. These latter vesicles aggregated and settled out of suspension—a property that was not observed with the initial preparation. Both the lyophilized PG/PC/chol:1/4/5 REV preparations and those stored at 4°C for 2 months exhibited similar antitumor activity when examined with an L1210 mouse leukemia model system (130). These results indicate that liposome preparations containing entrapped drugs can be stored for extended periods under appropriate conditions without undue loss of the encapsulated marker or alteration in the size of the liposomes; they also point out the feasibility of lyophilization and reconstitution for long-term storage of liposome-encapsulated drugs.

The use of liposomes for drug delivery necessitates the examination of the effects of various biological fluids on the stability of liposomes. It has been reported that liposomes recovered from plasma have altered electrophoretic mobilities, indicating possible adsorption of plasma proteins (14). The same study reported that α_2 macroglobulin seems to be specifically recovered with such liposomes. Albumin can also interact with SUV, bringing about a release of entrapped markers and mediating transfer of phospholipid from the vesicle to itself (222). There was no appreciable association of the albumin with the vesicle, and incubation of SUV and albumin does not lead to an alteration of the elution pattern of the vesicle as does incubation with plasma, suggesting that the SUV remain essentially intact, albeit leaky (222). The injection of liposomes in vivo or their incubation with plasma in vitro leads to a drastic alteration in their size and physical characteristics (105). Krupp and his colleagues pointed out that liposome-associated tracers are transferred to a particle similar to the high density lipoprotein (105). More detailed studies confirm that high density lipoproteins interact with liposomes, and these studies are in agreement on most points (21, 141, 183, 196, 197). SUV and MLV composed of phosphatidylcholine undergo a progressive transformation from spherical bilayer vesicles to discoidal particles with a diameter of approximately 150–180 Å and a thickness of 55 Å when incubated with high density lipoproteins (HDL) (196). The interaction of liposomes with HDL results in a disintegration of the vesicle and leads to substantial losses of the entrapped contents. This includes not only low molecular markers such as sucrose but also macromolecules such as albumin (183). Other nonlipoprotein serum proteins have also been implicated in the transfer of phospholipids from vesicles to a suitable acceptor such as mitochondria (16). This protein has an apparent molecular weight of 100,000, and its ability to catalyze the exchange of phospholipids is inhibited by albumin (16).

In order to effectively utilize liposomes as drug delivery systems, it is important to obtain a complete understanding of the serum components that interact with liposomes and how they bring about exchange of phospholipids and disintegration of vesicles. The influence of lipid composition and the effect of cholesterol on these processes have not as yet been studied in detail, but they appear to be important controlling factors (131). The information discussed in this section is of obvious importance for in vivo studies with liposomes where the selection of appropriate lipid mixtures and marker molecules could circumvent or minimize the problem of liposome breakdown and lipid exchange. Radiolabeled phospholipids or lipophilic compounds may exchange with the lipoprotein fraction—hence, the apparent distribution of liposomes may actually reflect a substantial contribution from lipid that has become associated with the lipoprotein fraction. If radiolabeled aqueous space markers are employed, the resulting distribution may be drastically distorted by markers that have been released from the vesicle.

SUMMARY

Liposomes have already proven to be an important system for studying basic biophysical and biochemical phenomena relating to cell membranes, and new methods will increase the utilization of liposomes as carrier systems in other areas of biology and medicine. However, it should be emphasized that liposomes are not a well-characterized system, but rather an evolving methodology—a situation that presents great advantages for the eventual utilization of liposomes in biological and medicinal applications, since basic information on the physicochemical properties of the liposome components can be used to develop a system that meets the specific requirements of a given application. Unfortunately, the flow of information between basic physicochemical research and pharmaceutical or medicinal applications has not been particularly rapid or effective in this field. Many of the drug delivery studies have utilized early methodologies and formulations for preparing liposomes, which were not designed or optimized for such applications. The design and optimization of new methodology for such applications requires interaction and appropriate feedback within a large interdisciplinary area, which is often difficult to achieve. We hope to catalyze such an interaction with this review by emphasizing both the wealth of methodological diversity and the problems associated with various liposome preparations.

The characterization of the specific liposomes prepared by any of the methods is always a tedious but unavoidable task since even small changes in lipid composition and methodological details can lead to substantial differences in the properties of the resultant vesicles. The minimal requirements for the characterization of lipid vesicles include: (a) assessment of the lipid purity both before and after vesicle preparations, (b) examination of the size distribution of the resulting vesicles, and (c) determination of the aqueous volume per mole of lipid or the molar ratio of the lipophilic compound that is liposome-associated to the lipid. To minimize lipid oxidation during preparation and storage, the lipids should be handled under an inert atmosphere such as argon or nitrogen. Furthermore, for those interested in using liposomes for drug delivery, the effect of the relevant biological fluid on the size, composition, and permeability properties of the liposomes should be determined. In this respect, there is an urgent need for further research in the area of tissue disposition of liposomes with well-defined size distribution and appropriate markers, long-term storage, stabilization against biological fluids, bioavailability, and targeting to specific cells and tissues. As the questions are answered, we are confident that appropriate utilization of the available liposome methodology combined with continuing advancing in the field of cellular membranes will establish liposomes as a valuable tool in biomedical applications.

ACKNOWLEDGMENTS

We are grateful for helpful discussions with Dr. E. Mayhew and Dr. S. Nir of Roswell Park Memorial Institute and for the secretarial assistance of Ms. H. Guillemin during the preparation of this review. Support for this work was provided in part by contract CM-77118 from NCI and grants CA25526 and GM26369.

Literature Cited

1. Abramson, M. B., Katzmann, R., Gregor, H. P. 1964. *J. Biol. Chem.* 239:70–76
2. Adrian, G., Huang, L. 1979. *Biophys. J.* 25:A292
3. Antanavage, J., Chien, T. F., Ching, Y. C., Dunlap, L., Mueller, P., Rudy, B. 1978. *Biophys. J.* 21:A122
4. Arrio, B., Chevallier, J., Jullien, M., Yon, J., Calbayrac, R. 1974. *J. Membr. Biol.* 18:95–112
5. Bangham, A. D. 1972. *Ann. Rev. Biochem.* 41:753–76
6. Bangham, A. D., DeGier, J., Greville, G. D. 1967. *Chem. Phys. Lipids* 1:225–46
7. Bangham, A. D., Hill, M. W., Miller, N. G. 1974. *Methods Membr. Biol.*

1:1-68
8. Bangham, A. D., Standish, M. M., Watkins, J. C. 1965. *J. Mol. Biol.* 13:238-52
9. Barenholz, Y., Amselem, S., Lichtenberg, D. 1979. *FEBS Lett.* 99:210-14
10. Barenholz, Y., Gibbes, D., Litman, B. J., Goll, J., Thompson, T. E., Carlson, F. D. 1977. *Biochemistry* 16:2806-10
11. Batzri, S., Korn, E. D. 1973. *Biochim. Biophys. Acta* 298:1015-19
12. Berden, J. A., Barber, N. W., Radda, G. K. 1975 *Biochim. Biophys. Acta* 375:186-208
13. Bergelson, L. D. 1979. *Methods Membr. Biol.* 9:275-335
14. Black, C. D. V., Gregoriadis, G. 1976. *Biochem. Soc. Trans.* 4:253-56
15. Blok, M. C., Van der Neut-Kok, E. C. M., Van Deenen, L. L. M., De Gier, J. 1975. *Biochim. Biophys. Acta* 406:187-96
16. Brewster, M. E., Ihm, J., Brainard, J. R., Harmony, J. A. 1978. *Biochim. Biophys. Acta* 529:147-59
17. Brunner, J., Skrabal, P., Hauser, H. 1976. *Biochim. Biophys. Acta* 455:322-31
18. Calhoun, W. I., Shipley, G. G. 1979. *Biochemistry* 18:1717-22
19. Chapman, D. 1975. *Q. Rev. Biophys.* 8:185-235
20. Chapman, D., Urbina, J., Keough K. M. 1974. *J. Biol. Chem.* 249:2512-21
21. Chobanian, J. V., Tall, A. R., Brecher, P. I. 1979. *Biochemistry* 18:180-86
22. Chong, C. S., Colbow, K. 1976. *Biochim. Biophys. Acta* 436:260-82
23. Chowhan, Z. T., Yotsuyanagi, T., Higuchi, W. I. 1972. *Biochim. Biophys. Acta* 266:320-42
24. Clejan, S., Bittman, R., Deroo, P. W., Isaacson, Y. A., Rosenthal, A. F. 1979. *Biochemistry* 18:2118-25
25. Craig, I. F., Boyd, G. S., Suckling, K. E. 1978. *Biochim. Biophys. Acta* 508:418-21
26. Cullis, P. R., De Kruijff, B., Richards, R. E. 1976. *Biochim. Biophys. Acta* 426:433-46
27. Cullis, P. R., De Kruijff, B. 1978. *Biochim. Biophys. Acta* 507:207-18
28. Cullis, P. R., Van Dijck, P. W. M., De Kruijff, B., De Gier, J. 1978. *Biochim. Biophys. Acta* 513:21-30
29. Darszon, A., Vandenberg, C. A., Ellisman, M. H., Montal, M. 1979. *J. Cell Biol.* 81:446-52
30. Dawson, R. M. 1967. *Biochem. J.* 102:205-10
31. Deamer, D. W. 1978. *Ann. NY Acad. Sci.* 308:250-58
32. Deamer, D., Bangham, A. D. 1976. *Biochim. Biophys. Acta* 443:629-34
32a. De Gier, J., Blok, M. C., Van Dijck, P. W. M., Mombers, C., Verkley, A. J., Van der Neut-Kok, E. C. M., Van Deenen, L. L. M. 1978. *Ann. NY Acad. Sci.* 308:85-99
33. De Gier, J., Haest, C. W. M., Mandersloot, J. G., Van Deenen, L. L. 1970. *Biochim. Biophys. Acta* 211:373-75
34. De Kruijff, B., Cullis, P. R., Radda, G. K. 1975. *Biochim. Biophys. Acta* 406:6-20
35. De Kruijff, B., Cullis, P. R., Radda, G. K. 1976. *Biochim. Biophys. Acta* 436:729-40
36. De Kruijff, B., Demel, R. A., Slotboom, A. J., Van Deenen, L. L. M., Rosenthal, A. F. 1973. *Biochim. Biophys. Acta* 307:1-19
37. Demel, R. A., Bruckdorf, K. R., Van Deenen, L. L. 1972. *Biochim. Biophys. Acta* 255:331-47
38. Dimitriadis, G. J. 1978. *FEBS Lett.* 86:289-93
39. Diplock, A. T., Lucy, J. A., Verrinder, M., Ziebniewski, A. 1977. *FEBS Lett.* 82:341-44
40. Dittmer, J., Lester, R. C. 1964. *J. Lipid Res.* 5:126-27
41. Enoch, H. G., Strittmatter, P. 1979. *Proc. Natl. Acad. Sci. USA* 76:145-49
42. Estep, T. N., Calhoun, W. I., Barenholz, Y., Biltonen, R. L., Shipley, G. G., Thompson, T. E. 1979. *Biophys. J.* 25:A172
43. Fahey, P. F., Webb, W. W. 1978. *Biophys. J.* 21:A124
44. Fendler, J. H., Romero, A. 1977. *Life Sci.* 20:1109-20
45. Fildes, F. J., Oliver, J. E. 1978. *J. Pharm. Pharmacol.* 30:337-42
46. Findlay, E. J., Barton, P. G. 1978. *Biochemistry* 17:2400-5
47. Finkelstein, M. C., Maniscalco, J., Weissmann, G. 1978. *Anal. Biochem.* 89:400-7
48. Finkelstein, M., Weissmann, G. 1978. *J. Lipid. Res.* 19:289-303
49. Fiske, C. H., Subbarow, Y. 1925. *J. Biol. Chem.* 66:375-400
50. Forge, A., Knowles, P. F., Marsh, D. 1978. *J. Membr. Biol.* 41:249-63
51. Franks, N. P. 1976. *J. Mol. Biol.* 100:345-58
52. Fry, D. W., White, J. C., Goldman, I. D. 1978. *Anal. Biochem.* 90:809-15
53. Fukuzawa, K., Ikeno, H., Tokumura, A., Tsukatani, H. 1979. *Chem. Phys. Lipids* 23:13-22
54. Galla, H. J., Sackmann, E. 1975. *J. Am. Chem. Soc.* 97:4144-20

55. Gebicki, J. M., Hicks, M. 1976. *Chem. Phys. Lipids* 16:142–60
56. Gebicki, J. M., Hicks, M. 1973. *Nature* 243:232–34
57. Gerritsen, W. J., Van Zoelen, E. J., Verkleij, A. J., De Kruijff, B., Van Deenen, L. L. 1979. *Biochim. Biophys. Acta* 551:248–59
58. Gerritsen, W. J., Verkleij, A. J., Zwaal, R. F. A., Van Deenen, L. L. 1978. *Eur. J. Biochem.* 85:255–61
58a. Geurts Van Kessel, W. S. M., Hax, W. M. A., Demel, R. A., De Gier, J. 1977. *Biochim. Biophys. Acta* 486:524–30
59. Goldman, R., Facchinetti, T., Bach, D., Raz, A., Shinitzky, M. 1978. *Biochim. Biophys. Acta* 512:254–69
60. Gray, J. I. 1978. *J. Am. Oil Chem. Soc.* 55:539–46
61. Gregoriadis, G. 1976. *N. Engl. J. Med.* 295:704–10; 765–70
62. Gregoriadis, G. 1978. *Ann. NY Acad. Sci.* 308:343–65
63. Gregory, D. W., Pirie, B. J. S. 1973. *J. Microsc. Oxford* 99:251–55
64. Grover, A. K., Forrest, B. J., Buchinski, R. K., Cushley, R. J. 1979 *Biochim. Biophys. Acta* 550:212 21
65. Guoit, P., Baudhuin, P., Gotfredsen, C. 1979. *Biochim. Biophys. Acta.* In press.
66. Gupta, C. M., Radhakrishnan, R., Khorana, H. G. 1977. *Proc. Natl. Acad. Sci. USA* 74:4315–19
67. Gutteridge, J. M. 1977. *Anal. Biochem.* 82:76–82
68. Hargreaves, W. R., Deamer, D. W. 1978. *Biochemistry* 17:3759–68
69. Harlos, K., Vaz, W. L. C., Kovatchev, S. 1977. *FEBS Lett.* 77:7–10
70. Hauser, H. 1971. *Biochem. Biophys. Res. Commun.* 45:1049–55
71. Hauser, H., Barratt, M. D. 1973. *Biochem. Biophys. Res. Commun.* 53:399–405
72. Helenius, A., Simons, K. 1975. *Biochim. Biophys. Acta* 415:29–79
73. Hertz, R., Barenholz, Y. 1975. *Chem. Phys. Lipids* 15:138–56
74. Hess, H. H., Derr, J. E. 1975. *Anal. Biochem.* 63:607–13
75. Hoffman, R. M., Margolis, L. B., Bergelson, L. D. 1978. *FEBS Lett.* 93:365–68
76. Holloway, P. W. 1973. *Anal. Biochem.* 53:304–8
77. Hopfer, U. 1978. *Am. J. Physiol.* 234:F89–96
78. Huang, C. H. 1969. *Biochemistry* 8:344–52
79. Huang, C. H., Charlton, J. P. 1971. *J. Biol. Chem.* 246:2555–60
80. Huang, C. H., Sipe, J. P., Chow, S. T., Martin, B. B. 1974. *Proc. Natl. Acad. Sci. USA* 71:359–62
81. Hwang, K. S. 1978. *J. Nucl. Med.* 19:1162–70
82. Inoue, K. 1974. *Biochim. Biophys. Acta* 339:390–402
83. Inoue, K., Kitagawa, T. 1976. *Biochim. Biophys. Acta* 426:1–16
84. Israelachvili, J. N. 1973. *Biochim. Biophys. Acta* 323:659–63
85. Israelachvili, J. N., Mitchell, D. J., Ninham, B. W. 1977. *Biochim. Biophys. Acta* 470:185–201
86. Jacobson, K., Papahadjopoulos, D. 1975. *Biochemistry* 14:152–61
87. Jacobson, K., Papahadjopoulos, D. 1976. *Biophys. J.* 16:549–60
88. Jain, M. K. 1975. *Curr. Top. Membr. Transp.* 6:1–57
89. Janiak, M. J., Loomis, C. B., Shipley, G. G., Small, D. M. 1974 *J. Mol. Biol.* 86:325–39
90. Johnson, S. M. 1973. *Biochim. Biophys. Acta* 397:27–41
91. Johnson, S. M., Bangham, A. D., Hill, M. W., Korn, E. D. 1971. *Biochim. Biophys. Acta* 233:820–26
92. Jonah, M. M., Cerny, E. A., Rahman, Y. E. 1978. *Biochim. Biophys. Acta* 541:321–33
93. Juliano, R. L., Stamp, D. 1978. *Biochem. Pharmacol.* 27:21–27
94. Jungalwala, F. B., Turel, R. J., Evans, J. E., Mc Cluer, R. H. 1975. *Biochem. J.* 145:517–26
95. Kagawa, Y., Racker, E. 1971. *J. Biol. Chem.* 246:5477–87
96. Kantor, H. L., Prestegard, J. H. 1975. *Biochemistry* 14:1790–95
97. Keough, K. M., Davis, P. J. 1979. *Biochemistry* 18:1453–59
98. Keough, K. M., Oldfield, E., Chapman, D., Beynon, P. 1973. *Chem. Phys. Lipids* 10:37–50
99. Kimelberg, H. K. 1976. *Mol. Cell. Biochem.* 10:171–90
100. Kimelberg, H. K., Mayhew, E. G. 1978. *CRC Crit. Rev. Toxicol.* 6:25–78
101. Kitagawa, T., Inoue, K., Nojima, S. 1976. *J. Biochem. Tokyo* 79:1123–33; 1147–55
102. Klein, R. A., Moore, M. J., Smith, M. W. 1971. *Biochim. Biophys. Acta* 233:420–33
103. Korenbrot, J. I. 1977. *Ann. Rev. Physiol.* 39:19–49
104. Kremer, J. M. H., Esker, M. W. J., Pathmamanoharan, C., Wiersema, P. H. 1977. *Biochemistry* 16:3932–35
105. Krupp, L., Chobanian, A. V., Brecher, P. I. 1976. *Biochem. Biophys. Res. Commun.* 72:1251–58
106. Kulpa, C. F., Tinghitella, T. J. 1976. *Life Sci.* 19:1879–88

107. Kunitake, T., Okanata, Y. *J. Am. Chem. Soc.* 99:3860–61
108. LaBelle, E. F., Racker, E. 1977. *J. Membr. Biol.* 31:301–15
109. Ladbrooke, B. D., Chapman, D. 1969. *Chem. Phys. Lipids* 3:304–19
110. Lammers, J. G., Liefkens, T. J., Bus, J., VanderMeer, J. 1978. *Chem. Phys. Lipids* 22:293–305
111. Larrabee, A. L., Babiarz, J., Laughlin, R. G., Gedder, A. D. 1978. *J. Microsc. Oxford* 114:319–27
112. Lawaczek, R., Kainosho, M., Chan, S. I. 1976 *Biochim. Biophys. Acta* 443:313–30
113. Lee, A. G. 1975. *Biochim. Biophys. Acta* 413:11–23
114. Lee, A. G. 1975. *Biochim. Biophys. Acta* 472:237–81; 285–344
115. Lentz, B. R., Litman, B. J. 1978 *Biochemistry* 17:5536–43
116. Levine, Y. K., Lee, A. G., Birdsall, N. J., Metcalfe, J. C., Robinson, J. D. 1973. *Biochim. Biophys. Acta* 291:592–607
117. Litman, B. J. 1973. *Biochemistry* 12:2545–54
118. Litman, B. J. 1974. *Biochemistry* 13:2844–48
119. London, E., Feigenson, G. W. 1978. *Anal. Biochem.* 88:203–11
120. Luna, E. J., McConnell, H. M. 1977. *Biochim. Biophys. Acta* 470:303–16
120a. Luna, E. J., McConnell, H. M. 1978. *Biochim. Biophys. Acta* 509:462–73
121. Lundberg, B. 1977. *Chem. Phys. Lipids* 18:212–20
122. Mabrey, S., Sturtevant, J. M. 1978. In *Methods Membr. Biol.* 9:237–74
123. Mannino, R. J., Allebach, E. S., Strohl, W. A. 1979. *FEBS Lett.* 101:229–32
124. Mason, J. T., Huang, C. 1978. *Ann. NY Acad. Sci.* 308:29–48
125. Massari, S., Pascolini, D. 1977. *Biochemistry* 16:1189–95
126. Matsumoto, S., Kohda, M., Murata, S. 1977. *J. Colloid Interface Sci.* 62:149–57
127. Mauk, M. R., Gamble, R. C. 1979. *Anal. Biochem.* 94:302–7
128. Mayhew, E., Papahadjopoulos, D., Rustum, Y. M., Dave, C. 1976. *Cancer Res.* 36:4406–11
129. Mayhew, E. G., Papahadjopoulos, D., Rustum, Y. M., Dave, C. 1978. *Ann. NY Acad. Sci.* 308:371–86
130. Deleted in proof
131. Mayhew, E. G., Szoka, F. C., Rustum, Y., Papahadjopoulos, D. 1979. *Cancer Treat. Rep.* In press
132. McIntosh, T. J. 1978. *Biochim. Biophys. Acta* 513:43–58
133. Melchoir, D. L., Steim, J. M. 1976. *Ann. Rev. Biophys. Bioeng.* 5:205–38
134. Michaelson, D. M., Horwitz, A. F., Klein, M. P. 1973. *Biochemistry* 12:2637–45
135. Milsmann, M. H., Schwendener, R. A., Weder, H. G. 1978 *Biochim. Biophys. Acta* 512:147–55
136. Morrison, W. R. 1964. *Anal. Biochem.* 7:218–24
137. Mortara, R. A., Quina, F. H., Chaimovich, H. 1978. *Biochem. Biophys. Res. Commun.* 81:1080–86
138. Mukherjee, A. B., Orloff, S., Butler, J. D., Triche, T., Lalley, P., Schulman, J. D. 1978. *Proc. Natl. Acad. Sci. USA* 75:1361–65
139. Murphy, T. J., Shamoo, A. E. 1978. *Biophys. J.* 21:A27
140. Naoi, M., Shimizu, T., Malviya, A. N., Yagi, K. 1977. *Biochim. Biophys. Acta* 471:305–10
141. Nichols, A. V., Gong, E. L., Forte, T. M., Blanche, P. J. 1978 *Lipids* 13:943–50
142. Nichols, J. W., Deamer, D. W. 1976. *Biochim. Biophys. Acta* 455:269–71
143. Nonaka, G., Kishimoto, Y. 1979. *Biochim. Biophys. Acta* 572:423–31
144. Ohnishi, S. I., Ito, T. 1974. *Biochemistry* 13:881–87
145. Oldfield, E., Chapman, D. 1972. *FEBS Lett.* 23:285–97
146. Olson, F., Hunt, C. A., Szoka, F. C., Vail, W. J., Papahadjopoulos, D. 1979 *Biochim. Biophys. Acta* 557:9–23
147. Ostro, M. J., Giacomoni, D., Dray, S. 1977. *Biochem. Biophys. Res. Commun.* 76:836–42
148. Pache, W., Chapman, D., Hillaby, R. 1972. *Biochim. Biophys. Acta* 255:358–64
149. Pagano, R. E., Sandra, A., Takeichi, M. 1978. *Ann. NY Acad. Sci.* 308:185–99
150. Pagano, R. E., Weinstein, J. N. 1978. *Ann. Rev. Biophys. Bioeng.* 7:435–68
151. Papahadjopoulos, D. 1976. In *Lipids*, ed. R. Paoletti, G. Porcellati, G. Jacini, 1:187–95. New York: Raven
152. Papahadjopoulos, D. 1977. *J. Colloid Interface Sci.* 58:459–67
153. Papahadjopoulos, D. 1978. *Ann. NY Acad. Sci.* 308:1–2
154. Papahadjopoulos, D. 1978. In *Light Transducing Membranes*, ed. D. Deamer, pp. 77–90. New York: Academic
155. Papahadjopoulos, D., Jacobson, K., Nir, S., Isac, T. 1973. *Biochim. Biophys. Acta* 311:330–48
156. Papahadjopoulos, D., Kimelberg, H. K. 1974. In *Progress in Surface Science*, ed. S. G. Davison, pp. 141–232. Oxford: Pergamon
157. Papahadjopoulos, D., Miller, N.

1967. *Biochim. Biophys. Acta* 135:624–38
158. Papahadjopoulos, D., Poste, G. A., Schaeffer, B. E., Vail, W. J. 1974. *Biochim. Biophys. Acta* 352:10–28
158a. Papahadjopoulos, D., Poste, G., Vail, W. J. 1979. In *Methods Membr. Biol.* 10:1–121.
159. Papahadjopoulos, D., Vail, W. J. 1978. *Ann. NY Acad. Sci.* 308:259–66
160. Papahadjopoulos, D., Vail, W. J., Jacobson, K., Poste, G. 1975. *Biochim. Biophys. Acta* 394:483–91
161. Papahadjopoulos, D., Vail, W. J., Newton, C., Nir, S., Jacobson, K., Poste, G., Lazo, R. 1977. *Biochim. Biophys. Acta* 465:579–98
162. Papahadjopoulos, D., Vail, W. J., Pangborn, W. A., Poste, G. 1976. *Biochim. Biophys. Acta* 448:265–83
163. Papahadjopoulos, D., Watkins, J. C. 1967. *Biochim. Biophys. Acta* 135:639–52
164. Petitou, M., Tuy, F., Rosenfeld, C. 1978. *Anal. Biochem.* 91:350–53
165. Phillips, M. C. 1972. In *Progress in Surface Science*, ed. J. F. Danielli, M. D. Rosenberg, D. A. Cadenhead, pp. 139–221. New York: Academic
166. Phillips, M. C., Ladbrooke, B. O., Chapman, D. 1970. *Biochim. Biophys. Acta* 186:35–44
167. Porter, N. A., Wolf, R. A., Nixon, J. R. 1979. *Lipids* 14:20–24
168. Poste, G., Papahadjopoulos, D. 1979. In *Uses of Liposomes in Biology and Medicine*, ed. G. Gregoriadis, A. Allison, pp. 101–51. London: Wiley
169. Poste, G., Papahadjopoulos, D., Vail, W. J. 1976. *Methods Cell Biol.* 14:33–71.
170. Prince, R. C., Crofts, A. R., Steinrauf, L. K. 1974. *Biochem. Biophys. Res. Commun.* 59:697–703
171. Rand, R. D., Tinker, D. O., Fast, P. G. 1971. *Chem. Phys. Lipids* 6:333–42
172. Razin, S. 1972. *Biochim. Biophys. Acta* 265:241–96
173. Reeves, J. P., Dowben, R. M. 1969. *J. Cell. Physiol.* 73:49–57
174. Reeves, J. P., Dowben, R. M. 1970. *J. Membr. Biol.* 3:123–41
175. Reiber, H. 1978. *Biochim. Biophys. Acta* 512:72–83
176. Robinson, N. 1960. *Faraday Discuss. Chem. Soc.* 56:1260–64
177. Robles, E. C., Van den Berg, D. 1969. *Biochim. Biophys. Acta* 187:520–26
178. Roseman, M., Litman, B. J., Thompson, T. E. 1975. *Biochemistry* 14:4826–30
179. Rothman, J. E., Dawidowicz, E. A. 1975. *Biochemistry* 14:2809–16
180. Ryman, B. E., Jeyasingh, J. K., Osborne, M. V., Patil, H. M., Richardson, V. J., Tuttersall, M. H. 1978. *Ann NY Acad. Sci.* 308:281–306
181. Sandra, A., Pagano, R. E. 1979. *J. Biol. Chem.* 254:2244–49
182. Saunders, L., Perrin, J., Gammack, D. B. 1962. *J. Pharm. Pharmacol.* 14:567–72
183. Scherphof, G., Roerdink, F., Waite, M., Parks, J. 1978. *Biochim. Biophys. Acta* 542:296–307
184. Schieren, H., Rudolph, S., Finkelstein, M., Coleman, P., Weissmann, G. 1978. *Biochim. Biophys. Acta* 542:137–53
185. Schullery, S. E., Garzaniti, J. P. 1973. *Chem. Phys. Lipids* 12:75–95
186. Schwarz, M. A., McConnell, H. M. 1978. *Biochemistry* 17:837–40.
187. Schwartz, F. T., Paltauf, F. 1977. *Biochemistry* 16:4335–39
188. Selser, J. C., Yeh, Y., Baskin, R. J. 1976. *Biophys. J.* 16:337–56
189. Sharma, P., Tyrrell, D. A., Ryman, B. E. 1977. *Biochem. Soc. Trans.* 5:1146–49
190. Shimshick, E. J., McConnell, H. M. 1973. *Biochemistry* 12:2351–60
191. Shinitzky, M., Barenholz, Y. 1974. *J. Biol. Chem.* 249:2652–57
192. Smolen, J. E., Shohet, S. B. 1974. *J. Lipid Res.* 15:273–80
193. Stockton, G. W., Polnaszek, C. F., Leitch, L. C., Tulloch, A. P., Smith, I. C. 1974. *Biochem. Biophys. Res. Commun.* 60:844–50
194. Szoka, F. Jr., Papahadjopoulos, D. 1978. *Proc. Natl. Acad. Sci. USA* 75:4194–98
195. Taber, R., Wilson, T., Papahadjopoulos, D. 1978. *Ann. NY Acad. Sci.* 308:268–74
196. Tall, A. R., Hogan, V., Askinazi, L., Small, D. M. 1978. *Biochemistry* 17:322–20
197. Tall, A. R., Small, D. M. 1977. *Nature* 265:163–64
198. Teige, B., McManus, T. T., Mudd, J. B. 1974. *Chem. Phys. Lipids* 12:153–71
199. Trauble, H., Eibl, H. J. 1974. *Proc. Natl. Acad. Sci. USA* 71:214–18
200. Trauble, H., Grell, E. 1971. *Neurosci. Res. Program Bull.* 9:373–80
201. Tyrrell, D. A., Heath, T. D., Colley, C. M., Ryman, B. E. 1976. *Biochim. Biophys. Acta* 457:259–302
202. Vail, W. J., Papahadjopoulos, D., Moscarello, M. 1974. *Biochim. Biophys. Acta* 345:463–67
202a. Vail, W. J., Stollery, J. G. 1979. *Biochim. Biophys. Acta* 551:74–84

203. Van Dijck, P. W., De Kruijff, B., Van Deenen, L. L., De Gier, J., Demel, R. A. 1976. *Biochim. Biophys. Acta* 455:576–87
204. See 58a
204a. Verkleij, A. J., De Kruijff, B., Ververgaert, P. H. J. Th., Tocanne, J. F., Van Deenen, L. L. M. 1974. *Biochim. Biophys. Acta* 339:432–37
205. Viktorov, A. V., Vasilenko, I. A., Barsukov, L. I., Evstigneeva, R. P., Bergelson, L. D. 1977. *Dokl. Akad. Nauk SSSR* 234:207–16
206. Warren, G. B., Toon, P. A., Birdsall, N. J., Lee, A. G., Metcalfe, J. C. 1974. *Biochemistry* 13:5501–7
207. Watts, A., Harlos, K., Maschke, W., Marsh, D. 1978. *Biochim. Biophys. Acta* 510:63–74
208. Watts, A., Marsh, D., Knowles, P. F. 1978. *Biochemistry* 17:1792–1801
209. Weinstein, J. N., Magin, R. L., Yatvin, M. B., Zaharko, D. S. 1979. *Science* 204:188–91
210. Weissmann, G., Bloomgarden, D., Kaplan, R., Cohen, C., Hoffstein, S., Collins, T., Gottlieb, A., Nagle, D. 1975. *Proc. Natl. Acad. Sci. USA* 72:88–92
211. Wicksell, S. D. 1925. *Biometrika* 17:84–99
212. Wilson, T., Papahadjopoulos, D., Taber, R. 1977. *Proc. Natl. Acad. Sci. USA* 74:3471–75
213. Wilson, T., Papahadjopoulos, D., Taber, R. 1979. *Cell* 17:77–84
214. Worcester, D. L., Franks, N. P. 1976. *J. Mol. Biol.* 100:359–78
215. Wreschner, D. H., Gregoriadis, G., Gunner, D. B., Dourmashkin, R. R. 1978. *Biochem. Soc. Trans.* 6:930–33
216. Wu, E.-S., Jacobson, K., Szoka, F., Portis, A. 1978 *Biochemistry* 17:5543–50
217. Yatvin, M. B., Weinstein, J. N., Dennis, H., Blumenthal, R. 1978. *Science* 202:1290–92
218. Yeagle, P. L., Hutton, W. C., Huang, C. H., Martin, R. B. 1975. *Proc. Natl. Acad. Sci. USA* 72:3477–81
219. Yeagle, P. L., Hutton, W. C., Martin, R. B., Sears, B., Huang, C. H. 1976. *J. Biol. Chem.* 251:2110–12
220. Yeagle, P. L., Martin, R. B. 1976. *Biochem. Biophys. Res. Commun.* 69:775–80
221. Young, S. P., Gomperts, B. D. 1977. *Biochim. Biophys. Acta* 469:281–91
222. Zborowski, J., Roerdink, F., Scherphof, G. 1977. *Biochim. Biophys. Acta* 497:183–91

DISPLAY AND ANALYSIS OF FLOW CYTOMETRIC DATA[1]

♦9155

J. W. Gray and P. N. Dean

Lawrence Livermore Laboratory, University of California, Biomedical Sciences Division, Livermore, California 94550

INTRODUCTION

A classical approach to understanding a complex biological organism is to study its components and through these studies attempt to understand the total organism. Flow cytometers assist the biologist in these efforts by allowing the high speed, accurate measurement of the distribution of cellular properties among the cells of an organism. Thus, distinct cell types can be recognized by flow cytometry, the relative amounts of some of their biochemical and physiological components can be measured, and their frequency of occurrence can be estimated.

In this manner a wealth of data on cell populations, which may be the source of new quantitative biological understanding, can be accumulated. However, full use of the data may require extensive use of analytical and computational techniques. In this paper, we review a three-step process for the utilization of flow cytometric data. The first step requires the presentation of the data in a comprehensible form and we compare several graphical procedures for this purpose. After the data are clearly presented, the next step in the understanding of related biological systems usually requires the comparison of one distribution with another; we review a variety of nonparametric analyses for this purpose. We use the term nonparametric to describe analyses that deal with the distributions directly, even though parametric tests such as t-tests may be used in the process. As understanding of the biological system increases, the comparison of the data with a mathematical model allows a quantitative test of the validity of the assumptions embodied in

---

[1]Reference to a company or product name does not imply approval or recommendation of the product by the University of California or the US Department of Energy to the exclusion of others that may be suitable.

the model. Assuming the model is valid, the extensive flow cytometric data can be reduced to a few biologically meaningful mathematical parameters. To illustrate this procedure we present several examples of the parametric analysis of flow cytometric data.

FLOW CYTOMETRY

The first step in flow cytometry is the reduction of the cell population of interest to a suspension of single cells stained with one or more fluorescent dyes specific to cellular components of interest. The cells are then passed, one by one, through a measuring region where the dye contents of thousands of cells per second are measured with accuracies approaching one percent. Other cellular properties such as cell volume, size as indicated by light scatter, and nuclear to cytoplasmic diameter ratios also may be measured. The measured values are variables (not parameters) and are referred to as such throughout this paper. DNA content is probably the most widely measured variable, but extensive measurements have also been made using probes that are specific to lipids, proteins, or cell surface antigens or are indicative of membrane or cytoplasmic mobility, molecular proximity, or enzyme activity (for detailed reviews see 3, 35, 47).

As each cell passes the measurement region an electrical pulse is produced, the height of which is proportional to the amount of the cellular component being measured. Its height is digitized and either added to the memory of a multichannel pulse-height analyzer or stored on magnetic tape or disk.

If only one or two variables are measured for each cell, the measurements from a large number of cells are accumulated in the multichannel analyzer memory to form univariate or bivariate distributions describing the frequency with which the measured properties occur among the cells of the population. For historical reasons each element of such a distribution is referred to as a channel. If more than two variables are measured for each cell, the data are usually written on disk or magnetic tape immediately following those for the previous cell so that measurements of a large number of cells form a list.

DATA DISPLAY

The first analytical step in the utilization of flow cytometric data is its display. This can be accomplished with varying degrees of sophistication depending on the computer graphics devices that are available. A review of these devices is beyond the scope of this paper. Instead, we illustrate what may be accomplished at several different levels of sophistication (see 49 for an excellent review of computer graphics techniques and devices).

We first discuss techniques for the display of the distribution of single variable data and then extend our discussion to include multivariable data. Generally, we speak of n-variable data as distributed in a n-dimensional hyperspace or n space. The display of the distribution of data in a n space requires a $n+1$ dimensional display, with the $n+1$ dimension being frequency.

Single Variable Data

LINEAR DISPLAY The predominant form of data in flow cytometric analyses is univariate; i.e. an array of numbers representing the distribution of one measured property, e.g. DNA content, among the members of the population. Thus, the data may be plotted in two dimensions (frequency vs variable magnitude). An example of the graphical display of a DNA content distribution is illustrated in Figure 1a in which frequency (or number of objects) is displayed linearly on the ordinate and the value of the measured variable (in this case DNA content) is displayed linearly on the abscissa.

A more satisfactory scale for the display of data with a large range of frequencies, when low-frequency events also are of interest, is to plot the frequencies semilogarithmically (semilog). For example, Figure 1b shows the use of this technique by Cram et al (13) to demonstrate the stimulation of peripheral bovine lymphocytes by protein from *Mycobacterium avium* (PPD-A) or *Mycobacterium bovis* (PPD-B).

It may also happen that the dynamic range of a measured variable is large and that both the large and small variable populations are of interest. In this case it is helpful to display the data with a logarithmic (log) abscissa. If the frequency distributions are obtained by linear analysis, the measured variable can still be plotted logarithmically by application of a log transformation in which a new abscissa is defined with equally spaced elements whose widths represent exponentially increasing measured variable ranges. For example, Pearlman (51) used a log transformation of the form

$$Y = a_1 + (a_2 - a_1)(\ln a_2 / a_1)^{-1} \ln(X / a_1) \qquad 1.$$

to define the boundaries of log abscissa elements into which he mapped the values from the linear elements. Y and X are the log and linear variables, and a_1 and a_2 are the beginning and ending channel numbers (in the log domain) of the log-transformed data.

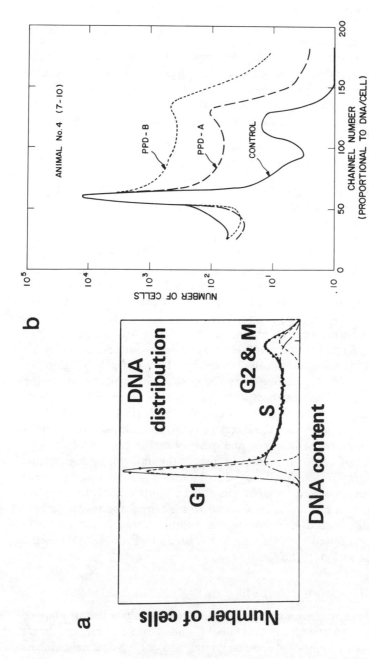

Figure 1 (Panel a) The DNA distribution measured for Chinese hamster ovary cells in asynchronous growth. The data points are shown as dots. The result of a least squares-best fit of the sum of two normal distributions and a broadened second order polynomial is shown as a solid line. (*Panel b*) The DNA distributions measured for bovine lymphocytes before (control) and after viral infection (PPD-A and PPD-B) (13). The ordinate is displayed logarithmically.

This form of data presentation has additional advantages. A "gain change" in the flow cytometer causes an expansion or contraction of the linear plot, thus changing the number of channels associated with each feature of the distribution and the frequency of objects associated with each element, and making channel-by-channel comparisons between distributions difficult. The same gain change causes only a shift along the abscissa in a semilog plot; the number of channels associated with each feature remains constant. Another advantage of the semilog plot is that heights of peaks with equal coefficients of variation are proportional to the areas under the peaks (assuming no peak overlap); this is of particular interest in flow cytometry since the measurements of homogeneous populations of cells tend to be normally distributed and

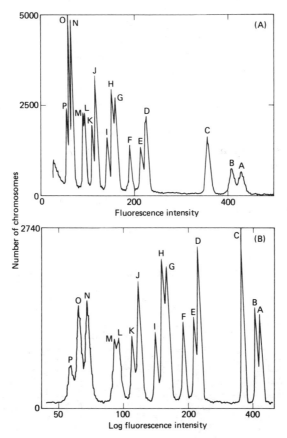

Figure 2 (*Panel A*) The fluorescence distribution from Chinese hamster M3-1 chromosomes stained with 33258 Hoechst. (*Panel B*) The same data plotted with a logarithmic abscissa (31).

to have constant coefficients of variation (19). A comparison of a linear and a semilog plot of the fluorescence distribution of isolated Chinese hamster chromosomes stained with 33258 Hoechst (31) is shown in Figure 2. Since the peaks have approximately constant coefficients of variation (actually increasing slightly with decreasing fluorescence intensity), the semilog plot in panel B reveals that peaks A, B, E, F, I, K, L, M, and P have about half the area of peaks C, D, G, H, J, N, and O. The smaller peaks are produced by chromosomes which are present in only one copy per cell (relative frequency 1), whereas the larger peaks are produced by chromosomes which are present in two copies per cell (relative frequency 2).

Two-Variable Data

SCATTERPLOTS The simplest way of displaying two-variable data, both electronically and computationally, is in a scatterplot. Figure 3 shows the use of this technique by Darzynkiewicz et al (15) for the linear representation of measurements of red ($\geqslant 600$ nm) and green (530 nm) fluorescence from lymphocytes stained with acridine orange. From the density of the points, it is possible to trace the evolution of the population from low red and green fluorescence prior to stimulation, to increased red fluorescence after stimulation, and to increased red and decreased green fluorescence after stimulation and growth in bromodeoxyuridine.

One disadvantage of scatterplots is their limited dynamic range. Regions where the frequency of cells is high may become completely black before a sufficient number of data points are plotted to properly define low-frequency features.

PERSPECTIVE PLOT A somewhat more quantitative method for the presentation of two-variable data is the perspective plot, an art tech-

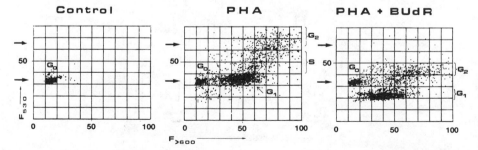

Figure 3 Scatterplots of red ($\geqslant 600$ nm) and green (530 nm) fluorescence from acridine orange-stained, phytohemagglutinin (PHA)-stimulated lymphocytes grown in the presence and absence of bromodeoxyuridine (15). Each point in the scatterplot represents the bivariate measurements of a single cell.

Figure 4 (*Panel a*) A perspective display of the green (protein content) and red (nucleic acid content) fluorescence from pooled human gynecologic specimens. (*Panel b*) The same data displayed as a contour plot (32).

nique used to give the viewer the illusion of three dimensions. The values of the two variables are presented as coordinates on a horizontal plane; the frequency of objects appears as distance above this plane, so that the distribution appears as a solid figure, e.g. as peaks and valleys. In the simplest technique used to accomplish this, the number of objects at each coordinate pair is represented by a single dot. Figure 4a shows the use by Habbersett et al (32) of a dot plot to represent the distribution of DNA and total protein content among cells from a gynecologic specimen. Displays of this type are generated by most two-variable

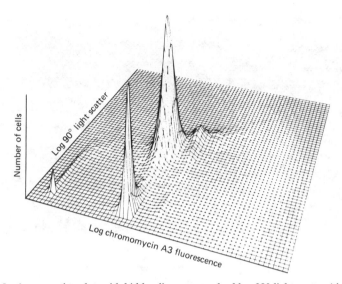

Figure 5 A perspective plot, with hidden lines removed, of log 90° light scatter (size) and log chromomycin A3 fluorescence (DNA content) from a human gynecologic specimen (38).

pulse-height analyzers found on flow cytometers. These plots can also be improved by graphic procedures that connect the dots with grid lines drawn parallel to the X and Y axes. Even with this improvement, however, the interpretation of these displays is difficult because all points are visible—both those in front of, and those behind, each feature. To overcome this defect, computer routines have been written which eliminate lines (49) for surfaces that would not be seen if the surface being viewed were solid and opaque. The limitation that some of the features of a distribution may be obscured by more prominent features can be removed by presenting the data in several different perspectives.

Figure 5 shows the perspective display (with hidden lines removed) of the distribution of log 90° light scatter and log chromomycin A3 fluorescence among cells from a gynecologic specimen (38). Note that log conversion techniques have been used in this bivariate data representation.

CONTOUR AND GRAY LEVEL DISPLAY Although perspective displays present two-variable data in an easily visualized form, with the relative frequencies of members of distinct populations readily apparent, they provide little quantitative topographic information. For example, it is almost impossible to determine bivariate modes or standard deviations from a perspective display. The contour plot (63) overcomes these limitations and allows a rough estimation of population magnitudes as well. Figure 4b shows the use of a contour plot to display the same data as in Figure 4a. The ease with which quantitative information about distribution features can be extracted is apparent. One can vary the display by selecting appropriate contour intervals. For example, linear spacing (contours at 100, 200...) emphasizes the most prominent features, whereas log spacing (contours at 10, 10^2, 10^3...) shows both low- and high-frequency populations. As discussed above, log conversions of the measured variables can enhance the utility of the contour plot.

In gray-level and pseudocolor displays of two-variable data, the variable axes are drawn orthogonally as in the contour plot. In the gray-level plot, the frequency in each distribution channel is plotted as a shade of gray, the darkness (or lightness) of which is usually proportional to the frequency. An example of a gray-level plot can be found in the section of this review on nonparametric analysis (Figure 8b). In the pseudocolor plot, the shades of gray are replaced by colors. The number of colors that can be distinguished is substantially greater than the number of shades of gray; therefore, the detail apparent in a pseudocolor display can be greater. The major disadvantages of the pseudocolor plot are the technical sophistication required to generate color displays and the expense of publishing colored plots.

Multivariable Data

When the number of measured variables exceeds two, frequency distributions are difficult to represent since the human mind has difficulty comprehending a world of more than three dimensions.

Probably the most common technique for the display of multivariable data involves its reduction to a set of one- or two-variable plots by a "gating" technique. For example, Salzman et al (54) simultaneously measured Coulter volume, small-angle light-scattering intensity, and DNA and RNA contents of Lewis lung carcinoma cells. They plotted univariate distributions of each variable alone and the same univariate distributions for those cells that exhibit a narrow range of DNA contents. This procedure can be extended to n-variable data and is available now in the more advanced commercial computer-based pulse-height analyzers, such as the Nuclear Data Model 6660.

Stöhr & Futterman (56) have presented a method of plotting three-variable data in two dimensions. Two of the variables are plotted along orthogonal X and Y axes while up to four intervals of the third variable are represented by colors. The frequency of objects is proportional to color intensity.

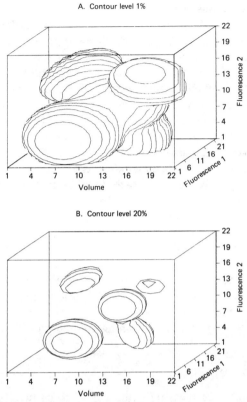

Figure 6 (*Panel A*) A three variable contour plot. The contour surface is drawn to connect channels whose values are 1% of the most frequent channel. (*Panel B*) The contour surface which connects channels whose values are 20% of the most frequent channel (58).

Valet (58) has developed a technique for the display of three-variable data which is the three-variable analog to the contour plot for two-variable data. In his procedure, a perspective display technique is used to show a constant frequency surface. A cluster of objects in the 3-space will appear as a "cloud." The perspective may be changed by rotating the data to reveal hidden clusters, and the frequency associated with the three-variable surface may be changed. This technique is illustrated in Figure 6.

If the variables being measured are adequate to separate the objects into sufficiently distinct subpopulations, it is possible to display the data so that the subpopulations are apparent. One such method, illustrated in Figure 7*A*, was used by Crowell et al (14) to display the intensity of light scattered from a mixture of 8-μm and 10-μm diameter microspheres into 28 different detectors. In this figure the scattered light

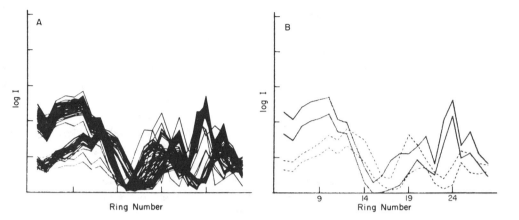

Figure 7 (*Panel A*) A multivariate plot of the light scattered from 8-μm and 10-μm diameter microspheres into 28 different angular ranges (14). The light scattered into each range was measured by a single detector ring; thus the light scatter intensity measurements are plotted as a function of ring number where larger ring numbers are associated with higher angles of scattering. (*Panel B*) The result of a cluster analysis of the data in panel *A*. Two clusters were found, corresponding to the two groups of particles. Each cluster was defined by a pair of lines that span two standard deviations around the center of each cluster.

intensities for each particle are plotted as a function of detector number (related to scattering angle) and points connected, so that each particle is represented as a line. Homogeneous classes of objects produce closely spaced groups of lines as is apparent from the figure. This type of plotting provides a direct, visual representation of the data and can show which variables best distinguish subpopulations of the distributions. As with scatterplots it has the disadvantage that data for only a small number of objects can be plotted, with the result that low-frequency subpopulations can not be detected.

NONPARAMETRIC COMPARISON

Methods for the direct comparison of distributions are also useful. Since the data are not parameterized in biologically meaningful terms, we term these analyses nonparametric. For example, given the DNA distribution for a population of cells, the type of nonparametric analysis presented here is not intended to yield the fraction of the population in each phase. Rather, this method of analysis attempts to detect changes by comparing individual distributions with each other or with a control distribution.

Univariate Comparison

Young (73) has used the Kolmogorov-Smirnov test to compare individual distributions with each other or with a mathematical function. Bagwell et al (4, 5) have extended this concept. They developed a nonparametric method for the comparison of one DNA distribution with an average DNA distribution or for the comparison of two average distributions. Figure 8a shows the application of their technique to DNA distributions from type A and B human peripheral lymphocytes. Type A cells were phytohemagglutinin (PHA)-stimulated human peripheral lymphocytes, and type B cells were the same as type A but treated with 200 μg of thymocyte extract. Average DNA distributions were produced by accumulating a set of type A distributions and a set of type B distributions. These distributions were then smoothed and translocated so that the modes of all G1 peaks fell in the same channel. The number of cells in each distribution was normalized to one by dividing the number in each channel by the total number of cells in the distribution. The replicate type A and B distributions were then averaged channel by channel to produce average distributions in which the means and standard deviations of the frequencies of cells in each channel were known. The two-tailed Student's (Gossett) t-test was used

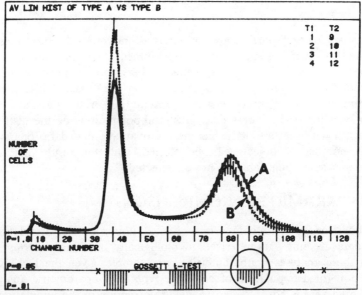

Figure 8a Average DNA distributions for type A and B peripheral lymphocytes following stimulation by phytohemagglutinin (PHA) and thymocyte extract. The lower portion shows regions of statistically significant difference according to the Gossett t-test (4, 5).

to determine areas of significant statistical difference between the type A and type B average distributions. Finally, regions of maximum difference were selected, and Bayes' theorem was applied to determine the probability that the distribution for an unknown cell population was produced by either type A or type B cells.

This method of analysis can yield information not available in some parametric methods of analysis, e.g. differences among distributions in the late S-phase region. However, it is very sensitive to changes in the coefficient of variation of the data which are due to variability in staining and/or fluctuations in flow cytometer performance, and it is imperative that close control be maintained over these variables and that standards be run frequently.

Bivariate Comparison

Another application of nonparametric comparison has been described by Bennett & Mayall (10). Their application compares bivariate distributions of cervical cells in which two variables were measured: log DNA content and log 90° light scatter. Their objective was to develop a method to detect differences between an average distribution for normal women and a distribution measured for a potential cancer patient. The average distribution was obtained by averaging distributions from normal patients with no history of dysplasia. Prior to averaging, these distributions were aligned so that a prominent feature in all distributions, e.g. a peak due to inflammatory cells, fell in the same channel in all distributions. Since the data were logarithmic, this required only a simple displacement of the distributions. Next, to give equal weight to all of the normal distributions, each was normalized so that the number of transition zone cells (i.e. cells with diploid DNA content and intermediate light scatter intensity) was constant. An average bivariate distribution was then formed in which the mean and the standard deviation of the number of cells in each channel was known. Unknown distributions were compared with the average by subtracting the average distribution from the unknown, channel by channel, and by computing the probability, p, of any positive difference occurring by chance; the one-sided tolerance test statistic was used, and it was assumed that the distribution values were distributed normally about some mean. An example of a bivariate distribution of probabilities of difference due to chance is plotted in Figure 8b. Although the p values are plotted as gray levels in this figure, Bennett & Mayall prefer a pseudocolor plot because of its larger dynamic range. As with the single variable method described above, this bivariate method of analysis also suffers from variation in flow cytometric resolution.

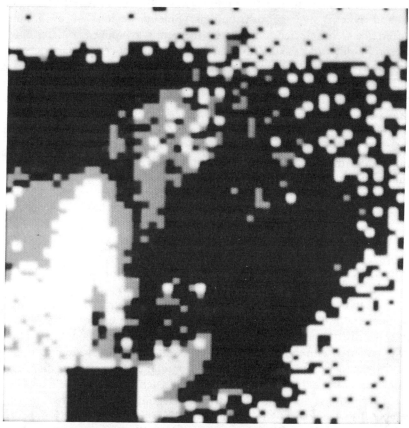

Figure 8b A gray level plot of regions of statistically significant difference between an average log DNA content vs log 90° light scatter distribution for gynecologic specimens and an unknown gynecologic distribution (38).

Multivariate Clustering

Nonparametric methods have also been used to analyze multivariable data. Crowell et al (14) have used cluster analysis methods to enable discrimination among particle types. They measured the intensity of light scattered at 28 different angles by a mixture containing 8-μm and 10-μm diameter microspheres as previously described. The 28-space data were then analyzed by a mathematical clustering algorithm (25) which searched for clusters of data points. The results of the analysis are shown in Figure 7*B*. Even without prior training on a learning set of data, two clusters were detected; the range of signal intensities at each ring (detector angle) is shown in the figure. However, no estimates of the risks of misclassification were made.

PARAMETRIC ANALYSIS

Increased understanding of flow cytometric distributions often leads to their approximation by mathematical functions. These functions may be chosen in a somewhat ad hoc fashion, as is the case when distributions are approximated by the sum of normal distributions, or they may be a description of the behavior of the biological system being studied. In either case, parametric analysis requires that the parameters of the mathematical function be adjusted such that the function is matched as closely as possible to the experimentally measured data. Suppose, for example, the data were to be approximated by the function $f(x) = a + bx$. The frequencies, $f(x)$, associated with the *variable* x would be matched to the data by adjusting the *parameters a* and *b*.

Best-Fit Procedures

The most simple technique for matching a mathematical function to experimental data is to vary parameters in the function by trial and error until the function matches the experimental data. This method is often the first step in the development of a parametric analysis, but it is unsatisfactory because there are no criteria with which to judge how well the function matches the data and because there is no order to the matching technique. Two people independently matching the same data will almost invariably arrive at different matches, and the parameters resulting from their analyses will be different. Thus, it is important to establish a criterion for defining the "best" fit and a procedure to arrive at this best fit. The criterion most commonly used is one in which the sum of the squares of the deviations of the data from the function is minimal. For example, consider the approximation of a fluorescence distribution, $Z = [z_1 z_2 ... z_n]$, where z_i is the frequency of objects with relative fluorescence, i, and n is the number of distribution elements, by a normal distribution

$$f(i; A, \mu, \sigma) = A(2\pi)^{-\frac{1}{2}} \sigma^{-1} \exp\left[-.5(i-\mu)^2 \sigma^{-2} \right]. \qquad 3.$$

The parameters A, μ, and σ, the area, mean, and standard deviation of the normal distribution, are adjusted until the sum of squares of the differences between the function and the data

$$E = \sum_{i=1}^{n} \left[f(i; A, \mu, \sigma) - z_i \right]^2 \qquad 4.$$

reaches a minimum value. A recent review article on fitting methods by Jennrich & Ralston (37) may be consulted for the various methods that can be employed. The interested reader also is directed to (1) and to the extremely useful book by Bevington (11), which presents a number of useful computer programs.

One-Parameter Analysis

A straightforward application of the fitting of normal distributions to flow cytometric data is the parameterization of the peaks in a fluorescence distribution of isolated Chinese hamster M3-1 chromosomes (29, 48). Here the parameters $\mathbf{p}=[A_1\mu_1\sigma_1 A_2\mu_2...]$ in the function

$$f(i;\mathbf{p}) = \sum_{j=1}^{K} A_j(2\pi)^{-1/2}\sigma_j^{-1}\exp\left[-.5(i-\mu_j)^2\sigma_j^{-2}\right] \qquad 5.$$

were adjusted until the sum of the squares of the deviations between the data and $f(i;\mathbf{p})$ was minimized. Examples of fits to parts of fluorescence distributions from human chromosomes are shown in Figure 9. Three

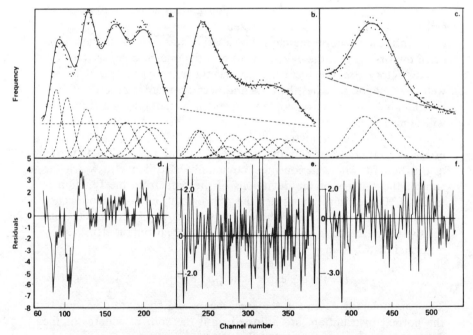

Figure 9 Least squares fit of portions of a fluorescence distribution of human chromosomes stained with ethidium bromide. (*Panels a, b, c*) Data shown as solid points, fitted functions as solid lines, and components of the fitted function as dashed lines. (*Panels d, e, f*) Deviations of the data from the fitted function (48).

parameters were determined for each peak: its mean, area, and standard deviation. These represent estimates of the mean fluorescence, frequency of occurrence, and measurement uncertainty or biological variability of the chromosome group producing that peak.

Perhaps the single most frequent use of flow cytometers is in the measurement of DNA content distributions as a way of characterizing the cell cycle traverse of cell populations. As a consequence, much effort has gone into developing methods of parameterizing DNA distributions in terms of the cell cycle (30). The cell cycle model that is the basis for parametric analysis was originally proposed by Howard & Pelc (36). In this model the cell cycle is divided into four phases, G1, S, G2, and M. After division in the mitotic phase, M, cells move into the G1 phase where they remain with constant DNA content until they enter the S phase and begin to synthesize DNA. After a cell has duplicated its DNA it enters the G2 phase where it remains with twice the G1 amount of DNA until mitosis. This model provides for a direct relationship between DNA content and phase of the life cycle. Figure 10 illustrates this relationship for asynchronous, exponentially growing cells (27). Panel A shows the cell maturity distribution (i.e. the distribution of the cells as a function of their position around the cycle). Panel B shows the relation between maturity and DNA content; the exact shape of the

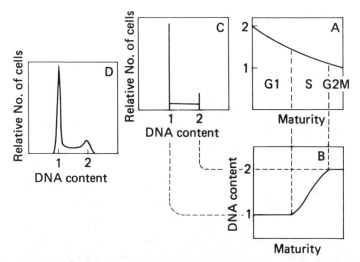

Figure 10 A model relating an experimentally measured DNA distribution to the cell cycle (27). (*Panel A*) The maturity distribution of an asynchronous cell population. (*Panel B*) The relation between DNA content and maturity. (*Panel C*) The DNA distribution that would be measured for the cell population with the properties specified in panels *A* and *B* if DNA content were measured with no error. (*Panel D*) The effect of broadening (produced by DNA content measurement variability) on the distribution in panel *C*.

S-phase portion is determined by the rate of DNA synthesis. The G2 and M phases are combined, since cells in these phases have the same DNA content. The curve in panel *B* allows the cell cycle distribution to be mapped into the idealized DNA distribution shown in panel *C*. All cells in G1 phase have one unit of DNA, cells in G2M have two units, and S-phase cells have intermediate amounts. If cellular DNA content could be measured perfectly, the experimentally measured DNA distributions might be similar to that shown in panel *C*. Unfortunately, neither the procedures for staining DNA with fluorescent dyes nor the flow cytometric methods of measurement are error free. These measurement errors are usually assumed to be normally distributed, with a standard deviation proportional to DNA content. Panel *D* shows the DNA distribution of panel *C* after broadening according to a normal distribution.

The model of the cell cycle just described suggests parameterization of DNA distributions into G1-, S-, and G2M-phase fractions. Thus, DNA distributions are usually represented as the sum of three functions. Following the assumption that measurement errors are normally distributed, the G1 and G2M phases are usually represented by normal distribution functions, although some authors (50, 74) have proposed the use of lognormal distributions. Thus the function representing the measured DNA distribution $Y(i)$ is written

$$Y(i) = A_1 (2\pi)^{-1/2} \sigma_1^{-1} \exp\left[-.5(i - \mu_1)^2 \sigma_1^{-2} \right]$$
$$+ A_2 (2\pi)^{-1/2} \sigma_2^{-1} \exp\left[-.5(i - \mu_2)^2 \sigma_2^{-2} \right]$$
$$+ S(i) \qquad\qquad 6.$$

where i is the channel number, A_1 and A_2 are the areas, μ_1 and μ_2 are the means, σ_1 and σ_2 are the standard deviations of the G1 and G2M peaks respectively, and $S(i)$ is the distribution of the S-phase cells. The selection of the S-phase function is crucial to the proper estimation of phase fractions since it is through this function that the overlap between the G1 and G2M peaks and the S phase is determined. The major difference in the many methods of DNA distribution analysis in use today is in the function used to approximate S phase. Functions used include constants, polynomials of varying degree (6, 8, 12, 19), exponentials (46), and a series of normal curves (20, 23, 24, 39). Each method has been reported by at least one author to provide an adequate fit to a set of test data. We describe here the two most frequently used models.

The S-phase function, $S(i)$, frequently has been represented by a series of normal curves so that

$$S(i) = \sum_{j=1}^{N} A_j (2\pi)^{-1/2} \sigma_j^{-1} \exp\left[-.5(i - \mu_j)^2 \sigma_j^{-2} \right] \qquad 7.$$

where the parameters A_j, μ_j, and σ_j are the area, mean, and standard deviation for the jth normal distribution, and N is the number of normal curves. Ideally one might like to use a separate normal curve for each channel of S phase. However for precise and unambiguous results, the number of normal curves must be kept low so that the total number of parameters is much lower than the number of channels. For a distribution containing 256 channels with $\mu_1 = 100$, S phase spanning 100 channels, and $N = 15$, there are a total of 51 parameters and about 130 channels involved in the calculation. The means and standard deviations of the S-phase peaks are often made functions of μ_1, μ_2, σ_1, and σ_2, thus reducing the number of parameters. To correctly partition a DNA distribution with this model it is necessary to constrain the first S-phase peak to remain at some DNA content greater than the G1 mean. In practice, the proper separation of the means of the G1 and first S-phase peaks (and the separation of the means of the last S-phase and G2M phase peaks) is usually determined in situations where a priori knowledge is available. These separations are then assumed to obtain in other situations.

The principal advantage of this method is that it does not require the S-phase distribution to be of low curvature. Thus, it has been widely used to determine phase fractions from DNA distributions for perturbed or partially synchronized cell populations for which the G1 and/or G2M peaks are often not distinct and for which the S-phase distribution contains significant structure. In these analyses the means and standard deviations of the normal curves usually are not free parameters. They are obtained from analyses of standard DNA distributions; for example, from asynchronously growing populations of the same type of cells. In fitting the perturbed population distributions, only the areas of the normal distributions are free parameters. Figure 11B shows the results of analyzing the DNA distribution from a synchronous Chinese hamster ovary (CHO)-line cell population six hours after release from synchrony in early G1 phase. Each normal curve used in the fit is plotted separately. Although the requirements on instrument stability and staining reproducibility are stringent, the method has been used successfully (23, 24).

For asynchronous, exponentially growing populations, constraints can be placed on the shape of the S-phase distribution, thereby reducing the difficulty in handling the overlap regions. For example, Dean & Jett (19) represented the S-phase distribution function $S(i)$ by a second degree polynomial. The actual formulation was more complicated than a simple polynomial since it took into account the broadening described earlier. The broadening was accomplished by replacing each channel, i, of S phase with a normal curve with mean j and standard deviation σ_j.

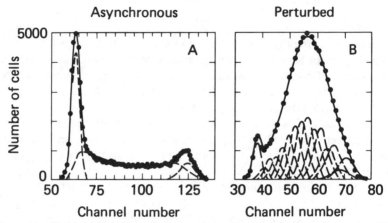

Figure 11 (*Panel A*) The DNA distribution of asynchronous exponentially growing CHO cells analyzed using a broadened polynomial to represent S phase. The solid line is the total fit to the distribution, the dashed lines show the two normal curves used to fit G1 and G2M phases and the broadened polynomial, and the circles are the data points. (*Panel B*) The DNA distribution of a synchronous CHO cell population six hours after release from synchrony in early G1 phase, analyzed using a series of normal curves. All of the normal curves are plotted as dashed lines, along with the total fit to the data (a solid line).

The area of each normal curve was the value of the polynomial at channel j. The number of S-phase cells in channel i was found by adding the contributions of all normal curves in S. The function was

$$S(i) = \sum_{j=\mu_1}^{\mu_2} (A + Bi + Ci^2)(2\pi)^{-1/2}\sigma_j^{-1}\exp\left[-.5(i-j)^2\sigma_j^{-2}\right] \qquad 8.$$

where A, B, and C were the coefficients of the polynomial. The standard deviation of each normal curve, σ_j, was scaled linearly between σ_1 and σ_2. The total function $Y(i)$ then had nine parameters, three for each phase, which were solved for by least squares nonlinear curve-fitting methods.

The result of this type of analysis is illustrated in Figure 11A, for CHO cells in exponential growth. The two normal curves, the broadened polynomial, and the total function are plotted along with the data. This method has been tested for a large variety of cell types and with synthesized distributions (16, 19), with very good results. This method of fitting DNA distributions of asynchronously growing cells has two principal advantages. The function has only nine parameters (there are usually over 100 data points) and the difficulty with the overlap regions, G1-S and S-G2M, is ameliorated since the parameters of the polynomial are determined mostly by the middle two thirds of S phase, where there is no overlap. It is important to note, however, that this method is not suitable for the analysis of DNA distributions from highly perturbed populations.

A simplified method that also fits S phase with a polynomial has recently been described by Dean (17). In this method the central part of S phase, which is free from contributions from either G1- or G2M-phase cells, is approximated by a second degree polynomial using a least squares fitting procedure. The polynomial is extrapolated to the G1- and G2M-peak means and integrated to obtain the S-phase fraction. The polynomial is subtracted from the total distribution to obtain the G1- and G2M-phase fractions. This method has been tested for a large number of synthesized distributions with a wide variety of phase fractions and coefficients of variation. For distributions with a coefficient of variation (CV) less than 8%, the average errors for the G1-, S-, and G2M-phase fraction estimates were 1.5%, 1.2%, and 4.1% respectively.

We have reviewed here only two methods for partitioning DNA distributions into G1, S, and G2M fractions. With the proliferation of methods and techniques proposed for such partitioning, selection of the proper method to be used with one particular set of data can be very difficult. In an effort to make selection of a method simpler and more informed, a number of studies have been made which compare different methods on a variety of data sets (7, 16, 30, 40, 45). Of particular interest is the study conducted by Baisch et al (7). In this study 11 researchers in 9 laboratories analyzed 50 DNA distributions with a variety of methods. The distributions include experimental as well as synthesized data for asynchronously growing and perturbed cell populations so that the performances of the various programs for a variety of data can be compared.

Using DNA distributions from cells in asynchronous exponential growth, the rate of DNA synthesis can be computed as a function of DNA content. Methods of computation have been described in detail by several authors (18, 41, 68, 71). Dean & Anderson (18), for example, made use of the principle that the amount of time spent in a given state (DNA content range in S phase) varies inversely with the rate of passage through the state. The distribution of cells in S phase then reflects the rate of DNA synthesis in S phase. The equation derived was

$$f(D) = [2M(1 - .5F_{G1}) - N(D)]\lambda n(D)^{-1} \qquad 9.$$

where $f(D)$ is the rate of DNA synthesis at DNA content D, λ is the exponential growth rate constant, M is the total number of cells in the distribution, F_{G1} is the fraction of cells in G1 phase and $N(D)$ and $n(D)$ are the integral and differential DNA distributions respectively. DNA synthesis rate calculations have been made for WI-38 cells and Chinese hamster ovary cells; in both cell lines the results have been confirmed using ^{3}H-TdR incorporation by synchronized cells (18).

Two-Variable Analysis

The parameterization of two-variable data is accomplished in much the same way as one-variable data. A mathematical function describing the data is formulated and the parameters therein are adjusted until the function matches the data as closely as possible. For example, Valet et al (59–62) approximated their bivariate measurements of erythrocyte volume and poly-l-ornithine content with a function which was the sum of three bivariate distributions, each of the form

$$G(\zeta,\eta) = A\exp\left[-.5(\zeta^2\sigma_\zeta^{-2} + \eta^2\sigma_\eta^{-2})\right], \qquad 10.$$

where A was the maximum amplitude of the two-variable distribution, and σ_ζ and σ_η were the standard deviations in the ζ and η directions respectively. Thus, each cell cluster in the experimental distribution was approximated by a normal distribution whose coordinates ζ and η were orthogonal but rotated relative to the experimental x-y coordinate system. The transformation between the x-y and ζ-η coordinates was

$$\zeta = (x-\bar{x})\cos\alpha + (y-\bar{y})\sin\alpha$$
$$\eta = -(x-\bar{x})\sin\alpha + (y-\bar{y})\cos\alpha$$

where $(\bar{x}, \bar{y})$ was the bivariate mean of the measured distribution and α was the angle of rotation between the $(\zeta\text{-}\eta)$ and $(x\text{-}y)$ coordinate systems. The entire experimental distribution was approximated as the sum of 3 bivariate distributions:

$$g(x,y) = \sum_{i=1}^{3} G_i[\zeta_i(x,y), \eta_i(x,y)]. \qquad 11.$$

The sum of squared differences between the experimentally measured values $Z(x_i, y_i)$ and the function $g(x_i, y_i)$ was minimized by varying the parameters $\bar{x}, \bar{y}, \sigma_\zeta, \sigma_\eta$, and A for each distribution. Figure 12, panel b shows the experimentally measured volume vs poly-l-ornithine distribution, panel a shows the fitted function consisting of the sum of 3 normal distributions, and panel c shows the difference between panels a and b.

Some of the more elaborate bivariate analyses involve the parameterization of time sequences of DNA distributions (variables: DNA content and time) in terms of cell cycle traverse. In these analyses, changes in the DNA distributions with time are presumed to reflect changes in the distribution of cells around the cell cycle and are parameterized in terms of a model of the cell cycle traverse which describes the evolution of the distribution of cells around the cell cycle. The G1, S, and G2M phases

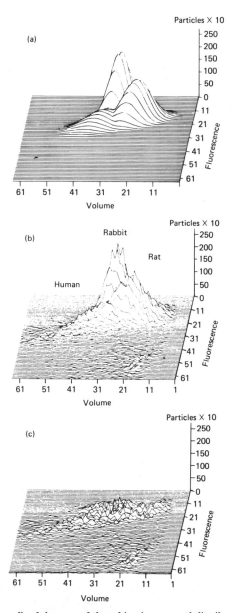

Figure 12 Least squares fit of the sum of three bivariate normal distributions to bivariate measurements of erythrocyte volume and poly-l-ornithine content (fluorescence). (*Panel a*) The fitted function. (*Panel b*) The experimental data. (*Panel c*) The deviations of the data from the function (60).

of the cell cycle are usually modeled as a series of compartments, called states, through which cells move as they mature. A number of mathematical procedures have been used to describe the movement of the cells through the states. One such technique describes the movement by a series of differential equations (26, 27). For example, if $N_i(t)$ is the number of cycling cells of maturity i at time t, and $B_i(t)$ is the number of out-of-cycle cells at that same maturity and time, then the changes in $B_i(t)$ and $N_i(t)$ are given by

$$\frac{dN_i(t)}{dt} = \lambda_{i-1} N_{i-1}(t) + \zeta_i(t) B_i(t) - [\lambda_i + \eta_i(t)] N_i(t) \qquad 12a.$$

and

$$\frac{dB_i(t)}{dt} = \eta_i(t) N_i(t) - [\zeta_i(t) + \mu_i(t)] B_i(t) \qquad 12b.$$

where the λ's are the transition probability densities for movement from one state to the next, the η's and ζ's are time-dependent probability densities for the arrest and resumption of cell cycle traverse, respectively, and the μ's are time-dependent probability densities for cell death. The division of each cell to form two at the end of mitosis is taken into account in the equation for the change in the number of cells in the first state as

$$\frac{dN_1(t)}{dt} = 2\lambda_k N_k(t) + \zeta_1(t) B_1(t) - [\lambda_1(t) + \eta_1(t)] N_1(t). \qquad 12c.$$

In the absence of perturbations, cell death, and cell decycling, the λ's within any phase are usually equated and the phase durations and dispersions for any phase can be written $T = n/\lambda$ and $CV = n^{-\frac{1}{2}}$ where n is the number of states in that phase. Given estimates for the initial distribution of cells around the cell cycle (i.e. $N_i(0)$, $B_i(0)$; $i = 1, 2, \ldots$) and estimates for the transition probability densities, the cell cycle distribution $Z_i(t) = N_i(t) + B_i(t)$ can be calculated for any later time by solving the differential equations.

The cell cycle distribution can be converted into a DNA distribution as illustrated in Figure 10. A number of states are assigned to each cell cycle phase, and a relation between maturity state number and total DNA content is defined so that the calculated cell cycle distributions can be converted into DNA content distributions whose G1-peak means and coefficients of variation are the same as those of the corresponding measured distribution. The model parameters (the λ's, η's, ζ's, μ's, number of states in each phase, DNA distribution peak means, and coefficients of variation) are estimated by varying them to minimize the sum-of-squares differences between the measured and calculated DNA distributions in the sequence. For example, Figure 13 shows the param-

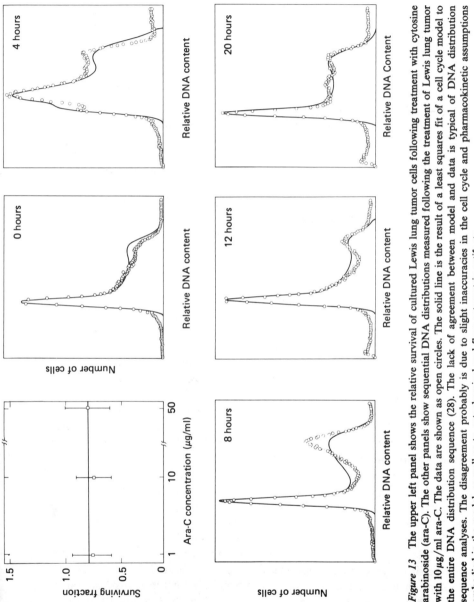

Figure 13 The upper left panel shows the relative survival of cultured Lewis lung tumor cells following treatment with cytosine arabinoside (ara-C). The other panels show sequential DNA distributions measured following the treatment of Lewis lung tumor with 10 µg/ml ara-C. The data are shown as open circles. The solid line is the result of a least squares fit of a cell cycle model to the entire DNA distribution sequence (28). The lack of agreement between model and data is typical of DNA distribution sequence analyses. The disagreement probably is due to slight inaccuracies in the cell cycle and pharmacokinetic assumptions embodied in the model, as well as to cytochemical and flow cytometric artifacts.

eterization of a sequence of DNA distributions measured for Lewis lung tumor cells (initially growing asynchronously) at various times after a 20-min exposure to 10-μg/ml cytosine arabinoside (28). The first panel, which shows the relative survival to various doses of cytosine arabinoside (ara-c), indicates that there was little cell death at 10 μg/ml so the μ's were assumed to be zero. The solid line shows the result of a least-squares best fit of the above model to the data; the λ's and the DNA distribution means and coefficients of variation were adjusted by a second order gradient technique, and the ζ's, η's, and number of states in each phase were varied by trial and error. The parameters thus extracted indicated that the Lewis lung cells were blocked over the last three fourths of S phase for four hours, after which time the cells resumed cycling with near normal G1-, S- and G2M-phase durations of 2.4, 9.4, and 0.6 hours respectively.

Of course, the DNA distribution sequences need not be parameterized in terms of the cell cycle model presented here. Discrete time (21, 33, 42–44, 52, 53, 57, 70, 71), stochastic (22), and probability density function models (34, 66) have also been used in the kinetic parameterization of DNA distribution sequences. Indeed, White & Thames (69) have shown that the discrete and continuous time models are closely related and they have derived relations between the two. Kim et al (42) have used a constrained least squares fitting procedure to parameterize DNA distributions from synchronous Chinese hamster ovary cells in terms of a discrete time cell cycle model. Woo & Enagonio (72) used a similar procedure to simultaneously parameterize DNA distributions from synchronous Chinese hamster ovary cells, as well as sequential measurements of spermidine content, thereby establishing a relation between cell maturity and spermidine content. Watson & Taylor (66) have extended the probability distribution function model of Hartmann & Pedersen (34) to the cell cycle parameterization of EMT6/M/CC cells; Donaghey et al (22) have parameterized sequential DNA distributions measured for T_2 human lymphoma cells following release from excess thymidine-induced block at the G1/S-phase boundary.

The cell cycle kinetic parameterization of time sequences of DNA distributions can be simplified somewhat by the reduction of the information used in the analysis. For example, each DNA distribution can be individually partitioned, as previously described, to yield fractions of cells in the G1, S, and G2M phases. The time variations in these fractions also can be parameterized in terms of the cell cycle models which allow calculation of the variation with time in the G1-, S-, and G2M-phase fractions. Alternatively, the DNA distribution sequence can be parameterized using only a portion of the information within each distribution (55, 74). In this approach, the fractions of the population at each of several DNA contents, i, are estimated for each DNA distribution (each value is designated an FP_i value). The FP_i values for each DNA content region, i, are plotted vs time to form curves that reflect the passage of synchronous cohorts of cells (55).

Flow cytometric data have also been used in quantitative studies of chemical reaction kinetics (64, 65, 67). For example, Watson et al have measured distributions of the fluorescence from 3-0 methyl fluorescein in EMT6 cells during treatment with the nonfluorescent substrate 3-0 methyl fluorescein phosphate. They postulate a model in which the substrate 3-0 methyl fluorescein phosphate, S, and a hydrolytic enzyme, E, combine with rate constant k_1 to form a complex, ES, which may dissociate to free enzyme and fluorescent product, 3-0 methyl fluorescein, P, with rate constant k_2 or revert to free enzyme and substrate with rate constant k_{-1}. Account was also made of the loss of the fluorescent product from the cells with rate constant k_3 so that the kinetic model took the form

$$E + S \underset{k_{-1}}{\overset{k_1}{\rightleftarrows}} ES \overset{k_2}{\rightarrow} \underset{\underset{E}{+}}{P} \overset{k_3}{\rightarrow} \text{loss}. \qquad 13.$$

The symbols [E], [S], [ES], and [P] represent the concentrations of the enzyme, substrate, enzyme substrate complex, and fluorescent product respectively.

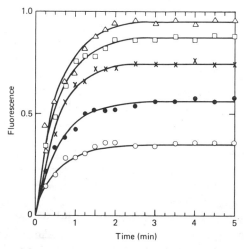

Figure 14 Sequential measurements of the fluorescence from 3-0 methyl fluorescein during treatment with various concentrations of 3-0 methyl fluorescein phosphate. The substrate concentrations were 48 μM(○), 95μM(●), 191μM(x), 286μm(□), and 381μM(△). The result of least squares fits of a kinetic model to the data are shown as lines (65; J. Watson, private communication).

Fluorescence distributions of the product P were measured every 15 seconds for the first two minutes and every 30 seconds thereafter. Figure 14 shows the median value of each fluorescence distribution plotted as a function of time after introduction of the substrate for several different concentrations of substrate.

In the limit when [ES] is nearly constant, the rate equations for the reaction shown in Equation 14

$$\frac{d[\text{ES}]}{dt} = k_1[\text{E}][\text{S}] - k_{-1}[\text{ES}] - k_2[\text{ES}] \qquad 14\text{a.}$$

$$\frac{d[\text{P}]}{dt} = k_2[\text{ES}] - k_3[\text{P}] \qquad 14\text{b.}$$

yield the expression

$$[\text{P}] = k_2[\text{E}_o]k_3^{-1}[\text{S}]([\text{S}] + (k_{-1} + k_2)/k_1)^{-1}(1 - \exp(-k_3 t)) \qquad 15.$$

for the variation of the median fluorescence with time, where E_o is the total enzyme concentration. Watson estimated the rate constant k_3 to be 1.75 min^{-1} by fitting the data in Figure 14 with an equation of the form $[\text{P}] = C[1 - \exp(-k_3 t)]$.

CONCLUSION

In this review we have illustrated computer techniques for data display, nonparametric, and parametric analyses. A number of examples have been used to illustrate this three-level approach. This group of examples was not intended to be exhaustive; the number and diversity of flow cytometric studies prevents this. Instead, we have chosen examples that suggest how the quantitative data provided by flow cytometry can be used to extend our understanding of biological systems. Current analytical techniques have developed as a result of close cooperation between scientists in diverse disciplines. Such cooperation is essential to future analytical developments as well. Without it, we may fall short of achieving our ultimate objective—the quantitative description of biological systems. The most sophisticated mathematical model is biologically useless if it is inconsistent with known biological principles. Thus, the mathematician is obligated to consult with his biological colleagues to insure that this does not happen. In addition, he may find that biological principles provide constraints which make the analyses more manageable. Of course, information must flow in both directions. The biological user of existing analytical procedures must insure that he is cognizant of the biological assumptions that may be hidden in the mathematical formulation. It is our hope that interdisciplinary cooperation will become even stronger and that the resulting improved flow cytometric display and analysis procedures will play an increasingly important role in quantitative biology and medicine.

ACKNOWLEDGMENT

We are indebted to numerous colleagues for critical evaluation of this manuscript and to many scientists for their permission to reproduce their published and unpublished work.

Work performed under the auspices of the US Department of Energy by the Lawrence Livermore Laboratory under contract number W-7405-ENG-48 with support of US Public Health Service Grant Number CA14533. We also thank Ms. Terry Cunningham for expert help in the preparation of this manuscript.

Literature Cited

1. Adby, P. R., Dempster, M. A. H. 1974. *Introduction to Optimization Methods.* London: Chapman & Hall. 204 pp.
2. Deleted in proof
3. Arndt-Jovin, D. J., Jovin, T. M. 1978. *Ann. Rev. Biophys. Bioeng.* 7:527–58
4. Bagwell, C. B. 1979. *Theory and application of DNA histogram analysis.* PhD thesis. Univ. Miami, Florida. 307 pp.
5. Bagwell, C. B., Hudson, J. L., Irvin, G. L. III. 1979. *J. Histochem. Cytochem.* 27:293–96.
6. Baisch, H. 1975. In *Mathematical Models in Cell Kinetics,* ed. A. J. Valleron, pp. 265–68. Ghent: European Press Medikon
7. Baisch, H., Beck, H.-P., Christensen, I. J., Hartmann, N. R., Fried, J., Dean, P. N., Gray, J. W., Jett, J. H., Johnston, D. A., White, R. A., Nicolini, C., Zietz, S., Watson, J. V. 1979. *Acta Pathol. Microbiol. Scand.* In press
8. Baisch, H., Gohde, W., Linden, W. A. 1975. *Radiat. Environ. Biophys.* 12:31–39
9. Barlogie, B., Drewinko, B., Johnston, D. A., Buchner, T., Hauss, W. H., Freireich, E. J. 1976. *Cancer Res.* 36:1176–81
10. Bennett, D. E., Mayall, B. H. 1979. *J. Histochem. Cytochem.* 27:579–83
11. Bevington, P. 1969. *Data Reduction and Error Analysis for the Physical Sciences.* New York: McGraw Hill
12. Christensen, I., Hartmann, N. R., Keiding, N., Larsen, J. K., Noer, H., Vindelov, L. 1978. In *Pulse-Cytophotometry,* P. III, ed. D. Lutz, pp. 71–78. Ghent, Belgium: European Press
13. Cram, L. S., Gomez, E. R., Thoen, C. O., Forslund, J. C., Jett, J. H. 1976. *J. Histochem. Cytochem.* 24:383–87
14. Crowell, J. M., Hiebert, R. D., Salzman, G. C., Price, B. J., Cram, L. S., Mullaney, P. F. 1978. *IEEE Trans. Biomed. Eng.* BME25:519–26
15. Darzynkiewicz, Z., Andreeff, M., Traganos, F., Sharpless, T., Melamed, M. R. 1978. *Exp. Cell. Res.* 115:31–35
16. Dean, P. N. 1978. See Ref. 12, pp. 63–69
17. Dean, P. N. 1979. *Cell Tissue Kinet.* In press
18. Dean, P. N., Anderson, E. C. 1975. In *Pulse-Cytophotometry,* P. I, ed. C. A. M. Haanen, H. F. P. Hillen, J. M. C. Wessels, pp. 77–86. Ghent, Belgium: European Press
19. Dean, P. N., Jett, J. H. 1974. *J. Cell. Biol.* 60:523–27
20. Dean, P. N., Jett, J. H. 1972. *Los Alamos Sci. Lab. Rep.* LA-5227-PR, Los Alamos, N. Mex., pp. 76–78
21. Dethlefsen, L. A., Riley, R. M., Roti Roti, J. L. 1979. *J. Histochem. Cytochem.* 27:463–69
22. Donaghey, C. E., Drewinko, B., Barlogie, B., Stubblefield, E. 1978. *BioSystems* 10:339–47
23. Fried, J. 1976. *Comput. Biomed. Res.* 9:263–76
24. Fried, J. 1977. *J. Histochem. Cytochem.* 25:942–51
25. Goad, C. A. 1978. *Los Alamos Sci.*

26. Gray, J. W. 1974. *J. Histochem. Cytochem.* 22:642–50
27. Gray, J. W. 1976. *Cell Tissue Kinet.* 9:499–516
28. Gray, J. W. 1979. *Acta Pathol. Microbiol. Scand.* In press
29. Gray, J. W., Carrano, A. V., Steinmetz, L. L., Van Dilla, M. A., Moore, D. H. II, Mayall, B. H., Mendelsohn, M. L. 1975. *Proc. Natl. Acad. Sci. USA* 72:1231–34
30. Gray, J. W., Dean, P. N., Mendelsohn, M. L. 1979. In *Flow Cytometry and Sorting*, ed. M. Melamed, P. Mullaney, M. Mendelsohn, pp. 383–407. New York: Wiley
31. Gray, J. W., Langlois, R. G., Carrano, A. V., Burkhart-Schultz, K., Van Dilla, M. A. 1979. *Chromosoma* 73:9–27
32. Habbersett, M. C., Shapiro, M., Bunnag, B., Nishiya, I., Herman, C. 1979. *J. Histochem. Cytochem.* 27:536–44
33. Hahn, G. M. 1970. *Math. Biosci.* 6:295–308
34. Hartmann, N. R., Pedersen, T. 1970. *Cell Tissue Kinet.* 3:1–11
35. Horan, P. K., Wheeless, L. L. 1977. *Science* 198:149–57
36. Howard, A., Pelc, S. R. 1953. *Heredity* 6:261–73
37. Jennrich, R. I., Ralston, M. L. 1979. *Ann. Rev. Biophys. Bioeng.* 8:195–238
38. Jensen, R. H., Mayall, B. H., King, E. B. 1979. In *The Automation of Cancer Cytology and Cell Image Analysis*, ed. N. J. Pressman, G. L. Wied, pp. 95–102. Chicago: Tutorials Cytol.
39. Jett, J. H. 1978. See Ref. 12, pp. 93–102
40. Johnston, D. A., White, R. A., Barlogie, B. 1978. *Comput. Biomed. Res.* 11:393–404
41. Kim, M., Perry, S. 1977. *J. Theor. Biol.* 68:27–42
42. Kim, M., Shin, K. G., Perry, S. 1978. *Math. Biosci.* 38:77–89
43. Kim, M., Wheeler, B., Perry, S. 1978. *Cell Tissue Kinet.* 11:497–512
44. Kim, M., Woo, K. B. 1975. *Cell Tissue Kinet.* 8:197–218
45. Lindmo, T., Aarnaes, E. 1979. *J. Histochem. Cytochem.* 27:297–304
46. Macdonald, P. D. M. 1975. In *Perspectives in Probability and Statistics*, ed. J. Gani, pp. 387–401. New York: Applied Probability Trust (Distributed by Academic Press)
47. Melamed, M. R., Mullaney, P. F., Mendelsohn, M. L., eds. 1979. *Flow Cytometry and Sorting.* New York: Wiley. 716 pp.
48. Moore, D. H. II 1979. *J. Histochem. Cytochem.* 27:305–10
49. Newman, W. M., Sproull, R. F. 1979. *Principles of Interactive Computer Graphics.* New York: McGraw-Hill. 607 pp.
50. Pearlman, A. L. 1976. *Quarterly No. 51, Lab. Chem. Biodyn., June–Sept.*, Lawrence Berkeley Lab., Berkeley, Calif., pp. 220–48
51. Pearlman, A. 1978. *Flow cytometric analysis of mitotic cycle perturbation by chemical carcinogens in cultured epithelial cells.* PhD thesis, Univ. California, Berkeley. 211 pp.
52. Roti Roti, J. L., Dethlefsen, L. A. 1975. *Cell Tissue Kinet.* 8:321–34
53. Roti Roti, J. L., Dethlefsen, L. A. 1975. *Cell Tissue Kinet.* 8:321–54
54. Salzman, G. C., Hiebert, R. D., Crowell, J. M. 1978. *Comput. Biomed. Res.* 11:77–88
55. Scherr, L., Zietz, S. 1976. *Radiat. Res.* 3:585
56. Stöhr, M., Futterman, G. 1979. *J. Histochem. Cytochem.* 27:560–63
57. Thames, H. D., White, R. A. 1977. *J. Theor. Biol.* 67:733–56
58. Valet, G. 1979. *Acta Pathol. Microbiol. Scand.* In press
59. Valet, G., Bamberger, S., Hofmann, H., Schindler, R., Ruhenstroth-Bauer, G. 1979. *J. Histochem. Cytochem.* 27:342–49
60. Valet, G., Fischer, B., Sundergeld, A., Hanser, G., Kachel, V., Ruhenstroth-Bauer, G. 1979. *J. Histochem. Cytochem.* 27:389–403
61. Valet, G., Hanser, G., Ruhenstroth-Bauer, G. 1978. See Ref. 12, pp. 127–36
62. Valet, G., Hofmann, H., Ruhenstroth-Bauer, G. 1976. *J. Histochem. Cytochem.* 24:231–46
63. Ward, S. A. 1978. *Comm. ACM* 21:788–90
64. Watson, J. V., Chambers, S. H. 1978. *Cell Tissue Kinet.* 11:415–22
65. Watson, J. V., Chambers, S. H., Workman, P., Horsnell, T. S. 1977. *FEBS Lett.* 81:179–82
66. Watson, J. V., Taylor, I. W. 1977. *Br. J. Cancer* 36:281–87
67. Watson, J. V., Workman, P., Chambers, S. H. 1978. *Br. J. Cancer* 37:397–402
68. White, R. A. 1979. *J. Theor. Biol.* In press
69. White, R. A., Rasmussen, S., Thames, H. D. 1979. *J. Theor. Biol.* 81:181–200

70. White, R. A., Thames, H. D. 1979. *J. Theor. Biol.* 77:141-60
71. Woo, K. B. 1979. *Cell Tissue Kinet.* 12:111-21
72. Woo, K. B., Enagonio, R. D. 1977. *Clin. Chem. NY.* 23:1409-15
73. Young, I. T. 1977. *J. Histochem. Cytochem.* 25:935-41
74. Zietz, S., Nicolini, C. 1978. In *Biomathematics and Cell Kinetics*, ed. A. -J. Valleron, P. D. M. Macdonald, pp. 357-94. New York: Elsevier/North Holland Biomedical

BIOMATHEMATICS IN ONCOLOGY: MODELING OF CELLULAR SYSTEMS

♦9156

Carol M. Newton

Department of Biomathematics, School of Medicine, University of California, Los Angeles, California 90024

INTRODUCTION

Several important trends conspire today to forecast an increasingly significant role for biomathematics in cancer research and treatment. First, there is rapid advancement of our knowledge about cellular mechanisms and responses to various agents. This enables more realistic modeling and, as the complexity of underlying biological systems is better revealed, it becomes more difficult for the unaided mind to project consequences of the detailed assumptions concerning these systems. Such projections are needed to design and interpret experiments and to transfer knowledge to applied areas such as medicine. The greater need for formal supports to our reasoning arises as our ability to produce better models advances. Second, while these new advances in knowledge enable more ambitious treatment strategies, some of the laboratory methods supporting them are being adapted for the more informative monitoring of patient responses that is required for model-aided optimizations of individualized therapy. For example, techniques are being improved to produce single-cell suspensions of solid tumors for flow microfluorometry monitoring of cellular DNA histograms during treatment. Gray's article in this issue (52) reviews new developments in theoretical supports to extracting information from these histograms, important now both to research and clinical application. Finally, the medical professional's receptiveness to technological supports has increased substantially in recent years. In addition, the clinical research trend in cancer today is toward aggressive therapy combining different modalities. There is hope that modeling will prove helpful to research on complex treatment strategies and their application to individual patients, and that there will be clinical acceptance of modeling supports if this proves to be true.

This review is designed to bring biomedical investigators abreast of the biomathematical work available to them in oncology and to provide theoreticians who might wish to enter this field an adequate background and reference resources. However, limited space requires selectivity. The decision has been made to emphasize cellular modeling while sketching the biomathematical fields with which it is likely to interact in the near research future; these also are fields that theoreticians attracted to this general area may find challenging to pursue. Advances in cellular modeling are needed to meet our rapid advances in knowledge about cells. Research on complex treatment strategies will center on exploitation of knowledge about how normal and malignant cells differ. However, this exploitation will be embedded theoretically in the context of other developments such as more efficient approaches to optimal control theory, more sophisticated aids to radiation treatment planning and optimization, improved pharmacokinetic models, etc. It will be seen that although each of these other developments has advanced considerably, for the most part they have yet to be linked to adequate cellular models. This linkage is a major research objective and a challenging invitation to strong theoreticians contemplating careers in oncology. Another consequence of selectivity is minimization of space devoted to formulas; they are presented only when believed to be more efficient than words in conveying a sense of a model's structure or complexity or if they are likely to be encountered frequently in the literature (e.g. the multitarget survival function).

General approaches to modeling cellular systems are presented first, together with references to some special systems. Then modeling of the effects on cells of various treatment modalities is considered, followed by a section on treatment optimization. Next, attention is directed to the importance of developing more effective systems for model exploration and presentation to biomedical investigators. Some final comments summarize impressions about important new research directions.

MODELING OF CELLULAR SYSTEMS

Several general mathematical modeling approaches to describing growing cellular systems can be traced from the early literature: ordinary differential-equation models of population numbers vs time, partial differential-equation models of numbers of cells and age distributions vs time, and stochastic models of population growth as a branching, renewal process. In addition, a general simulation approach that depicts cells as advancing stepwise through cycle provides a flexible basis for introducing age-dependent responses to therapeutic agents, uptake of labels, and progression delays. These techniques underly most of the cellular modeling applied to oncology. People contemplating biomathematical research in oncology should become familiar with them. Familiarity with models of special cellular systems also is recommended.

Ordinary Differential Equation Models of Growth of Cellular Systems

An assumption that growth rate is proportional to the number of cells present in the population,

$$dN/dt = kN \qquad \qquad 1.$$

leads to the familiar exponential growth law

$$N = N_0 e^{kt} \qquad \qquad 2.$$

where N is the population size at time t and N_0 is that at $t=0$. It has been known for some time that many tumors depart from this growth law, that their growth rate per cell decreases as N increases. Two empirical models have been widely used to depict such limited growth rates. The Verhulst equation (138)

$$dN/dt = kN(\theta - N)/\theta \qquad \qquad 3.$$

provides a growth rate that approaches zero as the population size approaches a limiting value, θ. The Gompertz function (48), which may be expressed as two simultaneous differential equations,

$$dN/dt = \gamma N$$
$$d\gamma/dt = -\alpha\gamma \qquad \qquad 4.$$

has been fitted rather successfully to tumor growth curves (79, 120, 127). The references cited explain the fitting methods, and that by Simpson-Herren & Lloyd (120) provides considerable data. Goel et al (47) observe that these are special cases of the more general expression

$$dN/dt = kNG(N/\theta), \qquad \qquad 5.$$

where G is any function that approaches zero monotonically as N approaches θ. This general case is developed and other theoretical aspects of population growth are well treated in that excellent reference.

Partial Differential Equation Models

The next important step toward realistic modeling is to describe a population in terms of its size and age distribution. Von Foerster (140) initiated this line of modeling, which received rigorous further development in a sequence of papers by Trucco (133, 134).

Von Foerster defines an "intrinsic life-function" $\phi(a)$ that represents the fraction of population elements that would survive up to age a in an ideal risk-free environment. There is a corresponding "intrinsic loss factor"

$$\theta_i(a) = -\phi^{-1}(a)\partial\phi(a)/\partial a. \qquad 6.$$

A "compound loss factor" $\theta(a)$ is the sum of intrinsic and environmental loss factors $\theta_i(a) + \theta_e(a)$. A conservation equation describes the progression of population elements in time and age, taking into account these losses.

$$N_a(t+\Delta t, a) = N_a(t, a-\Delta a) - N_a\theta. \qquad 7.$$

Expanding this expression around t and a, Von Foerster obtains the differential equation,

$$\partial N_a/\partial a + \partial N_a/\partial t = -N_a\theta, \qquad 8.$$

neglecting higher powers in Δa and Δt. An age-dependent birth process is then introduced whose "generation coefficient" $\gamma(a)$ is the expected number of offspring produced per element at a given age. In the steady state, the number of elements at age zero thus becomes

$$N_0 = \int_0^\infty N_a \gamma(a)\,da \qquad 9.$$

and a condition for realizing a population steady state can be derived from this, the definition of θ_e, and Equations 6 and 8:

$$\int_0^\infty \gamma(a)\phi(a)\exp\left[-\int_0^a \theta_e(a)\,da\right]da = 1. \qquad 10.$$

While Von Foerster's population model is general, he specializes its application to white and red blood cell populations, which are distinguished on the rather elementary basis of susceptibility to environmental factors.

Trucco allows that age should be treated independently as some cellular maturation index, but he initiates his modeling (133) by scaling age as time and thus arriving at an equation similar to Von Foerster's,

$$\partial n(t,a)/\partial t + \partial n(t,a)/\partial a = -\lambda(t, a, \cdots)n(t,a) \qquad 11.$$

where $n(t,a)$ is the number of cells of age a at time t and λ is a loss function corresponding to Von Foerster's θ. Trucco takes advantage of the relationship between a and t by transforming to variables

$$\xi = t - a \quad \text{and} \quad \eta = a, \qquad 12.$$

which facilitates the subsequent solution of Equation 11 with attention to boundary conditions on birth rate and initial age distribution, defined respectively as

$$\alpha(t) = n(t,0) \quad \text{and} \quad \beta(a) = n(0,a). \qquad 13.$$

The resulting expressions for $n(t,a)$ when $t > a$ and $t < a$ are:

$$n(t,a) = \alpha(t-a)\exp\left[-\int_0^a \lambda(t-a+x,x)\,dx\right] \quad t > a$$

$$n(t,a) = \beta(a-t)\exp\left[-\int_0^t \lambda(z,z+a-t)\,dz\right] \quad t < a. \qquad 14.$$

Trucco uses these equations to derive the steady-state solution for the cases where λ is a function of age only. The results are discussed for two forms of Von Foerster's $\phi(a)$, one of which corresponds to the incomplete gamma function assumed by Kendall (69) and Harris (56) in developing one of the most widely used stochastic models for population growth.

Trucco next applies this formulation (134) to a self-replenishing stem-cell system whose loss function consists of three terms representing respectively mitosis, death, and removal by differentiation into another cellular type.

$$\lambda(t,a) = \mu(t,a) + \rho(t,a) + \tau(t,a). \qquad 15.$$

Neglecting ρ, solutions are explored for several special assumptions concerning the forms of μ and τ. Trucco next attempts a solution for the Lajtha (80) stem-cell model, which depicts a large stem-cell pool from which cells are recruited into cycle at a rate required to counterbalance losses from the pool due to death and differentiation. He frankly discusses limitations of his approach, such as its inability to portray the time lag required for cycle transit. Concluding remarks leave no doubt that the domain of closed solutions must be left behind for more realistic explorations of cellular systems and their size-preserving feedbacks that require λ to have a functional dependence on n.

Weiss (142) adds an indicator of physiological age, p, modeling $n(t,a,p)$. Rubinow (111) replaces chronological age by a maturity variable μ whose rate of change ν may vary as a function of time, μ, and a vector of environmental factors, β. His conservation equation thus becomes

$$\partial n(\mu,t)/\partial t + \partial[\nu(\mu,t,\beta)\cdot n(\mu,t)]/\partial t = -\lambda n(\mu,t). \qquad 16.$$

He discusses the intrinsic difference between his and the preceding approaches, illustrating the solution of Equation 16 for when λ and ν are functions of μ only. Specializing to the case where the flux of cells per unit time completing mitosis is given by the gamma distribution, he derives a formula that agrees well with growth data for *Tetrahymena* cells. The Rubinow model assumes that cells enter mitosis as soon as μ reaches a critical value; dispersion of generation times, the time required to transit cycle, is ascribed to variations in ν. This approach supports Rubinow & Lebowitz's modeling of granulopoiesis and acute myeloblastic leukemia (114–116) and Himmelstein & Bischoff's chemotherapeutic model (61).

Stochastic Models

A number of textbooks on stochastic modeling treat birth-and-death processes. Bailey's (7) gives a good treatment of the extinction process, which relates to the problem of whether a cancer metastasis will survive. An article by Harris (56) provides a very useful entry to the literature on stochastic models of cellular systems, including both theoretical and experimental references. Among these, Kendall's classical article (69) introduces the gamma distribution formulation referred to above that leads to a distribution function for cell lifetimes of the form,

$$f(t) = (c^{k+1} t^k e^{-ct})/k! \qquad 17.$$

where t now takes on the meaning of cell age, time after birth. For $k=0$, this is an exponential function, and for all other values of k it has a peak that moves to higher values of t as k increases. It is somewhat unphysiological in not vanishing at $t=0$, but it is known to fit bacterial growth data rather well (109). Where $N(t)$ is the number of cells at time t and $p_i(t)$ is defined as the probability that $N(t)=i$, a generating function is defined,

$$G(s,t) = \sum_i p_i(t) s^i \qquad 18.$$

and an integral equation relating it to $f(t)$ is derived:

$$G(s,t) = \int_0^t G^2(s, t-u) f(u) du + s[1 - F(t)], \qquad 19.$$

where

$$F(t) = \int_0^t f(u) du. \qquad 20.$$

Assuming that a cell divides into two new cells at the end of its life, Harris derives means and standard deviations for $N(t)$ by the usual method of differentiating the generating function. For instance, $dG(s,t)/ds$ with s set equal to one yields the expected value of $N(t), m(t)$.

$$m(t) = 2 \int_0^t m(t-u) f(u) du + 1 - F(t). \qquad 21.$$

Results such as age distributions and existence of a constant coefficient of variation are discussed and the reader is referred to standard methods for solving renewal equations.

An advancement towards realistic modeling of cellular systems is contributed by Bronk et al (25, 26). Cell cycle is divided into G_1, S, G_2, and M phases, each of which may have a different mean transit time and transit-time variance. There is an excellent treatment of the decay of synchrony, age distributions, labeling, and response to some perturbations. They also use renewal equations, and they provide more information on their methods of solution. This work is highly recommended to people contemplating research on stochastic models of cellular systems.

All of the preceding models assume that the various steps through cycle are independent of one another and that generation times for daughters are independent of generation times for their parents. Powell (109) and others have noted findings suggestive of departures from the latter assumption, and it is not unreasonable to believe that some cells may proceed more slowly through all phases of cycle than others, especially if they are impaired. Schotz & Zelen (118) develop a model that permits correlations among transit times through the different cycle phases, evaluating the bias that could result from this when one is analyzing labeled-mitosis data. In addition to this fundamental sequence of papers on stochastic modeling of cellular systems, one can find some useful mathematical approaches in papers on carcinogenesis. King's book (70) is reported to give a very lucid coverage of stochastic modeling of carcinogenesis, and one also might be referred to Chapter 3 in Swan's book (128) or some recent reviews (59, 145). Except for factors such as possibly altered immune status, there are parallels between growth of tumor from a metastatic cell and from a cell that has undergone mutation.

Simulation Models

A number of the preceding mathematical papers must resort to computer methods to make their results explicit, to compare them to data. A number of the papers that lead more directly to simulation commence with mathematical descriptions of the basic modeling structure. Distinctions therefore are not nice. The models referred to in the following paragraphs have this in common: They proceed by a series of steps in time to follow an initial population as its component cells advance through cycle—e.g. as they undergo mitosis, risk death, are slowed in progress through G_2, become labeled in certain parts of cycle, differentiate by transfer to other subpopulations, reenter cycle, etc. Not all models have the features just alluded to; stepwise forward modeling is the primary common feature. This is an especially flexible framework within which one may introduce a rich variety of biological mechanisms, right down to the biochemical level if one wishes to step by small increments in time. However, a commitment to detailed modeling correlates with the unlikelihood that one can obtain closed-form mathematical solutions. Thus, one must be resigned to simulation and its well-known limitations. Useful results are seldom gained by carrying the mathematical development beyond the point where it contributes to efficient simulation, but it is important to carry it that far, as less efficient simulations of realistically complex systems can be prohibitively expensive.

Takahashi (129) depicts cell cycle as a sequence of compartments through which cells advance as they progress through cycle. This multicompartment model is described by a set of simultaneous differential equations. Because sojourn in any phase (G_1, S, etc.) is represented by advancement through a sequence of identical compartments, transit times in each phase, and for the cycle as a whole, follow the distribution in Kendall's model. At mitosis, a specified fraction of cells divides to reenter cycle. Labeled cells are treated as subpopulations of the compartments in which they reside. In later papers, cell death may occur throughout cycle (130) and an attempt is made to fit the model to data (131). Hahn (53) also models the advancement of cells through cycle as progression through a sequence of compartments, treating this from the outset as a discrete process, similar to the Leslie matrix approach encountered in ecological population modeling. The system is described by a cell-age state vector whose elements represent the number of cells in each compartment. At each step, this is acted upon by a dispersion matrix that determines what fraction of cells in each compartment will remain there, advance to the next compartment, or advance two compartments—and by a time-shift operator that carries out the advancement. Using this model to study the problem of obtaining measures of synchrony, Hahn obtains results similar to those established analytically by Bronk (25). He adds some capabilities for perturbing the system (54), e.g. labeling and killing cells while allowing for some repair. With Steward (125) he then uses the model to study optimal treatment strategies; he introduces a G_0 noncycling compartment that can contribute cells anywhere in cycle and provides some capability for introducing a block in G_2 as is seen in radiation therapy. He is able to exploit the synchronizing effect of a G_2 block to enhance killing. Our modeling, which also proceeds by discrete advancement of cells through a series of compartments, emphasizes a method of presentation for flexible exploration and for use by biomedical scientists (97–99, 101, 102).

Models of Special Cellular Systems

Modeling of hematopoiesis arose with active laboratory research in the late 1950s. Interpretation of laboratory research findings was an early motivating factor, and interests subsequently developed toward modeling the treatment of leukemia. Modeling of the immune system now responds to recent increased research activity in that area.

GRANULOPOIESIS Warner & Athens (141) sought to interpret labeling experiments and King-Smith & Morley (71) attempted to describe normal and impaired granulopoiesis as well as to explain the regulation of granulocyte levels in the peripheral blood. Rubinow and his associates have made major contributions to this area, commencing with an early paper describing population densities of neutrophil precursors in the bone marrow (112), followed by a comprehensive model of normal neutrophil production (114) and by a series of articles on modeling acute myeloblastic leukemia and its treatment (113, 115, 116). The basic modeling technique has been described. Neutrophil production is described as proceeding through a proliferative pool that has both cells in cycle and resting cells in G_0, followed by transit through nonproliferating maturation phases and by eventual release to the peripheral circulation, which includes marginal and circulating components. The stem-cell compartment that feeds this system is assumed to be self-maintaining. The rate at which cells in the proliferating pool are summoned from G_0, as well as the rate at which cells are released to the maturation sequence, is homeostatically controlled with respect to the total number of cells in the system. The rate at which mature cells are released from reserve into the circulation is controlled with respect to the number of cells in the blood. These investigators are aware of uncertainties concerning elucidation of feedback mechanisms, cycle parameters for the early proliferating phases, etc. Their interpretations of findings are appropriately cautious. Mauer et al and Evert (40, 41, 89) have modeled acute lymphoblastic leukemia, using difference equations describing the progression of cells through cycle. At mitosis, a cell may die, enter G_1, or rest in G_0. All cells are followed individually, enabling more detailed knowledge of what is going on at perhaps some additional expense in processing time and storage. Cycle-specific drug effects were studied.

ERYTHROPOIESIS An early venture into erythropoietic modeling (97, 98) was motivated by an apparent conflict between observations on responses to a sequence of erythropoietic challenges and the currently accepted hematopoietic model of Lajtha et al (80). The latter postulated that all death, differentiation, entries to cycle, and feedbacks to regulate the latter occurred in a large, well-mixed, resting stem-cell pool. The laboratory findings suggested that triggering into differentiation acted at some window through which cells progressed in time, rather than upon an entire resting pool. The new model kept the postmitotic location of Lajtha's pool, but permitted probabilities for death, differentiation, or entry to cycle within it to be age dependent. This model gave results for a wide variety of choices of parameters and forms of the age-dependent probabilities, that agreed qualitatively with the laboratory findings on erythropoiesis, whereas the old model could not. However, both models were susceptible to unphysiological oscillations when recovering from major, sudden challenges such as cell killing or depletion by large erythropoietic removals. Subsequent work investigated the stability of various postulated size-preserving feedback mechanisms for the stem-cell population and eventually led to a more complete hematopoietic model (99) that includes a sequence of proliferating and nonproliferating steps toward maturity for cells differentiating into erythropoietic or granulopoietic series. This model postulates an erythropoietin feedback, driven by the product of peripheral erythrocyte concentrations and oxygenation, that acts upon the stem-cell population to induce differentiation. The modeling technique is discrete forward simulation, similar to Hahn's, except that the maturation velocity can be placed under feedback control. A crude capability for labeling or killing cells is also present. Kretchmar's model (78) to investigate the action of erythropoietin is a rare example of cell-kinetic modeling on an analog computer. Vogel et al (139) are primarily concerned with a stochastic representation of the stem-cell system but extend and test their model against data by including erythropoietic differentiation and maturation.

Originators of the more detailed models of hematopoiesis are aware of gaps in our knowledge about some of the mechanisms that are depicted, e.g. various control mechanisms or sites where differentiation is triggered. However, if model exploration yields results that conflict with laboratory data, our alternative conceptualizations of underlying biological mechanisms can be narrowed down. One does not require quantitative disqualifications; the form of a model's responses may conflict with well-known qualitative experimental observations, e.g. with the absence of oscillations.

THE IMMUNE SYSTEM   Little need be added here to Swan's treatment of the immune system (128). Renewed interest in immune surveillance in oncology and recent rapid advances in the techniques of immunological research and experimental findings are just beginning to amass a level of knowledge that justifies support of experiment interpretation and planning by modeling of postulated mechanisms at the more detailed levels attempted by Bell (11–14) and by Bruni et al (27–29). Some interesting possible insights also have been developed on immune surveillance of malignancies (36) and on postulated immune mechanisms (108). A recent historical review (15) and the publication in which it appears should be of interest to serious researchers in this field.

OTHER SYSTEMS There are scattered papers on the modeling of other specialized systems, e.g. breast cancer (23), endometrial cancer (24), Ehrlich ascites tumor (66).

MODELING FOR LABORATORY ANALYSIS Insofar as some laboratory methods probe for cell-kinetic information, their data analyses are based on models the methods assume. If these assumptions differ for two methods, they may interpret the same set of data differently.

Analysis of labeling experiments Mendelsohn's (91) early derivation of the "growth fraction" from labeling data provided a valuable insight into why some tumors grow slowly, as did work by Steel (122). Mendelsohn applies this concept successfully to his analysis of the growth of C3H mammary tumors (92). Quastler was another early contributor (110). Barrett (9, 10) develops a statistical approach, based on distribution functions for transit times in the various phases of cycle, which ignores time spent in mitosis and adopts a maximum likelihood estimation procedure. He concedes Trucco's point (135) that his treatment may be somewhat inaccurate for expanding populations. Steel & Hanes (123) base the development of their widely used program on Barrett's model. It should be noted that their stepwise fitting procedure works from left to right, weighting earlier portions of the percent-labeled-mitoses curve more heavily and sometimes failing to fit well beyond the first peak or two. They concede that sometimes the analysis should be reported as meaningless and leave it to the user to decide when. Papers by Bronk (25, 26) and by Schotz & Zelen (118), discussed earlier, develop bases for more refined analysis. A number of papers in the National Cancer Institute's Monograph #30 discuss the analysis of labeling experiments; one of the most instructive is Mauer's comparison of problems in interpreting the various types of labeling experiments (88). Hartmann et al (57) provide a good recent reference that compares various methods for fitting percent-labeled mitoses curves. Subsequent to this, Thames & White (132) have developed a method that relates directly to the Hahn model, and White proposes an analysis (144) based on population projection matrix techniques that have been used for some time in population biology. The latter approach is especially attractive because of its numerical simplicity. It also can accommodate any initial age distribution or cycle transit-time distribution; the population is not required to be in a steady state. The requirement that distributions be discrete is examined. One must be cautious in adopting programs for the analysis of labeling data, both on the basis of fitting procedures used and because the model upon which the analysis is based may depart from what is appropriate for the cellular system being studied.

Analysis of flow microfluorometry data Flow microfluorometry (FMF) substantially increases the information that may be obtained about cell populations. Crissman et al provide a good early reference (35) for this technique, which enables the volume, scattering coefficient, and up to two fluorescent labels to be measured simultaneously for each cell passing through the flow system. Single-cell suspensions are required, a challenge that is being met by research on techniques for separating cells in solid tumors from surrounding stroma. There are encouraging early reports on FMF guidance of treatment (8, 34, 90). Analyses of FMF data may be based on modeling assumptions for label uptake, e.g. that the coefficient of variation of fluorescence is constant for all DNA levels. Tracing the development of methods for analyzing FMF data are Gray's early classic (50) and some good intermediate reviews (51, 93, 148). Most up-to-date is Gray's article in this issue (52).

MODELING EFFECTS OF VARIOUS TREATMENT MODALITIES

Preceding sections show that there is a considerable resource of theoretical work to draw upon to describe cellular systems, both in general and with regard to special systems such as the hematopoietic system. The extent to which models can be specific, detailed, and useful depends on the status of the knowledge we have derived from experiments. Laboratory research has now provided extensive information about cellular mechanisms for those animal or in vitro models that have been studied, and there is considerable information on responses to agents used in cancer therapy. Principles of cellular responses discovered in laboratory research are not necessarily applicable to humans, but knowledgeable investigators can make reasonable estimates about which ones might be most generalizable and about how they might be modified in their expression in humans. Some of these estimates generate hypotheses about how management of human disease might be improved, and these hypotheses in turn guide the design and analysis of investigations on experimental therapeutics in the laboratory and eventually in the clinic. By comparing the results of clinical investigations with what is estimated on the basis of extrapolations from laboratory research, we can advance our knowledge of how humans differ from models studied in the laboratory, building a progressively more adequate bridgehead for translating what we learn in the laboratory into the clinical domain. In order that extrapolations from the assumed mechanisms of damage and repair may be as precise as knowledge permits, it helps to describe those mechanisms in a way that can be treated quantitatively and that can be manipulated, i.e. a model. To be sure, much clinical research has progressed without the help of modeling, but to take major strides in aggressive, multimodal treatment we must exploit to the utmost the ways in which well-timed combinations of different treatment agents can probe significant response differentials between tumors and the normal tissues we seek most to spare. Such complexity taxes the unaided mind. Modeling of combined effects together with improved monitoring capabilities should contribute substantially to this clinical research effort as well as to the individualized application of its results to patient care.

Considerable laboratory research and modeling have been directed to the effects of individual treatment modalities, but much more remains to be done to assess ways in which the combined effects of different modalities differ from what might be predicted by simple additive models. Also, although there is an increasing fund of knowledge about effects of various treatment agents on intact animals, much of what we know has been derived from in vitro studies. Additional advances in experimentation and modeling are required to better characterize the nature and timing of acute repair responses in intact animals, as well as the long-term repair and degeneration processes that account for late effects.

As to modeling of responses to individual modalities, considerable attention has been directed to radiotherapy and chemotherapy. Modeling of immune mechanisms is somewhat recent, and only a few articles are found that might have direct relevance to cancer therapy. One does not model surgical management that intends complete excision, though regrowth from metastases or remaining tumor might be considered theoretically. To the extent that mechanisms of damage inflicted by chemotherapeutic agents are distinct from those for radiotherapy, additive models might be attempted as a first approximation for combined-modality therapy. Immunotherapy probably would not be simply additive, in that the other treatment modalities can be immunosuppressive. Thus, for now, multimodal modeling probably should be confined to chemotherapy and radiotherapy; much more remains to be learned about their interactions with basic and induced immune responses.

Modeling Radiotherapy

Radiotherapy has two major advantages. First, by implanting radiation sources in the tumor or by focusing a combination of beams from external sources on it, one can obtain some degree of localization of damage to the tumor area and selectively spare critical areas of normal tissue. Second, to implement combined-modality strategies whose timing exploits known cell-kinetic parameters, one can treat with a high-intensity radiation source within a few minutes. The time course of treatment to the tumor area is precisely known without the need for recourse to biopsies.

Theoretical work in support of radiotherapy was introduced very early by the physicists who have traditionally been associated with this medical specialty. Its primary clinical emphasis has been on the calculation of dose distributions throughout the treatment area and, more recently, on some semi-empirical estimates of the effects of fractionation, of delivering a dose in portions distributed over time rather than all at once. Radiological research has provided dose-response formulas that have been related grossly to mechanisms of radiation damage. Radiation effects also have been described by models that seek to take into account more precisely where damage occurs in cycle and what its effects are.

The calculation of radiation dose distributions rests upon well-known physical principles of interactions of radiation with matter. Until recently, the primary limitation has been precise anatomical information for each patient. With the advent of computerized axial tomography, this problem is reduced. For most practical purposes, tomographic densities are close to the tissue densities used in dosimetry, though one must be cautious about occasional artifacts. Some dosimetry programs now accept tomographic data on the patient's anatomy. Publications are abundant on classical radiation dosimetry (3). A series of international conferences on computers in radiotherapy provides up-to-date information on advances in this area (124).

It has long been known that damage to tissue is greater if the entire radiation dose is delivered at one time rather than by fractions that are administered over the course of a few weeks. Trade-offs between fractionation parameters have been studied for attaining a tolerance endpoint in various tissues. These parameters include the total dose delivered, the number of fractions into which it was divided (number of treatments), and the overall treatment time. At least two important processes are going on between the delivery of successive doses: tissues repair, and some previously hypoxic portions of the tumor become better oxygenated. The latter comes about because well-oxygenated portions of the tumor which are more susceptible to radiation damage are killed preferentially, allowing nearby hypoxic portions to be approached more closely by the vasculature. These hypoxic portions, in turn, become more radiosensitive.

From the outset, two basic theoretical approaches have been proposed to characterize the effective dose delivered in fractionated radiation therapy. Ellis (38, 39) derived an empirical formula,

$$D = (NSD) T^{0.11} N^{0.24} \qquad 22.$$

relating the tolerance dose for normal tissues (D rads) to total treatment time in days (T), number of fractions (N), and nominal standard dose (NSD). Conceptually, the NSD is the dose given in a single exposure that would treat connective tissue to the tolerance level, though one observes that data upon which the formula is based apply only over the range of 4 to 30 fractions. Cohen (31) based his calculation of effective dose on a model comprising types of damage inflicted and the regeneration of cells between treatments:

$$S = \left\{ e^{-JD_1} \cdot \left[1 - (1 - e^{-KD_1}) \right] \right\}^W \cdot e^{LT} \qquad 23.$$

where S is the fraction of cells surviving after W treatments of D_1 rads, each delivered over a total of T days. The first exponential term represents the irreversible single-event component of damage, the term within square brackets represents a reparable multi-target component of damage, and the final exponential factor represents regeneration attributable to proliferation of surviving cells. The last factor (32) has been replaced by a logistic function that decreases cell proliferation rate as a limiting population size is reached. Hall (55) compares the Ellis and Cohen approaches, noting that they agree closely for fractionation patterns commonly used in radiotherapy, as they should when one considers that they are based on essentially the same clinical data base. Differences arise when they are extrapolated beyond this range. Indeed, an approach originated by Dutreix et al (37) suggests that both of the above approaches may overestimate the sensitivity of tissue-sparing effects to changes in number of fractions when the number of fractions is large. Cohen continues the development of a model-based approach (33) that may come into wider use as combined effects from multimodal therapy must be assessed.

Simplicity has tended to favor use of empirical models, but they must be applied with attention to their underlying assumptions. Orton (105) found, for instance, that the NSD approach was misused by over 50% of experts solving a test problem, presumably because of their misunderstanding that NSD values can be added. Subsequently, Orton & Ellis (106) proposed a formulation, the TDF approach, that eliminates the rather meaningless NSD term. In a major series of papers, Kirk et al develop the Cumulative Radiation Effect (CRE), a generalized treatment of the Ellis approach; they extend and demonstrate the application of CRE to a variety of illustrative cases (72–76). Fractionated treatment regimens, constant-intensity sources, area and volume correction factors, and time gaps between regimens are treated. A unified treatment such as this for external-beam and constant-intensity (implanted) sources is an important step toward assessing the combined effects of two radiation treatment modalities. Such an assessment was also undertaken earlier by Liversage (85, 86) who modified Cohen's model to take better account of the buildup of sublethal damage.

Cohen's and other simple models are compared instructively in an article by Hethcote et al (60). It should be noted that only effects on healthy tissues are considered, that cycle dependence is ignored, and that factors such as tissue inhomogeneity with respect to oxygenation are not taken into account. The latter factor may be especially important for large tumors with poorly vascularized interiors. The time between administration of dose fractions should be designed to exploit differentials between tumor and surrounding normal tissues. First, proliferative repair of normal tissue with its injury-response feedback mechanisms may differ from that of tumor tissue. Second, the two tissues may differ to the extent that hypoxia reduces cell sensitivity to radiation and that formerly hypoxic cells are brought within range of good vascularization during the inter-treatment periods. Reporting on isotransplants of C3H mouse mammary tumors (63), Howes & Suit point to the potential for devising optimal fractionation schemes. Somewhat earlier, Fischer undertook theoretical studies (43, 44) on a model providing for reoxygenation of initially hypoxic tumor cells during a course of treatment. Only tumor cells were considered and an ad hoc formula related the fraction of cells that are oxygenated to the total number of cells remaining in the tumor. The multi-target expression referred to earlier was used to represent radiation effects, with different parameterizations for well-oxygenated and hypoxic cells. Although one may question some of the model's more detailed assumptions, the fact that local minima were found in the Strandqvist curves over a wide range of parameter choices points to the possibility for optimal treatment strategies and merits serious follow-up.

As discussed earlier, models exist that simulate cells' progression through cycle and that provide more flexible capabilities to portray phenomena such as cycle-specific responses to various treatment agents, Elkind repair, dose-dependent delays in G_2, etc. Two such models have been especially explored with radiation effects in mind (53, 54, 125, 101–103). The problem with more complex models is that part of their structure may represent postulates, though the remainder rests on good experimental grounds. Also, their parameters seldom are completely determined for application to specific systems. Nevertheless, they can be studied over a reasonable range of plausible postulates and parameters to see if they must be rejected because of gross conflict with a biological system's observed behavior or to discover important consequences that derive from their general nature. They lend themselves to exploring strategies for radiotherapy, if not application to individual cases.

Modeling Chemotherapy

Modeling chemotherapy has been a most active area of biomathematical research. Although some simple models have been instructive, modeling of chemotherapy is intrinsically complex. At the very least, one is combining a compartment-type pharmacokinetic model with a cell-kinetic model, and recent modeling research on the treatment of L1210 leukemia by ara-C (84, 96) points to the importance of intracellular pharmacological reactions such as conversion of ara-C to longer-acting ara-CTP, which is cytotoxic at high levels and which, at lower levels, causes progression delay in the S-phase. Few models have had the combined pharmacokinetic and cell-kinetic sophistication of this recent model. Most have emphasized either the pharmacokinetic or the cellular modeling. Yet, insights have been derived for general treatment strategies and, as Lincoln et al point out (84, pp. 37–41) in their analysis of research deriving from Skipper's early work (83, 137), even models based on imperfect or incomplete information can contribute significantly to research progress. They motivate deeper experimental investigation, direct attention to what assumed mechanisms may be at fault when experimentation and theoretical predictions disagree, and in general sharpen our focus on the interrelation of laboratory findings and the underlying biological mechanisms they purport to study.

Aroesty et al have written an excellent review article on early work in modeling chemotherapy (2), and there is a good brief survey by Lincoln et al (83).

MODELING THAT EMPHASIZES PHARMACOKINETICS Bellman, Jacquez, and Kalaba were among the earliest pharmacokinetic modelers; they developed a circulatory model of a number of parallel organ compartments, one of which could be tumor, into which the pharmacological agent can diffuse at different rates (16, 17, 64). This modeling was ahead of its time with respect to both its computational demands and the ability to determine model parameters. It has been somewhat forgotten because of consequent lack of early carry-through into practical applications. Bischoff & Dedrick's simpler two-compartment model (21) was more tractable, and subsequently a model was developed to describe methotrexate and ara-C distributions (22), taking into account such factors as biliary secretion of methotrexate and the rapid metabolism of ara-C in liver and other organs. Four branches of circulation are hepatic, gastrointestinal, renal, and "muscle," with the outlet of gastrointestinal feeding into the liver. Delays entailed in biliary secretion are modeled by imposing a sequence of three compartments before a sequence of four compartments that communicate with the gastrointestinal circulation. Fluid volumes, flows, organ volumes, tissue binding, and renal clearances are taken from the literature or independently measured. Model results compare satisfactorily with data from mice, rats, and man. After a rapid initial fall, blood levels of ara-C in man are assumed to fall off exponentially. A simple version of the Rubinow model is used to describe cells in exponential growth. In the one example of ara-C metabolism where pharmacokinetic and cellular models are coupled, much of the potential detail intrinsic to both models is lost by assumptions such as the simple exponential disappearance of ara-C from plasma and equal sensitivity throughout cell cycle to this agent, rather than sensitivity that predominates in S-phase. The latter would not be too restrictive an assumption in L1210 cells, whose cycle is dominated by S-phase, but it does evade the issue of cycle-specific therapy that would be applicable to other systems. Thus, the main contribution was formulation of what should be described in a more complete pharmacokinetic model and some checks to see that this did not contradict existing information. This work is described in some detail because it probably was the most influential early pharmacokinetic model for cancer chemotherapy researchers.

As mentioned earlier, more recent modeling of ara-C pharmacokinetics by Morrison et al (84, 96) includes an intracellular compartment within which ara-C converts to longer-acting ara-CTP. Nicolini et al model the interference of various drugs (e.g. ara-C, FUdR, ara-A, hydroxyurea, and methotrexate) with DNA metabolism (104), assuming a Michaelis-Menten formulation for enzyme reactions which illustrates different forms of drug interactions on cell growth.

One should be aware that there is an extensive literature on compartment theory that might be helpful to pharmacokinetic studies in cancer chemotherapy. A classical reference is Jacquez's textbook (65), and there have been developments since its publication. To aid specific clinical and laboratory studies of pharmacokinetic models, there are computer programs that determine underlying model parameters from those obtained from fitting the model to data. Well-tested computer software for fitting compartment models to data has been developed by Berman and others (20, 67, 77). These programs assume systems that can be linearized. As pharmacokinetic modeling becomes more sophisticated and drug interactions or other sources of nonlinearities cannot be avoided, direct fitting of differential equations to data may be required—a difficult theoretical area that is just beginning to be developed (95). Two recent articles in Annual Reviews of Biophysics and Bioengineering are especially helpful: First, when differential equations describing tightly-coupled systems give rise to very disparate time-constants, numerical solutions will require attention to "stiffness." Garfinkel et al (45) discuss where stiff differential equations arise in biomedical applications and present a variety of coping strategies, including references to available software. Second, Jennrich & Ralston provide a timely review of methods for fitting nonlinear models to data (68). If one contemplates individualizing chemotherapy to each patient's metabolism of the drugs to be given, an important objective will be to minimize the number and invasiveness of specimen removals required to assess that metabolism. This opens a rich area of research in optimal design. Those interested in pursuing it may be helped by Landaw's recent publication (81).

MODELING THAT EMPHASIZES CELLULAR RESPONSES A number of instructive studies have considered simple growth models that ignore cell-cycle-specific phenomena. Valeriote et al (136) assume exponential growth of cells, with constant generation times, that lose their proliferative capacity if they encounter drug concentrations above a threshold value during entry to mitosis. Berenbaum (19) compares recovery by proliferation, after sudden population depletion, for cell populations in exponential, Gompertzian, and steady-state growth. One may recall that the rate of regrowth between successive treatments will be greater for a Gompertz-type population the more its size has been depleted by previous treatments. Thus, one may expect that a sequence of equal doses which are equally spaced in time and are sufficient to effect initial successive reductions in tumor population size can eventually reach a plateau where inter-dose regrowth equals post-dose depletion. Aroesty et al (2) carry this type of study forward theoretically to suggest the strategy of an initial course of cycle-nonspecific therapy until the plateau is approached, followed by well-chosen cycle-specific therapy.

They cite evidence for the tumor regression plateauing effect found by Sullivan & Salmon (127) for IgG multiple myeloma.

Lloyd (87) compares a model that Chuang and he developed (30) with a simple exponential growth model for studying treatment strategies for B16 melanoma which employs cyclophosphamide, a relatively cycle-nonspecific agent. The Chuang-Lloyd model depicts passage of cells through G_1, S, G_2, and mitosis, after which a fraction enters G_0, with the remainder reentering G_1. The rate of reentry of cells from G_0 to G_1 is specified. Natural or drug-induced death can occur from any compartment, and only dead cells can be removed from the tumor site. The general form of the growth curve obtained from a trial-and-error parameterization that seeks to bring this model into correspondence with growth of B16 melanoma has a diminution in growth rate for large population size that is similar in nature but not identical in form to that of a Gompertz curve. The effects of a two-treatment schedule are compared for different choices of dose in both models of regrowth, assuming that the only effect of each dose is immediate death of a fraction of the population. Outcomes differ somewhat for the two models, though both agree in ranking the best and worst strategies with respect to tumor kill; tissue sparing is not considered.

Lincoln et al, in a most incisive analysis of some important issues that arise in chemotherapeutic modeling (84, pp. 31–41), trace research on ara-C treatment of L1210 leukemia in mice from Skipper's initial hypothesis that this is a good biological model for research on treatment of acute myelogenous leukemia (AML) in man (121). One strategy suggested by this model is to maintain a high level of ara-C for about 30 h to catch almost all of the L1210 cells as they enter S. Assuming that normal cells are relatively spared by their lower growth fractions, a resting period of several days before the next course of treatment permits them a higher degree of recovery by recruitment and proliferation than that possible for the few remaining leukemic cells. Schedules suggested by this model were attempted for treatment of patients with AML and gave good remission results. This success became somewhat hard to explain, however, when labeling experiments showed that human leukemia cells grow much more slowly than was assumed and studies on ara-C metabolism revealed the long-acting intracellular form, ara-CTP, mentioned earlier. Lincoln et al weigh possible explanations and seek improved strategies in view of the new information. This is an excellent example of a model-based discussion that does not require mathematical formulation but is complex enough to require these investigators to use a formal simulation model to check out the general ability of assumptions to explain phenomena, as well as to project improved treatment strategies for AML. Their model (84, 96) includes intracellular ara-C metabolism, a Bischoff-Dedrick type model for extracellular pharmacokinetics, a variation upon Rubinow's cell-kinetic model, cells with different maturation rates, drug-related cell killing and progression delay, etc. Inclusion of progression delay is found to be important for adequate simulation of divided-dose schedules. Earlier modelers of this system also should be mentioned. Stuart & Merkle were important pioneers (94, 126). As noted above, Bischoff et al (22) related their chemotherapeutic studies to this system for both ara-C and methotrexate. Shackney's (119) paper was contemporary with that of Skipper et al. Cells are depicted as passing through a sequence of compartments as they transit cycle. Cells in each compartment can undergo natural death or be partitioned into a pool that ceases to advance; both processes are modeled as assumed power functions of population size. There are two sub-populations of cells, one advancing faster than the other. Estimates of model parameters were attempted by trial-and-error on the basis of experimental data. The author is aware of and discusses some model limitations, and he is appropriately cautious in offering suggested treatment strategies as "exploratory proposals" rather than confident predictions. Model predictions and experimental results agree fairly well. Another contemporary model by Wilson & Gehan (146) should be noted. It includes a capability for modeling arrest in cycle.

Steward & Hahn (125) compare the effects of several cycle-specific drugs on malignant cells and normal stem cells, using experimentally determined age-response functions for each drug—vincristine, vinblastine, and hydroxyurea. These functions are assumed to be the same for both normal and malignant cells. All malignant cells are assumed to be in cycle, growing exponentially. Below 50% of their normal population size, normal stem cells are assumed to grow exponentially, but above this their per-cell growth rate decreases with population size, which tends asymptotically to an upper bound. Stem cells are assumed to be partitioned into cycling and noncycling pools, the latter increasing as total population size increases. Noncycling cells are assumed to be drug resistant. For each drug used, a specified fraction of the initial malignant cells are completely resistant; no cell is resistant to more than one drug. Exploration of the model suggests that, for two-drug therapy alternating vincristine and vinblastine, a very substantial therapeutic advantage may be gained by giving each dose in portions separated by an appropriate number of hours rather than all at once. The authors indicate that Skipper has observed a split-dose advantage in L1210 therapy. Adding hydroxyurea further attacks the resistant cells, and a three-drug, split-dose regimen appears quite promising. The authors are aware that their sharply-timed strategies depend on knowing the parameters of the tumor, which is much more feasible for the leukemias than for inaccessible solid tumors. They realize that this is not the only difficulty in determining model parameters for direct clinical application, but they quite correctly hold that the model is capable of making qualitative predictions about the likely efficacy of various types of protocols.

As described earlier, Rubinow models cells as progressing through cycle continuously in time, with respect to a maturation variable (112); Rubinow & Lebowitz specialize this formulation to a model for the neutrophil system. Based on the latter, they develop models for AML and its treatment by chemotherapy (115, 116). Two populations, normal neutrophils and leukemic cells, exist side by side. The leukemic cells are assumed to differ primarily from normal neutrophils by not exiting the proliferative pool to pass through stages of nonproliferative maturation before entering the blood. Instead, leukemic cells can transfer either way between a resting (G_0) and proliferative pool or they may exit to blood from the former. They die and leave the blood at a specified rate. As for the neutrophil population, parameters that control release of cells to the blood and shifting of cells between G_0 and the proliferative compartment are functions of total population size. But now one must take note of an important assumption that the investigators themselves realize requires further specific validation, even though it responds generally to some observations about interference between normal and leukemic cell populations: for both normal neutrophils and leukemic cells, the population-size control parameters depend on the total granulocyte population, normal plus malignant. In their investigation of the effects of chemotherapy applied to this model, Rubinow & Lebowitz simulate variations on the L-6 clinical research protocol (46) in which doses of ara-C and 6-thioguanine are given sufficiently close in time, so that it is unlikely that a leukemic cell will pass through S without encountering a high blood level of the drugs. The treatment continues until a rest period of several weeks is necessitated by toxicity, after which the treatment may be repeated. The authors conjecture whether this rather successful protocol, developed after much empirical trial and error, might have evolved more directly with the aid of modeling. Their present aim is the study of general strategies rather than individualization of each patient's therapy. Pharmacokinetics are not modeled per se; a fraction of all cells in S-phase is assumed to be killed at the time of drug administration, a simplification that should be reexamined if one shares Lincoln et al's concern for intracellular levels of long-acting ara-CTP (84). Each course of treatment is characterized by the number of times the drug combination is administered, the time between administrations, and the duration of the rest period after such a treatment course. The effects of varying these three parameters are examined, and outcomes are found to be quite sensitive to parameter choice. An optimum value of time between administrations appears to be duration of S-phase for the leukemic cells, which agrees with common sense, given the model's underlying assumptions. Calculations also suggest that longer rest periods should follow more intensive treatment courses. Their model agrees with the clinical finding that in cases where the treatment is favorable, remission is obtained after only one or two courses of treatment. Additional treatments may be detrimental. Also, the model and clinical experience agree that successful treatment of fast-growing leukemia entails larger toxic effects than that of slow-growing leukemia. Finally, the investigators predict improvements in treatment by taking advantage of differences in the durations of S-phase in normal and leukemic populations, a procedure that should be investigated when such determinations can be readily made for individual patients.

A recent paper by Nicolini et al (104) was referred to in the preceding pharmacokinetic section for its inclusion of a DNA metabolism model that depicts the sites of action of various chemotherapeutic agents. This model also is fairly sophisticated with respect to its cell-kinetic modeling, which permits effects such as temporary blocks to be produced. This modeling system is being developed in conjunction with a comprehensive laboratory research protocol that employs modern flow microfluorometry monitoring techniques, along with more traditional labeling methods. Some optimization features are automated, such as seeking combination of drugs, timing, and dosage that maximize synchrony.

TREATMENT OPTIMIZATION

There can be a variety of intents when one speaks of treatment optimization. The aim may be merely to find better general treatment principles, to point to the importance of a good choice of a certain parameter in a given treatment regimen, or to optimize application of a treatment regimen to each patient by monitoring or other informal means. On the other hand, some theoreticians would accept as "optimization" only a more rigorous demonstration that parameters and/or functions in a formal expression that characterizes some kind of strategy have been chosen so as to extremize some other function—the objective function that characterizes outcome of the undertaking guided by that strategy.

Let us refer to the latter undertaking as formal optimization and to the others as nonformal optimization.

Except for optimization of dose distributions in radiation treatment planning, research on formal optimization of cancer treatment strategies is quite recent. Chapter 7 in Swan's book (128) provides a review of some of this work, and some additional material can be found in a brief, earlier review (100).

Nonformal Optimization

Considerable research could be included in this broad category. Some characteristic examples already have been discussed. Among these are Steward & Hahn's (125) demonstrations that, in a basic multiple-drug protocol, appropriately timed split-doses should have a significant therapeutic advantage over administering equivalent single doses and that age-response differences of normal and malignant cells may be exploited by properly timed split-dose radiotherapy. One also recalls Aroesty et al's suggestion to initiate a sequence of doses of cyclenonspecific agents until a plateau is approached where brisker regrowth of the smaller tumor mass begins to counterbalance each kill, to be followed by a course of cycle-specific therapy (2).

Another study (100, 103) seeks to compare the effect of mixed-mode radiation therapy on two cellular systems, one growing at an exponential rate and the other capable of responding to injury by enhanced release of cells into cycle from G_0, as might a normal tissue with low growth-fraction and population-preserving feedbacks. An initial dose of neutrons is followed by a well-timed dose of conventional radiation. The first is assumed to damage cells throughout cycle, and the second is timed so that sparing in S-phase favors an injury-response wave of proliferating cells in the "normal" tissue. Note that attention here is upon timing to spare a chosen normal tissue rather than timing that requires a knowledge of tumor cell-kinetics. Since people are likely to differ far less with respect to the kinetics of normal tissues than to that of tumors of a given type, one should be better able to base fairly good estimates of normal tissue response on past studies, checking perhaps with flow microfluorometry monitoring when feasible. One then maximizes dose to the tumor by treating to tolerance of the normal tissue. An obvious limitation is that this method considers sparing of only one normal tissue; perhaps a second could be included by alternating treatment pairs. The general advantage of radiotherapy over chemotherapy for this type of strategy is that doses can be precisely timed and fairly well localized, so that sparing of a single type of affected tissue may be more feasible.

A good example of identifying parameters that are most important to optimize in an established treatment regimen is Rubinow & Lebowitz's studies on the L-6 protocol for treating AML (46). Times between doses turn out to be very important, and longer rest periods are needed between more vigorous courses of therapy.

Optimizing application of a treatment regimen to each patient by monitoring is well illustrated by Costanzi, May, Barranco et al (8, 34, 90). Bleomycin arrest in G_2 is used to obtain partial synchrony of cells. After release of the block, a sequence of tumor-tissue biopsies is taken and analyzed by flow microfluorometry until a significant rise of the DNA histogram in S-phase signals that it is time to administer a dose of some S-active agent. This does not require biomathematical or biostatistical supports, except to improve the analysis of FMF histograms or to suggest how to get the most information from a few well-timed biopsies.

Formal Optimization

WITH RESPECT TO SPACE   Except for local infusion or the administration of agents for which tumor uptake is especially high, strategies for localization tend not to be pursued for chemotherapy. On the other hand, one of the earliest treatment optimizations known and indeed widely used for many years has been Paterson & Parker's (107) pattern for the placement of implanted sources of radiation. With the advent of computer aids to radiation dosimetry about 20 years ago, it became feasible to investigate improvement of dose distributions for each patient for either implantation or external-beam therapy. With the improved tumor localization and tissue density estimates recently made possible by computerized axial tomography, there is renewed interest in formal spatial optimization methods in radiotherapy.

External-beam therapy   The score-function approach developed by Hope & Orr (62) has been in practical use for some years in the United Kingdom. Starting with a satisfactory arrangement of external beams, the dosimetrist instructs the computer to vary beam parameters stepwise over specified ranges and to calculate for each step a score function representing overall quality of the dose distribution. Subsequent refinements (137) automate decisions concerning choices of beams and beam modifiers such as wedges. An early linear programming approach by Bahr et al (4, 5) held doses to the tumor within acceptable bounds while minimizing doses to critical tissues. Linear programming is flexible but can be expensive when optimizations are complex. A simple system for optimizing beam weights runs on a minicomputer (82). A number of investigators reported current research on optimization at the Sixth

International Conference on the Use of Computers in Radiation Therapy, Göttingen, September 1977. A program developed by F. Gauwerky (private communication) and his associates at the University of Hamburg excels in the variety of treatment techniques it can muster, e.g. multicentered rotational therapy to optimize dose distributions to irregularly shaped tumors. Although some investigators have taken cellular responses into account in optimizing dose distributions (42, 49), much challenging work remains to be done on space-time optimization that exploits differences between the kinetics of tumor and various tissues.

Implantation therapy Implantation therapy has the natural advantage of introducing radiation sources directly into the tumor area. Thus, homogeneity of dose throughout the tumor area is a primary consideration in optimization. Paterson & Parker (107) provide an optimum pattern of radium needle placement that is applicable to a large number of cases. Another approach (100) is to accept the fact that tumors may be located where such optimum patterns may not be realizable or that, given tissue's natural elasticity, implants may not always go exactly where one wants them to. In this case, improved homogeneity of dose distribution throughout the tumor area can be achieved by withdrawing some sources earlier than others.

WITH RESPECT TO DOSE TIMING Research on formal methods for optimizing the timing doses for radiotherapy or chemotherapy is rather recent, though considerable attention has been directed to less formal studies. As yet, sophisticated optimization methods have not been applied to sophisticated models of biological systems. There is a good reason for this. The more realistic models of cellular systems, even without an interacting pharmacokinetic model, usually are nonlinear and require complex computer simulations. These models may be amenable to description by mathematical equations, but the solutions of those equations usually are not obtainable in closed form. This is particularly true for perturbed systems where size-preserving feedbacks are operating. Optimization methods applied to nonlinear systems where a number of parameters are to be optimized simultaneously tend to increase strongly in cost as the number of those parameters increases, even when operating on closed-form solutions. The cost and computer time can be prohibitive if, instead, a complex simulation is required for each iteration in the optimization process, when closed-form solutions are not available. Although there is an extensive literature on optimization and optimal control theory, much remains to be done to achieve low cost and efficiency for the types of systems we must consider. As our knowledge of the biology underlying cancer treatment grows and our ability to monitor patients being treated develops, the medical feasibility of optimizing an individual patient's course of treatment increases. The time may not be long before we shall have a compelling ability and motivation on the biomedical side to proceed with the development of clinically effective treatment-optimization programs. This in turn should motivate some very fundamental biomathematical research on optimization methods applied to complex, nonlinear systems. It will not be the first time that research in an applied area has motivated important mathematical advancements. The history of physics is rich with such examples.

In reading the present and future literature, it may help to have in mind a basic theoretical framework and definitions of some commonly used terms. The system to which treatment is to be directed is characterized by some mathematical or computational model. Where this is a system of equations, they are referred to as state equations. System variables include such things as time, numbers of cells at each location in cycle, total number of cells, number of cells dying at a given time, etc. Parameters include constants such as cell sensitivity to a given treatment agent as a function of location in cycle, maturation rates, numerical factors in size-preserving feedback functions, etc. The status of the system is assumed to be described at any given time by the values of a set of its variables, e.g. the numbers of cells in all relevant subdivisions of the system. Subdivisions can be partitioned on the basis of a cell's location in cycle, state of damage, state of labeling, etc. This set of variables is called the state vector of the system. Another function or computer simulation portrays external influences to be imposed on the system. We seek an optimum choice of this model's form and its parameters. In optimal control theory, the elements of this model to be chosen so as to optimize a certain outcome are called the control variables. Examples would be the time between each of a sequence of doses, size of each dose, total number of doses, etc. Finally, the outcome one wishes to optimize must be characterized. Optimization usually entails extremization (i.e. maximization or minimization) of a mathematical expression which characterizes that aspect of the total outcome one seeks to perfect. In the case of Hope & Orr's dose-distribution optimization, this was called the score function. In linear programming, the outcome function is a weighted linear sum of outcome variables and is called the objective function. In optimal control theory the outcome function to be extremized may be called the performance function or cost function and its elements, performance criteria or cost elements. These designations may be used interchangeably for different approaches.

Hethcote & Waltman (58) adopt a dynamic programming approach (18) for which performance criteria are elimination of the tumor and minimization of damage to normal tissues. Like the work described in the remainder of this section, this approach does not consider cycle-specific effects. For purposes of tractability, they use a simplified version of Fischer's (43, 44) model for treating cell oxygenation and the conventional one-hit multitarget model for cellular response to radiation. A logistic model depicts regrowth. More explicitly, the surviving fraction of oxygenated cells in the tumor after dose D of radiation is

$$S = 1 - [1 - \exp(-D/DO)]^{NO}, \qquad 24.$$

and that of the tumor's anoxic cells is

$$S = 1 - [1 - \exp(-D/DA)]^{NA}, \qquad 25.$$

where DO and DA are the characteristic doses for oxygenated and anoxic cells, respectively, and NO and NA are the extrapolation numbers that relate to cell repair. Partitioning of a total population of N tumor cells into anoxic and oxygenated cells is by way of Fischer's formula, where the oxygenated fraction of cells is given by $\exp(-\beta N)$. The correct functional form for this partitioning is not known and very likely would depend on tumor shape and vasculature. In the absence of better knowledge, any simple form that shifts the oxygenated fraction higher as tumor size regresses is reasonable. After the tumor has been damaged by radiation, only the surviving oxygenated cells regrow, and their growth is exponential at a rate α, giving the following expression for regrowth of the tumor as a whole:

$$dN/dt = \alpha N \exp(-\beta N). \qquad 26.$$

This corresponds to an assumption that those previously anoxic cells that are going to become oxygenated as the result of a treatment do so as soon as the treatment has been given rather than, more gradually, as dead cells are cleared away between them and better vasculature. Limitations of this assumption of sudden reoxygenation will be greater the more closely successive doses are spaced in time. Radiation-induced mitotic delay is assumed to be linear in the dose given, a rather well-established finding in laboratory research. This is applied to the model as a modification of t, the recovery time after a dose is given. An expression then is derived combining all of the above to give the total number of live tumor cells after irradiation by dose D followed by recovery time t. Surrounding normal tissues are assumed also to have a single-hit multitarget response to radiation, with constants DB and NB for characteristic dose and extrapolation number. All are assumed to be well oxygenated. Regrowth of surviving cells, $S(t)$, is given by a logistic function satisfying a Verhulst-type equation,

$$dS(t)/dt = rS(t)[H - S(t)], \qquad 27.$$

where H is the asymptotic limit of regrowth (e.g. the normal number of cells in that tissue area).

The optimization program seeks the best treatment strategy for a sequence of up to M doses. Each dose may be chosen from a finite set of possible dose sizes, each of which is followed by a recovery time that may be chosen from a finite set of possible recovery times. The optimization endpoint is reached when the tumor-cell population falls below one and the number of surviving nearby normal cells is maximized. A dynamic programming approach is applied to search for this maximum, given the restraint on final tumor size. In this example, then, the state vector includes the numbers of nearby normal cells, oxygenated tumor cells, and anoxic tumor cells. The control variables are number of doses, size of each dose, and recovery time after each dose. The cost function or objective function is the number of nearby normal cells surviving after the tumor has been eliminated. The authors find that best results are obtained when doses are increased as the treatment progresses, i.e. that larger doses with longer recovery times should be given after the tumor has become well oxygenated. The authors suggest that this strategy be checked experimentally.

Wheldon & Kirk (143) use the CRE system developed by Kirk and his associates (72-76) to characterize damage to connective tissue surrounding the tumor. The usual formula for single-hit multitarget cell survival after a single dose is assumed for the tumor as a whole, without partitioning into oxygenated and anoxic cells. Each treatment regimen considered consists of a sequence of K evenly-spaced doses of equal size. (Note that Hethcote & Waltman's conclusion would not be discovered under these constraints.) Regrowth of the tumor between doses is assumed to be exponential. Given parameters characterizing the connective tissue's response and tolerance level of damage (of which we have some approximate empirical knowledge), the schedule that minimizes the surviving fraction of tumor cells is sought. Given the foregoing simplifying assumptions, rather tractable expressions for determining optimal doses (d^*) and inter-dose time (t^*) result, though machine computation using a standard minimization technique (Newton-Raphson) is required to demonstrate the results. An approximately linear relationship is found to hold between d^* and the characteristic dose for the tumor cells (given a wide range of choice of values for the extrapolation number), as does a linear relationship between t^* and the doubling time for proliferation of the tumor cells. These authors, as those of the preceding paper, are fully aware of the limitations of their model with respect to adequate human data and portrayal of all relevant mechanisms. Tables of optimal values of t^* and d^* are given for various values of other model parameters, and the predicted outcomes of conventional treatment schedules are compared with those for optimal schedules in a number of cases. This comparison suggests that substantial gains should be realized by adhering to optimal schedules; the next step of laboratory validation seems to be warranted. The advantage appears to be greatest for high tumor doubling time and high extrapolation number for the single-dose survival function, i.e. for slow-growing tumors that repair readily. Unfortunately, the model's assumptions can be questioned most for tumors with long doubling times, in that the latter may relate to low growth fractions that might be expected to change during the course of treatment.

Almquist & Banks (1) note the disadvantages of dynamic programming for studying realistically complex systems. The method they use is known in optimal control theory as an optimal-control method with multistage system dynamics. This type of method has greater flexibility with respect to permissible cost functions, and its computing time and costs are less sensitive to model complexity. The method and appropriate references are well presented in the paper. As for Hethcote & Waltman, doses and time intervals between a sequence of doses need not be equal. The full Fischer model is used, where tumor cells are divided into four classes on the basis of being alive or dead, oxygenated or anoxic. The usual single-hit multitarget expression for fraction of cells surviving a single treatment is used, with different parameters for both anoxic and oxygenated alive tumor cells and for cells in the surrounding tissue. Dose-related mitotic delay and partitioning of tumor cells between oxygenated and anoxic fractions are handled in a manner similar to Hethcote & Waltman's, and only oxygenated cells in the tumor are assumed to proliferate. Dead cells are assumed to disintegrate at an exponential rate similar to that for growth of alive cells, an assumption that can be questioned. Repopulation of the surrounding connective tissue is ignored, an assumption that the authors realize to be questionable. Their focus appears to be largely on demonstrating an improved mathematical method, rather than on improving upon the Hethcote-Waltman formulation. They compare some variable-dose and equal-dose optimal treatments and come to the conclusion that one obtains better results for a given dosage for variable-dose methods. They corroborate Hethcote & Waltman's observations that larger doses should be given as treatment progresses, noting that in some cases there might be an initial dip after the first treatment. This agreement, considering the difference in how both papers treat recovery of surrounding tissue, offers the hope that the general strategy of increasing doses may become clinically useful before our knowledge of complex repair phenomena is perfect. An additional finding is that there may be a best choice of overall treatment time for a given number of treatments.

Bahrami & Kim (6) distinguish phases of cycle in their optimal-control study of chemotherapy. Zietz's dissertation (147) covers both cellular systems described by Gompertz functions and a more realistic model of cells in cycle, for which the effects of an S-specific agent are studied. His performance function seeks to minimize damage to normal tissues while maximizing damage to the tumor. The pharmacokinetic modeling is necessarily simple in both cases.

IMPORTANCE OF MODEL STRUCTURE AND PRESENTATION

The Need to Design Models for Use by Biologists

Let us remind ourselves once more of what modeling seeks to accomplish. As our biological knowledge advances, we begin to develop hypotheses about how unseen biological mechanisms work, and we modify those hypotheses as we learn more. Both the development of efficacious research strategies and the interpretation of new research findings seek reference to those postulated mechanisms. Sometimes we can negotiate such references in our heads, but when biological mechanisms are complex or highly interactive, or when more quantitative assessments are desired, we may need the help of a model that can be manipulated logically or numerically. Such models may also facilitate the transfer of new knowledge to medicine and other applied fields. Modeling should be a working aid to biomedical scientists, not a portrayal that stands apart as an end in itself.

If one believes this, it is mandatory to place modeling tools within the hands of biomedical investigators themselves. It is they who will know best if the formal representation proposed for a given biological system is behaving reasonably under a number of challenges representing known laboratory stresses applied to the system and who will suggest good challenges to test it. It is the well-trained biologist who is most likely to detect when a modeling outcome that reaches into an area, not yet examined experimentally, is particularly interesting and somewhat unexpected and who is able to propose explanations, relate this finding to others, or propose good follow-up experiments. A special effort must be made to provide biomedical investigators with graphical confirmation of the form of a formula that characterizes an assumed biological mechanism, to enable them to modify and explore the model without having to write down differential equations or other mathematical constructs with which they are not familar, and to describe the modeling system to them in a way they can understand and effectively criticize.

The Need to Design Flexible Models

Models must be able to change as our knowledge advances and must be presented to the investigator in a way that provides the greatest possible freedom for him to manipulate and revise them, to propose new treatment strategies, and to use them as a springboard to posing the next set of questions that should be pursued in laboratory and theoretical work. Ideally, the modeling system should be a neutral framework into which investigators may insert the assumed parameters and forms of the mechanisms they are studying, changing them at will. In practice, this ideal is not met; the demands of complete generality compete with demands of economy and ease of use. A realistic goal is to develop a modeling framework that imposes some well-understood limitations while permitting the user considerable freedom to manipulate both the parameters and forms of assumed biological mechanisms within those limits. If the basic modeling framework is built in a modular manner, such that extensive changes can be made in one part without necessitating more than minor revisions in the others, it is not difficult to remove some of the structural limitations as knowledge advances.

The Need for Comprehensive Models

The position has been taken that models are working tools to aid the design and interpretation of experiments, as well as investigations of how what is learned in the laboratory might be applied to medical care. One therefore is committed to designing systems that embed basic cell-kinetic models in a larger system that portrays the context in which the biomedical investigator will be studying or using these models. This requires providing flexible capabilities for imposing treatment agents, labels, etc. on preplanned or ad hoc schedules. Systems involved in experimental observations also must be portrayed, e.g. observations on percent-labeled mitoses or DNA histograms, as the simulation proceeds. This has been a central concern in the modeling investigation that is to be briefly described in the next section. Special techniques have been developed, such as a "clock processor" that coordinates the time-dependent exposure of cells to different treatment agents that have different dose-time functions and may be scheduled independently of one another. Much more remains to be done.

An Example of Research on a Comprehensive, Flexible Treatment Model for Use by Biologists

DESCRIPTION OF THE SYSTEM An interactive graphics approach was adopted because of its suitability for exploring complex models (101). First, it can simplify and reduce errors in the otherwise time-consuming task of specifying the model to be studied. Second, effective graphical and numerical summaries of the system's status can be presented while the model is running, giving the user a good feeling for what is happening and enabling him to stop or otherwise intervene at any time. Finally, when models are very complex, it may be important to provide the user with a variety of methods for viewing the model's outcomes. For instance, a relatively simple display can monitor the most important summary variables as the model is running, enabling the user to pause at any time for a deeper look at a larger set of variables. An interactive graphics system that permits programs to be written in FORTRAN and run on an IBM 370 series computer operating under TSO was used (117). Communication can be by way of ordinary dial-up telephone lines.

The modeling techniques employed (101, 102) evolved from earlier research on hematopoiesis (97–99). While certain aspects of the model are fixed, every effort is made to provide the user flexible capabilities for specifying both the form and parameters of other model components. Although some freedom to specify mathematical functions is provided, the guiding assumption in designing this system was that it should be accessible to biomedical investigators who are not experienced in specifying formulas or using computers. Thus, the emphasis is on permitting users to select from a list of commonly used functions. Program execution also is cheaper for this approach. Finally, the cell system is embedded in a more comprehensive model that handles imposition of treatments and labels and presents information (e.g. fraction of labeled mitoses) for monitoring.

Basic model features are as follows: two cellular systems are compared. One is growing exponentially and the other begins as a system of stable size. The first is referred to as "malignant" and the second as "normal." Such basic parameters as expected time in cycle, growth fraction, and normal cellular death are typed in for each system by the user as "fill-in"s in a text that is displayed to describe the basic system. The exponentially growing system is assumed not to be subject to size-preserving feedbacks, but recruitment of cells from G_0 into cycle (e.g. changing growth-fraction) regulates the size of the other system, and the user can specify parameters characterizing the strength of this feedback in a simple formula provided for that purpose. When cycle-specific parameters are to be specified, steps in cycle are displayed for both cellular systems, and the user may type in the appropriate letter or number for each step or merely revise some entries if the system has already been specified. The user may choose whether the effects of two treatment agents are to be additive or multiplicative at each point in cycle, or he may type in another formula for potentiation. Cell damage may be partitioned into instant death, death at mitosis, or damage that has a probability of repairing before mitosis. A gross dose-response function can be selected from among listed alternatives or typed in by the user. The user may request graphs of this function over specified dose ranges to see if its form is reasonable and if it is being properly computed over the desired dose range. For each treatment or label, the user indicates whether it is to be given according to a predetermined schedule or whenever the user pleases. For each administered dose or label, the user specifies whether exposure to it is to be instantaneous, constant over a specified period of time, or to fall off exponentially in time. For radiation, this modeling of exposure enables the user to simulate external-beam therapy and therapy with long-acting or short-acting sources. For chemotherapy, it enables only a first-order pharmacokinetic model with relatively rapid uptake to be considered. The user then selects which variables are to be displayed as the simulation proceeds. These include the number of cells in various subcategories with respect to malignancy, presence of a label, state of damage, and location in cycle. Examples are all cells, number of malignant cells repairing in S phase, etc. Ratios of these can be specified to yield derived variables such as the fraction of tumor cells in mitosis that are labeled.

After the task of initial model specification has been completed, the user is provided a menu, a list of options from which he may choose to guide what the computer does next. Options include termination, initiation of a simulation run, or any of the somewhat large number of possible model revisions. If a revision is indicated, the user is transferred directly to the display concerned with soliciting that revision and thereafter selectively to displays that permit adjustment of any other model features that might be affected by that revision. Every effort is made to minimize the user's burden of model specification; he need not go through the entire process again.

WHAT WAS LEARNED   On the whole, this interactive graphics approach appears to be satisfactory for investigators who are inexperienced in using computers. However, they must be serious investigators who want to exert control over the kind of model to be specified and who are willing to invest the initial model-specification effort that is required (about 15–30 min). Some might wish to settle for less freedom and less effort.

The most serious model limitations are felt to be the following: a Gompertz-type growth curve should be available for the "malignant" cells. The treatment of G_0 and G_1 should be more sophisticated. Much has been learned since this model was designed and even though much more remains to be learned, such basic features as delays after triggering from G_0 and variable partitioning of postmitotic cells between transfer to G_0 and G_1 should be better modeled. It may very well turn out that a major future source of therapeutic advantage will reside in exploitation of the difference between size-preserving mechanisms of normal and malignant populations. Tumor cells have not been partitioned into oxygenated and anoxic cells. A more satisfactory method for describing doses for combined radiation therapy and chemotherapy is required. At present, the chemotherapeutic dose is expressed as a rad equivalent, with respect to its overall dose-response curve, but can be weighted differently around cycle. When one gets down to modeling at this level a deeper consideration of the biological processes themselves is forced, a matter of value in itself. For instance, are all cells retarded equally in the dose-dependent G_2 delay for radiotherapy, or does it make a difference whether a cell is hit early or late in cycle or how many sublethal hits it receives? Another limitation of the model is that only the presence or absence of a label is recorded for cells.

Because of slow disappearance curves, impact on a wide range of normal cellular systems in the body, and absence of adequate monitoring procedures, explorations of more precisely timed chemotherapeutic strategies did not prove to be particularly promising, except for possible use of chemotherapeutic agents to produce partial synchrony by way of blocks, to be followed by well-timed radiotherapy. One finding with two radiotherapeutic modalities did seem to be promising—neutron irradiation followed by a well-timed treatment with gamma rays to spare normal tissue whose recovering wave of cells is well into S-phase (100, 103). Experimentation with guidance of flow microfluorometer assays should be the next step.

CONCLUSION

The time is right for intensifying cancer-related biomathematical research. Advancing biological knowledge enables both more realistic models and research on more ambitious strategies for treatment, which modeling can help. There is considerable past modeling research to build upon, but much remains to be done. There is the need to assemble chemotherapy models that are sophisticated in both their cell-kinetic and pharmacokinetic components. Space-time optimization for radiotherapy should be referred to more detailed cellular models and associated directly with the anatomical data computerized tomography provides. More efficient nonlinear optimization techniques must be developed so that more realistic models can be addressed cost-effectively in treatment-optimization research and its application to individual patients. Research also is needed on structuring models for easy modification and flexible exploration and for effective use by biomedical investigators.

Literature Cited

1. Almquist, K. J., Banks, H. T. 1976. *Math. Biosci.* 29:159–79
2. Aroesty, J., Lincoln, T. L., Shapiro, N., Boccia, G. 1973. *Math. Biosci.* 17:243–300
3. Attix, F. H., Roesch, W. C., Tochilin, E., eds. 1969. *Radiation Dosimetry*, Vols. I, II, III (see espec. Vol. III). New York/San Francisco/London: Academic
4. Bahr, G. K., Kereiakes, J. G., Horowitz, H., Finney, R., Galvin, J., Goode, K. 1968. *Radiology* 91:686–93
5. Bahr, G. K., Kereiakes, J. G., Horowitz, H., Compaan, P., Holt, J. G., Goode, K. 1970. *Phys. Med. Biol.* 15:210
6. Bahrami, I., Kim, M. 1975. *IEEE Trans. Autom. Control* AC20: 537–42
7. Bailey, N. T. J. 1967. *The Mathematical Approach to Biology and Medicine.* London/New York: Wiley. 296 pp.
8. Barranco, S. C., Luce, J. K., Romsdahl, M. M., Humphrey, R. M. 1973. *Cancer Res.* 33:882–87
9. Barrett, J. C. 1966. *J. Natl. Cancer Inst.* 37:443–50
10. Barrett, J. C. 1970. *Cell Tissue Kinet.* 3:349–53
11. Bell, G. I. 1970. *J. Theor. Biol.* 29:191–232
12. Bell, G. I. 1970. *Nature* 228:739–44
13. Bell, G. I. 1971. *J. Theor. Biol.* 33:339–78
14. Bell, G. I. 1971. *J. Theor. Biol.* 33:379–98
15. Bell, G. I., Perelson, A. S. 1978. In *Theoretical Immunology*, ed. G. I. Bell, A. S. Perelson, G. H. Pimbley. New York: Dekker. 646 pp.
16. Bellman, R., Jacquez, J. A., Kalaba, R. 1960. *Bull. Math. Biophys.* 22:181–98
17. Bellman, R., Jacquez, J. A., Kalaba, R. 1961. *Proc. 4th Berkeley Symp. Math. Stat. Probab.* 4:57–66. Berkeley: Univ. Calif. Press
18. Bellman, R., Dreyfus, S. 1962. *Applied Dynamic Programming.* Princeton, NJ: Princeton Univ. Press
19. Berenbaum, M. C. 1969. *Br. J. Cancer* 23:434–45
20. Berman, M., Weiss, M. F. 1977. *User's Manual for SAAM.* Bethesda, Md: Lab. Theor. Biol., Natl. Cancer Inst., Natl. Inst. Health
21. Bischoff, K. B., Dedrick, R. L. 1970. *J. Theor. Biol.* 29:63–83
22. Bischoff, K. B., Himmelstein, K. J., Dedrick, R. L., Zaharko, D. S. 1973. In *Chemical Engineering in Medicine and Biology, Adv. Chem. Ser. No. 118*, pp. 47–64
23. Blumenson, L. E., Bross, I. D. J. 1969. *Biometrics* 25:95–109
24. Blumenson, L. E., Bross, I. D. J. 1973. *J. Theor. Biol.* 38:397–411
25. Bronk, B. V., Dienes, G. J., Paskin, A. 1968. *Biophys. J.* 8:1353–98
26. Bronk, B. V. 1969. *J. Theor. Biol.* 22:468–92
27. Bruni, C., Giovenco, M. A., Koch, G., Strom, R. 1975. *Math. Biosci.* 27:191–211
28. Bruni, C., Giovenco, M. A., Koch, G., Strom R. 1975. In *Proc. 2nd US-Italy Semin. Var. Struct. Syst.*, ed. A. Ruberti, R. R. Mohler, pp. 244–64. New York: Springer
29. Bruni, C., Germani, A., Koch, G.,

30. Strom, R. 1976. *J. Theor. Biol.* 61:143–70
31. Chuang, S. N., Lloyd, H. H. 1975. *Bull. Math. Biol.* 37:147–60
32. Cohen, L. 1968. *Br. J. Radiol.* 41:522–28
33. Cohen, L., Scott, M. J. 1968. *Br. J. Radiol.* 41:529–33
34. Cohen, L. 1976. See Ref. 124, pp. 243–53
35. Costanzi, J. J. 1977. In *Growth Kinetics and Biochemical Regulation of Normal and Malignant Cells*, ed. B. Drewinko, R. M. Humphrey, pp. 879–84. Baltimore: William & Wilkins. 900 pp.
36. Crissman, H. A., Mullaney, P. F., Steinkamp, J. A. 1975. In *Methods in Cell Biology*, ed. D. M. Prescott, pp. 179–246. New York/San Francisco/London: Academic
37. Delisi, C. 1976. In *Environmental Health: Quantitative Methods*, ed. A. Whittemore, pp. 149–71. Philadelphia: SIAM Publ.
38. Dutreix, J., Wambersie, A., Bounik, C. 1973. *Eur. J. Cancer* 9:159–67
39. Ellis, F. 1967. In *Modern Trends in Radiotherapy*, ed. T. J. Deeley, C. A. P. Wood, 1:34–51. London: Butterworth
40. Ellis, F. 1968. *Curr. Top. Radiat. Res.* 4:359–97
41. Evert, C. F. 1975. *Simulation* 24:55–61
42. Evert, C. F., Palusinski, O. A. 1975. *Acta Haematol. Pol.* 6:175–84
43. Fischer, J. J. 1969. *Br. J. Radiol.* 42:925–30
44. Fischer, J. J. 1971. *Acta Radiol. Ther. Phys. Biol.* 10:73
45. Fischer, J. J. 1971. *Acta Radiol. Ther. Phys. Biol.* 10:267–78
46. Garfinkel, D., Marbach, C. B., Shapiro, N. Z. 1977. *Ann. Rev. Biophys. Bioeng.* 6:525–42
47. Gee, T. S., Yu, K.-P., Clarkson, B. D. 1969. *Cancer* 23:1019–32
48. Goel, N. S., Maitra, S. C., Montroll, E. W. 1971. *Rev. Mod. Phys.* 43(2):231–76
49. Gompertz, B. 1825. *Philos. Trans. R. Soc. London* 115:513–85
50. Graffman, S., Groth, T., Jung, B., Skollermo, B., Snell, J.-E. 1975. *Acta Radiol. Ther. Phys. Biol.* 14:54–62
51. Gray, J. W. 1974, *J. Histochem. Cytochem.* 22:642–50
52. Gray, J. W., Dean, T. N., Mendelsohn, M. L. 1979. See Ref. 93, pp. 383–407
53. Gray, J. W., Dean, P. N. 1979. *Ann. Rev. Biophys. Bioeng.* 9: In press
54. Hahn, G. M. 1966. *Biophys. J.* 6:275–90
55. Hahn, G. M. 1970. *Math. Biosci.* 6:295–304
56. Hall, E. J. 1969. *Br. J. Radiol.* 42:427–31
57. Harris, T. E. 1959. In *The Kinetics of Cell Proliferation*, ed. F. Stohlman, pp. 368–81. New York: Grune & Stratton
58. Hartmann, N. R., Gilbert, C. W., Jansson, B., Macdonald, P. D., Steel, G. G., Valleron, A.-J. 1975. *Cell Tissue Kinet.* 8:119–24
59. Hethcote, H. W., Waltman, P. 1973. *Radiat. Res.* 56:150–61
60. Hethcote, H. W. 1976. See Ref. 36, pp. 172–82
61. Hethcote, H. W., McLarty, J. W., Thames, H. D. Jr. 1976. *Radiat. Res.* 67:387–407
62. Himmelstein, K. J., Bishcoff, K. B. 1973. *J. Pharmacokinet. Biopharm.* 1:51–68
63. Hope, C. S., Orr, J. S. 1965. *Phys. Med. Biol.* 10:365–73
64. Howes, A. E., Suit, H. D. 1974. *Radiat. Res.* 57:342–48
65. Jacquez, J. A., Bellman, R., Kalaba, R. 1960. *Bull. Math. Biophys.* 22:309–22
66. Jacquez, John A. 1972. *Compartmental Analysis in Biology and Medicine*. Amsterdam/London/New York: Elsevier. 237 pp.
67. Jansson, B., Revesz, L. 1976. *Methods Cancer Res.* 13:228–90
68. Jennrich, R. I., Bright, P. B. 1976. *Technometrics* 18:385–92
69. Jennrich, R. I., Ralston, M. L. 1979. *Ann. Rev. Biophys. Bioeng.* 8:195–238
70. Kendall, D. G. 1948. *Biometrika* 35:316–30
71. King, T. J. 1974. *Developmental Aspects of Carcinogenesis and Immunity*. New York: Academic. 218 pp.
72. King-Smith, E. A., Morley, A. 1970. *Blood* 36:254
73. Kirk, J., Gray, W. M., Watson, E. R. 1971. *Clin. Radiol.* 22:145–55
74. Kirk, J., Gray, W. M., Watson, E. R. 1972. *Clin. Radiol.* 23:93–105
75. Kirk, J., Gray, W. M., Watson, E. R. 1973. *Clin. Radiol.* 24:1–11
76. Kirk, J., Gray, W. M., Watson, E. R. 1975. *Clin. Radiol.* 26:77–88
77. Kirk, J., Gray, W. M., Watson, E. R. 1975. *Clin. Radiol.* 26:159–76
78. Knott, G. 1975. *MLAB: An Online Modeling Laboratory*. Bethesda, Md: Comput. Res. Technol., Natl. Inst. Health

78. Kretchmar, A. L. 1966. *Science* 152:367–70
79. Laird, A. K. 1969. In *Human Tumor Cell Kinetics, Monogr. 30*, pp. 15–28. Washington DC: Natl. Cancer Inst.
80. Lajtha, L. G., Oliver, R., Gurney, C. W. 1962. *Br. J. Haematol.* 8:442–60
81. Landaw, E. M. 1979. *Optimal experimental design for biologic compartmental systems with applications to pharmacokinetics*. PhD thesis. Dep. Biomath. Univ. Calif., Los Angeles
82. Leavitt, D. D. Campbell, D. W., Stryker, J. A., Sherwood, G. 1975. *Radiology* 116:159–63
83. Lincoln, T. L., Aroesty, J., Meier, G., Gross, J. F. 1974. *Biomedicine* 20:9–16
84. Lincoln, T. L., Morrison, P. F., Aroesty, J., Carter, G. M. 1976. *Technical Report R-2001-HEW*. Santa Monica, Calif: RAND Corp. 46 pp.
85. Liversage, W. E. 1969. *Br. J. Radiol.* 42:153–54
86. Liversage, W. E. 1969. *Br. J. Radiol.* 42:432–40
87. Lloyd, H. H. 1977. In *Growth Kinetics and Biochemical Regulation of Normal and Malignant Cells*, ed. C. B. Drewinko, R. M. Hunter, pp. 455–469. Baltimore: William & Wilkins
88. Mauer, A. M. Saunders, E. F., Lampkin, B. C. 1969. See Ref. 79, pp. 63–78
89. Mauer, A. M., Evert, C. F., Lampkin, B. C., McWilliams, N. B. 1973. *Blood* 41:141–54
90. May, J. T. 1977. In *Proc. Southwest Sect. Am. Assoc. Cancer Res.* (Abstr.)
91. Mendelsohn, M. L. 1960. *Science* 132:1496
92. Mendelsohn, M. L. 1965. In *Cellular Radiation Biology*, pp. 498–513. Baltimore: William & Wilkins
93. Mendelsohn, M. L., Melamed, M. R., Mullaney, P., eds. 1979. *Flow Cytometry and Cell Sorting*. New York: Wiley. 716 pp.
94. Merkle, T. C., Stuart, R. N., Gofman, J. W. 1965. *Lawrence Radiat. Lab. Rep. No. UCRL-14505*, Livermore, Calif.
95. Milstein, J. 1975. *Estimation of the dynamical parameters of the Calvin photosynthesis cycle, optimization and ill conditioned inverse problem*. PhD thesis. Dep. Math., Univ. Calif., Berkeley. 241 pp.
96. Morrison, P. F., Lincoln, T. L., Aroesty, J. 1975. *Cancer Chemother. Rep.* 59:861–76
97. Newton, C. M. 1965. *Bull. Math. Biophys.* 27:275–90 (Spec. Issue)
98. Newton, C. M. 1966. *Ann. NY Acad. Sci.* 128:781–89
99. Newton, C. M. 1969. *Appl. Bio-med. Calcolo Elettron.* 2:73–84
100. Newton, C. M. 1971. *Br. J. Radiol.* Spec. Rep. No. 5, pp. 83–89
101. Newton, C. M. 1972. *Proc. San Diego Biomed. Symp.* pp. 189–99
102. Newton, C. M., Ryden, K. H. 1972. *Proc. 1st USA-Japan Comput. Conf., Tokyo, October 3–5*, pp. 277–86
103. Newton, C. M. 1977. *Symp. Proc. Inst. Lab. Anim. Resour., Div. Biol. Sci., Assem. Life Sci., Natl. Acad. Sci., Washington DC*, pp. 152–64
104. Nicolini, C., Milgram, E., Kendall, F., Giaretti, W. 1977. See Ref. 34, pp. 411–33
105. Orton, C. G. 1972. *Am. Assoc. Phys. Med. Q. Bull.* 6:13–175
106. Orton, C. G., Ellis, F. 1973. *Br. J. Radiol.* 46:529–37
107. Paterson, R., Parker, H. M. 1938. *Br. J. Radiol.* 11:252, 313
108. Perelson, A. S., Mirmirani, M., Oster, G. F. 1976. *J. Math. Biol.* 3:325–67
109. Powell, E. O. 1955. *Biometrika* 42:16–44
110. Quastler, H. 1963. In *Cell Proliferation*, ed. L. G. Lamerton, R. J. M. Fry. pp. 18–36. Oxford: Blackwell Scientific. 241 pp.
111. Rubinow, S. I. 1968. *Biophys. J.* 8:1055–72
112. Rubinow, S. I. 1969. *J. Cell Biol.* 43:32–39
113. Rubinow, S. I., Lebowitz, J. L., Sapse, A. 1971. *Biophys. J.* 11:175–88
114. Rubinow, S. I., Lebowitz, J. L. 1975. *J. Math. Biol.* 1:187–225
115. Rubinow, S. I., Lebowitz, J. L. 1976. *Biophys. J.* 16:897–910
116. Rubinow, S. I., Lebowitz, J. L. 1976. *Biophys. J.* 16:1257–71
117. Ryden, K. H., Newton, C. M. 1972. *Proc. Spring Jt. Comput. Conf.* 40:1145–56
118. Schotz, W. E., Zelen, M. 1971. *J. Theor. Biol.* 32:383–404
119. Shackney, S. E. 1970. *Cancer Chemother. Rep.* 54:399–429
120. Simpson-Herren, L., Lloyd, H. H. 1970. *Cancer Chemother. Rep. Pt. 1.* 54(3):143–74
121. Skipper, H. E., Schabel, F. M., Mellett, B. L., Montgomery, J. A. 1970. *Cancer Chemother. Rep.* 54:431–50
122. Steel, G. G. 1967. *Eur. J. Cancer*

123. Steel, G. G., Hanes, S. 1971. *Cell Tissue Kinet.* 4:93–105
124. Sternick, E. S., ed. 1976. *Computer Applications in Radiation Oncology; Proc. 5th Int. Conf. Use Comput. Radiat. Ther.* Hanover, NH: Univ. Press New Engl. 559 pp.
125. Steward, P. G., Hahn, G. M. 1971. *Cell Tissue Kinet.* 4:279–91
126. Stuart, R. N., Merkle, T. C. 1965. *Lawrence Radiat. Lab. Rep. No. UCRL-14505, Pt. II,* Livermore, Calif.
127. Sullivan, P. W., Salmon, S. E. 1972. *J. Clin. Invest.* 51:1697–1708
128. Swan, G. W. 1977. *Some Current Mathematical Topics in Cancer Research.* Ann Arbor, Mich: Publ. for Soc. Math. Biol. by Univ. Microfilms Int.
129. Takahashi, M. 1966. *J. Theor. Biol.* 13:202–11
130. Takahashi, M. 1968. *J. Theor. Biol.* 18:195–209
131. Takahashi, M., Hogg, J. D., Mendelsohn, M. L. 1971. *Cell Tissue Kinet.* 4:505–18
132. Thames, H. D. Jr., White, R. A. 1977. *J. Theor. Biol.* 67:733–56
133. Trucco, E. 1965. *Bull. Math. Biophys.* 27:285–304
134. Trucco, E. 1965. *Bull. Math. Biophys.* 27:449–71
135. Trucco, E., Brockwell, P. J. 1968. *J. Theor. Biol.* 20:321–37
136. Valeriote, F. A., Bruce, W. R., Meeker, B. E. 1966. *Biophys. J.* 6:145–52
137. Van der Laarse, R., Strackee, J. 1976. *Br. J. Radiol.* 49:450–57
138. Verhulst, P. F. 1845. *Mem. Acad. R. Med. Belg.* 18:1
139. Vogel, H., Niewisch, H., Matioli, G. 1969. *J. Theor. Biol.* 22:249–70
140. Von Foerster, J. V. 1959. See Ref. 56, pp. 382–407
141. Warner, H. R., Athens, J. W. 1964. *Ann. NY Acad. Sci.* 113:523
142. Weiss, D. W. 1968. *Bull. Math. Biol.* 30:427–35
143. Wheldon, T. E., Kirk, J. 1976. *Br. J. Radiol.* 49:441–49
144. White, R. A. *J. Theor. Biol.* 74:49–68
145. Whittemore, A., Keller, J. B. 1978. *SIAM Rev.* 20:1–30
146. Wilson, R. L., Gehan, E. A. 1970. *Comput. Programs Biomed.* 1:65–73
147. Zietz, S. 1977. *Cell cycle kinetic modeling and optimal control theory in the service of cancer chemotherapy.* PhD thesis. Dep. Math. Univ. Calif., Berkeley. 209 pp.
148. Zietz, S., Nicolini, C. 1978. In *Biomathematics in Cell Kinetics,* ed. A.-J. Valleron. Paris: Elsevier-North Holland

MEDICAL INFORMATION SYSTEMS ♦9157

Eugene M. Laska and Scott G. Abbey

Information Sciences Division, Rockland Research Institute, Orangeburg, New York 10962

INTRODUCTION

Socialized medicine and other related systems of social welfare, such as the Medicare and Medicaid programs in the United States, have helped make health care one of the major worldwide growth industries (62). International institutions, such as the World Health Organization, and national agencies, such as the US National Institutes of Health, have helped make generally acceptable and have propagated the concept of the individual's right to adequate medical care. Consequently, consumer use of health care providers has multiplied exponentially. Yet, with a few isolated but important exceptions, there is a shortage of well-trained physicians to meet the growing demand. Additionally, as medical knowledge expands and specialization becomes the norm, the number of people with access to the "proper" physician, particularly for treatment of the more esoteric illnesses, becomes a smaller fraction of the world's population. Although there are many programs designed to produce more physicians and to train more paramedical staff to function at higher medical levels, there is still not enough available expertise. Burgeoning costs exacerbated by enormous inflation rates have made it even more difficult to provide medical care for all who need it. Among the world's population, the continuing reality is one of many unmet medical needs.

For a number of years, scientists have dreamed of a care-giver, with a terminal by his side, in a fully computerized service environment. The computer is interposed to systematize and organize knowledge, to augment or control the collection of information, to provide logical support for decision making and, by monitoring progress, to act as an *aide memoire*. The concept, could it be implemented, would help put an end to the shortages of health care specialists. This chapter is an introductory survey of the use of computers in health care.

Computer hardware has evolved through three generations, with hundreds of millions of dollars spent on research and development of medical applications. The cost of hardware has plummeted, and the dream of a bedside terminal has been replaced by the dream of a bedside computer or even a pocket-sized computer gently tucked in next to the doctor's stethoscope. Computer hardware is at the stage where almost anything is possible (4, 12, 16, 17). Processing speeds and storage space pose no limitations. But medical logic needs to be developed; software needs to be written; and then the results need to be made broadly available. Progress toward this goal is clearly in evidence.

With the widespread commercial distribution of computers in the early 1960s, automation processes began to impact many facets of medical practice. The initial, commercially successful use of computers in medical facilities was in such well-defined business areas as accounts receivable, accounts payable, cost finding, rate setting, inventory, payroll, billing for services, and other typical "housekeeping" functions. Increasing patient loads, escalating medical costs, demands for higher standards of care and for expanded services, all served to raise the level of "front office" awareness of the benefits of automation. Increased revenues and decreased costs in many cases "justified" the expense of a large computer. Today, the number of researchers and commercial firms involved in computer applications is considerable.

SCOPE OF THIS SURVEY

The stage on which the health care drama is performed is usually the hospital, outpatient clinic, clinician's private office, or the patient's home. The star is the patient, and other players are the medical, paramedical, and support staff involved in endless information flow (30). Communications by and among staff contain information relating to the service process; these cover such areas as orders written by the physicians, patient progress, costs of care, payroll, bills, reimbursements, fiscal matters, statistics for review, control agencies, etc. Computer scientists and others have considered the total drama, and, wherever it has seemed reasonable, have attempted to automate clinical, administrative, and business processes and the information that reflects them. Systematic global approaches trying to include clinical, management and logistical activities associated with running a facility have been called automated medical record systems, hospital management systems, ambulatory patient care systems, and private practice systems. It was hoped that such a systems approach could, in time, fulfill the dream of the scientist and physician.

Much developmental activity has already begun. For example, projects trying to automate the medical record have attempted to provide a means towards systematizing both the record-keeping and the information-exchange process; instruments to monitor and control physiological functions under computer control have been developed; clinical decision support systems that "automate" medical knowledge and its use in the clinical situation are under development.

Virtually thousands of projects and investigators have made contributions to the field of computers in medicine which are worthy of discussion. This survey could not possibly review them exhaustively. Instead, it gives some background on automation efforts in the area of clinical information support. (There are several comprehensive bibliographies and many professional journals, some of which are oriented to specific medical specialties or other areas of technical interest not covered here. A sampling of these is listed in the Appendix at the end of this review.)

THE CLINICAL CARE PROCESS

In the clinical process, signs, symptoms, and test results are reviewed in logical sequence by the physician to diagnose the situation and to determine his next action. Whether it takes place in a hospital, in a clinic, or in a private office, this process is virtually the same. Figure 1 shows the flow of events. The physician must decide on an intervention or treatment (D) for a patient who has a particular problem or a set of problems (A and B). To do this, he must first evaluate the relative importance of each of the problems, signs, and symptoms to determine the most probable diagnosis (C). If his information is insufficient, he may order further tests (B). He must then review for each possible treatment intervention, one of which is taking no action, the possible outcomes and the probabilities of their occurrence. The basis for the selection is ultimately his own experience, his review of the medical literature, and his consulting with colleagues. Once he orders interventions, they are delivered by him, by supporting staff, or by the patient himself (E). The physician receives feedback by direct observation, from other staff, from the patient himself, and from the results of tests on the effectiveness of the interventions in terms of their impact on the patient's problems (B). The process of evaluating the patient continues until some resolution is reached.

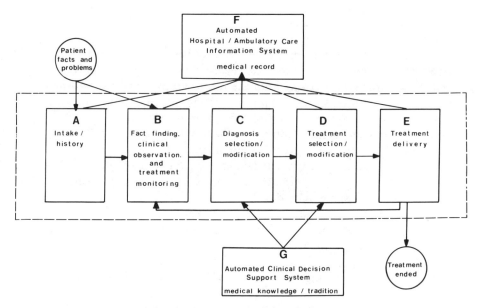

Figure 1 Automation, medical information, and the clinical process.

Facts about the patient—his problems, signs, symptoms (*A*), medical observations, the results of test evaluations (*B*), the diagnosis (*C*), therapeutic interventions ordered (*D*) and delivered (*E*), the rationale for these and reevaluations and assessments—are supposed to be entered in the medical record (*F*). Medical knowledge, giving a rationale for diagnostic and therapeutic decisions and providing, at least by implication if not specifically, the probabilities of outcomes for various interventions, is contained either in medical libraries or in the oral traditions of the specialists in the particular field (*G*). The arrows in Figure 1 show these relationships.

Computer technology is helping to increase the accuracy of assessments and diagnoses, to determine treatment choices, to monitor the course of therapy, and, in general, to achieve new levels of medical sophistication. A few brief descriptions of applications will serve to illustrate the efforts underway.

AUTOMATING THE CLINICAL PROCESS: SOME EXAMPLES

The organization of this section corresponds closely to the structure depicted in Figure 1.

Intake / History-Taking (A)

A great deal of a physician's time is spent obtaining information from the patient. The computer is a low-cost device capable of a reasonable job of obtaining at least the preliminary information (9, 55).

• In England, a micro-based medical interviewing system, with output available on a 40-character printer, is providing a low-cost approach to direct patient-computer communication (41).

• In Scotland, work is underway for computer questioning of a patient, where questions adapted for different medical problems and comprehensible to the patient are stored in a separate file. The patient's responses yield the data the computer analyzes statistically for medical diagnosis (41).

• In Utah, a self-administered, problem-oriented, patient-history-taking program calculates the probability of a disease based on Bayes' theorem. A printed output suggests diagnoses and potential problem areas. The history is administered as part of a multiphasic screening program (23, 65, 66).

• At a Veteran's Administration Hospital, a computer system has been designed to improve the "triage" process of referring individual patients to appropriate treatment services (68). An on-line system interviews potential clients, gathers information from the staff, and generates a tentative diagnosis and an initial problem list. The computer, after comparing patient characteristics with the specific admission criteria of each treatment unit in the mental health care delivery system, determines which is the best service for the patient.

• Canadian researchers are working on a computerized early-warning system for heart disease; one sensitive experimental model, which monitors and analyzes electrical activity and mechanical vibrations of the heart and detects usually unnoticeable cell damage, may be able to discover signs of heart disease years earlier than other methods (41).

• The foremost among the many facilities that have developed automated systems for use in multiphasic patient screening programs (2, 13) is the Kaiser Permanente Hospital in Oakland, California (23). These programs include history gathering from the patient, laboratory test results, and physical findings. In some facilities, an on-line computer administers the tests. In one case, a breathing test and an electrocardiogram are performed and immediately computer analyzed (47, 61).

Fact-Finding, Clinical Observation, and Treatment Monitoring (B)

Many patient observation devices are in various stages of development. In some cases, prototypes have been built; other devices are widely available. Many perform real-time monitoring of current vital functions (40) and can, at the same time, interface with the patient's medical record to document the assessments and services delivered.

• A microcomputer, worn by a patient during his daily activities, takes a continuous electrocardiogram (ECG). If it discovers an irregularity, it sounds a warning and displays a message. Sixteen seconds of the abnormal heartbeat, stored in a memory, coupled to a telephone, can be transmitted to a regional medical center for further interpretation. Instructions can then be given to the patient (41).

• In Japan, a Telemonitor provides a multi-channel patient-monitoring system measuring blood pressure, heartbeat, breath, temperature, and brain waves, with alarms at danger levels. A video screen displays output signals derived from a combination of the computer and the television cameras (41).

• In the United States, oscillometic measurement methods and programs processed through a microcomputer measure and monitor stroke volume and cardiac output and blood pressure, including arterial pressure, systolic pressure, diastolic pressure, and heart rate (64).

• Computerized spirometric and lung volume tests yield automated pulmonary function and blood-gas analysis results. The results of the data analysis are compared to established parameters for quality control and interpretation purposes (11, 23, 63).

Patients in the intensive care unit and in the operating room are monitored. Quantitative physiological data are input in real time, and alarms are generated when values go beyond established norms (26, 40, 49). Data are recorded for the medical record, for nurse's notes, for reports at shift changes, and for review.

Briefly, here are descriptions of some other devices in various stages of experimental and applied use.

• The Electro Retino-Oculo Graph, whose operations are automated through microcomputers, is a diagnostic device designed to detect various eye diseases (45).

• A miniature electronic pulse recorder has been devised (41).

• Through a scanner, a real-time ultrasonoscope permits viewing the heart and other vital organs on a cathode-ray tube (CRT) screen (25).

• A computer system for on-line decoding of ultrasonic Doppler signals from blood flow measurement is in development (10).

• In Japan a microcomputer is being used to monitor vital signs from four patients simultaneously, recording data related to heartbeat, blood pressure, venous pressure, respiration, temperature, and premature contraction (41).

• Changes in cardiac blood temperature can be detected (22).

• In Sweden experimenters are using computers in the continuous monitoring of the breathing of newborns who have respiratory ailments (1).

Diagnosis Selection (C)

Much work has been done in developing computer techniques to assist in the general areas of diagnosis, prognosis, and therapy selection (20, 24, 35, 36). Attempts are being made to organize patient data and patient classification so as to formalize the decision process, using both statistical methods and simulations of clinical reasoning (40).

• Using statistical models of growth stages of cells, estimates are made of the age of a specific cell. This technique is helpful in detecting uterine cancer (41).

• In Washington DC and in Colorado, ECG interpretative services provide for almost immediate computer diagnosis and analysis of electrocardiogram tracings (18, 23).

• Image-enhanced X rays help diagnostic interpretation and classification (5, 28). Such methods have been broadly applied, and there is great hope that they will be useful in early detection of breast cancer.

• An automated radiology system in a Missouri Medical Center (23) enables radiologists to construct an automated descriptive narrative of an X ray by selecting words and phrases from standardized tables (37).

• Automated procedures, developed at the Mayo Clinic and elsewhere, collect, score, interpret, and print out a narrative summary of the Minnesota Multiphasic Personality Inventory, a structured personality test used to diagnose psychiatric disorders (21, 58, 60). Other written programs also score standard psychological test batteries useful as diagnostic assists, such as the California Psychological History, Rorschach, the Wechsler Adult Intelligence Scale, and the Peabody Vocabulary Test (3, 40).

• In New York, a logical decision-tree program, DIAGNO III, produces a narrative and arrives at suggested psychiatric diagnoses based on data from a form, the Mental Status Examination Record (57, 59).

• In Bethesda, Maryland, new models of "synthetic" reasoning that simulate clinical behaviors relating to laboratory abnormalities are being used to study the complicated decision-reaching process in problems involving multiple disease entities (41).

• Researchers at the University of Utah are studying Mormon church genealogy records to trace patterns in genetic and other illnesses. To identify families and individuals at high risk, they are developing a computer system linking family genealogies with medical records (41).

Treatment Selection (D)

• A computerized information system at Duke University Hospital in North Carolina provides clinical, demographic, and laboratory data based on experiences with other patients. The information is for the clinician's use in decision making and prognostication for patients with transient cerebral ischemia (27).

• In Massachusetts at an anti-coagulant clinic, blood test values from the laboratory are entered into a computer that calculates and prints out for each patient the currently needed dosage of coumadin, the anti-coagulant (23).

• A private practitioner in Louisiana has developed an antibiotic dosing program that estimates the proper dosage for antibiotics based on endogenous creatinine clearance (41).

• In Japan a push-button optometer tells the user the exact eyeglass lenses required for both eyes in just four to five minutes (41).

Treatment Delivery (E)

• Experimentation in West Germany and elsewhere uses a microcomputer planted below the skin and an electronic control system to release amounts of insulin appropriate to maintain normal levels in a diabetic patient (41).

• A hospital in Massachusetts is using a simple computerized "Magic Light Pen" to help handicapped children develop motor skills (41).

• In Australia an electronic package implanted in a deaf patient's ear receives signals that are a transformation of the mechanical energy of speech sound waves into electrical impulses. These impulses stimulate hearing nerves and restore some hearing to the patient (41).

• A computer system developed in California tracks the motion of the eye and produces audible cues to eye position. This clues the amblyopic patient to "hold" the weak eye on center, using biofeedback techniques, to improve vision (41).

• In Minneapolis an electrical nerve stimulator that is under computer control and is applied directly to the skin is being used as a treatment for many types of acute and chronic pain (41).

Hospital and Ambulatory Care Information Systems

A Hospital Information System (HIS) or an Ambulatory Care Information System is usually set up to be patient-based, encompassing data related to the areas mentioned in Figure 1: Intake/History, Observation, Diagnosis Selection, Treatment Selection, and Treatment Delivery. Such systems customarily establish, through an intake or admission form, one or more files for each patient as he enters the hospital. At this point the information gathered might include presenting problems, demographic and identifying data, and, perhaps, referral sources, history of prior service, and financial status. Bed location and service delivery information on where in the facility the patient is being assigned and on what treatment he has received can also be stored. An entry noting the termination of the patient's involvement with the facility essentially closes his file and relegates it to the archives. The needs and the specialties of the facility determine what other clinical information is included in the computer file (42).

In addition to the patient data base, a set of business applications is usually included (8, 14, 43, 44). Personnel information files permit specialty identification, job and assignment control, budgetary planning, and time and attendance monitoring. Financial data files may facilitate billing, general accounting, budgeting, inventory, payrolls, cost finding, and rate setting. In an HIS, the patient data, personnel data, and the business data may be in one data base or in several.

• A system first installed at a hospital in California, where it has been operational since 1967, acts as a medical records communication system (23, 39, 56). Major modules in the system are Order Completion/Results Reporting for following up and recording the results of laboratory tests; Medical Functions for recording physician's drug orders and printing periodic administration schedules at nursing stations; and a Medical Information Library to provide clinical, diagnostic, and therapeutic knowledge, and the medical logic on which it is based, and to serve as a reminder and prompter to clinical staff. CRT's are used to display the information in an interactive mode. In particular, the Medical Information Library module allows the physician to step through a series of displays or to request a printout, which he can keep for permanent use. For example, each frame in a complete drug formulary contains information about the drug and its use in that hospital. The hospital may add to the frame knowledge it considers important to the local clinical staff. Standard sets of knowledge in the Medical Information Library include hyperemia workup assistance and respiratory therapy indications. Other aspects of medical data include X rays, diet, activity, and vital signs. Admission data, demographic background, insurance coverage, and other related business data are included. "Medalert" data can be retrieved instantly in the emergency room and in the outpatient department. The computer sends out appropriate laboratory, medication, X-ray, and dietary requisitions, prints medication labels, and routes orders to the appropriate department at the right time. Billing information and complete accounting entries are also prepared concurrently.

- A prepaid medical practice in Arizona uses a system that includes facilities for patient identification, registration, and appointment scheduling (23). Data are collected by computer mark-sense forms, and summaries, including demographic information, problem lists, most recent visits, and case history, are provided to the examining physician for use at the patient's next visit.
- In Boston an automated ambulatory care project was developed to guide paramedics in encounters with patients (6, 23). Input documents collect history, physical examination results, and laboratory data. After computer analysis, guidelines for the paramedic indicating whether the patient requires a further workup, a specific therapy, or a visit to a physician are printed out. The protocols and algorithms that have been developed are for the more common patient complaints usually seen in outpatient clinics.
- The Multi-State Information System (MSIS), a clinical and management information system (31, 32), serves mental health programs through a communication network operated from a host computer. Information is collected at the various service delivery facilities on structured forms, which are then either keyed or optically scanned. For each patient served, a computer record is created.

A variety of documents on which to input patient data are available. These can be used to collect admission data including demographic characteristics, presenting problems, psychiatric history, mental status, progress notes, medication orders, services received (15). MSIS also provides for the special data needs of facilities by allowing each to create its own data collection instruments; data are stored using general input programs.

Retrieval of reports from the data base is accomplished either through standard reports or through generalized retrieval programs that create lists or cross tabulations designed by systems users.

In addition to the patient-oriented systems, there are modules that assist service programs in personnel management, inventory control, cost finding and rate setting, and management by objective control. Quality assurance monitoring in the medication area is discussed below.

AUTOMATING THE MEDICAL RECORD

Since data-processing technology proved so effective in managing records in a business environment, it seemed only natural to apply similar techniques to the medical record (34). The earliest efforts at automation led to attempts to identify data elements that occur predictably at specific key points in the care process and to automate their collection and storage. Some information was already being collected in most hospitals on standardized forms such as the patient admission sheet. As may be inferred from the process illustrated in Figure 1, clinical assessments and examinations, diagnostic decisions, treatment plans, service delivery, and so on, occur in a somewhat predictable order with fairly consistent regularity across broad classes of patients. Items of information involved in the process at each of these key events fall more or less into broad classes. For each of these clinical events, automated systems designers have tried to classify, characterize, and structure the type of information needed, whether or not such data had, in the past, been recorded. Then, they created mechanisms, such as forms or screens or automated instrumentation, for data collection and entry into a computer record.

By standardizing the format for collecting information and by specifying that the data-processing documents be completed as part of the routine clinical process, facilities have tried to establish procedures that would impose a limited burden on the clinician while ensuring the availability of information needed, in any case. Once stored in a computer, easy retrieval of data in many formats, e.g. summaries, and in virtually unlimited combinations in response to many needs is possible.

This rationale as an argument for automating the medical record is natural and consistent with the growing demands of an assortment of regulatory, review, and funding agencies for more and more information from facilities falling under their jurisdiction. Often missing in such efforts is an adequate assessment of the impact such activity has on the physician who, after all, provides the bulk of the information. In spite of many laudable efforts, even the most sophisticated data collection vehicles are viewed by most clinicians as a drain on their time and an intrusion on their professionalism. Moreover, most subsequent uses of the file of clinical information have little to do with the day-to-day activities of the doctor. (Further discussion of this issue appears below.)

Forms capturing the specific details of various parts of the clinical process have been developed at virtually every facility making some serious effort to standardize clinical care and automate the record-keeping process. Most striking is the commonality of the forms used in all of these efforts. While forms developed by different groups present the material with different emphases in different orders and use different words to characterize the same ideas, the essential features of the particular part of the clinical care process involved are common to all. To the extent that the clinical event is clearly definable in some sense, the information content characterizing it tends to be well defined, too. However, when the structured entry characterizes the patient inadequately for clinical purposes, the computerized data are supplemented, in general, with a separate written note placed in the patient's chart.

Thus, the patient's medical record has now become a combination of completed forms, computer output, and manually entered idiosyncratic information.

Issues

Historically, the medical record is the information repository concerned with the patient and the care process afforded him including diagnostic assessments, treatment interventions, and outcome. Hence, this is a natural area in which to focus automation efforts. The primary purposes of the medical record are to serve as a clinical aid to the physician and to other practitioners and as a mode of communication among them, to document staff-patient interactions, and to accumulate data needed for treatment planning. In addition, the record has a clear value for research and it is useful in training, education, and staff supervision. Although the primary functions of the record are in the clinical realm, its use has been extended to such areas as administrative management, day-to-day operation, and utilization and peer review, i.e. monitoring of the appropriate amounts of service and assessing the quality of service.

Complaints about the time and effort expended in making entries in the record, and the frustrations resulting from the record's shortcomings when consulted for specific information, are well supported by critical articles in the professional literature (67). Records have been found to lack standardization, focus, or precise medical thinking; they often fail to contain information; and they are a less than perfect vehicle for effective communication. Entries are made without regard to how they will be used and frequently are not correlated with the clinical decision-making processes. Documents filed in records often appear in haphazard, random order. Whether used clinically or administratively, for audit or for research, records have been found to be unwieldy, unsystematic, fragmented, and illegible, with low validity and poor reliability.

Yet, vast expenditures of time, money, and personnel energies are invested in records. A recent study noted that within psychiatric facilities over ten percent of professional activities are devoted to recording and consulting medical records (54). Medical facilities have established medical records departments, which employ registered medical record librarians, assistants, and clerical personnel whose sole responsibility is the management of the medical record. Medical record librarians have emerged as a professional group with their own standards, organizational structure, and journal.

Despite the shortcomings noted, the medical record is regularly used by practitioners in clinical care both for inputting new data and for reviewing existing information (54). Interest in the automation of the information it contains derives from many facts. Practitioners recognize that information on the clinical events of the case cannot be exchanged orally. Medical knowledge and treatment strategies have become too broad and diversified for the practitioner to believe he can know and remember everything. The age of specialization brings many consultants and professional disciplines interacting with the same patient, but not always at the same time. The treatment decisions of a physician, once made, are often left to others to carry out, and the record documents whether or not such instructions have been implemented. Concern about malpractice, licensing, legal mandates, and judicial decisions also encourages the physician to regard the record as a legal safeguard.

In addition, several legal changes that carry considerable economic consequences have increased recognition of the importance of the medical record. These relate generally to the change in pattern of service delivery brought on by the involvement of the federal government in paying for care for the indigent and the aged. Accompanying these changes have been government regulations limiting costs by paying only for necessary services and defining minimum standards for care in environments in which federal dollars are spent. Other third-party payers have endorsed the regulations. Since virtually every major care-giving center in the country receives, to a lesser or greater extent, reimbursement from either the federal government or other third-party payers, the impact of the legislation is widespread.

Alternative Record Structures

The existing, traditional medical record is usually described as a "source-oriented" document, with entries organized along a time axis in the same sequence that they were made. The record thus resembles a diary more than a scientific journal. (The word "source" refers to the mode of organization—arranging entries according to class of recorders, e.g. doctors, nurses, laboratories, and so on.) Although medical records have existed for two millenia, once automation efforts began, it was natural to search for more clinically logical and meaningful alternative methods of organization.

Alternative record formats do not necessarily require the use of automation and have contributed by and in themselves to a systematization of the medical care process. Efforts to computerize the medical record have, wherever they have proven useful, integrated the best parts of these restructured forms of record-keeping practice.

The problem-oriented record (POR), pioneered by Dr. Lawrence Weed at the University of Vermont, is the principal new conceptualization of a structure of the medical record (29, 67). By referencing all collected information to clearly identified patient problems, the POR records the care process almost as if therapy were being administered within a framework of scientific investigation. A hypothesis is developed about the problems presented. This requires clear statements of the plans to alleviate each of the stated problems. The format of the POR is meant to parallel the structure of the care process by systematically collecting and organizing data on each problem in the same order that care is given. The patient provides information about his chief complaints; this and additional information are evaluated; and a treatment plan is formulated. Follow-up continues, step by step, until the problem is resolved.

The POR contains four major theoretical components: patient data base, problem list, therapeutic plans, and progress notes. The overall format and each subdivision of the POR are structured by this organization so that the collection and evaluation of data and the development and documentation of treatment and outcomes are all enumerated under specific problems. Weed's effort at automating the problem-oriented record is discussed below.

CLINICAL DECISION SUPPORT

Clinical decision making relies on knowledge that human beings obtain from memory or from records, such as a library, or from colleagues. For a computer to help, i.e. to evaluate the truth of an event in the light of information about relevant events, it must have available analogous resources, a source of information upon which to draw.

A clinical decision support system (CDSS) may be conceived of as an automated process by which the specifics about a particular patient are combined with formalized medical knowledge and translated into either a recommendation for a clinical action or an approval of a clinical action already taken (7, 40, 48). For example, a CDSS may provide a diagnostic suggestion to a clinician or may, on the basis of patient data, review a diagnosis recorded by him and agree that it is a realistic alternative. Automated clinical decision support generally takes one of two forms: in the first, using data stored in the system, the computer, comparing information on a particular patient with information on other patients, helps to place him within a population context; in the second, the computer performs some logical or numerical calculation, consults a reference, executes a series of rules that provides information that is difficult to obtain manually, or ensures that nothing is forgotten or overlooked. Examples of both types appear below.

The success of a CDSS is determined, first, by the extent to which facts about the patients are characterized by the entries in the patient data base; and, second, by the degree to which the characterizations of medical knowledge are complete and consistent with the current beliefs of the medical community. If a CDSS is to operate as part of the care process, it must relate to a patient information system that reflects the clinical situation reasonably well.

An idealized CDSS is composed of five parts:

1. a data base reflecting current knowledge or medical policy;
2. software used to organize, update, and review the knowledge policy data base;
3. procedures for collecting and/or reviewing the entries in a patient data base either to determine their consistency with, or to draw inferences from, the knowledge/policy data base;
4. a feedback mechanism to inform clinician of the findings; and
5. a mechanism to record and store the findings and the physician's reaction to them.

Computerization of the data base reflecting current knowledge or medical policy requires a formal structure. Although many medical concepts and policies are semantically compact in presentation, they are, in reality, complex logical structures whose subtleties, so easily stated in natural language, may not be representable in data structures amenable to computers.

There are four types of information found in a patient's medical record:

1. the patient's subjective judgments;
2. "hard facts" about the patient such as, address, sex, age, and items such as, "scar over patient's left eye";
3. judgmental medical observations about the patient such as, degree of palor or severity of distress; and
4. decisions made or actions taken with respect to the patient such as, drugs given, diagnosis made, tests ordered, and other therapies prescribed.

It must be assumed that any decision or action (the fourth type of data element) is a response to, or a consequence of, some patient complaint (first type), hard fact (second type), or judgmental information (third type).

Data items sufficient to justify the action or to provide reasonable clinical decision support may or may not be present in the patient data base. Absence of adequate data may imply an error in the data entry process, a failure to record something that should have been recorded, the absence of a mechanism for recording some salient information, or, in the case in which the CDSS is acting as a review mechanism, possibly a questionable action on the part of the treating staff. It may be presumed that those action elements relevant to billing will be contained in the data base because of their fiscal implications; however, it must not be assumed that the observational and other data elements necessary to the inference that must be made to justify the action of the therapeutic team will also be in the data base. Thus, as a CDSS develops, improvements to the associated automated patient-record system will be made. Actions or inferences that cannot be justified from patient facts may be considered exceptions, and, hence, feedback information is sent to the treating team.

The treating team can then act in one of several ways: they can indicate that data were entered erroneously; they can report that there is inadequate data entry (i.e. more information must be entered); they may produce justification for the action; and/or, possibly because of increased awareness of the views of the community of experts contributing to the knowledge/policy data base, they may change their action.

•MYCIN (51), a knowledge-based computer system, was developed at the Stanford University School of Medicine in California for use as a consultant to clinicians in the area of antimicrobial therapy for patients with severe infections. The system is meant to give clinically useful advice and the rationale behind decisions and to update its own data bank with additional knowledge in the process. Three interrelated parts make up the MYCIN system: a consultation system draws on data from the knowledge base and from information about the patient entered by the physician to provide therapeutic advice; an explanation system explains the reasoning used and the rationale for conclusions; and a knowledge acquisition system enables experts to add to the medical knowledge base. The program functions by asking the clinician for specific patient information, and, using general medical knowledge stored in its data bank, it outputs treatment advice with explanations.

Knowledge is stored in the data bank in a series of what computer scientists call "production rules" as opposed to the branching logic commonly used by many other programming systems. The production system generally has three main components: a data base with "facts" including the inferences which the system has made; rules or productions representing medical knowledge expressed in terms of antecedents and consequences, i.e. "theorems"; and a rule interpreter that selects which rules are to be used in which set of circumstances.

The MYCIN program (21a, 50, 52) is limited by its complex control structure and representational scheme and by its requirements for extensive hardware and memory. The system presently functions solely in a research environment, but the approach it represents is being experimented with in an increasing number of facilities. At this time, data bases on meningitis (6a, 51) and on mental disease (26a) are being established. A similar program to help localize damage to the central nervous system in comatose neurological patients is being developed at the University of Maryland where simple rules or predicates are being combined into macropredicates to improve compactness and provide a better representation of knowledge (43a).

• The Medication Order Review and Exception System is based on a medication order reporting system of the Multi-State Information System which allows for all drugs that are ordered to be recorded on a structured form (32). The data are entered into the patient's manual medical record and are also stored in the patient data base of the computer. A series of programs periodically reviews all outstanding orders and flags the exceptions to a set of generally accepted psychopharmacology practice rules determined by an expert committee of psychopharmacologists (38, 53). The physician is notified of the determination and may either retain his original prescription and document his reasons for so doing, or, based on the knowledge acquired from the printout, change his prescription. Prior to the existence of the system, the percentage of questionable prescriptions at one large psychiatric hospital was, by any standard, excessive; approximately 33% of all outstanding orders were in violation of polypharmacy or dosing rules. After a year of operation and continued feedback to the physicians, the percentage has been substantially reduced to approximately 10% (33).

This experience suggests that the likelihood of the computer's having an impact on clinical practice is dramatically increased when the assistance takes the form of *a posteriori* review. The doctor, in this case, must respond to the exception report because of legal, psychological, and professional implications. Consequently, he is in a reinforced learning situation. He must respond to the decision of his peers that, for a given set of clinical facts, there are reasonable actions that may be expected and, perhaps more importantly, there are actions that are questionable and that require careful review.

• The automated Mental Status Examination Record (MSER) provides an example of the use of a patient comparison to a standard population. The MSER is an automated instrument designed to allow a clinician to record the results of a detailed mental status examination. It was developed as part of the Multi-State Information System (32). Based on mental status data from a sample of a psychiatric inpatient population, the computer was programmed to derive twenty scaled scores. Two thousand Mental Status Examination Records were utilized in a principal component factor analysis, and twenty factors were identified. A new patient is evaluated on these factors, and his/her scores are related to those of the normative group. The MSER also illustrates the use of a computer for logical manipulation. The data on a completed MSER generate computer-suggested diagnoses for consideration by the clinician (59). The machine uses a rational, branching logic method of arriving at a differential diagnosis to classify each patient into one or more suggested diagnostic categories specified by the American Psychiatric Association (19). The logic of the analysis is based on abstractions of textbook classification rules and on the clinical experience of the developers.

• PROMIS is another clinical decision support system (23, 39, 46). The PROMIS laboratory at the University of Vermont Medical Center has been tackling the issues of knowledge representation in medical care by combining existing medical knowledge from two sources: knowledge about the patient recorded in a problem-oriented medical record, and knowledge about disease entities and medical practice represented by the content and structure of the hierarchical frames by which data are inputted and received. An automated problem-oriented medical record, the basis of the system, provides the set of rules, structure, and conditions to allow for problem solving and decision making by coupling what is known about the patient with what is known in the body of medical knowledge.

PROMIS provides a series of structured and branching logic displays on a CRT screen or console; these displays lead the practitioner inputting data through a sequence of entries, each of a greater level of detail, in a step-by-step search from among a vast number of possibilities for solutions to a patient's problem. As the practitioner is being led through this hierarchical structure, the automated problem-oriented record is concurrently reviewing and updating data. Rigorous use of the system demands common terminology and provides a logical approach to problem solving and decision making.

A typical beginning display might specify phases of medical action: Data Base, Problem List, Initial Plans, or Progress Notes. When any one option is touched, the screen displays a further series of choices related to that option. Each of these choices, when selected, branches further into additional options. The practitioner, responding to displayed questions and choices, is led further along the series of options, progressing from the basic structural displays to higher levels of detail whose choices represent specific medical knowledge contained within the wording and structure of the displays. Selected elements within certain frames, e.g. causes of initial concern about a particular medical problem, may initiate any of a number of actions, such as ordering a medical procedure, inputting information into patient's record, branching to a certain frame, requesting other information from patient's data bank, requesting medical information, requesting further investigation of potential causes of problems.

The "Initial Plan" option, when chosen, displays choices such as, "state aims for problem management"; or it issues instructions such as, "check how problem may be contributing to patient's sickness," "check for effects/disabilities produced by problem," "check function/status of systems involved with problem," "assess and follow cause," "investigate problem and etiology," "watch for complication," "institute and monitor treatment."

If needed, a user can request a further detailed definition of any of these areas before exercising the next option. Any option chosen produces a list of choices related to that option, and any number of options may be chosen. Further choices may produce specific medical knowledge about a particular problem from the medical libraries and/or information from the patient's own data base, thereby guiding the clinician to possible additional diagnostic or treatment procedures.

The content of each structural frame unit becomes the organizing structure for the next level or frame. The computer system proceeds to higher levels of abstraction and hypothesis formulation. The series of options the practitioner is led through for each patient is unique to the problems, needs, and characteristics of that patient since the branching route through any series has many possibilities.

This "coupling" of details of the patient's problems to medical knowledge libraries proceeds from the largest structure required for comprehensive care and most general questions to the smallest structure attuned to the finest level of detail. In this way, the medical knowledge programed into the computer system becomes available to the practitioner.

• HELP is a clinical decision support system based at the Latter Day Saints Hospital of the University of Utah (23, 40). The program applies decision logic to data to assist in medical decision making by examining the significance of all newly entered data. Automated patient records are established for all patients at admission or at preadmission screening. The data collected relate to patient history and physiological and laboratory findings; additional data generated by events occurring in the patient-care process, such as heart catherization procedures and the monitoring of vital life signs in operating rooms, intensive care units (ICU), and cardiac care units (CCU), are added. At discharge, final diagnosis and treatment data are also entered. Data are collected primarily through a variety of laboratory and monitoring systems and through terminals throughout the hospital. The system stores medical knowledge in the form of decision criteria called HELP sectors which have access to data from the patient's file. Also, upon request from any terminal, it is possible to assign the carrying out of a medical decision to a particular medical person. The system is data driven, i.e. when data are added to a patient's file, they are reviewed in the computer in conjunction with previously entered data, and output is automatically generated to alert the clinician whenever specific guidelines or decision criteria have been met. Conclusions from the HELP system are stored in the patient's file.

The system is quite extensive with many hundreds of HELP sectors which cover areas such as history, ECG interpretations, CCU and ICU monitoring, adverse drug reactions, X-ray interpretations, cardiovascular and pulmonary voluntary function testing, and outpatient management. The ultimate aim of HELP is to combine medical knowledge with a patient data base to provide guidance, consultation, and review in medical decision making.

CONCLUSION

Computer applications in medicine have advanced beyond the goal of replacing the manual aspects of the clinician's task in medical record keeping. The problem of the day is to develop clinical decision support systems. The former goal was an inadequate basis for the costly and invasive nature of structured data entry. Unfortunately, though the latter holds great promise, a considerable effort remains before the goal is reached.

Automated clinical decision support, to the degree that it has been developed so far, has not met with wide, general acceptance. The expert views such assistance as not necessary, not subtle enough, and typically, not addressed to the hard problems of knowing how and what to observe. He cites his superiority over computers at pattern recognition, the relatively high cost of computing, the economic consequence of spending his time filling out forms or at a terminal, the difficulty in formalizing and synthesizing the logical processes that a computer needs in order to be able to advise him, the intuitive rather than scientific character of the medical art, and so on. The nonexpert, not surprisingly, prefers to consult a human expert. Therefore it follows that this type of automation is not very useful to some clinicians.

However, the stage is set for substantial advances on several fronts. The maturation of computer hardware and software technology and new ideas and approaches to medical record keeping provide the means.

Federal legislation, the deficiencies of peer review in the face of social pressures and pressures from colleagues, and the aggressive review approach (relying on medical records) taken by the Joint Commission on Hospital Accreditation have increased the desirability of, and the demand for, methods to insure high quality of care. The examples described above provide persuasive evidence that the development of a comprehensive clinical decision support system is feasible. Because the adequacy of the consultation and/or medical review conducted by a CDSS is limited by the degree to which the patient data base reflects actual events, further development in automated patient-information systems that collect and store clinical data will be stimulated. Also, the value of consultations and/or review of clinical actions and decisions by a CDSS in determining medical appropriateness and in assuring quality care depends on the degree to which the medical knowledge data base represents accepted medical practice and current knowledge. The two types of systems are linked. Since both consumers and legislators need assurances that the medical industry is delivering the highest quality of care and is being adequately audited, there will be a simultaneous and rapid maturation of patient information systems and CDSS's.

There remain open many philosophical and practical questions:

1. Can medical knowledge be codified sufficiently so as to reflect clinical practice?
2. Can a consensus about what is acceptable practice be reached?
3. Can a mechanism be established to review and monitor medical research and the knowledge base and to change it as new knowledge is discovered and as practice evolves?
4. How can the contents of the knowledge data base be made accessible to the medical community?
5. Can such a system be developed and operated at an acceptable cost?
6. Could such systems win support from the medical profession, e.g. practitioners and educators?
7. What are the legal implications of clinical decision support systems, especially as they might affect malpractice cases?
8. Can we afford not to build clinical decision support systems?

ACKNOWLEDGMENTS

We gratefully acknowledge the contributions of our colleagues and friends who provided invaluable assistance through helpful discussions and critical comments. Most especially, we wish to thank Robert Schore, James Robinson, Jeffrey Crawford, Sal Vitale, Carole Siegel, Bram Cavin, Rheta Bank, and Barbara Weinbaum.

Appendix[1]

Annual Bibliography of Books on Computing and Computers, 1968, 1969, 1970...1979. Colorado Springs: Univ. Colorado

Brandys, J. E., Pace, G. 1979. *Physicians Primer on Computers—Private Practice*. Lexington, Mass: Lexington Books

Collen, M. D., ed. 1974. *Hospital Computer Systems*. New York: Wiley

Computer Applications in Medical Care, Ann. Proc. IEEE, New York, 1977, 1978, 1979

Computers and Medicine (a journal). Chicago: Am. Med. Assoc.

Computers and Biomedical Research, An International Journal. NY: Academic

Computers in Biology and Medicine (a journal). Elmsford, NY: Pergamon

Crawford, J. L., Morgan, D. W., Gianturco, D. T. 1974. *Progress in Mental Health Information Systems: Computer Applications*. Cambridge, Mass: Ballinger

Garrett, R. D. 1976. *Hospital Computer Systems and Procedures*, Vols. I, and II. New York: Van Nostrand Reinhold

Giebink, G. A., Hurst, L. L. 1975. *Computer Projects in Health Care*. Ann Arbor, Michigan: Health Admin. Press

International Journal of Biomedical Computing. Essex, England: Applied Science Publ.

Jacquez, J. A., ed. 1972. *Computer Diagnosis and Diagnostic Methods*. Springfield, Ill: Charles C. Thomas

Jensen, M. A., ed. 1975. *Patient Centered Health Systems, Proc. 5th Ann. Conf. Soc. Comp. Med.*, Minneapolis

Laska, E. M., Bank, R. 1975. *Safeguarding Psychiatric Privacy: Computer Systems and Their Uses*. New York: Wiley

McLachlan, G., Shegog, R. A. 1968. *Computers in the Service of Medicine*. New York: Oxford Univ. Press

Medinfo (annual publication of the World Conference on Medical Informatics). Uppsala, Sweden: Int. Fed. Inf. Proc.

Methods of Information in Medicine (a journal). Stuttgart, NY: F. K. Schattauer

Physicians Microcomputer Report (a journal). Lawrenceville, NJ: Gerald M. Orosz

Skofronick, J. G., Cameron, J. R. 1978. *Medical Physics*. New York: Wiley

Sondak, N., Schwartz, H. 1979. *Computers in Medicine*. New York: Artech House

Soucek, B., Carlson, A. D. 1976. *Computers in Neurobiology and Behavior*. New York: Wiley

Stacy, R. W., Waxman, B. D. 1965. *Computers in Biomedical Research*. New York: Academic

Tao, D. 1978. *Computer Applications in Medicine: A Survey of Vendors*. Lawrenceville, NJ: Physicians Microcomputer Rep./Gerald M. Orosz

Zimmerman, J., Rector, A. 1978. *Computers for the Physician's Office*. Forest Grove, Oregon: Research Studies Press

Literature Cited

1. Allen, L. P. et al. 1978. *Arch. Dis. Child.* 53(2):169-72
2. Aller, J. C., Ayers, W. R., Caceres, C. A., Cooper, J. K. 1953. *Proc. IEEE* 57(11)
3. Alltop, L. B., Wilson, I., Raby, W., Vernon, C. (no date). *State Psychiatric Record (Patient) Linkage System*, unpublished paper supported by NIMH grant I-DII MH 1297-1

[1] Any number of bibliographies on computers in medicine exist. These suggestions are by no means exhaustive, but are meant to guide the reader to other resources where extensive listings are available.

4. Amdahl, L. 1978. *Datamation* 24(12):18–20
5. Andrews, H. C., Hunt, B. R. 1977. *Digital Image Restoration*. Englewood Cliffs, NJ:Prentice-Hall
6. Barnett, G. 1979. *Hospital Computer Project*, NTIS #HS00240, Mass. Gen. Hosp.
6a. Blum, R. L. 1978. *Proc. 2nd Ann. Symp. Comput. Appl. Med. Care, IEEE*, pp. 303–7
7. Bordage, G., Elstein, A., Vinsonbaler, J., Wagner C. 1977. *Proc. 1st Ann. Symp. Comput. Appl. Med. Care, IEEE*, pp. 204–10
8. Brandejs, J. F., Pace, G. C. 1979. *Physicians Primer on Computers*. Lexington, Mass: Lexington Books. 178 pp.
9. Budd, M., Bleick, H., Sherman H. 1979. *Acquisition of Automated Medical Histories by Questionnaires*, NTIS #HSM 110-69-264, PB 233 784, Mass. Inst. Technol.
10. Callagan, D. A. 1967. In *Engineering in the Practice of Medicine*, ed. B. L. Segal, D. G. Kilpatrick, pp. 363–72. Baltimore, Md.:Williams & Wilkins
11. Cohen, M. L. 1969. *Comput. Biomed. Res.* 2:549
12. Cole, E., Gremillion, L., McKenney, J. 1979. *Commun. ACM* 22(4)
13. Collen, M. F., Rubin, L., Neyman, J., Dantzing, G. B., Baer, R. M., Siegeleub, A. B. 1964. *Am. J. Public Health* 54:741
14. Cooper, H. M. (no date). *Guidelines for a Minimum Statistical and Accounting System for Community Mental Health Centers*, US Dept. HEW NIMH Series C No. 7, US Gov. Print. Off.
15. Crawford, J., Conklin, G., McMahon, D., Vitale, S., Robinson, J. A., Geller, J., DeStefano, O. 1978. *2nd Ann. Symp. Comput. Appl. Med. Care. IEEE*, pp. 198–206
16. Cypser, R. J. 1978. *Communications Architecture for Distributed Systems*, pp. 35–36, 41. Reading, Mass:Addison-Wesley
17. Datapro Research Corp. 1977. *Processing Options: Large Centralized Computers vs. Minicomputers*, Datapro EDP Solutions Rep. E80-450-301
18. Dunn, R. A. 1977. *Proc. 1st Ann. Symp. Comput. Appl. Med. Care*, pp. 215–20
19. Feldman, S., Klein, D., Honigfeld, G. 1972. *Biometrics* 28:831–40
20. Fleiss, J. L., Spitzer, R. I., Cohen, J., Endicott, J. 1972. *Arch. Gen. Psychiatry* 27:643
21. Fowler, R. D. Jr. 1972. *Psychiatr. Ann.* 12(2):10–28
21a. Freiherr, G. 1979. *Res. Resour. Rep.* 111(12):1–6
22. Frolov, A. V., Sidorenko, G. I., Dubko, M. J., Drapeza, A. I., Sidorenko, E. R. 1977. *Biomed. Eng. USSR* 11(4):180–83
23. Giebink, G. A., Hurst, L. L. 1975. *Computer Projects in Health Care*. Ann Arbor: Health Administration Press. 211 pp.
24. Gleser, M. A., Collen, M. F. 1972. *Comput. Biomed. Res.* 5:180
25. Greenleaf, J. F., Johnson, S. A. 1978. *Ultrasound Med. Biol.* 3(4):327–40
26. Halloran, M. J. 1978. *Proc. 2nd Ann. Symp. Comput. Appl. Med. Care, IEEE*, pp. 629–33
26a. Heiser, J. F., Brooks, R. E., Ballard, J. P. 1977. *Abstr. 6th World Congr. Psychiatr., Honolulu, Hawaii*, p. 135
27. Heyman, A., Burch, J., Rosati, R., Haynes, C., Utley, C. 1979. *Neurology* 29(2):214–21
28. Huang, H. K. 1978. *Proc. 2nd Ann. Symp. Comput. Appl. Med. Care*, pp. 8–14
29. Hurst, J. W., Walker, H. K. 1972. *The Problem Oriented System*. New York: Medcom
30. Judstrup, R., Gross, M. 1966. *Health Serv. Res.* 1:235–71
31. Laska, E. 1974. In *Progress in Mental Health Information Systems: Computer Applications*, ed. J. Crawford, D. Morgan, D. Gianturco, pp. 231–51. Cambridge, Mass: Ballinger
32. Laska, E. M., Bank, R. 1975. *Safeguarding Psychiatric Privacy: Computer Systems and their Uses*. New York: Wiley. 452 pp.
33. Laska, E. M., Siegel, C., Simpson, G. 1978. *Abstr. World Conf. Psychiatry, 6th, Honolulu*, pp. 166–67
34. Laska, E. 1979. In *The Role of the Psychiatric Record in the Delivery of Mental Health Services*, C. Siegel, S. Fischer. NY: Bruner/Mazel. In press
35. Ledley, R. S. 1966. *J. Am. Med. Assoc.* 196:933
36. Ledley, R. S. 1969. *Proc. IEEE* 57:1900
37. Lodwick, G. 1974. *A Summary of Modular Computer Medicated Radiology Systems and an Economic Evaluation of Missouri Automated Radiology System, MARS*. Washington Univ., NTIS HS 00046, PB 248 341
38. *Manual for the Use of Psychotherapeutic Drugs in New York State Department of Mental Hygiene Facilities*. 1977. NY State Dept. Ment. Hyg. 23 pp.
39. Office of Technology Assessment,

Congress of the United States, 1977. *Policy Implication of Medical Information Systems*, pp. 21–26
40. Patrick, E. A. 1979. *Decision Analysis in Medicine: Methods and Application.* Boca Raton, Fla: CRC Press. 340 pp.
41. *Physicians Microcomputer Report*, 1979. Vols. 1, 2. Lawrenceville, NJ:G. Orosz
42. Price, J. 1978. *Comput. World*, March 27, pp. 59–61
43. Prince, T. R. 1970. *Information Systems for Management Planning and Control.* Homewood, Ill: Richard D. Irwin
43a. Reggia, J. A. 1978. *Proc. 2nd Ann. Symp. Comput. Appl. Med. Care, IEEE*, pp. 254–60
44. Ross, J. 1970. *Management by Information Systems.* Englewood Cliffs, NJ: Prentice Hall
45. Safir, A., Kashdan, N. R. 1977. *Proc. 1st Ann. Symp. Comput. Appl. Med. Care*, pp. 124–30
46. Schultz, J. R. 1976. *Proc. 3rd Illinois Conf. Med. Inf. Syst.*, pp. 1–4
47. Schumitzky, A., Rodman, J., Crone, J. 1977. *Proc. 1st Ann. Symp. Comput. Appl. Med. Care, IEEE*, pp. 154–61
48. Shane, L. L., Marler, E., Bauman, W. A. 1972. *NY State J. Med.*, June 1972, pp. 1636–44
49. Sheppard, L. C. et al. 1977. *Med. Instrum.* 11(5):296–301
50. Shortliffe, E., Davis, R., Axline, S., Buchanan, B., Green, C., Cohen, S. 1975. *Comput. Biomed.* Res. 8:303–20
51. Shortliffe, E. H. 1976. *Computer-based Medical Consultations: MYCIN.* New York: American Elsevier. 264 pp.
52. Shortliffe, E. H. 1977. *Proc. 1st Ann. Symp. Comput. Appl. Med. Care*, pp. 66–69
53. Siegel, C. 1976. *Proc. HEW-NIMH Region II Program Evaluation Conf.*, pp. 253–68
54. Siegel, C., Fischer, S. 1979. *The Role of The Psychiatric Record in the Delivery of Mental Health Services.* New York: Brunner/Mazel. In press
55. Slack, W. 1979. Computer-Based Medical Interviewing Project, NTIS #HS00283 Univ. Wisconsin
56. Sneider, R. M. 1978. *Proc. 2nd Ann. Symp. Comput. Appl. Med. Care, IEEE*, pp. 594–600
57. Spitzer, R., Endicott, J. 1971. *Arch. Gen. Psychiatry* 24:540–47
58. Spitzer, R., Endicott, J. 1974. *Am. J. Psychiatry* 131:523–30
59. Spitzer, R., Endicott, J. 1974. In *Progress in Mental Health Information Systems: Computer Applications*, ed. J. L. Crawford, D. Morgan, D. Gianturco, pp. 73–105. Cambridge, Mass:Ballinger
60. Spitzer, R., Endicott, J., Cohen, J., Fleiss, J. 1974. *Arch. Gen. Psychiatry* 31(8):197–201
61. Stiefel, M. L. 1978. *Mini Micro Syst.* 11(4):42–45
62. US Dept. Health, Education and Welfare. 1972 *The Size and Shape of the Medical Care Dollar*, Washington, DC
63. Vallbona, C., Penny, E., McMath, E. 1971. *Comput. Biomed. Res.* 4:623–33
64. Warner, H., Gardner, R. 1968. *Circulation* 3: Suppl. 2, pp. 68–74
65. Warner, H., Olmstead, C., Rutherford, B. 1972. *Comput. Biomed. Res.* 5(2):65–74
66. Warner, H. R. 1978. *Proc. 2nd Ann. Symp. Comput. Appl. Med. Care, IEEE*, pp. 401–4
67. Weed, L. 1969. *Medical Records, Medical Education and Patient Care.* Cleveland: Case Western Reserve. 297 pp.
68. Williams, T. A., Johnson, J. H., Bliss, E. L. 1975. *Am. J. Psychiatry* 132:1074–76

AUTHOR INDEX

(Names appearing in capital letters indicate authors of chapters in this volume.)

A

Aalberse, R. C., 343
Aarnaes, E., 529
Abbate, J., 206-9, 211
ABBEY, S., 581-604
Abdulaev, N. G., 283, 284
Abe, H., 274
Abelson, J., 182
Abraham, G. E., 333
Abramson, J. J., 274, 275
Abramson, M. B., 482
Adams, M. J., 45
Adams, P. R., 441
Adams, W. B., 454, 455, 461
Adby, P. R., 524
Addison, G. M., 339
Adelman, W. J., 149
Adler, A., 116, 120, 124
Adolfsen, R., 279, 280
Adrian, E. D., 150
Adrian, G., 483, 495, 496
Afromowitz, M. A., 49, 50
Aggarwal, J. K., 361
Aggarwal, R. K., 422
Aggerbeck, L., 5, 10, 11, 13, 14, 16-18, 27, 28
Aggerbeck, L. P., 16, 26
Agris, P. F., 206, 207
Akaike, N., 461
Akimenko, N. M., 116, 123, 127
Akimoto, K., 317
Akowitz, A., 273
Albani, M., 213
Albano, J., 59
ALBERS, R. W., 259-91; 261, 262, 264, 265, 286
Alger, J. R., 387, 388
Allebach, E. S., 490, 496
Allen, D. G., 98
Allen, G., 274
Allen, L. P., 586
Allen, S., 305
Allen, S. D., 304
Aller, J. C., 585
Allerhand, A., 206, 207
Alltop, L. B., 587
Almon, R. R., 249
Almquist, K. J., 570
Aloe, L., 228, 234, 236, 250
Altman, S., 182
Amberson, W. R., 152, 166
Amdahl, L., 582
Ames, B. N., 218
Amos, H., 227
Amselem, S., 486, 500

Amzel, L. M., 276, 277, 279, 280
Anan, F. K., 305, 310
Anbalagan, S., 421
Anderson, C. R., 441
Anderson, E. C., 529, 530
Anderson, P. M., 383
Anderson, R. A., 317
Anderson, W. A., 378
Andreeff, M., 514
Andres, R. Y., 232, 233, 249
Andrews, H. C., 587
Andrews, J. M., 169
Andrieu, J. M., 342
Angel, A., 171
Angeletti, P., 227, 228
Angeletti, P. U., 224, 225, 227, 228, 234, 236-38, 240, 249
Angeletti, R. H., 224-26, 232, 233
Antalis, T., 313
Antanavage, J., 491
Antoni, H., 447, 449, 450
Anundi, H., 249
Apgar, J., 181
Applequist, J., 115
Arch, S., 439
Argos, P., 226
Armando-Hardy, M., 77
Armitage, I. M., 386-90
Armstrong, C. M., 445
Armstrong, R. N., 377
Arnason, B. G. W., 227, 245, 250
Arnold, A. P., 155, 156
Arnott, S., 118
Aroesty, J., 557-61, 563, 564
Arrio, B., 480
Arter, D. B., 208
Arutyunyan, A. M., 303, 310, 315
Arvanitakl, A., 171
Asanuma, H., 155, 156
Ascher, P., 439, 440, 442-45, 455, 456
Asdourian, H., 227
Ash, D. E., 383
Ashby, J. P., 75
Ashcroft, S. J. H., 73, 75
Askinazi, L., 501, 502
Atanasov, B. P., 310
Aten, B., 340, 341
Athens, J. W., 549
Atkins, P. W., 129
Atkinson, D., 16
Attix, F. H., 554
Atwater, I., 74-76

Atwater, I. J., 73
Axelrod, D., 301
Axelrod, F., 244
Axline, S., 596
Ayers, W. R., 585
Ayres, M., 419
Azhderian, E., 206, 207, 209, 210

B

Baase, W. A., 118, 120
Babcock, G. T., 313
Babiarz, J., 477, 478
Bach, D., 498
Bacus, J. W., 399, 421, 422, 424
Badler, N. I., 361
Badoz, J., 293, 304, 307, 315, 316
Baer, R. M., 585
Bagwell, C. B., 520
Bahr, G. K., 565
Bahrami, I., 571
Bailey, D. B., 386, 390
Bailey, N. T. J., 546
Baird, B. A., 276-79
Baisch, H., 526, 529
Baker, P. F., 71, 72, 75
Balakrishnan, M. S., 368
Balazs, N. L., 113
Ballard, D. H., 400, 410, 411, 420
Ballard, J. P., 597
Bamberger, S., 530
Bane, J. L., 227
Banerjee, S. P., 231, 245
Bangham, A. D., 467-69, 471, 476, 477, 479-82, 486, 498, 499
Bank, R., 590, 597
Banks, H. T., 570
Banks, P., 71, 75
Banks, R. D., 285
Banthorpe, D. V., 249
Barakat, R. S., 342
Barber, N. W., 473, 478, 491
Barde, Y. A., 228
Barenholz, Y., 469, 471, 478, 483, 486, 493, 497, 500
Barker, J. L., 461
Barlogie, B., 529, 534
Barndt, R. Jr., 395
Barnes, W. M., 182
Barnett, G., 590
Barranco, S. C., 552, 565

Barratt, M. D., 478
Barrett, J. C., 551
Barron, D. H., 143, 166-68, 172
Barron, L. D., 129, 296
Barsukov, L. I., 473
Bartels, P. H., 395, 421
Barth, G., 303, 305, 308-10, 312, 315, 321
Bartlett, A., 343
Barton, P. G., 470-72
Baserga, R., 125
Baskin, R. J., 268, 478, 480
Bason, R., 315
Bastide, F., 265
Batts, A. A., 73, 75
Batzri, S., 446, 485
Baudhuin, P., 478
Bauer, P. J., 282, 286
Baukal, A., 335, 338
Bauman, A., 327
Bauman, W. A., 594
Bayev, A. A., 196
Bayley, P. M., 108, 110
Baylor, D. A., 163, 164, 169, 170, 445
Bazzone, T., 317
Beadle, B. C., 45
Bear, R. S., 86, 88
Bearden, J. Jr., 135
Beck, H. -P., 529
Beckenbach, E. S., 395
Becker, E., 459
Beckwith, J. B., 246
Beeler, G. W., 451, 458
Beeler, G. W. Jr., 449, 450, 458
Behnke, W. D., 317, 386, 390
Beigelman, P. M., 74, 75
Bekey, G. A., 394
Belanger, M. G., 422
Belforte, G., 394
Bell, G. I., 551
Bellman, R., 558, 568
Bendet, I. J., 135
Benigno, R., 211
Benkovic, S. J., 372, 374
Bennett, D. E., 521
Bennett, L. L., 73-75
Bennett, M. V. L., 150, 158, 173
Bennett, V., 458
Benovic, J. L., 375
Benowitz, L. I., 227, 241
Benson, E. S., 310, 367
Berden, J. A., 473, 478, 491
Berenbaum, M. C., 559
Beresewicz, A., 457
Bergelson, L. D., 473, 477-79, 481, 496
Berger, B. D., 243
Berger, D., 302
Berger, E. A., 230, 231

Bergmann, F., 165
Bergmans, J., 151
Berkinblit, M. B., 144, 146, 147
Berman, M., 264, 286, 559
Berridge, M. J., 55
Berry, M. S., 439
Berson, A. S., 394
Berson, S. A., 327-30, 333, 334, 340
Bertolini, L., 229
Besch, H. R., 457, 458, 460
Bessman, S. P., 74, 75
Bevington, P., 524
Beynon, P., 474
Bhagat, B., 240, 247
Bhoola, K. D., 59
Biales, B., 71, 72, 75
Bibbo, M., 395
Bidwell, D. E., 343
Bigazzi, M., 245
Biggins, R., 71, 75
Bihary, D. J., 354-56, 359, 360
Bill, A. H., 246
Billardon, M., 293, 304, 307
Biltonen, R., 191, 194, 196
Biltonen, R. L., 471
Bina-Stein, M., 191, 194-96, 211, 214
Binder, A., 277
Birdsall, N. J., 484, 500
Birdsall, N. J. M., 269
Bischoff, K. B., 546, 558, 561
Bishop, G. H., 165
Bishop, R., 71, 75
Bittman, R., 474
Bittner, G. D., 162
Bizzi, E., 171
Bjerre, B., 228, 243
Björklund, A., 228, 230, 243
Black, C. D. V., 501
Black, I. B., 228
Blackman, J. G., 62-64, 75, 440, 441, 444
Blair, E. A., 165
Blake, C. C. F., 285
Blake, R. D., 214
Blanchard, M. H., 227
Blanche, P. J., 501
Blankenhorn, D. H., 395
Blasi, F., 182
Blaurock, A. E., 282
Bleick, H., 584
Blinks, J. R., 83, 98
Bliss, E. L., 585
Bliss, T. V. P., 151
Blok, M. C., 472, 498
Bloomgarden, D., 496
Blostein, R., 264, 265
Blum, R. L., 590, 597
Blumenson, L. E., 551
Blumenthal, R., 264, 472
Boccara, A. C., 307
Bocchini, V., 224, 225

Boccia, G., 557, 559, 564
Bock, J. L., 367
Bodenstein, G., 394
Bogaard, M. P., 129
Bogart, B. I., 59
Bogomolni, R. A., 281
Bohren, C. F., 121, 128
Boiti, C., 340
Boland, R., 268, 269
Bolard, J., 310
Bolton, P. H., 199, 200, 206-9, 211, 214
Bolton, T. D., 437, 457
Bond, G. H., 275
Bonting, S. L., 263, 265
Booker, B., 224, 240
Booth, J. R., 361
Bordage, G., 594
Born, M., 137
Bornet, E. P., 275
Boschero, A. C., 73
Boschers, A. C., 76
Bothwell, M. A., 225
Bottenstein, J. E., 250
Bounik, C., 555
Bourne, J. R., 394
Bowditch, H. P., 81
Bowser-Riley, F., 439
Boyd, G. S., 475
Boyd, L. F., 228, 230-32, 237, 249, 250
Boyer, D. M., 228
Boyer, P. D., 274, 276, 279, 281, 367, 368
Boyer, S. H., 244, 245
Brace, C., 411
Braden, W. G., 229, 250
Bradley, D. F., 116
Bradshaw, R. A., 224-26, 228, 230-33, 237, 241, 245, 249, 250
Brady, A. J., 82, 97, 102
Brahms, J., 116-19
Brahms, S., 116-19
Brainard, J. R., 502
Bram, S. J., 118
Brand, B., 116, 120
Brandejs, J. F., 589
Brandt, B. L., 71, 72
Branton, W. D., 439
Brasfield, D. L., 340
Braunstein, A. I., 304
Brazkinkov, E. V., 262
Brecher, P. I., 501
Breddam, K., 317
Breeze, R. H., 304
Brenner, J. F., 421, 423, 429, 430, 432
Breton, J., 303
Brewer, C. F., 185
Brewster, M. E., 502
Briat, B., 293, 302, 304, 307, 310, 315, 316

AUTHOR INDEX 607

Bridgen, I., 283
Bright, P. B., 559
Brill, M. H., 146, 169
Brisson, G. R., 75
Brittain, H. G., 319
Brittain, T., 310, 313, 314
Brocki, T. R., 113
Brockwell, P. J., 551
Brodde, O. E., 458, 459
Bronk, B. V., 547, 548, 551
Brooker, G., 70
Brooks, J. C., 279
Brooks, R. E., 597
Brooks, S. H., 395
Bross, I. D. J., 551
Browett, W. R., 310
Brown, A. M., 461
Brown, G. L., 160
Brown, H. F., 447, 449, 454
Brown, K., 182
Brown, M. S., 16
Brown, R. M., 242
Brown, R. S., 181, 189, 191, 196
Brown, W., 85-88, 91, 92, 94
Brownell, P., 439, 455
Brownstein, M. J., 242
Bruce, W. R., 559
Bruckdorf, K. R., 474
Brücke, E. T. von, 166
Brudno, S., 116, 118, 119
Brunelli, M., 460
Bruni, C., 551
Bruni, C. B., 182
Brunner, J., 484
Brunso-Bechtold, J. K., 233
Brutsaert, D. L., 102
Bryant, F., 372, 374
Buchanan, B., 596
Buchinski, R. K., 498
Buckingham, A. D., 129, 296, 307
Budd, M., 584
Budzik, G. P., 185, 186, 190
Bueker, E. D., 223, 227, 236, 244
Bulbring, E., 446
Bull, T. E., 382
Bullock, T. H., 150, 154
Bunnag, B., 515
Bunnenberg, E., 293, 301, 303, 305, 308-10, 312, 315, 319-21
Burch, J., 588
Burdman, J. A., 245
Burgen, A. S. W., 449
Burgers, P. M. J., 372, 374
Burgess, G. M., 55
Burkhart-Schultz, K., 514
Burnham, P., 228
Burnham, P. A., 248
Burnstock, G., 236
Bürsch, J. H., 395

Burstein, D. E., 249
Burton, L. E., 225, 230
Bus, J., 469
BUSTAMANTE, C., 107-41
Butcher, F. R., 59, 65, 69, 70
Butcher, R. W., 457
Butler, J. D., 476, 482
Butler, V. P. Jr., 333
Byron, K. S., 227

C

Caceres, C. A., 585
Cahen, D., 45-47
Calabrese, L., 317, 319
Calbayrac, R., 480
Calhoun, W. I., 471, 475, 498
Calissano, P., 248, 249
Callagan, D. A., 586
Callender, A. B., 304-6
Calvert, T. W., 404, 406
Cambier, J. L., 395
Cammack, C., 304, 307, 315
Cammack, R., 304, 315, 316, 319
Campbell, D. W., 565
Campbell, I. D., 210
Campbell, J., 235
Campbell, K., 268, 271
Campbell, S. D., 49
Canfield, R. E., 245
Cantley, L. C., 264, 266
Cantley, L. G., 264
Cantor, C. R., 116-18, 120, 191
Cantor, K. P., 117, 118, 120, 121
Caplan, S. R., 46, 47
Carafoli, E., 275
Caramia, F., 228, 240, 249
Cardin, A. D., 386, 390
Carlson, F. D., 478, 483, 493
Carlton, S. G., 399
Carmeliet, E., 447
Carpenter, D. O., 439
Carrano, A. V., 514, 524
Carroll, D., 120, 121, 123
Carter, G. M., 557, 558, 560, 561, 563
Casciano, S., 245
Case, R. M., 55, 58, 61, 69
Cassim, J. Y., 109
Castel, M., 165
Castellucci, V., 460
Castleman, K. R., 351, 357, 360
Castro, J., 267
Castro, Z., 341
Catt, K. J., 335, 338
Catterall, W. A., 276
Cave, A., 385
Cavenett, B. C., 305
Cech, C. L., 115, 117, 119, 126
Cedar, H., 460

Cerny, E. A., 478
Chabay, I., 127
Chailakhyan, L. M., 144, 146, 147
Chaimas, N., 243
Chaimovich, H., 470, 483
Challis, T. W., 411
Chambers, R. W., 191
Chambers, S. H., 535
Chamley, J. H., 236
Chan, A., 116, 118-20
Chan, S. I., 483
Chance, B., 276, 279
Chandler, C., 235
Chandler, D. E., 57
Chandler, H. D., 285, 286
Chandrasekhar, S., 120, 128, 135
Chang, C., 116, 120
Chang, C. W., 394
Chang, C. Y., 405
Chang, S. H., 182, 184, 187
Chapman, D., 270, 470-74, 497
Charbonnel, B., 342
Charlemagne, D., 268
Charlton, J. P., 478, 479
Charpak, G., 100
Chase, M., 114
Chelbowski, J. F., 386, 387, 390
Chen, F. T., 267
Chen, J. S., 236, 249
Chen, M. G. M., 236, 249
Chen, Y. P., 421
Cheng, J. C., 304, 305, 310, 312
Cheng, S. M., 120
Chetverin, A. B., 262
Chevallier, J., 272, 480
Chew, H., 136
Chiang, R., 310, 312
Chien, T. F., 491
Chien, Y. P., 401, 409, 410
Childers, D. G., 361
Ching, Y. C., 491
Chirgadze, Y. N., 262
Chizmadzhev, Yu. A., 147
Chobanian, A. V., 501
Chobanian, J. V., 501
Chong, C. S., 480
Chow, S. T., 473
Chowhan, Z. T., 486
Christensen, I., 526
Christensen, I. J., 529
Christian, B., 71, 75
Christiansen, C., 262, 263
Christodoulakis, S., 411
Chu, L., 264
Chuang, S. N., 560
Chujo, M., 350, 351, 355, 356, 359
Chung, S., 167, 168, 174
Chung, S.-Y., 119

Church, G. M., 191, 196, 197
Churchill, L., 262
Cimino, G. D., 296, 305, 307
Cintrón, N., 276, 277, 279, 280
Cintrón, N. M., 280
Claret, M., 55
Clark, B. F. C., 181, 189
Clark, J. L., 340, 341
Clark, L. B., 320, 321
Clark, N. A., 282
Clarkson, B. D., 562, 565
Clayton, P. D., 361
Clejan, S., 474
Cleland, W. W., 373, 374
Clendenon, N., 246
Clingan, D., 267
Clough, D. L., 275
Cochran, W., 134
Cochrane, D. E., 71
Cohen, A., 420, 421
Cohen, C., 496
Cohen, I., 454
Cohen, J., 587
Cohen, L., 555
Cohen, M. L., 586
Cohen, P., 119
Cohen, S., 224, 237, 596
Cohen, S. A., 227
Cohn, M., 184, 191, 192, 194, 206, 207, 209, 363-67, 374, 377, 383
Colbow, K., 480
Cole, E., 582
Cole, P. E., 191, 192, 195, 199, 200, 202, 211, 214, 215
Coleman, J. E., 309, 317, 319, 386-90
Coleman, J. S., 192
Coleman, P., 486
Coleman, P. D., 350, 356, 357
Coleman, R. V., 317
Collen, M. F., 585, 587
Colley, C. M., 468
Collingwood, J. C., 305, 306, 319
Collins, T., 496
Collman, J. P., 310, 312
Compaan, P., 565
Conklin, G., 590
Connell, C. R., 310, 312
Connelly, C. M., 160, 170
Conners, J., 244
Conners, R. W., 398, 401, 403, 408, 411
Connolly, J. L., 237, 247
Cook, J. M., 249
Cook, J. R., 75
Cook, R. B., 121, 123, 321
Cooke, I. M., 164
Cooke, R., 91, 95
Cooper, D. B., 400
Cooper, H. M., 589
Cooper, J. K., 585
Cooper, R. H., 75

Coore, H. G., 73, 74
Coppin, C. M. L., 173
Coraboeuf, E., 447, 449, 457
Cornelius, R. D., 373, 374
Cortese, R., 218
Costanzi, J. J., 552, 565
Cotter, D. A., 422
Cottrell, G. A., 439
Coughlin, M. D., 228
Cover, T. M., 405
Cowman, M. K., 120, 124
Cozzari, C., 249
Craig, I. F., 475
Cram, L. S., 511, 512, 518, 519, 522
Cramer, F., 200, 207, 209, 211, 213
Cramer, S. P., 310, 313
Cranefield, P. F., 447
Crawford, D. W., 395
Crawford, J., 590
Crawford, J. L., 28
Crawford, W. O. Jr., 409, 413
Creed, K. E., 64
Crespi, H., 284
Crick, F. H. C., 134
Criddle, R. S., 280
Crissman, H. A., 552
Crofts, A. R., 496
Crone, J., 585
Crothers, D. M., 191, 192, 194, 195, 199, 200, 202, 211, 214, 215
Crowell, J. M., 517-19, 522
Cuatrecasas, P., 231, 444, 445, 458
Cullis, P. R., 473-75
Cunningham, L. W., 270
Currie, N., 71, 75
Curry, D. L., 73-75
Curry, K., 73, 75
Cushley, R. J., 185, 498
Cypser, R. J., 582
Czarnecka, M., 449

D

Dagorn, J. C., 61
Dahl, J. L., 260-63, 265, 273
Danchin, A., 191, 194
Dancis, J., 244
Daniel, W. E. Jr., 184, 206, 207, 209
Danielsson, S., 306
Dantzing, G. B., 585
Darszon, A., 488, 489
Darzynkiewicz, Z., 514
Das, G. C., 121
Daune, M., 121
Davanloo, P., 200, 207, 209, 211, 213
Davanloo-Malherbe, P., 209, 213

Dave, C., 499
Davis, B., 73, 75
Davis, P. J., 470, 471
Davis, R., 596
Davis, R. C., 317
Davis, W. J., 354-56, 359, 360
Davison, J. S., 60
Dawidowicz, E. A., 473
Dawson, C. M., 75, 76
Dawson, J. H., 305, 310, 312, 313, 315
Dawson, R. M., 469
Day, P., 305, 306, 319
Deamer, D., 486
Deamer, D. W., 268, 274, 470, 476, 486, 487, 492, 496, 500
Dean, P. M., 64, 73-76
DEAN, P. N., 509-39; 514, 525-30, 541, 552
Dean, T. N., 552
Dean, W. L., 270, 271, 286
DeBruin, O. A., 337
Deckelbaum, R. J., 16
Dedrick, R. L., 558, 561
Degani, C., 274
DeGennes, P. G., 120, 128, 135
de Gier, J., 263, 471, 472, 475, 478, 480, 498, 499
Deguchi, N., 261
De Iraldi, A. P., 240
De Kruijff, B., 469, 471, 473-75, 478, 479
Delabar, J. -M., 321
DeLarco, J. E., 229
del Castillo, J., 440, 442, 443
Delisi, C., 551
de Meis, L., 268
Demel, R. A., 471, 474, 475, 478
DeMori, R., 394
Dempsey, M. E., 367
Dempster, M. A. H., 524
DeMurcia, G., 121
Dencher, N. A., 282, 286
Denlinger, R., 246
Denning, R. G., 305, 306, 319
Dennis, H., 472
DePamphilis, M. L., 373
de Pont, J. J. H. H. M., 263, 265
De Robertis, E., 240
Deroo, P. W., 474
Derr, J. E., 468
Desmond, E. J., 305
DeStefano, O., 590
Dethlefsen, L. A., 534
Deupree, J. D., 261-63, 273
Deutsche, C. S., 116
Devis, G., 73
Devlin, F., 304, 315, 316
DeVoe, H., 114
De Weer, P., 264

AUTHOR INDEX 609

DeZoete, 337
Diamond, S. L., 394
Dichter, M., 71, 72, 75
Dichter, M. A., 72, 229, 230, 235
Dickman, S., 206, 207
Dienes, G. J., 547, 548, 551
DiFrancesco, D., 454
DiLauro, R., 182
Dimitriadis, G. J., 490
DiNocera, P. P., 182
Dinstein, I., 397, 399, 401
Diplock, A. T., 499
Di Polo, R., 268
Dirheimer, G., 214
Disch, R. L., 122
Dittmer, J., 469
Dixon, J. F., 261-63, 265, 267, 273
Djerassi, C., 293, 301, 303, 305, 308-10, 312, 315, 319-21
Dobson, C. M., 210
Dobson, J. G., 458
Dolce, G., 394
Dolinger, P. M., 305, 310, 315
Donaghey, C. E., 534
Donato, H. Jr., 320
Donatsch, P., 75
Donohue, J. F., 394
Dorkin, H., 245
Dorman, B. P., 116, 120-23, 125, 127, 128
Dormer, R. L., 58
Dorn, H., 174
Doskocil, J., 116, 123
Doty, P., 307, 308
Doty, R. W., 171
Dougherty, J. P., 273
Douglas, B. E., 319
Douglas, I. N., 304, 307, 315
Douglas, W. W., 55, 70, 71, 446
Dourmashkin, R. R., 490
Dowben, R. M., 476, 490
Downer, N. W., 184
Downie, D., 304
Downing, A. C., 152, 166
Drachev, L. A., 283, 284
Drakenberg, T., 385
Drapeza, A. I., 586
Dray, F., 342
Dray, S., 486
Drewinko, B., 534
Dreyfus, S., 568
Drezner, M. K., 228
Drucker, B., 92
Druckrey, H., 246
Duance, V. C., 272
Dubko, M. J., 586
Duda, R. O., 358, 400
Dudel, J., 170, 438, 449, 450
Duff, M. J. B., 361
Dugas, H., 199

Dulak, N. C., 262, 265
Dun, F. T., 166, 170
Duncan, G., 73, 76
Dunlap, K, 452
Dunlap, L., 491
Dunn, J. B. R., 304, 310, 315, 316
Dunn, M. F., 225
Dunn, R. A., 394, 587
Dupont, Y., 3, 4, 269, 273
Durie, B. G. M., 421
Dutreix, J., 555
Dutta, C. R., 152
Dutton, A., 275, 285
Duval, J. F., 303
Duysens, L. M. N., 121
Dwek, R. A., 375, 377, 390
Dwyer, S. J. III, 395, 401, 411
Dy, P., 245

E

Eargle, D. H., 367
Early, M., 166
Easty, D. M., 247
Easty, G. C., 247
Eaton, W. A., 310, 312, 315
Ebashi, S., 268
Ebner, E., 280
Eccles, J. C., 174
Eckstein, F., 372, 374
Edgar, D. H., 238
Edmonson, D. P., 117, 118, 121
Edwards, D. C., 228, 246
Ehrenberg, B., 107, 284
Eibl, H. J., 470
Eichhorn, G. L., 120, 123
Eimerl, S., 446
Eisenbarth, G. S., 228
Eisenbeis, S. A., 409
Eisner, D. A., 454
Ekins, R. P., 342
Elder, D. L., 293, 321
Eldridge, R., 244
Elfrim, L. G., 268
Ellerman, J., 73
Ellerman, J. E., 73
Elliott, G. F., 89, 91, 94
Elliott, H., 400
Ellis, F., 555, 556
Ellis, K. J., 319
Ellis, P. D., 386, 390
Ellisman, M. H., 488, 489
Ellory, J. C., 77
Elseviers, D., 182
Elstein, A., 594
Elvidge, J. A., 184, 186
Emler, C., 70
Enagonio, R. D., 534
Endicott, J., 587, 598
Endo, M., 82

Endoh, M., 83, 97, 98, 100-2, 457-59
Englander, S. W., 184, 214
Engstfeld, G., 447
Engstrom, H., 304
Enoch, H. G., 484, 500
Entman, M. L., 275
Epstein, P., 116, 120
Erard, M., 121
Erhard, K., 317
Erlanger, J., 165
Ernster, L., 276, 279
Esker, M. W. J., 485
Esmann, M., 262, 263
Estep, T. N., 471
Estival, A., 61
Evans, E. A., 184, 186
Evans, J. E., 469
Evans, P. R., 285
Evdokimov, Y. M., 116, 123, 127
Everett, G. A., 181
Evert, C. F., 549
Evstigneeva, R. P., 473
Eyring, H., 309, 320, 321
Eytan, E., 271, 280

F

Fabiato, A., 82
Fabiato, F., 82
Fabricant, R. N., 229, 244
Facchinetti, T., 498
Fahey, P. F., 491
Fahn, S., 265
Fain, J. N., 55
Farley, R. A., 267
Faruqi, A. R., 100
Fasman, G. D., 116, 118, 120, 121, 124
Fast, P. G., 472, 475
Fatt, P., 72
Favre, A., 191
Fee, J. A., 390
Feigenson, G. W., 493
Feigina, M. Yu., 283, 284
Feldman, S., 598
Feldtkeller, E., 305
Fellows, R. E., 242
Felsenfeld, G., 116, 121
Feltz, A., 445
Fendler, J. H., 468
Fenton, E. L., 246
Ferraris, F., 394
Ferreira, H. G., 77
Ferrendelli, J. A., 250
Fertel, R., 73
Fessenden-Raden, J. M., 279
Fidler, I. J., 247
Figueroa, N., 201-3, 215
Fildes, F. J., 498
Filimonov, V. V., 196
Finch, J. T., 181, 189
Findlay, E. J., 470-72

AUTHOR INDEX

Finkelstein, M., 468, 486
Finkelstein, M. C., 481
Finney, R., 565
Fischbach, G. D., 452, 453
Fischer, B., 530
Fischer, J. J., 556, 566, 568
Fischer, S., 592
Fisher, D. A., 230, 241
Fisher, K. A., 282
Fisher, V. J., 99
Fiske, C. H., 468
Fitzhugh, R., 149
Fleckenstein, A., 457
Fleckenstein, U. A., 447
Fleischer, S., 265, 268, 272
Fleiss, J., 587
Fleiss, J. L., 587
Fleminger, S., 77
Fless, G. M., 17
Flohe, L., 319
Foley, D. H., 408
Fomin, S. V., 144, 146, 147
Fong, C. P., 411, 416
Foote, A. M., 26
Forbes, A., 166
Forbush, B. III, 263
Ford, N. C. Jr., 191
Foreman, J. C., 69
Forge, A., 475, 476, 483
Forrest, B. J., 498
Forsén, S., 378, 380, 382
Forslund, J. C., 511, 512
Förster, T., 185
Forte, T. M., 501
Fortes, P. A. G., 265, 266
Fournier, M. J., 191
Fowler, R. D. Jr., 587
Fox, C. A., 152
Fox, J. J., 185
Fozzard, H. A., 394
Frankenhaeuser, B., 159, 169, 170
Frank-Kamenetskii, M. D., 117, 118
Franks, N. P., 474
Frazier, W. A., 226, 230-33, 241, 249, 250
Freiherr, G., 587, 596
Fresco, R., 214
Freund, A. M., 26
Frey, P. A., 372
Freygang, W., 169
Fric, I., 120, 121
Fried, J., 526, 527, 529
Frisch, H. L., 401
Fritz, S. L., 395, 401
Fritzinger, D. C., 191
Froehlich, J. P., 264, 273, 286
Frolov, A. V., 586
Frunzio, R., 182
Fry, D. W., 495
FU, K.-S., 393-436; 395-97,

399, 401, 404-10, 422-24, 428, 430, 431
Fujii, I., 117
Fujita, I., 310
Fujita, T., 250
Fujiwara, M., 458
Fukumura, T., 410
Fukunaga, K., 397, 404-6, 408
Fukuzawa, K., 499
Fulton, R. L., 115
Fung, C. H., 375, 376
Fuortes, M. G. F., 167
Furano, A. V., 237
Furukawa, S., 245
Futamachi, K. J., 171
Futterman, G., 517

G

Gabor, A. J., 171
Gabriel, A., 3, 4
Gabriel, M., 308, 317, 319
Gagnon, C., 228
Gainer, H., 461
Gakhova, E. N., 456
Gale, R., 296, 322
Galla, H. J., 472
Gallacher, D. V., 59
Galloway, M. M., 401
Gallus, G., 421, 423
Galvin, J., 565
Gamble, R. C., 184-90, 496
Gammack, D. B., 482
Gandini-Attardi, D., 237
Gangloff, J., 214
Garcia-Iniquez, L., 313
Gardner, J. D., 67
Gardner, J. F., 182
Gardner, R., 586
Gardner-Medwin, A. R., 157
Garfinkel, D., 559
Gargouil, Y. M., 447, 449
Garnier, A., 310
Garnier, D., 438, 447, 449, 450
Garrett, J. R., 58, 59
Garrett, R. A., 121, 127
Garty, H., 46, 47
Garvey, C. F., 350, 356, 357
Garzaniti, J. P., 494
Gassen, H. G., 213
Gasser, H. S., 151, 158, 165
Gebicki, J. M., 470, 492
Gedder, A. D., 477, 478
Gee, T. S., 562, 565
Geerdes, H. A. M., 208, 209, 211-13
Gehan, E. A., 561
Geissner-Prettre, C., 208
Geller, J., 590
Gelles, J., 264, 266
Gelsema, E. S., 421, 423, 429, 430, 432
Gennis, R. B., 117, 118

George, P. K., 16
George, W. J., 447
Ger, B. A., 439, 444
Gerber, G. E., 283, 284
Germani, A., 551
Gerritsen, W. J., 479, 484, 485, 489
Gerschenfeld, H. M., 439
Gersho, A., 36
Gevins, A. H., 394
Ghelis, C., 272
Giacomoni, D., 486
Giannoni, G., 116, 126, 127
Giaretti, W., 558, 563
Gibbes, D., 478, 483, 493
Giebink, G. A., 585-90, 598, 599
Giese, D. A., 394
Gilbert, C. W., 551
Gilbert, E. N., 401
Giles, W., 438, 449, 450, 453, 454, 457
Gilliatt, R. W., 151
GINSBORG, B. L., 55-80; 59-65, 72, 75, 439-41, 444, 446
Giotta, G. J., 262
Giovenco, M. A., 551
Girod, J. C., 116, 118
Gitler, C., 266, 280
Gitter-Amir, A., 121, 125
Givens, A. S., 394
Givol, D., 342
Glaser, L., 228, 237
Glaser, M., 121
Glasser, S., 361
Glaze, K. A., 228, 247
Gledhill, R. F., 169
Gleser, M. A., 587
Glitsch, H. G., 444
Glynn, I. M., 261, 264
Gnedenko, L. S., 144, 146, 147
Goad, C. A., 522
Godbillon, G., 319
Goebel, R., 264, 286
Goel, N. S., 543
Gofman, J. W., 561
Gohde, W., 526
Goldberg, N. D., 447
Goldfine, I. D., 232
Goldman, I. D., 495
Goldman, J. A., 70
Goldman, R., 498
Goldman, Y., 450, 458
Goldman, Y. E., 95
Goldsmith, L., 116, 120, 124
Goldstein, A., 343
Goldstein, D. A., 274
Goldstein, J. L., 16
Goldstein, L. D., 227
Goldstein, M. N., 245
Goldstein, S. S., 144-48, 161, 173

AUTHOR INDEX 611

Goll, J., 478, 483, 493
Goller, I., 236
Gomez, E. R., 511, 512
Gomez, M., 74
Gomperts, B. D., 69, 496
Gompertz, B., 543
Gong, E. L., 501
Gonzalez, R. C., 404
Goode, K., 565
Goodman, J. J., 272
Goodman, R., 229, 235, 238
Goodwin, B., 264
Goody, R. S., 374
Gopinath, R. M., 275
Gorden, P., 227, 230, 241
Gordon, D. G., 116, 121, 125, 128
Gordon, D. J., 121, 128
Gordon, W. A., 246
Gorin, P., 241
Gorry, G. A., 393
Gose, E. E., 395, 405, 421, 422
Gosule, L. C., 125
Gotfredsen, C., 478
Goto, M., 449, 450, 458, 460
Gottlieb, A., 496
Gotto, A. M., 16
Gotto, A. M. Jr., 16
Goupil, N., 449
Gouterman, M., 115, 310, 312, 322
Graf, T., 280
Graffman, S., 566
Graham, C. D., 449
Graham, D. E., 272
Graham, H. T., 150
Grand, R. J., 69, 70
Grange, J., 308, 317
Granot, J., 377
Grant, D. M., 206, 207
Grant, W. B., 304
Gratzer, W. B., 121, 127
Graubatz, G. L., 439
Gray, C. P., 283, 284
Gray, D. M., 116-19, 121, 123, 321
Gray, H. B., 317, 318
Gray, J. I., 468
GRAY, J. W., 509-39; 514, 524, 525, 529, 532, 534, 541, 552
Gray, L. J., 116
Gray, T. A., 59, 65-67, 70
Gray, W. M., 556, 569
Green, C., 596
Green, G., 116, 118, 119
Green, J. R., 422, 423
Green, N. M., 268-70, 273-75
Greenberg, R. E., 227
Greene, L. A., 72, 225, 227, 229-31, 235-37, 241, 242, 247-50

Greengard, P., 159, 169, 170, 447, 454, 457
Greenleaf, J. F., 586
Greenspan, C. M., 184
Greenwood, C., 310, 313, 314
Greer, J., 357
Gregor, H. P., 482
Gregoriadis, G., 468, 470, 478, 481, 490, 501
Gregory, D. W., 477
Gregory, R. P. T., 121, 125
Grell, E., 491, 492
Gremillion, L., 582
Greve, J., 118, 120
Greville, G. D., 480
Griffin, K. P., 303
Griffith, A. H., 246
Grisham, C. M., 266, 383
Grisham, G. C., 261, 266
Grodsky, G. M., 73-75
Gross, J. F., 557
Gross, K. P., 305
Gross, M., 582
Grosse, R., 261
Grossman, Y., 161, 163, 173, 175
Groth, T., 566
Groudinsky, O., 310, 315
Grover, A. K., 498
Grunberg-Manago, M., 191, 194
Grundfest, H., 151, 158, 169
Gruol, D. L., 439
Guéron, M., 191, 206, 207, 211
Guillery, R. W., 171
Guillory, R. J., 280
Guinier, A., 2, 7
Gulik, A., 26
Gulik-Krzywicki, T., 26
Gul'ko, F. B., 144, 161
Gulrajani, R. M., 454
Gunner, D. B., 490
Gunsalus, I. C., 312
Guo, S. M., 230, 241
Guoit, P., 478
Gupta, C. M., 469
Gupta, R. K., 375, 376
Gurney, C. W., 545, 549
GUROFF, G., 223-57; 228, 233, 234, 238, 239, 242, 249, 250
Guschlbauer, W., 321
Gutnick, M. J., 171
Gutteridge, J. M., 468

H

Haap, K., 460
Haar, F. V. D., 200, 207, 213
Haas, D. C., 250
Habbersett, M. C., 515
Haberditzl, W., 310
Habgood, J. S., 167

Hach, R., 192, 194, 195
Hachimura, K., 395
Hackney, J. F., 261-63, 265, 267, 273
Haest, C. W. M., 499
Hager, L. P., 310, 312
Hagiwara, S., 67, 71, 72, 76, 447
Hahn, G. M., 534, 548, 557, 561, 564
Haimovich, J., 342
Hakansson, R., 306
Hales, C. N., 74, 75, 338, 339
Hall, D. O., 304, 307, 315, 316
Hall, E. J., 555
Hall, E. L., 409, 413-15
Hall, K., 245
Hall, R. A., 59, 65-67, 70
Halloran, M. J., 586
Hamburger, V., 224, 227, 233-35
Hamill, W. D. Jr., 206, 207
Hamilton, B. L., 356, 359
Hamilton, F. D., 117
Hamilton, G. A., 386
Hammes, G. G., 276-79, 286
Hamori, J., 171
Hanes, S., 551
Hankinson, J. L., 394, 415
Hanlon, S., 116-20
Hanser, G., 530
Hanson, J., 81
Hanson, L. K., 310, 312
Hanson, O., 265, 286
Haralick, R. M., 397, 399, 401
Harding, M. J., 319
Hardwicke, P. M. D., 268, 269
Hardy, G. W., 285
Hargreaves, W. R., 470, 476, 492, 500
Harlos, K., 470, 497
Harlow, C. A., 395, 398, 401, 403, 408, 409, 411
Harmar, A. J., 454, 455, 461
Harmatz, D., 121
Harmony, J. A., 502
Harms, H., 305
Harper, J. F., 70
Harrington, W. F., 92
Harris, E. J., 438
Harris, L. D., 361
Harris, R. A., 114, 129
Harris, R. S., 99
Harris, T. E., 545, 546
Harrison, B. M., 169
Harrison, S. C., 269
Hart, P. A., 373
Hart, P. E., 358, 400
Hart, T., 242, 243
Hart, W. M. Jr., 265
Hartmann, J. R., 246
Hartmann, N. R., 526, 529, 534, 551

Hartzell, C. R., 386
Hartzell, H. C., 439-42, 444
Hasegawa, J. I., 410
Haselgrove, J. C., 88, 89, 91, 92, 94, 100
Haser, R., 285
Hashizume, H., 95, 117
Hasselbach, W., 268, 269
Hassner, D., 420, 421
Hastings, D. F., 262, 263
Hatanaka, H., 243, 250
Hatano, M., 293, 304, 310, 312, 313, 316, 322
Hatt, H., 162, 169, 170, 173
Hattan, D., 270, 273
Haugh, E. F., 112
Hauser, H., 478, 482, 484
Hauswirth, O., 454
Hawkins, E. R., 184
Haworth, R., 277
Hayashi, F., 206, 207
Hayashi, K., 245
Hayashi, Y., 264
Hayatsu, H., 185
Hayes, J. B., 259
Haymovits, A., 58, 60
Haynes, C., 588
Haynes, M., 121, 127
Heap, P. F., 59
Hearst, J. E., 117-23, 127
Heath, T. D., 468
Hebdon, G. M., 270
Hedeskov, C. H., 73
Hedlund, B. E., 310
Heilmeyer, L. M. G., 275
Heinemann, S., 230, 247
Heintzen, P. H., 395
Heiser, J. F., 597
Helenius, A., 485, 497
Heller, S. R., 185
Helmreich, E. J. M., 445
Helson, L., 244
Hemenger, R. P., 116
Henderson, R., 281-83
Hendry, I. A., 227, 228, 230, 233, 235, 241, 249
Henquin, J. C., 75, 76
Hepburn, H. R., 285, 286
Heppel, L. A., 267
Heppner, R. L., 450, 458
Herbert, V., 341
Herchuelz, A., 73
Herlands, L., 116, 120
Herman, C., 515
Herman, G. T., 357
Herrup, K., 231-33
Herschmann, H., 246
Herschmann, H. R., 229, 235, 238
Hertz, H. G., 380
Hertz, R., 497
Herzog, V., 60
Hesketh, T. R., 269

Hess, H. H., 468
Hethcote, H. W., 547, 556, 568
Heyman, A., 588
Heyn, M. P., 282, 286
Hicks, M., 470, 492
Hiebert, R. D., 517-19, 522
Hier, D. B., 250
Higuchi, W. I., 486
Hilaire, M., 303
Hilbers, C. W., 190, 199, 200, 202, 206, 207, 211, 212, 214, 216
Hilborn, D. A., 279, 286
Hilden, S., 260, 263, 264
Hill, A. V., 82
Hill, L. R., 118, 119
Hill, M. W., 468, 469, 471, 476, 477, 479-82, 498, 499
Hillaby, R., 497
Hille, B., 158
Hillman, D. E., 152, 350, 351, 355, 356, 359
Hill-Smith, I., 64, 440, 441, 444
Hilton, B. D., 184
Himmelstein, K. J., 546, 558, 561
Hino, N., 449
Hipps, K. W., 296
Hirabayashi, T., 87
Hirata, H., 276, 278, 279
Hirschfelder, J. O., 112
Hirsh, D., 218
Hishy, R. G., 384
HOBBS, A. S., 259-91
Hochstein, S., 145, 146, 149, 165, 170
Hodgkin, A. L., 71, 75, 144, 147, 150, 159, 169, 170, 445
Hodgson, K. O., 224, 226
Hoffer, M. M., 394
Hoffman, B. F., 449
Hoffman, H., 240
Hoffman, J. F., 263
Hoffman, R. M., 481, 496
Hoffstein, S., 496
Hofmann, H., 530
Hogan, V., 501, 502
Hogg, J. D., 548
Hogue-Angeletti, R., 228
Hogue-Angeletti, R. A., 226
Hokin, L. E., 260-65, 267, 273
Holasek, A., 16
Holbrook, S. R., 191, 196, 197
Holland, P. C., 268
Hollebone, B. R., 310, 322
Holley, R. W., 181
Holloway, P. W., 485
Holm, R. H., 310, 312
Holmes, K. C., 94
Holmes, O., 160
HOLMQUIST, B., 293-326;

293, 295, 303, 304, 307, 309, 310, 315-20
Holowka, D. A., 279
Holroyd, S. A., 270
Holt, J. G., 565
Holtwick, S., 240
Holy, A., 321
Holzwarth, G., 116, 119, 121, 125, 127, 128, 307
Holzwarth, N. A. W., 121, 127
Homareda, H., 264
Homer, R. B., 320
Honigfeld, G., 598
Hope, C. S., 565
Hopfer, U., 499
Hopfield, J. J., 310
Hopkins, B. E., 261, 262, 265, 267
Hopkins, D. A., 157
Horan, P. K., 510
Horemans, B., 75
Hori, S., 234, 250
Horii, Z. I., 237
Hörl, W. H., 275
Horn, J. P., 452
Horowitz, H., 565
Horowitz, J., 206
Horowitz, S. L., 394
Horrocks, W. D., 320
Horsnell, T. S., 535
Horstman, L. L., 279
Horton, W. J., 206, 207
Horwitz, A. F., 473
Hosoda, J., 118, 120
HOUSE, C. R., 55-80; 59-65, 75, 439-41, 444, 446
Houslay, M. D., 269
Hover, B. A., 73
Howard, A., 525
Howes, A. E., 556
Hoy, R., 411
Hoy, R. J., 401, 411, 416, 419
Hu, A., 363-67
Huang, C., 478, 479
Huang, C. H., 473, 474, 477-79, 482, 483, 493
Huang, H. K., 415, 587
Huang, L., 483, 495, 496
Huang, T. S., 398
Hubbard, J. I., 172
Hudson, B., 299
Hudson, J. L., 520
Hueckel, M. H., 398
Huff, K., 234, 250
Huff, K. R., 239, 242, 250
Hug, W., 117, 119, 126
Hughes, G. M., 162
Hullihen, J., 276, 277, 279, 280
Humphrey, R. M., 552, 565
Hunt, B. R., 587
Hunt, C. A., 477, 480, 481, 494
Huntington, S. K., 116, 118

AUTHOR INDEX 613

Hurd, R. E., 199, 201, 206-11, 216
Hurley, M. R., 265
Hurst, J. W., 593
Hurst, L. L., 585-90, 598, 599
Hutter, O. F., 438, 449, 450
Hutton, J. C., 73
Hutton, W. C., 266, 383, 473, 474
Huxley, A. F., 81, 82, 144, 150
Huxley, H. E., 81, 83, 85-92, 94, 95, 100, 101
Hwang, K. S., 496

I

Ibel, K., 3
Idahl, L. A., 73
Ierokomas, A., 271
Ifshin, M. S., 461
Ihm, J., 502
Iida, S., 185
Iizuka, T., 310
Ikehara, M., 321
Ikemoto, N., 272, 273
Ikemoto, Y., 449, 450, 458, 460
Ikeno, H., 499
Ikeno, T., 242
Iles, G. H., 271
Imahori, K., 117
Imai, Y., 310
Imamura, H., 459
Inesi, G., 269, 270, 272, 273
Ingebretsen, W. R. Jr., 459
Ingram, K., 421
Inoue, K., 468, 470, 497
Inskeep, W. H., 320
Inui, J., 459
Irvin, G. L. III, 520
Isaacson, Y. A., 474
Isac, T., 469, 472, 498
Ishii, D. N., 251
Ishimura, Y., 310
Israelachvili, J. N., 473, 483
Ito, F., 174
Ito, M., 166, 169
Ito, M. H., 95, 100
Ito, T., 472
Ivankovic, S., 246
Ivanov, V. I., 117-19
Iversen, L. L., 233, 249
Iwama, K., 171
Iwashi, H., 215
Iwatsuki, N., 60-62, 64, 65, 67, 68
Izen, E. H., 304, 316
Iziuka, T., 310
Izumi, T., 95, 100

J

Jack, A., 191, 196
Jack, J. J. B., 173

Jackson, R. L., 16
Jacobs, H. S., 339
Jacobs, J. W., 226
Jacobs, R., 299
Jacobs, S., 445, 458
Jacobson, G., 412, 416
Jacobson, K., 469-72, 489, 497, 498
Jacquez, J. A., 558
Jaffee, E. K., 374, 377
Jagendorf, A., 277
Jagoe, J. R., 414, 415
Jain, M. K., 473, 499
Jallon, J. H., 321
Jamieson, J. D., 58
Jansen, J. K. S., 163, 164
Jansson, B., 551
Jarrett, H. W., 275
Jasperson, J. N., 304
Jean, D.-H., 262
Jekowsky, E., 185-89
Jeng, I., 232, 249
Jenkinson, D. H., 55, 446
Jennrich, R. I., 524, 559
Jensen, D. E., 117, 120
Jensen, J., 265, 286
Jensen, R. H., 516, 522
Jesaitis, A. J., 265, 266
Jett, J. H., 511, 512, 514, 526-29
Jewell, B. R., 82, 83
Jeyasingh, J. K., 478
Ji, T. H., 123
Jilka, R. L., 270
Jimenez de Asua, L., 267
Jirakulsomchok, D., 59
Johnson, D. G., 227, 230, 241
Johnson, E. M. Jr., 233, 241, 249, 250
Johnson, J. H., 585
Johnson, J. P., 310, 313
Johnson, R. S., 116, 118, 120
Johnson, S. A., 586
Johnson, S. M., 475, 476, 479, 482, 483
Johnson, W. C., 116-20
Johnson, W. C. Jr., 108, 117
Johnston, D. A., 529
Johnston, P. D., 199, 201-3, 205, 207, 208, 210, 211, 214-16
Jolly, P. C., 264
Jonah, M. M., 478
Jones, C. R., 192, 211
Jones, E. G., 171
Jones, J. B., 304, 315, 316
Jones, J. R., 184, 186
Jordan, C. F., 116, 123, 127
Jørgensen, K. E., 273, 286
Jørgensen, P. L., 261, 262, 265, 266, 285, 286
Josephson, I., 458
Josephson, L., 264, 266

Jovin, T. M., 510
Joyner, R. W., 147, 148, 161, 171
Judstrup, R., 582
Julesz, B., 401
Juliano, R. L., 497
Jullien, M., 480
Jung, B., 566
Jungalwala, F. B., 469

K

Kachel, V., 530
Kaden, T. A., 317, 318
Kado, R. T., 441
Kadykov, V. A., 116, 123, 127
Kagawa, Y., 276-80, 484
Kagayama, M., 59
Kai, K., 185
Kainosho, M., 483
Kaiser, E. T., 377
Kaito, A., 322
Kajiyoshi, M., 305, 310
Kak, A. C., 358
Kalaba, R., 558
Kallenbach, N. R., 211, 214
Kalman, T. I., 185
Kamen, D. M., 310, 312
Kamiyama, A., 84, 89, 90, 95-97
Kan, L. S., 200, 207, 213
Kanal, L., 394
Kanamori, N., 174
Kandel, E. R., 451-53, 456, 460
Kaniike, K., 264, 275
Kanno, T., 57, 65, 66
Kantor, H. L., 468
Kao, R. L., 461
Kaplan, E., 395
Kaplan, J. H., 263
Kaplan, N. O., 458
Kaplan, R., 496
Kaplan, T., 116
Karlish, S. J. D., 261, 264, 266, 285
Karny, O., 279
Kasai, M., 272
Kasamatsu, T., 171
Kashdan, N. R., 586
Kassirer, J. P., 393
Kastrup, R. V., 210, 213
Katchman, A. N., 444
Kater, S. B., 55, 58
Kates, M., 282
Kato, H., 317
Kato, Y., 320
Katz, A., 273
Katz, A. M., 81, 274
Katz, B., 70, 171, 440-43
Katz, M., 420, 421
Katzmann, R., 482
Kaulen, A. D., 283, 284
Kavaler, F., 99

AUTHOR INDEX

Kawamura, H., 171
Kawazoe, Y., 185
Kawazu, S., 73, 76
Kay, C. M., 192
Kayalar, C., 279
Kayne, F. J., 383
Kayne, M. S., 192
Kazachenko, V. N., 456
Ke, B., 304
Kearns, D. R., 192, 199, 200, 206-11, 214
Kebabian, J. W., 457
Keeler, R. F., 71
KEHOE, J. S., 437-65; 439, 440, 446, 455, 456
Keiderling, T. A., 304, 305, 315, 316
Keiding, N., 526
Keller, J. B., 547
Keller, W., 121
Kellogg, J., 70
Kelly, R. C., 117, 120
Kemp, A., 279
Kempt, J. C., 304
Kendall, D. G., 545, 546
Kendall, F., 125, 558, 563
Kennedy, F. S., 317
Kenyon, G. L., 367
Keough, K. M., 470-72, 474
Kereiakes, J. G., 565
Kerker, M., 131, 136
Kersten, H., 213
Keth, H. D., 116, 126, 127
Kettler, R., 238
Khitrina, L. V., 283, 284
Khodorov, B. I., 144, 146, 161, 169, 173
Kholopov, A. V., 144, 146, 147
Khorana, H. G., 283, 284, 469
Kidd, A., 58, 59
Kidokoro, Y., 71, 72, 230, 247
Kidson, C., 119
Kielczewski, M., 310, 315
Kilkuskie, R., 119
Killingly, D., 182
Kim, M., 529, 534, 571
Kim, S. H., 181, 182, 189, 191, 196, 197
Kimberg, D. V., 69, 70
Kimelberg, H. K., 263, 468, 473, 474, 498, 499
Kimimura, M., 264
King, A. A., 45
King, E. B., 516, 522
King, T. J., 547
King-Smith, E. A., 549
Kinius, P., 394
Kinoshita, M., 395
Kirchberger, M. A., 274
Kirk, J., 556, 569
Kirkbright, G. F., 45
Kirkepar, S. M., 71

Kirkwood, J. G., 112
Kirsch, W. M., 245
Kirschberger, M. A., 275
Kirste, R. G., 10, 12
Kiselev, A. V., 283, 284
Kishimoto, Y., 469
Kitagawa, T., 468, 470, 497
Kleihues, P., 246
Klein, D., 598
Klein, M., 451-53, 460
Klein, M. P., 301-3, 305, 307, 315, 473
Klein, R. A., 500
Kling, D., 279
Klip, A., 268, 271, 274
Klodos, I., 266
Klug, A., 181, 189, 191, 196
Knott, G., 559
Knowles, A. F., 271, 280
Knowles, P. F., 475-79, 483, 493
Kobayashi, H., 310
Kobayashi, N., 310
Koch, G., 551
Kochupillai, N., 334, 336
KOCSIS, J. D., 143-79; 157
Koehler, K. A., 384
Koestner, A., 229, 236, 246, 247
Kohda, M., 492
Kohlhardt, M., 460
Koketsu, K., 437, 452, 457
Koles, Z. J., 146
Komoroski, R. A., 206, 207
Kondo, H., 377
Koning, R. E., 305, 306
Konishi, T., 284
Konstantinov, A. A., 303, 315
Kopin, I. J., 227, 230, 241
Korc, M., 58
Korenbrot, J. I., 484
Koreneva, L. G., 319
Korn, E. D., 482, 485
Kostner, G., 16
Kostyuk, P. G., 461
Kotler-Brajtburg, J., 73, 74
Koudelka, J., 123
Koval, G. J., 262, 264, 265, 286
Kovalev, S. A., 144, 146, 147
Kovatchev, S., 497
Kozlov, I. A., 276, 279
Krakauer, H., 196
Kratky, O., 3, 16
Kraus, J. D., 135
Kremer, J. M. H., 485
Kretchmar, A. L., 550
Kripke, M. L., 247
Krisch, B., 228, 236
Krishtal, O. A., 461
Krivacic, J., 121
Krnjević, K., 168, 170-72, 445, 460

Krueger, J. W., 96, 97
Kruger, R. P., 401, 403, 409, 413-15
Krupp, L., 501
Krzanowski, J., 73
Kuba, K., 437, 457
Kubota, M., 277
Kudo, I., 185
Kuffler, S. W., 439-42, 444
Kulick, J. H., 411
Kulikowski, A., 393
Kulpa, C. F., 481, 496
Kumar, S., 246
Kume, S., 267
Kundig, W., 259
Kunitake, T., 470, 492
Kunkel, G. A., 394, 419
Kunkel, H., 394
Kuroda, H., 174
Kushwaha, S. C., 282
Kuwahara, M., 395
Kwan, C. -Y., 317
Kwan, T., 310
Kyle, M. C., 394
Kyo, J. -Y., 317
Kyogoku, Y., 215
Kyte, J., 262, 263, 265, 267

L

LaBelle, E. F., 493, 499
LaCorbiere, M., 247, 248
Ladbrooke, B. D., 471, 472
Ladner, J. E., 191, 196
Laggner, P., 16, 272
Lagwinska, E., 268
Lainhart, W. S., 412, 416
Laird, A. K., 543
Lajtha, L. G., 545, 549
Lakshmanan, J., 228, 234, 237, 242, 250
Lalley, P., 476, 482
Lamb, T. D., 445
Lambert, A. E., 75
Lammers, J. G., 469
Lampkin, B. C., 549, 551
Lan, T. T. P., 118, 119
Landaw, E. M., 559
Lander, J. E., 181, 189
Landgraf, R., 73, 74
Lane, L. K., 261, 264, 275
Lang, D., 116-18, 121
Langlois, R. G., 514
Lanyi, J. K., 281
Larcher, D., 308, 317, 319
Larner, J., 70
Larrabee, A. L., 477, 478
Larrabee, M. G., 240, 249
Larsen, J. K., 526
LASKA, E. M., 581-604; 590, 597
Lastowecka, A., 71
Lau, C. Y., 317

AUTHOR INDEX 615

Laughlin, R. G., 477, 478
Laugier, R., 65
Lawaczek, R., 483
Lazarus, N. R., 73, 75
Lazdunski, M., 263
Lazo, R., 489
Leavitt, D. D., 565
Lebowitz, H. E., 228
Lebowitz, J. L., 546, 549, 562
Ledley, R. S., 415, 587
Lee, A. G., 470, 472, 484, 497, 500
Lee, D., 415
Lee, K., 378
Lee, K. S., 461
Lee, R. J., 99
Lee, S. C., 117
Leeman, S. E., 452, 453
Lefkowitz, R. J., 338, 458
Lehrich, J., 227
Leigh, J. S., 377
Leitch, L. C., 474
Leliboux, M., 302
Lelievre, L., 268
Le Maire, M., 13, 15, 27, 270, 273
Lemon, M. J. C., 59
Lenard, J., 121
Leng, M., 116, 121
Lentz, B. R., 472, 473
Lerman, L. S., 116, 119, 123, 127
Lerner, E. I., 45
Leroy, J.-L., 191
Leslie, B. A., 59, 60, 65, 66, 69, 70
Less, R. S., 16
Lester, R. C., 469
Lettvin, J. Y., 150, 151, 154, 158, 160, 167, 168, 174
Leute, R., 343
Lever, J. E., 267
Levialdi, S., 361
Levi-Montalcini, R., 224, 227, 228, 234-38, 240, 248, 249
Levin, A. I., 116
Levine, Y. K., 500
LEVINTHAL, C., 347-62; 347, 349, 355
Levitan, I. B., 454, 455, 461
Levitzki, A., 445
Levy, H. M., 264
Levy, J., 191, 194, 196
Lew, V. L., 77
Lewartowski, B., 449
Lewis, A., 284
Lewis, D. G., 117
Lewis, M. E., 242
LI, C.-C., 393-436; 394, 401, 411, 416, 419
Li, H. J., 116, 120
Li, T. J., 405
Li, T. M., 377, 378

Libby, D. B., 386
Libet, B., 457
Lichtenberg, D., 486, 500
Licko, V., 367
Liefkens, T. J., 469
Lightfoot, D. R., 207, 210
Lilja, H., 378, 385
Lin, Y. K., 395
Lincoln, T. L., 557-61, 563, 564
Lind, K. E., 273, 286
Linden, W. A., 526
Lindenmayer, G. E., 264
Linder, R. E., 305, 309, 310, 312, 315, 321, 322
Lindman, B., 378, 380, 385
Lindmo, T., 529
Ling, C. M., 339, 340
Lipscomb, W. N., 28
Lipsky, S. R., 185
Lipson, H., 83
Lisiewicz, A., 445
Lissack, T., 408
Litman, B. J., 472, 473, 478, 481, 483, 493, 500
Litt, M., 184
Liu, H. C., 422
Liu, H. H., 395
Liu, H. K., 357
Liu, S. C., 269
Liu, T. S., 394
Liuzzi, A., 237
Liversage, W. E., 556
Livingston, K., 245
Livramento, J., 185
Livshitz, M. A., 310, 315
Llinas, R., 350, 351, 355, 356, 359
Lloyd, H. H., 543, 560
Lobanov, N. A., 283, 284
Lobenstein, E. W., 108
Lodwick, G., 395, 587
Lodwick, G. S., 401, 403, 408, 411
Loeppky, R. N., 206, 207
London, E., 493
Long, R. A., 321
Longo, A. M., 230
Looney, Q., 319
Lorente de No, R., 150, 158, 171
Louis, C. F., 270
Lovenberg, W., 315
Low, H., 295, 305
Lowagie, C., 439
Lowe, D. A., 75
Lowe, G., 364, 371
Lowe, J. T., 296, 305, 307
Lowy, J., 89, 94
Loxom, F. M., 116
Lozier, R. H., 281
Lu, S. Y., 401
Luce, J. K., 552, 565

Luck, G., 118
Lucken, E. A. C., 380, 386
Lucy, J. A., 499
Lukins, H. H., 280
Luna, E. J., 471, 472, 478
Lundberg, B., 475
Lundquist, R., 206, 207
Luse, S. A., 249
Lutz, O., 364
Lux, H. D., 170, 171
LUZZATI, V., 1-29; 5, 7, 8, 10, 11, 13-18, 22, 26-28
Lymn, R. W., 90-92, 95, 100
Lynch, D. C., 191, 195, 196
Lysov, Y. P., 310, 315

M

Mabrey, S., 471
MACAGNO, E. R., 347-62; 347, 349, 355
Macdonald, P. D., 551
Macdonald, P. D. M., 526
MacDonnell, P., 238, 242, 250
MacDonnell, P. C., 234, 238, 239, 242, 250
MacLennan, D. H., 268, 271, 274, 279
Madden, T. D., 270
Madison, J. T., 181
Maeda, M., 185
MAESTRE, M. F., 107-41; 107, 116, 118-23, 125, 127, 128, 321
Magin, R. L., 472
Magleby, K. L., 441-43
Magni, F., 171
Mahler, H. R., 116, 118, 119
Maier, K. L., 280
Maitra, S. C., 543
Mäkelä, O., 342
Malaisse, W. J., 73, 75, 76
Malaisse-Lange, F., 75
Malkin, S., 45
Malviya, A. N., 500
Mamas, S., 342
Manabe, M., 310, 314
Mandal, P., 214
Mandersloot, J. G., 263, 499
Maniatis, T., 119
Maniscalco, J., 481
Manning, J., 321
Mannino, R. J., 490, 496
Marbach, C. B., 559
Marchiafava, P. L., 171
Marcus, M. A., 284
Marder, J., 439
Margolet, L., 245
Margoliash, E., 340, 341
Margolis, L. B., 481, 496
Marier, S. H., 66, 70
Markin, V. S., 147
Marler, E., 594

AUTHOR INDEX

Marquise, M., 181
Marrazzi, A. S., 171
Marsh, D., 470, 475-79, 483, 493
Martelli, A., 400
Martin, B. B., 473
Martin, R. B., 319, 320, 473, 474
Martin, W. G., 282
Martonosi, A., 268, 269, 271
Martonosi, A. N., 270
MARTY, A., 437-65; 442-45
Maschke, W., 470
Mason, J. T., 478, 479
Mason, S. F., 108, 319
Massaglia, A., 337
Massari, S., 472
Mateu, L., 5, 10, 16
Matioli, G., 550
Matschinsky, F. M., 73, 74
MATSUBARA, I., 81-105; 83-85, 87-93, 95-98, 100-2
Matsui, H., 264
Matsumoto, S., 246, 492
Matthews, B. H. C., 143, 166, 168, 172
Matthews, E. K., 64, 73-76
Matusbara, I., 92
Mauer, A. M., 549, 551
Mauk, M. R., 496
Maunsbach, A. B., 261
Max, S. R., 239
May, J. T., 552, 565
May, S. W., 317
Mayall, B. H., 516, 521, 522, 524
Mayer, S. E., 458, 459
Mayeri, E., 439, 455
Mayhew, E., 499
Mayhew, E. G., 468, 499, 502
Mazziotta, J., 356, 359
McAfee, D. A., 452, 460
McAllister, R. E., 454
McCaffery, A. J., 296, 305, 322
McCaman, M. W., 439
McCaman, R., 439
McCaman, R. E., 439
McCarty, R. E., 276
McClain, W. M., 129
Mc Cluer, R. H., 469
McColloch, W. S., 167
McCollum, L., 206-9, 211
McConnell, H. M., 471, 472, 478, 481
McDonald, G. G., 367
McDonald, R. L., 453
McDonald, T. F., 438, 449, 450
McDonald, W. I., 169
McFarland, T. M., 309
McGill, K. A., 269
McGinness, J. E., 116, 121, 125, 128
McGraw, J. P., 321

McGuire, J. C., 237, 242, 248, 250
McIntosh, T. J., 475
McKenna, C. E., 304, 315, 316
McKenna, D. G., 439
McKenna, M. -C., 304, 315, 316
McKenney, J., 582
McLarty, J. W., 556
McMahon, D., 590
McManus, T. T., 468
McMath, E., 586
McMillan, D. R., 317-19
McNaughton, P. A., 447, 449
McPherson, A., 181, 182, 189
McWilliams, N. B., 73, 75, 549
Meech, R. W., 76, 445, 446, 451, 453, 454
Meeker, B. E., 559
Meier, G., 557
Meinertz, T., 457
Meissner, G., 265, 268, 272
Meissner, H. P., 74-76
Meister, A., 369
Melamed, M. R., 510, 514, 552
Melchior, D. L., 470
Melga, G. K., 422
Mellett, B. L., 560
Mendelsohn, M. L., 510, 524, 525, 529, 548, 551, 552
Mennel, H. D., 246
Mercado, C. M., 185, 186
Meredith, A. L., 317
Merkle, T. C., 561
Merrell, R., 228, 237
Merrett, M., 285
Merrifield, R. E., 115
Merrill, E. G., 157
Merrill, S. H., 181
Merritt, E. A., 373
Merritt, I., 411
Metcalfe, J. C., 269, 484, 500
Meves, H., 71, 72, 75
Meyer, H., 224, 227
Meyer, H. W., 261
Meyer, R. B. Jr., 461
Michaelson, D. M., 473
Michalak, M., 268, 271
Midelfort, C. F., 368, 372
Migazaki, S., 71, 72
Mildvan, A. S., 266, 375-78
Miledi, R., 70, 168, 171, 172, 441
Miles, D. W., 309, 320, 321
Miles, L. E. M., 339
Milgram, E., 558, 563
Miller, A., 92
Miller, F., 60
Miller, J., 361
Miller, J. P., 461
Miller, M. N., 422
Miller, N., 469, 472, 482

Miller, N. G., 468, 469, 471, 476, 477, 479-82, 498, 499
Miller, R. A., 393
Millman, B. M., 84, 85, 87-90, 94
Milner, R. D. G., 74, 75
Milsmann, M. H., 484
Milstein, J., 559
Min, P. J., 405
Minchenkova, L. E., 117-19
Mineyev, A. P., 310, 315
Minota, S., 452
Minyat, E. E., 117, 118
Mirmirani, M., 551
Mitchell, D. J., 473
Mitchell, M. R., 60, 61, 65
Mitchell, O. R., 399
Mitchell, P., 259, 276, 279
Miura, K., 117
Miura, Y., 459
Miyamoto, H., 272
Miyazawa, T., 214
Mobley, W., 228, 243
Mobley, W. C., 225, 226
Moczydlowski, E. G., 266
Modine, F. A., 304, 316
Moffitt, W., 116
Mohr, S. C., 120
Moise, H., 118, 120
Mollenauer, L. F., 304
Moller, J. V., 270, 273, 286
Mombers, C., 472, 498
Mommaerts, W. F. H. M., 116-18
Monaco, H. L., 28
Mongar, J. L., 69
Monroy, G. C., 279, 280
Montague, W., 75
Montal, M., 488, 489
Montanari, U., 400
Montgomery, J. A., 560
Montgomery, P., 228, 242
Montroll, E. W., 543
Moody, M. F., 26
Moody, W. J. Jr., 171
Moore, D. H. II, 524
Moore, D. S., 117
Moore, G. E., 411
Moore, J. B. Jr., 225, 226
Moore, J. W., 147, 148, 161, 171
Moore, M. J., 500
Moore, R. Y., 243
Morad, M., 450, 458
Morales, M. F., 275, 285
Morel, C., 317
Morgan, W. T., 310, 314
Mori, W., 317, 319
Morimoto, K., 92
Morley, A., 549
Morowitz, H. J., 281, 284
Morris, M. E., 170
Morrisett, J. D., 16

AUTHOR INDEX 617

Morrison, P. F., 557, 558, 560, 561, 563
Morrison, W. R., 468
Mortara, R. A., 470, 483
Mortimer, B. D., 320
Morton, I. K. M., 446
Moscarello, M., 478
Moscowitz, A., 310
Motomura, S., 459
Moudrianakis, E. N., 277, 279, 280
Mouri, T., 71
Mowery, R. L., 302, 304
Moyle, M., 259
Mucciardi, A. N., 405
Mudd, J. B., 468
Mudge, A. W., 452, 453
Mueller, P., 491
Mugnaini, E., 234, 236
Mui, J. K., 399, 401, 404, 407, 408, 422-24, 428, 430, 431
Mukherjee, A. B., 476, 482
Mullaney, P. F., 510, 518, 519, 522, 552
Muller, J. L. M., 279
Müller, K., 16
Muraoka, T., 316
Murata, S., 492
Murphy, B. E. P., 342
Murphy, R. A., 227, 229, 245, 250
Murphy, T. J., 491
Musienko, V. S., 456
Mutani, R., 171
Myer, J. D., 393
Myers, M., 264

N

Nafie, L. A., 304, 305, 310, 312
Nagai, T., 320
Nagaiah, K., 228, 234, 242, 250
Nagle, D., 496
Nagle, J. F., 281, 284
Nakahara, A., 317, 319
Nakajima, H., 95, 100
Nakajima, S., 164, 170, 447
Nakamura, H., 271
Nakao, Y., 317, 319
Nakaya, T., 320
Nakazato, Y., 71
Naoi, M., 500
Narahashi, T., 71
Narendra, P. M., 405
Nargeot, J., 438, 447, 449, 450
Narumi, S., 250
Nathan, D., 451, 458
Nave, C., 117, 118, 121
Nawrath, H., 438, 447, 449, 450, 457, 460
Necheles, T. F., 421, 423, 429, 430, 432

Neelands, P., 411
Neet, K. E., 273
Negoro, T., 410
Neher, E., 437, 441-43
Nehrlich, S., 340
Nejdl, J. F., 395
Nelson, H., 279, 280, 286
Nelson, N., 279, 280, 286
Nelson, P. G., 453
Nelson, R. G., 118, 119
Nemerovski, M., 70
Nenquin, M., 75
Neufeld, H. H., 264
Neumann, J., 280, 286
Neurath, P. W., 421, 423, 429, 430, 432
Newerly, K., 327
Newman, E. A., 150, 154, 158, 167
Newman, J., 394
Newman, W. M., 511, 516
Newton, C., 489
NEWTON, C. M., 541-79; 548-50, 557, 564, 566, 573, 575
Neyman, J., 585
Ngo, E., 277
Nguyen, H. T., 304, 315, 316
Niall, H. D., 226
Nicholls, J. G., 163, 164, 169, 170
Nichols, A. V., 501
Nichols, J. W., 496
Nicolaieff, A., 8, 22
Nicolini, C., 125, 526, 529, 535, 552, 558, 563
Nieboer, E., 319
Niedergerke, R., 64, 447, 451, 458, 460
Niewisch, H., 550
Niggli, V., 275
Niitani, K., 422
Nikodijevic, B., 238, 250
Nikodijevic, O., 250
Ninham, B. W., 473
Nir, S., 469, 472, 489, 498
Nishigaki, I., 267
Nishimura, S., 214
Nishitani, H., 245
Nishiya, I., 515
Nishiyama, A., 59, 66
Nixon, J. R., 469
Noble, D., 447, 449, 454
Noble, S. J., 438, 447, 449, 450, 453, 454
Noebels, J. L., 171
Noer, H., 526
Nojima, S., 468, 470, 497
Noland, C., 340, 341
Nolle, A., 364
Noma, A., 441-44, 447
Nomura, J., 224, 225
Nonaka, G., 469

Norby, J. G., 265, 286
Norden, B., 110, 306
Norgren, P. E., 361
Notsani, B. E., 280
Nozawa, T., 293, 304, 310-14, 316, 322

O

Oakley, B., 59, 63
Obara, S., 158
Oble, D., 454
O'Brien, C., 184, 186
Ochi, R., 449
O'Connor, K. J., 73
Oesch, F., 236
Oesterhelt, D., 281-84
Ofengand, J., 206
Oger, J., 227
Ohlander, R. B., 354-56, 359, 360
Ohlendorf, C. E., 250
Ohmori, F., 274
Ohnishi, S. I., 472
Ohta, M., 71
Ohtsuka, O., 268
Ohtsuki, I., 268
Oikawa, K., 192
Ojeda, C., 438, 447, 449, 450
Okamoto, H., 276, 278, 279
Okanata, Y., 470, 492
Oldfield, E., 473, 474
Olender, E. J., 249
Oleszko, S., 277
Olins, A. L., 120, 121
Olins, D. E., 120, 121
Oliver, G. C. Jr., 340
Oliver, J. E., 498
Oliver, M., 268
Oliver, R., 545, 549
Olmstead, C., 585
Olson, F., 477, 480, 481, 494
Olson, G. B., 421
Olson, J., 185, 186
Olson, M. L., 115
Onoe, M., 395
Orii, Y., 310, 314
Orkand, R. K., 447
Orloff, S., 476, 482
Oron, Y., 70
Orr, J. S., 565
Ortner, P. B., 43
Orton, C. G., 555, 556
Osborne, G. A., 305
Osborne, M. V., 478
Oster, G. F., 551
Ostro, M. J., 486
Ostwald, T. J., 271
O'Toole, A. G., 447
Otten, U., 228, 236, 239, 243, 250
Ottolenghi, P., 265, 286
Otvos, J. D., 386-88, 390

AUTHOR INDEX

Ovchinnikov, Yu. A., 283, 284
Overby, L. R., 339, 340
Oyer, P. E., 340, 341

P

Pace, C. S., 73, 75
Pace, G. C., 589
Pache, W., 497
Packer, L., 280, 284
Padden, F. J. Jr., 116, 126, 127
Pagano, R. E., 468, 473, 486
Page, S., 64, 451, 458, 460
Page, S. G., 81
Paintal, A. S., 149, 150, 173
Palade, G. E., 58
Palmer, A. R., 390
Palmer, G., 310, 313, 314
Paltauf, F., 474
Palusinski, O. A., 549
Pangaro, P., 167
Pangborn, W. A., 489
Pantazis, N., 227
Pantazis, N. J., 227, 245
PAPAHADJOPOULOS, D., 467-508; 468-74, 477, 478, 480-82, 486-90, 494, 496, 498, 499, 501
Pappano, A. J., 447
Pappas, G. D., 173
Paradies, H. H., 277, 279, 280
Paraf, A., 268
Paravicini, U., 233, 249
Parello, J., 385
Parker, B. M., 340
Parker, C. W., 340
Parker, H. M., 565, 566
Parks, J., 501, 502
Parnas, H., 145, 146, 149, 165, 170
Parnas, I., 145, 146, 149, 161-63, 165, 170, 173, 175
Parod, R. J., 60, 65, 68
Partlow, L. M., 249
Pascolini, D., 472
Paskin, A., 547, 548, 551
Pastan, I., 338
Pastushenko, V. F., 147
Patel, D. J., 199, 200, 206
Paterson, R., 565, 566
Pathmamanoharan, C., 485
Patil, H. M., 478
Paton, K. A., 414, 415
Patrick, E. A., 393, 585-87, 594, 599
Patrick, J., 72
Pattison, S. E., 225
Pauker, S. G., 393
Paukstelis, J. V., 206, 207
Pauli, W., 131
Paupardin-Tritsch, D., 439
Pavlidis, T., 396, 399
Pearce, F. L., 249

Pearlman, A., 511
Pearlman, A. L., 526
Pearson, D., 246
Pearson, E. F., 305
Pearson, G. T., 60
Pecci-Saavedra, J., 171
Pedersen, G. L., 59, 66, 446
Pedersen, N. D., 310
Pedersen, P. L., 276, 277, 279, 280
Pedersen, T., 534
Pediconi, M., 229
Pelc, S. R., 525
Pellmar, T. C., 456
Penniston, J., 275
Penniston, J. T., 275
Penny, E., 586
Pentilla, T., 259
Penwick, J. R., 181
Pepe, F. A., 92
Peper, K., 441-43
Perdue, J. F., 261-63, 273
Perelson, A. S., 551
Perez-Polo, J. R., 225, 227, 230, 236, 241, 245, 246
Perrin, F., 130
Perrin, J., 482
Perrone, J. R., 261-63, 265, 267
Perry, J., 394
Perry, S., 529, 534
Perry, S. V., 87
Persoon, E., 398, 409, 410
Peters, A., 173
Petersen, O. H., 55, 59-62, 64-68, 70, 439, 446
Peterson, D. L., 109, 120, 307
Peterson, J. D., 340
Peterson, P. A., 249
Petitou, M., 468
Philipson, K. D., 118, 125
Phillips, A. W., 285
Phillips, M. C., 470, 472, 473, 499
Philpott, H. G., 60, 62
Picarelli, J., 59
Pick, U., 279
Pilarska, M., 268
Pilet, J., 117-19
Piltch, A., 225
Pilz, I., 3
Pimmel, R. L., 394
Pines, E., 51-53
Pinkerton, H., 240
Pinsker, H., 456
Pipberger, H. V., 394
Pirie, B. J. S., 477
Pitts, W. H., 167
Platonov, A. L., 116, 123
Pocchiari, F., 237
Podolsky, R. J., 95, 100
Poisner, A. M., 446
Poletayev, A. I., 117-19
Pollak, G. H., 96, 97

Pollard, H., 250
Polnaszek, C. F., 474
Polson, J. B., 447
Polyvtsev, O. F., 116, 123, 127
Pople, H. E., 393
Porod, G., 8
Porter, N. A., 469
Portis, A., 497
Post, R. L., 264, 265, 267
Poste, G., 468, 470, 478, 489
Poste, G. A., 471, 472
Pott, L., 441, 444
Potts, R. O., 191
Poulsen, J. H., 57-60, 63, 65
Powell, B. W., 421, 423
Powell, E. O., 546, 547
Powell, T. P. S., 171
Powers, S. G., 369
Praetorius, H. M., 394
Prat, J. C., 71
Pratt, W. K., 398
Pressman, H. J., 401
Prestegard, J. H., 468
Preston, K., 361
Preston, K. Jr., 395, 411, 413, 419, 421, 422
Preussmann, R., 246
Prewitt, J. M. S., 421
Price, B. J., 518, 519, 522
Price, J., 589
Price, S., 75
Prince, D. A., 171
Prince, R. C., 496
Prince, T. R., 589
Pringle, J. W. S., 82
Privalov, P. L., 196
Prokopczuk, A., 449
Pulliam, M. W., 228, 230, 232, 237
Pullman, B., 208
Pullman, M. E., 279, 280
Purves, R. D., 64, 440, 441, 444
Putney, J. W., 59, 60, 64-66, 68-70, 439, 446
Pyatigorskaya, T. L., 116, 123, 127

Q

Quastler, H., 551
Quested, P. N., 305, 306
Quick, D. C., 173
Quigley, G. J., 181, 182, 189, 191, 196-98
Quina, F. H., 470, 483
Quinn, P. J., 270

R

Raby, W., 587
Racker, E., 271, 275-77, 279, 280, 286, 484, 493, 499

Radda, G. K., 473, 478, 491
Radhakrishnan, R., 469
Rahman, Y. E., 478
Raiborn, C., 228, 248
RajBhandary, U. L., 182, 187
Rall, W., 144-48, 161, 173
Ralston, M. L., 524, 559
Rana, M. W., 240, 247
Rand, R. D., 472, 475
Randle, P. J., 73-75, 338
Randono, L., 228, 237
Rang, H. P., 160, 161, 164, 170
Rao, K. K., 304, 307, 315, 316
Raps, S., 121, 125
Rask, L., 249
Rasminsky, M., 146, 172
Rasmussen, S., 534
Ratliff, R. L., 117, 119
RAUSHEL, F. M., 363-92; 370, 383
Rawding, C., 383
Rawlings, J., 304, 310, 312, 315-18
Raymond, S. A., 150, 151, 154, 158, 160, 167, 168, 174
Raz, A., 498
Razin, S., 468, 484
Rechardt, L., 236
Records, R., 301, 309, 310, 320, 321
Reddy, D. R., 354-56, 359, 360
REDFIELD, A. G., 181-222; 199, 201-3, 205, 207, 208, 210, 215, 216
Redmann, S. M., 114
Reed, G. H., 383
Rees, E. D., 275, 285
Reeves, J. P., 476, 490
Reeves, R. H., 191
Reggia, J. A., 597
Reiber, H., 475
Reid, B. R., 199-201, 204, 206-11, 213, 214, 216
Reid, R. E., 206, 207, 209, 210
Rein, G., 72, 229, 230, 235
Rein, H., 310
Reithmeier, R., 268, 271
Renaud, L. P., 157
Renshaw, B., 171
Repke, D. I., 274
Repke, K. R. H., 261
Reuben, J., 380, 383, 386
Reuter, H., 447, 449, 450, 457, 458, 460
Revenko, S-V., 146, 169
Revesz, L., 551
Revoltella, R., 229, 245, 248
Reynolds, C. P., 227, 236
Reynolds, J. A., 262, 263, 273
Rhee, H. M., 262
Rhodes, D., 181, 189, 191, 196
Rhodes, W., 114, 116
Rialdi, G., 191, 194, 196, 337

Ribalet, B., 73-76
Ribas, J., 439
Ribiero, N. S., 206-9, 211
Rice, D. W., 285
Rich, A., 181, 182, 189, 191, 196-98
Richards, R. E., 474
Richardson, B. P., 75
Richardson, C. E., 317
Richardson, F. S., 319
Richardson, V. J., 478
Ridgway, E. B., 71, 72, 75
Riesner, D., 214
Riley, R. M., 534
Rink, T. J., 71, 75
Rinnert, H., 308, 319
Risler, J. -L., 310, 315
Risler, Y., 321
Risley, J. M., 367
Ritchie, A. K., 71, 72
Ritchie, J. M., 160, 161, 164, 170, 173
Ritman, E. L., 396
Rivoal, J. C., 304, 307, 315, 316
Rizzolo, L. J., 273-75
Robb, R. A., 396
Robbins, D. J., 319
Roberge, F. A., 454
Roberts, F. E., 409, 413
Roberts, M. L., 59, 61, 64
Robertson, P. Jr., 384
Robertus, J. D., 181, 189
Robillard, G. T., 199, 200, 204, 206-9, 211, 213, 214
Robins, R. K., 321
Robinson, D. R., 245
Robinson, J. A., 590
Robinson, J. D., 500
Robinson, N., 486
Robison, G. A., 457
Robles, E. C., 469
Roda, N., 458
Rodbard, D., 339
Rodman, J., 585
Roelofsen, B., 263
Roerdink, F., 501, 502
Roesch, W. C., 554
Rogart, R. B., 173
Roger, L. J., 242
Rogers, J. G., 245
Rogers, T. B., 263
Rohrer, H., 239
Roigaard-Petersen, H., 273, 286
Roisen, F. J., 229, 250
Rojas, E., 73-76
Rome, E. M., 87, 89
Römer, R., 192, 194, 195
Romero, A., 468
Romsdahl, M. M., 552, 565
Ronne, H., 249
Ronner, P., 275
Root, W., 420, 421

Rordorf, B. F., 192
Rosa, U., 337
Rosati, R., 588
Rose, I. A., 368, 369
Rose, S., 62, 246
Roseman, M., 500
Rosenberg, A., 310
Rosenberg, M. E., 151
Rosenberg, T., 259
Rosenbluth, J., 173
ROSENCWAIG, A., 31-54; 32, 34, 36-45, 48-53
Rosene, D. L., 154, 157, 158
Rosenfeld, A., 358, 399, 400
Rosenfeld, C., 468
Rosenfeld, L., 113, 137
Rosenheck, K., 121, 123, 125, 308
Rosenthal, A. F., 474, 475
Rosing, J., 279
Ross, J., 458, 589
Rossi, C. A., 337
Roth, J., 338, 342
Rothenberger, S. P., 341
Rothman, J. E., 473
Rothman, S. S., 61
Rothschild, K. J., 282
Rothschild, M. A., 327
Rotilio, G., 317, 319
Roti Roti, J. L., 526, 534
Rotmann, M., 449, 450
Rotolo, L. S., 415
Rotter, A., 116, 120
Rougier, O., 438, 447, 449, 450
Roveri, D. A., 279
Rowe, M. D., 322
Rowe, V., 250
Rowlands, J. R., 322
Rozengurt, E., 267
Rubenstein, A. H., 340, 341
Rubin, L., 585
Rubin, R. P., 70, 71
Rubinow, S. I., 545, 546, 549, 562
Ruckpaul, K., 310
Ruder, H. J., 339
Rudich, L., 59, 65, 69, 70
Rudolph, S., 486
Rudy, B., 491
Ruhenstroth-Bauer, G., 530, 531
Rumel, S. R., 361
Rustum, Y., 469, 499, 502
Rustum, Y. M., 499
Rutherford, B., 585
Ryden, K. H., 548, 557, 573
Ryman, B. E., 468, 478, 493

S

Saavedra, J. M., 439
Sabatini, M. T., 240
Sackmann, E., 472

AUTHOR INDEX

Safir, A., 393, 586
Saide, J. D., 227, 245
Saito, A., 57
Sakakura, H., 171
Sakamoto, Y., 73-76
Sakmann, B., 441
Sakurai, J. J., 113
Salemink, P. J. M., 206, 207, 212
Salmeen, I., 313, 315
Salmon, A., 310, 315
Salmon, S. E., 421, 543, 560
Salvi, M. L., 237
Salzman, G. C., 517-19, 522
Samejima, T., 117, 118, 310
Sanborn, S., 246
Sander, C., 192
Sanders, M. J., 270
Sandra, A., 473
Saneyoshi, M., 185
Sankar, P., 420, 421
Santella, R., 116, 120
Santella, R. M., 120
Santi, D. V., 185
Saori, H., 259
Sapse, A., 549
Sardet, C., 5, 15, 16, 26, 27
Saroste, M., 259
Sarton-Miller, I., 372
Sarzala, M. G., 268
Sashin, D., 411
Sato, R., 310
Sato, Y., 57
Sauer, K., 118, 125, 310, 311, 314, 315
Saunders, E. F., 551
Saunders, L., 482
Sauviat, M. P., 447, 449
Savol, A. M., 401, 411, 416
Saxon, D. S., 131
Sayer, P., 310, 312
Scales, D., 269, 270
Scanu, A. M., 16, 17
Scatchard, G., 192
Schabel, F. M., 560
Schaeffer, B. E., 471, 472
Schaffhausen, B., 116, 120, 124
Schanberg, S. M., 242
Schatz, P. N., 302, 304
Schatzmann, H. J., 268
Scheele, G. A., 58, 60
Scheibel, A. B., 167
Scheibel, M. E., 167
Schellman, J. A., 125
Schenkein, I., 227, 244
Schenker, A., 225, 226
Scheraga, H. A., 112, 192
Scherphof, G., 501, 502
Scherr, L., 535
Schieren, H., 486
Schimdt, P. G., 210
SCHIMMEL, P. R., 181-222; 182, 184-96

Schindler, R., 530
Schmähl, D., 246
Schmelz, H., 75, 76
Schmidt, P. G., 208, 213
Schmidt, U. D., 277, 279, 280
Schmitt, O. H., 171
Schnatterly, S. E., 304-6
Schneider, A. S., 116, 121, 123, 125
Schneider, B. S., 336-38
Schneider, C., 321
Schneider, J. A., 449
Schneider, M. J., 121, 123
Schnepp, O., 305
Schneyer, C. A., 55, 57, 59
Schneyer, L. H., 55, 57
Schoemaker, H. J. P., 185-90
Scholz, H., 447, 449, 457, 460
Schonfeld, M., 280, 286
Schoot, B. M., 265
Schoot-Uiterkamp, A. J. M., 386, 389, 390
Schotz, W. E., 547, 551
Schramm, M., 59, 61, 63, 65, 66, 69, 70, 446
Schreckenbach, T., 281
Schreier, A. A., 192-96
Schubert, D., 230, 237, 247, 248, 250
Schullery, S. E., 494
Schulman, J. D., 476, 482
Schultz, J. R., 62, 598
Schumann, H. J., 457-59
Schumann, L., 283
Schumitzky, A., 585
Schwab, M., 233, 249
Schwartz, A., 121, 261, 264, 275
Schwartz, A. A., 398
Schwartz, F. T., 474
Schwartz, J. H., 460
Schwartz, M. D., 395
Schwartz, T. L., 441
Schwartz, U., 213
Schwartz, W. B., 393
Schwartzkroin, P. A., 171
Schwarz, M. A., 472, 481
Schwendener, R. A., 484
Schyolkina, A. K., 117-19
Scobey, R. P., 171
Scott, M. J., 555
Scott, T. L., 274
Seaman, P., 271
Seamans, L., 310
Sears, B., 473
Sebald, W., 276, 280, 281
Seeman, N. C., 181, 182, 189
Sehlin, J., 76
Seibert, E. S., 246
Seigesmund, H. A., 152
Sela, M. J., 342
Selby, C. C., 86

Selinger, Z., 59, 61, 63, 65, 66, 69, 70, 446
Selles, W. D., 421, 423, 429, 430, 432
Selser, J. C., 478, 480
Selverston, A., 361
Selzer, R. H., 395
Sener, A., 73
Senior, A. E., 277, 279
Serbinova, T. A., 319
Serra, M., 117
Servis, R. E., 185
Shackney, S. E., 561
Shamoo, A. E., 264, 274, 275, 491
Shane, L. L., 594
Shanes, A. M., 159, 169, 170
Shanmugam, K., 397, 401
Shantz, M. J., 351, 356, 357
Shapiro, J. T., 116, 121
Shapiro, M., 515
Shapiro, N., 557, 559, 564
Shapiro, N. Z., 559
Shapiro, R., 185
Sharma, P., 493
Sharonov, Y. A., 293, 303, 310, 315
Sharonova, N. A., 310, 315
Sharp, T. R., 368
Sharpless, T., 514
Shashoua, V. E., 309
Shatwell, R. A., 296, 305
Shen, A. L., 192
Shepp, L. A., 401
Sheppard, H. C., 184, 186
Sheppard, L. C., 586
Sheridan, J. D., 68
Sherman, H., 584
Sherman, J. M., 59, 65, 66, 69
Sherman, R., 226
Sherwin, S. A., 227
Sherwood, G., 565
Shieh, P., 280
Shiek, C. F., 411
Shigekawa, M., 273
Shih, T. Y., 116, 118, 120, 121
Shillady, D. D., 309
Shimada, H., 310
Shimahara, T., 453, 460
Shimizu, M., 310
Shimizu, T., 310, 312, 313, 322, 500
Shimshick, E. J., 472
Shin, K. G., 534
Shin, Y. A., 120, 123
Shine, H. D., 227, 230, 241
Shinitzky, M., 469, 471, 498
Shinoda, Y., 155, 156
Shipley, G. G., 16, 471, 475, 498
Shipton, H. W., 394
Shohet, S. B., 468

AUTHOR INDEX 621

Shooter, E. M., 224-26, 230-32, 245, 249
Shortliffe, E. H., 393, 596, 597
Shu, M. J., 74, 75
Shue, K. F. R., 372
Shulman, R. G., 190, 202, 206, 207, 211, 214
Shure, M., 121
Sichel, K., 296
Sidorenko, E. R., 586
Sidorenko, G. I., 586
Siegel, B., 460
Siegel, C., 592, 597
Siegel, G. J., 264
Siegelbaum, S., 449
Siegeleub, A. B., 585
Sies, H., 60
Siggers, D. C., 244, 245
Sigrist, H., 280
Sigrist-Nelson, K., 280
Silinsky, E. M., 59, 61-63, 439
Simmons, R. M., 81, 82, 95
Simon, S. B., 439
Simon, W., 350, 356, 357
Simons, D. J., 172
Simons, K., 485, 497
Simpson, G., 597
Simpson, W. T., 109, 120, 307
Simpson-Herren, L., 543
Singer, C. E., 218
Singer, R. H., 227
Singer, S. J., 121, 275, 285
Singer, W., 170, 171
Sipe, J. P., 320, 473
Sipski, M. L., 121
Skaper, S. D., 237, 248, 250
Skinner, D. M., 117
Skinner, J., 245
Skipper, H. E., 560
Sklansky, J., 400, 410, 420, 421
Skollermo, B., 566
Skou, J. C., 262, 263, 265
Skrabal, P., 484
Skulachev, V. P., 276, 279, 283, 284
Slack, W., 584
Slater, E. C., 276, 279
Sligar, S. C., 312
Sliski, A. H., 227
Slotboom, A. J., 475
Small, D. M., 16, 501, 502
Smith, A. P., 225
Smith, B. E., 304, 315, 316
Smith, D. F., 73, 75
Smith, D. O., 162, 169, 170, 173
Smith, G. A., 269
Smith, G. J., 232
Smith, G. R., 218
Smith, I. C., 474
Smith, J. B., 279
Smith, L. F., 340
Smith, M. W., 500

Smith, R. K., 59
Smith, W. T., 261, 262, 265, 267
Smock, R., 439
Smolen, J. E., 468
Sneider, R. M., 589
Snell, J.-E., 566
Snellgrove, T. R., 305, 306
Snyder, S. H., 231
SOBEL, I., 347-62; 347, 349, 355
Soha, J. M., 398
Sokolove, P. G., 164
Söll, D., 182, 190
Solomon, F., 233, 249
Solomon, E. I., 317, 318
Somers, G., 73
Somoano, R. B., 48
Sone, N., 276, 278, 279
Soper, J. W., 276, 277, 279, 280
Soroka, B. I., 361
Sorrell, T. N., 310, 312
Sowersby, G., 305
Speake, R. N., 75
Sperelakis, N., 449, 458
Sperling, L., 26
Spiesberger, W., 395
Spira, M. E., 161, 163, 165, 170, 173, 175
Spire, J. P., 394
Spiro, D., 81
Spitsberg, V., 277
Spitzer, R., 351, 356, 359, 587, 598
Spitzer, R. I., 587
Sponar, J., 120, 121
Sprecher, C. A., 117
Springall, J., 310, 313, 314
Springall, J. P., 310, 313
Sprinkel, F. M., 309
Sprinzl, M., 200, 207, 209, 211, 213
Sproat, B. S., 364, 371
Sproull, R. F., 511, 516
Squire, J. M., 85, 92
Stach, R., 233
Stach, R. W., 225, 249
Stahn, R., 246
Stallcup, W., 247, 248
Stallcup, W. B., 72
Stamp, D., 497
Standish, M. M., 467, 480
Stanis, J. W., 394
Stark, H., 415
Staschewski, D., 364
Stauffer, G. V., 182
Stearns, S. D., 405
Steel, G. G., 551
Steim, J. M., 470
Stein, A., 191, 192, 194-96
Stein, D. G., 242, 243
Stein, L., 243

Steinberg, I. Z., 107, 108
Steiner, D. F., 340, 341
Steinkamp, J. A., 552
Steinmentz-Kayne, M., 211
Steinmetz, L. L., 524
Steinrauf, L. K., 496
Stenevi, U., 228, 230, 243
Stenger, R. J., 81
Stephens, E. M., 383
Stephens, P. J., 293, 296, 300-2, 304, 305, 307, 310, 312, 315-18
Sternick, E. S., 554
Sternweis, P. C., 278, 279
Stevens, C. F., 437, 441-43
Steward, J. K., 246
Steward, P. G., 548, 557, 561, 564
Stewart, J. E., 415
Stewart, M., 91, 92
Stewart, P. S., 274
Stickgold, R., 439-42, 444
Stiefel, M. L., 585
Stillings, S. N., 73
Stillman, J. S., 310
Stillman, M. J., 302, 310, 322
Stöhr, M., 517
Stöckel, K., 233, 249
Stockman, J., 394
Stockton, G. W., 474
Stoeckel, K., 228, 233, 249
Stoeckenius, W., 281, 282, 284
Stokes, B. O., 368
Stollery, J. G., 478, 489
St. Onge, R., 95, 100
Strackee, J., 557, 565
Strata, P., 171
Straub, R. W., 159, 160, 169, 170
Straus, E., 336-38
Strickland, R. W., 309
Strintzis, M. G., 394
Strittmatter, P., 484, 500
Strohl, W. A., 490, 496
Strom, R., 551
Stryker, J. A., 565
Stuart, R. N., 561
Stubblefield, E., 534
Stuhrmann, H. B., 3, 5, 10, 12, 16
Sturtevant, J. M., 471
Subbarow, Y., 468
Suckling, E. F., 449
Suckling, K. E., 475
Suda, K., 228
Suddath, F. L., 181, 182, 189
Suenaga, Y., 410
Suetaka, W., 302, 304
Suga, H., 84, 89, 90, 93, 95-97
Suit, H. D., 556
Sullivan, P. W., 543, 560
Sun, A. S. K., 315
Sundaralingam, M., 373

Sundberg, K. R., 115
Sundergeld, A., 530
Sussman, J. L., 181, 182, 189, 191, 196, 197
Sutherland, E. W., 457
SUTHERLAND, J. C., 293-326; 293, 295, 296, 301-3, 305-7, 310, 312, 315, 322
Sutton, R. N., 409
Suzuki, K., 169
Sverdlik, D. I., 122
SWADLOW, H. A., 143-79; 149, 150, 152-58, 173
Swan, G. W., 547, 550, 564
Swann, A. C., 264
Swann, J., 439
Swann, J. W., 439
Swarthof, T., 206, 207, 212
Sweadner, K. J., 262, 268
Swenberg, J. A., 246
Switzer, R. L., 377, 378
Sytkowski, A. J., 317, 318
Szczepaniak, A. C., 439
Sze, T. W., 411
Szentagothai, J., 171
Szoka, F. C., 469, 477, 480, 481, 494, 497, 499, 502
SZOKA, F. JR., 467-508; 487-89, 494, 496
Szutowicz, A., 230, 232, 241

T

Taber, R., 490, 496
Taborikova, H., 174
Tada, M., 268, 274, 275
Tai, T. H., 419
Takacs, P. Z., 305
Takagi, Y., 410
Takahashi, I., 166, 169
Takahashi, K., 164, 170
Takahashi, M., 548
Takeda, A., 310
Takeichi, M., 473
Takeuchi, A., 172
Takeuchi, N., 172
Taljidal, I. B., 76
Tall, A. R., 501, 502
Tanford, C., 270, 271, 273-75, 286
Tang, G. Y., 398
Tang, S. C., 310, 312
Taraskevich, P. S., 55
TARDIEU, A., 1-29; 5, 10, 11, 13-18, 26-28
Tarr, C. E., 200, 204, 207, 213
Tauc, L., 162, 453, 460
Taylor, C. W., 273, 286
Taylor, I. W., 534
Taylor, J., 395

Taylor, J. S., 270, 273
Taylor, P., 75
Taylor, T. N., 116-18, 121
Taylor, T. S., 317
Teeter, M. M., 191, 196-98
Teige, B., 468
Teitelbaum, H., 184
Ten Eick, R., 438, 449, 450
Tennent, D. L., 319
Tenu, J. P., 272
Terroux, K. G., 449
Terry, R. D., 169
Thames, H. D., 534
Thames, H. D. Jr., 551, 556
Thiery, J., 321
Thiery, J. M., 321
Thirion, C., 308, 319
Thoen, C. O., 511, 512
Thoenen, H., 228, 233, 236, 238, 239, 243, 249, 250
Thomas, D. D., 91, 95
Thomas, G., 191, 319
Thomas, G. J. Jr., 185
Thomas, L. J., 74, 75
Thomas, P. K., 169
Thomas, R. C., 164, 170, 264
Thompson, T. E., 471, 478, 483, 493, 500
Thompson, W. B., 401, 403, 413
Thomson, A. J., 302, 304, 307, 310, 313-16, 319, 322
Thorley-Lawson, D. A., 274, 275
Thulin, A., 58, 59
Thulin, E., 385
Tikhonenko, A. S., 116, 123
Tillack, T. W., 268-70
Timin, E. N., 144, 161, 173
Timir, Y. N., 146, 169
Tinghitella, T. J., 481, 496
Tinker, D. O., 472, 475
Tinoco, I., 320, 321
TINOCO, I. JR., 107-41; 107, 110, 113, 116-19, 121, 126
Tischler, A., 71, 72, 75
Tischler, A. S., 72, 229, 230, 235, 236
Titus, E. O., 265
Tobias, I., 113
Tocanne, J. F., 470, 478
Tochilin, E., 554
Todaro, G. J., 227, 229, 244
Todhunter, J. S., 419
Toennies, J. F., 167
Tokumura, A., 499
Tolkovsky, A. M., 445
Tolson, N., 234, 238, 239, 250
Tomasz, M., 185, 186
Tombol, T., 171
Tomita, J. T., 228

Tomita, T., 446
Tompson, J. G., 206, 207
Tonomura, Y., 268, 275, 285
Toon, P. A., 484
Toriwaki, J. I., 361, 410
Torten, M., 116, 121
Toschi, G., 237
Tou, J. T., 395, 404
Toussaint, G. T., 408
Towfig, F., 420, 421
Townsend, L. B., 320, 321
Traganos, F., 514
Trauble, H., 470, 491, 492
Trautwein, W., 438, 441-44, 447, 449, 450
Tregear, R. T., 92
Treistman, S. N., 461
Trentham, D. R., 367
Treu, J. I., 304-6, 310
Triche, T., 476, 482
Trifaró, J. M., 71
Tritthart, H., 457
Trotta, E. E., 268
Trube, G., 447
Trucco, E., 543-45, 551
Trudell, J. R., 310, 312, 315
Tsai, M. D., 371
Tsai, M. J., 394
Tsien, R., 449
Tsien, R. W., 447, 449, 450, 454, 457-59, 461
Ts'o, P. O. P., 192, 200, 207, 213
Tsukahara, N., 174
Tsukatani, H., 499
Tsvankin, D. Y., 116, 123, 127
Tulloch, A. P., 474
Tully, R. J., 401, 403, 408, 411
Tunis, M. J. B., 118, 119
Tunis-Schneider, M. J. B., 116, 118, 119, 122
Turel, R. J., 469
Turner, A. F., 401, 403, 409, 413-15
Turner, D. H., 107, 108, 295
Turner, D. L., 382
Turner, J. E., 228, 247
Tuttersall, M. H., 478
Tuy, F., 468
Tweedle, M., 313
Tyrrell, D. A., 468, 493
Tzagoloff, A., 279

U

Ueda, N., 59, 61, 65-67, 70
Uesugi, S., 262, 265
Ullman, E. F., 343
Ulmer, D. D., 315
Unsicker, K., 228, 236
Unwin, P. N. T., 282

AUTHOR INDEX 623

Urbina, J., 472
Urry, D. W., 121, 123
Utley, C., 588

V

Vachette, P., 13, 15, 26, 27
Vagnucci, A. H., 394
Vail, W. J., 468, 470-72, 477, 478, 480, 481, 489, 490, 494
Vaitukaitis, J., 339
Valeriote, F. A., 559
Valet, G., 518, 530
Vallbona, C., 586
Vallee, B. L., 293, 295, 303, 304, 307, 309, 310, 315-19
Valleron, A.-J., 551
Van Bemmel, J. H., 394
Van Boom, J. H., 211
Vand, V., 134
Van Deenen, L. L. M., 470-72, 474, 475, 478, 479, 484, 485, 489, 498, 499
Vandegrift, V., 117
Vandenberg, C. A., 488, 489
Van den Berg, D., 469
Vanderkooi, J. M., 271
van der Kraan, I., 279
Van der Laarse, R., 557, 565
VanderMaelen, C. P., 157
VanderMeer, J., 469
Van der Neut-Kok, E. C. M., 472, 498
van der Sluis, P. R., 279
VandeWalle, C. M., 60, 65, 66
Van Dijck, P. W., 471, 472, 475, 478, 498
Van Dilla, M. A., 514, 524
Van Divender, J. M., 383
Van Essen, D., 161, 163-65, 170, 175
Van Etten, R. L., 367
Van Holde K. E., 116-19
van Meter, W. G., 460
van Prooijen-van Eeden, A., 263, 265
Van Steelandt, J., 313
Van Winkle, W. B., 261, 275
Van Zoelen, E. J., 479
Varon, S., 224, 225, 228, 237, 248-50
Varshavsky, Y. M., 116, 123, 127
Vasilenko, I. A., 473
Vassort, G., 447, 449
Vastola, E., 421, 423, 429, 430, 432
Vaughn, M. R., 117, 118, 121
Vaught, L., 421
Vaz, W. L. C., 497

Vealla, C., 73, 75
Venable, J. H., 119
Venable, J. H. Jr., 116, 123, 127
Venkstern, T. V., 196
Venter, J. C., 458
Veprintsev, B. N., 456
Vereecke, J., 447
Verhulst, P. F., 543
Verjovski-Almeida, S., 273
Verkleij, A. J., 469, 479, 484, 485, 489
Verkley, A. J., 472, 498
Verniers, J., 75
Vernon, C., 587
Vernon, C. A., 249
Verrinder, M., 499
Verschoor, G. J., 279
Ververgaert, P. H. J. Th., 469
Vianna, A. L., 268
Vickery, L. E., 293, 304, 305, 307, 310, 311, 313-15
Vidal, J. J., 394
Vigneti, E., 228, 229, 245
Viktorov, A. V., 473
Vilenkin, S. Ya., 144, 161
VILLAFRANCA, J. J., 363-92; 368, 370, 383
Vincenzi, F. F., 275
Vindelov, L., 526
Vinograd, J., 121
VINORES, S., 223-57; 241, 246
Vinores, S. A., 229, 236, 246, 247
Vinsonbaler, J., 594
Viscarello, R. R., 237, 247
Vitale, S., 590
Vliek, R. M. E., 305, 306
Voelter, W., 301, 303, 308, 309, 319-21
Vogel, F., 261
Vogel, H., 550
Volkenstein, M. V., 310
Volkmann, R., 457
Voller, A., 343
Von Foerster, J. V., 543
von Hippel, P. H., 117, 120
Vosman, F., 200, 204, 207, 213
Vournakis, J. N., 192
Vvedenskaya, N. D., 144, 146, 147

W

Wachtel, H., 454, 455
Wachter, E., 280, 281
Waddell, W. R., 245
Wadhwani, B. J., 394
Wagenvoord, R. J., 279
Waghe, M., 246
Wagner, C., 594

Wagner, H. J., 261, 262, 265, 267
Wagner, T. E., 117, 121
Waite, M., 501, 502
Wakabayashi, T., 277
Walker, H. K., 593
Walker, I. D., 283
Walker, P., 230, 241
Wall, P. D., 157, 167
Wallick, E. T., 261, 264
Waltman, P., 568
Waltz, J. M., 169
Wambersie, A., 555
Wang, A. H. J., 181, 182, 189
Wang, C. C., 191
Wang, J., 120
Ward, J. W., 394
Ward, S. A., 516
Ware, F., 449
Waris, P., 236
Waris, T., 236
Warner, H., 585, 586
Warner, H. R., 361, 393, 549, 585
Warrant, W., 191, 196, 197
Warren, G. B., 269, 484
Warshaw, M. W., 116
Wasserman, L. R., 341
Watanabe, A. M., 457, 458, 460
Watanabe, K., 213
Watanabe, S., 272
Wataya, Y., 185
Watkins, J. C., 467, 480, 482, 486
Watson, E. R., 556, 569
Watson, J. V., 529, 534, 535
Watts, A., 470, 477-79, 493
WAXMAN, S. G., 143-79; 146, 149, 150, 152-58, 169, 173, 174
Webb, M. R., 367
Webb, P. K., 214
Webb, W. W., 491
Wechsler, H., 410
Wechsler, W., 246
Wechter, W. J., 185
Weder, H. G., 484
Weed, L., 592, 593
Weeks, T. A., 461
Wehner, H. D., 454
Wehrli, F. W., 380
Weichsel, M. E. Jr., 230, 241
Weidmann, S., 450, 458
Weiler-Feilchenfeld, H., 321
Weiner, C., 439
Weingart, R., 459
Weinreich, D., 439, 440
Weinstein, J. N., 468, 472, 486
Weinstein, M., 351, 357, 360
Weis, J. S., 227

AUTHOR INDEX

Weis, P., 236
Weiskopf, M., 116, 120
Weiss, B., 16
Weiss, D. W., 545
Weiss, L. C., 115
Weiss, M. F., 559
Weiss, R., 457
Weiss, S. G., 393
Weiss, S. J., 66, 70
Weissmann, G., 468, 481, 486, 496
Welcher, M., 185
Werman, R., 165
Weser, V., 319
Wessells, N. K., 249
West, G., 246
Westerfield, M., 147, 148, 161, 171
Westlund, K., 245
Weswig, P. H., 310
Weszka, J. S., 399
Weyand, T. G., 152
Whanger, P. D., 310
Wharton, D. C., 310, 314
Wheeler, B., 534
Wheeless, L. L., 395, 510
Wheldon, T. E., 569
White, J. C., 495
White, R. A., 529, 534, 551
Whiting, G. J., 460
Whitlock, C., 237, 247, 248, 250
Whittemore, A., 547
Wicksell, S. D., 478
Wide, L., 339
Wied, G. L., 395
Wielders, J. P. M., 279
Wiener, O., 135
Wiersema, P. H., 485
Wiklund, L., 228
Wikström, M., 259
Wilbrandt, W., 259
Wildenauer, D., 283, 284
Wilkie, D. R., 82
Willems, J. L., 394
Williams, J. A., 57, 58, 60, 65
Williams, R. J. P., 210, 281
Williams, T. A., 585
Willick, G., 192
Willis, W. D., 172
Willison, R. G., 151
Wilson, I., 587
Wilson, J. A. P., 64
Wilson, L. J., 313
Wilson, P. D., 171
Wilson, R. L., 561
Wilson, T., 490, 496
Wilson, W. A., 454-56

Wilson, W. H., 225, 230
Wimmer, M. J., 369
Winget, G. D., 280
Winick, M., 227
Wisdom, B. G., 343
Wise, C. D., 243
Wissler, R. W., 17
Witz, J., 8, 22
Wlodawer, A., 224, 226
Wojcik, J. D., 69, 70
Wolf, B., 116-20
Wolf, R. A., 469
Wolfson, J. M., 192
Wong, A. K. C., 394
Wong, K. L., 207, 209, 210
Woo, K. B., 529, 534
Wood, E. H., 396, 450, 458
Woodward, C. K., 184
Woody, R. W., 108, 116, 117
Woolf, C. J., 285, 286
Worcester, D. L., 474
Workman, P., 535
Worthington, C. R., 89
Worthington, D. R., 269
Wreschner, D. H., 490
Wright, P. E., 210
Wu, E.-S., 497
Wu, T. T., 116, 118, 119
Wundt, W., 165

X

Xuong, N. G., 361

Y

Yager, P., 285, 286
Yagi, K., 92, 93, 500
Yagi, N., 83, 92, 93, 95-98, 100-2
Yaksh, T. L., 157
YALOW, R. S., 327-45; 327-30, 333, 334, 336-38, 340, 341
Yamada, K. M., 249
Yamada, M., 274
Yamaizumi, Z., 214
Yamamoto, H., 310
Yamamoto, M., 65, 66
Yamamoto, T., 268, 310, 312, 322
Yamanaka, N., 274
Yamane, T., 207
Yamasaki, Y., 458
Yamauchi, O., 317, 319
Yang, G. J., 398
Yang, J. T., 109, 117, 118

Yang, S. K., 190
Yankner, B. A., 232, 249
Yanofsky, C., 182
Yarom, Y., 165, 170
Yarowski, P. J., 439
Yates, D. W., 266, 272, 285
Yates, M., 10, 11, 13, 17, 18, 26, 27
Yatvin, M. B., 472
Yau, K. W., 161, 163-65, 175
Yeager, C. L., 394
Yeagle, P. L., 473, 474
Yee, S. S., 49
Yeh, Y., 478, 480
Yip, C. C., 271
Yokoyama, S., 214
Yon, J., 272, 480
Yoneda, M., 310, 314
Yoshida, M., 276-79
Yoshida, S., 310
Yoshikami, D., 439-42, 444
Yotsuyanagi, T., 486
Young, I. T., 421, 520
Young, J. A., 55, 57
Young, J. N., 350, 356, 357
Young, J. Z., 161, 169
Young, M., 227, 245, 250
Young, S. P., 496
Young, T. Y., 404, 406
Younis, H., 280
Yu, K.-P., 562, 565
Yu, L., 95, 100
Yu, M. W., 228, 234, 238, 242, 250

Z

Zaharko, D. S., 472, 558, 561
Zamir, A., 181
Zandstra, P. J., 305, 306
Zborowski, J., 501
Zeimal, E. V., 439, 444
Zeitlin, G. M., 394
Zelen, M., 547, 551
Zenker, W., 169
Zhurkin, V. B., 310, 315
Ziebniewski, A., 499
Zietz, S., 526, 529, 535, 552, 571
Zimmer, C., 118
Zimmermann, J., 277
Ziskoven, R., 454
Zolin, V. F., 319
Zubieta, J. A., 315
Zubrzycka, E., 268
Zucker, R. S., 154, 159, 170
Zülch, K. J., 246
Zurawski, G., 182
Zwaal, R. F. A., 484, 485, 489

SUBJECT INDEX

A

Acetylcholine (ACh)
  cardiac action potential and, 449–50
  catecholamine secretion and, 70–71
  heart muscle and, 438–39, 449–50
  inducing calcium to enter adrenal chromaffin cells, 71–72
  pancreas and, 62
  synaptical potassium conductance and, 439–42
Acid phosphatase
  nuclear magnetic resonance (NMR) studies of, 367
Acinar cells
  electrical responses of, 61–64
  site of, 61–63
  fluid secretion of, 61
Acoustic neuroma
  bilateral
    nerve growth factor (NGF) and, 244
Actin
  arrangement in heart muscle, 87–88
  skeletal muscle contraction and, 81
Action potentials
  refractory period and, 150
  velocity of
    axonal diameter and, 144–46
Acute lymphoblastic leukemia
  modeling of, 549
Acute myeloblastic leukemia
  modeling of, 546–49
Acute myelogenous leukemia (AML)
  modeling of, 560–61
ADC-500 422–23
Adenine
  H-8 exchange rate constants for, 186
  magnetic circular dichroism (MCD) and, 320–21
Adrenal chromaffin cells, 70–72
  acetylcholine-induced calcium influx, 71–72
Adrenaline
  cardiac action potential and, 448–49
  sodium-potassium pump and, 454

Adrenal medullary cells
  nerve growth factor (NGF) and, 228
Adrenal medullary hyperplasia
  bilateral
    nerve growth factor (NGF) and, 245
α-Adrenergic agonists
  lacrimal gland response to, 57
  receptors for, 57
β-Adrenergic agonists
  receptors for, 57
Adrenergic compounds
  heart muscle and, 448–49
Alanine
  magnetic circular dichroism (MCD) and, 308
Alcohol dehydrogenase
  magnetic circular dichroism (MCD) and, 317
Algae
  marine
    photoacoustic spectroscopy (PAS) and, 42–43
Alkaline phosphatase
  magnetic circular dichroism (MCD) and, 317
  nuclear magnetic resonance (NMR) studies of, 367
Amino acids
  aromatic
    magnetic circular dichroism (MCD) and, 308–9
Aminoacyl tRNA synthetase
  tRNA interacting with, 190–91
γ-Aminobutyric (GABA)
  synaptical potassium conductance and, 439
Amylase
  salivary gland release of, 58–61
  secretion of
    cyclic nucleotides and, 69–70
Analytical ultracentrifugation
  liposome size distribution analysis and, 478, 493
Anthranoylouabain, 266
Antigens
  hepatitis B
    assay of, 339–40
    labeled
      in radioimmunoassay (RIA), 333–38
Antinerve growth factor antiserum, 228

cytotoxic effects of, 240–41
Aplysia
  axonal branch conduction in, 162
  neurons of
    potassium-dependent responses in, 440–46
Apple peel
  photoacoustic spectroscopy (PAS) and, 44–45
Ara-A
  DNA metabolism and, 558
Ara-C
  DNA metabolism and, 558
  Lewis lung tumor cell survival following treatment with, 533–34
  metabolism of, 560–61
Arginine-specific
  esteropeptidase activity
    in nerve growth factor (NGF), 225
Aromatic amino acids
  magnetic circular dichroism (MCD) and, 308–9
Arterial wall irregularity
  computer grading of, 395
ATP-ADP
  transphosphorylation reaction
    sodium ion-dependent, 265
ATPase
  F_1, 276–80
  molecular weights of, 276
  subunits of, 278–80
  ultrastructure of, 277
ATPase activity
  glycerol stabilizing, 270
Atropine
  agonist blockage and, 57
Attenuated total reflection spectroscopy, 32
Autocorrelation function
  for low density lipoprotein (LDL), 20–25
Axonal branching regions
  active impulse propagation through, 147–49
  supernormal period of, 154–57
Axonal excitability
  activity-dependent fluctuations in, 149–61
  membrane potential and, 158–61
  relative refractory period and, 150
  supernormal, 150–52
  strophanthidin and, 159

SUBJECT INDEX

Axonal impulse conduction
  through axonal branching
    regions, 147–49
  axonal diameter and, 144–46
  in invertebrates, 161–65
  low safety factor regions and,
    143–44, 161–68
  refractory period and, 150
  subnormal period of, 158
    mechanism of, 158–61
  supernormal period of,
    150–58
    mechanism of, 158–61
  threshold variations in,
    152–53
  velocity of
    activity-dependent
      fluctuations in, 149–61
    membrane potential and,
      158–61
    in vertebrates, 165–68
Axonal impulse conduction
  block
  mechanisms of, 168–75
Axons
  diameter of
    absolute refractory period
      and, 150
    action potential velocity
      and, 144–46
Azotobacter vinelandii
  "Mo-Fe" proteins from, 316
Azurin
  magnetic circular dichroism
    (MCD) and, 317

B

Bacillus cereus neutral protease
  magnetic circular dichroism
    (MCD) and, 317
Bacillus subtilis var. Niger
  photoacoustic spectroscopy
    (PAS) and, 48
Bacteria
  optical behavior of
    circular dichroism (CD)
      and, 121
    photoacoustic spectroscopy
      (PAS) and, 48
Bacteriophages
  circular dichroism (CD) and,
    121–23
  as a nonradioactive label, 342
Bacteriorhodopsin, 46, 281–84
  active site structure of, 284
  primary structure of, 283–84
  ultrastructure of, 282–83
Basophils (BAS), 423
  machine classification of,
    421–22
Bayes' theorem, 585
Beer's law, 294

Benzo(a)pyrene
  antinerve growth factor
    antiserum and tumor
    induction and, 246–47
Bessel functions, 133–34
Binary tree classifier
  of nucleated blood cells,
    428–32
Birth-and-death processes
  stochastic models of, 546–47
Blank scattering, 4
Bleomycin, 565
Blood
  whole
    photoacoustic spectroscopy
      (PAS) and, 39–40
Blood cells
  nucleated
    automated classification of,
      421–32
    binary tree classifier of,
      428–32
    segmentation of, 423–28
    tree classifier of, 407
Bombesin
  pancreatic activation and, 57
Born-Kirchoff approximation,
  131
Brain
  nerve growth factor (NGF)
    and, 241–43
N-Bromosuccinimide
  nerve growth factor (NGF)
    transport and, 233

C

Cadmium
  nuclear magnetic resonance
    (NMR) studies of,
    386–91
Caerulin
  pancreatic activation and, 57
Calcium
  adrenal chromaffin cells and,
    71–72
  conductance in neurons,
    451–53
  currents in heart muscle,
    457–60
  currents in neurons, 460–62
  electrical activity of
    pancreatic β-cells and,
    76
  exocrine gland responses and,
    64–69
  heart muscle contractile
    activity and, 82
  insulin secretion and, 73–75
  intracellular injection of
    imitating agonist
    application, 68–69

liposome preparation and,
  489–90
nuclear magnetic resonance
  (NMR) studies of,
  384–86
potassium permeability of
  neurons and, 445–47
Calcium ions
  membrane permeability of,
    448–49
Calcium pump, 274–75
Calcium transport
  ATP-dependent, 268–75
Calmodulin-ATPase
  interaction, 275
cAMP
  adrenal cortical
    nerve growth factor
    (NGF) and, 243
  calcium currents in neurons
    and, 460–62
  exocrine gland secretion and,
    69–70
  insulin secretion and, 75
  nerve growth factor (NGF)
    and, 234
Cancer
  see Oncology
Carbamyl phosphate synthetase
  nuclear magnetic resonance
    (NMR) studies of,
    369–71
Carbon
  nuclear magnetic resonance
    (NMR) applied to
    tRNA and, 204–6
Carbonic anhydrase
  magnetic circular dichroism
    (MCD) and, 317
Carboxypeptidase A
  magnetic circular dichroism
    (MCD) and, 317
Carcinogenesis
  stochastic modeling of, 547
Carcinoma
  thyroid medullary
    nerve growth factor
    (NGF) and, 245
Cardiac action potential
  acetylcholine (ACh) and,
    449–50
  adrenaline and, 448–49
Cardiac rhythm
  adrenergic activation and,
    454
Catalase
  spin-state determinations of
    magnetic circular
    dichroism (MCD)
    and, 310
Cataracts
  photoacoustic spectroscopy
    (PAS) and, 51

SUBJECT INDEX 627

Catecholamine secretion
  acetylcholine (ACh)
    inducing, 70–71
  ionic requirements for, 71–72
Cell cycle
  Chuang-Lloyd model of, 560
  divisions of, 547
  DNA distribution and,
    525–30
  multicompartment model of,
    548
Cells
  see specific type
Cellular system growth
  differential equation models
    of, 543–46
Central nervous system (CNS)
  absolute refractory period
    and axonal conduction
    velocity and, 150
  nerve growth factor (NGF)
    and, 230
  supernormal axonal
    conduction velocity in,
    152
cGMP
  exocrine gland secretion and,
    70
Chemotherapy
  modeling of, 557–63
Chest radiographs
  computer analysis of, 408–21
Chloride conductance
  cellular, 62–63
Chloroperoxidase
  magnetic circular dichroism
    (MCD) and, 310
Chloroplasts
  circular dichroism (CD)
    spectrum of, 125
Cholecystokinin (CCK)
  exocrine pancreas activation
    and, 57
Cholecystokinin-pancreozymin
  (CCK-PK)
  exocrine pancreas activation
    and, 57
Cholesterol
  liposome permeability and,
    499
Cholinergic agonists
  receptors for, 57
Chromatin
  optical behavior of
    circular dichroism (CD)
    and, 121
Chromatography
  column
    liposome size distribution
      analysis and, 493
  gel
    liposome size distribution
      analysis and, 478

high performance liquid
  (HPLC)
  pure lipid preparation and,
    469
sepharose
  liposome size distribution
    analysis and, 493
Chromosomes
  ethidium bromide stained
    fluorescence distribution
    of, 524
  fluorescence distribution of,
    513
Cicyclohexylcarbodiimide
  orientation of, 281
Circular birefringence, 109
Circular dichroism (CD),
  107–28
  of bacteriophages, 121–23
  of dinucleoside phosphate
    adenyl-3',5'-adenosine
    (ApA), 115
  fluorescence-detected, 107–8,
    125–26
  light-scattering suspensions
    and, 121–28
  of nucleohistone aggregates,
    123–24
  polymers and, 109–16
  scattering envelope of, 122
  see also Magnetic circular
    dichroism (MCD)
Circular dichroism (CD)
  spectrum
  of chloroplasts, 125
  of DNA, 117–20
  of nucleic acids, 120–21
  of RNA, 117–19
Circular intensity differential
  scattering (CIDS), 128–38
  absorptive phenomena and,
    135
  of helical molecules, 130
  for light incident along the z
    axis, 134–35
  measure of, 129
  theory of, 130–38
Clinical care process, 583–90
  automated, 584–90
Clinical decision support
  system, 594–600
Coal workers' pneumoconiosis
  (CWP)
  classification of, 412–16
Cobalt
  acetylcholine-induced
    epinephrine release from
    perfused rat adrenal
    glands and, 71
  proteins
    magnetic circular
    dichroism (MCD)
    and, 316–19

Coccolitus huxleyii
  photoacoustic spectroscopy
    (PAS) and, 44
Cockroach salivary gland
  calcium withdrawal and the
    electrical responses to
    agonists in, 66–68
Collagen
  heart muscle resting tension
    and, 88
Column chromatography
  liposome size distribution
    analysis and, 493
Competitive radioassay, 341–43
Complement fixation
  in nerve growth factor
    (NGF) assay, 228
Computer analysis
  of chest radiographs, 408–21
Computerized axial
  tomography (CAT)
  radiation dosimetry and, 554
Computers
  magnetic circular dichroism
    (MCD) spectroscopy
    and, 307
  in medical information
    systems, 581–601
  in tracing neuronal
    structures, 347–62
Concanavalin A (ConA)
  magnetic circular dichroism
    (MCD) and, 317
Copper
  nuclear magnetic resonance
    (NMR) studies of, 386
  proteins
    magnetic circular
    dichroism (MCD)
    and, 316–19
Coproporphyrin
  magnetic circular dichroism
    (MCD) and, 302–3
Corticosterone
  antibody specificity and, 340
Cortisol
  antibody specificity and, 340
  competitive radioassays for,
    342
Coulter diff3, 422
Coupling
  see Strong coupling, Weak
    coupling
Cryostats, 307
Cumulative radiation effect
  (CRE), 556
Cyclic nucleotides
  calcium currents in heart
    muscle and, 457–60
  neurotransmitter responses
    and, 447
  salivary gland secretion and,
    69–70

SUBJECT INDEX

Cyclophosphamide
  in treating B16 melanoma, 560
Cysteine
  magnetic circular dichroism (MCD) and, 308
Cytochrome a
  magnetic circular dichroism (MCD) and, 313–14
Cytochrome a_3
  magnetic circular dichroism (MCD) and, 313–14
Cytochrome b_2
  magnetic circular dichroism (MCD) and, 310
Cytochrome b_5
  Soret magnetic circular dichroism (MCD) and, 310
Cytochrome c
  fluorescence-detected magnetic dichroism (FDMCD) and, 295
  nerve growth factor (NGF) binding and, 232
  optical absorption spectrum of, 38–39
  Soret magnetic circular dichroism (MCD) and, 310
Cytochrome c'
  magnetic circular dichroism (MCD) and, 312
Cytochrome c oxidase
  magnetic circular dichroism (MCD) and, 313–14
  spin-state determinations of, 310
Cytochrome P-450
  magnetic circular dichroism (MCD) and, 312–13
  spin-state determinations of, 310
Cytochrome P-450
  magnetic circular dichroism (MCD) and, 312–13
  spin-state determinations of, 310
Cytoplasm
  white cell, 423–28
Cytosine
  magnetic circular dichroism (MCD) and, 321
Cytosine arabinoside (ara-C)
  DNA metabolism and, 558
  Lewis lung tumor cell survival following treatment with, 533–34
  metabolism of, 560–61

D

Decay catastrophe, 334–36
Dichrometers, 304–5
Dicyclohexylcarbodiimide (DCCD), 280
Differential equation models of cellular system growth, 543–46
Diffuse reflectance, 32
Digitoxin
  antibody specificity and, 340
Digoxin
  antibody specificity and, 340
Dimethylsubarimidate
  nerve growth factor (NGF) and, 225
Dinucleoside phosphate adenyl-3',5'-adenosine (ApA)
  circular dichroism (CD) and, 115
DNA
  cellular content of
    flow cytometric display of, 511–12
    nonparametric comparison of, 519
  chromosomal
    optical behavior of, 121
  circular dichroism (CD) scattering components of, 126
  circular dichroism (CD) spectrum of, 117–20
  in complexes and aggregates, 120–21
  distribution in cell cycle, 525–30
  distribution in gynecological specimens, 515–16
  distribution in peripheral lymphocytes, 520–21
  geometry of, 118
  liposome encapsulation of, 490, 495–96
  magnetic circular dichroism (MCD) and, 321–22
  metabolism
    chemotherapeutic agents and, 558
    nerve growth factor (NGF) and, 235
  synthesis
    as function of DNA content, 529–30
Dopamine
  cockroach salivary gland response to, 62
  synaptical potassium conductance and, 439–40
Dorsal rib contour detection, 411
Dunaliella tertiolecta
  photoacoustic spectroscopy (PAS) and, 44
Duysens flattening, 121

Dysautonomia
  familial
    nerve growth factor (NGF) and, 245

E

Electric dipole moment operator
  for polymers, 113
Electron microscopy
  negative stain
    liposome size distribution analysis and, 477–78, 494
Electron paramagnetic resonance (EPR), 266
Electro Retino-Oculo Graph, 586
Eledoisin
  salivary gland response to, 57
Endocrine gland cells, 70–77
Enzyme-immunoassay (EIA), 343
Enzymes
  nerve growth factor (NGF) and, 238
  phosphoryl transfer
    nuclear magnetic resonance (NMR) studies of, 364–78
Eosinophils (EOS), 423
  machine classification of, 421–22
Epidermal growth factor, 239
Erythrocytes
  photoacoustic spectroscopy (PAS) and, 39–40
Erythropoiesis
  modeling of, 549–50
Erythropoietin feedback, 550
Escherichia coli
  amino acid sequence of, 280–81
  tRNA
    tritium-labeling analysis of, 190
Ethanol
  in liposome preparation, 485
Ether
  in liposome preparation, 486–87
N-Ethylmaleimide (NEM)
  sodium-potassium-ATPase inhibited by, 265
Ethylnitrosourea (ENU)
  nerve growth factor (NGF) and tumor induction and, 246
Excitatory junctional potential (EJP), 438
Excitatory postsynaptical potential (EPSP), 437–38, 447–48

SUBJECT INDEX

Exocrine glands
  activation of, 57–58
  calcium-dependent responses of, 64–69
  exocytosis and protein secretion of, 58–61
  fluid secretion by sodium and, 57
  responses of, 58–61
  stimuli and, 56–64
  stimulus-response coupling in, 56
Exocytosis
  protein secretion and, 58–61
Eye lenses
  human
    photoacoustic spectroscopy (PAS) and, 49–51

F

Fatty acids
  free
    surface charge and permeability properties of liposomes and, 468
Ferrimyoglobin
  magnetic circular dichroism (MCD) spectra of, 311
Ferritin
  liposome encapsulation of, 490
Fibroblasts
  nerve growth factor (NGF) and, 227, 245
Flow cytometric data
  best-fit procedures and, 523–24
  bivariate comparison of, 521
  chemical reaction kinetics and, 535–36
  multivariate clustering of, 522–23
  nonparametric comparison of, 519–23
  one-parameter analysis of, 524–30
  parametric analysis of, 523–36
  two-variable analysis of, 530–36
  univariate comparison of, 520–21
Flow cytometric data display, 510–36
  multivariable, 517–36
  single variable, 511–14
  three-variable, 518
  two-variable, 514–17
Flow cytometry, 510
Flow microfluorometry (FMF)
  cellular systems and, 552
Fluorescence-detected circular dichroism (FDCD), 107–8, 125–26
Fluorescence-detected magnetic circular dichroism (FDMCD), 295
Free fatty acids
  surface charge and permeability procedures of liposomes and, 468
Freeze fracture
  liposome size distribution analysis and, 478
French press extrusion
  in liposome preparation, 485–86
FUdR
  DNA metabolism and, 558

G

D-Galactose
  insulin secretion and, 73–74
Gastrin
  radioimmunoassay (RIA) detection of
  standard curve for, 328
Gel chromatography
  liposome size distribution analysis and, 478
Glioblastoma
  nerve growth factor (NGF) and, 245
Glioma
  nerve growth factor (NGF) and, 227–29
Glucoreceptor
  in pancreatic β-cell, 73
Glucose
  oxidation of
    nerve growth factor (NGF) and, 237
D-Glucose
  insulin secretion and, 73–74
Glucose metabolism
  iodoacetate blocking, 73
Glucose-6-phosphate
  insulin secretion and, 73
Glutamate
  synaptical potassium conductance and, 439
Glutamine synthetase
  nuclear magnetic resonance (NMR) studies of, 368–69, 377–78
D-Glyceraldehyde
  insulin secretion and, 73
Glycerol
  ATPase activity and, 270
Gompertz function
  tumor growth and, 543
Granulopoiesis
  modeling of, 546–49
Guanine
  H-8 exchange rate constants for, 186
  residues
    kethoxal reacting with, 183–84
Guinier's law, 7
Gynecological specimens
  display of fluorescence from, 515–16

H

H-5 exchange
  of pyrimidines, 185
H-8 exchange
  kinetics of, 185–86
  of purines, 184
  rate constants for, 186
Halobacterium halobium
  bacteriorhodopsin and, 281–82
  photoacoustic spectroscopy (PAS) of photosynthesis in, 46–47
Hamiltonian polymer, 110
Heart
  parasympathetic neurons of, 439
Heart muscle
  acetylcholine (ACh), 438–39, 449–50
  adrenergic compounds and, 448–49
  arrangement of actin molecules in, 87–88
  collagen and the resting tension of, 88
  contractile apparatus of, 81
  contractility of
    calcium and, 82
  cyclic nucleotides and transmitter effects on calcium currents in, 457–60
  during cyclic contraction
    diffraction patterns from, 95–97
    time-resolved diffraction study of, 101–2
  mechanical properties of, 81–82
  myofilament lattice of, 93
  myosin-related diffraction patterns of, 85–87
  paired-pulse stimulation of, 102
  quiescent
    equatorial diffraction patterns from, 88–90
    meridional diffraction patterns from, 84–87
  in rigor
    equatorial diffraction patterns from, 90–94

630 SUBJECT INDEX

meridional diffraction
  patterns from, 87–88
  staircase phenomenon of,
    97–98
  X-ray diffraction of, 81–104
  quiescent and in rigor,
    84–94
HELP, 599–600
Hematopoiesis
  modeling of, 549–552
HEMATRAK, 422
Heme
  zero field splitting of, 301
Heme-hemopexin
  magnetic circular dichroism
    (MCD) and, 310
Heme proteins
  magnetic circular dichroism
    (MCD) and, 310
  photoacoustic
    spectroscopy (PAS) and,
    309–41
Hemocyanin
  magnetic circular dichroism
    (MCD) and, 319
Hemoglobin
  heme absorption spectrum of,
    39
  magnetic circular dichroism
    (MCD) and, 39–41
Hemopexin
  magnetic circular dichroism
    (MCD) and, 314
  Soret magnetic circular
    dichroism (MCD) and,
    310
Hemoproteins
  photoacoustic spectroscopy
    (PAS) and, 38–41
Hepatitis B antigen
  assay of, 339–40
High performance liquid
  chromatography (HPLC)
  pure lipid preparation and,
    469
Hirudo medicinalis
  axonal block and sensory
    fiber adaptation in, 163
Histamine
  synaptical potassium
    conductance and,
    439–40
Histidine
  magnetic circular dichroism
    (MCD) and, 308
Hormones
  see specific type
Horseradish peroxidase
  spin-state determinations of,
    310
Hospital and ambulatory care
  information systems,
    588–90

Hough accumulator array
  technique
  lung tumor detection and,
    420
Human eye lenses
  photoacoustic spectroscopy
    (PAS) and, 49–51
6-Hydroxydopamine
  nerve growth factor (NGF)
    and, 234–36
5-Hydroxytryptamine (5-HT)
  synaptical potassium
    conductance and, 439
  voltage-dependent calcium
    conductance in
    molluscan neurons and,
    451–52
Hydroxyurea
  DNA metabolism and, 558
  effects on malignant cells,
    560–61
Hypothalamic-pituitary-adrenal
  axis
  nerve growth factor (NGF)
    action on the brain and,
    242–43
Hypoxanthine
  magnetic circular dichroism
    (MCD) and, 320

I

Ideal solution
  X-ray scattering and, 6–8
Immune system
  in oncology, 550–51
Immunocytes (LYI), 423
Immunoenzymometric assay,
  343
Immunoradiometric assay
  (IMRA), 339–40
Immunosympathectomy, 224,
  240–41
Indole alkaloids
  indole chromophore in, 309
Inhibitory postsynaptic
  potential (IPSP), 437–38,
    447–48
Inosine
  magnetic circular dichroism
    (MCD) and, 320
Insulin secretion, 73–75
Internal reflection spectroscopy,
  32
Intrinsic life-function, 544
Invariant volume hypothesis
  low density lipoprotein
    (LDL) and, 18–19
  X-ray scattering and, 9–16
Iodoacetate
  glucose metabolism and, 73
Ionophoresis, 440
Ionophoric material

  isolated from
    264
Iron-sulfur proteins
  magnetic circular dichroism
    (MCD) and, 315–16
Isoprenaline
  mammalian salivary glands
    and, 62

K

Karhunen-Loéve expansion,
  404–5
Kethoxal
  guanine residues reacting
    with, 183–84
Klebsiella pneumoniae
  "Mo-Fe" proteins from, 316
Kolmogorov-Smirnov test, 520
Kramers degeneracy, 297, 304
Kronig-Kramers transforms,
  109

L

Lacrimal glands
  activation of, 57
Ladder-structured decision tree
  for lung tumor detection, 420
Lanthanides
  magnetic circular dichroism
    (MCD) and, 319–20
Lanthanum
  nuclear magnetic resonance
    (NMR) studies of,
    384–86
LARC, 422
Laser light scattering
  liposome size distribution
    analysis and, 478
Lasers
  magnetic circular dichroism
    (MCD) spectroscopy
    and, 305
  in photoacoustic
    spectroscopy (PAS), 37
L cells
  nerve growth factor (NGF)
    and, 227
Leaf matter
  photoacoustic spectroscopy
    (PAS) and, 41
Lettuce chloroplast membranes
  photoacoustic spectroscopy
    (PAS) of photosynthesis
    in, 45–46
Leukemia
  modeling of, 546, 549,
    560–61
Lewis lung carcinoma cells
  DNA and RNA content of,
    517
  survival following treatment

SUBJECT INDEX 631

with cytosine arabinoside
(ara-C), 533–34
Life function
intrinsic, 544
Light microscopy
liposome size distribution
analysis and, 476–80,
494
Light scattering
liposome size distribution
analysis and, 478–80,
493
Lipids
liposome composition and,
469–76
sodium-potassium-ATPase
interacting with, 263–64
synthesis of
nerve growth factor
(NGF) and, 237
Lipid vesicles
see Liposomes
Lipoproteins
fractionation of, 16
low density (LDL)
autocorrelation function
for, 20–25
invariant volume
hypothesis and, 18–19
morphological parameters
of, 20
X-ray scattering and,
16–26
Liposarcoma
nerve growth factor (NGF)
and, 245
Liposomes
composition of
lipids and, 469–76
drug delivery and, 501–2
lipid soluble compounds
incorporated into,
496–98
permeability properties of,
498–500
phospholipids used in, 471
preparation of, 469–76,
480–92
calcium-induced fusion
and, 489–90
detergent removal and,
483–85
ethanol injection and, 485
ether infusion and, 486–87
French press extrusion
and, 485–86
multilamellar, 480–82
reverse phase evaporation
and, 487–89
small unilamellar, 482–83
size distribution of, 476–80,
493–94

stability and storage of,
500–2
types of, 470
water-soluble compounds
encapsulated into,
494–95
Lithium isotope
nuclear magentic resonance
(NMR) studies of,
382–84
Low density lipoprotein (LDL)
see Lipoprotein
Lung boundary determination
algorithm for, 410
Lung carcinoma cells
Lewis
DNA and RNA content
of, 517
survival following
treatment with
cytosine arabinoside
(ara-C), 533–34
Lung diseases
classification of
computer analysis of chest
radiographs for,
408–21
Lung tumors
detection of, 419–21
Lymphoblasts (LYB), 423
Lymphocytes (LYM), 423
atypical (LYA), 423
machine classification of,
421–22
phytohemagglutin-stimulated
average DNA distributions
for, 520–21
scatterplots of fluorescence
from, 514
thymocyte-treated
average DNA distributions
for, 520–21
Lysophosphatides
surface charge and
permeability properties
of liposomes and, 468

M

Magnesium
nuclear magnetic resonance
(NMR) studies of,
384–86
Magnesium ions
tRNA interacting with,
191–98
Magnetically induced circular
differential scattering, 296
Magnetic circular dichroism
(MCD)
"A" type, 297–99
"B" type, 299
"C" type, 299–302

of discrete line spectra,
296–301
dispersion analysis in, 302–3
fluorescence-detected
(FDMCD), 295
moment analysis in, 302–3
on non-heme porphyrins, 322
of nucleic acids, 320–22
overlapping lines and
continuum band spectra
and, 301–2
of proteins, 307–20
theory of, 296–304
transmission detected,
293–95
Magnetic circular dichroism
spectroscopy
apparatus for, 304–7
Magnetic circular emission
(MCE), 295–96
Magnetic dipole moment
operator
for polymers, 113
Magnets
magnetic circular dichroism
spectroscopy and, 305–6
Manganese
calcium influx in excitable
cells inhibited in, 67
ions
tRNA interacting with,
191–98
Mannoheptulose
glucose-induced activity
inhibited by, 73
Marine algae
photoacoustic spectroscopy
(PAS) and, 42–43
Marine phytoplankton
photoacoustic spectroscopy
(PAS) and, 43–44
Maxwell model
for muscle contractility, 82
Medical image pattern
recognition, 396–408
classification strategies in,
406–7
classifier training and testing
in, 408
feature measurement in,
400–4
feature selection and
extraction in, 404–6
image segmentation in,
399–400
preprocessing in, 397–98
Medical imagery
classification of, 395–96
Medical information systems,
581–601
Medical records
automated, 590–94
problem-oriented, 593–94

632 SUBJECT INDEX

Melanoma
  B16
    treatment strategies for, 560
    nerve growth factor (NGF) and, 227–29
Membrane potential
  axonal conduction velocity or excitability and, 158–61
  extracellular potassium and, 158–59
Membrane transport
  ATP-dependent calcium, 268–75
  ATP-dependent sodium and potassium, 260–68
  proton ATP-dependent, 276–81
  systems structure and function related in, 285–86
Metal ions
  nuclear magnetic resonance (NMR) studies of, 378–91
Metamyelocytes (MET), 423
Methemoglobin
  magnetic circular dichroism (MCD) and, 312
Methotrexate
  DNA metabolism and, 558
Metmyoglobin
  Soret magnetic circular dichroism (MCD) and, 310–11
MgATP complexes
  stereochemistry of, 373–75
Mie-type scattering, 125
Monocytes (MON), 423
  machine classification of, 421–22
Monoiodoangiotensin II
  specific activity of, 336
Monosaccharides
  glucose-induced activity and, 73
Morowitz mechanism, 285
Mouse sarcomas 180 or 37
  nerve growth factor (NGF) and, 223–27
mRNA
  liposome encapsulation of, 490
Muscular dystrophy
  nerve growth factor (NGF) and, 245
MYCIN, 595–97
Myelinated nerve fibers
  conduction failure in, 146
Myeloblasts (MYB), 423
Myelocytes (MYC), 423
Myoblasts
  nerve growth factor (NGF) and, 227
Myoglobin
  magnetic circular dichroism (MCD) and, 310
  spin-state determinations of, 310
Myosin
  nuclear magnetic resonance (NMR) studies of, 367
  skeletal muscle contraction and, 81
  in the thick filament helical arrangement of, 86
Myosin projections
  during diastolic phase of heart muscle, 102–3

N

Negative state electron microscopy
  liposome size distribution analysis and, 477–78, 494
Nerve fibers
  myelinated
    conduction failure in, 146
Nerve growth factor (NGF), 223–52
  absence of, 240–41
  actions on the brain, 241–43
  actions on target cells, 234–40
  amino acid sequence of, 224–25
  assays of, 227–28
  biosynthesis of, 230–31
  cellular receptors for, 231–34
  forms, sources, and structure of, 224–27
  isolation from salivary glands, 225
  mechanisms of action of, 247–51
  molecular weight of, 224
  neurological disease and, 244–47
  peptides and, 226
  subunits of, 225–26
  target cells of, 228–30
  as therapeutic agent in humans, 246
  transcriptional effects of, 249
  transport of, 233–34
  tumors and, 244–47
Neurite outgrowth
  nerve growth factor (NGF) and, 234–36
Neuroblastoma
  nerve growth factor (NGF) and, 227–29, 245–46
Neurofibromatosis
  nerve growth factor (NGF) and, 244
Neurological disease
  nerve growth factor (NGF) and, 244–47
Neuroma
  bilateral acoustic
    nerve growth factor (NGF) and, 244
Neuronal structures
  computerized tracing of, 347–62
Neurons
  5-hydroxytryptamine-induced increase in voltage-dependent calcium conductance in, 451–52
Neurospora crassa
  amino acid sequence of, 280–81
Neutron scattering, 3, 10
  see also X-ray scattering
Neutrophils
  band (BAN), 423
  machine classification of, 421–22
  segmented (SEG), 423
Nickel proteins
  magnetic circular dichroism (MCD) and, 316–19
Nitrite reductase
  magnetic circular dichroism (MCD) and, 310
Nuclear Data Model 6600, 517
Nuclear magnetic resonance (NMR)
  applied to tRNA, 199–217
  liposome size distribution analysis and, 476–80
  of metal nuclei, 378–91
  of phosphoryl transfer enzymes, 364–78
  of yeast tRNA^{Phe}, 200
Nuclear Overhauser effect
  nuclear magnetic resonance (NMR) applied to tRNA and, 203–5, 210
Nucleated blood cells
  automated classification of, 421–32
  binary tree classifier of, 428–32
  segmentation of, 423–28
Nucleated red cell A (NRA), 423
Nucleated red cell B (NRB), 423
Nucleic acid bases
  exchange reactions associated with, 184–85
  optical activity of, 117
Nucleic acids

SUBJECT INDEX 633

circular dichroism (CD)
spectrum of, 120–21
magnetic circular dichroism (MCD) and, 320–22
primary structure effects of, 117
secondary structure effects of, 117–20
see also Polymers
Nucleohistone aggregates
circular dichroism (CD) and, 123–24
Nucleohistones
optical behavior of, 121
Nucleosides
H-8 exchange rate constants for, 186
Nucleotides
binding to sarcoplasmic reticulum vesicles, 272
cyclic
neurotransmitter responses and, 447
3'-Nucleotides
H-8 exchange rate constants for, 186

O

Octopamine
synaptical potassium conductance and, 439
Oligomycin, 280
sodium-potassium-ATPase inhibited by, 265
Oncology
chemotherapy modeling in, 557–63
differential equation models in, 543–46
laboratory analysis modeling in, 551–52
radiotherapy modeling in, 554–57
simulation models in, 547–48
stochastic models in, 546–47
treatment modalities in, 552–63
treatment optimization in, 563–571
Optical rotatory dispersion (ORD), 129
Optical spectroscopy, 31–32
Ornithine decarboxylase
nerve growth factor and, 239–42
Oubain
ATPase inhibited by, 260
blocking ion binding, 264
insulin secretion and, 74
post-tetanic hyperpolarization of mammalian C fibers blocked by, 160

P

Paget's disease of bone
nerve growth factor (NGF) and, 245
Pancreas
acetylcholine (ACh) and, 62
activation of, 56–57
calcium withdrawal and the electrical responses to agonists in, 66–68
peptide agonists and, 62
Pancreatic β-cells, 73–77
electrical activity of, 75–77
glucoreceptor in, 73
Pancreozymin (PZ)
exocrine pancreas activation and, 57
Papain
bacteriorhodopsin hydrolyzed by, 284
Patterson function
in X-ray diffraction, 109
Peptide agonists
pancreas and, 62
Peptide bonds
magnetic circular dichroism (MCD) and, 307–8
Peptide hormones
exocrine pancreas activation and, 57
viroimmunoassay and, 342
Peptidergic agonists
receptors, for, 57
Peptides
nerve growth factor (NGF) and, 226
Peptides substance P
salivary gland response to, 57
Peripheral nervous system
absolute refractory period and axonal conduction velocity and, 150
nerve growth factor (NGF) and, 241
supernormal axonal excitability or conduction velocity in, 151
Periplaneta americana
axonal impulse conduction in, 165
Peroxidase
salivary gland release of, 58–61
Perrin matrix, 130
Pharmacokinetics, 557–59
Phentolamine
agonist blockage and, 57
Phenylalanine
magnetic circular dichroism (MCD) and, 308–9
Phenylethanolamine
synaptical potassium conductance and, 439
Pheochromocytoma
multicentric
nerve growth factor (NGF) and, 245
rat
nerve growth factor (NGF) and, 229
stimulus secretion coupling and, 72
Phosphate analysis
liposome size distribution analysis and, 493
Phosphatidic acid
isolation of
yielding purified phospholipids, 469
Phosphatidylcholine, 263–64
isolation of
yielding purified phospholipids, 469
Phosphatidylethanolamine, 263–64
Phosphatidylglycerol
defined
yielding purified phospholipids, 469
Phosphatidylglycerophosphate, 282
Phosphatidylserine
isolation of
yielding purified phospholipids, 469
Phospholamban, 274
Phospholipids
purification of, 468–69
quantitation of, 468–69
sodium-potassium-ATPase and, 263–64
used in liposomes, 471
Phosphorus
nuclear magnetic resonance (NMR) applied to tRNA and, 204–6
Phosphoryl transfer enzymes
nuclear magnetic resonance (NMR) studies of, 364–78
Photoacoustic spectrometer
single-beam, 36
Photoacoustic spectroscopy (PAS), 31–54
advantages of, 33–34
biological studies using, 38–47
data acquisition in, 37–38
drugs in tissues and, 48–49
experimental chamber in, 37
experimental methodology of, 36–38
hemoproteins and, 38–41

SUBJECT INDEX

identifying bacterial states, 48
medical studies using, 48–53
photosynthesis and, 45–47
plant matter and, 41–45
radiation sources in, 36–37
theory of, 34–36
tissue studies using, 51–53
Photoelastic modulators (PEM), 304–5
Photosynthesis
  photoacoustic spectroscopy (PAS) and, 45–47
Phthalocyanine
  magnetic circular dichroism (MCD) and, 322
Phytoplankton
  marine
    photoacoustic spectroscopy (PAS) and, 43–44
Picornavirus RNA
  liposome encapsulation of, 490
Plant matter
  photoacoustic spectroscopy (PAS) and, 41–45
Plasmacytes (PLA), 423
Plastocyanin
  magnetic circular dichroism (MCD) and, 317
Platymonas sp.
  photoacoustic spectroscopy (PAS) and, 44
Pneumoconiosis
  coal workers' (CWP)
    classification of, 412–16
    detection of opacities in, 416–19
Poliovirus
  liposome encapsulation of, 490
Poly-L-lysine
  magnetic circular dichroism (MCD) and, 308
Polymers
  absorption bands of, 110
  circular dichroism (CD) and, 109–16
  electronic wavefunctions of, 110–12
  polarizability of, 114
  strong coupling of, 109–14
  symmetry of, 116
  weak coupling of, 114–15
  see also Nucleic acids
Polynucleotide phosphorylase
  nuclear magnetic resonance (NMR) studies of, 366–67
Polynucleotides
  magnetic circular dichroism (MCD) and, 321–22

Porphyrins
  magnetic circular emission (MCE) and, 295–96
  non-heme
    magnetic circular dichroism (MCD) and, 322
  in solution
    zero field splitting of, 301
  Zeeman splitting in, 293
Position-sensitive X-ray detectors, 3–4
Potassium
  extracellular
    resting membrane potential and, 158–59
    pancreatic activation and, 58
    synaptic responses and, 437–47
Potassium conductance
  cellular, 62–63
Potassium-dependent inhibitory potentials, 438–47
Potassium transport
  ATP-dependent, 260–68
Procarboxypeptidase A
  magnetic circular dichroism (MCD) and, 317
Proinsulin
  nerve growth factor (NGF) and, 226
PROMIS, 598
Promyelocytes (PRO), 423
Propranolol
  agonist blockage and, 57
Prostaglandins
  viroimmunoassay and, 342
Proteins
  exocytosis and exocrine gland secretion of, 58–61
  heart muscle contractility and, 82–83
  magnetic circular dichroism (MCD) and, 307–20
  synthesis of
    nerve growth factor (NGF) and, 237
Pulmonary artery boundaries
  algorithm for, 411
Pulmonary infiltrations
  classification of, 411–12
Purines
  H-8 exchange reaction of, 185
  magnetic circular dichroism (MCD) and, 320–21
Pyraminmonas sp.
  photoacoustic spectroscopy (PAS) and, 44
Pyrimidines
  H-5 exchange of, 185

  magnetic circular dichroism (MCD) and, 321
Pyrophosphatase
  inorganic
    nuclear magnetic resonance (NMR) studies of, 367
Pyruvate kinase
  nuclear magnetic resonance (NMR) studies of, 371–76

Q

Quinine
  insulin secretion and, 75–77

R

Radiation dosimetry
  computerized axial tomography (CAT) and, 554
Radioassay
  competitive, 341–43
Radioimmunoassay (RIA), 327–62
  gastrin detection by, 328
  labeled antigens in, 333–38
  non-equilibrium, 338–39
  sensitivity of, 329–33
  specificity of, 340–41
Radioiodine
  in peptide hormone assay, 333–38
Radioiodothyronines
  maximum specific activity of, 335
  production of, 336
  stability of, 335–36
Radiotherapy
  modeling of, 554–57
  optimizing, 565–71
Radiotracers
  liposome size distribution analysis and, 493
Radium implantation therapy, 566
Raman scattering, 32, 296
Random phase approximation, 114
Rayleigh-Debye scattering, 125
Rayleigh scattering, 136, 296
Red blood cells
  see Erythrocytes
Relaxin, 226
Retinylamine, 283
Retinylidene, 284
Reverse-phase evaporation vesicles (REV), 487–89

SUBJECT INDEX 635

Rib boundary determination
  algorithm for, 410–11
Ribosomes
  tRNA interacting with,
    190–91
Riley-Day syndrome
  nerve growth factor (NGF)
    and, 245
RNA
  circular dichroism (CD)
    spectrum of, 117–19
  geometry of, 118
  synthesis of
    nerve growth factor
      (NGF) and, 237
  see also mRNA, tRNA
RNA viruses
  circular dichroism (CD)
    spectra of, 125
Rubinow model, 546
Rubredoxin
  magnetic circular dichroism
    (MCD) and, 315–17

S

Saccharomyces cerevisiae
  amino acid sequence of,
    280–81
Salivary glands
  activation of, 57
  amylase release of, 58–61
  cockroach
    calcium withdrawal and
      the electrical
      responses to agonists
      in, 66–68
    dopamine and, 62
    isoprenaline and, 62
    nerve growth factor (NGF)
      isolated from, 225
  secretion of
    cyclic nucleotides in,
      69–70
Sarcoplasmic reticulum-ATPase
  primary structure of, 273–74
Sarcoplasmic reticulum vesicles
  nucleotide binding to, 272
Scanning transmission electron
  microscope (STEM), 349
Scatchard analysis, 192–96
Scatterplots, 514
Schizophrenia
  nerve growth factor (NGF)
    and, 245
Secretin
  exocrine pancreas activation
    and, 57
Sedimentation analysis
  of liposomes, 478–79
Sensory neurons

nerve growth factor (NGF)
  and, 228–30
Sepharose chromatography
  liposome size distribution
    analysis and, 493
Serotonin
  indole chromophore in, 309
Skeletal muscle
  arrangement of actin
    molecules in, 87
  contractile apparatus of, 81
  contracting
    diffraction patterns from,
      94–95
    time-resolved diffraction
      studies of, 100–1
  myofilament lattice in, 92
  myosin-related diffraction
    pattern of, 85–87
  quiescent
    equatorial diffraction
      patterns from, 88–90
    meridional diffraction
      patterns from, 84–87
  in rigor
    equatorial diffraction
      patterns from, 90–94
    meridional diffraction
      patterns from, 87–88
Snake venom
  nerve growth factor (NGF)
    and, 226–27
Sodium
  exocrine gland fluid secretion
    and, 57
  insulin secretion and, 75
Sodium conductance
  cellular, 62–63
Sodium ions
  tRNA interacting with,
    192–95
Sodium-potassium-ATPase
  amino acid sequence of, 267
  ionophoric material isolated
    from, 264
  lipid interactions of, 263–64
  molecular weights of
    subunits of, 262
  oligomeric structure of,
    261–63
  regulation of, 267–68
  stoichiometries for, 262
  ultrastructure of, 261
Sodium-potassium pump
  adrenaline and, 454
  binding sites of, 264–65
  kinetics of, 264–65
  regulation of, 267–68
  strophanthidin and, 159
Sodium transport
  ATP-dependent, 260–68

Spectrophotometers, 305
Spectroscopy
  attenuated total reflection,
    32
  internal reflection, 32
  magnetic circular dichroism
    (MCD)
    apparatus for, 304–7
  optical, 31–32
  see also Photoacoustic
    spectroscopy (PAS)
Sperm
  optical behavior of, 121
Spermidine
  tRNA interacting with,
    191–98
Spermine
  tRNA interacting with,
    191–98
Squalus acanthias
  sodium-potassium-ATPase
    pattern of, 261
Staircase phenomenon, 97–98
Stellacyanin
  magnetic circular dichroism
    (MCD) and, 317
Steroids
  cardiac
    ion binding sites of, 265
    radioimmunoassay (RIA) of,
      340–41
    viroimmunoassay and, 342
Stochastic model of
  birth-and-death processes,
    546–47
Storage rings
  X-ray scattering and, 3–4
Stratum corneum
  rat
    photoacoustic
      spectroscopic studies
      of, 51–53
Strong coupling
  energy matrix in, 111
  of polymers, 109–14
Strophanthidin
  ATPase inhibited by, 260
  supernormal axonal
    excitability and, 159
Superoxide dimutase
  magnetic circular dichroism
    (MCD) and, 317
Sympathetic nervous system
  nerve growth factor (NGF)
    and, 227
Sympathetic neurons
  nerve growth factor (NGF)
    and, 228–30
Synaptic current decay, 442
Synaptic responses
  slow

SUBJECT INDEX

potassium-dependent, 438–47
Synchrotrons
X-ray scattering and, 3–4

T

Tetrachlorosalicylanilide (TCSA)
photoacoustic spectroscopy (PAS) and, 48–49
Tetracycline (TCN)
photoacoustic spectroscopy (PAS) and, 49
Tetrahymena cells
growth data for, 546
Tetrodotoxin (TTX)
acetylcholine-induced epinephrine release from perfused rat adrenal glands and, 71
electrical activity of pancreatic β-cells and, 76
Thalassiosira pseudonana
photoacoustic spectroscopy (PAS) and, 44
Thermolysin
magnetic circular dichroism (MCD) and, 317
6-Thioguanine, 562
4-Thiouridine
in tRNA, 184
Thymine
magnetic circular dichroism (MCD) and, 321
Thyroid gland
medullary carcinoma of nerve growth factor (NGF) and, 245
Thyronines
substrates and products of iodination of, 335
Thyroxine
competitive radioassays for, 342
Time-dependent Hartree approximations, 114
Tocopherols
liposome permeability and, 499
Tolbutamide
insulin secretion and, 73–74
Tree classifier
of nucleated blood cells, 407, 428–32
Trinitrophenyl-ATP, 266
Tritium
labeling *Escherichia coli* tRNA, 190
probing specfic sites in tRNA, 183–91

Tritium exchange rates
for yeast tRNA^{Phe}, 187–88
tRNA
aminoacyl tRNA synthetase interacting with, 190–91
codon binding and, 211–13
Escherichia coli
tritium-labeling analysis of, 190
fragments of, 210
H-8 labeling reaction to, 185–90
ions interacting with, 191–98
X-ray crystallography of, 196–98
local stability and flexibility of, 213–14
nuclear magnetic resonance (NMR) and, 199–217
kinetic studies of, 201–3
nuclear Overhauser effect and, 203–5, 210
phosphorus and carbon nuclei used in, 204–5
resonance assignments of, 206–10
ribosomes interacting with, 190–91
ring current shifts in, 208–9
solvent exchange and, 214–17
tertiary resonances of, 207–8
4-thiouridine in, 184
tritium probes of specific sites in, 183–91
tRNA^{Phe}
yeast, 182
crystal structure of, 198
nuclear magnetic resonance (NMR) spectrum of, 200
sequence and cloverleaf structure of, 187
tritium exchange rates for, 187–88
Tropomyosin
in skeletal muscle, 87
Troponin
in skeletal muscle, 87
Tryptophan
cataract formation and, 51
fluorescence-detected magnetic circular dichroism (FDMCD) and, 295
magnetic circular dichroism (MCD) and, 308–9
T-testing, 520–21
Tumor cells
nerve growth factor (NGF) and, 229, 244–47
Tumors
growth of

Gompertz function and, 543
lung
detection of, 419–21
Tyrosine
cataract formation and, 51
magnetic circular dichroism (MCD) and, 308–9
Tyrosine hydroxylase
brain
nerve growth factor (NGF) and, 242
nerve growth factor (NGF) and, 229
specific activity of, 238

U

Ultracentrifugation
liposome size distribution analysis and, 478, 493
Uracil
magnetic circular dichroism (MCD) and, 321

V

Veratridine
calcium-dependent catecholamine release from perfused adrenal glands and, 71
Vinblastine
effects on malignant cells, 560–61
Vincristine
effects on malignant cells, 560–561
Viroimmunoassay, 342
Viruses
magnetic circular dichroism (MCD) and, 321–22
Voight model
for muscle contractility, 82
von Recklinghausen's disease
nerve growth factor (NGF) and, 244

W

Weak coupling
of polymers, 114–15
White blood cell cytoplasm
segmentation of, 423–28
White blood cell differential counters, 421–22

X

X-ray crystallography
of tRNA ion-binding sites, 196–98

SUBJECT INDEX

X-ray detectors
  position-sensitive, 3–4
X-ray diffraction
  of bacteriorhodopsin, 282
  of the heart, 81–104
    in the quiescent state and in rigor, 84–94
  Patterson function in, 109
X-ray DNA fiber analysis, 118–19
X-ray scattering
  collimation distortions in, 5
  distribution in, 1–2
  of DNA gels, 119–20
  electron density and, 8–9
  exposure times in, 3
  fluctuations in, 1
  ideal solution and, 6–8
  information content of, 14
  intensity in, 1–2, 6–7
  invariant volume hypothesis and, 9–16
  low density lipoprotein (LDL) and, 16–26
  of macromolecules in solution, 1–28
  scattering angles in, 2
  structure analysis and, 15–16
  techniques of, 4–5
  theory of, 5–16
  wavelengths used in, 2
X-ray scattering curves
  small-angle properties of, 2

Y

Yeast tRNA^{Phe}, 182
  crystal structure of, 198
  nuclear magnetic resonance (NMR) spectrum of, 200
  sequence and cloverleaf structure of, 187
  tritium exchange rates for, 187–88

Z

Zeeman splitting, 293, 297–99
Zinc
  nuclear magnetic resonance (NMR) studies of, 386

CUMULATIVE INDEXES

CONTRIBUTING AUTHORS, VOLUMES 5–9

A

Abbey, S. G., 9:581–604
Adrian, R. H., 7:85–112
Albers, R. W., 9:259–91
Allen, N. S., 7:469–95; 497–526
Allen, R. D., 7:469–95; 497–526
Antonini, E., 6:239–71
Arndt-Jovin, D. J., 7:527–58
Arnold, W., 6:1–6

B

Bárány, M., 8:1–25
Barrantes, F. J., 8:287–321
Barrington Leigh, J., 5:239–70
Bauer, W. R., 7:287–313
Beal, S., 5:561–87
Brehm, P., 8:353–83
Brinley, F. J. Jr., 7:363–92
Burt, C. T., 8:1–25
Bustamante, C., 9:107–41

C

Callender, R., 6:33–55
Caplan, S. R., 5:449–76
Caputo, C., 7:63–83
Careri, G., 8:69–97
Cervetto, L., 7:229–51
Chambers, J. L., 6:177–93
Chiancone, E., 6:239–71
Christiansen, G., 7:19–35
Cohen, J. S., 6:383–417
Cohen, S. M., 8:1–25
Colombetti, G., 7:341–61

D

Das, M., 8:165–93
Dean, P. N., 9:509–39
de Haas, G. H., 5:77–117

E

Eaton, W. A., 5:511–60
Eckert, R., 8:353–83

Egan, W., 6:383–417
Eisenberger, P., 7:559–78
Elashoff, R. M., 5:561–87
Elston, R. C., 7:253–86
Eyring, H., 6:111–33

F

Fasella, P., 8:69–97
Feldmann, R. J., 5:477–510
Fink, A. L., 6:301–43
Foster, K. W., 6:419–43
Fox, C. F., 8:165–93
Friesen, W. O., 7:37–61
Fu, K-S., 9:393–436
Fuortes, M. G. F., 7:229–51

G

Garfinkel, D., 6:525–42
Gennis, R. B., 6:195–238
Ginsborg, B. L., 9:55–80
Goldbeter, A., 5:449–76
Gratton, E., 8:69–97
Gray, H. B., 5:363–96
Gray, J. W., 9:509–39
Griffith, J. D., 7:19–35
Guroff, G., 9:223–57

H

Hagiwara, S., 8:385–416
Hambrecht, F. T., 8:239–67
Henderson, R., 6:87–109
Hilton, B. D., 8:99–127
Hobbs, A. S., 9:259–91
Hofrichter, J., 5:511–60
Holmquist, B., 9:293–326
Holten, D., 7:189–227
Holwerda, R. A., 5:363–96
Honig, B., 6:33–55
House, C. R., 9:55–80
Hudson, B. S., 6:135–50

J

Jaffe, L. A., 8:385–416
Jaffe, L. F., 6:445–76

Jennings, W. H., 5:177–204
Jennrich, R. I., 8:195–238
Johnson, F. H., 6:111–33
Jonas, A., 6:195–238
Jovin, T. M., 7:527–58

K

Katz, J. J., 7:393–434
Kearns, D. R., 6:477–523
Kehoe, J., 9:437–65
Kincaid, B. M. 7:559–78
Kocsis, J. D., 9:143–79
Kossiakoff, A. A., 6:177–93

L

Laska, E. M., 9:581–604
Latt, S. A., 5:1–37
Lechene, C. P., 6:57–85
Lenard, J., 7:139–65
Lenci, F., 7:341–61
Levinthal, C., 8:323–51
Levinthal, C., 9:347–62
Li, C-C., 9:393–436
Luzzati, V., 9:1–29
Lymn, R. W., 8:145–63

M

Macagno, E. R., 8:323–51
Macagno, E. R., 9:347–62
Maestre, M. F., 9:107–41
Makinen, M. W., 6:301–43
Marbach, C. B., 6:525–42
Marty, A., 9:437–65
Matsubara, I., 9:81–105
Meech, R. W., 7:1–18
Melchior, D. L., 5:205–38
Montal, M., 5:119–75

N

Neher, E., 6:345–81
Newton, C. M., 9:541–79
Norris, J. R., 7:393–434
Nuccitelli, R., 6:445–76

O

Ottensmeyer, F. P., 8:129–44

P

Pagano, R. E., 7:435–68
Pak, W. L., 5:397–448
Papahadjopoulos, D., 9:467–508
Pinto, L. H., 5:397–448
Pysh, E. S., 5:63–75

R

Ralston, M. L., 8:195–238
Rao, D. C., 7:253–86
Raushel, F. M., 9:363–92
Redfield, A. G., 9:181–221
Reite, M., 7:167–88
Richards, F. M., 6:151–76
Rosenbaum, G., 5:239–70
Rosencwaig, A., 9:31–54

S

Schimmel, P. R., 9:181–221
Schultz, A. R., 5:117–204

Shapiro, M. B., 5:177–204
Shapiro, N. Z., 6:525–42
Shindo, H., 6:383–417
Shipman, L. L., 7:393–434
Shulman, R. G., 7:559–78
Sobel, I., 8:323–51
Sobel, I., 9:347–62
Sobell, H. M., 5:307–35
Steim, J. M., 5:205–38
Steinberg, I. Z., 7:113–37
Stent, G. S., 7:37–61
Stephenson, J. L., 7:315–39
Stevens, C. F., 6:345–81
Stone, G. C., 8:27–45
Stover, B. J., 6:111–33
Stroud, R. M., 6:177–93
Sutherland, J. C., 9:293–326
Swadlow, H. A., 9:143–79
Szoka, F. Jr., 9:467–508

T

Tardieu, A., 9:1–29
Thurnauer, M. C., 7:393–434
Tinoco, I. Jr., 9:107–41

U

Ulbricht, W., 6:7–31

V

Verger, R., 5:77–117
Villafranca, J. J., 9:363–92
Vinores, S., 9:223–57

W

Waggoner, A. S., 8:47–68
Warner, R. R., 6:57–85
Warshel, A., 6:273–300
Wasielewski, M. R., 7:393–434
Waxman, S. G., 9:143–79
Weinstein, J. N., 7:435–68
Wetmur, J. G., 5:337–61
Wherland, S., 5:363–96
Wiederhold, M. L., 5:39–62
Wilson, D. L., 8:27–45
Windsor, M. W., 7:189–227
Wolfenden, R., 5:271–306
Woodward, C. K., 8:99–127

Y

Yalow, R. S., 9:327–45

Z

Zimmerman, J., 7:167–88

CHAPTER TITLES, VOLUMES 5–9

CONCEPTUAL AND PHYSICAL TOOLS FOR ANALYSIS
Mathematical Techniques

Computers in the Research Laboratory	M. B. Shapiro, A. R. Schultz, W. H. Jennings	5:177–204
The Design of Computing Systems for Molecular Modeling	R. J. Feldmann	5:477–510
Stiff Differential Equations	D. Garfinkel, C. B. Marback, N. Z. Shapiro	6:525–42
Fitting Nonlinear Models to Data	R. I. Jennrich, M. L. Ralston	8:195–238
The Microprocessor, A New Tool for the Biosciences	H. W. Shipton	8:269–86
Special Techniques for the Automatic Computer Reconstruction of Neuronal Structures	I. Sobel, C. Levinthal, E. R. Macagno	9:347–62
Machine-Assisted Pattern Classification in Medicine and Biology	C-C. Li, K-S. Fu	9:393–436
Biomathematics in Oncology: Modeling of Cellular Systems	C. M. Newton	9:541–79

Techniques for Characterizing Atoms, Molecules, Associations, and Processes Occurring Among Them

Optical Studies of Metaphase Chromosome Organization	S. A. Latt	5:1–37
Optical Activity in the Vacuum Ultraviolet	E. S. Pysh	5:63–75
Synchrotron X-Ray Sources: A New Tool in Biological Structural and Kinetic Analysis	J. Barrington Leigh, G. Rosenbaum	5:239–70
Linear Dichroism of Biological Chromophores	J. Hofrichter, W. A. Eaton	5:511–60
New Laser Techniques for Biophysical Studies	B. S. Hudson	6:135–50
Interpretation of Resonance Raman Spectra of Biological Molecules	A. Warshel	6:273–300
High-Resolution Nuclear Magnetic Resonance Studies of Double Helical Polynucleotides	D. R. Kearns	6:477–523
Electron Microscope Visualization of Chromatin and Other DNA-Protein Complexes	J. D. Griffith, G. Christiansen	7:19–35
Circular Polarization of Luminescence: Biochemical and Biophysical Applications	I. Z. Steinberg	7:113–37
X-Ray Absorption Spectroscopy of Biological Molecules	R. G. Shulman, P. Eisenberger, B. M. Kincaid	7:559–78
Hydrogen Exchange Kinetics and Internal Motions in Proteins and Nucleic Acids	C. K. Woodward, B. D. Hilton	8:99–127
Molecular Structure Determination by High Resolution Electron Microscopy	F. P. Ottensmeyer	8:129–44
Recent Developments in Solution X-Ray Scattering	V. Luzzati, A. Tardieu	9:1–29
Photoacoustic Spectroscopy	A. Rosencwaig	9:31–54
The Optical Activity of Nucleic Acids and Their Aggregates	I. Tinoco, Jr., C. Bustamante, M. F. Maestre	9:107–41

CHAPTER TITLES 641

Magnetic Circular Dichroism of Biological Molecules	J. C. Sutherland, B. Holmquist	9:293–326
Radioimmunoassay	R. S. Yalow	9:327–45
Biophysical Applications of NMR to Phosphoryl Transfer Enzymes and Metal Nuclei of Metalloproteins	J. J. Villafranca, F. M. Raushel	9:363–92
Techniques for Characterizing Organs, Organ Systems, and Organisms		
Statistical Modeling and Analysis in Human Genetics	R. C. Elston, D. C. Rao	7:253–86
Automated Cell Sorting with Flow Systems	D. J. Arndt-Jovin, T. M. Jovin	7:527–58
Analysis of Intact Tissue with ^{31}P NMR	C. T. Burt, S. M. Cohen, M. Bárány	8:1–25
Chemical Cross-Linking in Biology	M. Das, C. F. Fox	8:165–93
Electrical Properties of Egg Cell Membranes	S. Hagiwara, L. A. Jaffe	8:385–416
X-Ray Diffraction Studies of the Heart	I. Matsubara	9:81–105
Display and Analysis of Flow Cytometric Data	J. W. Gray, P. N. Dean	9:509–39
Medical Information Systems	E. M. Laska, S. Abbey	9:581–604

PARTICULAR CONSTITUENTS, ASSEMBLIES, AND RELATIONS AMONG THEM

Naturally Occurring Constituents		
Thermotropic Transitions in Biomembranes	D. L. Melchior, J. M. Steim	5:205–38
Symmetry in Nucleic Acid Structure and Its Role in Protein-Nucleic Acid Interactions	H. M. Sobell	5:307–35
Hybridization and Renaturation Kinetics of Nucleic Acids	J. G. Wetmur	5:337–61
Electron Transfer Reactions of Copper Proteins	R. A. Holwerda, S. Wherland, H. B. Gray	5:363–96
Oscillatory Enzymes	A. Goldbeter, S. R. Caplan	5:449–76
The Purple Membrane from *Halobacterium halobium*	R. Henderson	6:87–109
Mechanisms of Zymogen Activation	R. M. Stroud, A. A. Kossiakoff, J. L. Chambers	6:177–93
Reactivity and Cryoenzymology of Enzymes in the Crystalline State	M. W. Makinen, A. L. Fink	6:301–43
Carbon-13 Nuclear Magnetic Resonance Studies of Proteins	W. Egan, H. Shindo, J. S. Cohen	6:383–417
Enzyme Dynamics: The Statistical Physics Approach	G. Careri, P. Fasella, E. Gratton	8:69–97
Endogenous Chemical Receptors: Some Physical Aspects	F. J. Barrantes	8:287–321
Transfer RNA in Solution: Selected Topics	P. R. Schimmel, A. G. Redfield	9:181–221
Naturally Occurring Assemblies		
Assembly of Multisubunit Respiratory Proteins	E. Antonini, E. Chiancone	6:239–71
Virus Envelopes and Plasma Membranes	J. Lenard	
Structure and Reactions of Closed Duplex DNA	W. R. Bauer	7:287–313
Kinetic Analysis of Myosin and Actomyosin ATPase	R. W. Lymn	8:145–63
Comparative Properties and Methods of Preparation of Lipid Vesicles (Liposomes)	F. Szoka, Jr., D. Papahadjopoulos	9:467–508
Relational Features		
Interfacial Enzyme Kinetics of Lipolysis	R. Verger, G. H. de Haas	5:77–117
Areas, Volumes, Packing, and Protein Structure	F. M. Richards	6:151–76
Protein-Lipid Interactions	R. B. Gennis, A. Jonas	6:195–238

MOLECULAR MECHANISMS UNDERLYING PROCESSES THAT OCCUR IN CELLS OR AMONG CELLS

Gene Expression		
Genetic Approach to the Study of the Nervous System	W. L. Pak, L. H. Pinto	5:397–448

Transport

Experimental Membranes and Mechanisms of Bioenergy Transductions	M. Montal	5:119–75
Countercurrent Transport in the Kidney	J. L. Stephenson	7:315–39
Calcium Buffering in Squid Axons	F. J. Brinley Jr.	7:363–92
Axoplasmic Transport of Proteins	D. L. Wilson, G. C. Stone	8:27–45
Stimulus-Response Coupling in Gland Cells	B. L. Ginsborg, C. R. House	9:55–80
The Structure of Proteins Involved in Active Membrane Transport	A. S. Hobbs, R. W. Albers	9:259–91

Excitation and Movement

Ionic Channels and Gating Currents in Excitable Membranes	W. Ulbricht	6:7–31
Ultramicroanalysis: X-Ray Spectrometry by Electron Probe Excitation	C. P. Lechene, R. R. Warner	6:57–85
Conductance Fluctuations and Ionic Pores in Membranes	C. F. Stevens, F. Neher	6:345–81
Calcium-Dependent Potassium Activation in Nervous Tissues	R. W. Meech	7:1–18
Neural Circuits for Generating Rhythmic Movements	W. O. Friesen, G. S. Stent	7:37–61
Excitation and Contraction Processes in Muscle	C. Caputo	7:63–83
Charge Movement in the Membrane of Striated Muscle	R. H. Adrian	7:85–112
Excitation and Interactions in the Retina	L. Cervetto, M. G. F. Fuortes	7:229–51
Interactions of Liposomes with Mammalian Cells	R. E. Pagano, J. N. Weinstein	7:435–68
Three-Dimensional Computer Reconstruction of Neurons and Neuronal Assemblies	E. R. Macagno, C. Levinthal, I. Sobel	8:323–51
Modulation of Impulse Conduction Along the Axonal Tree	H. A. Swadlow, J. D. Kocsis, S. G. Waxman	9:143–79
Nerve Growth Factor: Mechanism of Action	S. Vinores, G. Guroff	9:223–57
Certain Slow Synaptic Responses: Their Properties and Possible Underlying Mechanisms	J. Kehoe, A. Marty	9:437–65

Interaction with Mechanical Forces

Transition State Analog Inhibitors and Enzyme Catalysis	R. Wolfenden	5:271–306

Interaction with Radiation

Delayed Light in Photosynthesis	W. Arnold	6:1–6
Resonance Raman Studies of Visual Pigments	R. Callender, B. Honig	6:33–55
Magnetic Phenomena of the Central Nervous System	M. Reite, J. Zimmerman	7:167–88
Picosecond Flash Photolysis in Biology and Biophysics	D. Holten, M. W. Windsor	7:189–227
Chlorophyll Function in the Photosynthetic Reaction Center	J. J. Katz, J. R. Norris, L. L. Shipman, M. C. Thurnauer, M. R. Wasielewski	7:393–434

Interaction and Communication Among Cells

Mechanosensory Transduction "Sensory" and "Motile" Cilia	M. L. Wiederhold	5:39–62
Electrical Controls of Development	L. F. Jaffe, R. Nuccitelli	6:445–76

Stability and Time-Dependent Instability

Reaction Rate Theory in Bioluminescence and Other Life Phenomena	F. H. Johnson, H. Eyring, B. J. Stover	6:111–33

Tropisms

Phototropism in Coprophilious Zygomycetes	K. W. Foster	6:419–43

Photobehavior of Microorganisms: A Biophysical Approach	F. Lenci, G. Colombetti	7:341–61
Cytoplasmic Streaming in Amoeboid Movement	R. D. Allen, N. S. Allen	7:469–95
Cytoplasmic Streaming in Green Plants	N. S. Allen, R. D. Allen	7:497–526

PROPERTIES AND REACTIONS OF ORGANS, ORGAN SYSTEMS, AND ORGANISMS

Dynamics of Systems of Interacting Elements

Dye Indicators of Membrane Potential	A. S. Waggoner	8:47–68
Special Aspects of Natural and Artificial Organ Systems		
Two-Stage Screening Designs Applied to Chemical-Screening Problems with Binary Data	R. M. Elashoff, S. Beal	5:561–87
Neural Prostheses	F. T. Hambrecht	8:239–67
Ionic Mechanisms of Excitation in *Paramecium*	R. Eckert, P. Brehm	8:353–83

ORDER FORM   ANNUAL REVIEWS INC.

Please list on the order blank on the reverse side the volumes you wish to order and whether you wish a standing order (the latest volume sent to you automatically upon publication each year). Volumes not yet published will be shipped in month and year indicated. Prices subject to change without notice. Out of print volumes subject to special order.

NEW TITLES FOR 1980

ANNUAL REVIEW OF PUBLIC HEALTH   ISSN 0163-7525
    Vol. 1 (avail. May 1980): $17.00 (USA), $17.50 (elsewhere) per copy

ANNUAL REVIEWS REPRINTS: IMMUNOLOGY, 1977–1979   ISBN 0-8243-2502-8
A collection of articles reprinted from recent *Annual Review* series
    Avail. Mar. 1980   Soft cover: $12.00 (USA), $12.50 (elsewhere) per copy

SPECIAL PUBLICATIONS

ANNUAL REVIEWS REPRINTS: CELL MEMBRANES, 1975–1977   ISBN 0-8243-2501-X
A collection of articles reprinted from recent *Annual Review* series
    Published 1978   Soft cover: $12.00 (USA), $12.50 (elsewhere) per copy

THE EXCITEMENT AND FASCINATION OF SCIENCE, VOLUME 1   ISBN 0-8243-1602-9
A collection of autobiographical and philosophical articles by leading scientists
    Published 1965   Clothbound: $6.50 (USA), $7.00 (elsewhere) per copy

THE EXCITEMENT AND FASCINATION OF SCIENCE, VOLUME 2: Reflections by Eminent Scientists
    Published 1978   Hard cover: $12.00 (USA), $12.50 (elsewhere) per copy   ISBN 0-8243-2601-6
                        Soft cover: $10.00 (USA), $10.50 (elsewhere) per copy   ISBN 0-8243-2602-4

THE HISTORY OF ENTOMOLOGY   ISBN 0-8243-2101-7
A special supplement to the *Annual Review of Entomology* series
    Published 1973   Clothbound: $10.00 (USA), $10.50 (elsewhere) per copy

ANNUAL REVIEW SERIES

Annual Review of ANTHROPOLOGY   ISSN 0084-6570
    Vols. 1–8 (1972–79): $17.00 (USA), $17.50 (elsewhere) per copy
    Vol. 9 (avail. Oct. 1980): $20.00 (USA), $21.00 (elsewhere) per copy

Annual Review of ASTRONOMY AND ASTROPHYSICS   ISSN 0066-4146
    Vols. 1–17 (1963–79): $17.00 (USA), $17.50 (elsewhere) per copy
    Vol. 18 (avail. Sept. 1980): $20.00 (USA), $21.00 (elsewhere) per copy

Annual Review of BIOCHEMISTRY   ISSN 0066-4154
    Vols. 28–48 (1959–79): $18.00 (USA), $18.50 (elsewhere) per copy
    Vol. 49 (avail. July 1980): $21.00 (USA), $22.00 (elsewhere) per copy

Annual Review of BIOPHYSICS AND BIOENGINEERING*   ISSN 0084-6589
    Vols. 1–8 (1972–79): $17.00 (USA), $17.50 (elsewhere) per copy
    Vol. 9 (avail. June 1980): $17.00 (USA), $17.50 (elsewhere) per copy

Annual Review of EARTH AND PLANETARY SCIENCES*   ISSN 0084-6597
    Vols. 1–7 (1973–79): $17.00 (USA), $17.50 (elsewhere) per copy
    Vol. 8 (avail. May 1980): $17.00 (USA), $17.50 (elsewhere) per copy

Annual Review of ECOLOGY AND SYSTEMATICS   ISSN 0066-4162
    Vols. 1–10 (1970–79): $17.00 (USA), $17.50 (elsewhere) per copy
    Vol. 11 (avail. Nov. 1980): $20.00 (USA), $21.00 (elsewhere) per copy

Annual Review of ENERGY   ISSN 0362-1626
    Vols. 1–4 (1976–79): $17.00 (USA), $17.50 (elsewhere) per copy
    Vol. 5 (avail. Oct. 1980): $20.00 (USA), $21.00 (elsewhere) per copy

Annual Review of ENTOMOLOGY*   ISSN 0066-4170
    Vols. 7–24 (1962–79): $17.00 (USA), $17.50 (elsewhere) per copy
    Vol. 25 (avail. Jan. 1980): $17.00 (USA), $17.50 (elsewhere) per copy

Annual Review of FLUID MECHANICS*   ISSN 0066-4189
    Vols. 1–11 (1969–79): $17.00 (USA), $17.50 (elsewhere) per copy
    Vol. 12 (avail. Jan. 1980): $17.00 (USA), $17.50 (elsewhere) per copy

Annual Review of GENETICS   ISSN 0066-4197
    Vols. 1–13 (1967–79): $17.00 (USA), $17.50 (elsewhere) per copy
    Vol. 14 (avail. Dec. 1980): $20.00 (USA), $21.00 (elsewhere) per copy

Annual Review of MATERIALS SCIENCE   ISSN 0084-6600
    Vol. 1–9 (1971–79): $17.00 (USA), $17.50 (elsewhere) per copy
    Vol. 10 (avail. Aug. 1980): $20.00 (USA), $21.00 (elsewhere) per copy

(continued on reverse)
*Price will be increased to $20.00 (USA), $21.00 (elsewhere) per copy effective with the 1981 volume.

Annual Review of MEDICINE: Selected Topics in the Clinical Sciences*   ISSN 0066-4219
    Vols. 1–3, 5–15, 17–30 (1950–52, 1954–64, 1966–79): $17.00 (USA), $17.50 (elsewhere) per copy
    Vol. 31 (avail. Apr. 1980): $17.00 (USA), $17.50 (elsewhere) per copy

Annual Review of MICROBIOLOGY   ISSN 0066-4227
    Vols. 15–33 (1961–79): $17.00 (USA), $17.50 (elsewhere) per copy
    Vol. 34 (avail. Oct. 1980): $20.00 (USA), $21.00 (elsewhere) per copy

Annual Review of NEUROSCIENCE*   ISSN 0147-006X
    Vols. 1–2 (1978–79): $17.00 (USA), $17.50 (elsewhere) per copy
    Vol. 3 (avail. Mar. 1980): $17.00 (USA), $17.50 (elsewhere) per copy

Annual Review of NUCLEAR AND PARTICLE SCIENCE   ISSN 0066-4243
    Vols. 10–29 (1960–79): $19.50 (USA), $20.00 (elsewhere) per copy
    Vol. 30 (avail. Dec. 1980): $22.50 (USA), $23.50 (elsewhere) per copy

Annual Review of PHARMACOLOGY AND TOXICOLOGY*   ISSN 0362-1642
    Vols. 1–3, 5–19 (1961–63, 1965–79): $17.00 (USA), $17.50 (elsewhere) per copy
    Vol. 20 (avail. Apr. 1980): $17.00 (USA), $17.50 (elsewhere) per copy

Annual Review of PHYSICAL CHEMISTRY   ISSN 0066-426X
    Vols. 10–21, 23–30 (1959–70, 1972–79): $17.00 (USA), $17.50 (elsewhere) per copy
    Vol. 31 (avail. Nov. 1980): $20.00 (USA), $21.00 (elsewhere) per copy

Annual Review of PHYSIOLOGY*   ISSN 0066-4278
    Vols. 18–41 (1956–79): $17.00 (USA), $17.50 (elsewhere) per copy
    Vol. 42 (avail. Mar. 1980): $17.00 (USA), $17.50 (elsewhere) per copy

Annual Review of PHYTOPATHOLOGY   ISSN 0066-4286
    Vols. 1–17 (1963–79): $17.00 (USA), $17.50 (elsewhere) per copy
    Vol. 18 (avail. Sept. 1980): $20.00 (USA), $21.00 (elsewhere) per copy

Annual Review of PLANT PHYSIOLOGY*   ISSN 0066-4294
    Vols. 10–30 (1959–79): $17.00 (USA), $17.50 (elsewhere) per copy
    Vol. 31 (avail. June 1980): $17.00 (USA), $17.50 (elsewhere) per copy

Annual Review of PSYCHOLOGY*   ISSN 0066-4308
    Vols. 4, 5, 8, 10–30 (1953, 1954, 1957, 1959–79): $17.00 (USA), $17.50 (elsewhere) per copy
    Vol. 31 (avail. Feb. 1980): $17.00 (USA), $17.50 (elsewhere) per copy

Annual Review of SOCIOLOGY   ISSN 0360-0572
    Vols. 1–5 (1975–79): $17.00 (USA), $17.50 (elsewhere) per copy
    Vol. 6 (avail. Aug. 1980): $20.00 (USA), $21.00 (elsewhere) per copy

*Price will be increased to $20.00 (USA), $21.00 (elsewhere) per copy effective with the 1981 volume.

To ANNUAL REVIEWS INC., 4139 El Camino Way, Palo Alto, CA 94306 USA
(Tel. 415-493-4400)

Please enter my order for the following publications:
(Standing orders: indicate which volume you wish order to begin with)

_____, Vol(s). ____ Standing order ____

_____, Vol(s). ____ Standing order ____

_____, Vol(s). ____ Standing order ____

_____, Vol(s). ____ Standing order ____

Amount of remittance enclosed $_____   California residents please add applicable sales tax.
Please bill me ☐   Prices subject to change without notice.

SHIP TO (include institutional purchase order if billing address is different)

Name _____

Address _____

_____

_____ Zip Code _____

Signed _____ Date _____

☐ Please add my name to your mailing list to receive a free copy of the current Prospectus each year.
☐ Send free brochure listing contents of recent back volumes for *Annual Review(s)* of